Algebras, Lattices, Varieties

Volume II

Mathematical
Surveys
and
Monographs

Volume 268

Algebras, Lattices, Varieties

Volume II

Ralph S. Freese
Ralph N. McKenzie
George F. McNulty
Walter F. Taylor

AMERICAN
MATHEMATICAL
SOCIETY
Providence, Rhode Island

2020 *Mathematics Subject Classification.* Primary 08-02, 08Bxx, 03C05, 06Bxx.

For additional information and updates on this book, visit
www.ams.org/bookpages/surv-268

Library of Congress Cataloging-in-Publication Data

Names: McKenzie, Ralph, author. | McNulty, George F., 1945- author. | Taylor, W. (Walter), 1940- author.
Title: Algebras, lattices, varieties / Ralph N. McKenzie, George F. McNulty, Walter F. Taylor.
Description: Providence, Rhode Island : American Mathematical Society : AMS Chelsea Publishing, 2018- | Series: | Originally published: Monterey, California : Wadsworth & Brooks/Cole Advanced Books & Software, 1987. | Includes bibliographical references and indexes.
Identifiers: LCCN 2017046893 | ISBN 9781470467975 (paperback) | 9781470471293 (ebook)
Subjects: LCSH: Algebra, Universal. | Lattice theory. | Varieties (Universal algebra) | AMS: General algebraic systems – Algebraic structures – Algebraic structures. | General algebraic systems – Varieties – Varieties. | Mathematical logic and foundations – Model theory – Equational classes, universal algebra. | Order, lattices, ordered algebraic structures – Lattices – Lattices. | Order, lattices, ordered algebraic structures – Modular lattices, complemented lattices – Modular lattices, complemented lattices.
Classification: LCC QA251 .M43 2018 | DDC 512–dc23
LC record available at https://lccn.loc.gov/2017046893

Volume 2 ISBN: 978-1-4704-6797-5
Volume 2 ISBN (Electronic): 978-4704-7129-3
Volume 3 ISBN: 978-1-4704-6798-2
Volume 3 ISBN (Electronic): 978-1-4704-7130-9

∞ The paper used in this book is acid-free and falls within the guidelines
established to ensure permanence and durability.
Visit the AMS home page at `https://www.ams.org/`

10 9 8 7 6 5 4 3 2 1 27 26 25 24 23 22

This volume is dedicated to our teachers—
Robert P. Dilworth, J. Donald Monk, Alfred Tarski, and Garrett Birkhoff

Contents

List of Figures

Preface for Volumes II and III

The two new volumes.

In the thirty or more years since *Algebras, Lattices, Varieties Volume I* appeared, the authors—and our readers—have seen the growth of many branches of our field of general algebra. Our first volume included five chapters—the first four laid the foundation for our enterprise, while the fifth provided a serious look at direct products and, particularly, at some circumstances under which a unique direct factorization result (like the familiar one for finite Abelian groups) might hold.

For presentation in the second and third volumes, following the reissuing of Volume I by the American Mathematical Society in their Chelsea series in 2018, we have selected six threads of development (three per volume), and have written six new chapters corresponding to these themes. (The three-volume set thus has eleven chapters, numbered consecutively.)

The six new chapters.

We begin Volume II with Chapter 6, *Classification of Varieties*, which describes the many classes (or properties) of varieties that arise from the existence (or not) of various term operations (as typified in the theory of Mal'tsev conditions). Chapter 7, *Equational Logic*, presents the study of equational theories through a detailed analysis of equational proofs—especially undecidability results and the investigation of finite axiomatizability of equational theories—and the small-scale analysis of subdirect representation. Chapter 8 *Rudiments of Model Theory*, provides tools and motivations from model theory, including logical compactness, reduced products, Jónsson's Lemma, and Baker's Finite Basis Theorem. (As noted in the Introduction to Volume 1, general algebra and model theory have become very distinct fields, but the more basic parts of model theory continue to play a strong role in our understanding of general algebra.)

In Chapter 9, *Finite Algebras and their Clones*, we examine, for a set A, the collection $\mathbf{Clo}\,A$ of all operations $A^n \longrightarrow A$ (with n ranging over ω, or sometimes over $\omega\smallsetminus\{0\}$). By a *clone of operations on A*— or simply *a clone*—we mean a subset C of $\mathbf{Clo}\,A$ that contains the coordinate projections and is closed under forming compositions. For example if C is a clone and $f \in C$, then $g(x_0,x_1,x_2,x_3) = f(f(x_0,x_1),f(x_2,x_3))$ must also belong to C. The *clone of an algebra* $\mathbf{A} = \langle A, F_t\rangle_{t\in T}$ is the smallest clone of operations on A that contains all the basic operations F_t. In many contexts, the interesting properties of $\mathbf{A}$ depend only on this clone. Landmark theorems in this chapter include: for A a two-element set, a description of all the clones on A (Emil L. Post, 1941); and for A an arbitrary finite set, a description of all the maximal proper subclones of $\mathbf{Clo}\,A$ (Ivo Rosenberg, 1970).

In Chapter 10, *Abstract Clone Theory*, the composition of operations in **Clo** A is regarded as forming an algebraic system in its own right. The family of clones $C \subseteq \mathbf{Clo}\, A$ (with A ranging over all sets) can be axiomatized in a way that makes no mention of the underlying sets A. (So abstract clones are related to clones of operations in the same way that abstract groups are related to permutation groups.) Abstract clones have also been known as *algebraic theories* (F. William Lawvere, 1963). Abstract clones support the study of varieties in the following way: we may attach to each variety $\mathcal{V}$ (of any signature) the clone of operations of the free algebra $\mathbf{F}_{\mathcal{V}}(\omega)$, which we may also designate as $\mathbf{C}(\mathcal{V})$. Now the correspondence $\mathcal{V} \longleftrightarrow \mathbf{C}(\mathcal{V})$ makes term-equivalence classes of varieties correspond to isomorphism classes of clones. And it makes interpretations of a variety $\mathcal{V}$ in a variety $\mathcal{W}$ correspond bijectively to clone homomorphisms $\mathbf{C}(\mathcal{V}) \longrightarrow \mathbf{C}(\mathcal{W})$. In this way many varietal properties can be seen as arising in tandem with algebraic properties at the clone level. Thus in Chapter 10 we reprise the subject of Mal'tsev conditions from Chapter 6. Varieties can be ordered by taking $\mathcal{V} \leq_{\text{int}} \mathcal{W}$ iff there is an interpretation of $\mathcal{V}$ in $\mathcal{W}$. The resulting ordered class is a lattice $\mathbf{L}_{\text{int}}$, and each Mal'tsev condition defines a filter on $\mathbf{L}_{\text{int}}$.

Our final chapter (11) is *The Commutator.* The commutator is a binary operation $\langle\alpha, \beta\rangle \longmapsto [\alpha, \beta] \in \mathbf{Con\, A}$, defined on **Con A**. Its theory was begun in the 1970's and 1980's, marking a new era in universal algebra. In that period important structural results were proved for the algebras in any congruence modular variety: results from group theory, ring theory and other classical algebraic systems were extended to this general framework. Chapter 11 revisits congruence-modular, congruence-distributive and Abelian varieties, as well as discussing nilpotence, solvability, finite axiomatizability, direct representability and residual smallness. In particular, the chapter represents a further study of congruence modularity. The chapter concludes with a proof of McKenzie's 1996 Finite Basis Theorem: *Every congruence-modular variety of finite signature that has a finite residual bound is finitely based.* (See Theorem 11.119 on page 367 of Volume III.) The commutator is defined in every variety and, in the last several years, it has been used to obtain substantial results for varieties that are not necessarily congruence modular. For example, Theorem 11.57 characterizes when an Abelian algebra is affine (polynomially equivalent to a module). Most of the results involving the commutator in nonmodular varieties are in §11.5 and §11.6.

Each of the six themes already appeared in Volume I. Chapter 6 really began, for example, in §4.7 and §4.9, with permutability and with distributivity of congruence lattices. The free-algebra material in §4.11 is a good basis for our Chapter 7. Birkhoff's ***HSP***-theorem (§4.11) was the earliest example of a preservation theorem, a theme that is developed more extensively in Chapter 8. The clones of term operations and of polynomial operations, which dominate Chapter 9, were set out in the first section of Chapter 4. Chapter 10 builds both on the category-theoretic axiomatization of clones introduced in Chapter 3 (pages 136–137), and on the interpretability notions of §4.12. Chapter 11 expands very thoroughly on the commutator material that was presented in §4.13. It is our hope that this continuity of subject matter will facilitate the readers' entry into these latest volumes.

Motifs.

There are leitmotifs that run throughout these two volumes. One example is the application of available *Mal'tsev conditions* to obtain insights about varieties, equational theories, and clones. Chapter 6 is largely devoted to laying the foundations of this motif. It re-emerges in later chapters to provide key hypotheses for decidability and finite axiomatizability results, for Jónsson's Lemma, and for many other results. Another leitmotif is the question, for a given (finite) algebra **A**, of *whether the equational theory* $\Theta\,\mathbf{A}$ *of the algebra* **A** *is finitely axiomatizable.* While this question does not occur in Chapter 6, in retrospect it is only because the concepts introduced in that chapter seem to be almost designed to provide methods for use in later chapters. The finite axiomatizability question is one of the central concerns of Chapter 7. Then in Chapter 8 we find Baker's Finite Basis Theorem according to which $\Theta\,\mathbf{A}$ is finitely axiomatizable provided **A** is of finite signature and the variety generated by **A** is congruence distributive. Some further results on finite axiomatizability occur in Chapters 8 and 11.

Yet a third motif is the consideration of the *subdirectly irreducible algebras in a variety*. When the subdirectly irreducible algebras in a variety can be completely understood, this leads to a deep understanding of the whole variety. This happens, for instance, when the signature is finite and there is a finite upper bound of the cardinalities of the subdirectly irreducible algebras. In Chapter 7 having such a finite residual bound is a key hypothesis for finite axiomatizability results, but it is also used to show that there is no computable procedure for determining, of a finite algebra of finite signature, whether the variety it generates has a finite residual bound. In Chapter 8 one great power of Jónsson's Lemma lies in its ability to describe all the subdirectly irreducible algebras in a congruence distributive variety.

We have written the six new chapters to be independent expositions of their respective topics. There are, nevertheless, important links between the chapters, especially where an earlier chapter supports a result in a later chapter. We mention here just a few examples of such dependency. Chapter 6 is largely devoted to a detailed development of particular Mal'tsev conditions. In Chapter 10 a general theory of Mal'tsev conditions is laid out in the categorical context of abstract clone theory. Then in Chapter 11 we show that certain properties connected with the commutator, such as having a difference term or a weak difference term, are definable by a Mal'tsev condition. Chapter 6 shows congruence meet semidistributivity is definable by a Mal'tsev condition, while Chapter 11 shows this property is equivalent to congruence neutrality, that is, the commutator equals the meet. The theory of directly representable varieties, introduced in Chapter 8, is expanded in Chapter 9 and in Chapter 11, using the properties of the commutator developed in Chapter 11. On a few occasions we have found it necessary to have forward references. For instance the Compactness Theorem is applied in Chapter 7, but not proved until Chapter 8; the characterization of congruence join semidistributivity, given in Chapter 6, requires substantial results from Chapter 11 in order to complete its proof.

The examples of such dependency found in the exercises are too numerous to list here. A final example in the main text is the theory of Boolean powers, found here in Chapter 8, which is used in Chapter 9 to demonstrate that varieties generated by primal algebras are all categorically equivalent.

Our selection of topics has also been attentive to our (rather personal) observations, over time, of the ideas and methods that have been found useful and productive in our own, and in others', further investigations. Here we might cite the *recursive solution* (or lack thereof) of the word problem for *presentations* of an algebra **A** in a variety $\mathcal{V}$; the *decidability* (or not) of the set $\Theta\mathbf{A}$ of equations valid in **A**; also *finite axiomatizability* of $\Theta\mathbf{A}$—especially for **A** a finite lattice with operators, and for **A** belonging to a variety with distributive congruence lattices; the strength of $\mathcal{V}$ (or of **A**) as measured by the family of Mal'tsev conditions that it satisfies (or not); the utility of the notion of a *system of definitions* of one variety $\mathcal{V}$ in a second variety $\mathcal{W}$; the *lattice of subvarieties* of a given variety; the *linearity* (*or not*) of various Mal'tsev conditions; the basic properties of *Taylor terms*.

Omissions and Inclusions.

The resulting conception of these two volumes has not, in fact, diverged far from what we envisioned thirty years ago. Meanwhile, of course, during this period the field has grown significantly, producing new themes (and many individual theorems) that we cannot accommodate in these volumes. Notable omissions are constraint satisfaction problems, tame congruence theory, natural duality theory (except for a departure point found in our section on Boolean powers, Chapter 8), the general theory of quasivarieties, topological algebra, substructural logics, higher-dimension commutators, the theory of free lattices, and others. Also missing are several significant theorems, like the Oates-Powell Theorem according to which the equational theory of each finite group is finitely axiomatizable, whose proofs use concepts and methods from parts of mathematics that could not be included in this volume, or whose proofs, like Jaroslav Ježek's profound study of subsets of lattices of equational theories definable in first-order logic, were so intricate and extensive that the space they would require was beyond what was available to us.

Nevertheless we have ventured to include a selection of results that were not available to us in 1986. Here we mention a few: congruence-semidistributive varieties (§6.4); the minimal idempotent variety of Olšák (§6.8); shift-automorphism algebras (§7.3); Willard's Finite Basis Theorem, etc. (§7.4); McKenzie's resolution of Tarski's Finite Basis Problem (§7.11); Nation's counterexample to the Finite Height Conjecture (§8.4). There are also a few results here that have not appeared in the literature—notably Don Pigozzi's representation of an algebraic lattice with an additional new top element as a principal filter in a lattice of equational theories.

Useful Background.

The best preparation a reader can bring to the study of the new volumes of *Algebras, Lattices, Varieties* is a command of the material in Volume I. Indeed, in the present volume we make reference to Volume I, usually giving the page number, for background material, the statements of particular definitions, theorems, and examples—although we have made an effort to restate most of these within the later volumes. But readers acquainted with *Universal Algebra* by Clifford Bergman (2012) should be comfortable continuing their study of our field in the present two volumes. The short volume *Allgemeine Algebra* by Thomas Ihringer (2003) or the older expositions *A Course in Universal Algebra* by S. Burris and H. P. Sankappanavar (1981) or the second edition of *Universal Algebra* by G. Grätzer (1979) would be useful background.

We have provided most sections in this volume with exercises, against which diligent readers might test their understanding of the material.

The beautiful edifice that we strive to portray in these volumes is the product of numerous mathematicians, who have worked over the last century to uncover and understand the fundamental structures of general algebra. During the course of writing we have gone again and again to the literature, coming away with a deep appreciation of this work. It is our hope that our readers will find here some of the beauty, joy, and power that we have found in this domain of mathematics.

Ralph S. Freese
Ralph N. McKenzie
George F. McNulty
Walter F. Taylor

Acknowledgments

Our colleagues Brian A. Davey, Heinz-Peter Gumm, Keith A. Kearnes, Peter Mayr, J.B. Nation, Ágnes Szendrei, Matthew Valeriote, and Ross Willard, kindly agreed to read earlier versions of *Algebras, Lattices, Varieties, Volumes II and III*. Their suggestions, including some simpler and more informative proofs, have improved these volumes. We appreciate their contributions.

Ralph S. Freese
Ralph N. McKenzie
George F. McNulty
Walter F. Taylor

CHAPTER 6

The Classification of Varieties

6.1. Introduction

Although varieties have brought some order to the diversity which we saw among individual algebras, we can again see a great diversity in looking over varieties themselves. Fortunately there have arisen various classifications of varieties which help to bring some order to this further diversity. In this chapter we will review and study some of the better known and more important properties which concern congruence relations on the algebras of a given variety.

The prototype for most of the properties treated in this chapter is that of congruence permutability, which we examined in §4.7 of Volume I. Recall that we proved in Theorem 4.141 that a variety $\mathcal{V}$ is congruence permutable (i.e., $\theta \circ \phi = \phi \circ \theta$ for all $\theta, \phi \in \operatorname{Con} \mathbf{A}, \mathbf{A} \in \mathcal{V}$) if and only if there exists a ternary term $p(x, y, z)$ such that $\mathcal{V}$ satisfies the equations

$$(6.1.1) \qquad p(x, z, z) \approx x \approx p(z, z, x).$$

In honor of this result of A. I. Mal'tsev (1954) varietal properties which can be defined in a similar manner, i.e., by the existence of terms satisfying a certain finite set Σ of equations, have come to be known as **strong Mal'tsev conditions**; we will also say that such a property is **strongly Mal'tsev definable**. In this chapter we will be examining properties P which are Mal'tsev definable; in most cases P will have a natural definition which is not of the Mal'tsev type, and our job will be to find a suitable set Σ of equations. In fact this Σ can be thought of as defining the 'most general' variety satisfying P, in that for P to hold, we require nothing more than the deducibility or **realization** of the equations Σ. This can be made precise by using the concept of the interpretation of one variety into another as defined in §4.12 and also reviewed at the end of this section. For example, Mal'tsev's result described above, can be stated as follows: a variety $\mathcal{V}$ is congruence permutable if and only if the variety $\mathcal{M}$ is interpretable into $\mathcal{V}$, where $\mathcal{M}$ is the variety with a single ternary operation symbol p satisfying (6.1.1). We also say $\mathcal{V}$ *realizes* Σ, where Σ is the set of the two equations of (6.1.1). This is Theorem 4.141 of §4.12. In general we say that an algebra $\mathbf{A}$ (or a variety $\mathcal{V}$) **realizes** a set of equations Σ if it is possible to interpret each operation symbol that occurs in Σ as a term of $\mathbf{A}$ (or $\mathcal{V}$) such that all the equations of Σ are satisfied by $\mathbf{A}$ (or $\mathcal{V}$).

Actually, most of the properties P that we study in this chapter fail to be strongly Mal'tsev definable, but satisfy the following more general situation instead. There exist strong Mal'tsev conditions $\mathrm{P}_1, \mathrm{P}_2, \ldots$ such that each P_i implies P_{i+1} and such that P is logically equivalent to the disjunction of all P_i. (Equivalently, there exist finite sets Σ_1, $\Sigma_2, \ldots$ of equations such that, if $\mathcal{V}_i$ is the variety defined by Σ_i, then $\mathcal{V}_{i+1}$ interprets into $\mathcal{V}_i$ and $\mathcal{V}$ satisfies P if and only if for some i, $\mathcal{V}_i$ interprets into $\mathcal{V}$.) We emphasize that the *signature*[1] of $\mathcal{V}$ need not be the same—nor even related in any way—to that of any of the Σ_i. We will call such a property **Mal'tsev definable**; sometimes also we refer to P, or to the sequence $\Sigma_1, \Sigma_2, \ldots$, as a **Mal'tsev condition**. A prototypical example of a Mal'tsev condition that is not a strong Mal'tsev condition (as will be shown in Corollary 6.84) is congruence distributivity. $\mathcal{V}$ is **congruence distributive** or CD if all the algebras of $\mathcal{V}$ have distributive congruence lattices. A variety $\mathcal{V}$ is CD if and only if, for some n, $\mathcal{V}$ realizes the set of equations Δ_n:

$$(\Delta_n)\qquad \begin{aligned} &d_0(x,y,z) \approx x; \\ &d_i(x,y,x) \approx x && \text{for all } i; \\ &d_i(x,x,z) \approx d_{i+1}(x,x,z) && \text{if } i \text{ is even}; \\ &d_i(x,z,z) \approx d_{i+1}(x,z,z) && \text{if } i \text{ is odd}; \\ &d_n(x,y,z) \approx z. \end{aligned}$$

The next theorem records this Mal'tsev condition for CD along with several other equivalent conditions which illustrate the various angles of looking at Mal'tsev conditions and some of the standard techniques for proving them. The theorem was proved for the most part in Theorem 4.144 of the first volume. A more detailed discussion will follow the proof.

THEOREM 6.1 (Bjarni Jónsson 1967). *The following are equivalent for a variety $\mathcal{V}$.*

(i) *$\mathcal{V}$ is congruence distributive.*

(ii) *$\mathcal{V}$ satisfies the congruence inclusion*

$$(*)\qquad \gamma \cap (\alpha \circ \beta) \subseteq (\gamma \cap \alpha) \vee (\gamma \cap \beta).$$

(iii) *For some n, $\mathcal{V}$ satisfies the congruence inclusion*

$$(**)\qquad \gamma \cap (\alpha \circ \beta) \subseteq (\gamma \cap \alpha) \circ^n (\gamma \cap \beta).$$

(iv) *For $\mathbf{A} = \mathbf{F}_{\mathcal{V}}(x,y,z)$ and $\alpha = \mathrm{Cg}^{\mathbf{A}}(x,y)$, $\beta = \mathrm{Cg}^{\mathbf{A}}(y,z)$ and $\gamma = \mathrm{Cg}^{\mathbf{A}}(x,z)$, the inclusion $(*)$ holds.*

(v) *For $\mathbf{A} = \mathbf{F}_{\mathcal{V}}(x,y,z)$ and $\alpha = \mathrm{Cg}^{\mathbf{A}}(x,y)$, $\beta = \mathrm{Cg}^{\mathbf{A}}(y,z)$ and $\gamma = \mathrm{Cg}^{\mathbf{A}}(x,z)$, $\langle x,z\rangle \in (\gamma \cap \alpha) \circ^n (\gamma \cap \beta)$, for some n.*

(vi) *$\mathcal{V}$ realizes Δ_n, for some n.*

(vii) ***Con** $\mathbf{F}_{\mathcal{V}}(x,y,z)$ is distributive.*

(viii) *Letting $\mathbf{F}_2 = \mathbf{F}_{\mathcal{V}}(a,b)$, $\mathbf{S}$ be the subalgebra of $\mathbf{F}_2^3$ generated by $\langle a,a,b\rangle$, $\langle a,b,a\rangle$ and $\langle b,a,a\rangle$, and $T \subseteq S$ be those elements whose middle component is a, there is a sequence of elements $\langle c_i,a,e_i\rangle \in T$ with*

$$(\dagger)\qquad \langle a,a,b\rangle = \langle c_0,a,e_0\rangle, \langle c_1,a,e_1\rangle, \ldots, \langle c_n,a,e_n\rangle = \langle b,a,a\rangle,$$

such that for each i, either $c_i = c_{i+1}$ or $e_i = e_{i+1}$.

(ix) (S. Burris, 1979)[2] *There is a finite set T of ternary terms so that*

$$\mathcal{V} \models t(x,u,x) \approx t(x,v,x) \textit{ for all } t \in T, \textit{ and}$$

$$\mathcal{V} \models \bigwedge_{t\in T} \big(t(x,x,y) \approx t(x,y,y)\big) \Rightarrow x \approx y.$$

[1] *Signature* is synonymous with *similarity type* as defined in §4.2 of Volume I. *Signature* has become common usage so we use it throughout this and the next volumes.

[2] Burris credits Kirby A. Baker (1977) for proving this condition implies congruence distributivity.

Proof. Clearly the congruence inclusion $(*)$ is implied by the distributive law:

$$\gamma \wedge (\alpha \vee \beta) \leq (\gamma \wedge \alpha) \vee (\gamma \wedge \beta)$$

so (i) ⇒ (ii).

That (ii) ⇒ (iv) is obvious. That (iv) ⇒ (v) we have by $(*)$

$$\langle x, z\rangle \in \gamma \cap (\alpha \circ \beta) \subseteq (\gamma \cap \alpha) \vee (\gamma \cap \beta) = \bigcup_n (\gamma \cap \alpha) \circ^n (\gamma \cap \beta),$$

which implies $\langle x, z\rangle \in (\gamma \cap \alpha) \circ^n (\gamma \cap \beta)$ for some n.

If (v) holds there are elements $x = u_0, u_1, \ldots, u_n = z$ in A such that u_i is related to u_{i+1} by $\gamma \wedge \alpha$ if i is even, and by $\gamma \wedge \beta$ if i is odd. Since $\mathbf{A} = \mathbf{F}_{\mathcal{V}}(x, y, z)$, there are terms d_i such that $u_i = d_i(x, y, z)$. So $x = u_0 = d_0(x, y, z)$ and $z = u_n = d_n(x, y, z)$. Note all the u_i's are γ-related so $x = u_0 \ \gamma \ u_i = d_i(x, y, z) \ \gamma \ d_i(x, y, x)$. But, since γ restricted to the subalgebra of $\mathbf{F}_{\mathcal{V}}(x, y, z)$ generated by x and y is trivial (we give a proof of this easy fact in Theorem 6.9 below), $x = d_i(x, y, x)$ for all i. If i is even then $\langle u_i, u_{i+1}\rangle \in \alpha$, and so $d_i(x, y, z) \ \alpha \ d_{i+1}(x, y, z)$. Since $\langle x, y\rangle \in \alpha$ we have $d_i(x, x, z) \ \alpha \ d_{i+1}(x, x, z)$. By Theorem 6.9 again, α is trivial on the subalgebra generated by x and z. Hence $d_i(x, x, z) = d_{i+1}(x, x, z)$. When i is odd a similar argument applies. Hence (vi) holds.

A proof that (vi) ⇒ (i) is given in the proof of Theorem 4.144 on page 248 of Volume I. Exercise 2 at the end of this section outlines the proof. With this we have established that (i), (ii), (iv), (v), and (vi) are pairwise equivalent.

To see (vi) ⇒ (iii) let $\mathbf{A}$ be an algebra in $\mathcal{V}$ with congruences α, β and γ. Suppose $\langle a, c\rangle \in \gamma \cap (\alpha \circ \beta)$. Then there is a $b \in A$ with $a \ \alpha \ b \ \beta \ c$. Using the equations (Δ_n) it is easy to establish $\langle d_i(a, b, c), d_{i+1}(a, b, c)\rangle \in \gamma \cap \alpha$, if i is even, and in $\gamma \cap \beta$, if i is odd. Since $a = d_0(a, b, c)$ and $c = d_n(a, b, c)$, we have $\langle a, c\rangle \in (\gamma \cap \alpha) \circ^n (\gamma \cap \beta)$, proving (iii). (iii) ⇒ (ii) is obvious, so we have that (i) to (vi) are pairwise equivalent.

Clearly (i) ⇒ (vii). Conversely, with α, β and γ as defined in (v) and using the distributivity assumption of (vii), we have $\langle x, z\rangle \in \gamma \wedge (\alpha \vee \beta) = (\gamma \wedge \alpha) \vee (\gamma \wedge \beta)$, which implies (v). So (i) to (vii) are all equivalent.

To see (vi) ⇒ (viii) just apply the d_i's to the three generators of $\mathbf{S}$:

$$\begin{aligned} d_i(\langle a, a, b\rangle, \langle a, b, a\rangle, \langle b, a, a\rangle) &= \langle d_i(a, a, b), d_i(a, b, a), d_i(b, a, a)\rangle \\ &= \langle d_i(a, a, b), a, d_i(b, a, a)\rangle \end{aligned}$$

Using d_{i+1} we get $\langle d_{i+1}(a, a, b), a, d_{i+1}(b, a, a)\rangle$, and by (vi) either the first or last components of these triples are equal, depending on the parity of i. This gives a sequence (†) of triples in $\mathbf{S}$. Conversely, if we have a sequence (†) in $\mathbf{S}$ then there are terms d_i in the generators of $\mathbf{S}$ giving this sequence. These d_i's satisfy (Δ_n). The details are left to the reader; see Exercise 1.

We can see (vi) ⇒ (ix) simply by letting $T = \{d_0, \ldots, d_n\}$.

To see (ix) ⇒ (v), let $\mathbf{A} = \mathbf{F}_{\mathcal{V}}(x, y, z)$ and $\alpha = \mathrm{Cg}^{\mathbf{A}}(x, y)$, $\beta = \mathrm{Cg}^{\mathbf{A}}(y, z)$ and $\gamma = \mathrm{Cg}^{\mathbf{A}}(x, z)$. Then, for all $t \in T$,

$$t^{\mathbf{A}}(x, x, z) \ \ \gamma \cap \alpha \ \ t^{\mathbf{A}}(x, y, z) \ \ \gamma \cap \beta \ \ t^{\mathbf{A}}(x, z, z)$$

So letting $\theta = (\gamma \wedge \alpha) \vee (\gamma \wedge \beta)$, we have $t^{\mathbf{A}}(x, x, z) \ \theta \ t^{\mathbf{A}}(x, z, z)$. Thus in $\mathbf{A}/\theta$

$$t^{\mathbf{A}/\theta}(x/\theta, x/\theta, z/\theta) = t^{\mathbf{A}/\theta}(x/\theta, z/\theta, z/\theta).$$

By (ix) this gives $\langle x, z\rangle \in \theta$ which implies (v) holds. ∎

Statement (ii) says that for all $\mathbf{A} \in \mathcal{V}$, and for all congruences α, β and γ of $\mathbf{A}$, the inclusion $(*)$ holds. Congruence inclusions like this are also referred to as Mal'tsev conditions. While the implication that (i) implies (ii) is trivial, it is not at all clear that (ii) implies (i) and this is the key to the theorem. While the proof that (ii) implies (i) was not too hard for congruence distributivity, see Exercise 2, we will see it is more involved for congruence modularity, as well as for other conditions.

As we said, the proof of much of this theorem is contained in the proof of Theorem 4.144 in Volume I. Interestingly, that theorem replaces Δ_n by Δ'_n, which is the same except 'even' and 'odd' are reversed. The Mal'tsev condition defined by the Δ'_n's is known as the **alvin variant**[3] of Jónsson's condition for congruence distributivity. The relation of Δ_n to Δ'_n are explored in the exercises. For example, if a variety realizes Δ'_n then it realizes Δ_{n+1} and if it realizes Δ_n then it realizes Δ'_{n+1}. Of course it follows from this that the class of varieties realizing Δ_n for some n is the same same as the class of varieties realizing Δ'_n for some n.

Statement (viii) is important computationally since it only requires finding the free algebra on 2 generators instead of 3.

Many of the varietal properties of this chapter arise naturally if one observes the behavior of congruence relations in vector spaces and in lattices. For instance, if θ is a congruence on a vector space $\mathbf{V}$ (over any field), then any two congruence blocks a/θ and b/θ have the same number of elements; in our terminology, the congruences are **uniform**. In §6.6 we will examine this and some related properties, especially the somewhat weaker notion of congruence **regularity** (if $|a/\theta| = 1$, then $|b/\theta| = 1$). Vector spaces are also congruence permutable, and, as we saw in Theorem 4.67, permutability of the congruence lattice implies its modularity. In §6.5 we present some Mal'tsev conditions for congruence modularity. Except for modularity, these properties all fail for the variety $\mathcal{L}$ of lattices, but $\mathcal{L}$ is even congruence distributive, as we saw in Chapter 2.

There is a geometric way of thinking about these sorts of problems which has proved quite helpful (see especially Rudolf Wille (1970) and H. P. Gumm (1983)). The traditional method (going back to Descartes) for constructing algebraic models of Euclidean geometry, i.e., analytic geometry, can be described as follows. Let $\mathbf{V}$ be a real vector space of dimension n. One then defines 'points' to be elements of $\mathbf{V}$ and 'lines' to be cosets of 1-dimensional subspaces of $\mathbf{V}$. In other words, lines are congruence blocks a/θ with $\mathbf{V}/\theta$ of dimension $n-1$ (and flats of dimension k are blocks a/θ with $\mathbf{V}/\theta$ of dimension $n-k$). If we extend this terminology from $\mathbf{V}$ to an arbitrary algebra $\mathbf{A}$, i.e., if we define all congruence blocks a/θ to be 'lines' or 'flats,' then some properties of this chapter have interesting geometric interpretations. For example congruence permutability simply asserts the existence of the fourth corner of a parallelogram (given $a\ \theta\ b\ \phi\ c$ there exists d such that $a\ \phi\ d\ \theta\ c$), as illustrated in Figure 6.1.

Congruence uniformity asserts that any two 'parallel' lines or flats have the same number of elements. (Here we are referring to a/θ and b/θ as 'parallel.') We refer the reader to the references above for a full treatment from this point of view; nevertheless various traces of these ideas will be seen in this chapter.

[3]This terminology, which was named after its connection with title of these volumes, was introduced in (Ralph Freese and Matthew Valeriote, 2009).

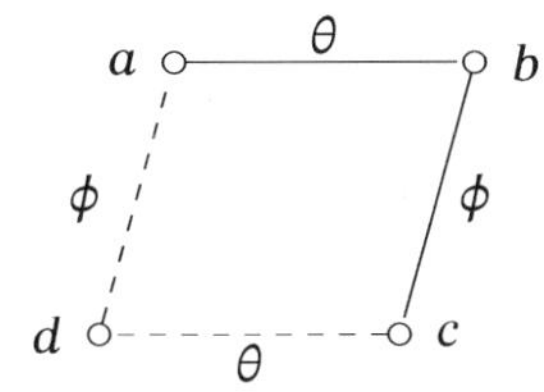

FIGURE 6.1. The Permutability Parallelogram

A lattice equation that holds identically in all the congruence lattices of the algebras in a variety $\mathcal{V}$ is called a **congruence identity** of $\mathcal{V}$. Theorem 6.1 gives a Mal'tsev condition for a variety to be congruence distributive. In §6.5 we give a Mal'tsev condition for a variety to be congruence modular and in §6.9 we take up the subject of congruence identities in general.

Theorem 4.143 shows that arithmeticity of a variety is strongly Mal'tsev definable. Recall that $\mathcal{V}$ is **arithmetical** if it is both congruence permutable and congruence distributive. In the presence of permutability, distributivity is equivalent to the following 'geometric' condition: if $\{B_1, \ldots, B_k\}$ is a finite set of congruence blocks such that no two have empty intersection, then $B_1 \cap \cdots \cap B_k$ is non-empty. (Compare with Helly's Theorem in geometry for convex sets.) This property is known as the **Chinese Remainder** property, since it generalizes the corresponding fact for the ring of integers, which is known as the Chinese Remainder Theorem (there the congruence blocks are simply arithmetic progressions).

The theory of varieties is essential for the results of this chapter: most of the results fail for algebras considered in isolation. For instance, any two element algebra obviously has permuting congruences, but there exist two element algebras with no Mal'tsev term p (such as we described at the beginning of this introduction). Nevertheless it is often the case that a stronger result than the result about varieties obtains. For example, by Theorem 6.119 if $\mathcal{V}$ is a regular variety then it is congruence modular. Although this result is not true for individual algebras, it is true that if every subalgebra of $\mathbf{A}^2$ is regular then $\mathbf{Con\ A}$ is modular; see Theorem 6.120. We loosely refer to these stronger results as *local* and call the results about varieties *global*. Another local result that will be proved in §6.9 is that if every subalgebra of $\mathbf{A}^2$ has a modular congruence lattice, $\mathbf{Con\ A}$ satisfies the arguesian equation.

Some Preliminaries

Terms and Signature. As we mentioned above, we will use the term *signature* in place of *similarity type* in this and the next volume. A **signature** is *a set I of operation symbols together with some method to determine the arity of each operation symbol.* Most commonly we have a function $\sigma : I \longrightarrow \omega$ which gives the arity of each operation symbol in I. We use σ to denote the signature in this case. (This makes sense since σ encodes the set I of operation symbols as its domain and, of course, gives the arity of each.) In less formal situations we will just mention the arity of the symbols; so, for example, we might say $\{h, f\}$ is a signature with h having arity 3 and f having arity 1. We still may refer to this signature as σ, where σ is the obvious function. Commonly used infix operation symbols such as $\cdot$, $+$, $\vee$ and $\wedge$ will naturally be assumed to have arity 2. So, for example, the signature of lattices is $\{\vee, \wedge\}$.

Definitions 4.113 and 4.114, Volume I, define the set $T_\sigma(X)$ of terms in signature σ and variables X, and the term algebra $\mathbf{T}_\sigma(X)$ of signature σ over a set X. A **term** of signature σ is a word or string on the alphabet $X \cup I$. $T_\sigma(X)$ is the smallest collection T of such words containing X with the property that if $p_0, \ldots, p_{n-1}$ are in T and $Q \in I$ is n-ary, then $Qp_0 \cdots p_{n-1}$ is in T. Note that if Q is a constant operation symbol, that is, it is 0-ary or nullary, then the word (consisting of the single token) Q is in T. As mentioned after Definition 4.113, we will often use the more common notation of terms: $Q(p_0, \ldots, p_{n-1})$ denoting $Qp_0 \cdots p_{n-1}$. (That we can identity the p_i's from the word $Qp_0 \cdots p_{n-1}$ follows from the unique readability of terms, Lemma 4.115.) Strictly speaking, if Q is nullary its common form should be $Q(\,)$, but we will almost always simply write Q.

T_σ and $\mathbf{T}_\sigma$ refer to $T_\sigma(X)$ and $\mathbf{T}_\sigma(X)$ where X is the canonical set of variables: $\{v_0, v_1, \ldots\}$.

Terms can be represented in various ways. For example they can be represented by a data structure called an (ordered) **term tree** or more generally by a **term digraph**. Digraph is short for directed graph, which we assume is acyclic. We illustrate these representations with an example. Let H and B be operation symbols of arities 3 and 2 respectively. Let $X = \{x, y, z\}$. The term tree for the term $HBxBxyzByBxy$ (in common notation $H(B(x, B(x, y)), z, B(y, B(x, y))))$ is given in Figure 6.2.

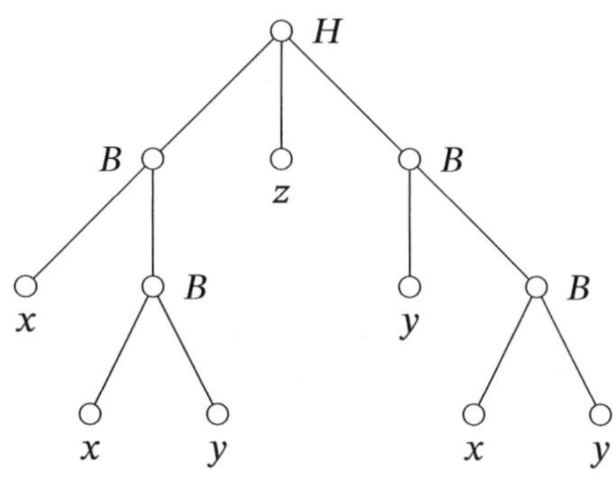

FIGURE 6.2. The term tree for $H(B(x, B(x, y)), z, B(y, B(x, y)))$

Notice the subterm $B(x, y)$ occurs twice. We can save space by employing a digraph; see Figure 6.3. This space savings can be important to algorithms and the savings can be exponential. For example, if we let $t_0 = x$, and $t_{n+1} = B(t_n, t_n)$, then the term tree of t_n is the complete binary tree of depth n. It has $2^{n+1} - 1$ nodes and $2^{n+1} - 2$ edges, while the digraph representation, given in Figure 6.4 for $n = 3$, has only $n + 1$ nodes and $2n$ edges.

A commonly encountered data structure to represent terms in computer investigations is very close to a digraph. It consists of a set of *nodes*. Each node has a *label* which is either an operation symbol or a variable. If the label of a node is an operation symbol f of arity n, then the node has an ordered list of (pointers to) the n children, each of which is a node. The same node can appear multiple times in a list of children and can be in more than one list. There is a special node called the *root*. Also, as with digraphs,

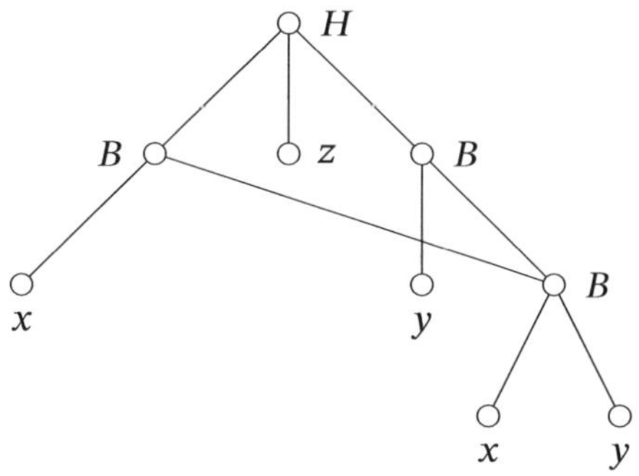

FIGURE 6.3. A digraph representation for $H(B(x, B(x, y)), z, B(y, B(x, y)))$

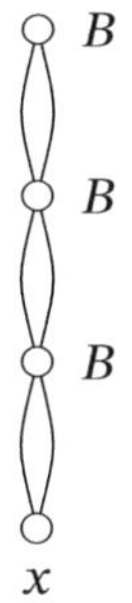

FIGURE 6.4. A digraph representation for t_3

we do not allow circularity in these structures. That is there cannot be a sequence of nodes $n_0, n_1, \ldots, n_k = n_0$ with n_{i+1} in the list of children of n_i.

Systems of definitions, reducts, interpretations

Interpretations. We begin by reviewing Definition 4.121 on page 232 of Volume I. For $\mathbf{A}$ an algebra of signature τ, p a τ-term (i.e., an element of T_τ), and $(x_0, x_1, \ldots, x_{n-1})$ an ordered list of distinct variables which contains every variable occurring in p, we define an n-ary operation on $\mathbf{A}$, $p^{\mathbf{A}}$—the **term operation of** p **on A**, or the **realization of** p **on A**—by recursion on the length of p, as follows. If p is a variable v_k, then $p^{\mathbf{A}}(a_0, \ldots, a_{n-1}) = a_k$. If p is formed as $p = Q(p_0, \ldots, p_{m-1})$, where Q is an m-ary operation symbol of τ, then

$$(6.1.2) \qquad p^{\mathbf{A}}(a_0, \ldots, a_{n-1}) = Q^{\mathbf{A}}(p_0^{\mathbf{A}}(a_0, \ldots .a_{n-1}), \ldots, p_{m-1}^{\mathbf{A}}(a_0, \ldots .a_{n-1})).$$

Thus each term p leads to multiple term operations on A. Which particular term operation is meant by $p^{\mathbf{A}}$ is usually clear from the context but if need be, we will write $p^{\mathbf{A}}(x_0, \ldots, x_{n-1})$.

As remarked in Volume I (*loc. cit.*), the operations $p^{\mathbf{A}}$ are the **term operations of A**. In situations where it is known that $a_0, \ldots, a_{n-1}$ are elements of an algebra $\mathbf{A}$, we often use a convention from Volume I and write $p(a_0, \ldots, a_{n-1})$ for $p^{\mathbf{A}}(a_0, \ldots, a_{n-1})$.

Observe that if p is an n-ary term and $n \leq n'$ for a natural number n', then p is also an n'-ary term. So we see that $p^{\mathbf{A}}$ should be regarded as a schema of infinitely many term operations. Which particular term operation is meant by $p^{\mathbf{A}}$ depends on the context.

In much of this chapter we will be concerned with the interpretation, or at least the interpretability, of one variety in another. Over the next three pages we provide an elementary review of the definition of interpretation that we have in Volume I, and various ancillary notions.

DEFINITION 6.2. We consider signatures $\sigma : I \longrightarrow \omega$ and $\tau : J \longrightarrow \omega$. A **system of definitions** of σ in τ is a map D from the operation symbols of σ into T_τ (the set of τ-terms) satisfying

(1) If Q is an operation symbol of σ with arity $r > 0$, then $D(Q)$ is a term of signature τ all of whose variables belong to $\{v_0, \ldots, v_{r-1}\}$. The term $D(Q)$ will also be denoted Q_D.
(2) If Q is a nullary operation symbol of σ, then $D(Q) = Q_D$ is a τ-term so that the only variables occurring in Q_D belong to the set $\{v_0\}$.

Observe that in (2) it is allowed that no variable occurs in the term Q_D.

In Exercise 16 on page 141 of Volume III, we will discuss the possibility of including nullary operations in the clone. We will say that such a system of definitions of σ in τ is **non-nullary** if $\sigma(Q) \neq 0$ for all $Q \in I$.

Now suppose that $\mathbf{A} = \langle A, F\rangle_{F\in J}$ is an algebra of signature $\tau : J \longrightarrow \omega$, and that $D : I \longrightarrow T_\tau$ is a non-nullary system of definitions of a signature σ in τ. We define an algebra $\mathbf{A}^D$ of signature σ, as follows. The universe of $\mathbf{A}^D$ is A. For $Q \in I$, the $\mathbf{A}^D$-operation corresponding to Q is the $\sigma(Q)$-ary term-operation $(Q_D)^{\mathbf{A}}$ (as defined just above). In symbols,

$$\mathbf{A}^D = \langle A, (Q_D)^{\mathbf{A}}\rangle_{Q\in I}. \tag{6.1.3}$$

A classical example—see Volume 1, p 245—has $\mathbf{A} = \langle A, \wedge, \vee, \neg\rangle$, $I = \{+, -\}$, and $D(+) = D(-) = (v_0 \wedge \neg v_1) \vee (v_1 \wedge \neg v_0)$ (the "symmetric difference"). The reader may verify that if $\mathbf{A}$ is a Boolean algebra, then $\mathbf{B} = \mathbf{A}^D$ turns out to be an Abelian group of exponent 2.

The construction of $\mathbf{A}^D$ is particularly simple when I is a subset of J and D is the function that maps each $Q \in I$ to the τ-term $Qv_0v_1\cdots$. In this case $\mathbf{A}^D$ is effectively the same as the algebra $\langle A, Q^{\mathbf{A}}\rangle_{Q\in I}$—which has the same universe as the given $\mathbf{A}$—but only a subset of its operations. This revised algebra is known as a **reduct** of $\mathbf{A}$ (Volume I, page 13); it may also be called **the *I*-reduct of A**—or other phrases when appropriate, such as, "Let $\mathbf{A}$ be a ring, and let $\mathbf{B}$ be its group reduct."

Generally speaking, the operations $(Q_D)^{\mathbf{A}}$ appearing in (6.1.3) are not fundamental operations of $\mathbf{A}$, so we will not call $\mathbf{A}^D$ a reduct of $\mathbf{A}$. But they are term operations; informally, the algebra $\mathbf{A}^D$ can be formed by first adjoining all term operations and then taking an appropriate reduct. Thus we sometimes call $\mathbf{A}^D$ a **term reduct of A**. To be more specific, it may be called **the *D*-reduct of A**. The algebra $\mathbf{A}^D$ may also be called the ***D*-interpretation of A (in signature** σ). One may define a **polynomial reduct** of $\mathbf{A}$ analogously. A frequent special case of term reduct is when we first adjoin all term operations of $\mathbf{A}$, and then retain only the term operations $Q^{\mathbf{A}}$ for which $\mathbf{A}$ satisfies $Q(x, x, \ldots) \approx x$. This is known as the **idempotent reduct of A**.

That concludes our discussion of interpretations for non-nullary signatures. If there is a nullary operation symbol $Q \in I$ with Q_D a unary term $t(v_0)$, then we must also be sure to have $t(v_0) \approx t(v_1)$ holding in the original algebra $\mathbf{A}$. In all cases where this identity needs to hold, we will make sure of the identity by including it among the τ-identities that we are imposing on $\mathbf{A}$. (See for instance condition (3) of Definition 6.3 just below. In such a case, of course, the nullary operation $(Q_D)^{\mathbf{A}}$ must be the constant value of the unary operation $t^{\mathbf{A}}$.) Thus for each system D arising in these volumes, $\mathbf{A}^D$ will be well defined.

Our principal application of systems of definitions, and D-reducts, especially in Chapter 6 (and also in Volume III), will be to facilitate a discussion of interpretation of one variety in another. Definition 6.3 recapitulates some of the definitions from Volume I, §4.12, that will be useful for this chapter.

DEFINITION 6.3. Let $\mathcal{V}$ and $\mathcal{W}$ be classes of algebras of signature σ and τ, respectively. An **interpretation of $\mathcal{V}$ in $\mathcal{W}$** is a system of definitions D of σ in τ that also satisfies:

(3) If Q is a nullary operation symbol of σ, and if $D(Q)$ is a unary τ-term $t(v_0)$, then $\mathcal{W} \models t(v_0) \approx t(v_1)$. (Hence $\mathbf{A}^D$ is well defined in the next condition.)
(4) If $\mathbf{A}$ is an algebra in $\mathcal{W}$, then $\mathbf{A}^D$ lies in $\mathcal{V}$.

By a **term equivalence**—we often say "equivalence"—of classes $\mathcal{V}$ and $\mathcal{W}$ of algebras, we mean a pair of interpretations, D of $\mathcal{V}$ in $\mathcal{W}$, and E of $\mathcal{W}$ in $\mathcal{V}$, such that $\mathbf{A}^{DE} = \mathbf{A}$ for all $\mathbf{A} \in \mathcal{W}$, and $\mathbf{B}^{ED} = \mathbf{B}$ for all $\mathbf{B} \in \mathcal{V}$. We say that $\mathcal{V}$ is **equivalent—or term equivalent—to $\mathcal{W}$** iff there exists an equivalence between them, and that $\mathcal{V}$ is **interpretable in $\mathcal{W}$** iff there exists an interpretation of $\mathcal{V}$ in $\mathcal{W}$. The relation of equivalence between varieties is written $\mathcal{V} \equiv \mathcal{W}$. Mostly in these volumes, $\mathcal{V}$ and $\mathcal{W}$ will be varieties.

When there is an interpretation of the variety $\mathcal{V}$ in the variety $\mathcal{W}$, we write $\mathcal{V} \leq_{\text{int}} \mathcal{W}$. This is a quasiorder[4] on the class of all varieties, whose induced order is a lattice. A **Mal'tsev condition**, in fact defines a filter of this lattice. We also refer to such a filter as a **Mal'tsev class** or a **Mal'tsev filter**. The lattice will be studied in detail in Volume III, Chapter 10. In terms of interpretability, a variety $\mathcal{W}$ **realizes** a set Σ of equations if there is an interpretation of the variety $\mathcal{V}_\Sigma = \mathbf{Mod}(\Sigma)$ defined by Σ in $\mathcal{W}$.

An interpretation D of $\mathcal{V}$ in $\mathcal{W}$ offers us the means to translate terms of signature σ into terms of signature τ. We define the **translation** $\mathrm{Tr}_D : T_\sigma \to T_\tau$ **between the sets of terms** by the following recursion:

(a) When t is variable, let $\mathrm{Tr}_D(t) = t$.
(b) When $t = Qt_0t_1\ldots$, let $\mathrm{Tr}_D(t) = Q_D(\mathrm{Tr}_D(t_0), \mathrm{Tr}_D(t_1), \ldots)$, for all $Q \in I$.

We apply Tr_D to equations by taking $\mathrm{Tr}_D(s \approx t)$ to be $\mathrm{Tr}_D(s) \approx \mathrm{Tr}_D(t)$. We apply Tr_D to sets Σ of equations by putting

$$\mathrm{Tr}_D[\Sigma] \;=\; \{\mathrm{Tr}_D(s \approx t) \mid s \approx t \in \Sigma\}.$$

The reader may easily observe that for any system D of definitions, the translation Tr_D is not very far from D itself. In the language of clone theory, D is the restriction of

[4]A **quasiorder** on a set S is a reflexive, transitive relation, $\leq$, on S. $x \equiv y$, defined by $x \leq y$ and $y \leq x$, is an equivalence relation on S. The relation $\leq$ induced on the equivalence classes (or blocks) is an order.

Tr_D to terms of depth 1, while Tr_D *is the unique clone morphism that extends D to the entire clone* T_σ. The clone-theoretic aspects of interpretation and interpretability will be explored at length in Volume III. In this Chapter we will give an elementary (correct but informal) argument that Tr_D is a clone morphism. In the first two pages of Chapter 4 (Volume I) it is made clear that the basic relation of clone theory is "$q = p(u_0, u_1, \dots)$," where q, p and the u_i represent (real or symbolic) functions. Now Exercise 4 tells us that this relationship is preserved upon going from q, p and all the u_i to their respective images under Tr_D. Hence Tr_D is a homomorphism of the clone-theoretic structure.

LEMMA 6.4. *Let $\mathcal{V}$ and $\mathcal{W}$ be classes of algebras, with D an interpretation of $\mathcal{V}$ in $\mathcal{W}$, and let* $\mathbf{A}$ *be any algebra in $\mathcal{W}$. For any term p in the signature of $\mathcal{V}$, the realization of p on* $\mathbf{A}^D$ *is the same as the realization of* $\mathrm{Tr}_D(p)$ *on* $\mathbf{A}$.

Proof. The proof is by induction on the complexity of p. Certainly these two realizations are equal if p is a single variable, so we move on to the case where p has the form $Qp_0p_1\cdots$.

Let $a_0, a_1, \dots$ lie in the universe of $\mathbf{A}$ (which is of course also the universe of the interpreted algebra $\mathbf{A}^D$). We begin with the definition of $\mathrm{Tr}_D(p)$:

$$(\mathrm{Tr}_D(p))(v_0, v_1, \dots) \;=\; (Q_D)(\mathrm{Tr}_D(p_0)(v_0, v_1, \dots), (\mathrm{Tr}_D(p_1)(v_0, v_1, \dots), \dots).$$

Therefore we have

$$\begin{aligned}(\mathrm{Tr}_D(p))^{\mathbf{A}}(a_0, a_1, \dots) &= (Q_D)^{\mathbf{A}}((\mathrm{Tr}_D p_0)^{\mathbf{A}}(a_0, a_1, \dots), (\mathrm{Tr}_D p_1)^{\mathbf{A}}(a_0, a_1, \dots), \dots) \\ &= Q^{(\mathbf{A}^D)}(p_0^{(\mathbf{A}^D)}(a_0, a_1, \dots), p_1^{(\mathbf{A}^D)}(a_0, a_1, \dots), \dots) \\ &= p^{(\mathbf{A}^D)}(a_0, a_1, \dots),\end{aligned}$$

where the first equation comes from (6.1.2), the second holds by induction, and the third comes from (6.1.2): This completes our proof that $(\mathrm{Tr}_D(p))^{\mathbf{A}} = p^{(\mathbf{A}^D)}$. ■

As a corollary of this lemma, for any equation $s \approx t$ of signature σ we have

$$\mathbf{A}^D \models s \approx t \text{ if and only if } \mathbf{A} \models \mathrm{Tr}_D(s \approx t). \tag{6.1.4}$$

From (6.1.4) one easily proves that if $\Sigma \models s \approx t$ in signature σ, then $\mathrm{Tr}_D(\Sigma) \models \mathrm{Tr}_D(s \approx t)$ in signature τ. In general, however, the converse fails without some further restriction on the interpretation D. One possible restriction will be seen in Theorem 7.42 on page 227.

We conclude this short introduction to interpretations with a fundamental, but easy, theorem on how equational axiom systems can be transported from one context to another using the translation function Tr_D. These results could be presented in the more elaborate context of Chapter 7 (making use of the deducibility relation $\vdash$, and the notion of an *equational theory*). But they also have a clear presentation in the simpler context (familiar from Volume I) of a variety $\mathcal{V}$ that is defined by a set Σ of equations: $\mathcal{V} = \mathbf{Mod}\,\Sigma$. When this holds, we may also say that Σ is an **axiom system,** or **equational base, for** $\mathcal{V}$.

Let us be given classes of algebras $\mathcal{V}$ and $\mathcal{W}$, of signatures $\sigma : I \longrightarrow \omega$ and $\tau : J \longrightarrow \omega$, and a pair of interpretations, $D : I \longrightarrow T_\tau$ and $E : J \longrightarrow T_\sigma$. By the **yoking equations of** D **and** E, we mean the set

$$\Gamma(D,E) = \{\mathrm{Tr}_E(\mathrm{Tr}_D(Qx_0x_1\cdots)) \approx Qx_0x_1x_2\cdots \ : Q \in J\}.$$

of equations in signature τ. We remark that if σ and τ are finite, then $\Gamma(D,E)$ is finite. Moreover, if finite, the set $\Gamma(D,E)$ is readily computable, using the recursion defined in conditions (a) and (b) on page 9.

LEMMA 6.5. *Let us be given classes $\mathcal{V}$ and $\mathcal{W}$ of algebras, of signatures $\sigma : I \longrightarrow \omega$ and $\tau : J \longrightarrow \omega$, and a pair of interpretations, $D : I \longrightarrow T_\tau$ and $E : J \longrightarrow T_\sigma$. An algebra* $\mathbf{A}$ *of signature τ satisfies the set $\Gamma(D,E)$ of equations if and only if $\mathbf{A}^{DE} = \mathbf{A}$.* ■

THEOREM 6.6. *Let $\mathcal{V}$ and $\mathcal{W}$ be classes of algebras, of any two signatures. Suppose that a pair of interpretations, D of $\mathcal{V}$ in $\mathcal{W}$, and E of $\mathcal{W}$ in $\mathcal{V}$, constitute a term equivalence between $\mathcal{V}$ and $\mathcal{W}$. If $\mathcal{V}$ is a variety with equational axiom set Σ, then $\mathcal{W}$ is a variety and*

$$\Lambda = \mathrm{Tr}_D[\Sigma] \cup \Gamma(D,E)$$

is an axiom system for $\mathcal{W}$.

Proof. We need to show that $\mathbf{Mod}\,\Lambda = \mathcal{W}$. We will consider an arbitrary τ-algebra $\mathbf{A}$, and prove that $\mathbf{A} \in \mathcal{W}$ if and only if $\mathbf{A} \models \Lambda$. First we consider $\mathbf{A} \in \mathcal{W}$. Since D and E together form an equivalence of varieties, we have $\mathbf{A}^{DE} = \mathbf{A}$, and by Lemma 6.5, $\mathbf{A} \models \Gamma(D,E)$, which is part of Λ. It remains to show that $\mathbf{A} \models \mathrm{Tr}_D[\Sigma]$, which is the other part of Λ. Since $\mathbf{A} \in \mathcal{W}$, we have $\mathbf{A}^D \in \mathcal{V}$. Thus $\mathbf{A}^D \models e$ for every equation $e \in \Sigma$. By (6.1.4), $\mathbf{A} \models \mathrm{Tr}_D(e)$ for every $e \in \Sigma$. By definition $\mathbf{A} \models \mathrm{Tr}_D[\Sigma]$.

We next consider an algebra $\mathbf{A}$ that models Λ, and prove that $\mathbf{A} \in \mathcal{W}$. In particular, $\mathbf{A} \models \Gamma(D,E)$, and so by Lemma 6.5 we have $\mathbf{A}^{DE} = \mathbf{A}$. And since $\mathbf{A} \models \mathrm{Tr}_D[\Sigma]$, we have that $\mathbf{A} \models \mathrm{Tr}_D(e)$ for each $e \in \Sigma$. By (6.1.4), $\mathbf{A}^D \models e$ for each $e \in \Sigma$, and so $\mathbf{A}^D \in \mathcal{V}$ (since Σ axiomatizes $\mathcal{V}$). Since E is an interpretation, we have $\mathbf{A}^{DE} \in \mathcal{W}$. So now $\mathbf{A} \in \mathcal{W}$. ■

COROLLARY 6.7. *Let $\mathcal{V}$ and $\mathcal{W}$ be varieties, each of finite signature. If $\mathcal{V}$ and $\mathcal{W}$ are term equivalent varieties, then $\mathcal{V}$ has a finite equational base iff $\mathcal{W}$ has a finite equational base.* ■

Note on equivalence relations

Also recall that $\mathbf{Eqv}\,A$ is the lattice of equivalence relations on the set A.

LEMMA 6.8. *If* $\mathbf{A}$ *is an algebra then* **Con A** *is a complete sublattice of* $\mathbf{Eqv}\,A$*, so the join of two congruences is their equivalence relation join. If* $\mathbf{A}$ *is a reduct of* $\mathbf{B}$*, then* **Con B** *is a sublattice of* **Con A**.

Proof. The first statement was observed in Volume I in §4.3, page 153. Since $\mathbf{A}$ is a reduct of $\mathbf{B}$, every congruence of $\mathbf{B}$ is a congruence of $\mathbf{A}$; that is, $\mathrm{Con}(\mathbf{B}) \subseteq \mathrm{Con}(\mathbf{A})$. That it is in fact a sublattice follows from the first statement. ■

In Theorem 6.1 we used the fact that $\mathrm{Cg}(x, y)$ of $\mathbf{F}_{\mathcal{V}}(x, y, z)$ restricted to the subalgebra generated by $\{x, z\}$, which is denoted $\mathrm{Sg}(\{x, z\})$, is trivial. Here we prove this.

THEOREM 6.9. *Let $\mathcal{V}$ be a variety and let $\mathbf{F} = \mathbf{F}_{\mathcal{V}}(X)$. Let Y be a subset of X and $\mathbf{S} = \mathrm{Sg}_{\mathbf{F}}(Y)$. Let π be a partition of X and let θ be the congruence on $\mathbf{F}$ generated by π. If each block of π contains at most one element of Y, then θ restricted to $\mathbf{S}$ is trivial.*

Proof. We may assume each block of π has exactly one element from Y (just add elements to Y to make this true; the conclusion with this enlarged Y implies it for the original Y.) Let $h : \mathbf{F}_{\mathcal{V}}(X) \twoheadrightarrow \mathbf{F}_{\mathcal{V}}(Y)$ be the retraction homomorphism extending the map from X to Y which sends $x \in X$ to the unique $y \in Y$ in the same block as x. Since h is the identity on Y, it is the identity on $\mathbf{S}$.

We claim $\theta = \ker h$. Clearly $\theta \subseteq \ker h$, so $\mathbf{F}_{\mathcal{V}}(X)/\theta$ maps homomorphically onto $\mathbf{F}_{\mathcal{V}}(Y)$. But the map that sends $y \in Y$ to the block of θ which contains y maps $\mathbf{F}_{\mathcal{V}}(Y)$ onto $\mathbf{F}_{\mathcal{V}}(X)/\theta$, and the claim follows easily from this.

Suppose a and $b \in \mathbf{S}$ and $\langle a, b\rangle \in \theta$. Then $h(a) = h(b)$. But h is the identity on $\mathbf{S}$. So $a = h(a) = h(b) = b$. ■

Exercises 6.10

1. Give the details of the proof that (vi) ⇔ (viii) in Theorem 6.1.

2. Prove that (vi) in Theorem 6.1 implies the distributive law,

$$\gamma \wedge (\alpha \vee \beta) = (\gamma \wedge \alpha) \vee (\gamma \wedge \beta) =: \theta.$$

To see this suppose $\langle a, c\rangle \in \gamma \wedge (\alpha \vee \beta)$. Then $\langle a, c\rangle \in \gamma$ and, for some k, there exists $a = b_0, \dots, b_k = c$ with $\langle b_i, b_{i+1}\rangle \in \alpha \cup \beta$ for $i < k$. Using (Δ_n) show $d_j(a, b_i, c)\ \theta\ d_j(a, b_{i+1}, c)$ for $j \leq n$. Then use the transitivity of θ to show $d_j(a, a, c)\ \theta\ d_j(a, c, c)$, and from this and (Δ_n) derive that

$$a = d_0(a, c, c)\ \theta\ d_n(a, c, c) = c.$$

3. The sets of equations Δ_n are defined with Theorem 6.1. Recall the *alvin variant* is given by the equations Δ_n' obtained from Δ_n by interchanging *even* and *odd*. Let $\mathcal{V}$ be a variety. Prove each of the following.
   (i) If $\mathcal{V}$ realizes Δ_n it realizes Δ_{n+1}'.
   (ii) If $\mathcal{V}$ realizes Δ_n' it realizes Δ_{n+1}.
   (iii) If n is odd, then $\mathcal{V}$ realizes Δ_n if and only if it realizes Δ_n'.
   (iv) $\mathcal{V}$ realizes Δ_2', if and only if it has a **Pixley term**; that is, one satisfying $p(x, z, z) \approx x$, $p(x, x, z) \approx z$ and $p(x, y, x) \approx x$.
   (v) $\mathcal{V}$ realizes Δ_2, if and only if it has a **majority term**; that is, one satisfying $m(x, x, z) \approx x$, $m(x, z, z) \approx z$ and $m(x, y, x) \approx x$).
   (vi) Show that if $\mathcal{V}$ has a Pixley term, it has a majority term.
   (vii) Show the variety of lattices has a majority term, but not a Pixley term.

   Note these results imply that if $\mathcal{V}$ realizes Δ_2', it realizes Δ_2. One might conjecture this extends to all even n, but it does not; see (Ralph Freese and Matthew Valeriote, 2009). Other connections of the parameters associated with various Mal'tsev conditions are explored in (Paolo Lipparini, 2020, 2021).

4. Let $p, u_0, u_1, \ldots$ be terms of signature $\sigma\colon I \longrightarrow \omega$, and let $D\colon I \to T_\tau$ be a system interpreting σ in τ. Let $q = p(u_0, u_1, \ldots)$ be the result of substituting u_i for v_i in p for all natural numbers i. Prove that

$$\mathrm{Tr}_D q = (\mathrm{Tr}_D p)(\mathrm{Tr}_D u_0, \ldots, \mathrm{Tr}_D u_{n-1}).$$

For this exercise one could use one's informal understanding of substitution, or look ahead to §7.6, where we have the tools for a formal proof by induction. Compare with Exercise 7.12.5 on page 166.

5. Prove that Tr_D is the unique clone morphism $T_\sigma \longrightarrow T_\tau$ extending $D\colon I \longrightarrow T_\tau$. (Use Exercise 4 and the argument outlined above.) This was remarked without formal proof in §4.12.

6.2. Permutability of Congruences

First we recall some definitions and basic results from §4.7. If α and β are binary relations on a set A, and $\alpha \circ \beta$ denotes their relational product (as defined in the preliminaries of Volume I), then we recursively define the relation $\alpha \circ^k \beta$ as follows: $\alpha \circ^1 \beta = \alpha$, $\alpha \circ^{k+1} \beta = \alpha \circ (\beta \circ^k \alpha)$. So $\alpha \circ^2 \beta = \alpha \circ \beta$, $\alpha \circ^3 \beta = \alpha \circ \beta \circ \alpha$, etc. We say that α and β k-**permute** if $\alpha \circ^k \beta = \beta \circ^k \alpha$. An algebra $\mathbf{A}$ is said to have k-**permutable congruences** if every pair of congruences of $\mathbf{A}$ k-permute. If $\mathbf{A}$ has k-permutable congruences then, by Lemma 4.66, $\alpha \vee \beta = \alpha \circ^k \beta$ for every pair of congruences of $\mathbf{A}$. In this situation, the terminology $\mathbf{A}$ has **type** $k-1$ joins is used since there are $k-1$ $\circ$'s in the expression $\alpha \vee \beta = \alpha \circ^k \beta$. Thus type 1 is the same as 2-permutability. In this case, we say that the algebra has permutable congruences. We saw in Theorem 4.67 that if an algebra $\mathbf{A}$ has 3-permuting congruences then **Con A** is modular. Later in this chapter we will construct a variety all of whose algebras have 4-permuting congruences, which is not congruence modular. Of course a variety is said to be k-permutable if each of its algebras is.

Permutability of congruences had a tremendous influence on the early development of universal algebra and its relation to lattice theory. The fact that permutability implies modularity goes back to Richard Dedekind (1900). He derives the modularity of the lattice of normal subgroups of a group from the fact that any two normal subgroups permute with each other, i.e., $\mathbf{AB} = \mathbf{BA}$ holds for normal subgroups. Essentially the same proof shows that any permutable variety is modular. The fact that a lattice of permuting equivalence relations (and hence of permuting congruence relations) is modular is explicit in (O. Ore, 1942). Birkhoff's application of Ore's Theorem, Corollary 2.48, which yielded the results on unique factorization of Chapter 5, is a prime example of the importance of permutability. Moreover various generalizations of the Jordan-Hölder theorem to general algebra relied on congruence permutability, since they required congruence modularity.

Our first new Mal'tsev condition of this chapter is for k-permutability of congruences, and is due to J. Hagemann and A. Mitschke (1973). The equivalence of (i), (ii) and (ii′) had been established earlier by E. T. Schmidt (1969). Other conditions equivalent to k-permutability are given in Exercise 12.

Deciding if $\mathcal{V}$ is k-permutable using condition (iv) has computational advantages over (iii). The latter involves a search in $\mathbf{F}_{\mathcal{V}}(x, y, z)$ while the former involves a search in a 3-generated subalgebra of $\mathbf{F}^2_{\mathcal{V}}(x, y)$. This is explored in Exercise 13.

THEOREM 6.11. *For a variety $\mathcal{V}$ and an integer $k \geq 2$ the following conditions are equivalent:*

(i) *$\mathcal{V}$ has k-permutable congruences.*
(ii) *$\mathbf{F}_{\mathcal{V}}(k+1)$ has k-permutable congruences.*
(ii′) *There exist terms $r_i(x_0, x_1, \dots, x_k)$, $i = 0, \dots, k$, for $\mathcal{V}$ such that the following are identities of $\mathcal{V}$: $x_0 \approx r_0(x_0, x_1, \dots, x_k)$, $r_k(x_0, x_1, \dots, x_k) \approx x_k$, and*

$$\begin{aligned} r_{i-1}(x_0, x_0, x_2, x_2, \dots) &\approx r_i(x_0, x_0, x_2, x_2, \dots) && \text{for } i \text{ even},\\ r_{i-1}(x_0, x_1, x_1, x_3, x_3, \dots) &\approx r_i(x_0, x_1, x_1, x_3, x_3, \dots) && \text{for } i \text{ odd}. \end{aligned}$$

(iii) *There exist terms $p_1, \dots, p_{k-1}$ for $\mathcal{V}$ such that the following are identities of $\mathcal{V}$:*

$$\begin{aligned} x &\approx p_1(x, z, z)\\ p_1(x, x, z) &\approx p_2(x, z, z)\\ &\ \vdots\\ p_{k-2}(x, x, z) &\approx p_{k-1}(x, z, z)\\ p_{k-1}(x, x, z) &\approx z. \end{aligned} \tag{6.2.1}$$

(iv) *The subalgebra of $\mathbf{F}^2_{\mathcal{V}}(x, y)$ generated by $\langle x, x\rangle$, $\langle x, y\rangle$ and $\langle y, y\rangle$ contains elements $\langle a_i, b_i\rangle$, $i = 0, \dots, k$, with $\langle y, y\rangle = \langle a_0, b_0\rangle$, $\langle x, x\rangle = \langle a_k, b_k\rangle$, and $b_i = a_{i+1}$.*

Proof. Clearly (i) implies (ii). Assume (iii) and let $\mathbf{A} \in \mathcal{V}$ and $\theta, \phi \in \operatorname{Con} \mathbf{A}$, and $a_0, \dots, a_k \in A$ satisfy

$$a_0 \ \theta \ a_1 \ \phi \ a_2 \ \theta \ a_3 \ \cdots \ a_k.$$

Let $b_i = p_i(a_{i-1}, a_i, a_{i+1})$, for $1 \leq i < k$. Then

$$\begin{aligned} a_0 &= p_1(a_0, a_1, a_1) \ \phi \ p_1(a_0, a_1, a_2) = b_1\\ b_1 &= p_1(a_0, a_1, a_2) \ \theta \ p_1(a_1, a_1, a_2) = p_2(a_1, a_2, a_2) \ \theta \ p_2(a_1, a_2, a_3) = b_2. \end{aligned}$$

Continuing in this way, we obtain

$$a_0 \ \phi \ b_1 \ \theta \ b_2 \ \cdots \ a_k,$$

showing that (i) holds.

Now suppose (ii) holds and let λ_0 and λ_1 be the endomorphisms of $\mathbf{F}_{\mathcal{V}}(x_0, \dots, x_k)$ such that

$$\begin{aligned} \lambda_0(x_i) &= x_{i-1} && \text{if } i \text{ is odd, and } x_i \text{ otherwise},\\ \lambda_1(x_i) &= x_{i-1} && \text{if } i > 0 \text{ and } i \text{ is even, and } x_i \text{ otherwise}. \end{aligned}$$

Let θ_0 and θ_1 be the kernels of λ_0 and λ_1. Then

$$x_0 \ \theta_0 \ x_1 \ \theta_1 \ x_2 \cdots x_{2i} \ \theta_0 \ x_{2i+1} \ \theta_1 \ x_{2i+2} \ \cdots \ x_k.$$

Since, by assumption, $\mathbf{F}_{\mathcal{V}}(k+1)$ has k-permutable congruences there are elements $q_1, \ldots, q_{k-1}$ in $\mathbf{F}_{\mathcal{V}}(x_0, \ldots, x_k)$ such that

$$x_0 \; \theta_1 \; q_1 \; \theta_0 \; q_2 \; \cdots \; q_{2i} \; \theta_1 \; q_{2i+1} \; \theta_0 \; q_{2i+2} \; \cdots \; x_k.$$

Let r_i be terms such that $q_i = r_i(x_0, \ldots, x_k)$. It is straightforward to use Theorem 6.9 to derive (ii′), but we give an argument using λ_0 and λ_1 instead. Since θ_j is the kernel of λ_j, we see

$$\begin{aligned} x_0 = \lambda_1(x_0) = \lambda_1(q_1) &= \lambda_1(r_1(x_0, \ldots, x_k)) \\ &= r_1(\lambda_1(x_0), \ldots, \lambda_1(x_k)) = r_1(x_0, x_1, x_1, x_3, \cdots) \end{aligned}$$

and, for $i < k/2$:

$$\begin{aligned} r_{2i}(x_0, x_1, x_1, x_3, \ldots) &= r_{2i}(\lambda_1(x_0), \lambda_1(x_1), \lambda_1(x_2), \lambda_1(x_3), \ldots) \\ &= \lambda_1(r_{2i}(x_0, x_1, x_2, x_3, \ldots)) \\ &= \lambda_1(r_{2i+1}(x_0, x_1, x_2, x_3, \ldots)) \\ &= r_{2i+1}(\lambda_1(x_0), \lambda_1(x_1), \lambda_1(x_2), \lambda_1(x_3), \ldots) \\ &= r_{2i+1}(x_0, x_1, x_1, x_3, \ldots). \end{aligned}$$

Similarly,

$$r_{2i+1}(x_0, x_0, x_2, x_2, \ldots) = r_{2i+2}(x_0, x_0, x_2, x_2, \ldots),$$

proving (ii′).

Now for $1 \leq i < k$, define

$$p_i(x_0, x_1, x_2) = r_i(\overbrace{x_0, \ldots, x_0}^{i \text{ times}}, x_1, \overbrace{x_2, \ldots, x_2}^{k-i \text{ times}}).$$

It easy to verify that $\mathcal{V}$ satisfies the equations (6.2.1), showing (ii′) implies (iii).

Thus we have shown that (i), (ii), (ii′) and (iii) are pairwise equivalent. They are equivalent to (iv) by a standard argument: if (6.2.1) holds, let $a_i = p_{k-i}(x, x, y)$ and $b_i = p_{k-i}(x, y, y)$. Then $\langle a_i, b_i \rangle = p_{k-i}(\langle x, x\rangle, \langle x, y\rangle, \langle y, y\rangle)$ and the properties of (iv) hold. Conversely if (iv) holds the terms giving the $\langle a_i, b_i\rangle$'s, indexed backwards, satisfy (6.2.1). ∎

There are natural examples of varieties which are 3-permutable but not permutable. In the exercises we present some of these examples. The next example presents a variety constructed by S. V. Polin (1977) (see also Alan Day and Ralph Freese (1980)) which is 4-permutable but not 3-permutable. In §6.9 we will see that this variety is not congruence modular but nevertheless its congruence lattices satisfy a nontrivial lattice equation.

EXAMPLE 6.12. For $i = 0,1$, let $\mathbf{A}_i = \langle \{0,1\}, \wedge, ', ^+ \rangle$ where $\wedge$ is the usual meet operation and $'$ and $^+$ are the unary operations given in the following tables.

$\mathbf{A}_0$	0	1
$'$	1	0
$+$	1	1

$\mathbf{A}_1$	0	1
$'$	0	1
$+$	1	0

Let $\mathbf{A} = \mathbf{A}_0 \times \mathbf{A}_1$ and let $\mathcal{P}$ denote the variety generated by $\mathbf{A}$. It is straightforward to verify that $\mathbf{Con\ A}$ is the lattice of Figure 6.5, where we have used juxtaposition to denote the elements of $\mathbf{A} = \mathbf{A}_0 \times \mathbf{A}_1$; see Exercise 19. The labels below are congruences as partitions with one-element blocks omitted.

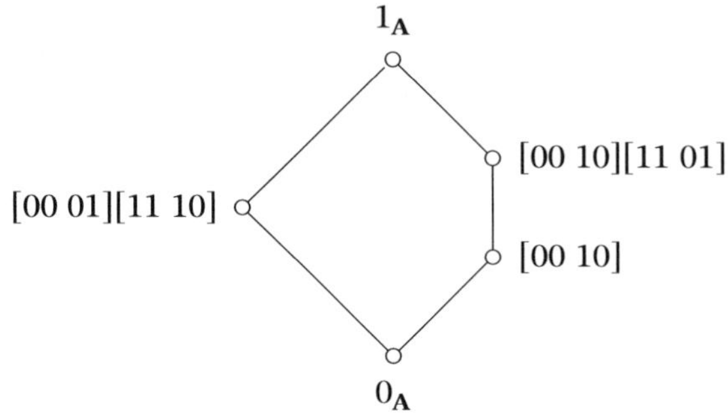

FIGURE 6.5. The Congruence Lattice of Polin's Algebra

Thus $\mathcal{P}$ is not congruence modular and hence, by Theorem 4.67, not 3-permutable. To see that it is 4-permutable, we define three terms:

$$p_1(x, y, z) = x \wedge (y \wedge z^+)^+$$
$$p_2(x, y, z) = [(x \wedge y')' \wedge (z \wedge y')' \wedge (x \wedge z)']'$$
$$p_3(x, y, z) = z \wedge (y \wedge x^+)^+.$$

We need to verify that the equations (6.2.1) of Theorem 6.11 for $k = 3$ are identities of $\mathcal{P}$. Since $p_1(x, y, z) = p_3(z, y, x)$ and $p_2(x, y, z) = p_2(z, y, x)$, we need only verify that $x \approx p_1(x, z, z)$ and $p_1(x, x, z) \approx p_2(x, z, z)$ are identities of $\mathbf{A}$. This is just a matter of checking on the algebras $\mathbf{A}_0$ and $\mathbf{A}_1$.

Since both $\mathbf{A}_0$ and $\mathbf{A}_1$ have the meet operation and each has a unary operation which is complementation, each is term equivalent to the two-element Boolean algebra. If $\mathcal{V}_i$ is the variety generated by $\mathbf{A}_i$ then

$$\mathcal{P} = \mathcal{V}_0 \vee \mathcal{V}_1.$$

(The set of all varieties of a fixed signature τ is closed under arbitrary intersection and has a greatest element, and thus is a complete lattice. See Section 4.5 of (Clifford Bergman, 2012) for the basic properties of this lattice.) Thus *the join of two congruence distributive varieties need not be congruence distributive, nor is it true that the direct product of two algebras with permuting congruences has permuting congruences.* Exercise 6.16.20 shows that $\mathcal{P}$ is residually large. Hence *the join of two residually finite varieties can be residually large.* No such counterexamples exist for congruence modular varieties. J. Hagemann and C. Herrmann (1979) have shown that *the join of two congruence distributive varieties in a congruence modular variety is congruence distributive.* Moreover, *the join of two residually small subvarieties of a congruence modular variety is residually small.* Furthermore, *the product of two algebras in a congruence modular variety with permuting congruences has permuting congruences.* These results can be found in (Ralph Freese and Ralph McKenzie, 1987) as Exercise 8.2, Theorem 11.1, and Exercise 6.8, respectively.

For $k \geq 5$ there do not seem to exist any naturally occurring k-permutable varieties which are not $(k-1)$-permutable. Of course, the equations (6.2.1) define a variety $\mathcal{V}_k$ such that an arbitrary variety $\mathcal{V}$ is k-permutable if and only if $\mathcal{V}_k$ is interpretable (see §4.12) into $\mathcal{V}$.

It is not difficult to find an individual congruence lattice which is k-permutable but not $(k-1)$-permutable (see Exercise 1), but finding a k-permutable variety which is not $(k-1)$-permutable is more difficult. The first example was given by (E. T. Schmidt, 1972). We present here an example due to Keith A. Kearnes (1993) which actually is closely related to Schmidt's original example. But first we note an interesting corollary to Theorem 6.11(iv).

COROLLARY 6.13 (Ralph Freese and Matthew Valeriote 2009). *If $\mathcal{V}$ is k-permutable but not $(k-1)$-permutable, then $k \leq |\mathbf{F}_{\mathcal{V}}(x, y)|$ and this bound is the best possible.*

Proof. Since $\mathcal{V}$ is k-permutable, but not $(k-1)$-permutable, the subalgebra of $\mathbf{F}^2_{\mathcal{V}}(x, y)$ generated by $\langle x, x\rangle$, $\langle x, y\rangle$ and $\langle y, y\rangle$ contains a sequence

$$\langle y, y\rangle = \langle a_0, b_0\rangle, \langle a_1, b_1\rangle, \langle a_2, b_2\rangle, \dots, \langle a_k, b_k\rangle = \langle x, x\rangle.$$

with $b_i = a_{i+1}$. We claim that, since this sequence cannot be shortened, $a_1, \dots, a_k$ are all distinct (as are $b_0, \dots, b_{k-1}$). Indeed, if $a_i = a_j$ for some $1 \leq i < j \leq k$, then the sequence obtained by deleting $\langle a_i, b_i\rangle, \dots, \langle a_{j-1}, b_{j-1}\rangle$ would witness that $\mathcal{V}$ is $k-(j-i)$ permutable. So $\mathbf{F}_{\mathcal{V}}(x, y)$ must have at least k elements.

In the next example we construct varieties showing this bound is the best possible. ■

EXAMPLE 6.14. Let $K_k = \{0, 1, \dots, k-1\}$. Let $\vee$ and $\wedge$ be the lattice operations on this chain. For each $a \prec b$ in this chain we define a unary operation which maps to b everything less than or equal to a and maps to a everything greater than or equal to b. This gives $k-1$ unary operations p_i, $i = 1, \dots, k-1$ whose precise definition is

$$p_i(x) = \begin{cases} i & \text{if } x < i \\ i-1 & \text{otherwise.} \end{cases}$$

Define ternary operations h_i, $i = 1, \dots, k-1$ by

$$h_i(x, y, z) = (x \wedge z) \vee (x \wedge p_{k-i}(y)) \vee (z \wedge p_i(y))$$

and let $\mathbf{K}_k$ be the algebra on K_k with basic operations $h_1(x, y, z), \dots, h_{k-1}(x, y, z)$. Let $\mathcal{V}_k$ be the variety generated by $\mathbf{K}_k$. We gather some facts about $\mathbf{K}_k$. It is convenient to define $h_0(a, b, c) = a$ and $h_k(a, b, c) = c$.

LEMMA 6.15. *Let $0 \leq v \leq u \leq k-1$ be elements of $\mathbf{K}_k$.*

(i) *$\mathbf{K}_k$ is generated by 0 and $k-1$. In fact,*

$$i = h_i(0, 0, k-1) = h_{i+1}(0, k-1, k-1), \qquad k = 0, \dots, k-1.$$

(ii) *The interval $I[v, u] = \{v, \dots, u\}$ is the subuniverse generated by v and u, and every non-empty subuniverse has this form.*

(iii) *The map which sends everything below v to v, everything above u to u, and fixes the elements in $I[v, u]$ is a homomorphism of $\mathbf{K}_k$ onto $I[v, u]$.*

(iv) *The map $\sigma(i) = (k-1) - i$ is an automorphism of* $\mathbf{K}_k$.

(v) *The extension of the map $x \mapsto 0$ and $y \mapsto k-1$ of $\mathbf{F}_{\mathcal{V}_k}(x,y) \twoheadrightarrow \mathbf{K}_k$ is an isomorphism.*

(vi) *The subalgebra of $\mathbf{K}_k^2$ generated by $\langle 0,0\rangle$, $\langle 0,k-1\rangle$ and $\langle k-1,k-1\rangle$ is $\{\langle a, b\rangle \in K_k^2 : b \geq a-1\}$. If one views K_k^2 as a $k \times k$ matrix, this is everything on or above the main diagonal, together with the $k-1$ elements just below the main diagonal.*

Proof. These can be proved with straightforward calculations. ■

By parts (iv) and (v) of the above lemma $\mathbf{K}_k$ is the two generated free algebra over $\mathcal{V}_k$ with generators 0 and $k-1$. Define a relation τ on pairs by $\langle a,b\rangle \;\tau\; \langle c,d\rangle$ if $b = c$ and call a sequence of τ-related pairs starting at $\langle k-1,k-1\rangle$ and ending with $\langle 0,0\rangle$ a *τ-sequence*. By (vi) the following τ-sequence lies in the subalgebra $\mathbf{S}$ of $\mathbf{K}_k^2$ generated by $\langle 0,0\rangle$, $\langle 0,k-1\rangle$ and $\langle k-1,k-1\rangle$:

$$(*) \qquad \langle k-1,k-1\rangle \;\tau\; \langle k-1,k-2\rangle \;\tau\; \langle k-2,k-3\rangle \;\tau\; \cdots \;\tau\; \langle 1,0\rangle \;\tau\; \langle 0,0\rangle.$$

By Theorem 6.11(iv) $\mathcal{V}_k$ is k-permutable.

We noted in the proof of Corollary 6.13 that the shortest τ-sequence must have distinct first coordinates except that the first two pairs will have equal first coordinates. We claim the above τ-sequence is the only one lying in $\mathbf{S}$ that has this property. The second element of any such sequence must have the form $\langle k-1,b\rangle$. But by (vi) of the lemma, b can only be $k-2$ or $k-1$. Only the former makes sense. The third element has the form $\langle k-2,c\rangle$ and there are only 3 choices for c and only the choice $c = k-3$ does not violate the rule that the first coordinates are distinct. Continuing in this way we see that $(*)$ is the unique shortest τ-sequence. It follows that $\mathcal{V}_k$ is not $k-1$ permutable. This example also completes the proof of Corollary 6.13. In addition the reader can verify that $h_0, h_1, \ldots, h_k$ are Hagemann-Mitschke terms.

Exercises 6.16

1. The variety of lattices is not congruence permutable. For every $k > 2$ there exists a lattice which is $(k-1)$-permutable but not k-permutable.

2. A **quasiprimal** algebra is a finite algebra whose clone of term operations contains the **ternary discriminator** operation:

$$t(a,b,c) = \begin{cases} c & \text{if } a = b \\ a & \text{if } a \neq b. \end{cases}$$

Prove that every quasiprimal algebra generates a congruence permutable variety.

3. If $\mathcal{V}$ has a 5-ary term q obeying the equations

$$x \approx q(x,y,y,z,z) \qquad q(x,x,y,y,z) \approx z$$

then $\mathcal{V}$ is congruence permutable.

4. The variety defined by the following equations is congruence permutable:

$$F(x,x,z) \approx z$$
$$H(u,u,x,y,w,z) \approx x$$
$$H(F(x,w,z),F(y,w,z),x,y,w,z) \approx y.$$

5. Implication algebras. (Aleit Mitschke, 1971) The variety of implication algebras has a single binary operation symbol, $\rightarrow$, and is defined by the equations

$$(x \rightarrow y) \rightarrow x \approx x$$
$$(x \rightarrow y) \rightarrow y \approx (y \rightarrow x) \rightarrow x$$
$$x \rightarrow (y \rightarrow z) \approx y \rightarrow (x \rightarrow z).$$

Prove that this variety is 3-permutable but not permutable. (Hint. Implication algebras can be interpreted in Boolean algebras by defining $x \rightarrow y$ as $y - x$; the reader should look for a nonrectangular subuniverse of $\langle 2, -\rangle^2$. To get 3-permutability, try something similar to the proof used in Example 6.14 with $k = 3$. In this case one column of the matrix will use only the Boolean operation $-$, and one can check that these same terms work for implication algebras.)

6. Right-complemented semigroups. (Bruno Bosbach, 1970) This variety has two binary operations, $\cdot$ and $*$, and is defined by the following equations:

$$x \cdot (x * y) \approx y \cdot (y * x)$$
$$(x \cdot y) * z \approx y * (x * z)$$
$$x \cdot (y * y) \approx x$$

Prove that right-complemented semigroups are 3-permutable but not permutable.[5] (Hint. This is similar to the previous exercise, except that now we need to notice that right-complemented semigroups can be interpreted in Boolean algebras by defining $x \cdot y$ to be $x \vee y$ and $x * y$ to be $y - x$. Now again look at Example 6.14.)

7. Heyting algebras. (S. Burris and H. P. Sankappanavar, 1981) A Heyting algebra has operations $\vee$, $\wedge$, $\rightarrow$, 0, 1 and is defined by the following equations, together with the equations of distributive lattice theory with 0 and 1:

$$x \rightarrow x \approx 1$$
$$(x \rightarrow y) \wedge y \approx y$$
$$x \wedge (x \rightarrow y) \approx x \wedge y$$
$$x \rightarrow (y \wedge z) \approx (x \rightarrow y) \wedge (x \rightarrow z)$$
$$(x \vee y) \rightarrow z \approx (x \rightarrow z) \wedge (y \rightarrow z)$$

[5] Actually only the first and third equations are needed to prove 3-permutability; see (J. Hagemann and A. Mitschke, 1973).

Prove that this variety is congruence permutable by verifying that

$$p(x, y, z) = (y \to x) \wedge (y \to z) \wedge (x \vee z)$$

is a Mal'tsev operation. (Hint: for a more conceptual approach, use the equations to verify that

$$(x \to y) \geq a \quad \text{if and only if} \quad y \geq a \wedge x,$$

i.e. $x \to y$ is the largest element a such that $a \wedge x \leq y$.)

8. Give an example of a variety $\mathcal{V}$ such that $\mathbf{F}_{\mathcal{V}}(k)$ has k-permutable congruences but $\mathcal{V}$ is not k-permutable.

9. (Rudolf Wille, 1970) Let $\theta \in \mathbf{Con\ A}$ and $f : \mathbf{A} \to \mathbf{B}$ be an onto homomorphism, with kernel ϕ. The relation

$$f(\theta) = \{\langle f(a), f(b)\rangle : \langle a, b\rangle \in \theta\}$$

may fail to be transitive, and thus to be a congruence. But if $\theta \circ \phi \circ \theta \subseteq \phi \circ \theta \circ \phi$ then $f(\theta)$ is a congruence. In fact, a variety $\mathcal{V}$ is congruence 3-permutable if and only if for every $\mathbf{A} \in \mathcal{V}$, every congruence $\theta \in \mathbf{Con\ A}$, and every onto homomorphism $f : \mathbf{A} \to \mathbf{B}$, $f(\theta)$ is a congruence. (Part of this exercise is worked out in Theorem 4.68.)

10. If $\theta \circ \phi \subseteq \phi \circ \theta$ for equivalence relations θ and ϕ on the same set, then $\theta \circ \phi = \phi \circ \theta = \theta \vee \phi$. But the corresponding assertion for triple products is false. Find an algebra $\mathbf{A}$ and congruences θ and ϕ on $\mathbf{A}$ such that $\theta \circ \phi \circ \theta < \phi \circ \theta \circ \phi$. (It still follows that $\theta \vee \phi = \phi \circ \theta \circ \phi$.) Is this situation possible inside a 3-permutable variety?

11. Give two Mal'tsev terms which differ on all nontrivial Boolean algebras.

12. (J. Hagemann, 1973a); see also (J. Hagemann and A. Mitschke, 1973). For a variety $\mathcal{V}$, the following three conditions are equivalent:
(i) $\mathcal{V}$ is congruence k-permutable;
(ii) for every $\mathbf{A} \in \mathcal{V}$ and every reflexive subuniverse S of $A \times A$,

$$S^{-1} \subseteq \overbrace{S \circ \cdots \circ S}^{k-1 \text{ times}}$$

(iii) for every $\mathbf{A} \in \mathcal{V}$ and every reflexive subuniverse S of $A \times A$,

$$\overbrace{S \circ \cdots \circ S}^{k \text{ times}} \subseteq \overbrace{S \circ \cdots \circ S}^{k-1 \text{ times}}$$

13. (Ralph Freese and Matthew Valeriote, 2009) This exercise explores the procedure suggested by Theorem 6.11 for testing if a variety $\mathcal{V}$ is k-permutable. (The procedure used by UACalc (Ralph Freese, Emil Kiss, and Matthew Valeriote, 2008) is based on this.) Let $\mathbf{S}$ be the subalgebra of $\mathbf{F}_{\mathcal{V}}^2(x, y)$ generated by $\langle x, x\rangle$, $\langle x, y\rangle$ and $\langle y, y\rangle$. Define a graph on S with edges $\langle a, b\rangle \to \langle c, d\rangle$ provided $b = c$. Starting at $\langle y, y\rangle$ do a breadth first search for $\langle x, x\rangle$: define levels starting with level 0, $S_0 = \{\langle y, y\rangle\}$. Inductively define level $m+1$ as

$$S_{m+1} = \{\langle c, d\rangle \in S : \langle a, b\rangle \to \langle c, d\rangle \text{ for some } \langle a, b\rangle \in S_m\} - \bigcup_{i=0}^{m} S_i.$$

a. Show that, if $\langle x, x\rangle \in S_k$, then $\mathcal{V}$ is k-permutable but not $(k-1)$-permutable.

b. Suppose

$$\langle y, y\rangle = \langle a_0, b_0\rangle \to \langle a_1, b_1\rangle \to \cdots \to \langle a_k, b_k\rangle = \langle x, x\rangle$$

and q_i is a term of $\mathcal{V}$ with $\langle a_i, b_i\rangle = q_i(\langle x, x\rangle, \langle x, y\rangle, \langle y, y\rangle)$. Show that $p_i = q_{k-i}$, $i = 0, \dots, k$ are Hagemann-Mitschke terms for $\mathcal{V}$.

14. (H. Lakser, 1982) If $\mathcal{V}$ is congruence k-permutable, $\mathbf{A} \in \mathcal{V}$, $a, b, c, d \in A$, and if $\langle c, d\rangle \in \mathrm{Cg}(a, b)$, then

$$\begin{aligned} c &= t_1(a, e_1, \dots, e_m) \\ t_1(b, e_1, \dots, e_m) &= t_2(a, e_1, \dots, e_m) \\ &\vdots \\ t_{k-1}(b, e_1, \dots, e_m) &= d \end{aligned}$$

for some $\mathcal{V}$-terms $t_1, \dots, t_{k-1}$ and for some $e_1, \dots, e_m \in A$. Conversely if all principal congruences in $\mathcal{V}$ can be expressed in this way, then $\mathcal{V}$ is congruence k-permutable. Compare this with Theorem 4.19 of §4.3. (Hint. Use the previous exercise.)

15. Suppose that every finite algebra in $\mathcal{V}$ has permutable congruences and that $\mathcal{V}$ is generated by its finite members. Must $\mathcal{V}$ be congruence permutable?

16. A subuniverse S of $\mathbf{A} \times \mathbf{B}$ is called **locally rectangular** if and only if whenever $\langle a, c\rangle, \langle b, c\rangle, \langle b, d\rangle \in S$, then $\langle a, d\rangle \in S$. Prove that $\mathcal{V}$ is congruence permutable if and only if every subuniverse of every product $\mathbf{A} \times \mathbf{B} \in \mathcal{V}$ is locally rectangular, and that this is equivalent to the property that every subuniverse of every direct square $\mathbf{A} \times \mathbf{A} \in \mathcal{V}$ is locally rectangular.

17. Show that if $\mathbf{A}$ is an algebra such that **Con A** contains a 0-1 sublattice isomorphic to $\mathbf{M}_3$ then $\mathbf{A}$ has at most one Mal'tsev term operation $p(x, y, z)$. The results of §4.13 are helpful.

18. **a.** Let $\mathbf{K}_k$ be the algebra from Example 6.14 and let $\mathcal{V}_k$ be the variety it generates. Show that $\mathcal{V}_k$ is congruence distributive with Jónsson terms $d_i(x, y, z) = h_i(x, h_i(x, y, z), z)$ when i is odd and $d_i(x, y, z) = h_i(x, y, z)$ when i is even, for $i = 0, \dots, k$.

b. Show that $\mathcal{V}_k$ does not have a shorter sequence of Jónsson terms.

c. Show that in any k-permutable, congruence distributive variety a minimal length sequence of Jónsson terms has length at most k (that is $k+1$ terms).

19. Let $\mathbf{A}$ be the algebra defined in Example 6.12. Show that **Con A** is the lattice given in Figure 6.5.

20. Polin's variety is residually large. Let $\mathcal{P}$ be Polin's variety given in Example 6.12 and let $\mathbf{A}_0$ and $\mathbf{A}_1$ be the algebras given in that example. Let I be a set and let $\mathbf{C} = \mathbf{A}_1^I$ and let $\mathbf{B} = \mathbf{A}_0 \times \mathbf{C}$. Let $\overline{1} \in C$ be the element all of whose coordinates are 1. Let θ be the equivalence relation on B which identifies all pairs whose second coordinates are equal except that $\langle 1, \overline{1}\rangle$ and $\langle 0, \overline{1}\rangle$ are not θ-related. Show that θ is a completely meet irreducible congruence on $\mathbf{B}$ and that $|B/\theta| = |I| + 1$. Thus Polin's variety $\mathcal{P}$ is residually large. That is, it has no cardinal bound on the size of its subdirectly irreducible members.

We now present five exercises in which congruence permutability is applied to develop the theory of topological algebras. All these results may be found in (Walter Taylor, 1977c). In all these exercises, $\mathbf{A} = \langle A, F_0, F_1, \ldots\rangle$ is an algebra, and $\mathcal{T}$ a topology on A such that each F_i is continuous as a function $A^{n_i} \to A$, where n_i is the arity of F_i. This algebra is assumed to lie in a congruence permutable variety $\mathcal{V}$. We also assume that F_0 is a Mal'tsev operation: $F_0(x, x, z) \approx z \approx F_0(z, x, x)$ are identities of $\mathcal{V}$. Thus, these exercises form a generalization of the theory of topological groups.

21. If θ is a congruence relation on $\mathbf{A}$, then the closure $\overline{\theta}$ (in the space $A \times A$) is again a congruence on $\mathbf{A}$.

22. If $U \subseteq A$ is open, and θ is any congruence on $\mathbf{A}$, then

$$\{v \in A : \langle u, v\rangle \in \theta, \text{ for some } u \in U\}$$

is also open.

23. For $\theta \in$ **Con A** there exists a unique topology on $\mathbf{A}/\theta$ so that all operations of $\mathbf{A}/\theta$ are continuous and $A \to A/\theta$ is an open continuous map.

24. If A is T_0, then A is Hausdorff.

25. In any case, $A/\overline{0}_{\mathbf{A}}$ is Hausdorff.

26. The claims of the last five exercises are false for topological algebras in general.

6.3. Generating Congruence Relations

Like the last section, the next several sections provide characterizations of certain properties of varieties that depend on the behavior of the congruences on every algebra in the variety. Understanding how congruences are generated is a key to this investigation. The Congruence Generation Theorem (Theorem 4.19 on page 155 of Volume I) is our starting point. In this section, we refine that theorem in several ways.

Let **A** be an algebra. A **basic translation** of **A** is a function $\lambda : A \to A$ such that there is an operation symbol Q of rank $r > 0$ and elements $a_0, \ldots, a_{r-1} \in A$ so that for some $i < r$

$$\lambda(x) = Q^{\mathbf{A}}(a_0, \ldots, a_{i-1}, x, a_{i+1}, \ldots, a_{r-1}) \text{ for all } x \in A.$$

Basic translations have **complexity** 1. A **translation** is just a composition of some finite sequence of basic translations. For a natural number ℓ, we say that a translation **has complexity at most** ℓ if it is the composition of a sequence of no more than ℓ of basic translations. Since a translation may arise as a composition of basic translations in several ways, we have framed the notion of complexity of a translation so that the set of complexities of a translation form an infinite interval of natural numbers, but naturally, we are most interested in the the least number in this interval. The identity map on A is a translation of complexity 0.

Each translation is associated with a term that is built from the terms associated to basic translations by means of repeated substitution. These terms have a special structure—they are sometimes called **slender** terms. Figure 6.6 below illustrates two slender terms that, for simplicity, happen to have only one operation symbol Q, which has rank 3.

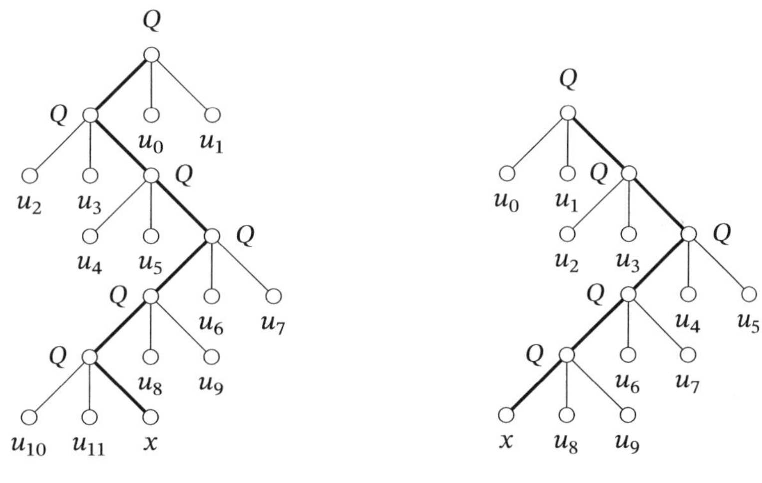

FIGURE 6.6. Two slender terms for representing translations

Observe that in the terms depicted in Figure 6.6

(a) each variable that occurs does so exactly once;
(b) the variable x occurs at the lowest level;
(c) all the leaves of the tree are labeled with variables;
(d) there is a branch beginning at the top node and ending at the leaf labeled with x that passes through all the internal nodes of the tree;
(e) each internal node is labeled with an operation symbol of positive rank.

Indeed, the tree depicting any slender term has all these properties. The term on the left in Figure 6.6 is

$$t(u_0.\ldots, u_{11}, x) := QQu_2u_3Qu_4u_5QQQu_{10}u_{11}xu_8u_9u_6u_7u_0u_1.$$

In an algebra **A** we can obtain a translation $\lambda(x)$ from t by specifying a 12-tuple $\langle a_0, a_1, \ldots, a_{11}\rangle$ of elements of A and putting

$$\lambda(x) := t^{\mathbf{A}}(a_0, a_1, \ldots, a_{11}, x).$$

In this way, the slender term t provides the *framework* for many translations, one for each 12-tuple of elements of A. Each such translation has complexity at most 6, a bound on the complexity being obtained simply by counting the occurrences of operation symbols in t. Likewise, given a translation $\lambda(x)$ of complexity at most ℓ there is at least one slender term t, in which there no more than ℓ occurrences of operations symbols, that is a framework for $\lambda(x)$.

It is sometimes convenient to expand the set of basic translations (and hence the set of translations) by allowing some finite set T of terms to also play the role of basic operation symbols in the definition above. We call the expanded set of basic translations obtained in this way T*-enhanced* basic translations, and we call their compositions T-*enhanced* translations. So, for each $t(v_0, \ldots, v_{r-1}) \in T$, where $v_0, \ldots, v_{r-1}$ includes all the variables occurring in t, and elements $a_0, \ldots, a_{r-1} \in A$ so that for some $i < r$

$$\lambda(x) = t^{\mathbf{A}}(a_0, \ldots, a_{i-1}, x, a_{i+1}, \ldots, a_{r-1}) \text{ for all } x \in A$$

is a T-enhanced basic translation. In this context, we understand the complexities of a T-enhanced translation to be measured against T-enhanced basic translations.

The following theorem, which has its roots in A. I. Mal'tsev (1954)(see also the translation (A. I. Mal'tsev, 1963)) refines the Congruence Generation Theorem 4.19 from Volume I:

THEOREM 6.17 (Mal'tsev's Congruence Generation Theorem). *Let* **A** *be an algebra,* $X \subseteq A \times A$, *and* p *and* q *be elements of* A. *Then*

$$\langle p, q\rangle \in \mathrm{Cg}^{\mathbf{A}}(X)$$

if and only if

there is a natural number n *and there are elements* $r_0, r_1, \ldots, r_n \in A$ *and translations* $\lambda_0, \lambda_1, \ldots, \lambda_{n-1}$ *of* **A** *and pairs* $\langle a_i, b_i\rangle \in X$ *so that*

(a) $p = r_0$ *and* $r_n = q$ *and*
(b) $\{r_i, r_{i+1}\} = \{\lambda_i(a_i), \lambda_i(b_i)\}$ *for all* $i < n$. ■

The set of unary polynomials appearing in the Congruence Generation Theorem has been replaced by a subset, namely the set of translations. The proof of the sharper version above differs in no important way from the proof we gave in Chapter 4. The only real change is a modification of Theorem 4.18 on page 154 in Volume I—basic to the proof of Theorem 4.19. There it was noted that an equivalence relation θ on A is a congruence of **A** if and only if it is invariant respect to every unary polynomial. Actually, more is true: an equivalence relation θ is a congruence if and only if it is invariant with respect to every basic translation. The proof given for Theorem 4.18 actually demonstrates this. We note that this modification of Theorem 4.18 is actually Theorem 1 in (A. I. Mal'tsev, 1954).

Here is another form of Mal'tsev's Congruence Generation Theorem that we find useful in the ensuing sections. Constructing a demonstration of this variant is a profitable exercise for our readers.

THEOREM 6.18. *For any algebra* $\mathbf{A}$ *and any* $X \subseteq A \times A$*, the congruence* $\mathrm{Cg}^{\mathbf{A}}(X)$ *generated by* X *consists of all pairs* $\langle c, d\rangle$ *such that for some natural numbers* m *and* k *and some* $\mathbf{e} \in A^k$*, and some* $(k+2)$*-ary terms* $t_0, \dots, t_{m-1}$*, and some pairs* $\langle a_i, b_i\rangle \in X$ *for* $i < m$*, we have*

$$\begin{aligned} c &= t_0^{\mathbf{A}}(a_0, b_0, \mathbf{e}) \\ t_0^{\mathbf{A}}(b_0, a_0, \mathbf{e}) &= t_1^{\mathbf{A}}(a_1, b_1, \mathbf{e}) \\ &\vdots \\ t_{m-1}^{\mathbf{A}}(b_{m-1}, a_{m-1}, \mathbf{e}) &= d. \end{aligned}$$

■

A stronger version of this theorem, in the case of k-permutable varieties, can be found in Exercise 6.16.14.

A useful special case of Mal'tsev's Congruence Generation Theorem is had by setting $X = \{\langle a, b\rangle\}$. This yields a characterization of the generation of principal congruences. Here are diagrams of the condition in Mal'tsev's Congruence Generation Theorem in the case of principal congruences. Arrows on the left are labeled with translations while on the right they are labeled with complexities.

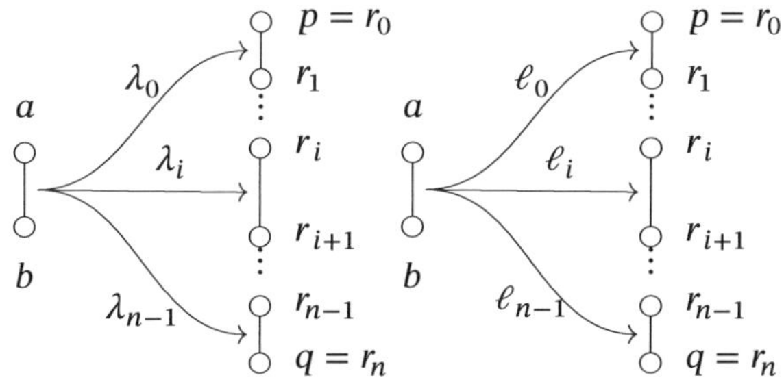

We use

$$\{a, b\} \looparrowright^n_\ell \{p, q\}$$

to denote that there is a sequence $r_0, \dots, r_n$ of no more than $n+1$ elements of A and a sequence $\lambda_0, \dots, \lambda_{n-1}$ of n translations of $\mathbf{A}$, each of complexity no more than ℓ, that witness $(p, q) \in \mathrm{Cg}^{\mathbf{A}}(a, b)$. The sequence $r_0, r_1, \dots, r_n$ is called a **principal congruence sequence**.

COROLLARY 6.19. *Let* $\mathbf{A}$ *be an algebra and let* $a, b, p,$ *and* q *be elements of* A*. Then*

$$\langle p, q\rangle \in \mathrm{Cg}^{\mathbf{A}}(a, b)$$
if and only if
there are natural numbers n *and* ℓ *such that* $\{a, b\} \looparrowright^n_\ell \{p, q\}$.

Moreover, if the signature of $\mathbf{A}$ *is finite, then* $\{x, y\} \looparrowright^n_\ell \{z, w\}$ *is expressible by a formula of first-order logic.*

Proof. Only the "moreover" statement requires proof, as the other part of the corollary just sets the content of Mal'tsev's Congruence Generation Theorem in the context when $X = \{\langle a, b\rangle\}$.

Since we assume the signature is finite, the set F of slender terms in which the number of occurrences of operation symbols is no more than ℓ is finite. For convenience, we also assume that distinct members of F have only the variable x in common and that all the other variables occurring in members of F are in the tuple $\mathbf{u} := \langle u_0, \dots, u_{m-1}\rangle$. The reader can check that $\{x, y\} \looparrowright^n_\ell \{z, w\}$ is expressed by the following formula of first-order logic.

$$\exists w_0, \dots, w_n \left[\bigvee_{t(x,\mathbf{u})\in F} \exists u_0, \dots, u_{m-1} \Psi_{t(x,\mathbf{u})}(x, y, z, w) \right]$$

where $\Psi_{t(x,\mathbf{u})}(x, y, z, w)$ is the formula

$$\big((t(x,\mathbf{u}) \approx z \wedge t(y,\mathbf{u}) \approx w\big) \vee \big(t(x,\mathbf{u}) \approx w \wedge t(y,\mathbf{u}) \approx z)\big)$$

While the full details of the syntax for first-order logic are available in §8.1, above the symbol $\vee$ denotes disjunction—that is "inclusive or" and $\wedge$ denotes conjunction—that is "and". ■

Exercises 6.20

1. Prove Theorem 6.18.

2. Suppose that $\mathbf{A}$ is an algebra belonging to a congruence permutable variety. For any elements a, b, p, and q in A. prove that

$$\langle p, q\rangle \in \mathrm{Cg}^{\mathbf{A}}(a, b)$$
if and only if
$$\{p, q\} = \{\mu(a), \mu(b)\} \text{ for some } \mu(x) \in \mathrm{Pol}_1(\mathbf{A})$$
if and only if
$$p = \mu(a), q = \mu(b) \text{ for some } \mu(x) \in \mathrm{Pol}_1(\mathbf{A}).$$

6.4. Congruence Semidistributive Varieties

A lattice is **meet semidistributive** if it satisfies the implication

$$a \wedge b = a \wedge c \text{ implies } a \wedge (b \vee c) = (a \wedge b) \vee (a \wedge c).$$

Join semidistributivity is defined dually:

$$a \vee b = a \vee c \text{ implies } a \vee (b \wedge c) = (a \vee b) \wedge (a \vee c).$$

A lattice is said to be **semidistributive** if it has both of these properties. Evidently, both are weakened forms of distributivity. For lattices in general, neither of these properties implies the other. However, when cast as properties of congruences lattices for varieties, there is a relationship. While there are varieties that are congruence meet semidistributive but that fail to be join semidistributive, congruence join semidistributive varieties must also be congruence meet semidistributive and hence also congruence semidistributive.

Let $\mathbf{A}$ be an algebra and α, β, and γ be congruences of $\mathbf{A}$. Consider the congruences defined by the recursion below:

$$\begin{aligned} \beta_0 &= \beta & \gamma_0 &= \gamma \\ \beta_{n+1} &= \beta \vee (\alpha \wedge \gamma_n) & \gamma_{n+1} &= \gamma \vee (\alpha \wedge \beta_n) \text{ for all } n \\ \beta_\omega &= \bigcup_{n<\omega} \beta_n & \gamma_\omega &= \bigcup_{n<\omega} \gamma_n \end{aligned} \tag{6.4.1}$$

Let $\mathcal{C}_2$ be the congruence inclusion

$$\alpha \wedge (\beta \circ \gamma) \subseteq \beta_\omega. \tag{$\mathcal{C}_2$}$$

$\mathcal{C}_2$ is one of several conditions we will present that characterize congruence meet semidistributivity of a variety.

Let $\mathcal{V}$ be a variety. As proved below, the following condition is also equivalent to the congruence meet-semidistributivity of $\mathcal{V}$.

The $\diamond$-Condition for $\mathcal{V}$

There exists a finite set T of terms such that, for all $\mathbf{A} \in \mathcal{V}$ and all finite sequences $p = r_0, \ldots, r_n = q \in A$, if $p \neq q$, there are $i < n$ and $e, f \in A$ with $e \neq f$ so that

$$\{r_i, r_{i+1}\} \looparrowright_1^2 \{e, f\} \text{ and } \{p, q\} \looparrowright_1^2 \{e, f\},$$

with respect to T-enhanced translations.

Here is a diagram of the $\diamond$-Condition.

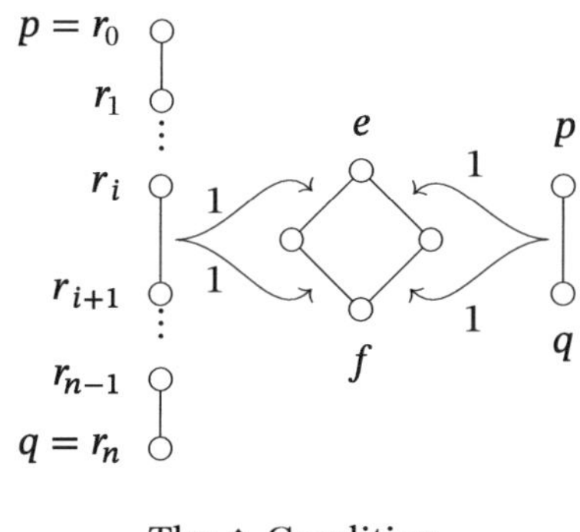

The $\diamond$-Condition

We also need the notion of a **bracket expression**. We define this notion by recursion: $\langle\,\rangle$ is a bracket expression of *nesting depth* 1, and for any natural number n, if $\mathbf{b_0}, \ldots, \mathbf{b_{n-1}}$ are bracket expressions, then $\langle \mathbf{b_0 b_1} \ldots \mathbf{b_{n-1}} \rangle$ is a bracket expression of nesting depth one more than the maximum of the nesting depths of the $\mathbf{b_i}$'s. Here is an example of a bracket expression of nesting depth 3: $\langle\langle\,\rangle\langle\langle\,\rangle\rangle\langle\,\rangle\rangle$. Each bracket expression consists of a string of brackets such that each left bracket $\langle$ is matched with a right bracket $\rangle$. We define a strong Mal'tsev condition, denoted $\mathcal{C}_2(\mathbf{b})$, for each bracket expression $\mathbf{b}$ by associating a ternary term $t_i(x, y, z)$ with each bracket, left and right, and imposing the following equations:

$$
\begin{aligned}
&t_0(x,y,z) \approx x\\
&t_i(x,x,z) \approx t_{i+1}(x,x,z) \text{ for each even } i < n\\
&t_i(x,z,z) \approx t_{i+1}(x,z,z) \text{ for each odd } i < n\\
&t_i(x,y,x) \approx t_j(x,y,x) \text{ for each matched pair of indices } i,j\\
&t_n(x,y,z) \approx z
\end{aligned}
\tag{6.4.2}
$$

Notice there are $n+1$ terms, so there must be $n+1$ brackets. Since there are an even number of brackets, n must be odd. Also notice that this condition bears a strong resemblance to the corresponding condition characterizing congruence distributivity—the distinction here is that the fourth equation listed above applies to all indices (and n is not required to be odd) in the congruence distributive case.

The next theorem gives several characterizations of congruence meet semidistributive varieties. The presentation primarily follows (Ross Willard, 2000), but the use of bracket expressions is from (Kirby A. Baker, George F. McNulty, and Ju Wang, 2004) and the $\Diamond$-condition is in (George F. McNulty and Ross Willard, 2018). Gábor Czédli (1983) had earlier characterizations and parts of the theorem are from (Keith A. Kearnes and Ágnes Szendrei, 1998) and (Paolo Lipparini, 1998).

In the proof of this theorem we shall first show that (i) to (v) are equivalent by showing each implies the next cyclically. We then show that (i) $\Rightarrow$ (vi). The proof (vi) $\Rightarrow$ (i) requires commutator theory so we postpone the proof until the commutator is developed in Chapter 11. Condition (vii) mentions **congruence neutrality**, which will be introduced and thoroughly studied in Chapter 11. The definition of the commutator for an arbitrary variety is given in Definition 4.150 and it is observed that $[\alpha,\beta] \leq \alpha \wedge \beta$. A variety is **congruence neutral** if this is equality; that is, if $[\alpha,\beta] = \alpha \wedge \beta$.

THEOREM 6.21. *The following are equivalent for a variety $\mathcal{V}$:*

(i) *$\mathcal{V}$ is congruence meet semidistributive.*
(iia) *The congruence inclusion $\mathcal{C}_2$ (page 27) holds in $\mathcal{V}$.*
(iib) *$(x,z) \in \beta_n$, for some n, where $\alpha = \mathrm{Cg}^{\mathbf{F}}(x,z)$, $\beta = \mathrm{Cg}^{\mathbf{F}}(x,y)$, $\gamma = \mathrm{Cg}^{\mathbf{F}}(y,z)$, and $\mathbf{F} = \mathbf{F}_{\mathcal{V}}(x,y,z)$. (See (6.4.1) for the definition of β_n.)*
(iii) *There is a bracket expression $\mathbf{b}$ such that $\mathcal{V}$ satisfies the Mal'tsev condition $\mathcal{C}_2(\mathbf{b})$.*
(iv) *There is a finite family $\{(\ell_p(x,y,z), r_p(x,y,z)) \mid p \in T\}$ of pairs of ternary terms such that*

$$\mathcal{V} \models \ell_p(x,y,x) \approx r_p(x,y,x) \text{ for all } p \in T,$$

$$\mathcal{V} \models \forall x,y \left[\bigwedge_{p\in T} (\ell_p(x,x,y) \approx r_p(x,x,y) \Leftrightarrow \ell_p(x,y,y) \approx r_p(x,y,y)) \Rightarrow x \approx y \right].$$

(v) *$\mathcal{V}$ satisfies the $\Diamond$-condition.*
(vi) *$\mathcal{V}$ satisfies an idempotent Mal'tsev condition that fails in every nontrivial variety of modules.*
(vii) *$\mathcal{V}$ is congruence neutral.*

Proof.

(i) $\Rightarrow$ (iia)
Let $\mathbf{A} \in \mathcal{V}$ and let α, β and γ be congruences of $\mathbf{A}$. Let β_n, β_ω, γ_n and γ_ω be the congruences defined by 6.4.1. Since $\bigvee_n \beta_n = \bigcup_n \beta_n$

$$\alpha \wedge \beta_\omega = \bigvee_{n<\omega} (\alpha \wedge \beta_n) \leq \bigvee_{n<\omega} (\alpha \wedge \gamma_{n+1}) = \alpha \wedge \gamma_\omega.$$

So $\alpha \wedge \beta_\omega = \alpha \wedge \gamma_\omega$, and hence by meet semidistributivity, $\alpha \wedge \beta_\omega = \alpha \wedge (\beta_\omega \vee \gamma_\omega)$. Thus

$$\alpha \cap (\beta \circ \gamma) \subseteq \alpha \wedge (\beta_\omega \vee \gamma_\omega) = \alpha \wedge \beta_\omega \leq \beta_\omega.$$

This proves (iia).

(iia) $\Rightarrow$ (iib)
By (iia) $\mathcal{C}_2$ holds for the α, β and γ of (iib). Clearly $\langle x, z\rangle \in \alpha \wedge (\beta \circ \gamma)$, and so in β_ω. Hence $\langle x, z\rangle$ is in β_n for some n.

(iib) $\Rightarrow$ (iii)
With α, β and γ as in (iib), we have $\langle x, z\rangle \in \beta_n$ for some n, and so $\langle x, z\rangle \in \alpha \wedge \beta_n$. We will show by induction on k that for any $\langle s, t\rangle \in \alpha \cap \beta_n$ each of the following hold.

(a) If $\langle s, t\rangle \in \alpha \cap \beta_k$, then there is a bracket expression $\mathbf{b}$ of nesting depth no more than k such that $\mathcal{V}$ realizes $\mathcal{C}_2(\mathbf{b})$ except for the equations $x \approx t_0$ and $t_n \approx z$. Instead we have $s = t_0^{\mathbf{F}}$ and $t = t_n^{\mathbf{F}}$.
(b) If $\langle s, t\rangle \in \alpha \cap \gamma_k$, then there is a bracket expression $\mathbf{b}$ of nesting depth no more than k such that $\mathcal{V}$ realizes $\mathcal{C}_2(\mathbf{b})$, but with the parities switched in the equations (6.4.2), except for the equations $x \approx t_0$ and $t_n \approx z$. Instead we have $s = t_0^{\mathbf{F}}$ and $t = t_n^{\mathbf{F}}$.

To start the induction suppose $\langle s, t\rangle \in \alpha \cap \beta_0$. Let $\mathbf{b}$ be $\langle\ \rangle$. We also need terms associated with the brackets. In this case there are only two, and since this bracket will (likely) be inside a bigger bracket expression, we will call the term $t_j(x, y, z)$ and $t_{j+1}(x, y, z)$ rather than $t_0(x, y, z)$ and $t_1(x, y, z)$. The index j will be determined later but it will be even. Let $t_j^{\mathbf{F}}(x, y, z) = s$ and $t_{j+1}^{\mathbf{F}}(x, y, z) = t$. The equations we need verify are $t_j(x, x, z) \approx t_{j+1}(x, x, z)$ and, since j and $j+1$ are matching, $t_j(x, y, x) \approx t_{j+1}(x, y, x)$. For the first

$$t_j(x, x, z)\ \beta\ t_j(x, y, z) = s\ \beta\ t = t_{j+1}(x, y, z)\ \beta\ t_{j+1}(x, x, z)$$

By Theorem 6.9, β is trivial on the subalgebra generated by x and z, so $t_j^{\mathbf{F}}(x, x, z) = t_{j+1}^{\mathbf{F}}(x, x, z)$, and consequently $t_j(x, x, z) \approx t_{j+1}(x, x, z)$. The other equation is handled similarly.

The case $\langle s, t\rangle \in \alpha \cap \gamma_0$ is similar, except j will be odd and γ will take the place of β.

For the induction step, first consider the case that $(s,t) \in \alpha \cap \beta_{k+1}$. Then $\langle s,t\rangle \in \beta \vee (\alpha \cap \gamma_k)$. Hence, there are $u_0, \ldots, u_{2m+1} \in \mathbf{F}_{\mathcal{V}}(x,y,z)$ such that

$$\begin{aligned} s &\equiv u_0 \mod \beta \\ u_0 &\equiv u_1 \mod \alpha \cap \gamma_k \\ u_1 &\equiv u_2 \mod \beta \\ &\vdots \\ u_{2m} &\equiv u_{2m+1} \mod \alpha \cap \gamma_k \\ u_{2m+1} &\equiv t \mod \beta \end{aligned}$$

Using the inductive hypothesis, for each even index $2j$ pick $\mathbf{b}_j$ to be a bracket expression witnessing $u_{2j} \equiv u_{2j+1} \mod \alpha \cap \gamma_k$. Then let $\mathbf{b}$ be the bracket expression $\langle \mathbf{b}_0 \ldots \mathbf{b}_m \rangle$. The case when $\langle s,t\rangle \in \alpha \cap \gamma_{k+1}$ is similar.

Finally, since $(x,z) \in \alpha \cap \beta_n$ we see that there is a bracket expression $\mathbf{b}$ so that $\mathcal{C}_2(\mathbf{b})$ holds in $\mathcal{V}$.

(iii) $\Rightarrow$ (iv)
Relabel the terms $t_0, \ldots, t_n$ occurring is $\mathcal{C}_2(\mathbf{b})$ so that the terms associated with the left bracket $\langle$ in $\mathbf{b}$ are denoted by ℓ_p ($p = 0, \ldots$ from left to right) while the matched right bracket $\rangle$ is denoted by r_p. Then T is an index set half the size of the one used in $\mathcal{C}_2(\mathbf{b})$. The first contention in (iv) follows immediately from (iii).

For the second contention in (iv), let $\mathbf{A} \in \mathcal{V}$ and let $a, b \in A$ so that

$$\mathbf{A} \models \ell_p(a,a,b) \approx r_p(a,a,b) \Leftrightarrow \ell_p(a,b,b) \approx r_p(a,b,b) \text{ for all } p \in T.$$

We prove by induction on the formation of bracket expressions that

$$\ell_p^{\mathbf{A}}(a,b,b) = r_p^{\mathbf{A}}(a,b,b) \text{ and } \ell_p^{\mathbf{A}}(a,a,b) = r_p^{\mathbf{A}}(a,a,b)$$

for all p involved in the bracket expression $\mathbf{b}$.

As a base step, consider the bracket expression $\langle\ \rangle$, possibly inside a larger bracket expression. This bracket is associated with the terms ℓ_p and r_p for some p. For example, in the bracket expression $\langle\langle\ \rangle\langle\langle\ \rangle\rangle\langle\ \rangle\rangle$, p could be 1, 3, or 4; that is $\ell_1 = t_1$ or $\ell_3 = t_4$ or $\ell_4 = t_7$. In any case, these will be consecutive brackets and so one of $\ell_p^{\mathbf{A}}(a,a,b) = r_p^{\mathbf{A}}(a,a,b)$ or $\ell_p^{\mathbf{A}}(a,b,b) = r_p^{\mathbf{A}}(a,b,b)$ must hold. So by the equivalence both must hold.

For the induction step, let the bracket expression be

$$\langle b_1 \cdots b_{n-1} \rangle$$

where the b_j's are simpler bracket expressions for which our induction hypothesis in assumed to hold. First observe that the rightmost bracket of b_j is followed by the leftmost bracket of b_{j+1} for all $j < n-1$. Take r_p to be the term associated with the rightmost bracket in b_j and ℓ_q to be the term associated with the leftmost bracket in b_{j+1}. Then we have

$$\begin{aligned} \ell_p^{\mathbf{A}}(a,a,b) &= r_p^{\mathbf{A}}(a,a,b) \qquad & \ell_p^{\mathbf{A}}(a,b,b) &= r_p^{\mathbf{A}}(a,b,b) \\ \ell_q^{\mathbf{A}}(a,a,b) &= r_q^{\mathbf{A}}(a,a,b) \qquad & \ell_q^{\mathbf{A}}(a,b,b) &= r_q^{\mathbf{A}}(a,b,b) \end{aligned}$$

But (6.4.2) provides us with

$$r_p^{\mathbf{A}}(a,a,b) = \ell_q^{\mathbf{A}}(a,a,b) \text{ or } r_p^{\mathbf{A}}(a,b,b) = \ell_q^{\mathbf{A}}(a,b,b)$$

Since each of the b_j has even length, all the b_j's are associated with i indices of the same parity. If the parity is even then the equations in the left column above yield

$\ell_{\leftarrow}^{\mathbf{A}}(a,a,b) = r_{\rightarrow}^{\mathbf{A}}(a,a,b)$, where $\ell_{\leftarrow}$ is the leftmost term associated with b_1 and $r_{\rightarrow}$ is the rightmost term associated with b_{n-1}. We also obtain $\ell_{\leftarrow}^{\mathbf{A}}(a,b,b) = r_{\rightarrow}^{\mathbf{A}}(a,b,b)$ by our inductive assumption.

Now let ℓ_* be the term associated with the leftmost bracket of our bracket expression and r_* be associated with the matched rightmost bracket. Suppose the index of the leftmost bracket (the one associated with ℓ_*) is odd. Then $\ell_{\leftarrow}$ has an even index and $\ell_*^{\mathbf{A}}(a,b,b) = \ell_{\leftarrow}(a,b,b)$ follows from $\mathcal{C}_2(\mathbf{b})$ as does $r_{\rightarrow}^{\mathbf{A}}(a,b,b) = r_*^{\mathbf{A}}(a,b,b)$. But we have just seen that $\ell_{\leftarrow}^{\mathbf{A}}(a,b,b) = r_{\rightarrow}^{\mathbf{A}}(a,b,b)$. So we obtain $\ell_*^{\mathbf{A}}(a,b,b) = r_*^{\mathbf{A}}(a,b,b)$. Unless we are at the outermost layer, one more application of the inductive hypothesis gives us $\ell_*^{\mathbf{A}}(a,a,b) = r_*^{\mathbf{A}}(a,a,b)$. For the outside layer, observe that we have by $\mathcal{C}_2(\mathbf{b})$

$$a = \ell_*^{\mathbf{A}}(a,a,b) = \ell_*^{\mathbf{A}}(a,b,b) \text{ and } r_*^{\mathbf{A}}(a,b,b) = r_*^{\mathbf{A}}(a,a,b) = b.$$

(iv) ⇒ (v)
The pairs of terms used for the ◇-condition are those found in (iv). Let $\mathbf{A} \in \mathcal{V}$ and let $p = u_0, \dots, u_n = q \in A$ with $p \neq q$. We must find $i < n$ and $e, f \in A$ so that the relations in the following diagram hold

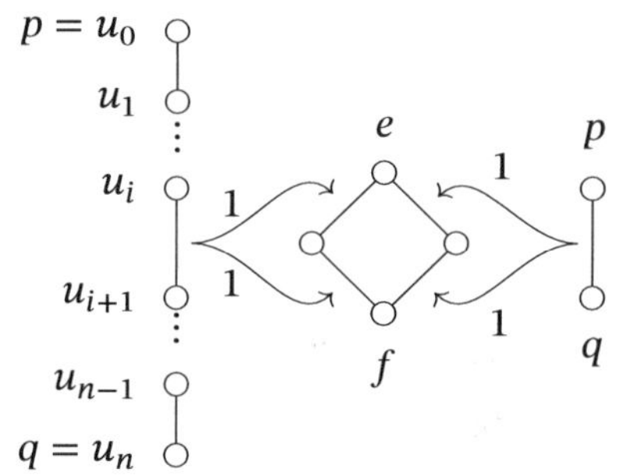

By the second condition of (iv) there is a $t \in T$ so that

$$\ell_t^{\mathbf{A}}(p,p,q) = r_t^{\mathbf{A}}(p,p,q) \Leftrightarrow \ell_t^{\mathbf{A}}(p,q,q) = r_t^{\mathbf{A}}(p,q,q) \text{ fails.}$$

We deal with the case

$$\ell_t^{\mathbf{A}}(p,p,q) = r_t^{\mathbf{A}}(p,p,q) \text{ and } \ell_t^{\mathbf{A}}(p,q,q) \neq r_t^{\mathbf{A}}(p,q,q).$$

The equation $\ell_t^{\mathbf{A}}(p,x,q) = r_t^{\mathbf{A}}(p,x,q)$ holds when $x = p$ and fails when $x = q$. So there must be $i < n$ so that

$$\ell_t^{\mathbf{A}}(p,u_i,q) = r_t^{\mathbf{A}}(p,u_i,q) \text{ and } \ell_t^{\mathbf{A}}(p,u_{i+1},q) \neq r_t^{\mathbf{A}}(p,u_{i+1},q).$$

Let $e = \ell_t^{\mathbf{A}}(p,u_i,q)$ and $f = r_t^{\mathbf{A}}(p,u_{i+1},q)$. Then the left corner of the ◇ can be $\ell_t^{\mathbf{A}}(p,u_i,q) = r_t^{\mathbf{A}}(p,u_i,q)$ and the right corner of the ◇ can be

$$\ell_t^{\mathbf{A}}(p,u_{i+1},p) = r_t^{\mathbf{A}}(p,u_{i+1},p).$$

(v) ⇒ (i)
To see that $\mathcal{V}$ is congruence meet-semidistributive, it is enough to show that for $\mathbf{A} \in \mathcal{V}$ and $\alpha, \beta, \gamma \in \operatorname{Con} \mathbf{A}$, if $\alpha \wedge \beta = 0_A = \alpha \wedge \gamma$, then $\alpha \cap (\beta \vee \gamma) = 0_A$. Suppose otherwise. Let $p, q \in A$ with $p \neq q$ so that $\langle p,q \rangle \in \alpha \cap (\beta \vee \gamma)$. This means $\langle p,q \rangle \in \alpha$ and $\langle p,q \rangle \in \beta \vee \gamma$. So there are $p = u_0, u_1, \dots, u_n = q$ in A so that $\langle u_i, u_{i+1} \rangle \in \beta \cup \gamma$ for all $i < n$. According to the ◇-condition there are $e, f \in A$ with $e \neq f$ so that $\langle e,$

$f\rangle \in \mathrm{Cg}^{\mathbf{A}}(p,q) \cap \mathrm{Cg}^{\mathbf{A}}(u_i, u_{i+1})$. But $\langle p,q\rangle \in \alpha$ and $\langle u_i, u_{i+1}\rangle \in \beta$ or else $\langle u_i, u_{i+1}\rangle \in \gamma$. So one of $\alpha \wedge \beta$ or $\alpha \wedge \gamma$ must be nontrivial.

(i) $\Rightarrow$ (vi)
Suppose $\mathcal{V}$ is congruence meet semidistributive. Then, since we know (i) $\Rightarrow$ (iii), there is a bracket expression **b** such that $\mathcal{V}$ satisfies the Mal'tsev condition $\mathcal{C}_2(\mathbf{b})$. Let $\mathcal{U}$ be a nontrivial variety of **R**-modules which satisfies this Mal'tsev condition. By (iii) $\Rightarrow$ (iia) we have that $\mathcal{C}_2$ holds in $\mathcal{U}$. Let **M** be a nontrivial module in $\mathcal{U}$. Then in $\mathbf{M} \times \mathbf{M}$ the two projections and the congruence generated by the diagonal form a copy of the lattice $\mathbf{M}_3$. Call these three congruences α, β and γ. Then it is easy to see from (6.4.1) that in $\mathbf{Con}(\mathbf{M}^2)$, $\beta_n = \beta$ and that $\alpha \wedge (\beta \circ \gamma) = \alpha \wedge (\beta \vee \gamma) = \alpha$. Hence, ($\mathcal{C}_2$) says $\alpha \subseteq \beta$. This contradiction shows that (vi) holds.

(vi) $\Rightarrow$ (i)
This is proved in Theorem 11.73 on page 329 of Volume III.

(i) $\Leftrightarrow$ (vii)
This equivalence is proved in Theorem 11.37 on page 300 of the next volume. ■

EXAMPLE 6.22. We illustrate the previous theorem by showing congruence lattices of semilattices are meet semidistributive (Dona Papert, 1964). Let $\mathcal{S}$ be the variety of semilattices with the single binary operation denoted by juxtaposition and let α, β and γ be the congruences of $\mathbf{F}_{\mathcal{S}}(x,y,z)$ given in (iib). We construct β_1, γ_1 and β_2 as indicated in (6.4.1) on page 27. The reader may easily verify that $\mathbf{F} = \mathbf{F}_{\mathcal{V}}(x,y,z)$ has universe $\{x, y, z, xy, xz, yz, xyz\}$, and that

$$\begin{aligned} \beta &= [x, y, xy][xz, yz, xyz][z] \\ \beta_1 &= [x, y, xy, xz, yz, xyz][z] \\ \beta_2 &= [x, y, z, xy, xz, yz, xyz] = 1_{\mathbf{F}} \end{aligned}$$

Since $\langle x, z\rangle \in \beta_2$, $\mathcal{V}$ satisfies (iib) and so is congruence meet-semidistributive.

To get the bracket expression and the corresponding Mal'tsev condition, we follow the proof above. We know $\langle x,z\rangle \in \beta_2 = \beta \vee (\alpha \wedge \gamma_1)$ and we look for a sequence of elements witnessing this with consecutive elements related by β or $\alpha \wedge \gamma_1$ and such that the first and last links are both β-links. We find

$$x \; \beta \; xy \; (\alpha \wedge \gamma_1) \; yz \; \beta \; xz \; (\alpha \wedge \gamma_1) \; z \; \beta \; z.$$

Since $\gamma_1 = \gamma \vee (\alpha \wedge \beta)$, we look for a sequence from xy to yz of elements that are related by γ or $\alpha \wedge \beta$:

$$xy \; \gamma \; xyz \; (\alpha \wedge \beta) \; yz \; \gamma \; yz.$$

Similarly,

$$xz \; \gamma \; xyz \; (\alpha \wedge \beta) \; yz \; \gamma \; z$$

This gives the bracket expression $\langle\langle\langle\ \rangle\rangle\langle\langle\ \rangle\rangle\rangle$ with terms that witness the corresponding Mal'tsev condition of (6.4.2):

$$x \quad xy \quad xyz \quad yz \quad yz \quad xz \quad xyz \quad yz \quad z \quad z.$$

In Exercise 1 you will prove that semilattices have meet semidistributive congruence lattices using condition (iv) of Theorem 6.21.

THEOREM 6.23 (Dona Papert 1964). *The congruence lattices of semilattices are meet semidistributive. More generally, if* $\mathbf{A}$ *is an algebra for which there is a binary term t such that* $\mathbf{A}$ *satisfies the semilattice identities for t, then* **Con A** *is meet semidistributive.*

Proof. The above example gives the first statement. An algebra that has a semilattice term clearly satisfies the same $\mathcal{C}_2(\mathbf{b})$ that semilattices do. ■

Now we turn to the characterization of congruence join semidistributivity; $\mathrm{SD}_\vee$, for short. Bjarni Jónsson (1980), Problem 2.18, asks if $\mathrm{SD}_\vee$ could be characterized even by a weak Mal'tsev condition. The next theorem shows it can in fact be characterized by a Mal'tsev condition, but getting there took some effort. Gábor Czédli (1981) gave a Mal'tsev-like characterization. That $\mathrm{SD}_\vee \Rightarrow \mathrm{SD}_\wedge$ was shown in Exercise 7.14(10) by David Hobby and Ralph McKenzie (1988). Their monograph also contains a proof of much of the next theorem for finite algebras in locally finite varieties. Condition (iv) of the next theorem, which has some advantages over the original formulation, is due to Ralph Freese and Ralph McKenzie (2017). We prove each of the conditions (i) to (v) implies the next. However, the proof that (v) $\Rightarrow$ (i) is a deep result of Keith A. Kearnes and Emil W. Kiss (2013) requiring commutator theory, and so will be postponed until Chapter 11. The difficult part boils down to proving that a variety which is congruence meet semidistributive and has a Hobby-McKenzie term is congruence join semidistributive; see Corollary 11.75.

THEOREM 6.24. *The following are equivalent for a variety* $\mathcal{V}$*:*

(i) $\mathcal{V}$ *is congruence semidistributive.*

(ii) $\mathcal{V}$ *is congruence join semidistributive.*

(iiia) *If* α, β *and* γ *are congruences on an algebra* $\mathbf{A} \in \mathcal{V}$*, then*

$$\gamma \cap (\alpha \circ \beta) \subseteq (\alpha \wedge \beta) \vee (\beta \wedge \gamma) \vee (\alpha \wedge \gamma).$$

(iiib) *If* $\alpha = \mathrm{Cg}^{\mathbf{F}}(x,y)$, $\beta = \mathrm{Cg}^{\mathbf{F}}(y,z)$ *and* $\gamma = \mathrm{Cg}^{\mathbf{F}}(x,z)$*, where* $\mathbf{F} = \mathbf{F}_{\mathcal{V}}(x,y,z)$*, then*

$$\langle x,z\rangle \in (\alpha \wedge \beta) \vee (\beta \wedge \gamma) \vee (\alpha \wedge \gamma).$$

(iv) *For some k,* $\mathcal{V}$ *has terms* $d_0(x,y,z), \dots, d_k(x,y,z)$ *satisfying*

$$\begin{aligned}
&d_0(x,y,z) \approx x;\\
&d_i(x,y,y) \approx d_{i+1}(x,y,y) \quad \text{if } i \equiv 0 \text{ or } 1 \pmod 3;\\
&d_i(x,y,x) \approx d_{i+1}(x,y,x) \quad \text{if } i \equiv 0 \text{ or } 2 \pmod 3;\\
&d_i(x,x,y) \approx d_{i+1}(x,x,y) \quad \text{if } i \equiv 1 \text{ or } 2 \pmod 3;\\
&d_k(x,y,z) \approx z.
\end{aligned}$$

(v) $\mathcal{V}$ *satisfies an idempotent Mal'tsev condition that fails in the variety of semilattices and in every nontrivial variety of modules.*

Proof. Clearly (i) implies (ii). To see (ii) implies (iiia) let $\langle a,c\rangle \in \gamma \cap (\alpha \circ \beta)$. Then $\langle a,c\rangle \in \gamma$ and, for some b, $\langle a,b\rangle \in \alpha$ and $\langle b,c\rangle \in \beta$. Let $\alpha' = \mathrm{Cg}^{\mathbf{A}}(a,b)$, $\beta' = \mathrm{Cg}^{\mathbf{A}}(b,c)$

and $\gamma' = \mathrm{Cg}^{\mathbf{A}}(a, c)$. Then $\alpha' \vee \beta' = \alpha' \vee \gamma' = \beta' \vee \gamma'$. Using join semidistributivity we have $\alpha' \vee (\beta' \wedge \gamma') = \beta' \vee (\alpha' \wedge \gamma') = \alpha' \vee \beta'$. So

$$\alpha' \vee (\beta' \wedge \gamma') \vee (\alpha' \wedge \gamma') = \alpha' \vee \beta' ;$$

$$\beta' \vee (\beta' \wedge \gamma') \vee (\alpha' \wedge \gamma') = \alpha' \vee \beta' ;$$

and finally applying join semidistributivity again we get

$$(\alpha' \wedge \beta') \vee (\beta' \wedge \gamma') \vee (\alpha' \wedge \gamma') = \alpha' \vee \beta' .$$

Since $\alpha' \subseteq \alpha$, etc., we see that $(a, c) \in (\alpha \wedge \beta) \vee (\beta \wedge \gamma) \vee (\alpha \wedge \gamma)$, proving (iiia).

Obviously (iiia) implies (iiib). And (iv) is an easy consequence of (iiib) using the arguments we have given for several of the previous Mal'tsev conditions.

To see (iv) implies (v) first suppose the variety of semilattices has terms $d_i(x, y, z)$ satisfying the equations of (iv). Since the free semilattice on $\{x, y, z\}$ has 7 elements, any term with three or less variables is equivalent to one of x, y, z, xy, xz, yz, xyz. We prove by induction on i that $d_i(x, y, z) \approx x$. Suppose $i \equiv 2 \pmod 3$ and that $d_i(x, y, z) \approx x$. By the third displayed equation of (iv) we see that $d_{i+1}(x, y, z)$ must be either x or z or xz, while the fourth equation implies it is either x or y or xy. Hence $d_{i+1}(x, y, z) \approx x$. A similar but easier argument shows this also holds for $i \equiv 0, 1 \pmod 3$. So we end up with $x \approx d_k(x, y, z) \approx z$, a contradiction.

Let $\mathbf{R}$ be a ring with $0 \neq 1$ and suppose $\{x, y, z\}$ are generators of a free $\mathbf{R}$-module $\mathbf{M}$. Suppose there are terms d_i such that condition (iv) holds in $\mathbf{M}$. Then there are ring elements r_i, s_i, t_i with $d_i^{\mathbf{M}}(x, y, z) = r_i x + s_i y + t_i z$, $i = 0, \ldots, k$. Now the first and second equation of (iv) with $i = 0$ show $x = r_1 x + s_1 y + t_1 y$. Since $\mathbf{M}$ is free, we have $r_1 = 1$ and $s_1 + t_1 = 0$. The third equation gives $r_1 + t_1 = 1$ and $s_1 = 0$. These combined to yield that $r_1 = 1$ and $s_1 = t_1 = 0$, so $d_1^{\mathbf{M}} = x$. We can continue in this manner to prove that $x = z$ in $\mathbf{M}$, contradicting that $\mathbf{R}$, and hence $\mathbf{M}$, is nontrivial. This completes the proof that (iv) implies (v). That (v) implies (i) is proved in Theorem 11.74 of Volume III. ■

Congruence Lattices of Semilattices

The main goal of this subsection is to show that, despite being a rather limited class, congruence lattices of semilattices do not satisfy any nontrivial lattice equation. As in Example 6.22 we let $\mathcal{S}$ be the variety of semilattices. And as in that example we use juxtaposition or $\cdot$ for the single operation which we view as meet so that $a \leq b$ is defined as $a = a \cdot b$. And as we saw from that example, the congruence lattice of a semilattice is meet semidistributive.

The two element semilattice is denoted $\mathbf{2}$ with universe $\{0, 1\}$ and whose operation is ordinary multiplication.

Let $\mathbf{S}$ be a semilattice and let D be a filter of $\mathbf{S}$ as defined in Definition 2.17 on page 48 of Volume I. The map

$$f(x) = \begin{cases} 1 & \text{if } x \in D \\ 0 & \text{otherwise} \end{cases}$$

defines a homomorphism from $\mathbf{S}$ to $\mathbf{2}$. Let $\psi_D \in \mathbf{Con}\,\mathbf{S}$ be the kernel of this homomorphism. When $D = I[a)$ is the principal filter generated by a, we denote this congruence by ψ_a. As long as D is a proper filter, that is $D \neq S$, ψ_D has two blocks and so is a coatom of $\mathbf{Con}\,\mathbf{S}$. Conversely, an easy argument shows every congruence with exactly two blocks has this form.

LEMMA 6.25. *Let* **S** *be a semilattice. Then the following statements hold.*

(i) $0_S = \bigwedge_{a \in S} \psi_a$.
(ii) *Every congruence* α *of* **S** *is the meet of the set of coatoms above it.*
(iii) *A congruence* θ *is completely meet irreducible iff* $\theta = \psi_D$, *for some proper filter* D.
(iv) *Up to isomorphism,* **2** *is the only subdirectly irreducible semilattice.*
(v) *Let* ψ_{D_1} *and* ψ_{D_2} *be distinct coatoms and suppose that* $D := D_1 \cap D_2$, *then the interval* $I[\psi_{D_1} \wedge \psi_{D_2}, 1_S]$ *is the lattice on the left of Figure 6.7 provided* $D \neq \emptyset$, $D \neq D_1$ *and* $D \neq D_2$*; otherwise it is the lattice on the right.*

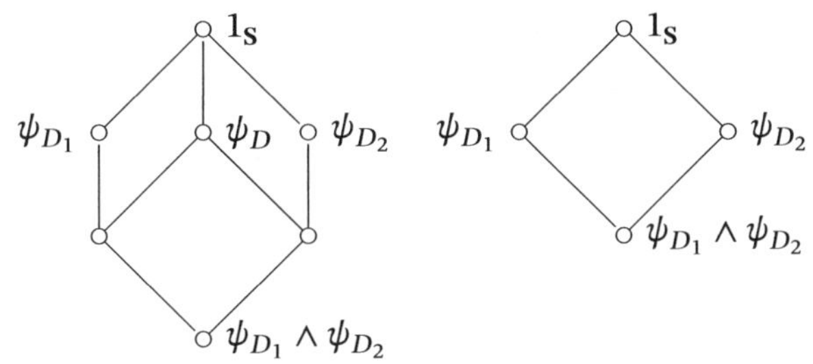

FIGURE 6.7. The Interval above the Meet of two Coatoms of **Con S**.

Proof. Suppose $\langle b, c \rangle \in \bigwedge_{a \in S} \psi_a$, with $b \neq c$. We may assume $b \not\leq c$. Then b and c are in different blocks of ψ_b so $\langle b, c \rangle \notin \psi_b$ and thus not in $\bigwedge_{a \in S} \psi_a$, contradicting our assumption. This shows $0_S = \bigwedge_{a \in S} \psi_a$ and thus (i) holds. Now using the Correspondence Theorem on page 151 of Volume I it follows that every congruence is the meet of the coatoms above it so (ii) holds. Moreover it shows that every congruences is the meet of coatoms with exactly two blocks. Consequently if ψ is completely meet irreducible, then ψ must be a coatom with two blocks. this proves the forward direction of (iii), and the other direction is obvious. (iv) follows from (iii) together with the obvious fact that **2** is, up to isomorphism, the only two-element semilattice,

To see (v), let $\mathbf{B} = \mathbf{S}/\psi_{D_1} \wedge \psi_{D_2}$, and note it is subdirectly embedded into the direct product of $\mathbf{S}/\psi_{D_1} \times \mathbf{S}/\psi_{D_2} = \mathbf{2} \times \mathbf{2}$. So either $\mathbf{B} = \mathbf{2} \times \mathbf{2}$ or **B** is a 3-element semilattice. There are two 3-element semilattices, the three element chain and the semilattice that looks like a V on $\{0, a, b\}$, with $a \cdot b = 0$. Both have the same congruence lattice; the one diagrammed on the right in Figure 6.7. For the other case, when $\mathbf{B} = \mathbf{2} \times \mathbf{2}$, the reader can verify **Con B** is the lattice on the left, and moreover that in **Con S**, $\psi_D \geq \psi_{D_1} \wedge \psi_{D_2}$, and that $\psi_D = (\psi_{D_1} \wedge \psi_D) \vee (\psi_{D_2} \wedge \psi_D)$. ■

At this point we assume **S** is finite. Then every filter is principal. Let a and b be distinct elements of **S**. If the set $\{a, b\}$ has an upper bound it has a least upper bound. This defines a partial operation denoted $a +^{\mathbf{S}} b$, which, following our convention, we shall write $a+b$. We let $\psi_a = \psi_{I[a)}$, etc. Then, if $a+b = c$ and a and b are incomparable, the interval $I[\psi_a \wedge \psi_b, 1^{\mathbf{S}}]$ in **Con S** is the lattice on the left in Figure 6.7 with ψ_D being ψ_c. In the other three cases, $a < b$, $b < a$, or a and b are incomparable and have no upper bound, $I[\psi_a \wedge \psi_b, 1^{\mathbf{S}}]$ is the lattice on the right.

LEMMA 6.26. *Let* **S** *be a finite semilattice with least element* 0*. Then* **Con S** *is dually isomorphic to* **Sub** $\langle S,+,0\rangle$.

Proof. For $d \in S$, let $S_d = \{0, d\}$. This is clearly an atom in **Sub** $\langle S,+,0\rangle$ and every element of **Sub** $\langle S,+,0\rangle$ is a join of atoms of this form. To prove the lemma we will show that the map $\psi_d \mapsto S_d$ extends to a dual isomorphism of **Con S** onto **Sub** $\langle S,+,0\rangle$. Since every element of **Con S** is the meet of coatoms and every element of **Sub** $\langle S,+,0\rangle$ is the join of atoms (the S_d's), to do this we need to show

$$\psi_c \geq \psi_{d_1} \wedge \cdots \wedge \psi_{d_k} \quad \text{if and only if} \quad S_c \leq S_{d_1} \vee \cdots \vee S_{d_k},$$

that is, we need to verify

$$\psi_c \geq \psi_{d_1} \wedge \cdots \wedge \psi_{d_k} \quad \text{if and only if} \quad c \in \mathrm{Sg}^{\langle S,+,0\rangle}(\{d_1,\ldots,d_k\}).$$

Note that the blocks of $\psi_{d_1} \wedge \cdots \wedge \psi_{d_k}$ are the nonempty subsets of S of the form

$$(*) \qquad \bigcap_{i\in K} I[d_i) - \bigcup_{j\in K^c} I[d_j)$$

for $K \subseteq \{1,\ldots,k\}$. We understand $\bigcap_{i\in K} I[d_i)$ to be S when $K = \emptyset$.

If $\psi_c \geq \psi_{d_1} \wedge \cdots \wedge \psi_{d_k}$ then we claim c is the least element of the $\psi_{d_1} \wedge \cdots \wedge \psi_{d_k}$-block containing it; for otherwise, since blocks are closed under $\cdot$, there would be $c' < c$ in that same block. But this would imply $\langle c', c\rangle \in \psi_c$, a contradiction. Let the block containing c be the one given by $(*)$ so $c \in \bigcap_{i\in K} I[d_i)$. Then c is an upper bound for d_i, $i \in K$. So $\sum_{i\in K} d_i$ exists and is the least element of this block. So $c = \sum_{i\in K} d_i$ and hence $c \in \mathrm{Sg}^{\langle S,+,0\rangle}(\{d_1,\ldots,d_k\})$.

For the converse suppose $c \in \mathrm{Sg}^{\langle S,+,0\rangle}(\{d_1,\ldots,d_k\})$. Then $c = \sum_{i\in K} d_i$ for some $K \subseteq \{1,\ldots,k\}$. So $\bigwedge_{i\in K} \psi_{d_i}$ has one block equal to $I[c)$ and the other blocks contained in the complement. Thus

$$\psi_c \geq \bigwedge_{i\in K} \psi_{d_i} \geq \psi_{d_1} \wedge \cdots \wedge \psi_{d_k},$$

proving the converse. ∎

It is worth noting that if **S**, which we still assume to be finite, has a greatest element 1, then our partial algebra is a total algebra and in fact is a join semilattice with 0. For each lattice term w with variables from X, we define a join semilattice with 0, denoted $\mathbf{S}(w)$. If $w = x \in X$ then $\mathbf{S}(x) = \langle \{0,1\},+,0\rangle$ is the two element semilattice with 0. Inductively define

$$\mathbf{S}(w_1 \vee w_2) = \mathbf{S}(w_1) \times \mathbf{S}(w_2)$$
$$\mathbf{S}(w_1 \wedge w_2) = \mathbf{S}(w_1) \times \mathbf{S}(w_2) - \Gamma,$$

where

$$\Gamma = \{\langle r,s\rangle \in \mathbf{S}(w_1) \times \mathbf{S}(w_2) : r = 1 \text{ or } s = 1, \text{ but not both}\}.$$

Of course in $\mathbf{S}(w_1 \wedge w_2)$ the join operation is given by

$$(**) \qquad \langle r_1, r_2\rangle + \langle s_1, s_2\rangle = \begin{cases} \langle r_1 + s_1, r_2 + s_2\rangle & \text{if } r_1 + s_1 \neq 1 \text{ and } r_2 + s_2 \neq 1 \\ \langle 1,1\rangle & \text{otherwise.} \end{cases}$$

Note $\mathbf{S}(w)$ is a join semilattice with a greatest element which we will denote by 1.

Let $\mathcal{L}$ be the variety of lattices. For each lattice term w we define by recursion on w a homomorphism

$$\varphi_w : \mathbf{F}_{\mathcal{L}}(X) \to \mathbf{Sub}(\mathbf{S}(w))$$

from the free lattice to the lattice of subalgebras of $\mathbf{S}(w)$. Let $y \in X$.

$$\varphi_x(y) = \begin{cases} \mathbf{S}(x) & \text{if } y = x \\ \{0\} & \text{otherwise} \end{cases}$$

$$\varphi_{w_1 \vee w_2}(y) = \{\langle r_1, r_2\rangle : r_1 \in \varphi_{w_1}(y),\ r_2 \in \varphi_{w_2}(y)\}$$

$$\varphi_{w_1 \wedge w_2}(y) = \{\langle r_1, r_2\rangle : r_1 \in \varphi_{w_1}(y) - \{1\},\ r_2 \in \varphi_{w_2}(y) - \{1\}\} \cup \Lambda$$

where Λ is $\{1\}$ if $1 \in \varphi_{w_1}(y)$ or $1 \in \varphi_{w_2}(y)$, and is $\emptyset$ otherwise. This defines φ_w on X and of course this extends to a homomorphism from $\mathbf{F}_{\mathcal{L}}(X)$ to $\mathbf{Sub}(\mathbf{S}(w))$.

LEMMA 6.27. *If u, w_1 and w_2 are lattice terms, then*

(i) $\varphi_{w_1 \vee w_2}(u) = \varphi_{w_1}(u) \times \varphi_{w_2}(u)$,
(ii) $\varphi_{w_1 \wedge w_2}(u) - \{1\} = \{\langle r, s\rangle \in \varphi_{w_1}(u) \times \varphi_{w_2}(u) : r \neq 1,\ s \neq 1\}$.

Proof. We shall prove the second statement and leave the first, which is similar but easier, to the reader. We induct on the length of u and w. If $u = y \in X$ then (ii) follows from the definition. Notice that if T_1 and T_2 are in $\mathbf{Sub}(\mathbf{S}(w))$, then, since $0 \in T_i$,

$$T_1 \vee T_2 = \{t_1 + t_2 : t_i \in T_i, i = 1, 2\}.$$

Suppose that $u = u_1 \vee u_2$. Then by $(**)$

$$\begin{aligned}\varphi_{w_1 \wedge w_2}(u) - \{1\} &= \varphi_{w_1 \wedge w_2}(u_1 \vee u_2) - \{1\} \\ &= \{\langle r_1 + r_2, s_1 + s_2\rangle : \langle r_i, s_i\rangle \in \varphi_{w_1 \wedge w_2}(u_i), i = 1, 2, r_1 + r_2 \neq 1, s_1 + s_2 \neq 1\}\end{aligned}$$

Now for $i = 0, 1$, induction gives

$$\langle r_i, s_i\rangle \in \varphi_{w_1 \wedge w_2}(u_i) - \{1\} = (\varphi_{w_1}(u_i) - \{1\}) \times (\varphi_{w_2}(u_i) - \{1\}),$$

and hence

$$\varphi_{w_1 \wedge w_2}(u) - \{1\} = \{\langle r, s\rangle \in \varphi_{w_1}(u) \times \varphi_{w_2}(u) : r \neq 1 \text{ and } s \neq 1\}$$

Now assume $u = u_1 \wedge u_2$. Then

$$\varphi_{w_1 \wedge w_2}(u) - \{1\} = (\varphi_{w_1 \wedge w_2}(u_1) - \{1\}) \cap (\varphi_{w_1 \wedge w_2}(u_2) - \{1\}),$$

and the lemma follows. ■

LEMMA 6.28. *If w is a lattice term, then $1 \in \varphi_w(w)$.*

Proof. As before we use induction on the length of w. The case $w = y \in X$ follows from the definitions. Suppose now $w = w_1 \vee w_2$. Then by induction $1 \in \varphi_{w_i}(w_i)$, $i = 1, 2$. By (i) of the last lemma, $\langle 1, 0\rangle \in \varphi_w(w_1)$ and $\langle 0, 1\rangle \in \varphi_w(w_2)$, so $\langle 1, 1\rangle \in \varphi_w(w_1) \vee \varphi_w(w_2) = \varphi_w(w)$. Of course $\langle 1, 1\rangle$ is the greatest element of $\mathbf{S}(w)$.

Now suppose $w = w_1 \wedge w_2$ and assume that $1 \in \varphi_{w_i}(w_i)$, $i = 0, 1$. We will show that $1 \in \varphi_w(w_i)$, $i = 0, 1$, which of course gives the desired result. We will prove more, namely,

$$\text{if } 1 \in \varphi_{w_1}(u) \text{ then } 1 \in \varphi_{w_1 \wedge w_2}(u) = \varphi_w(u).$$

If $u = y \in X$ then the definition of $\varphi_{w_1 \wedge w_2}(y)$ forces $\Lambda = \{1\}$ and so $1 \in \varphi_{w_1 \wedge w_2}(y)$.

Suppose $u = u_1 \vee u_2$. Then $1 \in \varphi_{w_1}(u) = \varphi_{w_1}(u_1) \vee \varphi_{w_1}(u_2)$, and so $1 = t_1 + t_2$ with $t_i \in \varphi_{w_1}(u_i)$. If $t_1 \neq 1$ and $t_2 \neq 1$, then $\langle t_1, 0\rangle \in \varphi_{w_1 \wedge w_2}(u_1)$ and $\langle t_2, 0\rangle \in \varphi_{w_1 \wedge w_2}(u_2)$. But in $\mathbf{S}(w_1 \wedge w_2)$, $1 = \langle 1, 0\rangle$. Consequently,

$$1 = \langle t_1, 0\rangle + \langle t_2, 0\rangle \in \varphi_{w_1 \wedge w_2}(u_1) \vee \varphi_{w_1 \wedge w_2}(u_2) = \varphi_{w_1 \wedge w_2}(u)$$

as claimed. If $t_i = 1$ for one of the i's, then by induction $1 \in \varphi_{w_1}(u_i)$ which gives $1 \in \varphi_{w_1 \wedge w_2}(u_i) \subseteq \varphi_{w_1 \wedge w_2}(u_i)$.

Now suppose $u = u_1 \wedge u_2$. Then $1 \in \varphi_{w_1}(u_1) \wedge \varphi_{w_1}(u_2)$. So by induction $1 \in \varphi_{w_1 \wedge w_2}(u_i)$ for both i's. So $1 \in \varphi_{w_1 \wedge w_2}(u)$. ■

For lattice terms u and w, $w \leq u$ means $w^{\mathbf{F}_{\mathcal{L}}(X)} \leq u^{\mathbf{F}_{\mathcal{L}}(X)}$ holds in the free lattice. This is a quasi-order on the set of terms.

LEMMA 6.29. *If u and v are lattice terms with $u \nleq v$, then $1 \notin \varphi_u(v)$.*

Proof. Suppose the statement of the lemma is false. Choose u of minimal length so that $1 \in \varphi_u(v)$ for some v with $u \nleq v$. Among such v's choose one of minimal length. The only hard case is $u = u_1 \wedge u_2$ and $v = v_1 \vee u_2$. Since $u \nleq v$ we have

$$(\dagger) \qquad u \nleq v_1, \quad u \nleq v_2, \quad u_1 \nleq v, \text{ and } u_2 \nleq v.$$

Since $1 \in \varphi_u(v) = \varphi_u(v_1) \vee \varphi_u(v_2)$, there are $t_i \in \varphi_u(v_i)$, $i = 1, 2$, with $1 = t_1 + t_2$. If $t_i = 1$ for one of the i's, then by the minimality of v, we have $u \leq v_i$, a contradiction. So $t_i \neq 1$ for $i = 1, 2$. By part (ii) of Lemma 6.27 we see that $t_i = \langle r_i, s_i\rangle$ with $r_i \in \varphi_{u_1}(v_i)$ and $s_i \in \varphi_{u_2}(v_i)$. So

$$1 = t_1 + t_2 = \langle r_1, s_1\rangle + \langle r_2, s_2\rangle = \langle r_1 + r_2, s_1 + s_2\rangle.$$

By the definition of $\mathbf{S}(w)$, this implies either $1 = r_1 + r_2$ or $1 = s_1 + s_2$ (or both). If $1 = r_1 + r_2$ in $\mathbf{S}(u_1)$, then $1 \in \varphi_{u_1}(v_1) \vee \varphi_{u_1}(v_2) = \varphi_{u_1}(v)$. By the minimality of the length of u, this implies $u_1 \leq v$, contracting (†). The equality $1 = s_1 + s_2$ leads to $u_2 \leq v$, again giving a contradiction. ■

Combining the last two lemmas we get the following theorem. For $\mathcal{K}$ a class of algebras let $\mathcal{K}_{\text{fin}}$ be the finite members and let $\mathbf{Con}(\mathcal{K})$ be $\{\mathbf{Con}\ \mathbf{A} : \mathbf{A} \in \mathcal{K}\}$. We also let $\mathbf{Sub}(\mathcal{K}) = \{\mathbf{Sub}\ \mathbf{A} : \mathbf{A} \in \mathcal{K}\}$.

THEOREM 6.30 (Ralph Freese and J. B. Nation 1973). *The class of congruence lattices of semilattices satisfies no lattice identity. In fact, $\mathbf{Con}(\mathcal{S}_{\text{fin}})$ and $\mathbf{Sub}(\mathcal{S}_{\text{fin}})$ satisfy no lattice identity.* ■

Exercises 6.31

1. Show that Theorem 6.21(iv) is satisfied in the variety of semilattices by using the family of pairs $\{\langle xyz, xy\rangle, \langle xyz, yz\rangle\}$, giving another proof that semilattices have meet semidistributive congruences.

2. Let $\mathbf{S}$ be a finite meet semilattice and $n = |S|$. An element $a \in S$ is *join irreducible* if a has a unique lower cover in the order on $\mathbf{S}$. We denote this lower cover by a_*.

a. Show the number of coatoms of **Con S** is $n-1$.

b. Show that if a is join irreducible then $\alpha = \mathrm{Cg}^{\mathbf{S}}(a, a_*)$ collapses a and a_* but no other pair; that is, all the blocks of α have only one element except for the block $[a, a_*]$. Of course α is an atom of **Con S**. Show every atom has this form.

c. Since **Con S** is finite and meet semidistributive each element of **Con S** is a unique, irredundant meet of meet irreducible congruences. Show that there is a unique set of k coatoms of **Con S** which irredundantly meet to 0, where k is the number of atoms of **Con S** (which by the previous part is the number of join irreducible elements of **S**). In fact this set is $\{\psi_a : a \text{ is join irreducible in } \mathbf{S}\}$.

d. Let T be a set with $n-1$ elements. Label each coatom of **Con S** with a unique element of T. The interval above the meet of two coatoms is one of the lattices in Figure 6.7. Define a partial operation $+^T$ on T as follows. If a and $b \in T$ and the interval above the meet of the coatoms labeled by a and b is

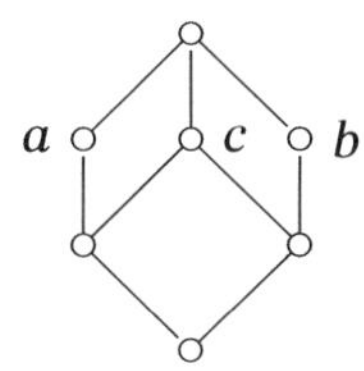

then $a +^T b = c$. Prove that $\langle T, \{+^T\}\rangle \cong \langle S - \{0\}, \{+'^{\mathbf{S}}\}\rangle$, where $+'^{\mathbf{S}}$ is the restriction of $+^{\mathbf{S}}$ to

$$\{\langle a, b\rangle : a \text{ and } b \text{ are incomparable and have an upper bound}\}.$$

3. Let **S** be a finite semilattice with n elements. Show the following are equivalent:

(i) **Con S** is distributive.
(ii) **Con S** is a Boolean algebra.
(iii) **Con S** $\cong \mathbf{2}^{n-1}$.
(iv) **S** has no pair of incomparable elements with an upper bound (**S** is a tree).

4. Can we recover S from Con S as an abstract lattice? No. As pointed out there are two semilattices with three elements and both have the same congruence lattice. However by the last exercise we can find $a+b$ provided a and b are incomparable and have an upper bound. Assume now that our finite meet semilattice has a greatest element 1 (as well as a least element 0). Then $+$ is everywhere defined and, of course, the join $+$ determines the meet $\cdot$. So it seems possible that if we restrict our attention to finite semilattices with a 1, **S** might be determined by **Con S**. The difficulty arises from the fact that if a and b are comparable, **Con S** determines that, but does not tell us which one is larger. Find two nonisomorphic semilattices, both with 1, which have isomorphic congruence lattices.

6.5. Congruence Modularity

The early importance of modularity of the congruence lattice is evident in various generalizations of the Jordan-Hölder theorem (see §2.4 of Volume I), and the Birkhoff-Ore theory of factorization in modular lattices, later refined by B. Jónsson to a theory of factorization of finite algebras with modular congruence lattice (see Chapter 5 of Volume I). The varietal theory of congruence-modularity came into its own with the advent of the **commutator** in modular varieties; we will devote an entire chapter to this topic in the next volume.

Here our purpose is again the modest one of describing congruence-modularity, especially by writing a Mal'tsev condition for it. It turns out that there are two significant Mal'tsev conditions for modularity, Day's Theorem (Theorem 6.32) and Gumm's Theorem (Theorem 6.44).

The modular law, which is a specialization of the distributive law, is:

$$(\alpha \vee \beta) \wedge \gamma \leq (\alpha \wedge \gamma) \vee \beta \qquad \text{if } \beta \leq \gamma. \tag{6.5.1}$$

This is equivalent to the following inequality:

$$(\alpha \vee (\beta \wedge \gamma)) \wedge \gamma \leq (\alpha \wedge \gamma) \vee (\beta \wedge \gamma). \tag{6.5.2}$$

Just as we did for distributivity, we need an approximate form of the modular law in which $\circ$ appears instead of $\vee$. It turns out that the most suitable approximation is

$$(\alpha \circ (\beta \wedge \gamma) \circ \alpha) \wedge \gamma \leq \underbrace{(\alpha \wedge \gamma) \circ (\beta \wedge \gamma) \circ (\alpha \wedge \gamma) \circ \cdots}_{n \text{ factors}}. \tag{Σ_n}$$

This approximation of modularity has the same shortcoming that we saw for Δ_n in our discussion of distributivity in the previous section, namely, Σ_n does not imply modularity of **Con A** when it is interpreted in the usual way. For a counterexample, see Exercise 6.46.6. However, we shall see from the proof of the next theorem that if we assume (Σ_n) holds for α any symmetric, reflexive subuniverse of $\mathbf{A}^2$, and for any congruences β and γ on **A**, then **Con A** is modular.

We can now present the main Mal'tsev condition for congruence modularity.

THEOREM 6.32 (Alan Day 1969). *For a variety $\mathcal{V}$, the following conditions are equivalent:*

(i) *$\mathcal{V}$ is congruence modular;*
(ii) **Con** $\mathbf{F}_{\mathcal{V}}(4)$ *is modular;*
(iii) *$\mathcal{V}$ satisfies some Σ_n (as a congruence identity);*
(iv) *There exist 4-ary terms $m_0, \ldots, m_n$ in the language of $\mathcal{V}$ such that the following identities hold in*

$$\begin{aligned} & m_0(x,y,z,u) \approx x && \\ & m_i(x,y,y,x) \approx x && \textit{for all } i \\ & m_i(x,x,u,u) \approx m_{i+1}(x,x,u,u) && \textit{for } i \textit{ even} \\ & m_i(x,y,y,u) \approx m_{i+1}(x,y,y,u) && \textit{for } i \textit{ odd} \\ & m_n(x,y,z,u) \approx u. && \end{aligned} \tag{M_n}$$

Moreover, $\mathcal{V}$ satisfies (iii) *for a fixed value of n if and only if $\mathcal{V}$ satisfies* (iv) *for the same value of n.*

Proof. We first prove that for fixed n, (iii) and (iv) are both equivalent to the auxiliary condition:

(v) If α, β, and γ are the principal congruences on $\mathbf{F}_{\mathcal{V}}(x, y, z, u)$ generated by $\{\langle x, y\rangle, \langle z, u\rangle\}$, $\{\langle y, z\rangle\}$, and $\{\langle x, u\rangle, \langle y, z\rangle\}$, then

$$\langle x, u\rangle \in \underbrace{(\alpha \wedge \gamma) \circ \beta \circ (\alpha \wedge \gamma) \circ \cdots}_{n \text{ factors}}. \tag{6.5.3}$$

We obviously have (iii) implies (v). To see that (v) implies (iv) note that (6.5.3) means that there are terms $m_0, \ldots, m_n$ such that

$$\begin{aligned} x = m_0^{\mathbf{F}}(x, y, z, u) \;\; (\alpha \wedge \gamma) \;\; m_1^{\mathbf{F}}(x, y, z, u) \;\; \beta \;\; m_2^{\mathbf{F}}(x, y, z, u) \\ (\alpha \wedge \gamma) \;\; m_3^{\mathbf{F}}(x, y, z, u) \;\; \beta \;\; \cdots \;\; m_n^{\mathbf{F}}(x, y, z, u) = u \end{aligned}$$

where $\mathbf{F} = \mathbf{F}_{\mathcal{V}}(x, y, z, u)$. Since $\beta \leq \gamma$ we have $x \; \gamma \; m_i^{\mathbf{F}}(x, y, z, u) \; \gamma \; m_i^{\mathbf{F}}(x, y, y, x)$. Since γ is trivial on the subalgebra of $\mathbf{F}$ generated by x and y by Theorem 6.9, $\mathcal{V}$ satisfies $x \approx m_i(x, y, y, x)$. Similar reasoning shows that the other equations of (iv) hold.

To see that (iv) implies (iii), let α, β, and $\gamma \in \mathbf{Con\,A}$ for some $\mathbf{A} \in \mathcal{V}$ and suppose that $\langle a, d\rangle \in \gamma \wedge (\alpha \circ (\beta \vee \gamma) \circ \alpha)$. Hence there are b and $c \in A$ so that the relations in the figure below hold.

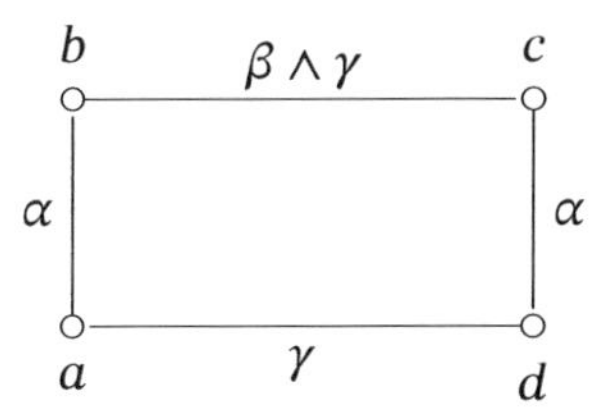

Let

$$e_i = m_i^{\mathbf{A}}(a, b, c, d) \qquad \text{for } 0 \leq i \leq n$$

Now (iv) yields

$$\begin{aligned} e_0 &= m_0^{\mathbf{A}}(a, b, c, d) = a \\ e_i &= m_i^{\mathbf{A}}(a, b, c, d) \;\; \gamma \;\; m_i^{\mathbf{A}}(a, b, b, a) = a && \text{for all } i \\ e_i &= m_i^{\mathbf{A}}(a, b, c, d) \;\; \alpha \;\; m_i^{\mathbf{A}}(a, a, d, d) = m_{i+1}^{\mathbf{A}}(a, a, d, d) \\ &\quad \alpha \;\; m_{i+1}^{\mathbf{A}}(a, b, c, d) = e_{i+1} && \text{for } i \text{ even} \\ e_i &= m_i^{\mathbf{A}}(a, b, c, d) \;\; \beta \;\; m_i^{\mathbf{A}}(a, b, b, d) = m_{i+1}^{\mathbf{A}}(a, b, b, d) \\ &\quad \beta \;\; m_{i+1}^{\mathbf{A}}(a, b, c, d) = e_{i+1} && \text{for } i \text{ odd} \\ e_n &= m_n^{\mathbf{A}}(a, b, c, d) = d \end{aligned}$$

Thus we have

$$a = e_0 \; (\alpha \wedge \gamma) \; e_1 \; (\beta \wedge \gamma) \; e_2 \; (\alpha \wedge \gamma) \; e_3 \; (\beta \wedge \gamma) \; e_4 \cdots e_n = d,$$

showing that (Σ_n) holds. Thus (iii), (iv), and (v) are equivalent.

Now the fact that (i) implies (ii), and that (ii) implies (v) are both trivial. We will show that (iv) implies (i), which will show that (i)–(v) are equivalent.

Let us first notice that if, in the above proof that (iv) implies (iii), we weaken the assumption that α is transitive and assume only that α is a reflexive symmetric subuniverse of $\mathbf{A}^2$, then we still obtain the slightly weaker conclusion that

$$\langle a,d\rangle \in (\alpha\wedge\gamma)\circ(\alpha\wedge\gamma)\circ(\beta\wedge\gamma)\circ(\alpha\wedge\gamma)\circ\cdots, \tag{6.5.4}$$

with the pattern of two $(\alpha\wedge\gamma)$–factors followed by one $(\beta\wedge\gamma)$–factor repeating as necessary, so the number of $(\alpha\wedge\gamma)$-factors divided by 2 plus the number of $(\beta\wedge\gamma)$-factors is n. (To the above argument, we need only add the observation that $m_i^{\mathbf{A}}(a,b,c,d)\ \gamma\ m_i^{\mathbf{A}}(a,a,d,d)$, which is true since both are γ-related to a.) In proving the modular law (6.5.2) from (iv) we will use (6.5.4) with α replaced by the reflexive symmetric subuniverse

$$\Gamma_k = \alpha\circ(\beta\wedge\gamma)\circ\alpha\circ(\beta\wedge\gamma)\circ\cdots\circ\alpha \qquad \text{with } 2k+1 \text{ factors}$$

Since $\bigcup_{k<\omega}(\Gamma_k\wedge\gamma) = (\alpha\vee(\beta\wedge\gamma))\wedge\gamma$, modularity will follow from the inclusions

$$\Gamma_k\wedge\gamma \leq (\alpha\wedge\gamma)\vee(\beta\wedge\gamma)$$

which we will prove by induction on k. The inclusion is obvious for $k=0$, and for the inductive step, we first observe that $\Gamma_{k+1} = \Gamma_k\circ(\beta\wedge\gamma)\circ\alpha$, and so if $\langle a,d\rangle \in \Gamma_{k+1}\wedge\gamma$ we have that for some $b,c\in A$ the following situation holds.

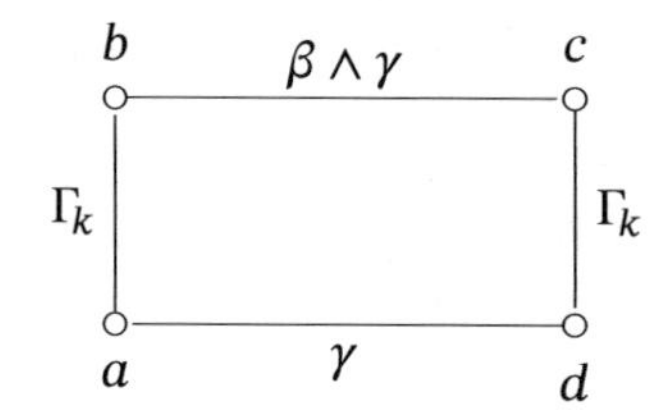

Thus by (6.5.4) we have $\langle a,d\rangle$ is in a finite relational product of the relations $\Gamma_k\wedge\gamma$ and $\beta\wedge\gamma$. Now by induction, we have $\langle a,d\rangle \in (\alpha\wedge\gamma)\vee(\beta\wedge\gamma)$. This completes the proof of the theorem. ■

It is easy to see that the sublattice of **Con** $\mathbf{F}_{\mathcal{V}}(x,y,z,u)$ generated by the congruences α, β, and γ defined in condition (v) is a homomorphic image of the lattice diagrammed in Figure 6.8 on page 42; see Exercise 3.

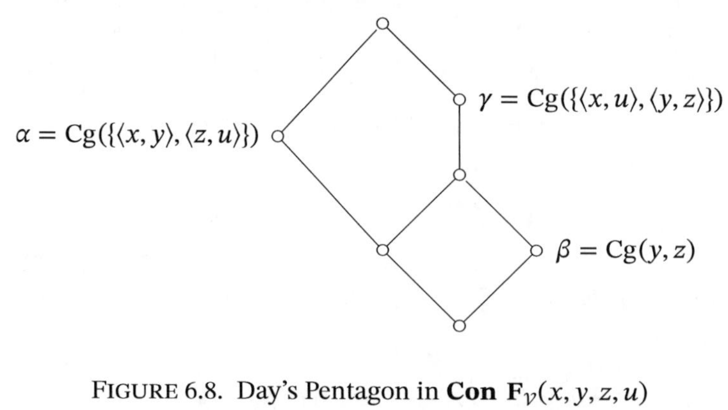

FIGURE 6.8. Day's Pentagon in **Con** $\mathbf{F}_{\mathcal{V}}(x,y,z,u)$

Obviously if $\mathcal{V}$ is congruence modular, the pentagon of Figure 6.8, which is know as **Day's pentagon**, must collapse. One of the surprising consequences of Day's result is that this is sufficient.

COROLLARY 6.33. *Let α, β, and γ be the congruences on $\mathbf{F}_{\mathcal{V}}(x, y, z, u)$ generated by $\{\langle x, y\rangle, \langle z, u\rangle\}$, $\{\langle y, z\rangle\}$, and $\{\langle x, u\rangle, \langle y, z\rangle\}$ (see Figure 6.8). Then $\mathcal{V}$ is congruence modular if and only if $\beta \vee (\alpha \wedge \gamma) = \gamma$, which is equivalent to*

$$\langle x, u\rangle \in \beta \vee (\alpha \wedge \gamma).$$

■

By Theorem 6.32, $\mathcal{V}$ is congruence modular if and only if **Con** $\mathbf{F}_{\mathcal{V}}(4)$ is modular. This statement is false if 4 is replaced by 3. Indeed, the variety of sets is not congruence modular, but **Con** $\mathbf{F}_{\mathcal{V}}(3)$ is the partition lattice on a three element set, which is $\mathbf{M}_3$.

Notice that the congruence α of the corollary is the kernel of the endomorphism of $\mathbf{F}_{\mathcal{V}}(x, y, z, u)$ which maps $x \mapsto x$, $y \mapsto x$, $z \mapsto u$, and $u \mapsto u$. By the same token, γ is also the kernel of a homomorphism from $\mathbf{F}_{\mathcal{V}}(4)$ onto $\mathbf{F}_{\mathcal{V}}(2)$. Since the copy of $\mathbf{N}_5$ of Figure 6.8 lies above $\alpha \wedge \gamma$, the $\mathbf{N}_5$ lies in the congruence lattice of a subdirect product of two copies of $\mathbf{F}_{\mathcal{V}}(2)$. Thus the next result follows from the last corollary.

COROLLARY 6.34. *A variety $\mathcal{V}$ is congruence modular if and only if the variety generated by $\mathbf{F}_{\mathcal{V}}(2)$ is.* ■

We will now use Theorem 6.32 to investigate other conditions equivalent to congruence modularity for varieties. We begin with an important lemma[6] due to H. P. Gumm (1978, 1980). Our proof follows Walter Taylor (1982).

LEMMA 6.35. *Let $\mathcal{V}$ be a congruence modular variety. Then there is a term $d(x, y, z)$ such that*

(i) *$\mathcal{V}$ satisfies $d(x, x, z) \approx z$.*
(ii) *If $\mathbf{A} \in \mathcal{V}$ and α, β, and $\gamma \in$ Con $\mathbf{A}$, and if $\langle a, b\rangle \in \alpha \wedge (\beta \vee \gamma)$ then*

$$a \ \ (\alpha \wedge \beta) \vee (\alpha \wedge \gamma) \ \ d(a, b, b).$$

Proof. Since $\mathcal{V}$ is modular there are terms such that $\mathcal{V}$ satisfies (M_n). Define terms $q_i(x, y, z)$ inductively by letting $q_0(x, y, z) = z$ and

$$q_{i+1}(x, y, z) = \begin{cases} m_{i+1}(q_i(x, y, z), x, y, q_i(x, y, z)) & i \text{ even} \\ m_{i+1}(q_i(x, y, z), y, x, q_i(x, y, z)) & i \text{ odd.} \end{cases}$$

Now define $d(x, y, z) = q_n(x, y, z)$. A straightforward induction shows $\mathcal{V}$ satisfies $q_i(x, x, z) \approx z$ from which (i) follows.

Let $\theta = (\alpha \wedge \beta) \vee (\alpha \wedge \gamma)$. Suppose that $\langle x, y\rangle \in \alpha \wedge (\beta \vee \gamma)$. Then $x \ \alpha \ y$ and there are elements $x_j \in A$ such that $x = x_0 \ \beta \ x_1 \ \gamma \ x_2 \ \cdots \ x_k = y$. We will show by induction on i that

$$q_i(x, y, y) \ \theta \ \begin{cases} m_i(y, y, x, x) & i \text{ even} \\ m_i(y, y, y, x) & i \text{ odd} \end{cases}$$

[6]In §11.5 we will strengthen condition (ii) to say *if $\langle a, b\rangle \in \alpha$ then $a \ [\alpha, \alpha] \ d(a, b, b)$.*

This is trivial for $i = 0$. Assume i is even. Then

$$\begin{aligned} m_{i+1}(q_i(x,y,y),x,x,q_i(x,y,y)) &= q_i(x,y,y) \\ &\ \theta\ m_i(y,y,x,x) \\ &= m_{i+1}(y,y,x,x). \end{aligned}$$

Hence we have the following relations:

$$\begin{array}{ccc} m_{i+1}(q_i(x,y,y),x,x,q_i(x,y,y)) & \theta & m_{i+1}(y,y,x,x) \\ \beta & & \beta \\ m_{i+1}(q_i(x,y,y),x,x_1,q_i(x,y,y)) & \alpha & m_{i+1}(y,y,x_1,x) \\ \gamma & & \gamma \\ m_{i+1}(q_i(x,y,y),x,x_2,q_i(x,y,y)) & \alpha & m_{i+1}(y,y,x_2,x) \\ \beta & & \beta \\ \vdots & & \vdots \\ m_{i+1}(q_i(x,y,y),x,y,q_i(x,y,y)) & \alpha & m_{i+1}(y,y,y,x) \end{array}$$

Since $\theta = \alpha \wedge (\beta \vee \theta) = \alpha \wedge (\gamma \vee \theta)$ by modularity, the first α can be replaced with θ. Then the second α can be replaced with θ. Finally, using the definition of q_i and the last row, we obtain

$$q_{i+1}(x,y,y) = m_{i+1}(q_i(x,y,y),x,y,q_i(x,y,y))\ \theta\ m_{i+1}(y,y,y,x).$$

For i odd the proof is similar (although somewhat more awkward since you use the sequence $x\ \beta\ x_1\ \gamma\ x_2\ \cdots\ y$ backwards).

Thus when $i = n$, we obtain $x\ \theta\ d(x,y,y)$ as desired. ■

The next corollary shows that in an algebra in a modular variety, the congruences $\alpha \wedge (\beta \vee \gamma)$ and $\gamma \vee (\alpha \wedge \beta)$ always permute. This is a surprising result about modular varieties. But notice that in a distributive variety $\alpha \wedge (\beta \vee \gamma) \leq \gamma \vee (\alpha \wedge \beta)$, and so these congruences obviously permute. This phenomenon of a result which is nontrivial for modular varieties but trivial for distributive varieties is really part of the commutator theory introduced in Chapter 4 of the first volume. We will take up a thorough study of the commutator in Volume III.

COROLLARY 6.36. *Let α, β, and $\gamma \in$ Con **A**, where **A** is an algebra in a congruence modular variety $\mathcal{V}$. Then $\alpha \wedge (\beta \vee \gamma)$ and $\gamma \vee (\alpha \wedge \beta)$ permute.*

Proof. As before let $\theta = (\alpha \wedge \beta) \vee (\alpha \wedge \gamma)$ and assume that

$$x\ \ \alpha \wedge (\beta \vee \gamma)\ \ y\ \ \gamma \vee (\alpha \wedge \beta)\ \ z$$

Then we calculate

$$x\ \theta\ d(x,y,y)\ \gamma \vee (\alpha \wedge \beta)\ d(x,y,z)\ \alpha \wedge (\beta \vee \gamma)\ d(x,x,z) = z.$$

Since $\theta \leq \gamma \vee (\alpha \wedge \beta)$, this shows that $\gamma \vee (\alpha \wedge \beta)$ and $\alpha \wedge (\beta \vee \gamma)$ permute. ■

COROLLARY 6.37. *Suppose that **A** lies in a congruence modular variety and that **L** is a sublattice of **Con A** isomorphic to $\mathbf{M}_3$. Then all the members of L permute with each other. If in addition the least and greatest elements of **L** are the least and greatest elements of **Con A**, then the term $d(x,y,z)$ of Lemma 6.35 is a Mal'tsev term for **A**, and thus $\mathbf{V}(\mathbf{A})$ is permutable.*

Proof. The first part follows directly form the previous corollary. To see the rest, let β, γ, and δ be three congruences pairwise meeting to 0_A and pairwise joining to 1_A. By the first part, these congruences permute with one another and so $\mathbf{A} \cong \mathbf{B} \times \mathbf{C}$, where $\mathbf{B} = \mathbf{A}/\beta$ and $\mathbf{C} = \mathbf{A}/\gamma$. If $b \in B$ and $c, c' \in C$, then $\langle b, c\rangle \; \beta \wedge (\gamma \vee \delta) \; \langle b, c'\rangle$ and so by Lemma 6.35

$$\langle b, c\rangle \;\; (\beta \wedge \gamma) \vee (\beta \wedge \delta) \;\; d(\langle b, c\rangle, \langle b, c'\rangle, \langle b, c'\rangle).$$

Since this congruence is 0_A, $c = d(c, c', c')$, showing that $d(x, y, z)$ is a Mal'tsev term for **C**. Similarly it is a Mal'tsev term for **B**, and thus for **A**. ∎

A short introduction to the commutator, a subject that will be taken up in earnest in Chapter 11, was given in §4.13 of Volume I and the notion of an Abelian algebra was defined. That definition is repeated in (6.11.1) of §6.11. Abelian algebras play an important role in many aspect of our field. One of the fundamental achievements of the commutator theory is Herrmann's theorem, which was proved in Theorem 4.155 but using the stronger hypothesis that the variety is congruence permutable; see also Satz 8.3.4 in (Thomas Ihringer, 2003) and Theorem 7.35 in Clifford Bergman (2012). However, combining Theorem 4.155 with the theory developed above, we are able to obtain the full result.

THEOREM 6.38 (Christian Herrmann 1979). *Let* **A** *be an algebra in a congruence modular variety. Then* **A** *is Abelian if and only if there is a ring* **R** *such that* **A** *is polynomially equivalent to an* **R***-module with universe A.*

Proof. This follow from Theorem 4.155 and Corollary 6.39 below. ∎

COROLLARY 6.39. *Let* **A** *be an Abelian algebra in a congruence modular variety, then* $V(\mathbf{A})$ *is congruence permutable.*

Proof. Definition 4.151 defines $\Delta(\mathbf{A})$, a congruence on $\mathbf{A}^2$, by

$$\Delta(\mathbf{A}) = \mathrm{Cg}^{\mathbf{A}^2}(\{\langle\langle x, x\rangle, \langle y, y\rangle\rangle : x, y \in A\}).$$

Let $\Delta = \Delta(\mathbf{A})$. We claim that the projection kernels, η_0 and η_1, both join with Δ to $\mathbf{1}_{\mathbf{A}^2}$ and meet with it to $\mathbf{0}_{\mathbf{A}^2}$. The calculations for the join are straightforward. For the meet suppose $\langle\langle x, y\rangle, \langle x, z\rangle\rangle \in \eta_0 \wedge \Delta$, then $\langle\langle y, y\rangle, \langle y, z\rangle\rangle \in \eta_1 \circ (\eta_0 \wedge \Delta) \circ \eta_1$. Hence by modularity

$$\langle\langle y, y\rangle, \langle y, z\rangle\rangle \in \eta_0 \wedge (\eta_1 \vee (\eta_0 \wedge \Delta)) = (\eta_0 \wedge \eta_1) \vee (\eta_0 \wedge \Delta) = \eta_0 \wedge \Delta$$

Since **A** is Abelian, Theorem 4.152(iii) implies $y = z$ and so $\eta_0 \wedge \Delta = \mathbf{0}_{\mathbf{A}^2}$.

The result now follows from Corollary 6.37. ∎

While Herrmann's theorem does not hold in every variety, it does extend beyond congruence modular varieties. This topic will be covered in §11.5.

In the introduction we pointed out that congruence permutability could be viewed as completing a parallelogram, see Figure 6.1. In modular varieties such parallelograms can be completed provided we have certain additional points. We now investigate some of these geometric conditions which are equivalent to congruence modularity. The first condition is due to H.-P. Gumm, who initiated this type of condition.

We call a term $d(x, y, z)$ a **modular difference term** or a **Gumm difference term**[7] for $\mathcal{V}$ if for any $\mathbf{A} \in \mathcal{V}$, $\alpha, \beta, \gamma \in \mathrm{Con}\ \mathbf{A}$, and x, y, z, u, and $v \in \mathbf{A}$, the implication of Figure 6.9 holds, see (H. P. Gumm, 1978, 1980). (Points lying on a horizontal line are α related and points lying on a vertical line are β related.)

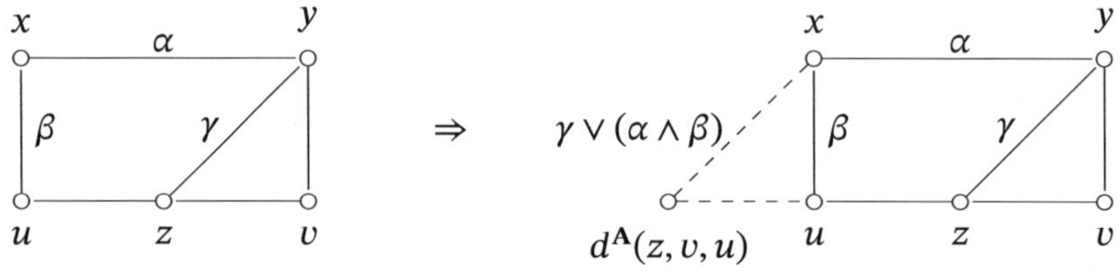

FIGURE 6.9. Gumm (Modular) Difference Term

THEOREM 6.40. *A variety $\mathcal{V}$ is congruence modular if and only if $\mathcal{V}$ has a Gumm difference term. Moreover, if there is a term $d(x, y, z)$ such that $\mathcal{V}$ satisfies conditions (i) and (ii) of Lemma 6.35, then $d(x, y, z)$ is a Gumm difference term for $\mathcal{V}$.*

Proof. Assume $\mathcal{V}$ is congruence modular. Then it satisfies condition (iv) of Theorem 6.32. Let $d(x, y, z)$ be a term satisfying conditions (i) and (ii) of Lemma 6.35. Assume that $\mathbf{A} \in \mathcal{V}$ and that the relations of the left side of Figure 6.9 hold in $\mathbf{A}$. It follows easily from condition (i) that $d(z, v, u)\ \alpha\ u$. The following relations are easily checked.

$$d(z, v, u)\ \gamma\ d(y, v, u) \tag{6.5.5}$$

$$d(y, v, u)\ \beta\ d(y, y, x) = x \tag{6.5.6}$$

$$d(y, v, u)\ \alpha\ d(x, u, u) \tag{6.5.7}$$

Since $\langle x, u\rangle \in \beta \wedge (\alpha \vee \gamma)$, condition (ii) implies that $d(x, u, u)\ (\alpha \wedge \beta) \vee (\beta \wedge \gamma)\ x$. Hence by (6.5.7), $d(y, v, u)\ \alpha \vee (\beta \wedge \gamma)\ x$. So by (6.5.6), $d(y, v, u)\ \beta \wedge (\alpha \vee (\beta \wedge \gamma))\ x$. But $\beta \wedge (\alpha \vee (\beta \wedge \gamma)) = (\alpha \wedge \beta) \vee (\beta \wedge \gamma)$ and thus by (6.5.5), $d(z, v, u)\ \gamma \vee (\alpha \wedge \beta)\ x$, as desired.

For the converse, see Exercise 6.46.1. ■

A useful variant of Gumm's difference term is a four variable term discovered by Emil Kiss (1992). This four place term is known as a **Kiss 4-ary difference term** and is studied in more detail in §11.5 of Volume III.

THEOREM 6.41. *The following are equivalent for a variety $\mathcal{V}$.*

(i) *$\mathcal{V}$ is congruence modular.*

(ii) *If α, β, and $\gamma \in$ **Con A**, $\mathbf{A} \in \mathcal{V}$, then*

$$(\alpha \vee \gamma) \wedge (\beta \vee \gamma) \subseteq (\gamma \vee (\alpha \wedge \beta)) \circ \alpha.$$

[7] In these volumes we distinguish between a **Gumm difference term** and a **difference term**, which is a term $d(x, y, z)$ satisfying $d(x, x, z) \approx z$ and $a\ [\theta, \theta]\ d^{\mathbf{A}}(a, c, c)$ whenever $a\ \theta\ c$ holds in the variety. The relationship between these two kinds of difference terms is explained in Theorem 11.25 of Chapter 11. Suffice it to say they are equivalent in a congruence modular variety. This terminology differs from that of (Ralph Freese and Ralph McKenzie, 1987).

(iii) *There is a term* $q(x, y, z, u)$ *such that if* α, β, *and* $\gamma \in \mathbf{Con\ A}$, $\mathbf{A} \in \mathcal{V}$, *then the implication of Figure* 6.10 *holds.*

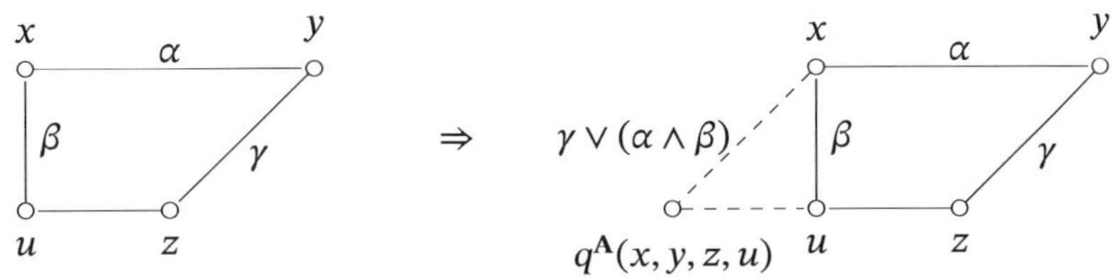

FIGURE 6.10. Kiss's 4-Variable Difference Term

Proof. Let α, β, and $\gamma \in \mathbf{Con\ A}$, $\mathbf{A} \in \mathcal{V}$. By modularity and Corollary 6.36 we have

$$\begin{aligned}(\alpha \vee \gamma) \wedge (\beta \vee \gamma) &= (\alpha \wedge (\beta \vee \gamma)) \vee \gamma \\ &= (\alpha \wedge (\beta \vee \gamma)) \vee \gamma \vee (\alpha \wedge \beta) \\ &= [\gamma \vee (\alpha \wedge \beta)] \circ [(\alpha \wedge (\beta \vee \gamma))] \\ &\subseteq [\gamma \vee (\alpha \wedge \beta)] \circ \alpha,\end{aligned}$$

proving that (i) implies (ii).

Assume (ii) and let α, β, and γ be the congruences on $\mathbf{F} = \mathbf{F}_{\mathcal{V}}(x, y, z, u)$ generated by $\{\langle x, y\rangle, \langle z, u\rangle\}$, $\{\langle x, u\rangle\}$, and $\{\langle y, z\rangle\}$, respectively. The inclusion of (ii) immediately implies that there is an element $q^{\mathbf{F}}(x, y, z, u) \in \mathbf{F}_{\mathcal{V}}(x, y, z, u)$ satisfying the indicated relations of Figure 6.10. It is easy to see that $q(x, y, z, u)$ satisfies (iii).

To see that (iii) implies (i), let α, β, and γ be the congruences in Figure 6.8. Now the relations of the first part of Figure 6.10 hold with the roles of β and γ reversed. Thus by (iii) we have

$$x \ \ (\beta \vee (\alpha \wedge \gamma)) \ \ q(x, y, z, u) \ \ \alpha \ \ u.$$

Since $\beta \le \gamma$, we have that $u \ \gamma \ x \ \gamma \ q(x, y, z, u)$ and hence, $u \ \alpha \wedge \gamma \ q(x, y, z, u)$. Thus $x \ \beta \vee (\alpha \wedge \gamma) \ u$ and the modularity of $\mathcal{V}$ now follows from Corollary 6.33. ■

LEMMA 6.42. *Let* $q(x, y, z, u)$ *be a term which satisfies condition* (iii) *of the last theorem. Then*

(i) $q(x, x, u, u) \approx u$ *is an identity of* $\mathcal{V}$.
(ii) $q(x, y, y, x) \approx x$ *is an identity of* $\mathcal{V}$.
(iii) $r(x, y, z, u) = q(z, y, x, u)$ *also satisfies condition (iii) of Theorem* 6.41.

Proof. If we let $\beta = \gamma = 0$ then $\gamma \vee (\alpha \wedge \beta) = 0$, and so the implication of Figure 6.10 with $x = u$ and $y = z$ gives $q(x, y, y, x) = x$. A similar argument proves (ii).

To see (iii), suppose that the relations indicated in the left side of Figure 6.10 hold for some $\mathbf{A} \in \mathcal{V}$. Note

$$r(x, y, y, u) \ \beta \ r(x, y, y, x) = q(y, y, x, x) = x$$

$$q(x, y, y, u) \ \beta \ q(x, y, y, x) = x$$

Moreover, $r(x, y, y, u) \ \alpha \ r(x, x, x, u) = q(x, x, x, u) \ \alpha \ q(x, y, y, u)$. Thus $r(x, y, y, u)$ $\alpha \wedge \beta$ $q(x, y, y, u)$. Since $r(x, y, z, u) \ \gamma \ r(x, y, y, u)$ and $q(x, y, z, u) \ \gamma \ q(x, y, y, u)$, we have that

$$r(x, y, z, u) \ \gamma \vee (\alpha \wedge \beta) \ q(x, y, z, u).$$

Moreover, $r(x, y, z, u) \ \alpha \ r(x, x, u, u) = q(u, x, x, u) = u$. From this it follows that $q(x, y, z, u)$ satisfies condition (iii) of Theorem 6.41. ■

The next theorem shows that we can form a modular difference term from a 4-ary difference term.

THEOREM 6.43. *If $q(x, y, z, u)$ is a term satisfying condition (iii) of Theorem 6.41, then $d(x, y, z) = q(x, y, z, z)$ is a modular difference term.*

Proof. Suppose that $\mathbf{A}$ is an algebra in $\mathcal{V}$ which has elements and congruences satisfying the relations of the left side of Figure 6.9. By part (iii) of the last theorem, the relations of the right side of Figure 6.9 hold if we replace $d^{\mathbf{A}}(z, w, u)$ by $q^{\mathbf{A}}(z, y, x, u)$. But $q^{\mathbf{A}}(z, y, x, u)$ and $d^{\mathbf{A}}(z, w, u) = q^{\mathbf{A}}(z, w, u, u)$ are both α-related to u, and clearly they are β-related. This implies that the relations of Figure 6.9 hold. ■

Now we come to Gumm's Mal'tsev condition for congruence modularity. A sequence of terms p, $d_1, \ldots, d_k$ satisfying equations (G_k) in (iii) below will be called a sequence of Gumm terms for $\mathcal{V}$.

THEOREM 6.44. *For a variety $\mathcal{V}$ the following conditions are equivalent.*

(i) *$\mathcal{V}$ is congruence modular.*

(ii) *For any $\mathbf{A} \in \mathcal{V}$ and any α, β, and $\gamma \in \operatorname{Con} \mathbf{A}$,*

$$\gamma \cap (\alpha \circ \beta) \subseteq [(\alpha \wedge \gamma) \vee (\beta \wedge \gamma)] \circ \beta \circ \alpha.$$

(iii) *For some k, there exist terms p, $d_0, d_1, \ldots, d_k$ in the language of $\mathcal{V}$ such that the following equations hold identically in $\mathcal{V}$:*

$$(G_k) \qquad \begin{aligned} d_0(x, y, z) &\approx x, \\ d_i(x, y, x) &\approx x && \textit{for all } i, \\ d_i(x, z, z) &\approx d_{i+1}(x, z, z) && \textit{for } i \textit{ even}, \\ d_i(x, x, z) &\approx d_{i+1}(x, x, z) && \textit{for } i \textit{ odd}, \\ d_k(x, z, z) &\approx p(x, z, z), \\ p(x, x, z) &\approx z. \end{aligned}$$

Proof. To see (i) implies (ii), suppose that α, β, and $\gamma \in \operatorname{Con} \mathbf{A}$ for some $\mathbf{A} \in \mathcal{V}$ and that $\langle a, c\rangle \in \gamma$, $\langle a, b\rangle \in \alpha$, and $\langle b, c\rangle \in \beta$. Since $\mathcal{V}$ is modular, it has a term $d(x, y, z)$ satisfying conditions (i) and (ii) of Lemma 6.35. Using this we calculate

$$(*) \qquad a \ (\alpha \wedge \gamma) \vee (\beta \wedge \gamma) \ d(a, c, c) \ \beta \ d(a, b, c) \ \alpha \ d(a, a, c) = c.$$

Now assume (ii) holds. In $\mathbf{F}_{\mathcal{V}}(x, y, z)$ let $\alpha = \operatorname{Cg}(x, y)$, $\beta = \operatorname{Cg}(y, z)$, and $\gamma = \operatorname{Cg}(x, z)$. Since $\langle x, z\rangle \in \gamma \cap (\alpha \circ \beta)$, there are three variable terms t and p such that

$$x \ (\alpha \wedge \gamma) \vee (\beta \wedge \gamma) \ t(x, y, z) \ \beta \ p(x, y, z) \ \alpha \ z.$$

By the first relation there are terms d_i, for $i = 0, \dots, k$, such that $d_0(x, y, z) = x$, $d_i(x, y, z)\ \beta\ d_{i+1}(x, y, z)$ for i even, $d_i(x, y, z)\ \alpha\ d_{i+1}(x, y, z)$ for i odd, $d_i(x, y, z)\ \gamma\ d_{i+1}(x, y, z)$ for all i, and $d_k(x, y, z) = t(x, y, z)$. These facts imply that $\mathcal{V}$ satisfies the equations (G_k) by the usual arguments. For example, since $d_k(x, y, z)\ \beta\ p(x, y, z)$ and $y\ \beta\ z$, we have $d_k(x, z, z)\ p\ (x, z, z)$. Now Theorem 6.9 gives $d_k(x, z, z) = p(x, z, z)$. Since this holds in the free algebra, $\mathcal{V}$ satisfies the equation $d_k(x, z, z) \approx p(x, z, z)$. Similar arguments prove the other equations of (G_k).

Given Gumm terms satisfying (iii), we can define Day terms by:

$$\begin{aligned} m_{2i-1}(x, y, z, u) &= d_i(x, y, u) && \text{for } i \text{ even,} \\ m_{2i-1}(x, y, z, u) &= d_i(x, z, u) && \text{for } i \text{ odd,} \\ m_{2i}(x, y, z, u) &= d_i(x, z, u) && \text{for } i \text{ even,} \\ m_{2i}(x, y, z, u) &= d_i(x, y, u) && \text{for } i \text{ odd.} \end{aligned}$$

Also let $m_{2k+1}(x, y, z, u) = p(y, z, u)$ and $m_{2k+2}(x, y, z, u) = u$. It is straightforward to verify that these terms satisfy (M_{2k+1}). ■

The $p(x, y, z)$ given in the proof of the equations (G_k) is in fact a Gumm difference term, and conversely every Gumm difference term can be part of such a set of equations; see Exercise 8.

We close this section with Gumm's Shifting Lemma (H. P. Gumm, 1978), which plays an important role in commutator theory.

THEOREM 6.45 (THE SHIFTING LEMMA). *If* **Con A** *is modular and* α, β, $\gamma \in$ Con **A**, *then the implication of Figure* 6.11 *holds for* x, y, z, *and* $u \in A$. *Moreover, if the implication of Figure* 6.11 *holds for all algebras in a variety* $\mathcal{V}$, *then* $\mathcal{V}$ *is congruence modular.*

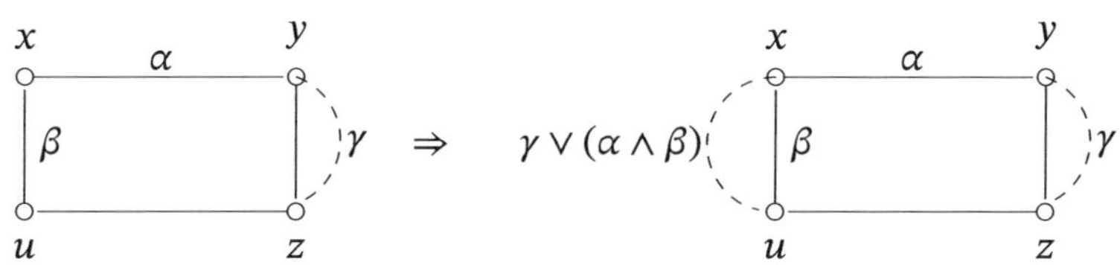

FIGURE 6.11. The Shifting Lemma

Proof. To see the first statement, note that $\langle x, u\rangle \in \beta\wedge(\alpha\vee(\beta\wedge\gamma))$ and so by modularity $\langle x, u\rangle \in (\alpha \wedge \beta) \vee (\beta \wedge \gamma)$. For the second statement, see Exercise 6.46.7. ■

Exercises 6.46

1. Use Corollary 6.33 to show that if a variety $\mathcal{V}$ has a modular difference term, then $\mathcal{V}$ is modular.

2. (Alan Day, 1969) Show how to derive terms satisfying (M_2) from a Mal'tsev term $p(x, y, z)$. Conversely, given terms satisfying (M_2) derive a Mal'tsev term.

3. Show that the sublattice generated by the congruences α, β and γ from condition (v) of Theorem 6.32 is a homomorphic image of the lattice diagrammed in Figure 6.8.

4. Show that if $\mathcal{V}$ is 3-permutable with terms p_1 and p_2 which satisfy condition (iii) of Theorem 6.11 (with $k = 3$), then $\mathcal{V}$ satisfies (M_3) as given in Theorem 6.32.

5. Show that the variety of lattices satisfies (M_3). Thus the converse of the previous exercise is not true.

6. Show that Polin's algebra, given in Example 6.12, satisfies (Σ_3) on page 40, but is not modular.

7. Prove that if the implication of Figure 6.11 holds for all algebras in a variety $\mathcal{V}$, then $\mathcal{V}$ is congruence modular.

8. Show that the term $p(x, y, z)$ from the proof of (ii)$\Rightarrow$(iii) of Theorem 6.44 is a Gumm difference term. (Hint: use Theorem 6.40.) Also show that if $p(x, y, z)$ is a Gumm difference term then there are terms $d_i(x, y, z)$, $i = 0, \ldots, k$, which together with p satisfy the equation of (G_k) given in Theorem 6.44.

6.6. Congruence Regularity and Uniformity

A family F of equivalence relations on a set A is called **regular** if each $\theta \in F$ is determined (within F) by each of its blocks, i.e., if $\theta, \phi \in F$ and $a/\theta = a/\phi$ for some $a \in A$, then $\theta = \phi$. This property is actually familiar, when we consider a non-zero linear map $\varphi : \mathbb{R}^3 \to \mathbb{R}$. The congruence relation $\theta = \ker \varphi$ partitions $\mathbb{R}^3$ into an infinite family F of parallel planes. Any single plane P of F determines the entire family F, and hence determines θ in an obvious way: F is simply the family of all planes parallel to P. Thus, in general, we call an algebra **A** **congruence regular** if and only if Con **A** is a regular family of equivalence relations, and as usual, we will call a variety congruence regular if and only if each of its algebras is congruence regular. As we will see in §6.7, congruence regularity of a variety $\mathcal{V}$ is stronger than modularity, but independent of distributivity; it also implies that $\mathcal{V}$ is congruence k-permutable for some k. Later in this section we will consider the even stronger properties of congruence uniformity and coherence.

Our first lemma shows that regularity can be established for an entire variety by checking the above condition only for $\theta = 0$.

LEMMA 6.47. *$\mathcal{V}$ is congruence regular if and only if the following condition holds for all $\mathbf{A} \in \mathcal{V}$, $\phi \in$ Con $\mathbf{A}$, and all $a \in A$:*

$$a/\phi = \{a\} \Rightarrow \phi = 0.$$

Proof. The condition is a special case of the definition (namely $\theta = 0$), and hence follows from regularity. Conversely, let us suppose that the condition holds for $\mathcal{V}$. To prove regularity, let $\mathbf{A} \in \mathcal{V}$, and $\theta, \phi \in$ Con $\mathbf{A}$ and suppose that $a/\theta = a/\phi$ for some

$a \in A$. It easily follows that $a/(\theta \wedge \phi) = a/(\theta \vee \phi)$. Thus $\theta \vee \phi$ considered as a congruence on $\mathbf{A}/\theta \wedge \phi$ has a singleton block. The condition tells us that $\theta \vee \phi$ is the zero congruence on $\mathbf{A}/\theta \wedge \phi$. Thus $\theta \vee \phi = \theta \wedge \phi$, so $\theta = \phi$, establishing regularity. ■

Before presenting a Mal'tsev condition for regularity, let us remark that well before the theory of Mal'tsev conditions, Robert L. Vaught (1961) made some general model-theoretic observations on regularity. His remarks, which were based on Beth's Theorem (Theorem 8.129), held that, under regularity, there must exist some formulas of first order logic yielding one congruence block in terms of another one; see Exercise 8.132.6 for the details. Of course condition (iii) of the following Theorem yields such formulas very explicitly (interpreting = as an arbitrary congruence).

G. Grätzer (1970a) and Rudolf Wille (1970) showed congruence regularity is definable by a Mal'tsev condition; see also (J. Hagemann, 1973b).

THEOREM 6.48. *For a variety $\mathcal{V}$, the following conditions are equivalent:*

(i) *$\mathcal{V}$ is congruence regular;*
(ii) *$\mathbf{F}_{\mathcal{V}}(3)$ is congruence regular;*
(iii) *there are ternary terms $g_1, \dots, g_n$ in the language of $\mathcal{V}$ such that $\mathcal{V}$ satisfies the implication*

$$\left[\bigwedge_{i=1}^{n} g_i(x, y, z) \approx z \right] \Longleftrightarrow x \approx y.$$

(iv) *$\mathcal{V}$ has ternary terms $g_1, \dots, g_n$ and 5-ary terms $f_1, \dots, f_n$ such that the following are identities of $\mathcal{V}$:*

$$\begin{aligned} g_i(x, x, z) &\approx z \qquad 1 \le i \le n \\ x &\approx f_1(x, y, z, z, g_1(x, y, z)) \\ f_1(x, y, z, g_1(x, y, z), z) &\approx f_2(x, y, z, z, g_2(x, y, z)) \\ f_2(x, y, z, g_2(x, y, z), z) &\approx f_3(x, y, z, z, g_3(x, y, z)) \\ &\vdots \\ f_n(x, y, z, g_n(x, y, z), z) &\approx y. \end{aligned}$$

REMARK 6.49.

(a) In the Corollary to Theorem 6.120 in §6.10 we will be able to derive a slightly simpler version of this Mal'tsev condition, one in which the f_i's do not depend on their fifth variables.
(b) A congruence regular variety is congruence modular and n-permutable for some n, see Theorem 6.71.

Proof. The implications (iv)$\Rightarrow$(iii) and (i)$\Rightarrow$(ii) are immediate. For (iii)$\Rightarrow$(i), we will assume (iii) and derive the condition of Lemma 6.47. Given a singleton congruence block a/ϕ, we will use the terms g_i supplied by (iii) to prove that ϕ is $0_{\mathbf{A}}$. For any $\langle b, c\rangle \in \phi$ we have, for each i,

$$g_i(b, c, a) \;\phi\; g_i(b, b, a) = a$$

and so, by assumption

$$g_i(b, c, a) = a$$

for each i. Now, by (iii), $b = c$, and thus $\phi = 0_{\mathbf{A}}$.

To complete the proof we will show that (ii) implies (iv). We will apply regularity to two congruences θ and ϕ on the free algebra $\mathbf{F}_{\mathcal{V}}(x, y, z)$, which we now define. First, θ is the kernel of the endomorphism extending the map

$$x, y \mapsto x \qquad z \mapsto z.$$

And then ϕ is the congruence generated by

$$Z = \{\langle z, u\rangle : u \in z/\theta\}.$$

It is obvious that $\phi \subseteq \theta$ and $z/\theta \subseteq z/\phi$; hence $z/\theta = z/\phi$. By regularity $\theta = \phi$ and hence in particular, $\langle x, y\rangle \in \phi$. By Theorem 6.18 on page 25, we have

$$\begin{aligned} x &= t_1(z, g_1, e_1, \ldots, e_m) \\ t_1(g_1, z, e_1, \ldots, e_m) &= t_2(z, g_2, e_1, \ldots, e_m) \\ &\vdots \\ t_n(g_n, z, e_1, \ldots, e_m) &= y \end{aligned}$$

for some $e_1, \ldots, e_m, g_1, \ldots, g_n \in \mathbf{F}_{\mathcal{V}}(x, y, z)$ such that $\langle z, g_i\rangle \in \theta$ for each i. From the definition of θ it is clear that for each i, $\mathcal{V}$ identically satisfies $g_i(x, x, z) \approx z$. Now if we define

$$f_i(x, y, z, u, v) = t_i(u, v, e_1(x, y, z), \ldots, e_m(x, y, z)),$$

then we obtain the equations (iv) by the usual reasoning of this chapter. ■

Groups of course are congruence regular; in fact the equations (iv) obtain with $n = 1$, by defining

$$g_1(x, y, z) = x^{-1}yz \qquad f_1(x, y, z, u, v) = xuz^{-1}.$$

For some subtler examples of regularity, the reader is referred to Exercises 3, 4, 15, 18, 19, 20, 22, 23, and Example 6.52 below.

Notice that although condition (iii) of the last theorem is simpler than (iv), condition (iii) involves implications while (iv) is expressed in terms of equations (and thus (iv) is a Mal'tsev condition). The next theorem shows that it is always possible to convert a varietal condition expressed with implications into a Mal'tsev condition. It is an immediate consequence of Theorem 6.18.

THEOREM 6.50. *Let $\mathcal{V}$ be a variety and let s_i and t_i, $i = 0, \ldots, n$, be k-ary terms. Then the following are equivalent:*

(i) *$\mathcal{V}$ satisfies the implication*

$$\left[\bigwedge_{i=1}^{n} s_i(\overline{x}) \approx t_i(\overline{x})\right] \Rightarrow s_0(\overline{x}) \approx t_0(\overline{x})$$

(ii) *For some m, $\mathcal{V}$ has $(k+2)$-ary terms $f_1, \ldots, f_m$ satisfying*

$$\begin{aligned} s_0(\overline{x}) &\approx f_1(\overline{x}, s_{i_1}(\overline{x}), t_{i_1}(\overline{x})) \\ f_1(\overline{x}, t_{i_1}(\overline{x}), s_{i_1}(\overline{x})) &\approx f_2(\overline{x}, s_{i_2}(\overline{x}), t_{i_2}(\overline{x})) \\ &\vdots \\ f_m(\overline{x}, t_{i_m}(\overline{x}), s_{i_m}(\overline{x})) &\approx t_0(\overline{x}) \end{aligned}$$

■

We now turn our attention to the closely related topic of congruence uniformity. As we mentioned in the introduction to this chapter, a variety $\mathcal{V}$ is **congruence uniform** if and only if each congruence θ on every algebra $\mathbf{A} \in \mathcal{V}$ is a **uniform equivalence relation** in the sense that all blocks of θ have the same cardinality. Now it is readily apparent from Lemma 6.47 that every congruence uniform variety is congruence regular. As we will see in the exercises, uniformity itself is not Mal'tsev definable, and in fact remains rather mysterious. Thus it turns out to be more fruitful, in the present context, to study a slightly weaker, but Mal'tsev definable, property which is nevertheless still strong enough to imply regularity.

We define a variety $\mathcal{V}$ to be **weakly congruence uniform** if and only if there exists a cardinal function f, mapping the class of cardinals into itself, such that $|b/\theta| \leq f(|a/\theta|)$ whenever $\mathbf{A} \in \mathcal{V}$, a, $b \in A$, and $\theta \in \operatorname{Con} \mathbf{A}$. Uniformity, of course, corresponds to the special case $f(\kappa) = \kappa$. Let us define the **iterated exponential** $\beth_n(\kappa)$, $n \in \omega$, by

$$\beth_0(\kappa) = \kappa \qquad \beth_{n+1}(\kappa) = 2^{\beth_n(\kappa)}.$$

We call a variety **congruence n-uniform** if it satisfies the definition of weak congruence uniformity with $f(\kappa) = \beth_n(\kappa + \aleph_0)$. We shall show that a variety is weakly congruence uniform if and only if it is congruence n-uniform for some n. Exercise 6.58.17 shows that if one block of a congruence of an algebra in a weakly congruence uniform variety is finite, then all blocks are finite. Hence the strongest of these conditions, congruence 0-uniformity, guarantees that all the blocks of a congruence with an infinite block will have the same size. Notice that even this is weaker than congruence uniformity. The next theorem and the example following it show that these concepts are stronger than congruence regularity.

THEOREM 6.51. *Every weakly congruence uniform variety is congruence regular.*

Proof. Suppose that $\mathcal{V}$ is weakly congruence uniform with f as in the definition. By way of contradiction, we will assume that $\mathcal{V}$ is not congruence regular. Thus by Lemma 6.47 above, $\mathcal{V}$ has an algebra $\mathbf{A}$ which has a congruence ϕ with both a one-element block and a block with more than one element: $|a/\phi| = 1$ and $|b/\phi| = \mu \geq 2$. Let ν be a cardinal satisfying $2^{\nu} > f(|A|)$. Let $\mathbf{C} = \mathbf{A}^{\nu}$ and let $\overline{\phi}$ be the congruence on $\mathbf{C}$ that identifies $\overline{x}$ and $\overline{y}$ if and only if $x_i \;\phi\; y_i$ for $1 \leq i < \nu$ (no restriction on the 0 coordinate). If we let $\overline{a}$ and $\overline{b}$ be the ν-tuples all of whose coordinates are a and b, respectively, then, since $\overline{a}$ is congruent modulo $\overline{\phi}$ only to ν-tuples which differ in at most the zeroth coordinate, $|\overline{a}/\overline{\phi}| = |A|$ and

$$|\overline{b}/\overline{\phi}| \geq \mu^\nu \geq 2^\nu > f(|A|) = f(|\overline{a}/\overline{\phi}|),$$

in contradiction to the definition of weak uniformity. ■

EXAMPLE 6.52. This example shows that there is a congruence regular variety which is not weakly congruence uniform.

Let $\mathcal{V}$ be the variety with operation symbols h, g_1, and g_2 defined by the equations

$$\begin{aligned} g_1(x, x, z) \approx g_2(x, x, z) &\approx z \\ h(g_1(x, y, z), x, y, z) &\approx x \\ h(g_2(x, y, z), x, y, z) &\approx y. \end{aligned}$$

It is obvious that $\mathcal{V}$ satisfies (iii) of Theorem 6.48, and hence is congruence regular. For the failure of weak uniformity, we let μ be any ordinal ≥ 2, and define ternary operations g_1 and g_2 on $\mu \times \{0, 1\}$ as follows:

$$g_1(\langle a, i\rangle, \langle b, j\rangle, \langle c, k\rangle) = \begin{cases} \langle c, k\rangle & \text{if } \langle a, i\rangle = \langle b, j\rangle \\ \langle 0, i + j + k\rangle & \text{otherwise} \end{cases}$$

$$g_2(\langle a, i\rangle, \langle b, j\rangle, \langle c, k\rangle) = \begin{cases} \langle c, k\rangle & \text{if } \langle a, i\rangle = \langle b, j\rangle \\ \langle 1, k\rangle & \text{otherwise.} \end{cases}$$

The addition here is modulo 2, of course. Notice that every set B with $2 \times 2 \subseteq B \subseteq \mu \times 2$ is closed under g_1 and g_2. Given such a B, we consider the following four conditions on a quaternary operation h,

$$\begin{aligned} h(\langle 0, i + j + k\rangle, \langle a, i\rangle, \langle b, j\rangle, \langle c, k\rangle) &= \langle a, i\rangle \\ h(\langle 1, k\rangle, \langle a, i\rangle, \langle b, j\rangle, \langle c, k\rangle) &= \langle b, j\rangle \\ h(\langle c, k\rangle, \langle a, i\rangle, \langle a, i\rangle, \langle c, k\rangle) &= \langle a, i\rangle \\ h(\langle *, s\rangle, \langle *, i\rangle, \langle *, j\rangle, \langle *, k\rangle) &= \langle *, s + j + k\rangle \end{aligned}$$

where the $*$'s represent arbitrary elements.

The first three conditions specify h on some special elements of B^4. It is easy to check that where these specifications overlap with one another they agree, and so we easily obtain a partial operation obeying the first three conditions. But the fourth condition holds for this partial operation (as one easily checks), and so it is evidently possible to find a total operation h satisfying all four conditions.

It is now trivial to check that the algebra $\mathbf{B} = \langle B, h, g_1, g_2\rangle$ satisfies the equations defining $\mathcal{V}$. Moreover it is apparent from the fourth condition that second coordinate projection defines a congruence θ on $\mathbf{B}$. Since the image of the second projection has only two elements, this congruence has exactly two blocks,

$$B_i = B \cap (\mu \times \{i\}) \qquad \text{for } i = 0, 1.$$

By construction, the cardinalities of these blocks are completely arbitrary except for the obvious constraint that $2 \leq |B_i| \leq \mu$. Since μ itself is arbitrary, we have an obvious failure of weak uniformity.

We remark that any congruence regular variety which is not weakly congruence uniform must be residually large; see Exercise 6.58.14.

The few examples of uniformity which we know (see, for example, Exercises 6.58.18, 22 and 23) seem to have a general flavor reminiscent of group theory, especially the cancellation available in group theory. But weak uniformity, in its most general form, has a surprisingly different flavor, namely that of infinitary combinatorics. In order to develop a Mal'tsev condition for weak uniformity, we need a couple of lemmas from Ramsey theory. This theory began with the famous theorem of F. P. Ramsey (1930), and was later developed by P. Erdős, A. Hajnal and R. Rado.

Let $X^{(n)}$ denote the set of all n-element subsets of X. For finite n and cardinals κ, λ, μ, the symbolic expression

$$\kappa \to (\lambda)^n_\mu$$

abbreviates the following assertion: if $\kappa^{(n)} = \bigcup \mathcal{M}$ with $|\mathcal{M}| \leq \mu$, then there exists $D \in \mathcal{M}$ and $Y \subseteq \kappa$ with $|Y| = \lambda$, such that $Y^{(n)} \subseteq D$. The famous theorem of F. P. Ramsey (which appeared as a lemma to a theorem much less remembered) was that for all finite l, m, n, there exits a finite k such that

$$k \to (l)^n_m.$$

In fact, $m^{ml^{\cdot^{\cdot^{ml}}}}$ (ml appearing n times) is large enough (see, e.g., pages 7–9 of (Ronald L. Graham, Bruce L. Rothschild, and Joel H. Spencer, 1980)).

THEOREM 6.53 (P. Erdős and R. Rado 1956).

$$(\beth_n(\mu))^+ \to (\mu^+)^{n+1}_\mu$$

for every $n \in \omega$ and every infinite μ.

THEOREM 6.54 (P. Erdős, A. Hajnal, and R. Rado 1965). *The assertion*

$$\beth_n(\mu) \to (n+2)^{n+1}_\mu$$

fails for all $n \in \omega$ and all μ.

The next theorem will use Theorem 6.53 to show that the class of weakly congruence uniform varieties is defined by a Mal'tsev condition. Rather than explicitly presenting this Mal'tsev condition, which would be rather messy, we will show that a variety $\mathcal{V}$ is weakly congruence uniform if and only if it satisfies certain implications. Since, by Theorem 6.50 these implications are equivalent to a Mal'tsev condition, we will be content to present the more perspicuous implications.

For $m \geq n \geq 0$ and $0 \leq i < j \leq m$, terms $F_1, \ldots, F_p$ of rank $n+2$ will be called (a set of weak) **uniformity terms for $\mathcal{V}$ of type $\langle n, m, p, i, j\rangle$** if $\mathcal{V}$ obeys the equations

$$F_t(z, x, \ldots, x) \approx z$$

and the (universal closure of the) implication

$$\left[\bigwedge_{t=1}^{p} \bigwedge F_t(z, \overline{x}) \approx F_t(z, \overline{y})\right] \Rightarrow x_i \approx x_j,$$

where the second conjunction is over all of the $(n+1)$-sequences $\overline{x}$ and $\overline{y}$ from $\{x_0, \dots, x_m\}$ with strictly increasing subscripts. For instance, if $n = 0$ or $n = m$, then the only varieties having uniformity terms of type $\langle n, m, \dots \rangle$ are the trivial ones (those satisfying $x_0 \approx x_1$). For $n = 1$, $m = 2$, $p = 1$, $i = 0$ and $j = 2$, our implication is simply

$$[F_1(z, x_0, x_1) \approx F_1(z, x_0, x_2) \approx F_1(z, x_1, x_2)] \Rightarrow x_0 \approx x_2.$$

THEOREM 6.55 (Walter Taylor 1974). *The following statements are equivalent for a variety $\mathcal{V}$.*

(i) *$\mathcal{V}$ is weakly congruence uniform.*
(ii) *For some $0 \le n \le m$, $0 \le i < j \le m$, and p, $\mathcal{V}$ has uniformity terms of type $\langle n, m, p, i, j \rangle$.*
(iii) *For some n, $\mathcal{V}$ has n-uniform congruences.*

Moreover, if (ii) *holds for some n, then* (iii) *holds with that same n.*

Proof. We prove the last assertion first. Thus we are given uniformity terms $F_1, \dots, F_p$. Let $\mathbf{A} \in \mathcal{V}$ and let $\theta \in \operatorname{Con} \mathbf{A}$. By way of contradiction, let us suppose that $|a/\theta| \le \mu$, where $\mu \ge \aleph_0$, and that $|b/\theta| > \beth_n(\mu)$. Let $<$ be a total order on b/θ. We divide $(b/\theta)^{(n+1)}$ into $\mu = \mu^p$ classes

$$(b/\theta)^{(n+1)} = \bigcup \{D_\alpha : \alpha \in (a/\theta)^p\} \tag{6.6.1}$$

where

$$D_\alpha = \big\{\{b_0, \dots, b_n\} : b_0 < \cdots < b_n,\ F_t(a, b_0, \dots, b_n) = \alpha_t, \text{ for } 1 \le t \le p\big\}.$$

(And thus (6.6.1) follows easily from the fact that $F_t(z, x, \dots, x) = z$ for each t.) Now Lemma 6.53 yields the existence of an infinite set, and hence a set Y with $|Y| = m+1$, such that $Y^{(n+1)} \subseteq D_\alpha$ for a fixed α. Taking $Y = \{b_0, \dots, b_m\}$ with $b_0 < \cdots < b_m$, we see by definition of D_α that, for each t and for each $\{b_{i_0}, \dots, b_{i_n}\} \subseteq \{b_0, \dots, b_m\}$, with $b_{i_0} < \cdots < b_{i_n}$, all values $F_t(a, b_{i_0}, \dots, b_{i_n})$ are equal. Therefore by the defining implications for uniformity terms, $b_i = b_j$, in contradiction to the previously established relation that $b_i < b_j$. This contradiction establishes the desired conclusion that $|b/\theta| \le \beth_n(\mu)$.

Clearly (iii) implies (i). To complete the proof of the theorem, we assume that $\mathcal{V}$ is weakly congruence uniform and show that $\mathcal{V}$ has uniformity terms. Let f be the function defined on the cardinals which witnesses the weak uniformity of $\mathcal{V}$. Clearly we may assume that f is monotone. Let μ be an infinite cardinal at least as large as the cardinality of the set of operation symbols for $\mathcal{V}$ and let $\nu = f(\mu)$.

Let $\mathbf{F} = \mathbf{F}_{\mathcal{V}}(x_i : i < \nu^+)$ be the free algebra on ν^+ generators. We then define θ to be the kernel of the endomorphism of $\mathbf{F}$ which is specified as follows:

$$x_0 \mapsto x_0 \quad \text{and} \quad x_i \mapsto x_1 \quad \text{if } i > 0.$$

We further define $Z \subseteq F^2$ to be the set of all pairs

$$\langle g^{\mathbf{F}}(x_0, x_{i_1}, \dots, x_{i_s}), g^{\mathbf{F}}(x_0, x_{j_1}, \dots, x_{j_s}) \rangle,$$

where g is any $\mathcal{V}$-term such that $\mathcal{V}$ identically satisfies $g(x_0, x_1, \dots, x_1) \approx x_0$, and where the subscripts are as follows: $0 < i_1 < \cdots < i_s$ and $0 < j_1 < \cdots < j_s$. Defining ψ to be the congruence on $\mathbf{F}$ generated by Z, it is obvious that $Z \subseteq \theta$ and hence that $\psi \subseteq \theta$. Therefore θ may be regarded as a congruence on $\mathbf{F}/\psi$.

Let a and b denote the images of x_0 and x_1 in $\mathbf{F}/\psi$. Clearly $|a/\theta|$ is the number of ψ-blocks contained in the θ-block x_0/θ. Every member of x_0/θ is clearly given by a term $g(x_0, x_{i_1}, \cdots)$ with $0 < i_1 < \cdots$ and with $g(x_0, x_1, \ldots, x_1) \approx x_0$ holding in $\mathcal{V}$. Regardless of the sequence $i_1, i_2, \ldots$, formally identical g's yield the same ψ-class; therefore the number of ψ-blocks in x_0/θ is limited by the cardinality of the language of $\mathcal{V}$. Hence $|a/\theta| \leq \mu$, and thus $|b/\theta| \leq \nu$.

Now the θ-class of x_1 in $\mathbf{F}$ obviously contains the ν^+ distinct elements x_i, for $0 < i < \nu^+$ (since we may assume $\mathcal{V}$ nontrivial). As we just saw, the image of this class in $\mathbf{F}/\psi$ has at most ν elements, and so we obviously have $x_i \ \psi \ x_j$ for some i, j with $0 < i < j < \nu^+$. Thus $\langle x_i, x_j \rangle$ is in the congruence generated by a finite subset of pairs from Z. One may add fictitious variables so that the finitely many g appearing in these pairs all have the same arity, which we may take to be $n + 1$ (thus defining n). Taking $F_t(z, x_1, \ldots, x_n)$ to be $g(z, x_1, \ldots, x_n)$ (with one F_t for each of the formally distinct g's appearing), and considering, without loss of generality, each of the finitely many subscripts appearing to be a *finite* ordinal, it is routine to verify that the desired implications hold in $\mathcal{V}$. ∎

In the case of a countable set of operations, the implication (i) implies (iii) follows from a general model-theoretic result of Robert L. Vaught (1965). (See also Theorem 7.2.6 of C. C. Chang and H. J. Keisler (1990).) Exercises 19 and 20 will show that (ii) of Theorem 6.55 does not imply (iii) for a smaller n.

We round out this section with a look at D. Geiger's notion of coherence. It is a Mal'tsev definable property which implies 0-uniformity, the strongest of our weak uniformity conditions. An algebra $\mathbf{A}$ will be called **coherent** if the following condition holds for all $\phi \in \operatorname{Con} \mathbf{A}$ and all subalgebras $\mathbf{B}$ of $\mathbf{A}$: if B contains a ϕ-block, then B is a union of ϕ-blocks. Naturally, a variety $\mathcal{V}$ is called coherent if each $\mathbf{A} \in \mathcal{V}$ is coherent.

THEOREM 6.56 (David Geiger 1974). *A variety $\mathcal{V}$ is coherent if and only if there exist terms H, $F_1, \ldots, F_n$ such that the following are identities of $\mathcal{V}$:*

$$F_i(x, x, z) \approx z \quad \text{for } 1 \leq i \leq n$$
$$H(y, F_1(x, y, z), \ldots, F_n(x, y, z)) \approx x.$$

Proof. See Exercise 24. ∎

THEOREM 6.57. *Every coherent variety is weakly congruence uniform and permutable; in fact, it is congruence* 0*-uniform.*

Proof. Let $\mathcal{V}$ be a coherent variety. To prove that it is congruence 0-uniform, we need to show that if $a, b \in \mathbf{A} \in \mathcal{V}$ and $\theta \in \operatorname{Con} \mathbf{A}$ with a/θ infinite, then $|b/\theta| \leq |a/\theta|$. This follows from the fact that if $F_1, \ldots, F_n$ are the terms of Theorem 6.56 then

$$x \mapsto \langle F_1(x, b, a), \ldots, F_n(x, b, a) \rangle$$

is a bijection from b/θ into $(a/\theta)^n$, which is easily verified.

To see that $\mathcal{V}$ is permutable, let

$$p(x, y, z) = H(z, F_1(x, y, z), \ldots, F_n(x, y, z)).$$

It is easy to verify that $p(x, y, z)$ is a Mal'tsev term for $\mathcal{V}$. ∎

Exercise 25 gives a version of the last theorem for individual algebras.

Exercises 6.58

1. The variety of distributive lattices is not congruence regular.

2. The variety of implication algebras (see Exercise 6.16.5) is not congruence regular. (It helps to remember that every family of sets closed under set difference forms an implication algebra by defining $x \to y$ to mean $y - x$.)

3. If $\mathcal{V}$ has terms F and H obeying the equations

$$F(x, x, z) \approx z$$
$$H(F(x, y, z), y, z) \approx x,$$

then $\mathcal{V}$ is both coherent and congruence uniform. Thus, for example, the variety of quasigroups (defined in §3.4) is both coherent and congruence uniform. Hint: Look at the proof of Theorem 6.57.

4. (J. Varlet, 1972) The variety of three-valued Lukasiewicz algebras is congruence regular. This is the variety generated by $\langle 3, \wedge, \vee, ^+, ^*, 0\rangle$ where $\wedge$ and $\vee$ are the ordinary infimum and supremum on $3 = \{0, 1, 2\}$ ordered as usual, and $^+$ and * are unary operations defined by the following table

	0	1	2
$*$	2	0	0
$+$	2	2	0

5. (Junji Hashimoto, 1963; G. Grätzer, 1970b) **A** is congruence regular if and only if for all $a, b, c \in A$ there exists a finite sequence $\langle d_0, \dots, d_m\rangle$ of elements of A such that

$$\operatorname{Cg}(a, b) = \bigvee_{i=0}^{m} \operatorname{Cg}(c, d_i).$$

6. Give a direct proof that (iii) implies (iv) in Theorem 6.48.

7. For the $\mathcal{V}$ given in Example 6.52 find 5-ary terms $f_1, f_2, \dots$ satisfying condition (iv) of Theorem 6.48.

8. (J. Hagemann, 1973b; G. Grätzer, 1970b), see also K. Fichtner (1968) A variety $\mathcal{V}$ with constant terms $c_1, \dots, c_n$ is said to be **weakly congruence regular** with respect to $c_1, \dots, c_n$ if and only if each congruence θ on arbitrary $\mathbf{A} \in \mathcal{V}$ is determined by its n blocks $c_1/\theta, \dots, c_n/\theta$. That is, if $\theta, \psi \in \operatorname{Con} \mathbf{A}$ and $c_i/\theta = c_i/\psi$, for $i = 1, \dots, n$, then $\theta = \psi$. Prove that $\mathcal{V}$ is weakly congruence regular with respect to $c_1, \dots, c_n$ if and only if for some m, there exist binary terms $s_1, t_1, s_2, t_2, \dots, s_m, t_m$ and ternary terms $r_1, \dots, r_m$ such that the following

equations hold identically in $\mathcal{V}$:

$$\begin{aligned}
s_i(x,x) &\approx t_i(x,x) \approx c_j \qquad \text{for } i \equiv j \pmod m\\
r_1(s_1(x,y),x,y) &\approx x\\
r_i(t_i(x,y),x,y) &\approx r_{i+1}(s_{i+1}(x,y),x,y)\\
r_m(t_m(x,y),x,y) &\approx y.
\end{aligned}$$

Some earlier work on varieties weakly regular with respect to one constant symbol has been carried out by Robert Vaught, see Exercise 8.132.6.

In the Exercises 6.16.17–22, we presented some exercises indicating how some features of the theory of topological groups could be replicated for topological algebras in permutable varieties. Here we continue this development for topological algebras in congruence regular varieties. These results are due to Walter Taylor (1974, 1977a). Our assumptions on the algebra **A** and the topology are exactly as stated before Exercise 6.16.17, except that we are not assuming permutability. All topological notions in use here are standard, and may be found in any general topology text.

9. If **A** lies in a congruence regular variety and $\langle A, \mathcal{T}\rangle$ has an isolated point, then $\langle A, \mathcal{T}\rangle$ is discrete.

10. If **A** lies in a congruence regular variety and $\langle A, \mathcal{T}\rangle$ is an infinite compact metric space, then $|A| = 2^{\aleph_0}$. Hint: Use the Baire category theorem. It is known that every uncountable compact metric space has power at least $2^{\aleph_0}$.

11. If **B** is a topological algebra in a congruence regular variety and **A** is a topological subalgebra of $\mathbf{B}^I$ such that **A** has an isolated point, then there exists a finite set $J \subseteq I$ such that the projection of $\mathbf{B}^I$ onto $\mathbf{B}^J$ is one-to-one on A.

***12.** If **A** lies in a congruence regular variety and **A** is embeddable in some algebra which carries a compact Hausdorff topology, then $|B_2| \leq 2^{B_1}$ for B_1 and B_2 any two infinite blocks of any congruence $\theta \in \operatorname{Con} \mathbf{A}$. (See page 351 of Walter Taylor (1974) for a solution.)

13. If **A** is a congruence regular algebra, $\theta \in \operatorname{Con} \mathbf{A}$ and $a, b \in A$, then any set of congruences on **A** which separates points of a/θ will also separate points of b/θ.

14. Let $\mathcal{V}$ be a residually small, congruence regular variety in which each subdirectly irreducible algebra has power $\leq \mu$. If $\mathbf{A} \in \mathcal{V}$ and $a, b \in A$, then $|b/\theta| \leq \mu^{|a/\theta|}$ for a/θ infinite. Hint: use the previous exercise together with the method of proof for the Subdirect Representation Theorem (4.44).

15. Prove that the variety $\mathcal{V}$ presented in Example 6.52 has the following property which is stronger than what we required there. For any family of cardinals $\mu_i \geq 2$, for $i \in I$, there exists an algebra $\mathbf{B} \in \mathcal{V}$ and a congruence θ on **B**

which partitions B into blocks B_i with $|B_i| = \mu_i$ for all i. Prove moreover that the above property still holds if we enrich $\mathcal{V}$ by adding binary operations $\vee$ and $\wedge$ and add the axioms of lattice theory. Thus regularity and distributivity together do not imply weak uniformity.

16. Give a syntactic proof of Theorem 6.51. That is, deduce (iii) of Theorem 6.48 from the existence of uniformity terms.

17. If $\mathcal{V}$ is weakly congruence uniform, then there exists a function $f : \omega \to \omega$ such that $|b/\theta| \leq f(|a/\theta|)$ whenever $\mathbf{A} \in \mathcal{V}$, a, $b \in A$, $\theta \in \operatorname{Con} \mathbf{A}$, and a/θ is finite. However this condition does not guarantee uniformity. A first proof arises more or less directly from the definition, by using the compactness theorem. A more sophisticated argument will yield an explicit formula for f depending on the n, m and p appearing in uniformity terms for $\mathcal{V}$. Here one will need the explicit form of Ramsey's Theorem mentioned prior to Theorem 6.53.

18. A **discriminator variety** is a variety that has a ternary term $t(x, y, z)$ and is generated by a set of algebras such that t acts as a ternary discriminator (as defined in Exercise 2 of Section 6.2) on each. Prove that every discriminator variety is coherent. (Hint: there exist very simple Geiger terms in the ternary discriminator, with $n = 2$.) Then prove that every discriminator variety is congruence uniform. (In the case of a congruence having a finite block, a special argument will be required.) Discriminator varieties are studied in more detail in §9.5.

19. Let $\mathcal{V}$ be defined by the equations:

$$F(x, x, z) \approx z$$
$$G(w, w, w, x_0, x_1) \approx x_0$$
$$G(F(x_0, x_1, z), F(x_0, x_2, z), F(x_1, x_2, z), x_0, x_1) \approx x_1$$

Clearly $\mathcal{V}$ has a uniformity term of type $\langle 2, 3, 1, 0, 1\rangle$, and hence $|b/\theta| \leq 2^{|a/\theta|}$ whenever $\mathbf{A} \in \mathcal{V}$, $\theta \in \operatorname{Con} \mathbf{A}$, and a,$b \in A$. The task here is to show that the above estimate is best possible for this variety, i.e., for all infinite μ, there exist $\mathbf{A}$, θ, a,and b as above such that $|a/\theta| = \mu$ and $|b/\theta| = 2^{\mu}$. Hint: Let A be the subset of $\{0, 1\}^{\mu}$ consisting of sequences α such that either α has finite support or $\alpha_0 = 0$. Now define

$$F(\alpha, \beta, \gamma) = \begin{cases} \text{the majority value} & \text{if } |\{\alpha, \beta, \gamma\}| < 3\text{; otherwise:} \\ \alpha + \beta + \gamma \pmod 2 & \text{unless exactly two of } \alpha_0, \beta_0, \text{ and } \gamma_0 \\ & \text{are 0, in which case:} \\ \{0, m\} + \gamma & \text{where } m \text{ is the first place the two} \\ & \text{that agree at 0 differ.} \end{cases}$$

(Here, by $\{0, m\}$ we mean the characteristic function, i.e. δ such that $\delta(0) = \delta(m) = 1$ and $\delta(x) = 0$ otherwise.) One may check that the kernel of first coordinate projection is a congruence θ for this operation, and that a corresponding G may be found so that θ remains a congruence.

***20.** For $2 \leq n < \omega$, there exists a variety $\mathcal{V}_n$ which has a uniformity term of type $\langle n, n+1, 1, 0, 1\rangle$, and such that for each infinite μ, $\mathcal{V}_n$ has an algebra $\mathbf{A}$ with a congruence θ which has blocks of power μ and $\beth_n(\mu)$. Hint: The equations involved form an obvious generalization of those of the previous exercise. The construction of $\mathbf{A}$ uses the Erdős-Hajnal-Rado result (6.54); it seems to be necessarily more complicated than it was for $n = 2$ in the previous exercise. For a detailed solution, see pages 346–348 of Walter Taylor (1974).

21. Weak uniformity is not definable by a strong Mal'tsev condition. (Use the previous exercise.)

***22.** Let $\mathcal{V}$ be defined by the following infinite collection of equations:

$$\begin{aligned} G_1(x,x,z) &\approx G_2(x,x,z) \approx z \\ K(G_1(x,y,z), G_2(x,y,z), y, z) &\approx x \\ F_i(x,x,z) &\approx z \qquad \text{for } i \in \omega \\ H_i(F_i(x,y,z), x, y, z) &\approx x \qquad \text{for } i \in \omega \\ H_i(F_j(x,y,z), x, y, z) &\approx y \qquad \text{for } i \neq j \end{aligned}$$

Prove that $\mathcal{V}$ is coherent and congruence uniform, but that every finite subset of these equations defines a variety which is not congruence uniform. (For a proof of the final assertion, see pages 353–355 of Walter Taylor (1974).) Thus, although coherence implies 0-uniformity (any congruence with an infinite block must be uniform), it does not imply uniformity. Moreover, uniformity is not equivalent to a Mal'tsev condition. Hint: we readily satisfy the Mal'tsev condition for coherence by taking F_1 (of the coherence condition) to be G_1, F_2 to be G_2 and F_3 to be z.

23. The variety axiomatized by the following equations is congruence uniform and congruence permutable but not coherent:

$$\begin{aligned} F(x,x,z) &\approx z \\ H(u,u,x,y,w,z) &\approx x \\ H(F(x,w,z), F(y,w,z), x, y, w, z) &\approx y. \end{aligned}$$

Hint: show that $H(F(z,z,y), F(x,y,y), z, x, z, y)$ is a Mal'tsev term. (See page 356 of (Walter Taylor, 1974) for the proof that this variety is uniform and not coherent.)

24. Prove Theorem 6.56.

25. Prove that if $\mathbf{A}$ has a congruence with blocks of power μ and ν, where μ is an infinite cardinal at least as large as the cardinality of the basic operations of $\mathbf{A}$ and $\nu > \mu$, then $\mathbf{A}$ is not coherent.

26. Prove that the variety of Heyting algebras (defined in Exercise 18 of §4.5) is not congruence regular. Hint: Show that there is a three-element Heyting algebra which has a two-element homomorphic image.

6.7. Linear Mal'tsev Conditions, Derivations, Strong Mal'tsev Conditions

Linear Mal'tsev Conditions

In this section we will define linear terms, but first we will define a slightly more general concept, that of a basic term. A term t is **basic** provided there is at most one occurrence of an operation symbol of positive arity in t. If c is a 0-ary operation symbol in the signature, we call the associated term $c()$, which we usually denote simply by c, a **basic constant**. So a basic term is a variable, a basic constant, or an operation symbol of positive arity applied to variables and basic constants. A basic term without constants is called **linear**. For example, if p is a ternary operation symbol, then $p(x, y, x)$ is a linear term but $p(x, p(x, y, x), x)$ is not.[8] An equation is **linear** (or **basic**) if each side of the equation is linear (or basic). Most of the Mal'tsev conditions we have studied and will study are given by linear equations. These include congruence permutability, congruence k-permutability, congruence distributivity, congruence meet (join) semidistributivity, as well as congruence modularity and having a Taylor term, etc. Walter Taylor (2009) characterizes when a variety has an equational basis consisting of linear equations. This is presented as Theorem 6.59.

Since, as we will see, derivations with linear equations can be much more tractable, this section uses the ideas of Taylor's result, Theorem 6.59 below, to show some Mal'tsev classes cannot be characterized with linear equations; see Theorem 6.60.

What Mal'tsev conditions cannot be defined by linear equations? One example is the (idempotent) Mal'tsev condition of having a semilattice term. That is, a term t satisfying

$$\begin{aligned} t(x, x) &\approx x \\ t(x, y) &\approx t(y, x) \\ t(x, t(y, z)) &\approx t(t(x, y), z) \end{aligned} \tag{6.7.1}$$

(By Theorem 6.23 an algebra with such a term has meet semidistributive congruences.) While the third equation above is not linear, how does one show that there isn't a set of linear equations defining this property?

Let A and B be sets and suppose we have a set-retraction of B onto A. This means there are maps

$$f : B \twoheadrightarrow A \text{ and } g : A \rightarrowtail B \text{ with } f(g(x)) = x. \tag{6.7.2}$$

[8]The terminology *linear term* and *linear equation* is not perfect but it has become standard. An alternate terminology is *simple*. Another approach is to define a linear term as one of depth at most 1 (depth is defined on page 86).

Now suppose q is an n-ary operation on B. We can use q (and f and g) to define an operation on A, which we denote $q_{f,g}$ and define it by

$$q_{f,g}(a_1, \dots, a_n) = f(q(g(a_1), \dots, g(a_n))) \tag{6.7.3}$$

Now suppose that $\mathbf{B}$ is an algebra (on B of course). We will form an algebra $\mathbf{A}$ employing (6.7.3), using two different sets of operations on B.

First we will use the basic operations of $\mathbf{B}$ in our construction of $\mathbf{A}$. In this case $\mathbf{A}$ will have the same signature as $\mathbf{B}$ and we call $\mathbf{A}$ a **basic set-retract algebra** of $\mathbf{B}$. Explicitly, if p is an m-ary operation symbol of the signature of $\mathbf{B}$ and (6.7.2) holds, then the operation on A is given by

$$p^{\mathbf{A}}(a_1, \dots, a_m) = f(p^{\mathbf{B}}(g(a_1), \dots, g(a_m))). \tag{6.7.4}$$

We say that a class of algebras $\mathcal{K}$ is *closed under basic set-retracts* if every basic set-retract $\mathbf{A}$ of $\mathbf{B}$ lies in $\mathcal{K}$, for every $\mathbf{B} \in \mathcal{K}$.

The following theorem of Walter Taylor, which characterizes when a variety has a linear basis for its equations, is the genesis of our development of linear Mal'tsev conditions. While this theorem is not required for our examples below showing certain Mal'tsev conditions cannot be given by linear equations, it is used in the proof of the characterization of linear Mal'tsev conditions given in Theorem 6.63.

See Exercise 10.144.19 on page 238 of Volume III for a related application.

THEOREM 6.59 (Walter Taylor 2009). *Let $\mathcal{V}$ be a variety. Then $\mathcal{V}$ has an equational basis consisting of linear equations if and only if $\mathcal{V}$ is closed under basic set-retracts.*

Proof. Suppose Σ_0 is an equational basis for $\mathcal{V}$ consisting of linear equations. Let $\mathbf{B} \in \mathcal{V}$ and let A, f and g satisfy (6.7.2) and let $\mathbf{A}$ be the basic set-retract on A defined by f and g. A linear equation has either the form

$$p(x_{j_1}, \dots, x_{j_m}) \approx q(x_{k_1}, \dots, x_{k_n}), \tag{6.7.5}$$

where p is an m-ary operation symbol and q is an n-ary operation symbol; or it has the form $q(x_{k_1}, \dots, x_{k_n}) \approx x$. Suppose $\mathbf{B}$ satisfies (6.7.5). Then, by (6.7.4), we have in $\mathbf{A}$

$$\begin{aligned} p^{\mathbf{A}}(a_{j_1}, \dots, a_{j_m}) &= f(p^{\mathbf{B}}(g(a_{j_1}), \dots, g(a_{j_m}))) \\ &= f(q^{\mathbf{B}}(g(a_{k_1}), \dots, g(a_{k_n}))) = q^{\mathbf{A}}(a_{k_1}, \dots, a_{k_n}). \end{aligned}$$

This shows that (6.7.5) holds in $\mathbf{A}$. A similar argument handles the other form of a linear equation. Thus $\mathbf{A}$ satisfies Σ_0 and so $\mathbf{A} \in \mathcal{V}$, as desired.

Conversely, suppose $\mathcal{V}$ is closed under basic set-retracts. Letting Σ_0 be the linear equations that are true in $\mathcal{V}$, we have

$$\mathcal{V} \subseteq \mathbf{Mod}\ \Sigma_0. \tag{6.7.6}$$

To prove the theorem, we need to show the opposite containment holds. So let $\mathbf{A} \in \mathbf{Mod}\ \Sigma_0$. Let $\mathbf{F} = \mathbf{F}_{\mathcal{V}}(A)$ be the free algebra in $\mathcal{V}$ with free generating set A (the universe of $\mathbf{A}$). We will define a map $f : F \to A$. First, we require $f(a) = a$ for $a \in A$. If $\mathbf{A}$ were in $\mathcal{V}$ then f would naturally extend to a homomorphism. We only know $\mathbf{A} \in \mathbf{Mod}\ \Sigma_0$ but this is enough to define f at the first level of generation of $\mathbf{F}$ by A. What we mean by the first level is

$$A \cup \{q^{\mathbf{F}}(a_1, \dots, a_n) : q \text{ is a basic operation symbol}, a_i \in A\}. \tag{6.7.7}$$

We extend the definition of f to this set by defining

$$f(q^{\mathbf{F}}(a_1,\dots,a_n)) = q^{\mathbf{A}}(a_1,\dots,a_n). \tag{6.7.8}$$

To see that this extension of f is well defined, let p be another basic operation symbol and suppose $p^{\mathbf{F}}(a_{j_1},\dots,a_{j_m}) = q^{\mathbf{F}}(a_{k_1},\dots,a_{k_n})$. Then, since the elements of A freely generate $\mathbf{F}$, $\mathcal{V}$ satisfies the equation $p(x_{j_1},\dots,x_{j_m}) \approx q(x_{k_1},\dots,x_{k_n})$. Since this is a linear equation it is in Σ_0, and so $\mathbf{A}$ satisfies it. Hence $p^{\mathbf{A}}(a_{j_1},\dots,a_{j_m}) \approx q^{\mathbf{A}}(a_{k_1},\dots,a_{k_n})$. This, and a similar argument for the case $q^{\mathbf{F}}(a_1,\dots,a_n) = a$, shows that f is well defined on the set given by (6.7.7). For elements of F not in this set, we define f arbitrarily.

Now if we let $g : A \to F$ be the identity on A, then f and g satisfy the conditions of (6.7.2) with B replaced by F. Using (6.7.8) one can see that $\mathbf{A}$ is the basic set-retract of $\mathbf{F}$ determined by f and g. Since we are assuming $\mathcal{V}$ is closed under basic set retracts, we have $\mathbf{A} \in \mathcal{V}$. This shows that equality holds in (6.7.6), and the theorem follows. ■

In order to show that certain Mal'tsev conditions cannot be defined by linear equations, we use the set of *all* term operations on $\mathbf{B}$ in our construction of $\mathbf{A}$. In this case $\mathbf{A}$ and $\mathbf{B}$ will have different signatures. Let σ be the signature of $\mathbf{B}$. For each σ-term t, we introduce a new operation symbol $\bar{t}$. The arity of $\bar{t}$ is the same as that of t, which is the number of distinct variables occurring in t. On the other hand t is (usually) a composition of simpler terms, while $\bar{t}$ is an operation symbol not connected with the decomposition of t. We let $\overline{\sigma}$ be the signature $\{\bar{t} : t \text{ is a } \sigma\text{-term}\}$.

Given set retraction maps satisfying (6.7.2) and using (6.7.3) with $q = t^{\mathbf{B}}$, we get

$$\bar{t}^{\mathbf{A}}(a_1,\dots,a_n) = (t^{\mathbf{B}})_{f,g}(a_1,\dots,a_n) = f(t^{\mathbf{B}}(g(a_1),\dots,g(a_n))). \tag{6.7.9}$$

We call the algebra $\mathbf{A}$ with universe A and basic operations $\bar{t}^{\mathbf{A}}$, t a σ-term, a **full set-retract** of $\mathbf{B}$.

THEOREM 6.60. *If* $\mathbf{B}$ *satisfies a Mal'tsev condition given by linear equations and* $\mathbf{A}$ *is a full set-retract of* $\mathbf{B}$ *(or more generally, if* $\mathbf{A}$ *is an expansion of a full set-retract of* $\mathbf{B}$*), then* $\mathbf{A}$ *also satisfies this Mal'tsev condition.*

Proof. Suppose $\mathbf{A}$ is a full set-retract of $\mathbf{B}$ with maps f and g satisfying (6.7.2), and that $\mathbf{B} \models s \approx t$ for terms s and t in the signature of $\mathbf{B}$. Then by (6.7.9)

$$\begin{aligned}\bar{s}^{\mathbf{A}}(a_1,\dots,a_n) &= f(s^{\mathbf{B}}(g(a_1),\dots,g(a_n)))\\ &= f(t^{\mathbf{B}}(g(a_1),\dots,g(a_n))) = \bar{t}^{\mathbf{A}}(a_1,\dots,a_n),\end{aligned}$$

for all $a_1,\dots,a_n \in A$. It follows that $\mathbf{A}$ (or any expansion of $\mathbf{A}$), satisfies any linear Mal'tsev condition satisfied by $\mathbf{B}$. ■

By way of example suppose $\mathbf{B}$ has a Mal'tsev term $p(x,y,z)$ and $\mathbf{A}$ is a full set-retract of $\mathbf{B}$. We claim $\bar{p}(x,y,z)$ is a Mal'tsev term for $\mathbf{A}$. To see this suppose a and $c \in A$. Then

$$\bar{p}^{\mathbf{A}}(a,a,c) = (p^{\mathbf{B}})_{f,g}(a,a,c) = f(p^{\mathbf{B}}(g(a),g(a),g(c))) = f(g(c)) = c.$$

Similarly, $\bar{p}^{\mathbf{A}}(a,c,c) = a$ showing $\bar{p}(x,y,z)$ is a Mal'tsev term for $\mathbf{A}$.

The next theorem was proved by Keith Kearnes and John Snow in the late 1990's.

THEOREM 6.61. *Having a semilattice term cannot be defined by a linear Mal'tsev condition.*

Proof. To see this we will use Theorem 6.60. Let **B** be the direct product of two copies of the two element join semilattice, with elements $0 < a,\ b < 1$. Let $A = \{0, a, b\}$, let g be the identity on A, and let $f(1) = 0$ and $f(x) = x$ for $x \in A$. Each term in the signature of semilattices (that is, one binary operation symbol $\vee$) is equivalent to a term of the form $x_1 \vee \cdots \vee x_n$, associated left to right. Let $t_n(x_1, \ldots, x_n) = x_1 \vee \cdots \vee x_n$. If $a_1, \ldots, a_n$ are in A then by its definition

$$\bar{t}_n^{\mathbf{A}}(a_1, \ldots, a_n) = f(\bigvee a_i) = \begin{cases} 0 & \text{if } \bigvee a_i = 1 \\ \bigvee a_i & \text{otherwise} \end{cases}$$

Consequently if both a and b occur in $\{a_1, \ldots, a_n\}$ then $\bar{t}_n^{\mathbf{A}}(a_1, \ldots, a_n) = 0$; otherwise it is the join. These are the basic operations of **A**. Note this definition gives that $\bar{t}_n^{\mathbf{A}}(a_1, \ldots, a_n)$ is totally symmetric and idempotent. It also yields that the equation

$$\bar{t}_{n+1}(x, x, x_2, \ldots, x_n) \approx \bar{t}_n(x, x_2, \ldots, x_n) \tag{6.7.10}$$

holds in **A**.

Note the table for $\bar{t}_2^{\mathbf{A}}(x, y)$ is

$\bar{t}_2^{\mathbf{A}}$	0	a	b
0	0	a	b
a	a	a	0
b	b	0	b

We claim $\mathbf{F}_{V(\mathbf{A})}(x, y)$ has only 5 elements:

$$x, \quad y, \quad \bar{t}_2(x, y), \quad \bar{t}_2(x, \bar{t}_2(x, y)) \quad \text{and} \quad \bar{t}_2(y, \bar{t}_2(x, y)).$$

To prove this recall what is sometimes called the Birkhoff construction of the free algebra. To find $\mathbf{F}_{V(\mathbf{A})}(x, y)$ we consider all ordered pairs $\langle c, d\rangle$ from A. We view these pairs as maps from $\{x, y\}$ into A with $x \mapsto c$ and $y \mapsto d$. Since $|A| = 3$, there are 9 pairs. The free algebra is the subalgebra of $\mathbf{A}^9$ generated by

$$\begin{aligned} x &= \langle 0, 0, 0, a, a, a, b, b, b\rangle \\ y &= \langle 0, a, b, 0, a, b, 0, a, b\rangle. \end{aligned} \tag{$*$}$$

If $\langle c, d\rangle$ is one of our pairs we let $\eta_{\langle c,d\rangle}$ be the kernel of the homomorphism of $\mathbf{F}_{V(\mathbf{A})}(x, y)$ into **A**. Of course $\bigwedge_{\langle c,d\rangle \in A^2} \eta_{\langle c,d\rangle} = 0$. Since **A** is idempotent, $\eta_{\langle 0,0\rangle} = \eta_{\langle a,a\rangle} = \eta_{\langle b,b\rangle} = 1_{\mathbf{F}_{V(\mathbf{A})}(x,y)}$, and this means that, if we omit these three congruences, the meet will still be 0.

Now suppose that $\langle c', d'\rangle$ is another pair and suppose that the map $c \mapsto c'$, $d \mapsto d'$ extends to a homomorphism on the subalgebra generated by $\{c, d\}$. Then the homomorphism on $\mathbf{F}_{V(\mathbf{A})}(x, y)$ sending $x \mapsto c'$ and $y \mapsto d'$ factors through the one sending $x \mapsto c$ and $y \mapsto d$, and from this it follows that $\eta_{\langle c,d\rangle} \leq \eta_{\langle c',d'\rangle}$. Using this principle and the easily proved fact that **A** has an automorphism interchanging a and b, we see $\eta_{\langle 0,a\rangle} = \eta_{\langle 0,b\rangle}$, $\eta_{\langle a,0\rangle} = \eta_{\langle b,0\rangle}$, and $\eta_{\langle a,b\rangle} = \eta_{\langle b,a\rangle}$. Hence $\eta_{\langle 0,a\rangle} \wedge \eta_{\langle a,0\rangle} \wedge \eta_{\langle a,b\rangle} = 0$. Consequently we only need the second, fourth and sixth coordinates in $(*)$. That is, the map $x \mapsto \langle 0, a, a\rangle$, $y \mapsto \langle a, 0, b\rangle$ extends to an embedding of $\mathbf{F}_{V(\mathbf{A})}(x, y)$ into $\mathbf{A}^3$.

The claim is that the subalgebra generated by x and y is

$$\begin{aligned} x &= \langle 0, a, a\rangle \\ y &= \langle a, 0, b\rangle \\ \bar{t}_2(x,y) &= \langle a, a, 0\rangle \\ \bar{t}_2(x, \bar{t}_2(x,y)) &= \langle a, a, a\rangle \\ \bar{t}_2(y, \bar{t}_2(x,y)) &= \langle a, a, b\rangle. \end{aligned}$$

So we need to show this set of five three-tuples is closed under $\bar{t}_k^{\mathbf{A}^3}(x_1, \dots, x_k)$. By (6.7.10) we may assume $k \le 5$. We can also exploit the symmetry of these operations. With this it is not hard to show the above five elements form a subuniverse. For example,

$$\begin{aligned} &\bar{t}_5^{\mathbf{A}^3}(\langle 0,a,a\rangle, \langle a,0,b\rangle, \langle a,a,0\rangle, \langle a,a,a\rangle, \langle a,a,b\rangle) \\ &\qquad = \langle \bar{t}_5^{\mathbf{A}}(0,a,a,a,a), \bar{t}_5^{\mathbf{A}}(a,0,a,a,a), \bar{t}_5^{\mathbf{A}}(a,b,0,a,b)\rangle = \langle a,a,0\rangle. \end{aligned}$$

(This gives the identity $\bar{t}_5(x,y,\bar{t}_2(x,y),\bar{t}_2(x,\bar{t}_2(x,y)),\bar{t}_2(y,\bar{t}_2(x,y)) \approx \bar{t}_2(x,y)$.)

So, up to equivalence, these are the only binary term operations of $\boldsymbol{V}(\mathbf{A})$. Now $\bar{t}_2(x,y)$ is not a semilattice term since it is not associative, and $r(x,y) = \bar{t}_2(x,\bar{t}_2(x,y))$ and $s(x,y) = \bar{t}_2(y,\bar{t}_2(x,y))$ are not semilattice terms since they are not commutative. In fact, $r(y,x) = s(x,y)$. ■

Most of the above facts were discovered with the aid of The Universal Algebra Calculator (see Freese et al. (2008)). If we let $\mathbf{A}_k$ be the reduct of $\mathbf{A}$ to the signature $\{\bar{t}_2, \dots, \bar{t}_k\}$ then the above arguments show that $\mathbf{F}_{\boldsymbol{V}(\mathbf{A}_k)}(x,y)$ has 5 elements for all k. In particular $\mathbf{F}_{\boldsymbol{V}(\mathbf{A}_2)}(x,y) \cong \mathbf{F}_{\boldsymbol{V}(\mathbf{A}_3)}(x,y)$. This situation does not hold generally: using the Calculator one sees that $|\mathbf{F}_{\boldsymbol{V}(\mathbf{A}_2)}(x,y,z)| = 96$, while $|\mathbf{F}_{\boldsymbol{V}(\mathbf{A}_3)}(x,y,z)| = 97$.

For information on Birkhoff's construction of free algebras, see (Joel Berman, 2005; Ralph Freese, 2015).

Recall from the last section that an algebra $\mathbf{B}$ is *congruence regular* if $a/\theta = a/\phi$ implies $\theta = \phi$ for congruences θ and ϕ and $a \in B$. Theorem 6.48(iv) of §6.6, gives a (nonlinear) Mal'tsev condition for the class of varieties whose members are congruence regular. Here we will show that congruence regularity cannot be defined by a linear Mal'tsev condition.

THEOREM 6.62. *Congruence regularity cannot be defined by a linear Mal'tsev condition.*

Proof. To prove this we give a four element algebra $\mathbf{B}$ lying in a congruence regular variety having a full set-retract which is not congruence regular. Let

$$B = \{\langle 0,0\rangle, \langle 0,1\rangle, \langle 1,0\rangle, \langle 1,1\rangle\}$$

with the 3-place operation $x + y + z$ modulo 2. As $\mathbf{B}$ is the idempotent reduct of a vector space, the variety generated by $\mathbf{B}$ is congruence regular. Let $A = \{0,1,2\}$ and define maps $f : B \to A$ and $g : A \to B$ by $f(\langle x,y\rangle) = x + y$ (so $f(\langle 1,1\rangle) = 2$) and $g(0) = \langle 0,0\rangle$, $g(1) = \langle 1,0\rangle$ and $g(2) = \langle 1,1\rangle$.

The terms for $\mathbf{B}$ have the form $t(x_1, \dots, x_n) = y_1 + \cdots + y_m$ modulo 2, where the y_j's are a subset of the x_i's and m is odd (so that t is idempotent).

We claim that the partition θ with blocks $[0,2]$ and $[1]$ is a congruence of $\mathbf{A}$. Since every term operation of $\mathbf{A}$ is totally symmetric, it is enough to show $\bar{t}^{\mathbf{A}}(0, a_2, \ldots, a_n)\ \theta\ \bar{t}^{\mathbf{A}}(2, a_2, \ldots, a_n)$ for $a_2, \ldots, a_n \in A$. We may assume $\bar{t}^{\mathbf{A}}$ depends on its first variable. Let $b_2, \ldots, b_n \in B$. If $t^{\mathbf{B}}(\langle 0,0\rangle, b_2, \ldots, b_n) = \langle u, v\rangle$, then $t^{\mathbf{B}}(\langle 1,1\rangle, b_2, \ldots, b_n) = \langle u+1, v+1\rangle$ modulo 2. Since $f(\langle 0,1\rangle) = f(\langle 1,0\rangle)$ it follows that θ is a congruence on $\mathbf{A}$. We leave the details for the reader. This means θ and the zero congruence have a block in common and so $\mathbf{A}$ is not congruence regular. Thus congruence regularity cannot be defined by a linear Mal'tsev condition. ■

Now we turn to the characterization of Mal'tsev conditions definable by linear equations.

THEOREM 6.63 (W. Taylor, 2009, and K. Kearnes, L. Sequeira, and Á. Szendrei, unpublished). *A Mal'tsev condition* C *can be defined by linear equations if and only if, for every $\mathcal{V}$ satisfying* C *and every $\mathbf{B} \in \mathcal{V}$, every full set-retract of* $\mathbf{B}$ *also satisfies* C.

Proof. The forward direction is proved by Theorem 6.60.

To see the converse, let $\mathcal{V}$ be a variety with signature σ. As above, we let $\overline{\sigma}$ be the signature whose basic operation symbols are $\{\bar{t} : t \text{ is a } \sigma\text{-term}\}$ where, if t is a σ-term of arity k, then $\bar{t}$ is an operation symbol of arity k. We claim that the collection $\overline{\mathcal{V}}$ of all full set-retracts of members of $\mathcal{V}$ is a variety. We will show that it is closed under homomorphic images and leave it to the reader to show it is closed under subalgebras and direct products. Suppose $\mathbf{A}$ is a full set-retract of $\mathbf{B}$ as given in (6.7.2) and (6.7.9). Without loss of generality we may assume $A \subseteq B$ and that the map g is the identity. Let θ be a congruence of $\mathbf{A}$. Let $R \subseteq A$ be a transversal of the blocks of θ; that is, each block contains exactly one member of R. Let $r : A \twoheadrightarrow R$ be the retraction map that sends each $a \in A$ to the unique $x \in R$ with $a\ \theta\ x$. Of course R is a subset of B and $r \circ f$ is a retraction of B onto R. Let $\mathbf{R} \in \overline{\mathcal{V}}$ be the corresponding full set-retract of $\mathbf{B}$ onto R. We claim $\mathbf{R} \cong \mathbf{A}/\theta$. Let $r_1, \ldots, r_k \in R$. Of course these elements are in A and hence in B. Let t be a k-ary σ-term and let $b_0 = t^{\mathbf{B}}(r_1, \ldots, r_k)$ and $a_0 = f(b_0)$, and let r_0 be the unique element of R with $r_0\ \theta\ a_0$ so that $r(a_0) = r_0$. From these facts we see $\bar{t}^{\mathbf{A}}(r_1, \ldots, r_k) = a_0$, and hence

$$\bar{t}^{\mathbf{A}/\theta}(r_1/\theta, \ldots, r_k/\theta) = a_0/\theta = r_0/\theta.$$

On the other hand,

$$\bar{t}^{\mathbf{R}}(r_1, \ldots, r_k) = r(f(t^{\mathbf{B}}(r_1, \ldots, r_k))) = r(f(b_0)) = r(a_0) = r_0.$$

These combine to show the map $r_i \mapsto r_i/\theta$, $r_i \in R$, is an isomorphism of $\mathbf{R}$ and $\mathbf{A}/\theta$. Hence $\overline{\mathcal{V}}$ is closed under homomorphic images.

Thus $\overline{\mathcal{V}}$ is a variety of signature $\overline{\sigma}$. We claim $\overline{\mathcal{V}}$ is closed under basic set-retracts and thus, by Theorem 6.59, has a linear basis for its equations. To see this suppose $\mathbf{C}$ is a basic set-retract of $\mathbf{A}$ which is a full set-retract of $\mathbf{B} \in \mathcal{V}$. Then there are maps f, f', g and g$'$ such that

$$f : B \twoheadrightarrow A \text{ and } g : A \rightarrowtail B \text{ with } f(g(x)) = x,$$
$$f' : A \twoheadrightarrow C \text{ and } g' : C \rightarrowtail A \text{ with } f'(g'(x)) = x.$$

Using (6.7.9) and (6.7.4), we see that, if t is a term in the signature of $\mathbf{B}$ and $\bar{t}$ is a basic operation symbol for $\mathbf{A}$ (and for $\mathbf{C}$, since $\mathbf{A}$ and $\mathbf{C}$ have the same signature), we have,

$$\begin{aligned}\bar{t}^{\mathbf{C}}(c_1,\dots,c_m) &= f'(\bar{t}^{\mathbf{A}}(g'(c_1),\dots,g'(c_m)))\\ &= f'(f(t^{\mathbf{B}}(g(g'(c_1)),\dots,g(g'(c_m)))))\\ &= f''(t^{\mathbf{B}}(g''(c_1),\dots,g''(c_m))),\end{aligned}$$

where $f'' = f' \circ f : B \twoheadrightarrow C$ and $g'' = g \circ g' : C \rightarrowtail B$. Thus $\mathbf{C}$ is a full set-retract of $\mathbf{B}$, and hence $\mathbf{C} \in \overline{\mathcal{V}}$. This shows $\overline{\mathcal{V}}$ is closed under basic set-retracts and so has a linear basis for its equations.

We also claim $\overline{\mathcal{V}} \leq_{\text{int}} \mathcal{V}$. ($\leq_{\text{int}}$ is defined on page 9.) To see this we first need a system of definitions D, as defined in Definition 6.3, mapping the basic operation symbols of $\overline{\mathcal{V}}$ into terms in the signature of $\mathcal{V}$. This is easy: we define $D(\bar{t}) = t$. Now, as explained on page 8 and (6.1.3), we need to associate to each $\mathbf{B} \in \mathcal{V}$ an algebra $\mathbf{B}^D \in \overline{\mathcal{V}}$. The universe of $\mathbf{B}^D$ should be the same as that of $\mathbf{B}$ and its operations should be $D(\bar{t})^{\mathbf{B}} = t^{\mathbf{B}}$, for t a term in the signature σ of $\mathbf{B}$. To achieve this, we let $\mathbf{A} \in \overline{\mathcal{V}}$ be a particularly simple full set-retract: take $A = B$ and let f and g from (6.7.2) both be the identity map. Then (6.7.9) simplifies to

$$\bar{t}^{\mathbf{A}}(x_1,\dots,x_n) = t^{\mathbf{B}}(x_1,\dots,x_n).$$

From this it follows that $\mathbf{B}^D = \mathbf{A}$ has the required properties, showing $\overline{\mathcal{V}} \leq_{\text{int}} \mathcal{V}$.

Now suppose C is a Mal'tsev class closed under forming full set-retracts. Then C is a filter in $\mathbf{L}_{\text{int}}$ with a co-initial (the order-theoretical dual of co-final) descending chain of varieties

$$\mathcal{V}_0 \geq_{\text{int}} \cdots \geq_{\text{int}} \mathcal{V}_{n-1} \geq_{\text{int}} \mathcal{V}_n \geq_{\text{int}} \mathcal{V}_{n+1} \geq_{\text{int}} \cdots,$$

each finitely based. Now $\overline{\mathcal{V}_n} \leq_{\text{int}} \mathcal{V}_n$ and, by assumption, $\overline{\mathcal{V}_n}$ is in C. This implies that, for some $k > n$, $\mathcal{V}_k \leq_{\text{int}} \overline{\mathcal{V}_n} \leq_{\text{int}} \mathcal{V}_n$. By repeated use of this fact we can obtain an infinite subsequence of the $\mathcal{V}_n$'s such that, after renumbering, we have

$$\mathcal{V}_0 \geq_{\text{int}} \overline{\mathcal{V}_0} \geq_{\text{int}} \mathcal{V}_1 \geq_{\text{int}} \overline{\mathcal{V}_1} \geq_{\text{int}} \mathcal{V}_2 \geq_{\text{int}} \cdots.$$

As we noted above, $\overline{\mathcal{V}_n}$ has a basis Γ_n consisting of linear equations. Since $\mathcal{V}_{n+1}$ is finitely based and $\mathcal{V}_{n+1} \leq_{\text{int}} \overline{\mathcal{V}_n}$, there is a finite subset of Γ_n (by the compactness theorem, Theorem 8.30) such that the variety $\mathcal{U}_n$ it defines satisfies

$$\mathcal{V}_n \geq_{\text{int}} \overline{\mathcal{V}_n} \geq_{\text{int}} \mathcal{U}_n \geq_{\text{int}} \mathcal{V}_{n+1}.$$

It follows that the $\mathcal{U}_n$'s witness that C is a Mal'tsev condition defined by linear equations. ■

Other examples of Mal'tsev condition classes not definable by linear equations are given in the exercises.

Derivatives

Let Σ be a finite set of equations. We say a variety $\mathcal{V}$ **realizes** Σ if it satisfies Σ as a strong Mal'tsev condition; that is, for each operation symbol occurring in Σ there is a term for $\mathcal{V}$ such that the equations of Σ are true for $\mathcal{V}$.

We will introduce some special methods to prove implications between sets of equations: when is it the case that $\mathcal{V}$ realizes Σ_1 implies it realizes Σ_2? This topic will be covered more systematically in §10.8.

Σ is **inconsistent** if $\Sigma \models x \approx y$. Otherwise it is **consistent**. Σ *idempotent* means $\Sigma \models f(x, \dots, x) \approx x$ for all operation symbols f occurring in Σ. We introduce two kinds of derivatives of Σ, which are certain augmentations of Σ. Namely the *derivative* and the *order derivative*.

If $\Sigma \models x \approx f(\mathbf{w})$, where $\mathbf{w}$ is a sequence of not necessarily distinct variables and f is an operation symbol occurring in Σ, then f is **weakly independent** of its ith place under Σ for each i with $w_i \neq x$. (So if $\Sigma = \{p(x, y, y) \approx x, p(y, y, x) \approx x\}$, then p is weakly independent of all of its places.) The **derivative**, Σ', is the augmentation of Σ by equations that say for each operation symbol f occurring in Σ that f is **independent** of its ith place whenever Σ implies f is weakly independent of its ith place. This means that if f is weakly independent of its ith place under Σ, then we add

$$f(x_0, \dots, x_{n-1}) \approx f(x_0, \dots, x_{i-1}, y_i, x_{i+1}, \dots, x_{n-1})$$

to Σ'. We shall show that if Σ is a set of idempotent equations and the derivative, Σ', is inconsistent, then any variety that realizes Σ is congruence modular. The converse holds when Σ is linear.

The order derivative is related to congruence k-permutability. To define it, again let Σ be an idempotent set of equations. The **order derivative** of Σ, denoted Σ^+, is the augmentation of Σ by additional equations as follows. If

$$\Sigma \models x \approx f(\mathbf{w})$$

where $\mathbf{w}$ is a vector of not necessarily distinct variables and f is an operation symbol occurring in Σ, then Σ^+ contains all equations

$$x \approx f(\mathbf{w}')$$

where, for each i, $w'_i = x$ or w_i. To motivate this notion, suppose $\mathcal{V}$ is an idempotent variety and that $\mathbf{F}_{\mathcal{V}}(x, y)$ supports a compatible (partial) order in which x is the least element and y is the greatest. Now if $x = f(\mathbf{w})$, where $\mathbf{w}$ is a sequence of x's and y's, then it is easy to see $x = f(\mathbf{w}')$ must also hold. However, as we shall see in Theorem 6.68, having a compatible linear order is inconsistent with k-permutability. We shall show that if some iterated order derivative of Σ is inconsistent then Σ implies congruence n-permutability, for some n. The converse holds when Σ is linear.

THEOREM 6.64 (Topaz Dent, Keith A. Kearnes, and Ágnes Szendrei 2012).

(i) *Let Σ be an idempotent set of equations. If Σ' is inconsistent then any variety that realizes Σ is congruence modular.*

(ii) *If $\mathcal{V}$ is a variety with modular congruences, then $\mathcal{V}$ realizes some idempotent Σ such that Σ' is inconsistent.*

Proof. Suppose $\mathcal{V}$ is a variety that realizes Σ. Suppose also that $\mathcal{V}$ is not congruence modular. We will show that Σ' is consistent. By Lemma 6.8 the variety defined by Σ is not congruence modular and so we may assume $\mathcal{V}$ is presented by Σ; that is, $\mathcal{V}$ is the variety whose signature consists of the operation symbols occurring in Σ, and is defined by Σ.

Let a, b, c and d be the free generators of $\mathbf{F} = \mathbf{F}_{\mathcal{V}}(4) = \mathbf{F}_{\mathcal{V}}(a, b, c, d)$ and let

$$\alpha = \operatorname{Cg}^{\mathbf{F}}(\langle a, b\rangle, \langle c, d\rangle), \quad \beta = \operatorname{Cg}^{\mathbf{F}}(\langle b, c\rangle), \quad \gamma = \operatorname{Cg}^{\mathbf{F}}(\langle a, d\rangle, \langle b, c\rangle).$$

Since $\mathcal{V}$ is not congruence modular, Corollary 6.33 implies that

$$\beta \vee (\alpha \wedge \gamma) < \gamma, \tag{6.7.11}$$

see Figure 6.8 on page 42.

Let $\mathbf{T} = \mathbf{F}_{\mathcal{V}}(r, s)$ be the free algebra on two generators. The homomorphisms φ_α and $\varphi_\gamma : \mathbf{F} \twoheadrightarrow \mathbf{T}$ determined by

$$\varphi_\alpha : a \mapsto r, \quad b \mapsto r, \quad c \mapsto s, \quad d \mapsto s,$$

and

$$\varphi_\gamma : a \mapsto r, \quad b \mapsto s, \quad c \mapsto s, \quad d \mapsto r,$$

have kernels α and γ, respectively. Indeed, clearly $\alpha \leq \ker \varphi_\alpha$ and using the universal mapping property of Definition 4.107, one can show the opposite inclusion. Define a map $\varphi : \mathbf{F} \to \mathbf{T}^2$:

$$a \mapsto (r, r), \quad b \mapsto (r, s), \quad c \mapsto (s, s), \quad d \mapsto (s, r).$$

This map is onto: let $p(r, s)$ and $q(r, s)$ be in $\mathbf{T}$ and let $m = p(q(a, b), q(d, c)) \in F$. Using idempotence one can show $\varphi(m) = (p, q)$.

Again the reader can show the kernel of φ is $\alpha \wedge \gamma$, and thus

$$\mathbf{F}/\alpha \wedge \gamma \cong \mathbf{T}^2.$$

Since Day's pentagon lies in the filter of $\mathbf{Con\ F}$ above $\alpha \wedge \gamma$ this pentagon maps under φ to a pentagon in $\mathbf{Con\ T}^2$ by the Correspondence Theorem (4.12) on page 151 of Volume I. Under this correspondence α maps to η_0, γ maps to η_1 and $\beta \vee (\alpha \wedge \gamma)$ maps to $\delta = \mathrm{Cg}^{\mathbf{T}^2}(\langle\langle r, s\rangle, \langle s, s\rangle\rangle)$, where η_0 and η_1 are the projection kernels. Corollary 6.33 says $\langle a, d\rangle \notin \beta \vee (\alpha \wedge \gamma)$. Hence $\langle\langle r, r\rangle, \langle s, r\rangle\rangle \in \eta_1 - \delta$.

Let G be the η_1-class of $\langle r, r\rangle$. Since $\mathcal{V}$ is idempotent, there is a subalgebra $\mathbf{G}$ of $\mathbf{T}^2$ with universe G. Let $\overline{\mathbf{G}}$ be the quotient under δ restricted to G. $\overline{\mathbf{G}}$ is nontrivial and satisfies Σ since it is in $\mathcal{V}$. We will show that $\overline{\mathbf{G}}$ satisfies Σ', which will prove the result.

Let f be a function symbol of arity n occurring in Σ, and assume that Σ implies it is weakly independent of its first argument:

$$\Sigma \models f(y, \mathbf{w}) \approx x, \tag{6.7.12}$$

where x and y are distinct variables and $\mathbf{w}$ is a sequence of not necessarily distinct variables. Let $\langle g_0, r\rangle, \langle g_1, r\rangle, \dots, \langle g_{n-1}, r\rangle \in G$. We claim

$$f^{\mathbf{T}^2}(\langle g_0, r\rangle, \dots, \langle g_{n-1}, r\rangle) = f^{\mathbf{T}^2}(\langle g_0, r_0\rangle, \dots, \langle g_{n-1}, r_{n-1}\rangle), \tag{6.7.13}$$

where $r_i = r$ if there is an x in the ith place of $f(y, \mathbf{w})$ in (6.7.12), and $r_i = s$ otherwise. (6.7.13) is clear for the first coordinate. For the second coordinate the left side is $f^{\mathbf{T}}(r, \dots, r) = r$ and the right side is $f^{\mathbf{T}}(r_0, \dots, r_{n-1})$. But the latter is r by (6.7.12).

Let $w = w(r, s) \in \mathbf{T}$ be arbitrary. Then $\langle w, s\rangle \ \delta \ \langle s, s\rangle$. Indeed,

$$\langle w, s\rangle = \langle w(r, s), w(s, s)\rangle = w(\langle r, s\rangle, \langle s, s\rangle) \ \delta \ w((\langle s, s\rangle, \langle s, s\rangle) = \langle s, s\rangle$$

We need to show that f is independent of its first argument modulo δ on $\mathbf{G}$. Let $\langle g_0, r\rangle, \dots, \langle g_{n-1}, r\rangle$ and $\langle h, r\rangle$ be arbitrary elements of $\mathbf{G}$. Then (6.7.13) holds and by the same token

$$f^{\mathbf{T}^2}(\langle h, r\rangle, \langle g_1, r\rangle, \dots, \langle g_{n-1}, r\rangle) = f^{\mathbf{T}^2}(\langle h, r_0\rangle, \langle g_1, r_1\rangle, \dots, \langle g_{n-1}, r_{n-1}\rangle).$$

Notice $r_0 = s$ by (6.7.12). By our claim above, $\langle g_0, s\rangle \ \delta \ \langle h, s\rangle$. Hence,

$$\begin{aligned} f^{\mathbf{T}^2}(\langle g_0, r\rangle, \langle g_1, r\rangle, \dots, \langle g_{n-1}, r\rangle) &= f^{\mathbf{T}^2}(\langle g_0, r_0\rangle, \langle g_1, r_1\rangle, \dots, \langle g_{n-1}, r_{n-1}\rangle) \\ &\ \delta\ f^{\mathbf{T}^2}(\langle h, r_0\rangle, \langle g_1, r_1\rangle, \dots, \langle g_{n-1}, r_{n-1}\rangle) \\ &= f^{\mathbf{T}^2}(\langle h, r\rangle, \langle g_1, r\rangle, \dots, \langle g_{n-1}, r\rangle), \end{aligned}$$

proving f is independent of its first argument modulo δ on $\mathbf{G}$. Thus,

$$\mathbf{G}/\delta|_{\mathbf{G}} \models f(x, \mathbf{z}) \approx f(y, \mathbf{z}),$$

where x, y and all the variables in $\mathbf{z}$ are distinct.

For (ii) see Exercise 4. ■

EXAMPLE 6.65. Day's Mal'tsev condition for congruence modularity, Theorem 6.32, contains the equation

$$m_i(x, y, y, u) \approx m_{i+1}(x, y, y, u).$$

Unlike most Mal'tsev conditions we are studying, this one involves three variables. However, if we replace this equation by either

$$(6.7.14) \qquad m_i(x, x, x, u) \approx m_{i+1}(x, x, x, u) \quad \text{or} \quad m_i(x, y, y, y) \approx m_{i+1}(x, y, y, y),$$

the resulting Mal'tsev condition still characterizes congruence modularity; see Exercise 4. (The equation $m_0(x, y, z, u) \approx x$ can be replaced with its two variable consequences and the resulting Mal'tsev condition still defines congruence modularity.) The first proof that congruence modularity could be defined by two variable identities is due to J. B. Nation (1974). His condition uses 5-place terms.

EXAMPLE 6.66. Let Σ be a finite set of equations and let f be an n-ary operation symbol in the signature of Σ, with $n \geq 3$, and suppose Σ implies that f is idempotent and is weakly independent of each of its places. If $\mathcal{V}$ is a variety which realizes Σ, we say that $f(x_0, \dots, x_{n-1})$ is a **cube term** for $\mathcal{V}$. Of course, in any model of Σ', f is constant. But also $f(x, \dots, x) \approx x$ which means models of Σ' have only one element. So by Theorem 6.64 we have the following corollary.

COROLLARY 6.67. *If $\mathcal{V}$ realizes a cube term, it is congruence modular.* ■

Cube terms were introduced in (Joel Berman, Paweł Idziak, Petar Marković, Ralph McKenzie, Matthew Valeriote, and Ross Willard, 2010). They are important in the study of Constraint Satisfaction Problems (CSP's), as described in that paper.

Let $\mathbf{A}$ be an algebra. An order relation $\leq$ on A is a **compatible order** on $\mathbf{A}$ if $f(a_0, \dots, a_{n-1}) \leq f(b_0, \dots, b_{n-1})$ whenever $a_i \leq b_i$, $i = 0, \dots, n-1$, for all basic (equivalently term) operations f. The next theorem was proved by J. Hagemann (1973a); see Ralph Freese (2013) for an explicit proof. M. Valeriote and R. Willard (2014) show that if $\mathcal{V}$ is an idempotent variety and is not n-permutable for any n, then $\mathcal{D} \leq_{\text{int}} \mathcal{V}$, where $\mathcal{D}$ is the variety of distributive lattices, and where $\leq_{\text{int}}$ is the relation of interpretability for varieties (see page 9).

THEOREM 6.68. *The following are equivalent for a variety $\mathcal{V}$.*

(i) *$\mathcal{V}$ is not congruence n-permutable for any n.*

(ii) *$\mathcal{V}$ contains a nontrivial member **A** which has a compatible order which is not an antichain.*

*If $\mathcal{V}$ is idempotent, then the compatible order on the algebra **A** in clause* (ii) *can be taken to have least and greatest elements* 0 *and* 1 *with* $0 \neq 1$.

Proof. Assume $\mathcal{V}$ is not n-permutable for any n. Let $\mathbf{F}$ be the $\mathcal{V}$-algebra freely generated by x and y. Let $\mathbf{R}$ be the subalgebra of $\mathbf{F}^2$ generated by $\{(x,x),(x,y),(y,y)\}$ and let $\mathbf{T}$ be the transitive closure of $\mathbf{R}$. $\mathbf{T}$ is also a subalgebra of $\mathbf{F}^2$ and is reflexive and transitive so is a compatible quasiorder on $\mathbf{F}$. Clearly $(x,y) \in T$.

Suppose $(y,x) \in T$. Then there are elements $w_j \in F$, $j = 0,\dots,n$, such that $w_0 = y$, $w_n = x$ and $(w_j, w_{j+1}) \in R$, $j < n$. Thus

$$(y,y),(y,w_1),(w_1,w_2),\dots,(w_{n-1},x),(x,x)$$

are all in $\mathrm{Sg}_{\mathbf{F}^2}((x,x),(x,y),(y,y))$. By (iv) of Theorem 6.11, $\mathcal{V}$ is n-permutable, a contradiction. Thus there is a compatible quasiordering $\leq$ on $\mathbf{F}$ with $x \leq y$ and $y \not\leq x$.

Let

$$\theta = \{(a,b) \in \mathbf{F}^2 \,:\, a \leq b \text{ and } b \leq a\}$$

be the equivalence relation associated with $\leq$. Since $\leq$ is compatible, θ is a congruence. Let $\mathbf{A} = \mathbf{F}/\theta$. $\mathbf{A}$ is nontrivial since $(x,y) \notin \theta$. And $\mathbf{A}$ has a compatible ordering which we also denote $\leq$. Thus, if $a = x/\theta$ and $b = y/\theta$, then $a < b$. This shows the order on $\mathbf{A}$ is not an antichain and so proves (i) $\Rightarrow$ (ii).

For the other implication suppose $\mathbf{A} \in \mathcal{V}$ is nontrivial and has a compatible order which is not an antichain. Let $a < b$ in this order. Suppose $\mathcal{V}$ is k-permutable with a set Σ of Hagemann-Mitschke terms p_i as in Theorem 6.11. Then

$$b = p_1^{\mathbf{A}}(b,a,a) \leq p_1^{\mathbf{A}}(b,b,a) = \cdots = p_{k-1}^{\mathbf{A}}(b,a,a) \leq p_{k-1}^{\mathbf{A}}(b,b,a) = a,$$

a contradiction.

For the final remark, assuming $\mathcal{V}$ is idempotent, and $t^{\mathbf{F}}(\mathbf{w})$ is an element of $\mathbf{F}$, for some $\mathbf{w}$ with $w_i \in \{x,y\}$, we have

$$x = t(x,\dots,x) \leq t(\mathbf{w}) \leq t(y,\dots,y) = y.$$

It follows that $0 = x/\theta$ and $1 = y/\theta$ are the least and greatest elements in the order on $\mathbf{A}$. ∎

THEOREM 6.69. *Let $\mathcal{V}$ be a variety. Then*

(i) *If Σ is an idempotent set of equations such that $\mathcal{V}$ realizes Σ and some iterated order derivative of Σ is inconsistent, then $\mathcal{V}$ is congruence k-permutable for some k.*

(ii) *If $\mathcal{V}$ is congruence k-permutable for some k then $\mathcal{V}$ realizes some Σ whose n^{th} iterated order derivative is inconsistent for some n.*

Proof. First we note that (i) holds for $\mathcal{V}$ if and only if it holds for $\mathcal{V}_\Sigma = \mathbf{Mod}(\Sigma)$, the variety presented by Σ, by a standard argument. Hence we may assume $\mathcal{V}$ is idempotent. To see (i) assume $\mathcal{V}$ realizes Σ and is not congruence k-permutable for any k. By Theorem 6.68 there is an $\mathbf{A} \in \mathcal{V}$ having a bounded compatible order with 0 as the least element and 1 as the greatest.

Suppose Σ implies

$$(+) \qquad x \approx f(\mathbf{u}, y, \mathbf{v}),$$

where $\mathbf{u}$ and $\mathbf{v}$ are vectors of not necessarily distinct variables. For each variable z_i occurring in $\mathbf{u}$ or $\mathbf{v}$ except x and y we substitute an element $c_i \in A$. For x we substitute an element $a \in A$. Let $\mathbf{u}(b)$ be $\mathbf{u}$ with the above substitution and substituting b for y. $\mathbf{v}(b)$ is defined similarly. By (+)

$$\begin{aligned} a &= f(\mathbf{u}(0), 0, \mathbf{v}(0)) \\ &\leq f(\mathbf{u}(b), a, \mathbf{v}(b)) \\ &\leq f(\mathbf{u}(1), 1, \mathbf{v}(1)) = a \end{aligned}$$

showing $\mathbf{A}$ satisfies $x \approx f(\mathbf{u}, x, \mathbf{v})$. Repeated applications of this argument show that $\mathbf{A}$ is a model of Σ^+. Hence Σ^+ is consistent and, by Theorem 6.68, does not imply k-permutability for any k. Repeating this argument we see that every order derivative of Σ is consistent.

For (ii) suppose that $\mathcal{V}$ is congruence k-permutable. Then $\mathcal{V}$ realizes the Hagemann-Mitschke terms of Theorem 6.11(iii). An easy inductive arguments shows that the jth order derivative of Σ implies that $x \approx p_j(x, x, y)$. Thus the kth order derivative implies $x \approx p_{k-1}(x, x, y) \approx y$. ■

EXAMPLE 6.70. Theorem 6.48(iv) on page 51 gives a set Σ of equations for congruence regularity. Σ includes the equations $g_i(x, x, z) \approx z$ for $1 \leq i \leq n$. Clearly Σ^+ includes $g_i(z, x, z) \approx z$. Making the substitution $x \mapsto y$ and $z \mapsto x$, we see $\Sigma^+ \models g_i(x, y, x) \approx x$. In the other equations of Theorem 6.48(iv) make the substitution $z \mapsto x$ (fixing the other variables). Using these equations on the right side of the ith equation gives

$$(*) \qquad f_i(x, y, x, x, g_i(x, y, x)) \approx f_i(x, y, x, g_i(x, y, x), x).$$

Now using $(*)$ on the equations of Theorem 6.48(iv) after substituting $z \mapsto x$ easily yields $x \approx y$. Thus Σ^+ is inconsistent. Hence congruence regular varieties are k-permutable for some k by the previous theorem.

Since Σ^+ inconsistent implies Σ' is inconsistent, Theorem 6.64 implies congruence regular varieties are also congruence modular. Both these facts are results of J. Hagemann (1973b). A direct derivation of Hagemann-Mitschke terms from terms for congruence regularity is given in Theorem 6.121 in the Relationships section, §6.10.

THEOREM 6.71. *If a variety $\mathcal{V}$ is congruence regular, then it is congruence modular and congruence k-permutable, for some k.* ■

EXAMPLE 6.72. The converses of Theorem 6.64(i) and Theorem 6.69(i) fail. Let Σ be the usual equations defining lattices in $\vee$ and $\wedge$. It is easy to see that neither of these is independent of either of their places. So $\Sigma' = \Sigma$. But of course the variety of lattices is congruence distributive, and hence congruence modular, showing the converse of Theorem 6.64(i) is false.

For an example showing that the converse of Theorem 6.69(i) is not true we can take Σ to be the equations defining the variety of idempotent quasigroups. (The variety of all quasigroups was discussed on page 123 of Volume I.) The operation symbols are $\cdot$, $/$, and $\backslash$. Σ has the following 7 equations:

$$x \cdot x \approx x \qquad x/x \approx x \qquad x \backslash x \approx x$$
$$x \cdot (x \backslash y) \approx y, \qquad (x/y) \cdot y \approx x,$$
$$x \backslash (x \cdot y) \approx y, \qquad (x \cdot y)/y \approx x.$$

This variety is nontrivial: for $a \neq 0$ or 1 in a field, let $x \cdot y = ax + (1-a)y$ and $x/y = y \backslash x = a^{-1}x + (1-a^{-1})y$. It is also congruence permutable. For example,

$$p(x, y, z) = (x/(y \backslash y)) \cdot (y \backslash z) \approx (x/y) \cdot (y \backslash z)$$

is a Mal'tsev term; see (Ralph Freese and Ralph McKenzie, 1987) for a proof.[9] However, since the operation symbols are all binary it is easy to see that $\Sigma^+ = \Sigma$.

Despite these examples, the converses of Theorem 6.64(1) and Theorem 6.69(1) are true when Σ is linear. We begin with the converse of Theorem 6.69(1) which is slightly easier.

THEOREM 6.73. *The following are equivalent for an idempotent set Σ of linear equations.*

(i) *Some iterated order derivative is inconsistent.*
(ii) *Any variety that realizes Σ is congruence n-permutable for some n.*
(iii) *The variety $\mathcal{V}_\Sigma$ = $\mathbf{Mod}\,\Sigma$ axiomatized by Σ is congruence n-permutable for some n.*

Proof. (i) implies (ii) follows from Theorem 6.69(i) and (ii) implies (iii) is clear. To see (iii) implies (i) suppose (i) fails. So assume every iterated order derivative is consistent. Let Ω be the union of all the order derivatives. Then $\Omega^+ = \Omega$.

Let $\mathbf{V}$ be the algebra on $\{0, 1\}$ such that for each operation symbol f occurring in Ω we have $f^{\mathbf{V}}(v_1, \ldots, v_n) = 1$ if and only if

$$\Omega \models f(x_{v_0}, \ldots, x_{v_{n-1}}) \approx x_1. \tag{6.7.15}$$

We claim that $\mathbf{V}$ is in $\mathcal{V}_\Omega = \mathbf{Mod}\,\Omega$. The equations in Ω are all linear so they have the form

$$f(\mathbf{w}) \approx g(\mathbf{u}) \quad \text{or} \quad f(\mathbf{w}) \approx x$$

where $\mathbf{w} = (w_0, \ldots, w_{n-1})$ and $\mathbf{u} = (u_0, \ldots, u_{m-1})$ are vectors of not necessarily distinct variables. Let σ map the variables occurring in $\mathbf{w}$ and $\mathbf{u}$ into $V = \{0, 1\}$. Suppose $f^{\mathbf{V}}(\sigma\mathbf{w}) = 1$. Then $\Omega \models f(x_{\sigma w_0}, \ldots, x_{\sigma w_{n-1}}) \approx x_1$. Since $f(\mathbf{w}) \approx g(\mathbf{u})$ is in Ω,

$$\Omega \models f(x_{\sigma w_0}, \ldots, x_{\sigma w_{n-1}}) \approx g(x_{\sigma u_0}, \ldots, x_{\sigma u_{m-1}}).$$

Hence $\Omega \models g(x_{\sigma u_0}, \ldots, x_{\sigma u_{m-1}}) \approx x_1$, and thus $g^{\mathbf{V}}(\sigma\mathbf{u}) = 1$. It follows from symmetry that if either $f^{\mathbf{V}}(\sigma\mathbf{w}) = 1$ or $g^{\mathbf{V}}(\sigma\mathbf{u}) = 1$ then the other is, and from this we conclude that $f^{\mathbf{V}}(\mathbf{w}) = g^{\mathbf{V}}(\mathbf{u})$ holds under every substitution.

The argument showing $\mathbf{V}$ satisfies an equation of the form $f(\mathbf{w}) \approx x$ is similar and left to the reader. One approach is to add the equation $h(x) \approx x$ to Ω, where h is a unary operation symbol not occurring in Ω, and using the above argument on $f(\mathbf{w}) \approx h(x)$.

---

[9]That the variety of quasigroups is congruence permutable is a result of A. I. Mal'tsev (1954).

Using $\Omega^+ = \Omega$, it is easy to see that the operations of $\mathbf{V}$ preserve the order on $\mathbf{V}$. Thus by Theorem 6.68 $\mathcal{V}_\Omega$ is not congruence n-permutable for any n. Since $\mathcal{V}_\Omega \subseteq \mathcal{V}_\Sigma$ it is also not congruence n-permutable for any n. ■

THEOREM 6.74. *The following are equivalent for an idempotent set Σ of linear equations.*

(i) *Σ' is inconsistent.*
(ii) *Any variety that realizes Σ is congruence modular.*
(iii) *The variety $\mathcal{V}_\Sigma = \mathbf{Mod}\,\Sigma$ axiomatized by Σ is congruence modular.*

Proof. As in the last theorem, we need to show that (iii) implies (i). So suppose Σ' is consistent. We define algebras $\mathbf{V}$ and $\mathbf{V}'$ both on $\{0,1\}$ with the signature of Σ (which is the same as the signature of Σ'). $f^{\mathbf{V}}(v_0,\dots,v_{n-1}) = 1$ if and only if (6.7.15) holds with Ω replaced by Σ. Similarly $f^{\mathbf{V}'}(v_0,\dots,v_{n-1}) = 1$ if and only if (6.7.15) holds with Ω replaced by Σ'. Just as in the previous proof $\mathbf{V} \in \mathcal{V}_\Sigma$ and $\mathbf{V}' \in \mathcal{V}_{\Sigma'}$. Since $\Sigma \subseteq \Sigma'$, $\mathbf{V}'$, and so $\mathbf{V}' \times \mathbf{V}$, are in $\mathcal{V}_\Sigma$. We will complete the proof by showing $\mathbf{V}' \times \mathbf{V}$ has a nonmodular congruence lattice.

We use juxtaposition to denote the elements of $\mathbf{V}' \times \mathbf{V}$; so its universe is $\{00, 01, 10, 11\}$. We claim $\mathbf{Con}(\mathbf{V}' \times \mathbf{V})$ contains the pentagon lattice of Figure 6.5 on page 16 as a sublattice. Clearly the projection kernels as well as the least and greatest elements of that lattice are congruences. Let θ be the partition whose only block with more than one element is $[00\ 10]$. We need to show that θ is a congruence. If it is not, there is a basic translation $\lambda(x)$ such that $\langle \lambda(00), \lambda(10)\rangle \notin \theta$. Now $\theta \subseteq \eta$, the kernel of the projection onto $\mathbf{V}$ (the second projection kernel). So $\lambda(00)$ and $\lambda(10)$ are equal in their second coordinate. The only possibilities are

(1) $\lambda(00) = 01$ and $\lambda(10) = 11$, or
(2) $\lambda(00) = 11$ and $\lambda(10) = 01$.

Now $\lambda(ab) = f^{\mathbf{V}'\times\mathbf{V}}(ab, \mathbf{00}, \mathbf{10}, \mathbf{01}, \mathbf{11})$, where, for example, $\mathbf{10}$ is a sequence, possibly of length 0, of the form $(10, 10, \dots, 10)$, and f is an operation symbol of Σ. The second coordinate of both cases above yields $f^{\mathbf{V}}(0, \mathbf{0}, \mathbf{0}, \mathbf{1}, \mathbf{1}) = 1$. By the definition of $f^{\mathbf{V}}$, $\Sigma \models f(x_0, \mathbf{x_0}, \mathbf{x_0}, \mathbf{x_1}, \mathbf{x_1}) \approx x_1$, so Σ entails that $f(p, \mathbf{q}, \mathbf{r}, \mathbf{s}, \mathbf{t})$ is weakly independent of its first three blocks of variables. So relative to Σ', $f(p, \mathbf{q}, \mathbf{r}, \mathbf{s}, \mathbf{t})$ is independent of its first three blocks of variables. But now looking at the first coordinate, if either (1) or (2) holds we get

$$f^{\mathbf{V}'}(0, \mathbf{0}, \mathbf{1}, \mathbf{0}, \mathbf{1}) \neq f^{\mathbf{V}'}(1, \mathbf{0}, \mathbf{1}, \mathbf{0}, \mathbf{1}).$$

This contradicts that $f^{\mathbf{V}'}$ is independent of its first argument. Thus θ is a congruence of $\mathbf{V}' \times \mathbf{V}$, showing that its congruence lattice is not modular, demonstrating the contrapositive of (iii) implies (i). ■

Basic equations were defined at the beginning of this section. David Kelly (1972) and (1973) gave a calculus for basic equations which shows that deciding $\Sigma \models \varphi$, where φ and the equations of Σ are basic, is recursive; see Exercise 7.12.4. In particular this recursive decision method applies to linear equations.

COROLLARY 6.75. *The following two problems are decidable. For a finite, idempotent set Σ of linear equations,*

(i) *does the realization of Σ imply congruence modularity?*
(ii) *does the realization of Σ imply k-permutability, for some k?* ■

Without the assumption that Σ is a set of linear equations, these problems are undecidable as we will see in Chapter 7; see Theorem 7.71 and the exercises of that section.

Strong Mal'tsev Properties

As mentioned at the beginning of this chapter several interesting properties are strongly Mal'tsev definable. These include congruence permutability, having a majority term, having a Pixley term. Other properties such as congruence distributivity are Mal'tsev definable and appear not to be strongly Mal'tsev definable. But is that really the case? Are there any Mal'tsev definable properties that are not strongly Mal'tsev definable? In this subsection we will give a construction of Marcin Kozik, Andrei Krokhin, Matthew Valeriote, and Ross Willard (2015) showing that several important Mal'tsev properties are in fact not strongly Mal'tsev definable. The fact that congruence distributivity and congruence modularity are not strongly Mal'tsev definable was first proved by K. Fichtner (1972) and David Kelly (1973). Another example of a Mal'tsev class that is not a strong Mal'tsev class is the class of varieties with weakly uniform congruences; see Exercise 6.58.21.

An n-ary **near unanimity term** t is defined by

$$t(y, x, \dots, x) \approx t(x, y, x, \dots, x) \approx \cdots \approx t(x, \dots, x, y) \approx x.$$

These will be studied in §6.10. By Theorem 6.125, a variety with a near unanimity term is congruence distributive.

We will be constructing several algebras on a two element set so it is convenient to let $\mathbf{2} = \{0, 1\}$. For $N > 2$ and $1 \le r \le N - 2$ let $\varphi_{N,r} : \mathbf{2}^N \to \mathbf{2}$ be given by

$$\varphi_{N,r}(\mathbf{x}) = \begin{cases} 1 & \text{if } |\{i : x_i = 1\}| > r\,, \\ 0 & \text{otherwise} \end{cases}$$

The reader can easily verify that $\phi_{N,r}$ is a near unanimity operation on $\mathbf{2}$.

Let $n > 1$ and define $N = n(n-1)^{n-1} + 1$. Define $r_i = (n-1)^i$. Let σ_n be a signature with n ternary symbols, h_i, $i = 0, \dots, n-1$, and one N-ary symbol q. For each n we define n algebras $\mathbf{D}_{n,i}$, $i = 0, \dots, n-1$, on $\mathbf{2}$ with signature σ_n. Define the operations by

$$q^{\mathbf{D}_{n,i}} = \varphi_{N,r_i}$$

$$h_j^{\mathbf{D}_{n,i}}(x, y, z) = \begin{cases} x & \text{if } j < i \\ x \vee (\overline{y} \wedge z) & \text{if } j = i \\ x \vee z & \text{if } j > i, \end{cases}$$

where $\overline{y}$ is the complement of y. Let $\mathcal{V}_n$ be the variety generated by $\mathbf{D}_{n,0}, \dots, \mathbf{D}_{n,n-1}$.

LEMMA 6.76. *For each $n > 1$, $\mathcal{V}_n$ is finitely generated, is $(2n+1)$-permutable and has an N-ary near unanimity term.*

Proof. Let $p_0(x, y, z) = x$, $p_{2n+1}(x, y, z) = z$ and

$$p_{i+1}(x, y, z) = h_i(x, y, z) \qquad i = 0, \dots, n-1$$
$$p_{2n-i}(x, y, z) = h_i(z, y, x) \qquad i = 0, \dots, n-1$$

In Exercise 6 the reader is asked to verify that these p_i's satisfy the conditions of Theorem 6.11 on page 14. We already noted the other properties hold. ■

For $m > 1$ let $\mathcal{F}_m$ be the set of all functions $\mathbf{2}^m \to \mathbf{2}$ and let $\leq$ denote the pointwise order on $\mathcal{F}_m$. Let $\pi_1, \dots, \pi_m$ be the m-ary projection functions on $\mathbf{2}$. If $f \in \mathcal{F}_m$ then f *dominates a projection* if $f \geq \pi_i$ for some i. For $S \subseteq \{1, \dots, m\}$, let $\chi_S \in \mathbf{2}^m$ be defined by

$$\chi_S(k) = \begin{cases} 1 & \text{if } k \in S, \\ 0 & \text{otherwise} \end{cases}$$

We define $\rhd$ to be the binary relation on $\mathcal{F}_m$ that satisfies the following condition: for all $f, g \in \mathcal{F}_m$, we have $f \rhd g$ if and only if for all $S \subseteq \{1, \dots, m\}$, if $g(\chi_S) = 1$, then there exists $k \in S$ such that $f \geq \pi_k$.

LEMMA 6.77. (i) $\pi_k \rhd \pi_k$, $k = 1, \dots, m$.

(ii) *If* $f_i \rhd g_i$, *for* $i = 0, 1$, *then* $(f_0 \vee f_1) \rhd (g_0 \vee g_1)$.

(iii) *If* $f \rhd g$ *then* $f \geq g$.

(iv) *If* $f_1 \geq f_0 \rhd g_0 \geq g_1$ *then* $f_1 \rhd g_1$.

(v) $f \in \mathcal{F}_m$ *is a term operation of* $\langle \mathbf{2}, \vee \rangle$ *if and only if* f *is idempotent and* $f \rhd f$.

Proof. We leave the proofs of (i) to (iv) as an exercise. To see (v) suppose $f \in \mathcal{F}_m$ is a term operation of $\langle \mathbf{2}, \vee \rangle$. Then $f = \bigvee_{j \in J} \pi_j$ for some $J \subseteq \{1, \dots, m\}$. Of course f is idempotent. Suppose $S \subseteq \{1, \dots, m\}$ with $f(\chi_S) = 1$. This implies $J \cap S \neq \emptyset$. If $k \in J \cap S$, then $f \geq \pi_k$, which shows $f \rhd f$.

To see the converse suppose f is idempotent and $f \rhd f$. Let J be the set of $j \in \{1, \dots, m\}$ with $f(\chi_{\{j\}}) = 1$. Let $g = \bigvee_{j \in J} \pi_j$. Using that f is idempotent, we see that $f(\chi_{\{1,\dots,m\}}) = 1$ and so $f \rhd f$ implies $f \geq \pi_i$, for some i. So J is not empty and hence g is well defined. We want to show $f = g$. That $f \geq g$ follows from the definition of J and $f \rhd f$. Now suppose that $S \subseteq \{1, \dots, m\}$ and $f(\chi_S) = 1$. Then $f \geq \pi_k$ for some k. From this we get that $k \in J$, and hence $g(\chi_S) = 1$ showing $g \geq f$. ■

LEMMA 6.78. *Let* $n \geq 1$ *and suppose* $1 \leq r \leq N - 2$ *with* $r < N/n$. *Then for all* $1 \leq m \leq n$, *if* $f_1, \dots, f_N \in \mathcal{F}_m$ *and each* f_i *dominates a projection, then the composition* $\varphi_{N,r} \circ (f_1, \dots, f_N)$ *also dominates a projection.*

Proof. There are N f_i's each dominating at least one of the m projections, so there is a projection π_i dominated by at least $\lceil N/m \rceil$. Since $r < N/n \leq \lceil N/m \rceil$, the composition $\varphi_{N,r} \circ (f_1, \dots, f_N)$ also dominates π_i. ■

LEMMA 6.79. *Let* $n \geq 2$ *and suppose* $1 \leq s \leq r \leq N-2$ *with* $r < N/n$ *and* $s \leq r/(n-1)$. *If* $1 \leq m \leq n$ *and if* $f_1, \dots, f_N, g_1, \dots, g_N \in \mathcal{F}_m$ *satisfy*

(i) *each* g_i *dominates a projection, and*

(ii) $f_i \rhd g_i$ *for all* i,

then

$$\varphi_{N,s} \circ (f_1, \dots, f_N) \rhd \varphi_{N,r} \circ (g_1, \dots, g_N).$$

Proof. Let $\hat{f} = \varphi_{N,s} \circ (f_1, \dots, f_N)$, $\hat{g} = \varphi_{N,r} \circ (g_1, \dots, g_N)$, and let S be such that $\hat{g}(\chi_S) = 1$. By Lemma 6.78, $\hat{g}$ dominates a projection. Since $\hat{f} \geq \hat{g}$, $\hat{f}$ dominates the same projection. Note this proves the result in the case $|S| = m$. Thus we assume $|S| \leq m - 1$.

Since $\hat{g}(\chi_S) = 1$, there are at least $r + 1$ g_i's with $g_i(\chi_S) = 1$, and for each such g_i, f_i dominates a projection on a coordinate in S. Hence we have at least $\lceil (r+1)/|S| \rceil$ f_i's dominating a common projection in S. Since $s \leq r/(n-1) \leq r/|S| < \lceil (r+1)/|S| \rceil$, $\hat{f}$ dominates this projection as well. ∎

LEMMA 6.80. *For all $n \geq 2$ and $1 \leq m \leq n$, if t is an m-ary term in the signature σ_n and $f_i = t^{\mathbf{D}_{n,i}}$ for $i = 0, \dots, n-1$, then*

(i) *each f_i is idempotent,*
(ii) *f_{n-1} dominates a projection,*
(iii) *$f_0 \rhd f_1 \rhd \cdots \rhd f_{n-1}$.*

Proof. Since $\mathbf{D}_{n,i}$ is idempotent, (i) holds. For (ii) and (iii) we use induction on the complexity of t. If t is a variable then (ii) is clear and (iii) follows from Lemma 6.77(i).

So $t = h_j(r, s, w)$ or $t = q(u_1, \dots, u_N)$ and we may assume r, s, w and $u_1, \dots, u_N$ are m-ary terms in σ_n which satisfy the claims of the lemma. First consider $t = h_j(r, s, w)$. Let $r_i = r^{\mathbf{D}_{n,i}}$, $s_i = s^{\mathbf{D}_{n,i}}$ and $w_i = w^{\mathbf{D}_{n,i}}$, $i = 0, \dots, n-1$. Now $f_{n-1} \geq r_{n-1}$ because $h_j^{\mathbf{D}_{n,i}}(x, y, z) \geq x$ by its definition above. Now, since r_{n-1} dominates a projection, f_{n-1} dominates the same projection. This proves (ii).

To see (iii), fix i with $0 < i < n$. If $i < j$, $(f_{i-1}, f_i) = (r_{i-1} \vee w_{i-1}, r_i \vee w_i)$ and so $f_{i-1} \rhd f_i$ by Lemma 6.77(ii) and the fact that r and w satisfy (iii). If $i = j$ then $(f_{i-1}, f_i) = (r_{i-1} \vee w_{i-1}, r_i \vee (\overline{s_i} \wedge w_i))$ and by the above argument $f_{i-1} \rhd r_i \vee w_i$. Since $r_i \vee w_i \geq f_i$, we can use Lemma 6.77(iv) to conclude $f_{i-1} \rhd f_i$. If $i = j + 1$ a similar argument works. If $i > j + 1$ then $(f_{i-1}, f_i) = (r_{i-1}, r_i)$, and so $f_{i-1} \rhd f_i$ follows since (iii) holds for r.

For the case $t = q(u_1, \dots, u_N)$ use induction and Lemma 6.78 to prove (ii) and 6.79 to prove (iii); see Exercise 8. ∎

LEMMA 6.81. *Suppose $n \geq 2$ and t is an m-ary σ_n term. If $m \leq n$ then there is an $i < n$ such that $t^{\mathbf{D}_{n,i}}$ is a term operation of $\langle \mathbf{2}, \vee \rangle$.*

Proof. Let t be an m-ary σ_n term. Define $f_i = t^{\mathbf{D}_{n,i}}$ for $i = 0, \dots, n-1$. Assume f_i is not a term operation of $\langle \mathbf{2}, \vee \rangle$ for any i. For $i < n$, define $T_i = \{k : \pi_k \leq f_i\}$. Note $T_0 \supseteq T_1 \supseteq \cdots \supseteq T_{n-1} \neq \emptyset$ by Lemma 6.77(iii) and Lemma 6.80 parts (ii) and (iii).

For $i < n$, we have $f_i \not\rhd f_i$ by Lemma 6.77(5) and our assumption. So pick $S_i \subseteq \{1, \dots, m\}$ with $f_i(\chi_{S_i}) = 1$ but $S_i \cap T_i = \emptyset$. Since f_i is idempotent $S_i \neq \emptyset$. Also, if $0 < i < n$, then $S_i \cap T_{i-1} \neq \emptyset$ because $f_{i-1} \rhd f_i$ by Lemma 6.80(iii). Thus $T_{i-1} \supsetneq T_i$ for all $0 < i < n$, showing $|T_0| \geq n$. Hence $m \geq |T_0 \cup S_0| = |T_0| + |S_0| \geq n+1$, contradicting $m \leq n$. ∎

THEOREM 6.82 (Marcin Kozik, Andrei Krokhin, Matthew Valeriote, and Ross Willard 2015). *Let $\mathcal{V}_n$ be the varieties defined before Lemma* 6.76. *Any strong Mal'tsev class containing $\mathcal{V}_n$ for all $n \geq 2$, contains the variety of semilattices.*

Proof. Let $\mathcal{S}$ be the variety of semilattices. Let $\mathcal{U}$ be a finitely presented variety that is interpretable into every $\mathcal{V}_n$. This means there is a finite set Σ of equations such that each $\mathcal{V}_n$ realizes Σ. Since $\mathcal{S}$ and each $\mathcal{V}_n$ is idempotent, we can add equations to Σ expressing that each operation symbol of Σ is idempotent and it will still be the case that each $\mathcal{V}_n$ realizes this expanded Σ, and $\mathcal{S}$ realizes the expanded Σ if and only if it realized the original Σ. Hence we assume that Σ is idempotent; that is, it implies each of its operation symbols is idempotent.

We pause to define the star product of terms.

DEFINITION 6.83. If $r(x_1, \dots, x_n)$ and $s(x_1, \dots, x_m)$ are terms, define the **star product** $r \star s$ by

$$\begin{aligned} t(x_1, x_2, \dots, x_{nm}) &= (r \star s)(x_1, \dots x_{nm}) \\ &= r(s(x_1, \dots, x_m), s(x_{m+1}, \dots, x_{2m}), \dots, s(x_{(n-1)m+1}, \dots, x_{nm})). \end{aligned}$$

Suppose r and s are terms in the signature of Σ. Since Σ is idempotent, we can recover r and s from t:

$$r(y_1, y_2, \dots, y_n) = t(y_1, \dots, y_1, y_2, \dots, y_2, \dots, y_n, \dots, y_n)$$

where the blocks of repeated variables each have length m, and

$$s(z_1, \dots, z_m) = t(z_1, \dots, z_m, \dots z_1, \dots, z_m)$$

where the block $z_1, \dots, z_m$ is repeated n times. Using the $\star$ product we can assume the signature of Σ consists of a single operation symbol $h(x_1, \dots, x_m)$.

We wish to show that $\mathcal{U}$ is interpretable into $\mathcal{S}$. Choose $n \geq m$. Let $t(x_1, \dots, x_m)$ witness that $\mathcal{U}$ interprets into $\mathcal{V}_n$. By the previous lemma there is an i such that $t^{\mathbf{D}_{n,i}}$ is a term operation of $\langle \mathbf{2}, \vee \rangle$. Thus the 2 element semilattice $\langle \mathbf{2}, \vee \rangle$ has a term that also satisfies the equations in Σ. Thus $\mathcal{U}$ interprets into the variety generated by $\langle \mathbf{2}, \vee \rangle$, which is $\mathcal{S}$. ■

COROLLARY 6.84. *The following properties cannot be defined by a strong Mal'tsev condition:*

(i) *having a near unanimity term,*
(ii) *congruence distributivity,*
(iii) *congruence modularity,*
(iv) *congruence $\models$ for a nontrivial lattice equation,*
(v) *k-permutability for some k,*
(vi) *having a Hobby-McKenzie term (defined in §6.9).*

Proof. This follows from the theorem once we show each of these Mal'tsev classes contains the $\mathcal{V}_n$'s but does not contain the variety of semilattices. Since, as noted in §6.9, a near unanimity term is a Hobby-McKenzie term, the Mal'tsev class (vi) contains the $\mathcal{V}_n$'s. And, as we noted earlier, a variety with a near unanimity term is congruence distributive so the classes (ii), (iii) and (iv) each contain the $\mathcal{V}_n$'s.

By the result of Ralph Freese and J. B. Nation (1973), Theorem 6.30, congruence lattices of semilattices satisfy no nontrivial lattice identity. Hence (iv) does not contain the variety of semilattices. Theorem 6.113 shows that (vi) does not contain the variety of semilattices. (Actually K. Kearnes, E. Kiss and Á. Szendrei have shown that the classes (iv) and (vi) are the same; see Theorem 6.113.) The other classes clearly do not contain the variety of semilattices. ■

Exercise 6.58.21 shows that the (Mal'tsev) class of varieties with weakly uniform congruences is not a strong Mal'tsev class.

On the other hand, in the next section we will prove the surprising result that the class of varieties having a Taylor term is a strong Mal'tsev class. For the class of all varieties having meet semidistributive congruence lattices the question is open. However there is a strong Mal'tsev condition such that a finitely generated variety is congruence meet semidistributive if and only if it satisfies this condition (Marcin Kozik, Andrei Krokhin, Matthew Valeriote, and Ross Willard, 2015).

Exercises 6.85

1. Show that if $\mathcal{V}$ is a nontrivial variety defined by a set of linear equations, then $\mathcal{V}$ has algebras of every cardinality (except 0).

2. Modify Theorem 6.61 to show the property of having a binary, commutative, idempotent term t satisfying $t(x, t(x, y)) \approx t(x, y)$ cannot be defined by a linear Mal'tsev conditions.

3. Consider the property $\mathcal{V}$ does not have a two element algebra. Use Theorem 6.60 to show this property is not definable by a linear Mal'tsev condition. Hint: the variety generated by the three element group does not contain the two element group. [There is a (nonlinear) Mal'tsev condition for this property; see Theorem 5.15 in (Walter Taylor, 1973)].

4. Show that if Σ is the set of Day's equations given in Theorem 6.32(iv), or if Σ is this set of equations with $m_i(x, y, y, u) \approx m_{i+1}(x, y, y, u)$ replaced with either of the equations of (6.7.14), then Σ' is inconsistent. Also show that if Σ is the set of Gumm's equations given in Theorem 6.44(iii), then Σ' is inconsistent.

5. Let Σ be the usual equations defining lattices in the signature $\{\vee, \wedge\}$. Prove $\Sigma' = \Sigma$ even though lattices have distributive, and hence modular, congruence lattices. This shows the converse of Theorem 6.64(i) fails.

6. Show that the p_i's of Lemma 6.76 satisfy the conditions of Theorem 6.11.

7. Prove statements (i) to (iv) of Lemma 6.77.

8. Complete the proof of Lemma 6.80. Hint: review the interpretation of q in $\mathbf{D}_{n,i}$.

9. Show the varieties $\mathcal{V}_n$ defined before Lemma 6.76 are not congruence regular.

6.8. Taylor Classes of Varieties

Let M and N be two $m \times n$ arrays, or matrices, of terms in a signature $\rho : I \longrightarrow \omega$, and F, G two n-ary operation symbols of ρ. The (formal) matrix equation $F(M) \approx G(N)$ is taken as shorthand for the following tuple of m equations:

$$F(M_{i,0}, \dots, M_{i,n-1}) \approx G(N_{i,0}, \dots, N_{i,n-1}), \tag{6.8.1}$$

where i ranges from 0 to $m-1$. By **the ith row of** M, we mean the tuple $\langle M_{i,0}, \dots, M_{i,n-1}\rangle$; thus Equation (6.8.1) can be summarized as

$$F(i^{\text{th}} \text{ row of } M) \approx G(i^{\text{th}} \text{ row of } N).$$

Until further notice, we restrict this notation to the special case that $F = G$ and the matrix entries M_{ij} and N_{ij} are variables.

DEFINITION 6.86. As above, let $\rho : I \longrightarrow \omega$ be a signature possessing an n-ary operation symbol F. Let M and N be two $n \times n$ matrices of variables, with x's along the diagonal of M and y's along the diagonal of N. The (M, N, F)-**Taylor equations** are

$$F(x, \dots, x) \approx x; \qquad F(M) \approx F(N). \tag{6.8.2}$$

(This is an $(n+1)$-tuple of equations.) If a variety $\mathcal{V}$ realizes these equations, we say that $\mathcal{V}$ **has an** (M, N)-**Taylor term**. In greater detail, let us suppose that the signature of $\mathcal{V}$ is σ and that $D : I \longrightarrow T_\sigma$ is a system of definitions of ρ in σ (or an interpretation of ρ in σ, see page 8). If the equations (6.8.2) are realized by $\mathcal{V}$ under the system of definitions $D : I \longrightarrow T_\sigma$ (i.e. if $\mathbf{A}^D$ satisfies (6.8.2) for each $\mathbf{A} \in \mathcal{V}$), then the σ-term F^D will be called[10] an (M, N)-**Taylor term for** $\mathcal{V}$.

A **Taylor term for** $\mathcal{V}$ is an (M, N)-Taylor term for $\mathcal{V}$ for some appropriate M and N. A **Taylor term for an algebra A** is a Taylor term for $\boldsymbol{HSP}\mathbf{A}$.

We first note that we obviously can admit M and N under the apparently weaker condition that $M_{i,i} \neq N_{i,i}$ for each $i < n$. (For Clifford Bergman (2012), this is the definition.) For such M and N an easy change of variables on the ith row renders $M_{i,i} = x$ and $N_{i,i} = y$. In fact we can utilize M and N under the even weaker conditions that M and N each have n columns (and any number of rows) and that

$$(\forall j)(\exists i(j))\, M_{i(j),j} \neq N_{i(j),j}. \tag{6.8.3}$$

Under this condition we may simply build new $(n \times n)$ matrices M' and N' where the $i(j)^{\text{th}}$ row of M is used as the jth row of M' (and similarly for N and N').

On the other hand, if (6.8.3) fails for M and N, i.e. if there exists j such that M and N agree on their j^{th} columns, then (6.8.2) is satisfied for F equal to x_j, the jth coordinate projection. In other words if we allowed this situation every variety would have a Taylor term. To exclude that possibility, we require condition (6.8.3) to hold, at the bare minimum. We often deal with M and N that have been revised (as described above) so that $\{M_{jj}, N_{jj}\} = \{x, y\}$ for all j.

[10] As we have seen for other Mal'tsev conditions in these volumes, common usage often refers to the symbol F and its interpretation F^D by the same letter. We will continue to allow this practice. Thus in Example 3 below, Equation (6.8.5) really should say that $F^D = p(q(x_0, \dots), \dots)$.

It is now apparent that for any two fixed appropriate $n \times n$ matrices M, N, the existence of an (M, N)-Taylor term on $\mathcal{V}$ is a (strong) Mal'tsev condition on $\mathcal{V}$, as described at the start of §6.1. This may be called the **(M, N)-Taylor condition.** As noted there, the validity of the (M, N)-Taylor condition for $\mathcal{V}$ is equivalent to the interpretability of the equations (6.8.2) in $\mathcal{V}$. The major result of this section—see Theorem 6.92 below—exhibits two 12×12 matrices of x's and y's, M and N, such that the (M, N)-Taylor condition is the weakest of all idempotent Mal'tsev conditions. This means that if $\mathcal{W}$ is a variety that satisfies any nontrivial, idempotent Mal'tsev condition, then the equations (6.8.2) are interpretable in $\mathcal{W}$, which is to say that $\mathcal{W}$ realizes (6.8.2).

EXAMPLES OF TAYLOR TERMS

1. Congruence permutability. It is not hard to see, in practice, that many of the Mal'tsev conditions discovered in this chapter are either Taylor conditions or closely related to such conditions. We present one example now, while reserving the full statement to Theorem 6.87. Consider the equations

$$F(x.\dots,x) \approx x; \qquad F(M) = F\begin{pmatrix} x & x & y \\ x & x & y \\ y & x & x \end{pmatrix} \approx F\begin{pmatrix} y & y & y \\ y & y & y \\ y & y & y \end{pmatrix} = F(N).$$

It is easy to check that each equation here is one of Mal'tsev's original two equations for congruence permutability, and that both of his equations are included. Therefore the Taylor condition here is equivalent to the original condition of Mal'tsev. (The second rows of M and N repeat the first rows; this is allowed by our definition.) Thus every congruence-permutable variety has this Taylor term.

2. Near unanimity. One may make a very similar construction of M and N for the strong Mal'tsev condition of an n-ary **near unanimity term** which is defined by the equations

$$\begin{aligned} F(x, y, y, \dots, y) &\approx y \\ F(y, x, y, \dots, y) &\approx y \\ &\vdots \\ F(y, y, \dots, y, x) &\approx y, \end{aligned}$$

for a single n-ary operation symbol F. It is not hard to see that these n equations can be packaged as a single $n \times n$ matrix equation $F(M) \approx F(N)$, so that the matrix equation, together with idempotence, defines near-unanimity. Thus n-fold near unanimity is both a Mal'tsev condition and a Taylor condition.

3. Congruence 3-permutability. Let $\mathcal{V}$ be a congruence-3-permutable variety. By Theorem 6.11 there are ternary $\mathcal{V}$-terms p and q that satisfy these equations in $\mathcal{V}$:

(6.8.4) $\qquad x \approx p(x, z, z); \qquad p(x, x, z) \approx q(x, z, z); \qquad q(x, x, z) \approx z.$

We will see that

(6.8.5) $\qquad F(x_0, \dots, x_8) \;=\; p(q(x_0, x_1, x_2), q(x_3, x_4, x_5), q(x_6, x_7, x_8))$

is a Taylor term for $\mathcal{V}$. We claim that F is an (M, N)-Taylor term with M and N taken as the following two matrices:

$$M = \begin{pmatrix} x & x & x & y & y & y & y & y & y \\ x & x & x & x & x & x & y & y & y \\ x & x & y & x & x & y & x & x & y \end{pmatrix};$$

$$N = \begin{pmatrix} x & x & x & x & x & x & x & x & x \\ x & y & y & x & y & y & x & y & y \\ y & y & y & y & y & y & y & y & y \end{pmatrix}.$$

We first ask the reader (see Exercise 3) to verify that the matrices M and N satisfy (6.8.3). This establishes that M and N do form a Taylor condition. It is not hard to convert M (resp. N) to a 9×9 matrix that has x's (resp. y's) on the diagonal, in the manner described above. This conversion, however, is not necessary for establishing that F is a Taylor term for $\mathcal{V}$. It is a simple matter (see Exercise 3) to verify that the equations $F(M) \approx F(N)$ follow from (6.8.5) and the 3-permutability equations (6.8.4). Thus the F defined in (6.8.5) is an (M, N)-Taylor term for $\mathcal{V}$. Thus every 3-permutable variety has a Taylor term.

The method described here for 3-permutability carries over with minimal changes to k-permutability for arbitrary finite k. Note that in (6.8.5), $F = p \star q$, where the $\star$-operator was defined in Definition 6.83. Similar naïve methods can be applied to many of the Mal'tsev conditions in this chapter, to show that they also yield Taylor conditions. The theorem that follows (Theorem 6.87) assures us that this is possible for any non-trivial idempotent Mal'tsev condition.

Active idempotent varieties

Let us call an algebra **trite**[11] if it has more than one element and each of its operations is a projection operation. One easily sees that every term operation in a trite algebra is a projection as well. A variety $\mathcal{V}$ will be called **passive** (see (Clifford Bergman, 2012)) if $\mathcal{V}$ contains a trite algebra. (In this case, $\mathcal{V}$ has trite algebras of every cardinality.) If $\mathcal{V}$ has no trite models, then we call $\mathcal{V}$ **active**. Note a variety having only one-element algebras is active. One easily sees that the varieties of interest in this chapter—with permutable, distributive, modular, regular, etc., congruences, and other classes as well—never have trite models, and hence are active varieties. This observation is the starting point for the following theorem. The theorem was stated in (Walter Taylor, 1977b).

THEOREM 6.87. *$\mathcal{V}$ has a Taylor term if and only if $\mathcal{V}$ has a reduct that is active and idempotent. In particular, if $\mathcal{V}$ has a Taylor term, then $\mathcal{V}$ is active.*

Proof. **($\Rightarrow$)** First let us suppose that $\mathcal{V}$ has a Taylor term; more precisely, that there exist an idempotent n-ary term F and square matrices M (resp. N) with x's (resp. y's) on the diagonal, such that $\mathcal{V}$ satisfies $F(M) \approx F(N)$. We will prove by contradiction that $\mathcal{V}$ has no trite models. Suppose $\mathbf{A}$ is a trite algebra in $\mathcal{V}$, so $\mathbf{A}$ satisfies $F(x_0, \ldots, x_{n-1}) \approx x_j$ for some j. Now we have $F(j^{\text{th}}$ row of $M) \approx F(j^{\text{th}}$ row of $N)$ as one identity of $\mathcal{V}$. But this equation is synonymous with $x \approx y$, which contradicts the assumption that $\mathbf{A}$ has

[11] Terminology suggested by Bernhard Banaschewski.

more than one element. Thus $\mathcal{V}$ has no trite models, and hence is active. Clearly the F-reduct of $\mathcal{V}$ is active and idempotent.

($\Leftarrow$ linear identities) For the converse, it will suffice to assume that $\mathcal{V}$ is idempotent and active, and to prove that $\mathcal{V}$ has a Taylor term. We will first give the proof for a special case: $\mathcal{V}$ has finite signature and is defined by a finite set of linear equations (see §6.7). This special argument requires little more than a systematic version of Example 3 above; it is of some interest in this chapter, because many of the Mal'tsev conditions of the chapter are defined by linear equations. In this case, the method is so simple that for many of the smaller Mal'tsev conditions of the chapter a Taylor condition can be computed by hand. Later we shall return to the general case.

Let s and t be terms with arities m and n, respectively. Recall the definition of $s \star t$, given in Definition 6.83:

$$s \star t(x_0, \ldots, x_{mn-1}) \;=\; s(t(x_0, \ldots, x_{n-1}), \ldots, t(x_{(m-1)n}, \ldots, x_{mn-1})), \tag{6.8.6}$$

with the right-hand side denoting substitution—see e.g. page 151 of §7.2. (The $\star$-product also played a role in the previous section.) We can write the right side of (6.8.6) as

$$s(t(X_0), \ldots, t(X_{m-1}))$$

where X_0 is the sequence of variables $x_0, \ldots, x_{n-1}$, etc. Now assume s and t are both idempotent. Then

$$t(x_0, \ldots, x_{n-1}) = t(X_0) \approx s(t(X_0), \ldots, t(X_0)),$$

and taking X_i to be the constant sequence $x_i, \ldots, x_i$, we see

$$s(x_0, \ldots, x_{m-1}) \approx s(t(X_0), \ldots, t(X_{m-1})).$$

Thus both s and t can be recovered from $s \star t$.

The reader may easily check that $\star$-multiplication is an associative operation on the set of all terms. Supposing the operation symbols of $\mathcal{V}$ to be $F_0, F_1, \ldots F_{m-1}$, we define H to be the term $F_0 \star F_1 \star \cdots \star F_{m-1}$. We claim H is a Taylor term for $\mathcal{V}$.

Let us assume that all F_i have the same arity, N. (This assumption simplifies the notation, but does not essentially change any reasoning.) We may now use idempotence and the above arguments to derive two simple facts about the values of H on tuples that contain some special kinds of repetition. Note that the arity of H is N^m.

LEMMA 6.88. *Let s and t be idempotent terms in a variety $\mathcal{V}$.*

(i) *Suppose that s has arity N and t has arity N^r for some r. If X is an input N^{r+1}-sequence of variables, and if X has period N^r (under the cyclic shift operator), then $\mathcal{V}$ satisfies $s \star t(X) \approx t(X')$, where X' is one period, of length N^r.*

(ii) *Suppose that t has arity N and s has arity N^r for some r. If X is an input N^{r+1}-sequence of variables, and if X consists of N^r constant blocks of length N, then $\mathcal{V}$ satisfies $s \star t(X) \approx s(X')$, where X' is formed by collapsing each constant block of length N.*

Proof. See Exercise 6.91.4. ∎

COROLLARY. *If $F_0, \dots, F_{m-1}$ and H are idempotent terms as described above, then for each i,*

$$\mathcal{V} \models F_i(x_0, \dots, x_{N-1}) \approx H(X_i), \tag{6.8.7}$$

where X_i is the sequence of variables x_j, whose subscripts appear as the ith digit in the base-N expansion of the sequence $0, 1, \dots, N^m - 1$. ■

Now let us consider the set Σ of equations defining $\mathcal{V}$. By assumption, each equation in Σ has the form

$$F_j(x_{\sigma(0)}, \dots, x_{\sigma(N-1)}) \approx F_k(x_{\tau(0)}, \dots, x_{\tau(N-1)}), \tag{6.8.8}$$

where σ and τ are selfmaps of $\{0, \dots, N-1\}$, and where j and k may be the same or different elements of $\{0, \dots, m-1\}$. Now applying the substitution σ to Equation (6.8.7) we obtain

$$\mathcal{V} \models F_j(x_{\sigma(0)}, \dots, x_{\sigma(N-1)}) \approx H(\sigma X_j),$$

with X_j as indicated after (6.8.7), and σX_j the result of applying the substitution σ to the sequence X_j. (It is equal to $\sigma \circ X_j$.) By transitivity of $\approx$, we now have

$$\mathcal{V} \models H(\sigma X_j) \approx H(\tau X_k) \tag{6.8.9}$$

(one such equation for each linear $\mathcal{V}$-identity of type (6.8.8)). Now, to show that H is a Taylor term for $\mathcal{V}$, we will show that condition (6.8.3) holds for the matrices M and N whose rows come from the (left and right sides of) Equations (6.8.9).

We prove (6.8.3) by contradiction. If it fails then there exists $q \in \{0, \dots, N^m - 1\}$ such that $M_{i,q} = N_{i,q}$ for all i. In terms of our matrices M and N, this means that for each Equation (6.8.9), the two sides of (6.8.9) have the same variable in the qth position of H. This means that if we define $\overline{H} : A^{M^n} \longrightarrow A$ via $\overline{H}(x_0, x_1, \dots) = x_q$ (the qth projection operation), then the algebra $\langle A, \overline{H} \rangle$ satisfies each Equation (6.8.9). Now let us augment the algebra $\langle A, \overline{H} \rangle$ with operations $\overline{F}_i$ defined as follows. Expand q in base-N notation, and let d_i be the ith digit in this expansion. Then define $\overline{F_i} : A^N \longrightarrow A$ to be the ${d_i}^{\text{th}}$ projection operation. It is not hard to check—directly from the definitions—that the algebra $\langle A; \overline{H}, \overline{F}_i \rangle_{i<m}$ models Equation (6.8.7).

We wish to prove that the trite algebra $\langle A; \overline{H}, \overline{F}_i, \rangle_{i<m}$ models Equations (6.8.8), in other words that its $\overline{F}_i$-reduct lies in $\mathcal{V}$. So we suppose that $a_0, a_1, \dots \in A$, and we compute

$$\begin{aligned} \overline{F}_j(a_{\sigma(0)}, \dots, a_{\sigma(N-1)}) &= \overline{H}(\sigma A_j) \\ &= \overline{H}(\tau A_k) = \overline{F}_k(a_{\tau(0)}, \dots, a_{\tau(N-1)}). \end{aligned}$$

(Here A_i is defined analogously to the X_i found in (6.8.7): it is the sequence of length N^m of elements of $\{a_0, \dots, a_{N-1}\}$ whose subscripts appear as the ith digit in the base-N expansion of the sequence $0, 1, \dots, N^r - 1$.) Here the first and last equalities come from (6.8.7), and the middle equality comes from (6.8.9). Thus $\mathcal{V}$ has the trite model $\langle A; \overline{F}_i \rangle_{i<N}$, in contradiction to our assumption that $\mathcal{V}$ is an active variety. Thus H is an (M, N)-Taylor term for $\mathcal{V}$, where M and N arise from Equations (6.8.9) above.

($\Leftarrow$ general case.) In fact it will be enough to prove this implication in the special case that $\mathcal{V}$ is an idempotent variety of finite signature, defined by a finite set of equational axioms. For, given an active variety $\mathcal{V} = \mathbf{Mod}\,\Sigma$, with Σ possibly infinite, we may first apply the Compactness Theorem 7.30 to prove that there is a finite subset Σ_0 of Σ such that $\mathbf{Mod}\,\Sigma_0$ is active. (Details are left to the reader in Exercise 5.)

So now we assume, as before, that $\mathcal{V}$ has the signature of operation symbols $F_0, \ldots,$ F_{m-1}, each N-ary, and that $\mathcal{V}$ is defined by a finite set Σ of equations in the operations F_i. The equations in Σ are taken to be $\mu_j \approx \nu_j$ for $0 \leq j < k$.

Depth. The depth $h(\sigma)$ of a term σ (in any signature ρ) is defined recursively as follows. Each variable has depth 0. If $\sigma = F(\tau_0, \ldots, \tau_{n-1})$ for an n-ary ρ-operation F and for ρ-terms τ_i $(0 \leq i < n)$, then $h(\sigma) = 1 + \max\{h(\tau_i) : i < n\}$. The depth of a formal equation is the larger of the depths of its two sides. (Thus a linear Mal'tsev condition—as presented in §6.7—is one whose equations are of depth 0 or 1.) We will take d to be the maximum depth of the equations $\mu_j \approx \nu_j$. We then define D to be the set of all terms in this signature, in variables $x_0, \ldots, x_{N-1}$, and of depth d or less. Clearly D is a finite set.

Definition of t and s. Now, letting $\hat{F}_i$ denote the term $F_i(x_0, \ldots, x_{N-1})$ (for $i < m$), we define t_0 to be the term $\hat{F}_0 \star \hat{F}_1 \star \cdots \star \hat{F}_{m-1}$, and then we let $t = t_0^{\star d}$ (the $\star$-product of d copies of t_0). This is a $\mathcal{V}$-term with $n = N^{dm}$ variables. Finally we define $s = t \star t$, a $\mathcal{V}$-term with n^2 variables.

Substitution into t and s. Let X denote an n-tuple of variables selected from the set $\{x_0, \ldots, x_{N-1}\}$; we may write X as $\langle x_{\phi(i)} : i < n\rangle$ for an appropriate function $\phi \colon n \longrightarrow N$. Then if u is a term in $x_0, \ldots, x_{n-1}$, we may use either $u(X)$ or $u(\langle x_{\phi(i)} : i < n\rangle)$, or simply $u(x_\phi)$ or $u(\phi)$ to denote the substituted term $u(x_{\phi(0)}, \ldots, x_{\phi(n-1)})$. If $0 \leq q < N^2$, then q has a unique representation as $q = a + bn$ with $0 \leq a, b < n$. We define $\eta_0(q) = a$ and $\eta_1(q) = b$. Then we have $\eta_0, \eta_1 : n^2 \longrightarrow n$. We therefore may substitute each η_i into s, forming $s(\eta_0)$ and $s(\eta_1)$, each a term in n variables. The reader may check that $\mathcal{V}$ satisfies $s(\eta_0) \approx s(\eta_1) \approx t$ (see Exercise 6).

LEMMA 6.89. *If α is any term in D, there exists a function $\phi \colon n \longrightarrow N$ so that $\mathcal{V} \models t(\langle x_{\phi(i)} : i < n\rangle) \approx \alpha$.*

Proof. By induction on i (for $1 \leq i \leq d$), one may prove that every α in D of depth $\leq i$ has the required form. (In fact the only $\mathcal{V}$-equations used in this argument are the idempotent laws. See Exercise 7.) ■

Now for each $\alpha \in D$, we select one ϕ obeying the conclusion of Lemma 6.89, and designate it ϕ_α. In other words, we now have $\mathcal{V} \models t(x_{\phi_\alpha}) \approx \alpha$ for each $\alpha \in D$.

The variety $\mathcal{W}$. We specify the signature of $\mathcal{W}$ to have a single operation S, which is n^2-ary. The axioms come in three forms:

i. If ϕ and ψ are any maps $n^2 \longrightarrow n^2$, and if $\mathcal{V} \models s(\phi) \approx s(\psi)$, then $S(\phi) \approx S(\psi)$ is an axiom of $\mathcal{W}$. Let Δ_0 stand for this set of equations.
ii. $S(\eta_0) \approx S(\eta_1)$ is an axiom of $\mathcal{W}$. (This in fact repeats one of the equations from Part i.) Let Δ_1 stand for this equation.
iii. Define T to be the $\mathcal{W}$-term $S(\eta_0)$. Then $S \approx T \star T$ is an axiom of $\mathcal{W}$. Let Δ_2 stand for this equation.

It is not hard to see that if $\mathbf{A} = \langle A; F_i \rangle_{i<m} \in \mathcal{V}$, then $\langle A; s^{\mathbf{A}} \rangle \in \mathcal{W}$, and thus that $\mathcal{W}$ is interpretable in $\mathcal{V}$. We are interested, however, in establishing the reverse: an interpretation of $\mathcal{V}$ in $\mathcal{W}$. For $0 \leq i < m$, we define $\delta_i(x_0, \ldots, x_{N-1})$ to be the substituted $\mathcal{W}$-term $T(\phi_{\hat{F}_i})$, where the $\mathcal{V}$-terms $\hat{F}_i$ and the substitutions ϕ_α are as defined above. Now for $\mathbf{B} = \langle B; S \rangle \in \mathcal{W}$, we define $\mathbf{B}'$ to be $\langle B, \delta_i^{\mathbf{B}} \rangle_{i<m}$.

LEMMA 6.90. *For all $\mathbf{B} \in \mathcal{W}$, we have $\mathbf{B}' \in \mathcal{V}$. In other words the system of definitions $\langle \delta_i : i < m \rangle$ is an interpretation of $\mathcal{V}$ in $\mathcal{W}$.*

Proof. We first proceed by induction on $i < d$ to show that for every $\alpha \in D$ of depth at most i, and any $\mathbf{B} \in \mathcal{W}$, the operations $\alpha^{\mathbf{B}'}$ and $T(\phi_\alpha)^{\mathbf{B}}$ are equal. If α is one of the basic terms $\hat{F}_i$, this holds by definition of δ_i. Then we consider the case where $\alpha = \beta(\gamma_0, \ldots, \gamma_{N-1})$, and β and all γ_i have depth $< i$. By induction, we have

$$\begin{aligned} \alpha^{\mathbf{B}'} &= \beta^{\mathbf{B}'}(\gamma_0^{\mathbf{B}'}, \ldots, \gamma_{N-1}^{\mathbf{B}'}) \\ &= T(\phi_\beta)(T(\phi_{\gamma_0}), \ldots, T(\phi_{\gamma_{N-1}})) \\ &= S(\phi_\lambda) \text{ for some } \lambda. \end{aligned}$$

Now working in the equational theory of $\mathcal{V}$, we calculate

$$\begin{aligned} s(\phi_\alpha \circ \eta_0) \approx t(\phi_\alpha) &\approx \alpha \\ &= \beta(\gamma_0, \ldots, \gamma_{N-1}) \\ &\approx t(\phi_\beta)(t(\phi_{\gamma_0}), \ldots, t(\phi_{\gamma_{N-1}})) \\ &\approx s(\phi_\lambda) \text{ for the same } \lambda. \end{aligned}$$

Remembering Δ_0 (see Part i of our axioms for $\mathcal{W}$), we see that $\mathcal{W}$ satisfies the equation $S(\phi_\alpha \circ \eta_0) \approx S(\phi_\lambda)$. Now we may say that $\alpha^{\mathbf{B}'} = S(\phi_\lambda) = S(\phi_\alpha \circ \eta_0) = T(\phi_\alpha)$, as was to be proved. This finishes the inductive argument.

To complete the proof of Lemma 6.90, we let $\mu_i \approx \nu_i$ be one of the defining equations of $\mathcal{V}$. In $\mathcal{V}$ we have that $\mu_i = t(\phi_{\mu_i})$ and $\nu_i = t(\phi_{\nu_i})$, by Lemma 6.89; hence the equation $t(\phi_{\mu_i}) \approx t(\phi_{\nu_i})$ holds identically in $\mathcal{V}$, and is equivalent, via $t = s(\eta_0)$ to a linear equation in s, so that $T(\phi_{\mu_i}) \approx T(\phi_{\nu_i})$ holds identically in $\mathcal{W}$. From our recent inductive argument, we see that $\mathbf{B}' \models \mu_i \approx \nu_i$. ■

Completion of the proof of Theorem 6.87. We next wish to prove that $\mathbf{Mod}\,\Delta_0$ is active, i.e. contains no trite algebra. Let us assume, for a contradiction, that $\mathbf{Q} = \langle Q; S^{\mathbf{Q}} \rangle$ is a trite algebra satisfying Δ_0 (Part (i) of the axioms for $\mathcal{W}$). This means that $S^{\mathbf{Q}}$ is, say, the qth coordinate projection, for some $q < n^2$. The equation $S(\eta_0) \approx S(\eta_1)$ implies that $q = p + pn$ for some $p < n$. From this one easily verifies that $\mathbf{Q} \models S \approx T \star T$, and so in fact $\mathbf{Q} \in \mathcal{W}$. Considering now the algebra $\mathbf{Q}'$, it is clear that $\mathbf{Q}'$ is trite. It follows moreover from Lemma 6.90 that the trite algebra $\mathbf{Q}'$ lies in $\mathcal{V}$, contrary to our assumption that $\mathcal{V}$ is active. This contradiction establishes that $\mathbf{Mod}\,\Delta_0$ is active.

Considering the linear equations Δ_0 as they apply to the $\mathcal{V}$-term s, we again see that they rule out the existence of a trite algebra. This can happen only if (6.8.3) holds for the term s. Thus s is a Taylor term for $\mathcal{V}$, and Theorem 6.87 is proved. ■

One significant conclusion that we may draw from Theorem 6.87 is that the class of all varieties possessing a Taylor term is a *Mal'tsev class.* (Equivalently, to have a Taylor term is a *Mal'tsev property*, as defined near the start of §6.1.) To prove this requires knowing that the Taylor class is closed under the formation of product varieties, which are defined in Chapter 10. Theorem 6.87 is just what is needed to prove closure under product varieties. See Exercise 10.153.8 on page 258 of Volume III, where we return to this subject and describe the details. But in fact that proof can be seen as redundant, since we have (since 2016) a proof that the Taylor class is a *strong Mal'tsev class*. This will be a corollary of Theorem 6.92 just below.

Exercises 6.91

1. Show that a 2-place term is a Taylor term if and only if it is idempotent and commutative.

2. Prove the associativity of the $\star$ operation on terms.

3. Given matrices M and N as defined in Example 2, prove that they satisfy (6.8.3). Verify that, for this M and N, the equations $F(M) \approx F(N)$ follow from (6.8.5) and the 3-permutability equations (6.8.4).

4. Prove Lemma 6.88 and its corollary. To understand the statements it is useful to give a concrete example. So, in the case $N = 2$ and $r = 3$ (so s has arity $N = 2$ and t has arity $N^r = 8$), part (i) says
$$s(t(x_0, \dots, x_7), t(x_0, \dots, x_7)) \approx t(x_0, \dots, x_7),$$
which follows from idempotence of s.

5. Give a detailed argument, based on the Compactness Theorem (Theorem 8.30), for the finiteness conclusion in the proof of Theorem 6.87. It may help to note that if G is an r-ary operation symbol and $0 \leq g < r$, then there is a first-order sentence that says, "G projects each r-tuple to its gth coordinate."

6. Prove that $\mathcal{V}$ satisfies $s(\eta_0) \approx s(\eta_1) \approx t$. (As asserted just before Lemma 6.89.)

7. Complete the inductive argument for the proof of Lemma 6.89.

8. Let $\mathbf{A}$ be a finite unary algebra.
   - **a.** Show that $\boldsymbol{V}(\mathbf{A})$ is active if and only if there is a unary term t in the signature of $\mathbf{A}$ satisfying $t(x) \approx t(y)$. Hint: If no such term exists, then the subalgebra generated by x is disjoint from the one generated by y in the 2-generated free algebra over $\mathbf{A}$. These two subalgebras form a partition which is a congruence. The quotient by this congruence is a two element trite algebra.
   - **b.** Show that if $|A| > 1$ and each basic operation of $\mathbf{A}$ is a permutation, then $\boldsymbol{V}(\mathbf{A})$ is passive.
   - **c.** Show that if $A = \{0, 1, 2\}$ and has only one operation which interchanges 0 and 1 and sends 2 to 1 then $\boldsymbol{V}(\mathbf{A})$ is passive. But if the single operation sends 2 to 1, 1 to 0 and fixes 0, then $\boldsymbol{V}(\mathbf{A})$ is active.

9. Derive Taylor terms and equations for congruence regularity. More concretely, let $\mathcal{V}$ be the variety defined by the equations of Theorem 6.48(iv), page 51, say with $n = 3$. Exhibit a term T in the $f_n()$ and the $g_n()$, and exhibit matrices M and N, so that $\mathcal{V}$ satisfies $T(M) \approx T(N)$, and so that (6.8.3) holds. (It may be useful to consider the relationships that are diagrammed in Figure 6.19 (page 120). Note also that Olšák's Theorem 6.92 gives an immediate answer for M and N, but does not readily yield a concrete expression for T.

** **10.** (Alexandr Kazda, 2019) A **weak near-unanimity term** for a variety $\mathcal{V}$ is a $\mathcal{V}$-term $F(x_0, \ldots, x_{n-1})$ such that $\mathcal{V}$ satisfies the laws

$$\begin{aligned} F(x, \ldots, x) &\approx x, \\ F(y, x, \ldots, x) \approx F(x, y, x, \ldots, x) &\approx \cdots \approx F(x, \ldots, x, y). \end{aligned}$$

Prove that these equations yield no single-equation Taylor condition. A single-equation Taylor condition for this context would comprise a term T whose only operation symbol is F, together with single-row matrices M, N such that the given equations imply $T(M) \approx T(N)$, and such that (6.8.3) holds for M and N.

** **11.** (Miroslav Olšák, private communication.) Any Taylor condition with n variables implies a Taylor condition with two equations and n^2 variables. ("Two equations" means two equations plus the equation of idempotence.)

Olšák's Theorem

For M and N two $m \times n$ matrices of variables, satisfying (6.8.3), let us define $\mathcal{T}_{M,N}$ to be the class of all varieties $\mathcal{V}$ with an (M, N)-Taylor term, i.e. those with an idempotent n-ary term T satisfying $T(M) \approx T(N)$. As remarked above, $\mathcal{T}_{M,N}$ is the strong Mal'tsev class defined by $T(M) \approx T(N)$ plus idempotence. Now the class of all Taylor varieties, $\mathcal{T} = \bigcup \mathcal{T}_{M,N}$ (the union over all appropriate M and N), is ostensibly not a Mal'tsev class. Nevertheless, it turns out that $\mathcal{T}$ is a strong Mal'tsev class. According to a recent—and surprising—theorem of M. Olšák there are 2×12 matrices M_0, N_0 such that $\mathcal{T} = \mathcal{T}_{M_0,N_0}$. Equivalently, $\mathcal{T}_{M,N} \subseteq \mathcal{T}_{M_0,N_0}$ for all appropriate M and N. Another way to say it is that if a variety $\mathcal{V}$ has any Taylor term, then $\mathcal{V}$ has an (M_0, N_0)-Taylor term. Thus $T(M_0) \approx T(N_0)$ may be said to define the **weakest Taylor condition**. We begin our discussion of Olšák's Theorem with a description of M_0 and N_0.

Let K be the 4×12 matrix whose columns are all the four-tuples $\langle a, b, c, d\rangle \in \{x, y\}^4$ with $a \neq b$ or $c \neq d$, organized in some reasonable list from left to right (there are twelve such columns). Let r_0, r_1, r_2, r_3 be the first, second, third and fourth rows of K, respectively. Olšák's equations are then

$$S(r_0) \approx S(r_1); \qquad S(r_2) \approx S(r_3).$$

Organizing the columns alphabetically, the equations become

$$S(x, x, x, x, x, x, y, y, y, y, y, y) \approx S(x, x, y, y, y, y, y, y, x, x, x, x) \tag{6.8.10}$$

$$S(x, y, x, x, y, y, x, y, x, x, y, y) \approx S(y, x, x, y, x, y, y, x, x, y, x, y). \tag{6.8.11}$$

Or, to describe the matrices directly, we have

$$M_0 = \begin{pmatrix} x & x & x & x & x & x & y & y & y & y & y & y \\ x & y & x & x & y & y & x & y & x & x & y & y \end{pmatrix};$$

$$N_0 = \begin{pmatrix} x & x & y & y & y & y & y & y & x & x & x & x \\ y & x & x & y & x & y & y & x & x & y & x & y \end{pmatrix}.$$

THEOREM 6.92 (Miroslav Olšák 2017). *Every Taylor condition implies the (M_0, N_0)-Taylor condition.*

Proof. Let us be given (M, N, F)-Taylor equations $F(M) \approx F(N)$ as defined in Definition 6.86, where M and N are $n \times n$ matrices with x's on the diagonal of M and y's on the diagonal of N. We shall prove the existence of an F-term S satisfying Equations (6.8.10)–6.8.11. In fact, we shall prove that (6.8.10) is a consequence of the idempotence of F, and (6.8.11) is a consequence of $F(M) \approx F(N)$.

We begin with the construction of some algebras in the signature of a single n-ary operation F. Let $\mathbf{A}$ be the free idempotent algebra of this signature freely generated by $\{x_A, y_A\}$. Let $\mathbf{B}$ be the free algebra of the same signature, freely generated by $\{x_B, y_B\}$ in the variety axiomatized by the equations $F(M) \approx F(N)$. Let $\mathbf{C} = \mathbf{A} \times \mathbf{B}$.

We define a subalgebra of $\mathbf{C}$. Take

$$Q = \{\langle a, b, c, d\rangle \in \{x_A, y_A\}^2 \times \{x_B, y_B\}^2 : a \neq b \text{ or } c \neq d\},$$

and define $\mathbf{P}$ as the subalgebra of $\mathbf{A}^2 \times \mathbf{B}^2$ generated by Q. Our goal is to show that P contains a 4-tuple $\langle a, a, b, b\rangle$ (what Olšák calls a "double loop"). This will mean there is an F-term S such that

$$\begin{aligned} \mathbf{A} \models S(x,x,x,x,x,x,y,y,y,y,y,y) &\approx S(x,x,y,y,y,y,y,y,x,x,x,x) \\ \mathbf{B} \models S(x,y,x,x,y,y,x,y,x,x,y,y) &\approx S(y,x,x,y,x,y,y,x,x,y,x,y). \end{aligned}$$

Theorem 6.92 is an immediate corollary of this fact.

The proof now divides into two parts. In the first part, we construct a special subalgebra $\mathbf{R}$ of $\mathbf{A}^2$; see Lemma 6.93. In the second part, we shall prove that the graph $\langle A, R\rangle$ of Lemma 6.93 has a loop: $\langle a, a\rangle \in R$ for some a (see Lemma 6.95).

LEMMA 6.93. *There is a subuniverse R of $\mathbf{A}^2$ satisfying*

(i) *R is symmetric: $\langle a, b\rangle \in R$ implies $\langle b, a\rangle \in R$.*
(ii) *For all $\langle a, b\rangle \in R$ there exists $c \in B$ so that $\langle a, b, c, c\rangle \in P$.*
(iii) *The graph $\langle A, R\rangle$ has an odd cycle.*
(iv) $R \lhd_T \mathbf{A}^2$ (R absorbs $\mathbf{A}^2$ with respect to T). *This means that whenever $\overline{a}_0, \dots, \overline{a}_{n-1} \in A^2$ and $\overline{a}_i \in R$ for all but at most one i, then $T(\overline{a}_0, \dots, \overline{a}_{n-1}) \in R$.*

Proof. To construct R to satisfy Lemma 6.93, we recursively define a sequence of elements σ_j (for $j \in \omega$) in B as follows. To begin, we note that there is a unique automorphism of $\mathbf{B}$, $b \longmapsto \overline{b}$, that satisfies $x_B \longmapsto y_B$ and $y_B \longmapsto x_B$. We take $\sigma_0 = x_B$. Now suppose that $j \geq 0$ and $j = mn + i$ with $0 \leq i < n$. We define σ_{j+1} as follows. We write our i^{th} Taylor equation (encoded in the ith rows of M and N) as

$$L_i(x,y) \;=\; F(\text{row } i \text{ of } M) \;=\; F(\text{row } i \text{ of } N) \;=\; R_i(x,y).$$

We now put $\sigma_{j+1} = L_i(\sigma_j, \overline{\sigma}_j)$.

Now we put

$$R_j \;=\; \{\langle a,b\rangle \in A^2 : \langle a,b,\sigma_j,\sigma_j\rangle \in P\}; \quad R \;=\; \bigcup_{j\in\omega} R_j.$$

Note that R_0 contains the pair $\langle x_A, y_A\rangle$. In due course we will see that $R_j \subseteq R_{j+1}$ for all j, and that R has the properties listed in Lemma 6.93.

We now consider three automorphisms of $\mathbf{A}^2 \times \mathbf{B}^2$. First, we have $\langle a,b,c,d\rangle \longmapsto \langle a,b,\overline{c},\overline{d}\rangle$ (with $\overline{c}$ and $\overline{d}$ as defined above). Second is $\langle a,b,c,d\rangle \longmapsto \langle b,a,c,d\rangle$, and third is $\langle a,b,c,d\rangle \longmapsto \langle a,b,d,c\rangle$.

LEMMA 6.94. (i) *The algebra* $\mathbf{P}$ *is invariant under each of the three automorphisms of* $\mathbf{A}^2 \times \mathbf{B}^2$ *defined above.*

(ii) *For all* $a,b \in A$, *there exists* m *such that for all* $k \geq m$ *we have* $\langle a,b,\sigma_k,\overline{\sigma_k}\rangle \in P$ *and* $\langle a,b,\overline{\sigma_k},\sigma_k\rangle \in P$.

Proof. Statement (i) follows from the observation that the generating set Q is invariant under each of the three automorphisms.

For (ii) we will use the $\star$-product defined by Definition 6.83 and used in Theorem 6.87. For $m \geq 1$ we will consider the term $T^{\star m}$, which is the m-fold $\star$-product of T with itself. Its arity is n^m. In particular, we write $T^{*m}(X)$ for the set of all values $T^{\star m}(u_0, \ldots, u_{n^m-1})$ where $\overline{u} \in X^{m^n}$. Here we shall have X contained in either A, A^2, B, B^2 or $C = A^2 \times B^2$, with the term to be evaluated in the appropriate algebra.

It is easily proved by induction on m that

$$\sigma_{m+1}, \overline{\sigma_{m+1}} \in T^{\star m}(\{x_B, y_B\}) \qquad \text{for all } m \in \omega,$$

and in fact

$$\langle \sigma_{m+1}, \overline{\sigma_{m+1}}\rangle \in T^{\star m}(\{\langle x_B, y_B\rangle, \langle y_B, x_B\rangle\}) \qquad \text{for all } m \in \omega.$$

Moreover, due to the idempotence of $\mathbf{A}$, the set $\{x_A, y_A\}^2$ generates $\mathbf{A}^2$ and every $\langle a,b\rangle \in A^2$ belongs to $T^{\star m}(\{\langle x_A, y_A\rangle\}^2)$ for m sufficiently large. For m that large, we have that

$$\begin{aligned}\{\langle a,b,\sigma_{m+1},\overline{\sigma_{m+1}}\rangle, \langle a,b,\overline{\sigma_{m+1}},\sigma_{m+1}\rangle\} &\subseteq \\ \subseteq T^{\star m}(\{x_A,y_A\}^2 \times \{\langle x_B,y_B\rangle, \langle y_B,x_B\rangle\} &\subseteq P.\end{aligned}$$

This proves (ii) and completes the proof of Lemma 6.94 ■

Returning to the proof of Lemma 6.93, we note that each R_j is symmetric, by Part (i) of Lemma 6.94. Therefore R, the union of the R_j's is itself symmetric (and Part (i) is proved). For Part (ii) of Lemma 6.93, we refer only to the definition of R_j, where to have $\langle a,b\rangle \in R_j$ we required that $\langle a,b,\sigma_j,\sigma_j\rangle$ belong to P. Now Part (ii) holds because R is the union of the sets R_j.

As promised, we show that $R_j \subseteq R_{j+1}$ for each j. Let us say that $j = mn + i$ with $m \in \omega$ and $0 \leq i < n$. For $\langle a,b\rangle \in R_j$, we have $\langle a,b,\sigma_j,\sigma_j\rangle \in P$. By Lemma 6.94(i) we have $\langle a, b, \overline{\sigma_j}, \overline{\sigma_j}\rangle \in P$. Now

$$\langle a, b, \sigma_{j+1}, \sigma_{j+1}\rangle = L_i(\langle a, b, \sigma_j, \sigma_j\rangle, \langle a, b, \overline{\sigma_j}, \overline{\sigma_j}\rangle) \in P.$$

Thus $\langle a, b\rangle \in R_{j+1}$.

We now turn our attention to Part (iv) of Lemma 6.93. To show that R absorbs A^2 with respect to T, we will let $0 \leq i < n$ and assume that

$$\langle\langle a_0, b_0\rangle, \ldots, \langle a_{n-1}, b_{n-1}\rangle\rangle = \langle \overline{a}, \overline{b}\rangle \in (A^2)^n,$$

with $\langle a_u, b_u\rangle \in R$ for $u \neq i$. We need to show that $\langle T(\overline{a}), T(\overline{b})\rangle \in R$. By Part (ii) of Lemma 6.94 and the fact that R_j is an increasing sequence of sets, we can choose m so that $m = pn + i$ and

$$\begin{aligned}\bigcup_{u\neq i}\{\langle a_u, b_u, \sigma_m, \sigma_m\rangle\} \cup \bigcup_{0\leq u<n} \{\langle a_u, b_u\rangle\} \times \{\langle \sigma_m, \overline{\sigma_m}\rangle, \langle \overline{\sigma_m}, \sigma_m\rangle\} \\ = \Gamma \subseteq P.\end{aligned}$$

It should be obvious that

$$\begin{aligned}\langle T(\overline{a}), T(\overline{b}), \sigma_{m+1}, \sigma_{m+1}\rangle &= \langle T(\overline{a}), T(\overline{b}), L_i(\sigma_m, \overline{\sigma_m}), R_i(\sigma_m, \overline{\sigma_m})\rangle \\ &\in T^{*m}(\Gamma) \subseteq P.\end{aligned}$$

This is because at position i, expressed as instances of T, $L_i(x, y)$ has an x and $R_i(x, y)$ has a y. Thus, indeed, $T(\langle a_0, b_0\rangle, \ldots, \langle a_{n-1}, b_{n-1}\rangle) \in P$.

We now turn our attention to Part (iii) of Lemma 6.93. We will show that the graph $\langle A, R\rangle$ has an odd cycle (actually a $(2n+1)$-cycle). We first observe that $\langle x_A, y_A, \sigma_0, \sigma_0\rangle \in Q \subseteq P$, and hence $\langle x_A, y_A\rangle \in R$. To simplify notation, we shall finish the proof of Lemma 6.93 writing x for x_A and y for y_A. We have $\langle T(x, \ldots, x), T(x, y, \ldots, y)\rangle \in R$, since $\langle x, y\rangle \in R$ and R absorbs $\langle x, x\rangle$. Thus

$$x = T(x, \ldots, x) \;\; R \;\; T(x, y, \ldots, y) \;\; R \;\; T(y, x, \ldots x).$$

To continue another two steps, with similar justification, we have

$$T(y, x, \ldots, x) \;\; R \;\; T(x, x, y, \ldots, y) \;\; R \;\; T(y, y, x, \ldots x).$$

After $2n + 1$ steps we have

$$x = T(x, \ldots, x) \;\; R^{2n} \;\; T(y, \ldots, y) = y \;\; R \;\; x,$$

an R-cycle of odd length. This ends the proof of Lemma 6.93. ■

We now enter the second (and final) part of the proof of Theorem 6.92. We will prove that R contains a loop, in other words, that $a\,R\,a$ for some a. According to Part (ii) of Lemma 6.93, this means that there exists $c \in C$ such that $\langle a, a, c, c\rangle \in P$. And according to the remarks just before Lemma 6.93, the existence of such a double loop will guarantee the existence of the desired Olšák-term S, thereby completing the proof of Theorem 6.92. We begin with some remarks and a lemma on graph theory.

Suppose $g : K^n \longrightarrow K$, and $L \subseteq K$. We say L **absorbs K with respect to** g, and write $L \lhd_g K$, to mean the following: if $k_j \in K$ (for $0 < j < n$), and if $k_j \in L$ for all but at most one value of j, then $g(k_0, \ldots, k_{n-1}) \in L$.

Let $\langle X, E\rangle$ be a symmetric graph. For $x \in X$, write

$$N(x) = \{y \in X : \langle x, y\rangle \in E\}$$

(the set of neighbors of x). Write $N(\langle X, E\rangle)$ for the set of $x \in X$ such that $N(x) \neq \emptyset$ (the domain of E). Suppose that $g: X^k \longrightarrow X$ for some $k \geq 1$. We say that E **semi-absorbs with respect to** g iff for all $x \in X$, $N(x) \lhd_g N(\langle X, E\rangle)$. Notice that for the R of Lemma 6.93, R semi-absorbs with respect to T. (This follows easily from Lemma 6.93 using $R \lhd_T A^2$ and the idempotence of T.) Thus the next lemma completes the proof of our main result, Theorem 6.92.

LEMMA 6.95. *Suppose that $\langle A, R\rangle$ is a symmetric graph with an odd cycle and that $g: A^n \longrightarrow A$ for some $n \geq 1$. Suppose further that R is a subuniverse of $\langle A, g\rangle^2$ and R semi-absorbs with respect to g. Then R has a loop: $a\,R\,a$ for some $a \in A$.*

Proof. Note that we are not assuming idempotence for g in this lemma. Notice also that A is non-empty (because of the cycle), and by replacing A with $N(A)$ (which is a g-subuniverse), we may assume that $N(A) = A$.

For $n \geq 1$ and ℓ an odd number ≥ 1, we let $P(n, \ell)$ stand for the validity of the Lemma in the case of n-ary g and the existence of an odd cycle of size ℓ or smaller. The proof will be a double induction on n and ℓ, organized as follows. We begin with an easy direct argument for $P(1, 1)$. Then in proving $P(n, \ell)$ we allow ourselves to assume the truth of $P(n, \ell')$ for any smaller ℓ', and the truth of (n', ℓ') for any ℓ' and any smaller n'. (This is induction over the well-ordered set $\omega \cdot \omega$.)

If $n = 1$, then semi-absorption amounts to this: If $x, y \in A$, then $g(y) \in N(x)$. Taking $x = g(y)$, we have a loop $g(y)\,R\,g(y)$. Thus we have a basis for our induction. Our inductive proof now divides into three cases: (a) $\ell = 1$, (b) $\ell > 3$, and (c) $\ell = 3$.

In Case (a), we have a cycle of length 1, i.e. a loop, and the lemma holds. For the next case, $\ell > 3$, we proceed as follows. We replace R by $R' = R^3$. Note that $R \subseteq R'$ because R is symmetric: $a, b \in R$ yields a path $a\,R\,b\,R\,a\,R\,b$, giving $\langle a, b\rangle \in R'$. Now let $a_0\,R\,a_1\,R\,\cdots\,R\,a_{\ell-1}\,R\,a_\ell = a_0$ be a cycle in R of length ℓ. Then

$$a_0\,R'\,a_3\,R'\,a_4\,R'\,\cdots\,R'\,a_\ell = a_0$$

is an R'-cycle of length $\ell - 2$. Moreover, it is easy to see that g respects R' and R' semi-absorbs with respect to g. By induction, it follows that R' has a loop: $a\,R'\,a$. This means that there exist b, c such that $a\,R\,b\,R\,c\,R\,a$, i.e. that R has a 3-cycle. Now by induction R has a loop.

We now come to Case (c), the proof of $P(n, \ell)$ when $\ell = 3$. So $\langle A, R\rangle$ has a triangle $a\,R\,b\,R\,c\,R\,a$. We call $\langle a, b, c\rangle$ a **weak 3-clique**. ("Weak" because it is possible that some two of a, b, c are equal—of course, if that is the case we have a loop and we are done.) In general, a **weak k-clique in** $\langle A, R\rangle$ is a sequence $\langle a_0, \ldots, a_{k-1}\rangle$ with $a_i\,R\,a_j$ whenever $0 \leq i < j < k$. We now have a weak $(0 + 3)$-clique.

Claim. If $\langle A, R\rangle$ has a weak $(i + 3)$-clique, then it has a weak $(i + 1 + 3)$-clique.

To prove the claim, we begin by writing the given $(i + 3)$-clique as $a_0, a_1, \ldots, a_{i+2}$, where $a_j\,R\,a_k$ when $0 \leq j < k < i + 3$. We will write this also as

$$a_0, \ldots, a_{i-1}, a, b, c.$$

Now let $A' = N(a_0) \cap N(a_1) \cap \cdots \cap N(a_{i-1}) \cap N(a)$. Note that A' is a subuniverse of $\langle A, g\rangle$ due to semi-absorption, and $A' \supseteq \{b, c\}$ and $b\,R\,c$. Put $R' = (R \cap (A')^2)^3$ (3-fold relational product), and $g_a(x_0, \cdots, x_{n-2}) = g(x_0, \cdots, x_{n-2}, a)$ for $x_0, \ldots, x_{n-2} \in A'$. In subsequent paragraphs, we will show that A', R', g_a satisfy the hypotheses of $P(n-1, \ell')$

for some ℓ'. Then by induction, we know that R' has a loop, i.e. there are $a', b', c' \in A'$ such that $a' \, R \, b' \, R \, c' \, R \, a'$. Thus

$$a_0, a_1, \dots, a_{i-1}, a', a, b', c'$$

is a weak (i+1+3)-clique in $\langle A, R\rangle$. This finishes the Claim, modulo the hypotheses of $P(n-1, \ell)$.

First, we need to show that R has an odd cycle that lies completely in A'—which will be a cycle for R', since R restricted to A' is a subset of R'. Indeed we have

$$\begin{aligned} g(b, \dots, b) \, R \, g(a, c, \dots, c) \, R \, g(c, b, \dots) \, R \, g(b, a, c, \dots, c) \\ R \, g(c, c, b, \dots, b) \, R \, \cdots \, R \, g(c, \dots, c) \, R \, g(b, \dots, b). \end{aligned}$$

This is a cycle of length $2n+1$, and each of its elements is in A', by semi-absorption.

That A' is closed under the action of g_a follows from semi-absorption with respect to g. That R' semi-absorbs with respect to g_a is seen as follows. Suppose $p, a_1, \dots a_{n-2}, u$ are elements of A', and that $a_1, \dots, a_{n-2}$ are R'-neighbors of p. We can exhibit these neighbor relations as $a_i \, R \, b_i \, R \, c_i \, R \, p$ (for $1 \le i < n-1$), with $b_i, c_i \in A'$. We also have $u \, R \, a \, R \, b \, R \, c$ and $a \, R \, b \, R \, a \, R \, p$—these follow from our choice of a, b, c and from the definition of A'. Thus

$$g(\overline{a}, u, a) \, R \, g(\overline{b}, a, b) \, R \, g(\overline{c}, b, a) \, R \, p,$$

by semi-absorption, since $c_1, \dots, c_{n-2}$ are all neighbors of p. All four of the displayed elements belong to A', by semi-absorption. Thus the displayed relations show $g_a(a_1, \dots, a_{n-2}, u)$ is an R'-neighbor of p. Thus R' semi-absorbs with respect to g_a.

That g_a respects R' is shown like this: Suppose that $\langle a_i, d_i\rangle \in R'$ for $0 \le i < n-1$. Thus we have $b_i, c_i \in A'$ with $a_i \, R \, b_i \, R \, c_i \, R \, d_i$. We also have $a \, R \, b \, R \, c \, R \, a$. Now consider

$$g_a(\overline{a}) \;=\; g(\overline{a}, a) \; R \; g(\overline{b}, b) \; R \; g(\overline{c}, c) \; R \; g(\overline{d}, a) \;=\; g_a(\overline{d}).$$

All elements named in the displayed relations belong to A' by semi-absorption. Thus $\langle g_a(\overline{a}), g_a(\overline{d})\rangle \in R'$ as desired. This concludes the proof of the Claim.

Continuing with the proof of Lemma 6.95, we observe from the Claim (and induction) that $\langle A, R\rangle$ has a weak n-clique: $a_0, \dots, a_{n-1}$ with $a_i \, R \, a_j$ for $0 \le i < j < n$. Put $p = g(a_0, \dots, a_{n-1})$. For each i, it follows by semi-absorption that p is a neighbor of a_i. But then $p = g(a_0, \dots, a_{n-1})$ is a neighbor of p (a trivial consequence of semi-absorption). So we have $p \, R \, p$, the desired loop, and the proof of Lemma 6.95 is complete. ■

As remarked above, this loop in R yields a double loop in P, which in turn completes the proof of Olšák's Theorem 6.92. ■

COROLLARY 6.96. *The class $\mathcal{T}$ of all varieties that satisfy some Taylor condition (see page 89) is a strong Mal'tsev class.*

Proof. $\mathcal{T}$ is the class of all varieties in which one can interpret the equations (6.8.10)–6.8.11 on page 89, along with idempotence. ■

Exercises 6.97

1. Prove that, as claimed early in the proof of Theorem 6.92, the discovery of a double loop leads to the existence of an idempotent S satisfying (6.8.10)–6.8.11. In particular, confirm that the pattern of x's and y's is exactly as determined by the twelve elements of Q.

6.9. Congruence Identities

In this section we give a brief overview of the subject of lattice equations satisfied by the congruence lattices of all the algebras in a variety. As mentioned at the beginning of this chapter, a lattice equation valid in all the congruence lattices of every member of a variety $\mathcal{V}$ is called a **congruence identity for** $\mathcal{V}$. Modularity and distributivity are the two most important examples. We have seen, by the results of B. Jónsson and A. Day, that both of these equations have a Mal'tsev condition associated with them. After these results were proved, there was hope that, for any lattice equation ϵ, the class of varieties whose congruence lattices satisfied ϵ would be Mal'tsev definable. This question is still open. There are some results in the congruence modular case. Chapter XIII of (Ralph Freese and Ralph McKenzie, 1987) gives infinitely many inequivalent (in the strong sense that they correspond to distinct congruence varieties) lattice identities all stronger than the modular law, all of which have a Mal'tsev characterization. In (Gábor Czédli and Eszter K. Horváth, 2003) the authors extended this to every lattice identity that implies congruence modularity. We mention that (Alan Day and Ralph Freese, 1980) and (Gábor Czédli and Ralph Freese, 1983) give algorithms to decide the following problems: given lattice equation ϵ is it the case that every variety that satisfies ϵ as a congruence identity is congruence modular? (congruence distributive?)

The only general result we have is the result of Alden F. Pixley (1972) and Rudolf Wille (1970) that such a class is weakly Mal'tsev definable (weakly Mal'tsev definable classes are discussed in §10.8, page 232 of Volume III).

However, it is still possible that every lattice equation has a Mal'tsev condition associated with it; that is, we know of no lattice equation which is only weakly Mal'tsev definable.

For a variety $\mathcal{V}$ we define the **congruence variety** associated with $\mathcal{V}$ to be variety of lattices generated by the congruence lattices of the members of $\mathcal{V}$; that is,

$$\mathbf{V}\{\mathbf{Con\ A} : \mathbf{A} \in \mathcal{V}\}.$$

J. B. Nation (1974) showed in his thesis that there are lattice equations ϵ satisfied by some nonmodular lattices, which nevertheless have the property that any variety for which ϵ is a congruence identity is congruence modular. Let $s \approx t$ be a lattice equation and let Σ be a set of lattice equations. Then the notation

$$\Sigma \vDash_{\text{con}} s \approx t$$

indicates that if $\mathcal{V}$ is any variety of algebras, such that all of the congruence lattices of all the members of $\mathcal{V}$ satisfy the equations in Σ, then these congruence lattices will also satisfy $s \approx t$. In this case we say that $s \approx t$ is a **congruence consequence** of Σ. Thus, in this terminology, Nation's result was that there exists a lattice equation ϵ such that modularity is a congruence consequence of ϵ, but not a consequence of ϵ. Of course the statement ϵ_2 is a congruence consequence of ϵ_1 will mean $\{\epsilon_1\} \vDash_{\text{con}} \epsilon_2$. We call two sets Σ_0 and Σ_1 of lattice equations **congruence equivalent** if each element of Σ_0 is a congruence consequence of Σ_1 and vice versa. At the time of Nation's thesis only four congruence inequivalent lattice equations were known: $x \approx x$, $x \approx y$, distributivity and modularity. However, using modular lattice theory, it is possible to construct a lattice equation for each prime number p which holds in the congruence lattice of a vector space over a field **F** if and only if the characteristic of **F** is p; see Exercise 6.117.6. Thus all of these equations are congruence inequivalent.

E. Gedeonová 1972 and P. Mederly 1975 were able to show that for certain lattice equations ϵ the class of varieties having ϵ as a congruence identity is Mal'tsev definable. Moreover, there are nonmodular lattices which satisfy ϵ, but not every lattice satisfies ϵ. However Day, using the techniques of Nation, was able to show that each such ϵ is congruence equivalent to modularity, and thus his Mal'tsev condition for modularity would also define the class of varieties having ϵ as a congruence identity. The question of the existence of a lattice equation ϵ, not congruence equivalent to distributivity or modularity, such that the class of varieties whose congruence lattice satisfy ϵ is Mal'tsev definable was settled in (Ralph Freese and Ralph McKenzie, 1987). Using the commutator theory it was shown that there are infinitely many congruence inequivalent lattice equations ϵ such that the class of varieties satisfying ϵ is Mal'tsev definable.

Another surprising result in this area of congruence identities is the fact that if a variety is congruence modular its congruence lattices must also satisfy certain stronger equations as well. In this section we will introduce the Arguesian equation and prove that if a variety is congruence modular then it satisfies the Arguesian equation; see Theorem 6.109 below.

Hobby-McKenzie terms are covered at the end of this section.

Congruences as Algebras and their Congruence Lattices

Here we present a simple but useful construction. The reader may recognize it as a special kind of pullback, also known as a fiber product.

Let **A** be an algebra and α a congruence on **A**. Then it is easy to check that α is a subuniverse of $\mathbf{A}^2$. We denote the algebra on this subuniverse as $\mathbf{A}(\alpha)$ or by $\mathbf{A} \times_\alpha \mathbf{A}$.

For any congruence θ on **A** we define two congruences on $\mathbf{A}(\alpha)$ by

$$\theta_i = \{\langle\langle a_0, a_1\rangle, \langle b_0, b_1\rangle\rangle \in A(\alpha) \times A(\alpha) : \langle a_i, b_i\rangle \in \theta\}, \tag{6.9.1}$$

for $i = 0, 1$. We denote the kernels of the projections by η_0 and η_1 (rather than the awkward 0_0 and 0_1).

LEMMA 6.98. *Let α, θ and ψ be congruences of an algebra* **A**. *With notation as above we have:*

(i) *The map $\langle a_0, a_1\rangle \mapsto \langle a_1, a_0\rangle$ defines an automorphism of $\mathbf{A}(\alpha)$ which interchanges θ_0 and θ_1.*
(ii) *The map $\theta \mapsto \theta_i$ is a lattice isomorphism of* **Con A** *onto the interval $\mathbf{I}[\eta_i, 1_{\mathbf{A}(\alpha)}]$ of* **Con** $\mathbf{A}(\alpha)$, *for $i = 0, 1$. So $(\theta \vee \psi)_i = \theta_i \vee \psi_i$ and dually.*
(iii) *If $\psi \geq \alpha$ then $\psi_0 = \psi_1$.*
(iv) $\eta_0 \vee \eta_1 = \alpha_0$ $(= \alpha_1)$.
(v) *η_0 and η_1 permute.*
(vi) $(\theta_0 \wedge \theta_1) \vee \eta_0 = \theta_0$; *in fact* $\theta_0 = \eta_0 \circ (\theta_0 \wedge \theta_1) \circ \eta_0$.

Proof. The first statement is obvious and the second statement follows from the Correspondence Theorem (Theorem 4.12). The other statements follow from elementary calculations with elements. For example, to see (iii) suppose $\langle a_0, a_1\rangle \, \psi_0 \, \langle b_0, b_1\rangle$. Then $a_0 \; \alpha \; a_1$, $b_0 \; \alpha \; b_1$, and $a_0 \; \psi \; b_0$. Since $\psi \geq \alpha$, all four elements are in the same ψ-class. Hence $\langle a_0, a_1\rangle \, \psi_1 \, \langle b_0, b_1\rangle$. To see (vi) suppose $\langle a_0, a_1\rangle \, \theta_0 \, \langle b_0, b_1\rangle$. Then

$$\langle a_0, a_1\rangle \; \eta_0 \; \langle a_0, a_0\rangle \; (\theta_0 \wedge \theta_1) \; \langle b_0, b_0\rangle \; \eta_0 \; \langle b_0, b_1\rangle$$

So $\theta_0 \subseteq \eta_0 \circ (\theta_0 \wedge \theta_1) \circ \eta_0$. The other containment is clear. The other parts are proved similarly. ∎

The above construction is useful for proving theorems about congruence identities and congruence varieties. It is used in the proof of the theorem (Theorem 6.120) of S. Bulman-Fleming, Alan Day, and Walter Taylor (1974) giving a local version of Theorem 6.71. In fact this appears to be one of the first uses of this construction. As we shall see below it plays a role in showing that congruence modular implies congruence Arguesian. We will now use the above construction to give a proof of a version of Nation's result described in the introduction to this section on page 95. His original proof took the weak Mal'tsev condition associated with a certain lattice equation and derived the Mal'tsev condition for modularity, much in the same spirit as several of the proofs of this chapter.

Let $\mathbf{L}_{14}$ be the lattice diagrammed in Figure 6.12 below.

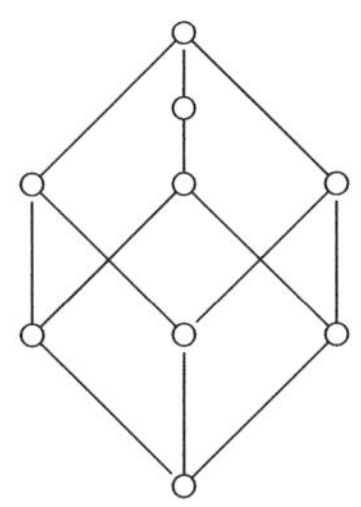

FIGURE 6.12. The Lattice $\mathbf{L}_{14}$

THEOREM 6.99. *Let* $\mathbf{A}$ *be an algebra with a nonmodular congruence lattice. Then there is an algebra* $\mathbf{B} \in S(\mathbf{A}\times\mathbf{A})$ *such that* **Con B** *has a sublattice isomorphic to the lattice* $\mathbf{L}_{14}$.

Proof. Since **Con A** is nonmodular, it has a sublattice isomorphic to $\mathbf{N}_5$, which we label as indicated in Figure 6.13.

We claim the sublattice of **Con** $\mathbf{A}(\alpha)$ generated by α_i, β_i and γ_i, $i = 0, 1$, is the lattice given by Figure 6.14. Indeed, using Lemma 6.98 it is straightforward to verify all the joins and meets are correct. ∎

It is easy to check that $\mathbf{N}_5$ satisfies the following equation, see Exercise 6.117.8.

$$(x \wedge y) \vee (x \wedge z) \approx x \wedge [(x \wedge y) \vee (x \wedge z) \vee (y \wedge z)] \tag{6.9.2}$$

COROLLARY. *If* $\mathcal{V}$ *is a variety all of whose congruence lattices satisfy* (6.9.2), *then* $\mathcal{V}$ *is congruence modular.*

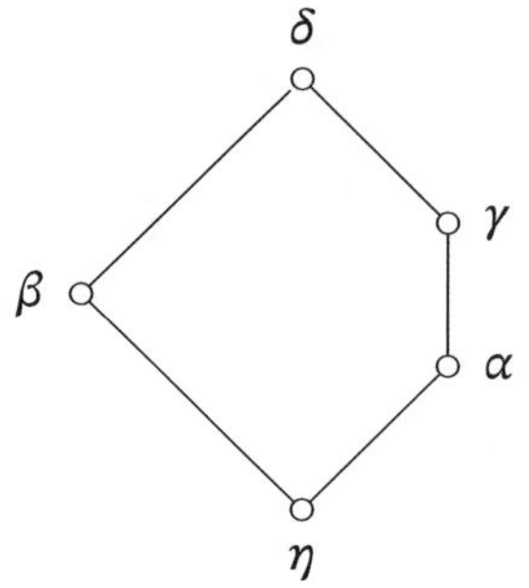

FIGURE 6.13. The Pentagon, $\mathbf{N}_5$, Labeled.

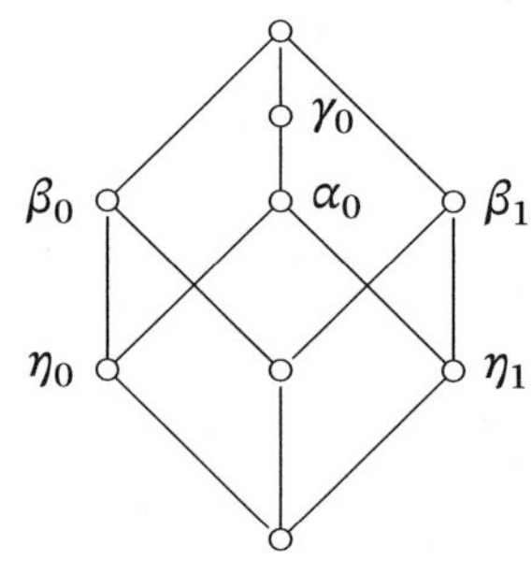

FIGURE 6.14. The Lattice $\mathbf{L}_{14}$, Labeled

Proof. If $\mathcal{V}$ is not congruence modular, then by the last theorem there is an algebra $\mathbf{B} \in \mathcal{V}$ whose congruence lattice has a sublattice isomorphic to the lattice pictured in Figure 6.12. It is easy to check that this lattice fails (6.9.2). ■

As Theorem 6.99 shows if **Con A** contains a copy of $\mathbf{N}_5$ labeled as in the diagram just before Figure 6.14, then in $\mathbf{A}(\alpha) = \mathbf{A} \times_\alpha \mathbf{A}$ the sublattice generated by α_i, β_i and γ_i, $i = 0, 1$, is exactly the lattice of Figure 6.14 in all circumstances. As we saw in Theorem 6.99, if $\mathcal{V}$ is not congruence modular, then $\mathbf{L}_{14}$ is a sublattice of **Con A**, for some $\mathbf{A} \in \mathcal{V}$.

This nice situation does not occur for $\mathbf{A} \times_\beta \mathbf{A}$ and $\mathbf{A} \times_\gamma \mathbf{A}$. To get some more insight into this, let **P** be Polin's algebra as described in Example 6.12 on page 15. Exercise 2 of this section shows that the sublattice generated by α_i, β_i and γ_i in $\mathbf{Con}(\mathbf{P} \times_\beta \mathbf{P})$ is the lattice of Figure 6.15. (Actually this lattice is all of $\mathbf{Con}(\mathbf{P} \times_\beta \mathbf{P})$.)

The lattice $\mathbf{L} = \mathbf{Con}(\mathbf{P} \times_\beta \mathbf{P})$ of Figure 6.15 is subdirectly irreducible by Exercise 2 (page 115). It is also a bounded epimorphic image of a free lattice. More precisely, if $h : \mathbf{F} = \mathbf{F}_{\mathcal{L}}(a_0, a_1, b, c_0, c_1) \twoheadrightarrow \mathbf{L}$ maps $a_i \mapsto \alpha_i$, $b \mapsto \beta_0$, and $c_i \mapsto \gamma_i$, then each element $w \in \mathbf{L}$ has a least preimage $\mu(w)$ and a greatest preimage $\sigma(w)$. (Normally the least preimage is denoted $\beta(w)$ and a greatest preimage $\alpha(w)$ but to avoid confusion we are using $\mu(w)$ and $\sigma(w)$.) If r denotes the upper solid point in Figure 6.15 and s is the

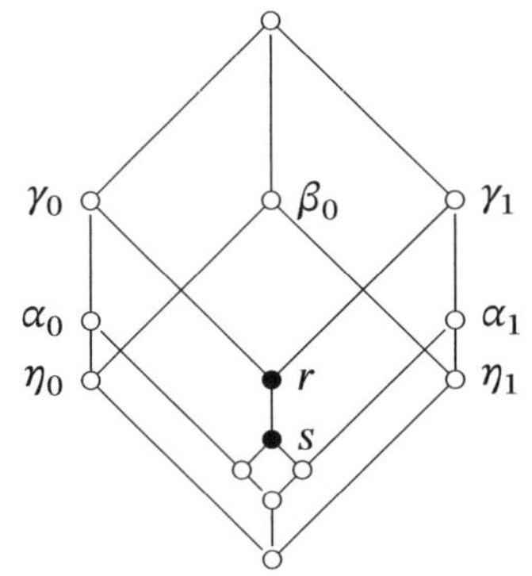

FIGURE 6.15. $\mathbf{Con(P \times_\beta P)}$

lower solid point, then

$$\sigma(s) = [(c_0 \vee a_0) \wedge (a_1 \vee (b \wedge (c_1 \vee a_1)))] \vee [(c_1 \vee a_1) \wedge (a_0 \vee (b \wedge (c_0 \vee a_0)))]$$

(6.9.3)

$$\begin{aligned}\mu(r) = {} & [(b \wedge c_0 \wedge a_0) \vee (c_0 \wedge ((b \wedge c_1 \wedge a_1) \vee (c_0 \wedge a_0 \wedge c_1 \wedge a_1)))] \\ & \wedge [(b \wedge c_1 \wedge a_1) \vee (c_1 \wedge ((b \wedge c_0 \wedge a_0) \vee (c_0 \wedge a_0 \wedge c_1 \wedge a_1)))].\end{aligned}$$

For the straightforward but somewhat tedious method to calculate $\sigma(s)$ and $\mu(r)$, see (Ralph Freese, Jaroslav Ježek, and J. B. Nation, 1995). The concepts of splitting lattice and bounded homomorphism originated in (Ralph McKenzie, 1972).

In the free lattice **F**, we obviously have $\mu(r) \wedge \sigma(s) \prec \mu(r)$. Now if $v \prec u$ in any lattice then there is a unique largest congruence, $\psi(u, v)$ that separates u and v and the quotient by this congruence is subdirectly irreducible. For our free lattice **F** this means that, for any homomorphism $f : \mathbf{F} \twoheadrightarrow \mathbf{K}$, either **K** satisfies the relation $\mu(r) \leq \sigma(s)$ or the homomorphism h factors through f, so that, in this latter case, **L** is a homomorphic image of **K**.

THEOREM 6.100. *If $\mathcal{V}$ is not congruence modular then $\mathbf{L} = \mathbf{Con(P \times_\beta P)}$ is a sublattice of* **Con A** *for some* $\mathbf{A} \in \mathcal{V}$.

Proof. By Corollary 6.33 we can find an $\mathbf{A} \in \mathcal{V}$ and a, b, c and $d \in A$ such that $\alpha = \mathrm{Cg}^{\mathbf{A}}(\langle a, b\rangle)$, $\beta = \mathrm{Cg}^{\mathbf{A}}(\langle a, c\rangle, \langle b, d\rangle)$, and $\gamma = \mathrm{Cg}^{\mathbf{A}}(\langle a, b\rangle, \langle c, d\rangle)$ generate the pentagon of Figure 6.13. If $\mu(r) \leq \sigma(s)$ in $\mathbf{Con(A \times_\beta A)}$ then $\alpha = \gamma$ in **Con A**, a contradiction; see Exercise 3. As remarked above this means that the sublattice of $\mathbf{Con(A \times_\beta A)}$ generated by α_i, β_0, and γ_i, $i = 0, 1$, has **L** as a homomorphic image. But **L** satisfies Whitman's condition (W) (see **W2.** on page 242 of Volume I), and is a bounded epimorphic image of a free lattices. These facts show that **L** is a projective lattice (see Corollary 5.9 of (Ralph Freese, Jaroslav Ježek, and J. B. Nation, 1995)) and hence is a sublattice of $\mathbf{Con(A \times_\beta A)}$. ■

The ideas above are the basis of the theorem of Alan Day and Ralph Freese (1980). Let $\mathcal{P}$ be variety generated by Polin's algebra.

THEOREM 6.101. *If $\mathcal{V}$ is not congruence modular then* $\mathbf{V}(\mathbf{Con}(\mathcal{P})) \subseteq \mathbf{V}(\mathbf{Con}(\mathcal{V}))$.

While we will not be able to give a full proof of this result, the idea is based on the examples above. The congruence lattices of the finitely generated free algebras in $\mathcal{P}$ are in fact splitting lattices and it can be shown that if **Con A** is not modular, then a certain subdirect power of **A** fails the associated splitting equation. Some of these ideas are explored in the exercises.

It turned out to be difficult to construct a variety which was not congruence modular, but for which some nontrivial lattice equation is a congruence identity. S. V. Polin (1977) constructed his algebra for exactly this reason. In Example 6.12 we showed the variety $\mathcal{P}$ generated by **P** has 4-permutable congruences and that it is not congruence modular. Moreover the signature of $\mathcal{P}$ has a binary operation symbol $\wedge$ whose interpretation on each algebra in $\mathcal{P}$ is a semilattice operation.

THEOREM 6.102. *The following equation is a congruence identity for Polin's variety $\mathcal{P}$.*

$$\begin{aligned} x \wedge (y \vee z) \leq {} & [x \wedge (y \vee (x \wedge (z \vee (x \wedge (y \vee (x \wedge z))))))] \\ & \vee [x \wedge (z \vee (x \wedge (y \vee (x \wedge (z \vee (x \wedge y))))))]. \end{aligned} \tag{6.9.4}$$

$\mathbf{M}_3$ *fails this equation.*

Proof. It is easy to see that this equation fails in $\mathbf{M}_3$. Let α, β, and γ be congruences of an algebra $\mathbf{B} \in \mathcal{P}$, and let $\langle a, b\rangle \in \alpha \wedge (\beta \vee \gamma)$. Since $\mathcal{P}$ is 4-permutable, there are elements c, d, and $e \in B$ such that the congruences indicated in the figure below hold.

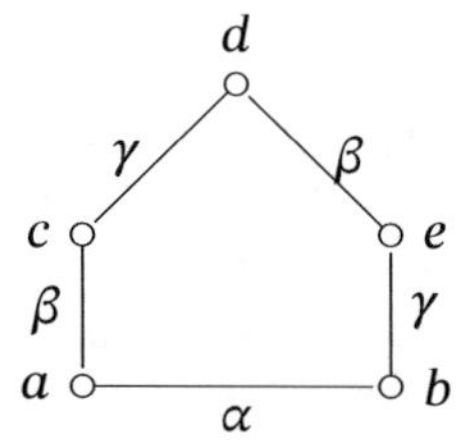

The following relations hold.

$$\begin{gathered} a \wedge b \quad \alpha \quad b \\ a \wedge b \wedge c \quad (\alpha \wedge \beta) \quad b \wedge c \\ a \wedge b \wedge d \quad (\alpha \wedge (\gamma \vee (\alpha \wedge \beta))) \quad b \wedge d \\ a \wedge b \wedge e \quad (\alpha \wedge (\beta \vee (\alpha \wedge (\gamma \vee (\alpha \wedge \beta))))) \quad b \wedge e \\ a \wedge b = a \wedge b \wedge b \quad (\alpha \wedge (\gamma \vee (\alpha \wedge (\beta \vee (\alpha \wedge (\gamma \vee (\alpha \wedge \beta))))))) \quad b \wedge b = b. \end{gathered}$$

The first relation is obvious since $a \wedge b \;\; \alpha \;\; b \wedge b = b$. Each of the others follows easily from the previous ones.

By symmetry we have

$$a \quad (\alpha \wedge (\beta \vee (\alpha \wedge (\gamma \vee (\alpha \wedge (\beta \vee (\alpha \wedge \gamma))))))) \quad a \wedge b,$$

Thus $\langle a, b\rangle$ is in the right side of the equation, proving the theorem. ■

Actually $\mathcal{P}$ satisfies a simpler congruence identity, namely:

$$x \wedge (y \vee z) \leq y \vee (x \wedge (z \vee (x \wedge y))), \tag{6.9.5}$$

see Theorem 7.1 in (Alan Day and Ralph Freese, 1980). The dual of this equation is also a congruence identity of $\mathcal{P}$. Moreover, they show that

$$\begin{aligned} &y \wedge ([x \wedge (z \vee (x \wedge y))] \vee (y \wedge z)) \\ &\qquad \leq [x \wedge (y \vee (z \wedge (x \vee y)))] \vee (z \wedge (x \vee y)) \end{aligned} \tag{6.9.6}$$

is a congruence identity of $\mathcal{P}$ but that the dual of (6.9.6) is not a congruence identity of $\mathcal{P}$. Thus the set of congruence identities of a variety need not be self dual. Gábor Czédli (1994) gave a modular example.

The subspaces of a vector space form a complemented, algebraic, atomic modular lattice, i.e., a projective geometry. These lattices actually satisfy an equation stronger than modularity known as the Arguesian equation. As we shall see, this equation is closely related to Desargues' Law of projective geometry, and thus the name. Let $\mathbf{L}$ be a lattice. A **triangle** in $\mathbf{L}$ is an element of L^3. Associated with any two triangles $\overline{a} = \langle a_0, a_1, a_2 \rangle$ and $\overline{b} = \langle b_0, b_1, b_2 \rangle$ are the elements

$$\begin{aligned} p &= (a_0 \vee b_0) \wedge (a_1 \vee b_1) \wedge (a_2 \vee b_2) \\ c_i &= (a_j \vee a_k) \wedge (b_j \vee b_k) \quad \text{for} \quad \{i, j, k\} = \{0, 1, 2\}. \end{aligned} \tag{6.9.7}$$

Triangles $\overline{a}$ and $\overline{b}$ are said to be **centrally perspective** if

$$(a_0 \vee b_0) \wedge (a_1 \vee b_1) \leq a_2 \vee b_2 \tag{6.9.8}$$

and are said to be **axially perspective** if $c_2 \leq c_0 \vee c_1$. **Desargues' implication** is the statement that any two centrally perspective triangles are axially perspective. When the triangles $\overline{a}$ and $\overline{b}$ are centrally perspective, we call the element p the **center of perspectivity**. A lattice $\mathbf{L}$ is said to be **Arguesian** if for all $\overline{a}$ and $\overline{b} \in L^3$ we have

$$(a_0 \vee b_0) \wedge (a_1 \vee b_1) \wedge (a_2 \vee b_2) \leq a_0 \vee (b_0 \wedge (b_1 \vee (c_2 \wedge (c_0 \vee c_1)))), \tag{6.9.9}$$

where $\overline{c} = \langle c_0, c_1, c_2 \rangle \in L^3$ is as defined in (6.9.7). Of course this is equivalent to a lattice equation, which is known as the **Arguesian equation**.

LEMMA 6.103. *Every Arguesian lattice is modular but not conversely.*

Proof. See Exercises 6.117.9 and 6.117.10. ■

In Theorem 4.67 we saw Jónsson's result that a lattice of 3-permuting equivalence relations is modular. The next result, also due to Jónsson, shows that a lattice of permuting equivalence relations is Arguesian. Thus a permutable variety is congruence Arguesian. In Theorem 6.109 we will prove the much stronger result that if a variety is congruence modular then it is congruence Arguesian. Along the way we will develop some of the theory of Arguesian lattices.

THEOREM 6.104. *Every lattice of permuting equivalence relations is Arguesian. If an algebra* $\mathbf{A}$ *has permuting congruences, then* $\mathbf{Con\ A}$ *is Arguesian.*

Proof. Suppose that $\mathbf{L}$ is a lattice of permuting equivalence relations on a set S, and that a_i and b_i, $i = 0, 1, 2$, are elements of L. Let $\langle u, v\rangle$ be in the left side of (6.9.9). Then $\langle u, v\rangle \in a_i \vee b_i = a_i \circ b_i$ for $i = 0, 1, 2$. Hence there are elements $z_i \in S$ such that the relations of the figure below hold. Note that $\langle z_0, z_1\rangle \in c_2 \wedge (c_0 \vee c_1)$. Using this and the relationships of the figure it is easy to verify that $\langle u, v\rangle$ is in the right side of (6.9.9). ■

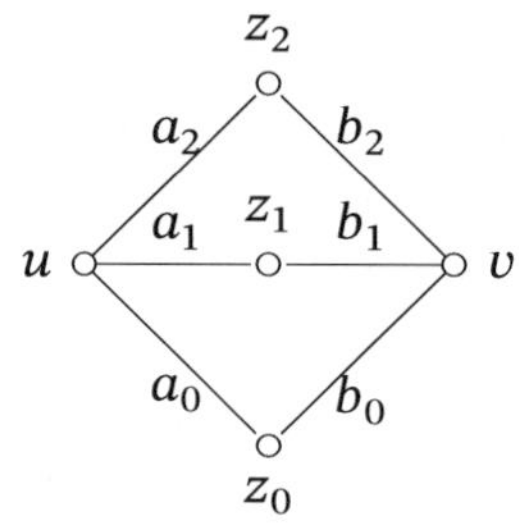

THEOREM 6.105. *Let* $\mathbf{L}$ *be a modular lattice. Then the following are equivalent.*

(i) $\mathbf{L}$ *is Arguesian,*
(ii) *any two centrally perspective triangles of* $\mathbf{L}$ *are axially perspective,*
(iii) *any two centrally perspective triangles* $\overline{a}$, $\overline{b}$ *of* $\mathbf{L}$ *which satisfy*

$$p \vee a_i = p \vee b_i = a_i \vee b_i \quad i = 0, 1, 2 \tag{6.9.10}$$

are axially perspective. (See (6.9.7) *for the definition of* p.*)*

Proof. Clearly (ii) implies (iii). We will first show that (iii) implies (i). Assume (iii). We need to show that (6.9.9) holds for arbitrary triangles $\overline{a}$ and $\overline{b}$ in $\mathbf{L}$. It suffices to show that (6.9.9) holds for triangles that satisfy (6.9.10). To see this, suppose that (6.9.9) holds whenever $\overline{a}$ and $\overline{b}$ satisfy (6.9.10), and let $\overline{a}$ and $\overline{b}$ be arbitrary triangles of $\mathbf{L}$ and let

$$a_i' = a_i \wedge (b_i \vee p), \quad b_i' = b_i \wedge (a_i \vee p), \quad i = 0, 1, 2.$$

Successive applications of modularity will establish that $a_i' \vee b_i' \geq p$ for each i. From this it readily follows that the new p', defined like p, but using the primed triangles, satisfies $p' \geq p$. On the other hand, the primed quantities are evidently all below their unprimed counterparts, and so in fact $p' = p$. Thus the primed triangles satisfy (6.9.10) and so, by our assumption, they satisfy (6.9.9). Hence

$$p \leq a_0' \vee (b_0' \wedge (b_1' \vee m'))$$

where $m' = c_2' \wedge (c_0' \vee c_1')$. Since the primed elements are smaller than their unprimed counterparts, it follows that (6.9.9) holds for the original (unprimed) triangles.

Now assume that our centrally perspective triangles $\overline{a}$ and $\overline{b}$ satisfy (6.9.10). Let $\overline{b'} = \overline{b}$ and $\overline{a'} = \langle a_0, a_1, a_2 \vee (a_0 \wedge (a_1 \vee b_1))\rangle$. Using (6.9.10) (for the unprimed triangles) it is easy to show that the primed triangles are centrally perspective. Now notice that by modularity

$$\begin{aligned} p' &= (a_0 \vee b_0) \wedge (a_1 \vee b_1) \wedge (a_2 \vee b_2 \vee (a_0 \wedge (a_1 \vee b_1))) \\ &= p \vee (a_0 \wedge (a_1 \vee b_1)). \end{aligned}$$

From this it is straightforward to verify that the primed triangles satisfy (6.9.10). Thus we can apply (iii). Since $c_2 = c_2'$ and $c_1 = c_1'$, we have

$$\begin{aligned} c_2 &\le c_1 \vee ((b_1 \vee b_2) \wedge (a_1 \vee a_2 \vee (a_0 \wedge (a_1 \vee b_1)))) \\ &= c_1 \vee ((b_1 \vee b_2) \wedge (a_2 \vee [(a_0 \vee a_1) \wedge (a_1 \vee b_1)])) \\ &= c_1 \vee ((b_1 \vee b_2) \wedge (a_1 \vee a_2 \vee (b_1 \wedge (a_0 \vee a_1)))) \\ &= c_1 \vee ((b_1 \vee b_2) \wedge (a_1 \vee a_2)) \vee (b_1 \wedge (a_0 \vee a_1)) \\ &= c_0 \vee c_1 \vee (b_1 \wedge (a_0 \vee a_1)). \end{aligned}$$

Now if we meet both sides with $a_0 \vee a_1$ and then join both sides with b_1 we obtain $b_1 \vee c_2 = b_1 \vee (c_2 \wedge (a_0 \vee a_1)) \le b_1 \vee ((a_0 \vee a_1) \wedge (c_0 \vee c_1))$. But by modularity $b_1 \vee c_2 = (a_0 \vee a_1 \vee b_1) \wedge (b_0 \vee b_1)$. By two applications of (6.9.10) we have $a_0 \vee a_1 \vee b_1 = a_0 \vee p \vee b_1 = a_0 \vee b_0 \vee b_1 \ge b_0$. Thus $b_1 \vee c_2 = b_0 \vee b_1$ and hence $b_0 \vee b_1 \le b_1 \vee ((a_0 \vee a_1) \wedge (c_0 \vee c_1))$. Meeting with b_0 we get

$$\begin{aligned} b_0 &\le b_0 \wedge [b_1 \vee ((a_0 \vee a_1) \wedge (c_0 \vee c_1))] \\ &\le b_0 \wedge (b_0 \vee b_1) \wedge [b_1 \vee ((a_0 \vee a_1) \wedge (c_0 \vee c_1))] \\ &= b_0 \wedge [b_1 \vee ((b_0 \vee b_1) \wedge (a_0 \vee a_1) \wedge (c_0 \vee c_1))] \\ &= b_0 \wedge [b_1 \vee (c_2 \wedge (c_0 \vee c_1))] \\ &= b_0 \wedge [(a_0 \wedge b_0) \vee b_1 \vee (c_2 \wedge (c_0 \vee c_1))] \\ &= (a_0 \wedge b_0) \vee (b_0 \wedge [b_1 \vee (c_2 \wedge (c_0 \vee c_1))]) \end{aligned}$$

Thus

$$p \le a_0 \vee b_0 \le a_0 \vee (b_0 \wedge [b_1 \vee (c_2 \wedge (c_0 \vee c_1))])$$

as desired.

To see that (i) implies (ii), let $\overline{a}$ and $\overline{b}$ be centrally perspective triangles of L. Then it follows from (6.9.9) and the central perspectivity that

$$(a_0 \vee b_0) \wedge (a_1 \vee b_1) \le a_0 \vee b_1 \vee (c_2 \wedge (c_0 \vee c_1)).$$

Now we join $a_0 \vee b_1$ to both sides and then we meet both sides with c_2. The left side becomes

$$c_2 \wedge (a_0 \vee b_0 \vee b_1) \wedge (a_0 \vee a_1 \vee b_1) = c_2$$

Thus we have

$$\begin{aligned}
c_2 &\le c_2 \wedge (a_0 \vee b_1 \vee (c_2 \wedge (c_0 \vee c_1))) \\
&= [c_2 \wedge (a_0 \vee b_1)] \vee [c_2 \wedge (c_0 \vee c_1)] \\
&= c_2 \wedge (c_0 \vee c_1 \vee (c_2 \wedge (a_0 \vee b_1))) \\
&\le c_0 \vee c_1 \vee [a_0 \wedge (b_0 \vee b_1)] \vee [b_1 \wedge (a_0 \vee a_1)] \\
&= [(a_1 \vee a_2) \wedge (b_1 \vee b_2)] \vee [(a_0 \vee a_2) \wedge (b_0 \vee b_2)] \\
&\qquad \vee [a_0 \wedge (b_0 \vee b_1)] \vee [b_1 \wedge (a_0 \vee a_1)] \\
&= [(b_1 \vee b_2) \wedge (a_1 \vee a_2 \vee (b_1 \wedge (a_0 \vee a_1)))] \\
&\qquad \vee [(a_0 \vee a_2) \wedge (b_0 \vee b_2 \vee (a_0 \wedge (b_0 \vee b_1)))] \\
&= [(b_1 \vee b_2) \wedge (a_2 \vee ((a_1 \vee b_1) \wedge (a_0 \vee a_1)))] \\
&\qquad \vee [(a_0 \vee a_2) \wedge (b_2 \vee ((a_0 \vee b_0) \wedge (b_0 \vee b_1)))] \\
&= [(b_1 \vee b_2) \wedge (a_1 \vee a_2 \vee (a_0 \wedge (a_1 \vee b_1)))] \\
&\qquad \vee [(a_0 \vee a_2) \wedge (b_0 \vee b_2 \vee (b_1 \wedge (a_0 \vee b_0)))]
\end{aligned}$$

Since $\overline{a}$ and $\overline{b}$ are centrally perspective, $a_0 \wedge (a_1 \vee b_1) \le a_0 \wedge (a_2 \vee b_2)$ and $b_1 \wedge (a_0 \vee b_0) \le b_1 \wedge (a_2 \vee b_2)$. Thus we have

$$\begin{aligned}
c_2 &\le [(b_1 \vee b_2) \wedge (a_1 \vee a_2 \vee (a_0 \wedge (a_2 \vee b_2)))] \\
&\qquad \vee [(a_0 \vee a_2) \wedge (b_0 \vee b_2 \vee (b_1 \wedge (a_2 \vee b_2)))] \\
&= [(b_1 \vee b_2) \wedge (a_1 \vee a_2 \vee (b_2 \wedge (a_0 \vee a_2)))] \\
&\qquad \vee [(a_0 \vee a_2) \wedge (b_0 \vee b_2 \vee (a_2 \wedge (b_1 \vee b_2)))] \\
&= c_0 \vee (b_2 \wedge (a_0 \vee a_2)) \vee c_1 \vee (a_2 \wedge (b_1 \vee b_2)) \\
&= c_0 \vee c_1,
\end{aligned}$$

completing the proof. ∎

The last theorem has two interesting corollaries. The first was discovered by A. Day and D. Pickering 1984, and the second by B. Jónsson 1972.

COROLLARY 1. *A lattice is Arguesian if and only if it satisfies*

$$(a_0 \vee b_0) \wedge (a_1 \vee b_1) \wedge (a_2 \vee b_2) \le a_0 \vee b_1 \vee (c_2 \wedge (c_0 \vee c_1)). \tag{6.9.11}$$

Proof. By Exercise 6.117.9 each of (6.9.9) and (6.9.11) imply modularity. Clearly (6.9.9) implies (6.9.11). Now our proof in the last theorem that (i) implies (ii) only used (6.9.11), not (6.9.9). Thus (6.9.11) implies (ii) which is equivalent to (i), the Arguesian equation. ∎

COROLLARY 2. *The class of Arguesian lattices is self-dual.*

Proof. Suppose that **L** is an Arguesian lattice, and let $\overline{a}$ and $\overline{b}$ be centrally perspective triangles in the dual of **L**. Then, $(a_0 \wedge b_0) \vee (a_1 \wedge b_1) \ge a_2 \wedge b_2$. Let

$$\overline{a'} = \langle a_0 \wedge a_2, b_0 \wedge b_2, a_0 \wedge b_0 \rangle \qquad \overline{b'} = \langle a_1 \wedge a_2, b_1 \wedge b_2, a_1 \wedge b_1 \rangle$$

Then $(a'_0 \vee b'_0) \wedge (a'_1 \vee b'_1) \leq a_2 \wedge b_2$, and by the dual central perspectivity, this is less than or equal to $a'_2 \vee b'_2$, showing that $\overline{a'}$ and $\overline{b'}$ are centrally perspective in **L**. Thus, by hypothesis and Theorem 6.105, they are axially perspective. Hence

$$\begin{aligned}
[(a_0 \wedge a_2) \vee (b_0 \wedge b_2)] \wedge [(a_1 \wedge a_2) \vee (b_1 \wedge b_2)] \\
&= (a'_0 \vee a'_1) \wedge (b'_0 \vee b'_1) \\
&\leq [(a'_0 \vee a'_2) \wedge (b'_0 \vee b'_2)] \vee [(a'_1 \vee a'_2) \wedge (b'_1 \vee b'_2)] \\
&\leq (a_0 \wedge a_1) \vee (b_0 \wedge b_1).
\end{aligned}$$

Thus $\overline{a}$ and $\overline{b}$ are axially perspective in the dual of **L**. Hence, by the previous theorem, the dual of **L** is Arguesian, as was to be shown. ■

Projective geometries and the lattices associated with them are defined in §4.8 of the first volume. Exercise 6.117.10 outlines the construction of a projective plane whose lattice is not Arguesian. On the other hand the lattice associated with any non-planar projective geometry is Arguesian, see Exercise 6.117.11. Roughly the argument runs as follows. It is easy to see that if $\overline{a}$ and $\overline{b}$ are centrally perspective triangles which do not lie in the same plane, then they are axially perspective. In fact c_0, c_1 and c_2 lie in the intersection of the planes determined by $\overline{a}$ and $\overline{b}$. If $\overline{a}$ and $\overline{b}$ do lie in the same plane, but there is a point of the geometry not in that plane, then it is possible to construct a third triangle $\overline{d}$ not in the plane but centrally perspective to both $\overline{a}$ and $\overline{b}$. Thus $\overline{d}$ is axially perspective to both $\overline{a}$ and $\overline{b}$ and from this one can show that $\overline{a}$ and $\overline{b}$ are axially perspective.

The next lemma, which is due to Jónsson, uses some of these ideas. In fact the triangle $\overline{d}$, constructed in the proof, can be shown to be centrally perspective to both $\overline{a}$ and $\overline{b}$, see Exercise 6.117.12.

LEMMA 6.106. *Let* **L** *be a modular lattice and let $\overline{a}$ and $\overline{b}$ be centrally perspective triangles in* **L** *which satisfy* (6.9.10). *Let $u = a_0 \vee a_1 \vee a_2 \vee b_0 \vee b_1 \vee b_2$, and suppose that there exist $q, r \in L$ such that*

$$p \vee q = p \vee r = q \vee r, \qquad u \wedge q = p \wedge a_2, \qquad u \wedge r = p \wedge b_2.$$

Then $\overline{a}$ and $\overline{b}$ are axially perspective.

Proof. Let $\overline{d}$ be the triangle with $d_i = (a_i \vee q) \wedge (b_i \vee r)$. Figure 6.16 gives a geometric representation of this situation. We begin by showing that

$$a_0 \wedge (b_0 \vee b_1) \leq d_0 \vee d_1. \tag{6.9.12}$$

This inclusion is proved with the following calculations.

$$\begin{aligned}d_0 \vee d_1 &\geq [(a_0 \vee q) \wedge (b_0 \vee r)] \vee (q \wedge (b_1 \vee r)) \\ &= (a_0 \vee q) \wedge [b_0 \vee r \vee (q \wedge (b_1 \vee r))] \\ &= (a_0 \vee q) \wedge [b_0 \vee r \vee (b_1 \wedge (q \vee r))] \\ &\geq (a_0 \vee q) \wedge (b_0 \vee r \vee (b_1 \wedge p)) \\ &= (a_0 \vee q) \wedge (b_0 \vee r \vee (b_1 \wedge (a_0 \vee b_0))) \\ &\geq (a_0 \vee q) \wedge (b_0 \vee b_1) \wedge (a_0 \vee a_1) \\ &\geq a_0 \wedge (b_0 \vee b_1).\end{aligned}$$

Now we show that

$$(a_0 \vee q) \wedge (b_0 \vee b_1 \vee r) \leq d_0 \vee d_1. \tag{6.9.13}$$

To see this note

$$\begin{aligned}(a_0 \vee q) \wedge (b_0 \vee b_1 \vee r) &\leq (a_0 \vee p \vee q) \wedge (b_0 \vee b_1 \vee r) \\ &= [(a_0 \vee p \vee q) \wedge (b_0 \vee b_1)] \vee r\end{aligned}$$

Using the fact that $b_0 \vee b_1 \leq u$ and $u \wedge q \leq p \leq a_0 \vee b_0$, we see that the right side is equal to $[(a_0 \vee b_0) \wedge (b_0 \vee b_1)] \vee r = (a_0 \wedge (b_0 \vee b_1)) \vee b_0 \vee r$. Thus

$$\begin{aligned}(a_0 \vee q) \wedge (b_0 \vee b_1 \vee r) &\leq (a_0 \vee q) \wedge [(a_0 \wedge (b_0 \vee b_1)) \vee b_0 \vee r] \\ &= [a_0 \wedge (b_0 \vee b_1)] \vee [(a_0 \vee q) \wedge (b_0 \vee r)] \\ &= [a_0 \wedge (b_0 \vee b_1)] \vee d_0.\end{aligned}$$

Now (6.9.13) follows from (6.9.12).

Using (6.9.13), we derive

$$\begin{aligned}d_0 \vee d_1 &\geq [(a_0 \vee q) \wedge (b_0 \vee b_1 \vee r)] \vee d_1 \\ &= (a_0 \vee q \vee d_1) \wedge (b_0 \vee b_1 \vee r) \\ &= (a_0 \vee q \vee a_1) \wedge (b_0 \vee b_1 \vee r) \\ &\geq c_2.\end{aligned} \tag{6.9.14}$$

By an argument similar to the proof of (6.9.13) we obtain:

$$(a_1 \vee a_2 \vee q) \wedge (b_2 \vee r) \leq c_0 \vee d_2.$$

Thus

$$\begin{aligned}c_0 \vee d_2 &\geq [(a_1 \vee a_2 \vee q) \wedge (b_2 \vee r)] \vee c_0 \\ &= (a_1 \vee a_2 \vee q) \wedge (b_2 \vee c_0 \vee r) \\ &= (a_1 \vee a_2 \vee q) \wedge (b_1 \vee b_2 \vee r) \\ &\geq d_1.\end{aligned} \tag{6.9.15}$$

Similarly we have

$$d_0 \leq c_1 \vee d_2. \tag{6.9.16}$$

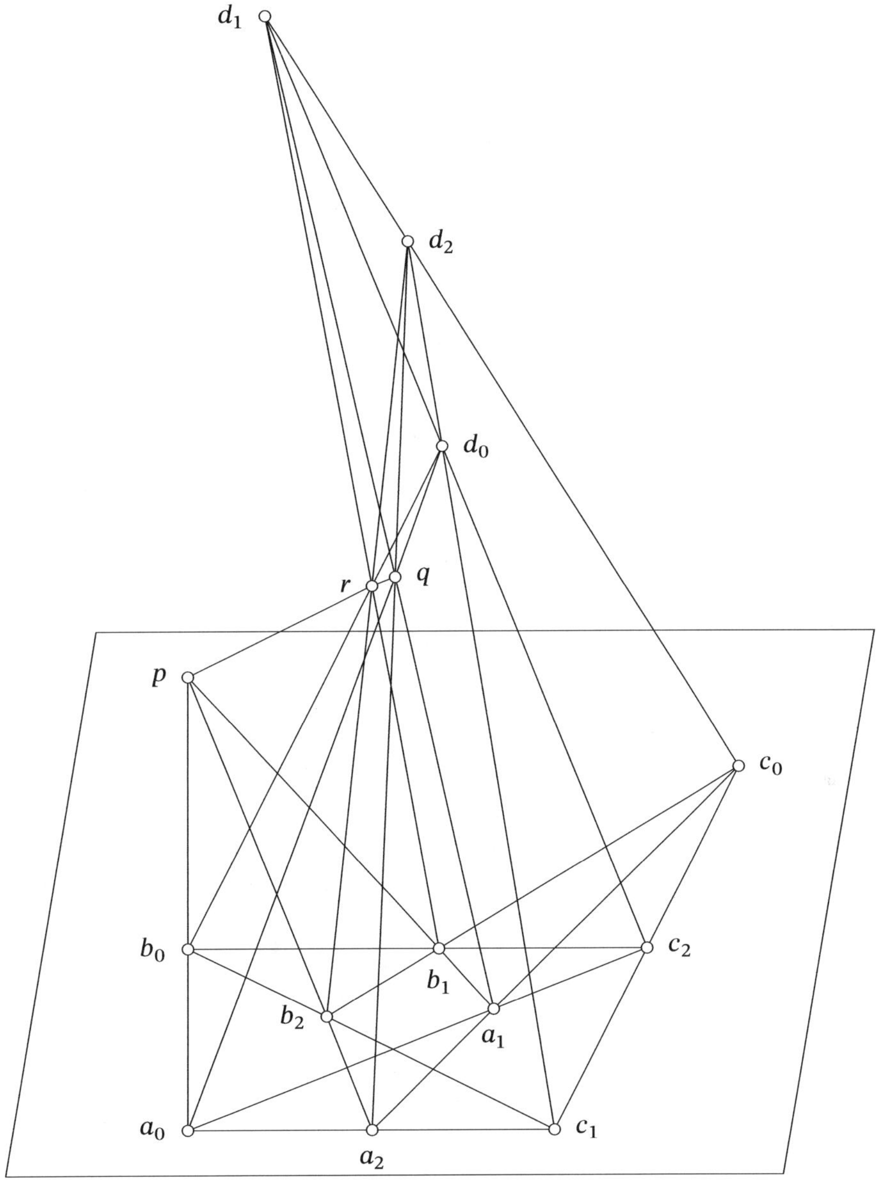

FIGURE 6.16. Desargues' Configuration in 3 Dimensions

Now by (6.9.14), (6.9.15), and (6.9.16) we have

$$\begin{aligned} c_2 &\leq u \wedge (d_0 \vee d_1) \\ &\leq u \wedge (c_0 \vee c_1 \vee d_2) \\ &= c_0 \vee c_1 \vee (d_2 \wedge u). \end{aligned}$$

Since we have $q \wedge u \leq a_2$ and $r \wedge u \leq b_2$, $d_2 \wedge u = (a_2 \vee q) \wedge u \wedge (b_2 \vee r) \wedge u = a_2 \wedge b_2$. Thus $c_2 \leq c_0 \vee c_1$, completing the proof. ∎

Recall from the beginning of this section that if α is a congruence on $\mathbf{A}$ then $\mathbf{A}(\alpha)$ is the subalgebra of $\mathbf{A}^2$ whose universe is α. Also recall that if θ is any congruence of $\mathbf{A}$, then, for $i = 0, 1$, $\theta_i = \{\langle\langle a_0, a_1\rangle, \langle b_0, b_1\rangle\rangle : \langle a_i, b_i\rangle \in \theta\}$ is a congruence of $\mathbf{A}(\alpha)$; see Lemma 6.98.

LEMMA 6.107. *Let α, β, γ, and δ be congruences of $\mathbf{A}$ satisfying*

$$\delta \le \alpha \wedge \beta = \alpha \wedge \gamma = \beta \wedge \gamma \tag{6.9.17}$$

then in $\mathbf{Con}\ \mathbf{A}(\alpha)$ *we have*

$$\alpha_0 \wedge \beta_1 = \alpha_0 \wedge \gamma_1 = \beta_1 \wedge \gamma_1 \tag{6.9.18}$$

$$\beta_1 \vee \delta_0 = \alpha_0 \vee \beta_0 \tag{6.9.19}$$

$$\gamma_1 \vee \delta_0 = \alpha_0 \vee \gamma_0. \tag{6.9.20}$$

Proof. By Lemma 6.98(iv) $\eta_0 \vee \eta_1 = \alpha_0 = \alpha_1$, where η_0 and η_1 are the kernels of the projection homomorphisms. If in formula (6.9.18) we change all the subscripts to 1 then the formula is true by (6.9.17) and Lemma 6.98(ii). Since $\alpha_0 = \alpha_1$ it follows that (6.9.18) holds.

To see (6.9.19) first note that if θ and ψ are congruences of $\mathbf{A}$ then $\theta_0 \vee \psi_1 \ge \eta_0 \vee \eta_1 = \alpha_0 = \alpha_1$, and by Lemma 6.98(vi) $\theta_0 \le \eta_0 \vee \theta_1 \le \delta_0 \vee \theta_1$. Thus, using part (iii) of the lemma,

$$\beta_1 \vee \delta_0 \ge \beta_0 \vee \alpha_0 = \beta_1 \vee \alpha_1 = \beta_1 \vee \alpha_0 \ge \beta_1 \vee \delta_0$$

proving (6.9.19). Similar arguments prove (6.9.20). ∎

LEMMA 6.108. *Let $\mathcal{V}$ be a variety and let $\mathcal{K}$ be the class of all lattices which can be embedded into the dual of* $\mathbf{Con}\ \mathbf{A}$ *for some $\mathbf{A} \in \mathcal{V}$. Then, for any $\mathbf{L} \in \mathcal{K}$, and any $p, s, t, u \in L$ which satisfy $p \vee s = p \vee t = s \vee t \le u$, $\mathbf{L}$ has an extension $\mathbf{L}' \in \mathcal{K}$ such that, for some $q, r \in L'$,*

$$p \vee q = p \vee r = q \vee r, \qquad q \wedge u = p \wedge s, \qquad r \wedge u = p \wedge t.$$

Proof. This is immediate from the last lemma. ∎

With these lemmas we can prove the following theorem of Ralph Freese and Bjarni Jónsson (1976).

THEOREM 6.109. *Let $\mathcal{V}$ be a congruence modular variety. Then $\mathcal{V}$ is congruence Arguesian.*

Proof. Let $\mathbf{A} \in \mathcal{V}$ and let $\mathbf{L}$ be the dual of $\mathbf{Con}\ \mathbf{A}$. Let $\overline{a}$ and $\overline{b}$ be centrally perspective triangles in $\mathbf{L}$ which satisfy (6.9.10). Applying Lemma 6.108, with $s = a_2$ and $t = b_2$, we embed $\mathbf{L}$ into a modular lattice $\mathbf{L}'$ possessing elements q and r satisfying the condition stated in Lemma 6.108. Now by Lemma 6.106 $\overline{a}$ and $\overline{b}$ are axially perspective. By Theorem 6.105, $\mathbf{L}$ is Arguesian, and so, by the second corollary to that theorem, $\mathbf{Con}\ \mathbf{A}$ is Arguesian. ∎

The proof of the last theorem shows that it is really a local theorem:

COROLLARY 6.110. *If all the subalgebras of* $\mathbf{A} \times \mathbf{A}$ *have modular congruence lattices, then* **Con A** *is Arguesian.* ∎

The variety of Arguesian lattices is not a congruence variety. Indeed, Ralph Freese (1994), using the identities of Mark Haiman (1991), shows that *no modular congruence variety, other than the trivial variety and the variety of distributive lattices, can have a finite basis for its identities.*[12]

A related result: the congruence variety of the variety of groups is distinct from that of Abelian groups. This result, which solves an old problem of B. Jónsson, is due to P. P. Pálfy and C. Szabó (1995).

The Pixley-Wille algorithm

Let ϵ be a pure lattice equation. Alden F. Pixley (1972) and Rudolf Wille (1970) showed that the class of varieties that satisfy ϵ as a congruence identity, can be defined by a **weak Mal'tsev condition**, that is, a conjunction of Mal'tsev conditions. (For more on weak Mal'tsev conditions, see §10.8, page 232.) They developed a procedure to produce this weak Mal'tsev condition called the Pixley-Wille algorithm which we describe now.

We may assume that ϵ has the form $u \leq v$ for lattice terms u and v since $u \approx v$ in a lattice if and only if $u \leq v$ and $v \leq u$. Fix $n \geq 1$ and let u°^n} denote the term in the signature $\{\circ^n, \cap\}$ obtained from u by replacing $\wedge$ with $\cap$ and $\vee$ with $\circ^n$, where $\circ^n$ is a binary operation symbol. For example, if $u = x \wedge (y \vee z)$ then $u^{\circ^n} = x \cap (y \circ^n z)$. If $x_0, \dots, x_{k-1}$ are the variables of u and $\alpha_0, \dots, \alpha_{k-1}$ are congruences of an algebra $\mathbf{A}$, then $u(\alpha_0, \dots, \alpha_{k-1})$ is of course also a congruence. Interpreting $\circ^n$ as the iterated relation product as defined at the beginning of §6.2, $u^{\circ^n}(\alpha_0, \dots, \alpha_{k-1})$ need not be a congruence but will be a reflexive, compatible relation of $\mathbf{A}$. Clearly

$$u^{\circ^n}(\alpha_0, \dots, \alpha_{k-1}) \subseteq u(\alpha_0, \dots, \alpha_{k-1}).$$

Moreover, if $\langle a, b\rangle \in u(\alpha_0, \dots, \alpha_{k-1})$ then, for some n, $\langle a, b\rangle \in u^{\circ^n}(\alpha_0, \dots, \alpha_{k-1})$. So

$$\bigcup_{n=1}^{\infty} u^{\circ^n}(\alpha_0, \dots, \alpha_{k-1}) = u(\alpha_0, \dots, \alpha_{k-1}). \tag{6.9.21}$$

Consequently, $u(\alpha_0, \dots, \alpha_{k-1}) \leq v(\alpha_0, \dots, \alpha_{k-1})$ holds throughout a variety if and only if

$$u^{\circ^n}(\alpha_0, \dots, \alpha_{k-1}) \subseteq v(\alpha_0, \dots, \alpha_{k-1}), \tag{6.9.22}$$

holds for all n. We will show that the class K_n of varieties that satisfy (6.9.22) forms a Mal'tsev class. Combining this with (6.9.21) we see that the class of varieties that satisfy the congruence inequality $u \leq v$ is the intersection of the K_n's, and so is defined by a weak Mal'tsev condition.

The proof that the class of varieties satisfying (6.9.22) can be defined by a Mal'tsev condition is similar to some of our earlier arguments. Let S, whose size will be determined later, be a set containing a and b. Let $\mathbf{F} = \mathbf{F}_{\mathcal{V}}(S)$ and let $n \geq 1$. In the first half of the procedure we associate a congruence with each variable, and hence each subterm, of u. We begin by associating with each subterm of u a list of pairs from S. First we

[12] R. Freese and P. Lipparini, in unpublished work, have extended this result to all varieties: *if* $\mathcal{V}$ *is a variety of algebras with a finitely based proper congruence variety, then* $\mathcal{V}$ *is congruence semidistributive.*

add $\langle a,b\rangle$ to the list associated by u. If $u = u_0 \wedge u_1$, we put $\langle a,b\rangle$ on the lists associated with both u_0 and u_1. If $u = u_0 \vee u_1$ we add new elements $a_1,\ldots,a_n$ to S and put $\langle a,a_1\rangle$ and $\langle a_i,a_{i+1}\rangle$ onto the list for u_0 for i even, and $\langle a_i,a_{i+1}\rangle$ onto the list for u_1 for i odd. $\langle a_n,b\rangle$ is on the list for u_0 or u_1 depending on the parity of n. We continue in this way until we reach the (lattice) variables $x_0,\ldots,x_{k-1}$. In this way each variable has a list of pairs from S associated with it. Let π_i be the partition on S generated by the pairs associated with x_i and let $\overline{x}_i$ be the congruence on $\mathbf{F}$ generated by π_i. For a lattice term $w(x_0,\ldots,x_{k-1})$ we let $\overline{w} = w(\overline{x}_0,\ldots,\overline{x}_{k-1})$. By construction, $\langle a,b\rangle \in \overline{u}^{\circ n}$ and so in $\overline{u}$. An example of this procedure is given after the next theorem.

Now if $\mathcal{V}$ satisfies the congruence identity $u^{\circ n} \leq v$, we must have $\langle a,b\rangle \in \overline{v}$. If $v = v_0 \vee v_1$ there must be terms $t_1,\ldots,t_m$ with variables S such that

$$a\ \overline{v}_0\ t_1\ \overline{v}_1\ t_2 \cdots t_m\ \overline{v}_1\ b, \tag{6.9.23}$$

if m is odd (and ending with $t_m\ \overline{v}_0\ b$ if m is even). Continue in this way until we get down to the lattice variables. At this point each pair t_i and t_j, $i<j$, is associated with a list of lattice variables such that $t_i\ \overline{x}\ t_j$ for each x in this list. By its construction $\overline{x}$ is generated by the partition π on S. Linearly order S as $\{a,b,s_1,s_2,\ldots\}$. From $t_i\ \overline{x}\ t_j$ we associate the equation $t_i^\pi \approx t_j^\pi$, where in t_i^π we substitute for each variable in S the first variable in S equivalent to it under π. Let $\Sigma_{n,m}$ be the set of equations that arise in this way, where m comes from (6.9.23) and is the largest over all such sequences. If $\mathcal{V}$ satisfies (6.9.22) then, using Theorem 6.9, one can show the equations of $\Sigma_{n,m}$ hold in $\mathbf{F}_{\mathcal{V}}(S)$ and hence in all of $\mathcal{V}$. Conversely, by construction, if $\mathcal{V}$ realizes $\Sigma_{n,m}$ for some m, then $\mathcal{V}$ satisfies (6.9.22).

Summing this up, we have:

THEOREM 6.111. *Let $u \leq v$ be a lattice inequality. Then, for each n, the class of varieties satisfying the inclusion $u^{\circ n} \subseteq v$ can be defined by a Mal'tsev condition K_n. A variety satisfies the congruence identity $u \leq v$ if and only if it satisfies every K_n.* ■

As an example consider the distributive law written as

$$u = x \wedge (y \vee z) \leq (x \wedge y) \vee (x \wedge z) = v.$$

If we take $n=1$ then $u^{\circ 1} = x \cap (y \circ z)$. In doing the first half only one element needs to be added to S so we take $S = \{a,b,c\}$. The associated partitions are $\pi_x = [a,b][c]$, $\pi_y = [a,c][b]$, and $\pi_z = [a][b,c]$. The equations of $\Sigma_{1,m}$ are (after permuting b and c and changing t_i to d_i)

$$\begin{aligned} d_0(a,b,c) &\approx a; \\ d_i(a,b,a) &\approx a && \text{for all } i; \\ d_i(a,a,c) &\approx d_{i+1}(a,a,c) && \text{if } i \text{ is even;} \\ d_i(a,c,c) &\approx d_{i+1}(a,c,c) && \text{if } i \text{ is odd;} \\ d_{m+1}(a,b,c) &\approx c. \end{aligned} \tag{6.9.24}$$

Our arguments show that the congruence containment $x \cap (y \circ z) \leq (x \wedge y) \vee (x \wedge z)$ is defined by the above Mal'tsev condition. Bjarni Jónsson (1967), however, has shown that this Mal'tsev condition actually defines congruence distributivity; see Theorem 6.1. Similarly, congruence modularity can be defined by a Mal'tsev condition, not just a weak Mal'tsev condition; see Theorem 6.32. It is possible that every congruence identity can be defined by a Mal'tsev condition, but that is not currently known. However, Gábor Czédli and Eszter K. Horváth (2003) have shown that if a congruence identity implies congruence modularity, then it can be defined by a Mal'tsev condition.

See Exercise 13 for another example of the use of the Pixley-Wille algorithm.

Hobby-McKenzie terms

The (M, N, F)-Taylor equations, namely $F(x, \dots, x) \approx x$ and $F(M) \approx F(N)$, were introduced in Equations (6.8.2) on page 81. There M (resp. N) was an $n \times n$ square matrix of x's and y's with all x's (resp. all y's) along the diagonal. Here we will restrict the matrix M to have all entries on or below the diagonal to be x, not just the diagonal itself. For N we only require there be y's on the diagonal, just as we did for Taylor equations. Under these restrictions, Equations (6.8.2) may be called **Hobby-McKenzie equations for** F. If a variety $\mathcal{V}$ realizes these equations, we say that $\mathcal{V}$ **has a Hobby-McKenzie term** (or **has an** (M, N)**-Hobby-McKenzie term**). If D is a system of definitions realizing (6.8.2) in $\mathcal{V}$, then F^D will be called a **Hobby-McKenzie term** (or an (M, N)-**Hobby-McKenzie term) for** $\mathcal{V}$.

In fact a variety $\mathcal{V}$ has a Hobby-McKenzie term if and only if it realizes a set Γ of equations in a single operation symbol G, with the following properties: Γ implies $G(x, \dots, x) \approx x$, and, for each $\emptyset \neq I \subseteq \{0, \dots, n-1\}$, Γ implies an equation of the form $G(\text{variables}) \approx G(\text{variables})$, where the set of variables occurring at the indices in I on the left side is different from the set on the right side. The next lemma shows these two formulations are equivalent.

LEMMA 6.112. *Let F be a function symbol with arity n and let Σ be a finite set of equations in the signature $\{F\}$. Consider the following properties of Σ:*

(A) *There are $n \times n$ matrices M and N such that Σ entails the (M, N)-Hobby-McKenzie equations for F.*
(B) *Σ entails $F(x, \dots, x) \approx x$ and, for each nonempty $I \subseteq \{0, \dots, n-1\}$, an equation of the form $F(\text{variables}) \approx F(\text{variables})$, where the variables occurring at the indices in I on the left side is different from the set on the right side.*

If (A) *holds then* (B) *holds. If* (B) *holds then there is a permutation σ of $\{0, 1, \dots, n-1\}$ such that* (A) *holds with F replaced by*

$$G(x_0, \dots, x_{n-1}) = F(x_{\sigma(0)}, \dots, x_{\sigma(n-1)}). \qquad (*)$$

Proof. To see that (A) $\Rightarrow$ (B), let $I \subseteq \{0, \dots, n-1\}$ be nonempty. Then I has a largest element k. The equation $F(M_{k,0}, \dots, M_{k,n-1}) \approx F(M_{k,0}, \dots, M_{k,n-1})$ guaranteed by (A) witnesses the desired property of (B).

To see the other implication we first note that if (B) holds then it holds with equations in $\{x, y\}$. Indeed, the set of variables at the positions in I on the left side of the equation for I is distinct from that of the right. By the left-right symmetry of equations, we may assume there is a variable v_i at position i, for some $i \in I$, on the right side that does not occur at any positions in I on the left. Replace all occurrences of v_i on both sides with y and replace all other variables with x. Clearly the derived equations still has the desired property for I.

Now consider $I = \{0, \dots, n-1\}$. Since the only nonempty subsets of $\{x, y\}$ are $\{x\}$, $\{y\}$ and $\{x, y\}$, (B) implies there is an equation where every variable is x on one side and there is at least one y occurring on the other side. If this y occurs at position k on the right side then we can define G as in $(*)$ with σ as the permutation interchanging k and $n-1$. Then G satisfies an equation of the form

$$G(x, \dots, x) \approx G(*, \dots, *, y).$$

This gives the equation for the last row of the matrices. A similar argument with $I = \{0, \dots, n-2\}$ gives the next to last equation, and so on. ■

It is easy to see that if a variety has a near unanimity term then this same term is a Hobby-McKenzie term for that variety. Indeed, we can take M to be the matrix with all x's and N to be the matrix with y's on the main diagonal and x's elsewhere. Similarly a Mal'tsev term for a variety is a Hobby-McKenzie term. Using the $\star$ product introduced in §6.7 one can show congruence distributive varieties have a Hobby-McKenzie term. In Exercise 15 you are asked to show if a variety is 3-distributive, that is, satisfies (Δ_3) on page 2, then the star product $d_1 \star d_2$ is a Hobby-McKenzie term with arity 9.

Similarly congruence modular varieties and varieties with k-permutable congruences for some k have Hobby-McKenzie terms. The variety of semilattices does *not* have a Hobby-McKenzie term even though it has a Taylor term, namely $t(x, y) = x \cdot y$. These facts are consequences of the next theorem.

Let $\mathbf{A}$ be an algebra with signature σ. Let T_σ be the set of all σ-terms; see §6.1. The **idempotent reduct** of $\mathbf{A}$ is the algebra on A whose operations are $t^{\mathbf{A}}$ for all $t \in T_\sigma$ such that $\mathbf{A} \models t(x, \dots, x) \approx x$. For a variety $\mathcal{V}$ we define the **idempotent reduct** of $\mathcal{V}$, denoted $\mathcal{V}_{\text{idem}}$, to be the variety generated by the idempotent reduct of $\mathbf{F}_{\mathcal{V}}(\omega)$.

THEOREM 6.113. *The following are equivalent for a variety $\mathcal{V}$.*

(i) *$\mathcal{V}$ has a Hobby-McKenzie term.*
(ii) *$\mathcal{V}$ satisfies an idempotent Mal'tsev condition that fails in the variety of semilattices.*
(iii) *$\mathcal{V}_{\text{idem}} \not\leq_{\text{int}} \mathcal{S}$, where $\mathcal{S}$ is the variety of semilattices.*
(iv) *$\mathcal{V}$ satisfies a nontrivial (pure lattice) congruence equation.*

Proof. We will first show (i), (ii), and (iii) are equivalent. Then we will show (iv) implies (ii). That (i) implies (iv) is a deep result of Keith A. Kearnes and Emil W. Kiss (2013), Theorem 7.15 and A.2, and Keith A. Kearnes and Ágnes Szendrei (1998). The the proof will not be given here.

(i) ⇒ (ii): Suppose $F(x_0, \dots, x_{n-1})$ is a Hobby-McKenzie term for the variety of semilattices. Since semilattice terms are meets of variables, we may assume $F(x_0, \dots, x_{n-1}) = \bigwedge_{i \in I} x_i$, for some $I \subseteq \{0, \dots, n-1\}$. It is easy to see the definition of the Hobby-McKenzie property fails for this I, and (ii) follows.

(ii)⇒(iii): Since Mal'tsev conditions are disjunctions of strong Mal'tsev conditions, there is a finitely presented idempotent variety, $\mathcal{V}'$ such that $\mathcal{V}' \leq_{\text{int}} \mathcal{V}$ and $\mathcal{V}' \not\leq_{\text{int}} \mathcal{S}$. Since $\mathcal{V}'$ is idempotent, $\mathcal{V}' \leq_{\text{int}} \mathcal{V}_{\text{idem}}$. It follows that $\mathcal{V}_{\text{idem}} \not\leq_{\text{int}} \mathcal{S}$.

(iii) ⇒ (i): If the idempotent reduct, $\mathcal{V}_{\text{idem}}$, of $\mathcal{V}$ has a Hobby-McKenzie term, then of course $\mathcal{V}$ does. Thus we may assume $\mathcal{V}$ is idempotent and so $\mathcal{V} \not\leq_{\text{int}} \mathcal{S}$.

As in the proof Theorem 6.87 we can reduce to the case the signature of $\mathcal{V}$ has two operation symbols, h with arity n and k with arity n^2. The equations defining $\mathcal{V}$ are

$$k \approx h \star h \quad \text{and} \quad h(x, \dots, x) \approx x$$

together with linear equations only involving k; that is, equations of the form $k(\text{variables}) \approx k(\text{variables})$.

Let $h^{\star 1} = h$, $h^{\star 2} = k = h \star h$, and $h^{\star r} = h \star \cdots \star h$ (with r h's). So the arity of $h^{\star r}$ is n^r. Let Γ be the set of all linear equations $h^{\star i}(\text{variables}) = h^{\star j}(\text{variables})$ which are true in $\mathcal{V}$. There are two alternatives, depending on whether Γ can be realized by the variety of semilattices.

Alternative I: Γ cannot be realized by the variety of semilattices.
Then there is a finite subset Γ' of Γ that cannot be realized in semilattices. By Lemma 6.88, $h^{\star i}$ can be obtained from $h^{\star j}$ by identifying certain variables, provided $i < j$. Using this we can assume our finite subset Γ' of Γ consists of equations of the form $h^{\star r}(\mathit{variables}) \approx h^{\star r}(\mathit{variables})$ for a sufficiently large r. We claim, in this situation, that h^r is a Hobby-McKenzie term. If this fails there is an I with $\emptyset \neq I \subseteq \{0, \dots, n^r - 1\}$, such that for all the equations in Γ' the set of variables at the positions in I are the same for each side of the equation. But then Γ' can be realized in semilattices by using $\bigwedge_{i \in I} x_i$ to realize $h^{\star r}(x_0, \dots, x_{n^r-1})$, which is impossible. So $h^{\star r}$ is a Hobby-McKenzie term, as desired.

Alternative II: Γ can be realized by semilattices
In this case, for each i, there is a semilattice term $\overline{h^{\star i}}$ so that these terms satisfy the equations in Γ.

Now the arity of $h^{\star m}$ is n^m, which equals the number of sequences of length m with entries in $\{0, \dots, n-1\}$. We denote the set of such sequences by Q_m and we linearly order them with the lexicographic order. Thus we can use these sequences to index the positions and variables in $h^{\star m}$: $h^{\star m}(\cdots, x_\alpha, \cdots, x_\beta, \cdots)$, where α, $\beta \in Q_m$. Since semilattice terms are just products of variables,

$$\overline{h^{\star m}}(\cdots, x_\alpha, \cdots, x_\beta, \cdots) = \bigwedge_{\gamma \in D} x_\gamma,$$

where D consists of those γ's such that $\overline{h^{\star m}}$ depends on x_γ. Let $Q = \bigcup_m Q_m$. If α and $\beta \in Q$ we let $\alpha^\frown\beta$ denote the *concatenation* of these sequences.

The reader can verify each of the following, see Exercise 16.

(6.9.25) If $\alpha \notin D$ then $\alpha^\frown\beta$ and $\beta^\frown\alpha \notin D$ for all $\beta \in Q$.

(6.9.26) If $\alpha^\frown\langle j\rangle \notin D$ then $\alpha^\frown\langle i, j\rangle \notin D$.

(6.9.27) $\alpha \in D$ if and only if $D[\alpha] \neq \emptyset$.

where $D[\alpha] := \{i < n : \alpha^\frown\langle i\rangle \in D\}$.

LEMMA 6.114. *For every $\alpha \in D$ there is an $i < n$ such that $\alpha^\frown\langle i\rangle \in D$ and $|D[\alpha^\frown\langle i\rangle]| < |D[\alpha]|$.*

Proof. Since $\alpha \in D$, $D[\alpha]$ is not empty by (6.9.27). Let

$$p(x_0, \cdots, x_{n-1}) = \bigwedge_{i \in D[\alpha]} x_i.$$

Since $\mathcal{V} \not\leq_{\text{int}} \mathcal{S}$, $p \star p$ cannot satisfy all the equations satisfied by $h^{\star 2} = h \star h$. Thus there is an equation

$$h^{\star 2}(u_{0,0}, u_{0,1}, \cdots, u_{n-1,n-1}) \approx h^{\star 2}(v_{0,0}, v_{0,1}, \cdots, v_{n-1,n-1})$$

where $u_{i,j}$ and $v_{i,j}$ are variables, satisfied by $\mathcal{V}$ but failing for p. This means

$$\{u_{i,j} : i, j \in D[\alpha]\} \neq \{v_{i,j} : i, j \in D[\alpha]\}$$

Substitute these variables into $h^{\star m+2}$, where $\alpha \in Q_m$, replacing the subterm $h^{\star 2}(\langle x_{\alpha^\frown\langle i,j\rangle} : i, j < n\rangle)$ to obtain an equation of Γ. Since $\overline{h}^{\star m+2}$ satisfies this equation, there are $i, j \in D[\alpha]$ such that $\alpha^\frown\langle i, j\rangle \notin D$. So $\alpha^\frown\langle i\rangle \in D$ and $j \in D[\alpha] - D[\alpha^\frown\langle i\rangle]$. But (6.9.26) and (6.9.27) imply $D[\alpha^\frown\langle i\rangle] \subseteq D[\alpha]$, proving the lemma. ■

The lemma implies $D = \emptyset$. So $\overline{h} = \overline{h^{\star}1}$ is independent of its variables. But then by idempotence $x \approx h(x, \dots, x) \approx h(y, \dots, y) \approx y$ holds in $\mathcal{V}$. But a variety of one-element algebras has a Hobby-McKenzie term (it is x with the equation $x \approx y$ supporting the Hobby-McKenzie property).

So, under either alternative we have a Hobby-McKenze term and the demonstration of (iii)⇒(i) is complete.

(iv) ⇒ (ii): Suppose $\mathcal{V}$ satisfies a nontrivial congruence identity ϵ. By the Pixley-Wille algorithm, Theorem 6.111, there are Mal'tsev conditions K_n such that a variety satisfies the congruence identity ϵ if and only if it satisfies each K_n. By the theorem of Ralph Freese and J. B. Nation (1973) (Theorem 6.30) the variety of semilattices satisfies no nontrivial congruence identity; in particular, it does not satisfy ϵ. Hence it must fail K_n for some n. Thus (ii) holds. ■

COROLLARY 6.115. *If a variety $\mathcal{V}$ has any of the following properties then it has a Hobby-McKenzie term.*

- *distributive congruence lattices;*
- *modular congruence lattices;*
- *join-semidistributive congruence lattices;*
- *k-permutable congruence lattices.*

Proof. Since each of these properties is definable by an idempotent Mal'tsev condition and the variety of semilattices is not congruence modular, is not congruence join semidistributive, nor is it k-permutable, the corollary follows from the equivalence of (i) and (ii) in Theorem 6.113. ■

The following lemma is used §7.8.

LEMMA 6.116. *Let $\mathcal{V}$ be a nontrivial variety having a Hobby-McKenzie term. Then there is a term t in the signature of $\mathcal{V}$ where both the variables x and y occur, such that $\mathcal{V} \models x \approx t$.*

Proof. Let $t(x_1, \dots, x_n)$ be a (M, N)-Hobby-McKenzie term for $\mathcal{V}$. If t does not depend on any of its arguments, then it is constant in every member of $\mathcal{V}$. But then $x \approx t$ implies $\mathcal{V}$ is trivial, contrary to hypothesis.

Let i be the largest index such that t depends on x_i. The ith Hobby-McKenzie equation has the form

$$t(x, \dots, x, *, \dots, *) \approx t(*, \dots, *, y, *, \dots, *)$$

where each $*$ is either x or y. Since t is independent of its variables beyond i, we have

$$x \approx t(x, \dots, x) \approx t(x, x, \dots, x, *, *, \dots, *) \approx t(*, \dots, *, y, *, \dots, *).$$

Of course y occurs in the right-most term. If x does not also occur we get $x \approx y$, again a contradiction. ■

Exercises 6.117

1. In this and the next few exercises $\mathbf{P}$ denotes Polin's four-element algebra described Example 6.12 and $\mathcal{P}$ denotes the variety it generates. Show that the congruence lattice of $\mathbf{P} \times_\alpha \mathbf{P}$ is the lattice of Figure 6.14.

2. Show that the congruence lattice of $\mathbf{P} \times_\beta \mathbf{P}$ is the lattice of Figure 6.15. Show that this lattice is subdirectly irreducible with the two solid elements as the (only) critical pair. (This means that this pair of elements belong to every nonzero congruence of this lattice.)

3. (Ralph Freese, 1995) Let $\mathbf{A}$ be an algebra with x, y, z and $u \in A$ such that $\alpha = \mathrm{Cg}^{\mathbf{A}}(\langle x, y\rangle)$, $\beta = \mathrm{Cg}^{\mathbf{A}}(\langle x, z\rangle, \langle y, u\rangle)$, and $\gamma = \mathrm{Cg}^{\mathbf{A}}(\langle x, y\rangle, \langle z, u\rangle)$ generate the pentagon of Figure 6.13. Show that in $\mathbf{A} \times_\beta \mathbf{A}$, if $\beta(r) \leq \alpha(s)$ [$\alpha(s)$ and $\beta(r)$ are defined in (6.9.3)], then $\alpha = \gamma$.

4. Show that $\mathbf{Con}(\mathbf{P} \times_\gamma \mathbf{P})$ is the lattice diagrammed in Figure 6.17. Show that this lattice is subdirectly reducible and, in fact, lies in the variety generated by $\mathbf{N}_5$.

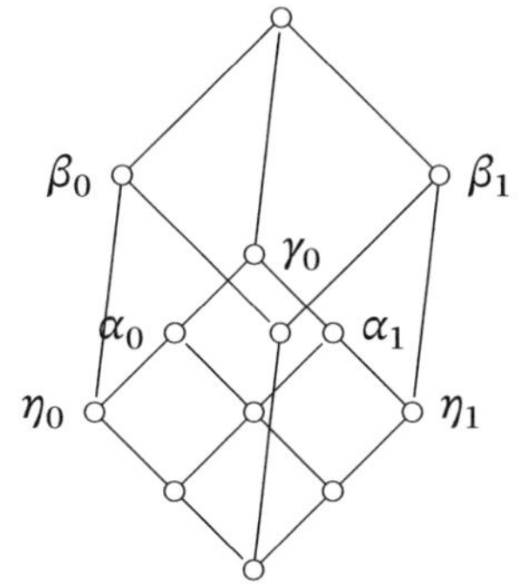

FIGURE 6.17. The Congruence Lattice of $\mathbf{P} \times_\gamma \mathbf{P}$

***5.** The congruence lattice of **P** is given in Figure 6.5. Let α, β and γ be the nontrivial congruences with $\alpha < \gamma$. Show that the universe of $\mathbf{F}_{\mathcal{P}}(1)$ consists of the 12 elements of $\mathbf{P}^4$ which, when written as 2×2 matrices, have the form $\begin{bmatrix} a_{00} & a_{01} \\ a_{10} & a_{11} \end{bmatrix}$ where the elements of each column are β related, the elements of the first row are α related, and the elements of the second row are γ related. Show that $\mathbf{Con}(\mathbf{F}_{\mathcal{P}}(1))$ is the lattice of Figure 6.18.

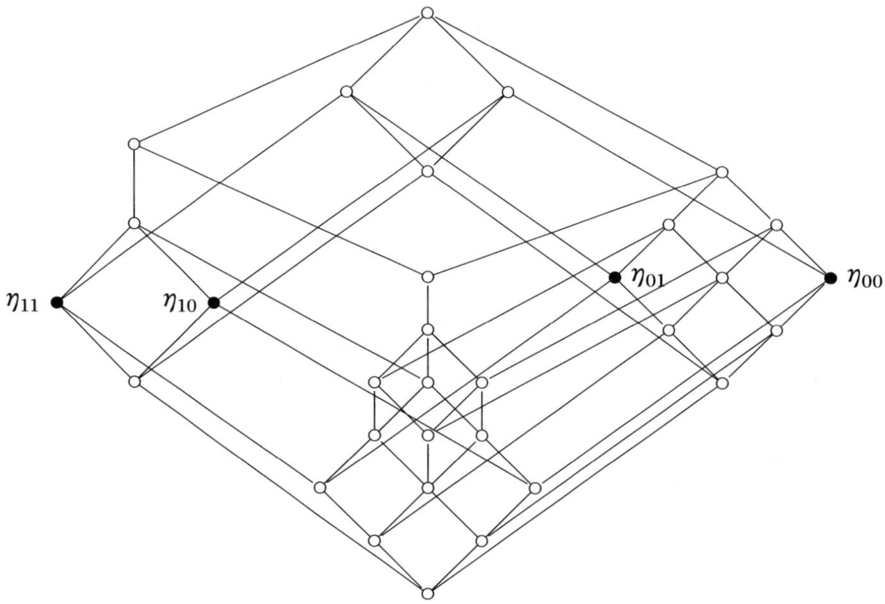

FIGURE 6.18. $\mathbf{Con}(\mathbf{F}_{\mathcal{P}}(1))$

6. This exercise shows that there are infinitely many congruence inequivalent lattice equations.

a. The concept of a spanning n-frame in a lattice is defined in Exercise 4.89.15 in the first volume. Let $x_0, \dots, x_n$ be elements of a modular lattice and define

$$u = \bigwedge_{i=0}^{n} \bigvee_{j\neq i} x_j$$

$$v = \bigvee_{i=0}^{n} \bigwedge_{j\neq i} \bigvee_{k\neq i,j} x_k$$

$$a_i = (x_i \vee v) \wedge u = (x_i \wedge u) \vee v. \tag{6.9.28}$$

Show that $\langle a_0, \dots, a_n\rangle$ is an n-frame spanning the interval sublattice $\mathbf{I}[v, u]$.

b. Let $\langle A_0, \dots, A_n\rangle$ be a spanning n-frame in the lattice of subspaces of a vector space $\mathbf{V}$. Show that

$$\mathbf{V} \cong \mathbf{A}_0 \times \cdots \times \mathbf{A}_{n-1}.$$

Let $\{f_\lambda : \lambda \in \Lambda\}$ be a basis for $\mathbf{A}_n$. By the above there are elements $e_{i,\lambda} \in A_i$ such that $f_\lambda = \sum_{i=0}^{n-1} e_{i,\lambda}$. Show that $\{e_{i,\lambda} : \lambda \in \Lambda\}$ is a basis for $\mathbf{A}_i$. *Hint*: Show that $\mathbf{A}_n$ and $\sum_{i=0}^{n-1} \mathbf{A}_i$ are complements.

c. Let $\langle a_0, a_1, a_2, a_3\rangle$ be a 3-frame in a modular lattice. Let

$$c = [(a_1 \vee a_2) \wedge (a_0 \vee a_3)] \vee [(a_0 \vee a_1) \wedge (a_2 \vee a_3)]$$

and define $r_0 = a_2$ and inductively

$$r_{k+1} = (a_2 \vee a_3) \wedge (a_0 \vee (c \wedge (a_1 \vee r_k))).$$

Now let $\langle \mathbf{A}_0, \mathbf{A}_1, \mathbf{A}_2, \mathbf{A}_3\rangle$ be a spanning 3-frame in the lattice of subspaces of a vector space $\mathbf{V}$ over a field $\mathbf{F}$. Using the previous part, show that the subspace corresponding to r_k as defined above is spanned by

$$\{ke_{0,\lambda} + ke_{1,\lambda} + e_{2,\lambda} : \lambda \in \Lambda\}.$$

Hint: By the last part we can find a basis, $\{e_{i,\lambda} : \lambda \in \Lambda\}$ of $\mathbf{A}_i$, for $i = 0, 1, 2$, so that $\{e_{0,\lambda} + e_{1,\lambda} + e_{2,\lambda} : \lambda \in \Lambda\}$ is a basis of $\mathbf{A}_3$. Using this show that the subspace corresponding to c defined above is

$$\Big\{ \sum_{\lambda \in \Lambda} c_\lambda e_{0,\lambda} + (c_\lambda + d_\lambda) e_{1,\lambda} + d_\lambda e_{2,\lambda} : c_\lambda, d_\lambda \in \mathbf{F}, \lambda \in \Lambda \Big\}.$$

Now use induction to prove the result.

d. Let $a_0, \dots, a_n$ be the lattice terms in the variables $x_0, \dots, x_n$ defined by (6.9.28). Then r_k can be viewed as a term in these variables. If $\mathbf{F}$ is a field and $\mathcal{V}_\mathbf{F}$ is the variety of all vector spaces over $\mathbf{F}$, then $\mathcal{V}_\mathbf{F}$ satisfies the congruence equation

$$r_k \approx a_2$$

if and only if the characteristic of $\mathbf{F}$ divides k. Thus there are infinitely many congruence inequivalent congruence equations.

7. (Alan Day and J. B. Nation, see (Bjarni Jónsson, 1980)) Let x, y, and z be variables and let $y_1 = y$, $z_1 = z$, $y_{n+1} = y \vee (x \wedge z_n)$, and $z_{n+1} = z \vee (x \wedge y_n)$. Let δ_n be the inclusion

$$x \wedge (y \vee z) \leq (x \wedge y_n) \vee (x \wedge z_n).$$

Show that if $\mathbf{L}$ is a $2n$-permutable sublattice of the congruence lattice of a semilattice then $\mathbf{L}$ satisfies δ_{2n}. Notice that Theorem 6.102 follows from this exercise.

8. Verify that (6.9.2) holds in $\mathbf{N}_5$. Hint: First show that it holds in any distributive lattice and thus one can assume that in any failure x, y, and z are assigned to a generating set of $\mathbf{N}_5$.

9. Show that every Arguesian lattice is modular. Also show that any lattice satisfying (6.9.11) is modular. Hint: Derive the modular law by applying either (6.9.9) or (6.9.11) to the triangles $\overline{a} = \langle x \wedge y \wedge z, x, x \rangle$ and $\overline{b} = \langle y \vee z, y \wedge z, y \rangle$.

10. Recall from Definition 4.78 and Exercise 4.89.6 that a projective plane is a pair $\boldsymbol{\pi} = \langle \mathrm{P}, \Lambda \rangle$ such that P is a set and Λ is a set of subsets of P and the following axioms are satisfied.
(i) Any two distinct points belong to one and only one line.
(ii) Any two distinct lines contain one and only one point.
(iii) There exist four points, no three of which are collinear.
A **partial projective plane** is a pair $\boldsymbol{\pi} = \langle \mathrm{P}, \Lambda \rangle$ satisfying
i′. Any two distinct points belong to at most one line.
ii′. Any two distinct lines contain at most one point.
(Notice that **ii.**′ follows from **i.**′) Let $\boldsymbol{\pi}_0 = \langle \mathrm{P}_0, \Lambda_0 \rangle$ be a partial projective plane satisfying (iii). This partial plane can be completed to a projective plane using the free plane construction of Marshall Hall described below. Inductively define $\boldsymbol{\pi}_{n+1} = \langle \mathrm{P}_{n+1}, \Lambda_{n+1} \rangle$ from $\boldsymbol{\pi}_n = \langle \mathrm{P}_n, \Lambda_n \rangle$ as follows. If n is even, enlarge Λ_n by adding a new line for each pair of points of P_n not contained on a line. At this stage, each of these new lines contains only two points. If n is odd, the dual construction is used: a new point is added for each pair of nonintersecting lines. Let $\mathrm{P} = \bigcup \mathrm{P}_n$ and $\Lambda = \bigcup \Lambda_n$ and $\boldsymbol{\pi} = \langle \mathrm{P}, \Lambda \rangle$.

a. Show that $\boldsymbol{\pi}$ is a projective plane.

b. Let $\boldsymbol{\pi}$ be the plane obtained from the above construction starting with P_0 a four element set and $\Lambda_0 = \emptyset$. Show that the lattice $\mathbf{L}^{\pi}$ associated with this plane (defined in §4.8) is modular but not Arguesian. Hint: A *confined configuration* in a (partial) projective plane is a pair $\langle \mathrm{Q}, \Gamma \rangle$ where Q is a finite, nonempty subset of the set of points and Γ is a finite, nonempty subset of the set of lines such that each element of Q lies on at least three lines in Γ and each line in Γ contains at least three points of Q. Show that $\boldsymbol{\pi}$ contains no confined configuration. Then show that if $\langle a_0, a_1, a_2 \rangle$ and $\langle b_0, b_1, b_2 \rangle$ are triangles in a projective plane which are centrally perspective from p, and which are also axially perspective, then the ten points p, a_0, a_1, a_2, b_0, b_1, b_2, c_0, c_1, and c_2 (the c_i are defined by (6.9.7)) are contained in a confined configuration which also has ten lines.

11. The lattice associated with a projective geometry is defined in §4.8. Show that the lattice associated with a projective geometry, which is not a projective plane, is Arguesian. Hint: See the remarks before Lemma 6.106.

12. Show that the triangle $\overline{d}$ defined in the proof of Lemma 6.106 is centrally perspective with both $\overline{a}$ and $\overline{b}$ of that lemma.

13. (George Hutchinson and Gábor Czédli, 1978) The lattice identity

$$(x_0 \vee x_1) \wedge (x_2 \vee x_3) \leq [(x_0 \vee x_2) \wedge (x_1 \vee x_3)] \wedge [(x_0 \vee x_3) \wedge (x_1 \vee x_2)] \tag{Fano}$$

is known as the Fano identity; see (Rudolf Wille, 1969).

a. Let **R** be a ring with unit. Show the variety of all unitary **R**-modules satisfies (Fano) as a congruence identity if and only if the characteristic of **R** is 2. *Hint:* Look at $\mathbf{R}^3$ as an **R**-module.

b. Show that the Pixley-Wille algorithm using $n = m = 1$ gives

$$\begin{aligned}
a_0 &\approx t_0(a_0, a_1, a_2, a_3)\\
a_1 &\approx t_1(a_0, a_1, a_2, a_3)\\
t_0(a_0, a_1, a_0, a_3) &\approx t_3(a_0, a_1, a_0, a_3)\\
t_2(a_0, a_1, a_2, a_0) &\approx t_3(a_0, a_1, a_2, a_0)\\
t_0(a_0, a_1, a_1, a_3) &\approx t_4(a_0, a_1, a_1, a_3)\\
t_2(a_0, a_1, a_2, a_1) &\approx t_4(a_0, a_1, a_2, a_1)\\
t_2(a_0, a_1, a_0, a_3) &\approx t_5(a_0, a_1, a_0, a_3)\\
t_1(a_0, a_1, a_2, a_1) &\approx t_5(a_0, a_1, a_2, a_1)\\
t_2(a_0, a_1, a_1, a_3) &\approx t_6(a_0, a_1, a_1, a_3)\\
t_1(a_0, a_1, a_2, a_0) &\approx t_6(a_0, a_1, a_2, a_0)
\end{aligned}$$

Show that a congruence permutable variety satisfies the Fano identity if and only if it realizes this strong Mal'tsev condition.

14. Let Σ be a set of Hobby-McKenzie equations of the form $F(M) \approx F(N)$ and $F(x, \dots, x) \approx x$, where M and N are as described on page 111 Show $\Sigma^{(n-1)} \models x \approx y$, where $\Sigma^{(n-1)}$ is the $(n-1)$ iterated derivative of Σ.

15. Let $\mathcal{V}$ be a 3-distributive variety; that is, $\mathcal{V}$ realizes (Δ_3) on page 2. Show that $d_1 \star d_2$ is a 9-place Hobby-McKenzie term for $\mathcal{V}$.

16. Prove (6.9.25), (6.9.26) and (6.9.27).

6.10. Relationships

In this section we examine briefly the relative strengths of some of the more important classes which comprise our classification scheme for varieties. The varietal properties under consideration will include several from earlier in this chapter as well as some new properties. The implications we will establish are indicated in Figure 6.19.

K_n is the class of varieties in which no algebra has a universe which can be written as the union of n proper subuniverses. The table below is a key for the other properties.

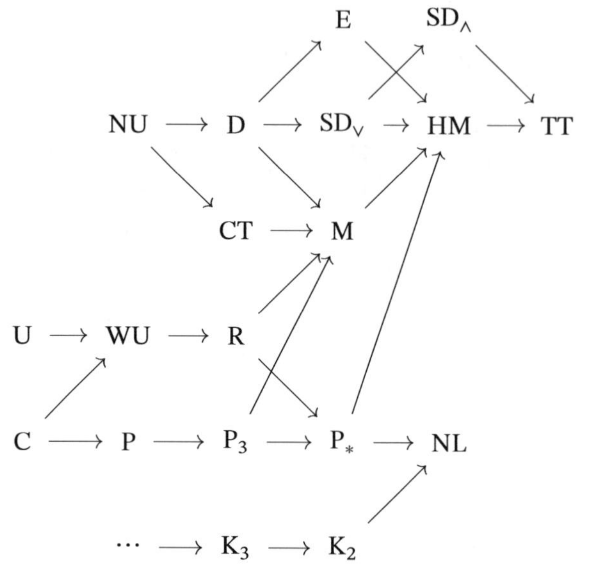

FIGURE 6.19. Relationships between Classes of Varieties

U :	congruence uniformity	C :	coherence
WU :	weak uniformity	R :	regularity
P_3 :	3-permutability	P :	permutable congruences
P_* :	k-permutability for some k	D :	distributive congruences
HM :	has a Hobby-McKenzie term	M :	modular congruences
$SD_\wedge$:	congruence meet semidistributive	NL :	no compatible linear order
$SD_\vee$:	congruence join semidistributive	TT :	has a Taylor term
E :	congruences satisfy (6.9.4)	NU :	has a near unanimity term
		CT :	has a cube term

We use these same letters to denote the class of all varieties satisfying the corresponding property. Thus R is also used to denote the class of all regular varieties. Each of these classes except U is a Mal'tsev class.[13] Some of the properties are new and will be explained below. E, for example, is the class of all varieties which have (6.9.4) on page 100 as a congruence identity. We propose to establish the implications indicated in Figure 6.19. Moreover, we will see that none of the arrows can be reversed. It may be possible that there are some true relationships not implied by this diagram; for example, it is possible that $K_2 \to P_*$. However, from our examples it will be clear that very few further implications could hold.

Several implications are trivial:

$$U \to WU \qquad D \to M \qquad D \to E \qquad D \to SD_\vee \qquad P \to P_3 \to P_*$$

The implication $K_{n+1} \to K_n$ is also clear. Jónsson's implication $P_3 \to M$ was established in Theorem 4.67. $SD_\vee \to SD_\wedge$ by Theorem 6.24. By Theorem 6.51, WU $\to$ R. Theorem 6.57 gives C $\to$ WU and C $\to$ P, and Theorem 6.71 gives R $\to$ M and R $\to$ P_*.

[13]That U is not a Mal'tsev class is proved in Theorem 4.1 of (Walter Taylor, 1974).

Of course these implications yield that U $\to$ M and it is worth mentioning that varieties which are congruence modular and nilpotent are congruence uniform; see Theorem 11.90 in Volume III.

Near unanimity terms and their connection with the Baker-Pixley Theorem are developed later in this section. Theorem 6.125 shows NU $\to$ D. Cube terms are defined in Example 6.66, and Corollary 6.67 proves CT $\to$ M. It is also clear from the definition of a cube term that a near unanimity term is a cube term and so NU $\to$ CT.

Corollary 6.115 gives the implications:

$$\mathrm{E} \to \mathrm{HM} \qquad \mathrm{SD}_\vee \to \mathrm{HM} \qquad \mathrm{M} \to \mathrm{HM} \qquad \mathrm{P}_* \to \mathrm{HM}$$

The next theorem is a corollary of Theorem 6.87. A Mal'tsev condition is trivial if it is satisfied by every variety.

THEOREM 6.118. *If a variety satisfies a nontrivial, idempotent Mal'tsev condition, then it has a Taylor term.* ■

From this result HM $\to$ TT and $\mathrm{SD}_\wedge \to$ TT follow.

NL is simply the class of varieties in which no algebra with more than one element admits a linear ordering; see (S. Bulman-Fleming and W. Taylor, 1976). (We say an algebra **A** *admits an order* if there is a compatible order on **A**, as defined above Theorem 6.68 on page 71.) The implication $\mathrm{P}_* \to$ NL follows easily from Theorem 6.68 on page 71.

As mentioned before, K_n is the class of varieties in which no algebra has a universe which can be written as the union of n proper subuniverses. Thus K_{n+1} implies K_n *a fortiori*. To see $\mathrm{K}_2 \to$ NL, suppose an algebra **A** admits a linear order $\leq$. Then

$$A^2 = \{\langle x, y\rangle \,:\, x \leq y\} \cup \{\langle x, y\rangle \,:\, y \leq x\}.$$

Moreover, each of these two sets is a subuniverse of $\mathbf{A}^2$. If $|A| \geq 2$, each is a proper subuniverse.

This completes the proof of all of the implication of Figure 6.19.

In the next theorem we reprove Hagemann's Theorem (Theorem 6.71). We show regularity implies modularity by deriving Day terms from the terms for congruence regularity. We then give a local version.

THEOREM 6.119. *Every congruence regular variety is congruence modular.*

Proof. If $\mathcal{V}$ is congruence regular, then $\mathcal{V}$ has terms f_i, g_i, for $1 \leq i \leq n$, satisfying condition (iv) of Theorem 6.48. Now we define new $\mathcal{V}$-terms $m_0, \dots, m_{2n+1}$ as follows:

$$\begin{aligned} m_0(x,y,z,w) &= x \\ m_{2i-1}(x,y,z,w) &= f_i(x,w,w,w,g_i(y,z,w)) \quad \text{for } 1 \leq i \leq n \\ m_{2i}(x,y,z,w) &= f_i(x,w,w,g_i(y,z,w),w) \quad \text{for } 1 \leq i \leq n \\ m_{2n+1}(x,y,z,w) &= w. \end{aligned}$$

We claim that these terms satisfy Day's equations for congruence modularity given in Theorem 6.32. To verify that these equations hold, we make the following deductions from the regularity equations of Theorem 6.48.

$$\begin{aligned} m_{2i-1}(x,y,y,x) &\approx f_i(x,x,x,x,g_i(y,y,x)) \\ &\approx f_i(x,x,x,x,x) \\ &\approx x. \end{aligned}$$

A similar calculation yields $m_{2i}(x,y,y,x) \approx x$. Now,

$$\begin{aligned} m_0(x,x,w,w) &\approx x \\ &\approx f_1(x,w,w,w,g_1(x,w,w)) \\ &\approx m_1(x,x,w,w) \\ m_{2n}(x,x,w,w) &\approx f_n(x,w,w,g_n(x,w,w),w) \\ &\approx w \\ &\approx m_{2n+1}(x,y,z,w). \end{aligned}$$

Finally

$$\begin{aligned} m_{2i-1}(x,y,y,w) &\approx f_i(x,w,w,w,g_i(y,y,w)) \\ &\approx f_i(x,w,w,g_i(y.y.w),w) \\ &\approx m_{2i}(x,y,y,w) \\ m_{2i}(x,x,w,w) &\approx f_i(x,w,w,g_i(x,w,w),w) \\ &\approx f_{i+1}(x,w,w,w,g_{i+1}(x,w,w)) \\ &\approx m_{2i+1}(x,x,w,w). \end{aligned}$$

■

The above theorem tells us that if **Con A** is nonmodular, then some $\mathbf{B} \in \boldsymbol{HSP}(\mathbf{A})$ has nonregular congruences. The following local result tells us that such a **B** may be found in $\boldsymbol{S}(\mathbf{A}^2)$.

THEOREM 6.120 (S. Bulman-Fleming, Alan Day, and Walter Taylor 1974). *If every subalgebra of* $\mathbf{A}^2$ *is congruence regular, then* **Con A** *is modular.*

Proof. Let α, β, and γ be elements Con **A** with $\beta \le \gamma$. We will show by induction that

$$(\alpha \circ^n \beta) \wedge \gamma \le (\alpha \wedge \gamma) \vee \beta. \tag{6.10.1}$$

We extend the definition of $\alpha \circ^n \beta$ given in §6.2 to include the case $n = 0$ by letting $\alpha \circ^0 \beta = 0_{\mathbf{A}}$. Since $\alpha \circ^1 \beta = \alpha$, the cases $n = 0$ and 1 are trivial. Now assume that (6.10.1) holds for a particular n, where $n > 1$. Of course $\alpha \circ^n \beta$ is a subuniverse of $\mathbf{A}^2$. We let **B** be the corresponding subalgebra. Using the construction of Lemma 6.98, for any $\psi \in$ Con **A**, and for $i = 0, 1$, we let $\psi_i \in$ Con **B** be defined by $\langle a_0, a_1\rangle\ \psi_i\ \langle b_0, b_1\rangle$ if $a_i\ \psi\ b_i$. Let $\theta = (\alpha \wedge \gamma) \vee \beta$, and note that $\theta \le \gamma$. Thus $\theta_0 \wedge \theta_1 \le \gamma_0 \wedge \theta_1$. Now if

$$\langle a, a\rangle\ \gamma_0 \wedge \theta_1\ \langle b_0, b_1\rangle$$

then $a\ \theta\ b_1$. Since $\langle b_0, b_1\rangle \in B$, (6.10.1) implies that $\langle b_0, b_1\rangle \in \theta$. Thus $a\ \theta\ b_0$ and hence

$$\langle a, a\rangle\ \theta_0 \wedge \theta_1\ \langle b_0, b_1\rangle.$$

This shows that the $\langle a, a\rangle$-blocks of $\theta_0 \wedge \theta_1$ and $\gamma_0 \wedge \theta_1$ are the same. Thus $\theta_0 \wedge \theta_1 = \gamma_0 \wedge \theta_1$ by regularity.

Now suppose that $\langle a, c\rangle \in (\alpha \circ^{n+1} \beta) \wedge \gamma$. Then there are elements b and d such that the relations of the figure hold.

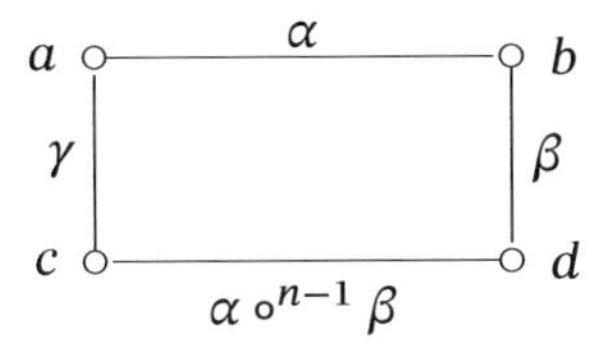

Clearly $\langle a, b\rangle$ and $\langle c, d\rangle$ are in B and $\langle a, b\rangle\ \gamma_0 \wedge \theta_1\ \langle c, d\rangle$. Hence $\langle a, b\rangle\ \theta_0 \wedge \theta_1\ \langle c, d\rangle$. Thus $a\ \ \theta\ \ c$, completing our inductive argument. Of course modularity follows from (6.10.1). ■

COROLLARY. *If every finite algebra in a variety $\mathcal{V}$ is congruence regular, then every finite algebra in $\mathcal{V}$ is congruence modular.* ■

Of course, Hagemann's Theorem 6.119 is also a corollary of the last theorem; in fact the proof of the theorem, when applied to the appropriate free algebra, yields Hagemann's terms seen in the proof of Theorem 6.119. (See Exercise 6.128.1.) The exponent 2 appearing in the theorem is best possible in the sense that there exists an eight element algebra **A** with **Con A** nonmodular, but such that every subalgebra of **A** is congruence regular (see Exercise 6.128.2).

If $\mathcal{V}$ is congruence regular, then $\mathcal{V}$ has terms f_i, g_i, for $i = 1, \ldots, n$, satisfying the equations in condition (iv) of Theorem 6.48. If we let

$$p_i(x, y, z) = f_i(x, z, z, g_i(y, z, z), g_i(x, y, z))$$

then it is easy to verify that the Hagemann-Mitschke equations of Theorem 6.11 (on page 14) hold, giving another proof of Theorem 6.71.

We still do not know any local proof for this last result. Nevertheless, it has an interesting corollary which simplifies the equations for regularity given in Theorem 6.48.

THEOREM 6.121. *A variety $\mathcal{V}$ is congruence regular if and only if there exist ternary terms $g_1, \ldots, g_n$ and 4-ary terms $f_1, \ldots, f_n$ such that the following equations hold identically in $\mathcal{V}$.*

$$\begin{aligned} g_i(x, x, z) &\approx z \qquad \text{for } 1 \le i \le n \\ x &\approx f_1(x, y, z, z) \\ f_1(x, y, z, g_1(x, y, z)) &\approx f_2(x, y, z, z) \\ &\ \vdots \\ f_n(x, y, z, g_n(x, y, z)) &\approx y \end{aligned}$$

Proof. We return to the proof of Theorem 6.48, especially the part (ii) implies (iv). As before, we take ϕ to be the congruence generated by Z. But instead of applying Theorem 6.18, we apply H. Lakser's special description of congruence generation for congruence n-permutable varieties given in Exercise 6.16.14 of §6.2. The details are left to the reader. ■

Now we will give examples showing that none of the implications of Figure 6.19 can be reversed. In most cases either well known or previous examples show that the reverse implication fails. For example the variety of groups shows that K_2 does not imply K_3. The idempotent reduct of groups shows that NL does not imply K_2. The variety of lattices shows that M does not imply NL and hence M implies neither R nor P_3.

Let $\mathcal{V}_k$ be the varieties given Example 6.14. $\mathcal{V}_3$ and the variety of implication algebras (see Exercise 6.16.5 and Exercise 6.58.2) show that P_3 implies neither R nor P. It follows that neither M nor P_* imply R. Heyting algebras show that P does not imply R, see Exercise 6.58.26, and hence P does not imply C. Polin's variety (Example 6.12) and $\mathcal{V}_4$ show that P_* does not imply P_3. Example 6.52 on page 54 shows that R does not imply WU; Exercise 6.58.19 shows that WU does not imply U. Exercise 6.58.23 shows that U does not imply C and so WU does not imply C. This same exercise gives another proof that P does not imply C. HM does not imply E since otherwise M would imply E which is not the case by Theorem 6.102.

The variety of semilattices combined with Theorem 6.113 shows TT does not imply HM. Semilattices also show that $SD_\wedge$ does not imply $SD_\vee$. The variety of groups refutes TT implies $SD_\wedge$ and HM implies $SD_\vee$. Polin's variety refutes HM implies M and $SD_\vee$ implies D. The variety of lattices shows HM does not imply P_*. Theorem 6.125 proves D does not imply NU. Exercise 8 shows that M does not imply CT.

The only possible reverse implication remaining for consideration is NL implies P_*. For an example blocking this implication, take $\mathcal{V}$ to be the variety of algebras $\langle A, \vee, \wedge, 0, 1, f\rangle$, where $\langle A, \vee, \wedge, 0, 1\rangle$ is a lattice with least element 0, and greatest element 1, and f is a unary operation obeying $f(0) \approx 1$ and $f(1) \approx 0$. Using finite chains, it is easy to see that $\mathcal{V}$ does not have n-permutable congruences for any n, that is $\mathcal{V} \notin P_*$. Now suppose $\mathbf{A} \in \mathcal{V}$, with $|A| \geq 2$, admits a linear order $\leq$ (not to be confused with the lattice order). Thus $1 \neq 0$, and so, without loss of generality, $0 < 1$ in the linear order on $\mathbf{A}$. Hence, $1 = f(0) < f(1) = 0$. This contradiction establishes that such an $\mathbf{A}$ cannot admit a linear order, and hence that $\mathcal{V} \in$ NL. In Exercise 6.128.18 you get to show that P_* and NL *are equivalent* for idempotent varieties.

Although we do not propose to consider every possibility, we will conclude this section with examples blocking two possible further implications in Figure 6.19. The first example is Polin's variety which we saw in Example 6.12 is 4-permutable but not modular. Thus we conclude that P_* does not imply M. Below we shall present the result of E. T. Schmidt that regularity does not imply permutability, but first we shall discuss the implications of uniformity. Examples showing C $\nrightarrow$ U and that U $\nrightarrow$ C are given in Theorem 4.2 of (Walter Taylor, 1974).

Perhaps the most important remaining problem in this area is

(6.10.2) *Does uniformity imply permutability?*

As we noted earlier, by Theorem 11.90, congruence modularity together with nilpotence implies congruence uniformity. For locally finite varieties we shall prove uniformity does imply permutability. In fact we shall show that a locally finite variety in U is in C, and we know C $\rightarrow$ P.

THEOREM 6.122 (David M. Clark and Peter H. Krauss 1976). *Let $\mathcal{V}$ be a variety and let $\mathcal{V}_{\text{fin}}$ be its finite members.*

(i) *If $\mathcal{V}_{\text{fin}}$ is congruence uniform, then $\mathcal{V}_{\text{fin}}$ is congruence coherent.*
(ii) *If $\mathbf{F}_{\mathcal{V}}(3)$ is finite and $\mathcal{V}_{\text{fin}}$ is congruence coherent, then $\mathcal{V}$ is congruence coherent.*

Proof. For part (i), let $\mathbf{A} \in \mathcal{V}_{\text{fin}}$, let $\mathbf{B}$ be a subalgebra of $\mathbf{A}$, and let $\theta \in \operatorname{Con} \mathbf{A}$ such that some θ-class C is contained in B. We have to show that B is a union of θ-classes. So let $b \in B$. Then $b/\theta|_B = (b/\theta) \cap (B \times B) \subseteq b/\theta$. Now by the uniformity of $\mathcal{V}_{\text{fin}}$ we know that both $\mathbf{A}$ and $\mathbf{B}$ have uniform congruences. By the first we see that $|C| = |b/\theta|$ and by the second that $|C| = |b/\theta|_B|$. This gives

$$|C| = |b/\theta|_B| \leq |b/\theta| = |C|.$$

Hence $|b/\theta|_B| = |b/\theta|$. But $b/\theta|_B \subseteq b/\theta$, while these sets are finite. So $b/\theta|_B = b/\theta$. So we conclude that B is a union of θ-classes. Hence $\mathcal{V}_{\text{fin}}$ is congruence coherent.

For part (ii), let $\mathbf{A} \in \mathcal{V}$, let $\mathbf{B}$ be a subalgebra of $\mathbf{A}$, and let $\theta \in \operatorname{Con} \mathbf{A}$ such that some θ-class C is contained in B. It suffices to show that if $a \in A$ and $b \in B$ and $a \equiv b \mod \theta$, then $a \in B$. Pick $c \in C$ and let $\mathbf{A}'$ be the subalgebra of $\mathbf{A}$ generated by $\{a, b, c\}$. Let $\mathbf{B}'$ be the subalgebra with universe $A' \cap B$ and let $\theta' = \theta|_{A'}$. Since $\mathbf{F}_{\mathcal{V}}(3)$ is finite, so is the algebra $\mathbf{A}'$ as well as its subalgebra $\mathbf{B}'$. Notice that $C \cap A'$ is a θ'-class and $C \cap A' = C \cap B'$. By congruence coherence of $\mathbf{A}'$ we see that $a \equiv b \mod \theta'$. Thus, $a \in B$, as desired. ∎

In §5.2 in Volume I, we observed a connection between the direct factorizations of an algebra $\mathbf{A}$ and its congruence lattice. In particular, for $\theta, \varphi \in \operatorname{Con} \mathbf{A}$ we took $\theta \times \varphi$ to be $\theta \wedge \varphi$ provided $\theta \circ \varphi = 1_A$. Then $\mathbf{A}/(\theta \times \varphi) \cong \mathbf{A}/\theta \times \mathbf{A}/\varphi$ by the natural map. Congruence permutability will allow us the replace the proviso with $\theta \vee \varphi = 1_A$. Thus congruence permutability is a key for unlocking an understanding of direct products and of directly indecomposable algebras. We exploit this in §11.9 in Volume III.

Now we give the result of E. T. Schmidt that regularity does not imply permutability. We extend this result by showing that weak uniformity of congruences does not imply permutability. Perhaps, as noted above, the most important remaining problem in this area is whether uniformity implies permutability. As we noted earlier, by Theorem 11.90 congruence modularity together with nilpotence implies congruence uniformity. If the converse of this were true (and we already know uniformity implies modularity) then, using Corollary 11.81, congruence uniformity would imply permutability. We do know that congruence uniformity implies permutability for locally finite varieties.

THEOREM 6.123 (E. T. Schmidt 1970). *There is a variety which is weakly congruence uniform (and hence regular) but not permutable.*

Proof. We take $\mathcal{V}$ to be the variety generated by two four-element algebras $\mathbf{A}_0$ and $\mathbf{A}_1$. Each $\mathbf{A}_i$ is a lattice (as depicted below) with six unary operations, μ_0, μ_1, λ_0, λ_1, c_0, and c_1 and four constants 0, 1, a_0, and a_1. We define the lattice structure of $\mathbf{A}_0$ by the figure:

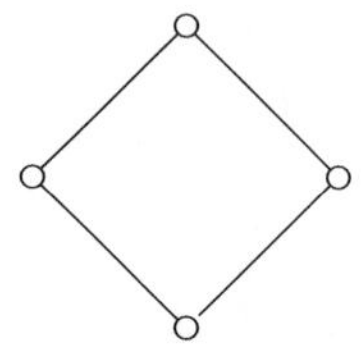

The unary operations are defined by

$$\mu_0(x) = x$$
$$\mu_1(x) = \begin{cases} 1 & \text{if } x = 1 \\ a_0 & \text{otherwise} \end{cases}$$
$$\lambda_0(x) = x \vee a_0$$
$$\lambda_1(x) = \begin{cases} 0 & \text{if } x = a_0 \\ x & \text{otherwise} \end{cases}$$
$$c_1(x) = \neg x$$
$$c_0(x) = (\neg x) \vee a_0.$$

Here $\neg x$ denote the complement of x. For the purposes of this proof only, we define the *dual* of a term (or formula) to be the usual lattice theoretic dual term except that we also interchange the subscripts 0 and 1 of a_0 and a_1 and the six unary operations. Now $\mathbf{A}_1$ is the lattice diagrammed in the figure below.

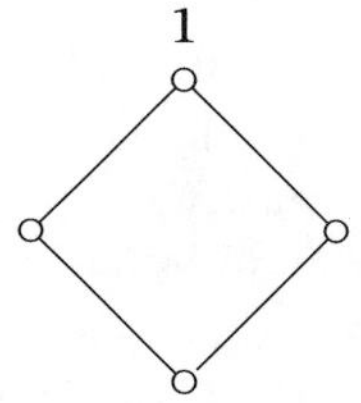

The unary operations are defined by duality as described above. Thus, for example, in the formula in $\mathbf{A}_1$ we have

$$\lambda_0(x) = \begin{cases} 1 & \text{if } x = a_1 \\ x & \text{otherwise} \end{cases}.$$

Obviously, the variety $\mathcal{V}$ defined in this way is self-dual: its set of identities is closed under the formation of dual equations.

In order to see that $\mathcal{V}$ is not congruence permutable, we leave it to the reader to check that the sublattice of $\mathbf{A}_0 \times \mathbf{A}_1$ given in Figure 6.20 is in fact a subalgebra. Now the failure of permutability is evident from the fact that this subuniverse contains $\langle 0, a_0 \rangle$, $\langle a_0, a_0 \rangle$, and $\langle a_0, 1 \rangle$ but not $\langle 0, 1 \rangle$, and thus the two coordinate projections do not permute.

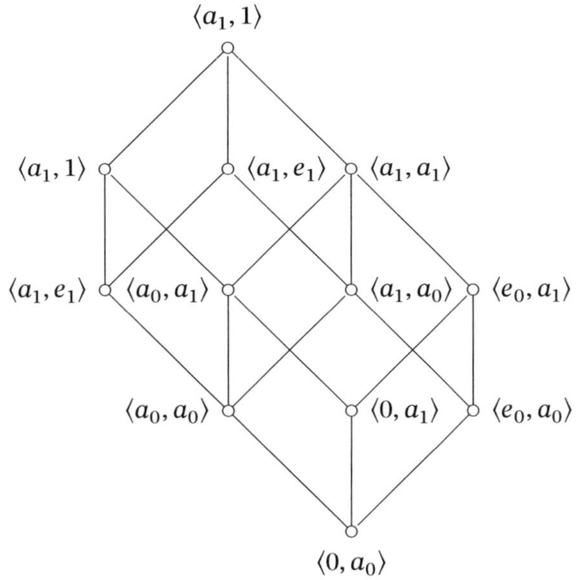

FIGURE 6.20. A Subalgebra of $\mathbf{A}_0 \times \mathbf{A}_1$

By Jónsson's Lemma, Lemma 8.57, $\mathcal{V}$ is residually finite. Therefore, by Exercise 6.58.14 weak uniformity of $\mathcal{V}$ will follow from the regularity of $\mathcal{V}$. Now to prove regularity, we will establish condition (iii) of Theorem 6.48. That is, we will describe ternary $\mathcal{V}$-terms g_i, for $1 \leq i \leq 4$, such that, for any $x, y \in \mathbf{A} \in \mathcal{V}$, $x = y$ if and only if $g_i(x, y, z) = z$ for all $z \in A$. We will in fact continue to use our notion of duality; for this reason we will prefer to denote our four $\mathcal{V}$-terms by g_i and g_i' for $i = 0, 1$, giving explicit definitions only for $i = 0$, and allowing duality to take care of the case $i = 1$. We proceed by first defining some auxiliary terms.

$$\begin{aligned}
t_0(x, y) &= [c_0(x \vee a_0) \wedge (y \vee a_0)] \vee [(x \vee a_0) \wedge c_0(y \vee a_0)]\\
s_0(x, y) &= \mu_0(t_0(x, y))\\
\hat{z} &= a_1 \wedge (z \vee a_0)\\
q_0(x, y, z) &= s_0(x, y) \vee \hat{z}\\
r_0(x, y, z) &= s_0(x, y) \wedge \hat{z}\\
g_0(x, y, z) &= z \vee (q_0(x, y, z) \wedge c_1(\hat{z}))\\
g_0'(x, y, z) &= z \wedge c_0(r_0(x, y, z))
\end{aligned}$$

Before proving that the terms g_0, g_0', g_1, and g_1' are as required, we first leave it to the reader to verify that $\mathbf{A}_0$ satisfies the following equations and their duals.

$$\begin{aligned}
c_1(x) &\leq a_1 & x \vee c_1(x) &\approx x \vee a_1\\
c_1(a_1) &\approx 0 & x \wedge c_1(x) &\approx 0\\
\lambda_0(\mu_0(x)) &\geq x & \mu_0(x) &\leq a_1\\
a_0 \approx \mu_0(a_0) &\approx \lambda_0(a_0) & \mu_0(x \vee a_0) &\geq a_0
\end{aligned}$$

and thus these equations and their duals hold on $\mathcal{V}$. It is now easy to verify the following identities of $\mathcal{V}$.

$$\begin{aligned}
t_0(x,y) &\geq a_0\\
s_0(x,y) &\geq a_0\\
r_0(x,y,z) &\geq a_0\\
t_0(x,x) &\approx a_0\\
s_0(x,x) &\approx a_0\\
q_0(x,x,z) &\approx a_0 \vee \hat{z} \approx \hat{z}\\
r_0(x,x,z) &\approx a_0 \wedge \hat{z} \approx a_0\\
g_0(x,x,z) &\approx z \vee (\hat{z} \wedge c_1(\hat{z})) \approx z\\
g_0'(x,x,z) &\approx z \wedge c_0(a_0) \approx z
\end{aligned}$$

Thus, if $x = y$ we have (by duality)

$$g_0 = g_0' = g_1 = g_1' = z. \tag{6.10.3}$$

To complete the proof we will assume (6.10.3) and use the identities of $\mathcal{V}$ to prove that $x = y$. We first deduce that $q_0(x,y,z) = \hat{z}$ and dually, as follows.

$$\begin{aligned}
z \wedge a_1 &= g_0(x,y,z) \wedge a_1\\
&= [(q_0(x,y,z) \wedge c_1(\hat{z})) \vee z] \wedge a_1\\
&= (q_0(x,y,z) \wedge c_1(\hat{z})) \vee (z \wedge a_1)
\end{aligned}$$

In the last step we have used the distributivity of $\mathcal{V}$ and the fact that

$$q_0(x,y,z) \wedge c_1(\hat{z}) \leq a_1.$$

And now, continuing,

$$\begin{aligned}
\hat{z} &= (z \wedge a_1) \vee a_0\\
&= (q_0(x,y,z) \wedge c_1(\hat{z})) \vee (z \wedge a_1) \vee a_0\\
&= (q_0(x,y,z) \wedge c_1(\hat{z})) \vee \hat{z}\\
&= q_0(x,y,z) \wedge (c_1(\hat{z}) \vee \hat{z})\\
&= q_0(x,y,z) \wedge a_1 = q_0(x,y,z)
\end{aligned}$$

In the last equation we have used the identity $\mu_0(x) \leq a_1$ and the definition of q_0. We next deduce that $r_0(x,y,z) = a_0$, and dually, as follows.

$$\begin{aligned}
a_0 \vee z = a_0 \vee g_0'(x,y,z) &= a_0 \vee (z \wedge c_0(r_0(x,y,z)))\\
&= (a_0 \vee z) \wedge c_0(r_0(x,y,z)),
\end{aligned}$$

since $a_0 \leq c_0(u)$ for all u. Now, since $a_0 \leq r_0(x,y,z)$, $a_0 = (z \vee a_0) \wedge a_0 \wedge r_0(x,y,z)$. Using one of our first identities $x \wedge c_0(x) \approx x \wedge a_0$, we have

$$\begin{aligned}
a_0 &= (z \vee a_0) \wedge c_0(r_0(x,y,z)) \wedge r_0(x,y,z)\\
&= (z \vee a_0) \wedge r_0(x,y,z) = r_0(x,y,z).
\end{aligned}$$

Now, recalling that $q_0(x,y,z)=\hat{z}$ from above, we have

$$\begin{aligned} s_0(x,y) &= s_0(x,y)\wedge(s_0(x,y)\vee\hat{z}) = s_0(x,y)\wedge q_0(x,y,z)\\ &= s_0(x,y)\wedge\hat{z} = r_0(x,y,z) = a_0\end{aligned}$$

By duality, we also have $s_1(x,y)=a_1$. Now observe that

$$\begin{aligned} a_0 &= \lambda_0(a_0) = \lambda_0(a_0)\wedge t_0(x,y)\\ &= \lambda_0(s_0(x,y))\wedge t_0(x,y) = \lambda_0(\mu_0(t_0(x,y)))\wedge t_0(x,y) = t_0(x,y).\end{aligned}$$

And so, by duality, we have $a_0 = t_0(x,y)$ and $a_1 = t_1(x,y)$. Since t_0 is merely the symmetric difference for the (Boolean) interval $\mathbf{I}[a_0,1]$, we deduce that $x\vee a_0 = y\vee a_0$. Likewise, by using t_1 on $\mathbf{I}[0,a_1]$, we deduce that $x\wedge a_1 = y\wedge a_1$, and hence also that $x\wedge a_0 = y\wedge a_0$. Now by distributivity it follows that $x=y$, see Theorem 2.51**v**. This completes the proof. ■

Near Unanimity Terms and the Baker-Pixley Theorem

Let $\mathcal{V}$ be a variety. On page 76 we said that $\mathcal{V}$ has a **near unanimity term** if, for some n, there is a term $t(x_0,\dots,x_{n-1})$ in the signature of $\mathcal{V}$ satisfying the equations

$$t(y,x,x,\dots,x)\approx t(x,y,x,\dots,x)\approx\cdots\approx t(x,x,x\dots,x,y)\approx x. \tag{6.10.4}$$

LEMMA 6.124. *A variety $\mathcal{V}$ has an n-ary near unanimity term if and only if the subalgebra of $\mathbf{F}_{\mathcal{V}}(x,y)^n$ generated by*

$$\langle x,\dots,x,y\rangle,\langle x,\dots,x,y,x\rangle,\dots,\langle y,x,\dots,x\rangle$$

contains $\langle x,x,\dots,x\rangle$.

Proof. If we apply an n-ary near unanimity term t to the n-tuples above we get $\langle x,\dots,x\rangle$. Conversely, if $\langle x,\dots,x\rangle$ is in the subalgebra generated by the above n-tuples, then there is a term t which, when applied to these tuples, gives $\langle x,\dots,x\rangle$. Using that x and y are free generators it follows that this t is a near unanimity term for $\mathcal{V}$. ■

Next we prove the result of Aleit Mitschke (1978) that a variety with a near unanimity term is congruence distributive. The proof we give, which does not involve nesting of terms, is an unpublished result H. P. Gumm.

THEOREM 6.125 (Aleit Mitschke 1978). *If a variety $\mathcal{V}$ has a near unanimity term then it is congruence distributive, but not conversely.*

Proof. Let $t(x_1,\dots,x_n)$ be a near unanimity term for $\mathcal{V}$ and, for $0\le i\le(n-2)$, define terms $d_0,\dots,d_{2n-3}$ as follows:

$$\begin{aligned} d_{2i} &= t(z,\dots,z,y,x,\dots x) && i\ z\text{'s}\\ d_{2i+1} &= t(z,\dots,z,z,x,\dots x) && (i+1)\ z\text{'s}\end{aligned}$$

It is straightforward to verify that the d_j's satisfy the equations (Δ_{2n-3}) of Theorem 6.1, and so by that theorem $\mathcal{V}$ is congruence distributive.

For an example of a congruence distributive variety without a near unanimity term, let $A = \{0, 1, 2\}$ and $\mathbf{A}$ have a single ternary operation given by

z	0	1	2
$h(0,0,z)$	0	1	1
$h(0,1,z)$	0	0	0
$h(0,2,z)$	0	0	0
$h(1,0,z)$	1	1	1
$h(1,1,z)$	1	1	1
$h(1,2,z)$	1	1	1
$h(2,0,z)$	2	2	2
$h(2,1,z)$	2	2	2
$h(2,2,z)$	1	1	2

Let $\mathcal{V}$ be the variety generated by $\mathbf{A}$. The reader can verify that $d_0(x, y, z) = x$, $d_3(x, y, z) = z$ and

$$d_1(x, y, z) = h(x, h(x, y, z), z), \quad d_2(x, y, z) = h(z, y, x)$$

are Jónsson terms, showing $\mathcal{V}$ is congruence distributive.

We use the above lemma to show that $\mathcal{V}$ does not have a near unanimity term. First we note that $\mathbf{A}$ is the two generated free algebra over $\mathcal{V}$ with generators 0 and 2. This is just a matter of showing that if $a, b \in A$, then $0 \mapsto a$, $2 \mapsto b$ can be extended to homomorphism. To see that $\langle 0, 0, \ldots, 0\rangle$ is not in the subalgebra generated by $\langle 0, 0, \ldots, 2\rangle, \ldots, \langle 2, 0, \ldots, 0\rangle$ it suffices to show $A^n - \{\langle 0, 0, \ldots, 0\rangle\}$ is a subuniverse of $\mathbf{A}^n$. This in turn follows from the fact that $h(a, b, c) = 0$ implies $a = 0$, which is obvious from the table defining h. ■

Another example of a congruence distributive variety that does not have a near unanimity term is the variety of implication algebras, which are defined in Exercise 6.16.5 on page 19.

The next theorem comes from (Kirby A. Baker and Alden F. Pixley, 1975). The equivalence of parts (i), (ii) and (iii) was proved earlier by A. P. Huhn (1972); see also (A. P. Huhn, 1983). A *partial operation* on A is a partial function from A^m to A, for some m. A *finite partial operation* on A is a partial operation whose domain is finite.

THEOREM 6.126. *Let $\mathcal{V}$ be a variety and n be a natural number with $2 \le n$. The following are equivalent.*

(i) *$\mathcal{V}$ realizes an $n+1$-ary near unanimity term.*
(ii) *For any $\mathbf{A} \in \mathcal{V}$, any natural number $m \ge n$, any $a_0 \ldots, a_{m-1} \in A$, and any $\theta_0 \ldots, \theta_{m-1} \in \operatorname{Con} \mathbf{A}$, if the m congruences*

$$(*) \qquad \begin{aligned} x &\equiv a_0 \pmod{\theta_0} \\ x &\equiv a_1 \pmod{\theta_1} \\ &\vdots \\ x &\equiv a_{m-1} \pmod{\theta_{m-1}} \end{aligned}$$

are solvable n at a time, then they are solvable simultaneously.

(iii) *For all* $\mathbf{A} \in \mathcal{V}$ *and all* $\varphi, \theta_0, \ldots, \theta_{n-1} \in \operatorname{Con} \mathbf{A}$,

$$\varphi \circ \Big(\bigwedge_{i<n} \theta_i\Big) = \bigcap_{i<n} \Big(\varphi \circ \bigwedge_{j \neq i} \theta_j\Big) \tag{$**$}$$

(iv) *For any* $\mathbf{A} \in \mathcal{V}$*, a finite partial operation* f *has an interpolating term operation of* $\mathbf{A}$ *if and only if* f *preserves all subuniverses of* $\mathbf{A}^n$.

Proof. To see (i) $\Rightarrow$ (ii) first assume $m = n + 1$. (If $m = n$, then (ii) clearly holds.) The hypothesis of (ii) implies there are elements $b_i \in A$, $i = 0, \ldots, n$ such that $b_i \ \theta_j \ a_j$ for $j \neq i$. Let $c = t(b_0, \ldots, b_n)$, where t is the $(n+1)$-ary near unanimity term given by (a). Then

$$c = t(b_0, b_1, \ldots, b_n) \equiv t(b_0, a_0, \ldots, a_0) = a_0 \pmod{\theta_0}.$$

Similar calculations show $c \ \theta_i \ a_i$, showing $x = c$ is a solution of $(*)$. Since an $(n+1)$-ary near unanimity term is a k-ary near unanimity term for $k \geq n + 1$, we can repeat this process to prove $(*)$ can be solved for any $m > n$.

For (ii) $\Rightarrow$ (iii), suppose $\mathbf{A} \in \mathcal{V}$, and $\varphi, \theta_0, \ldots, \theta_{n-1} \in \operatorname{Con} \mathbf{A}$. Suppose $\langle a, c \rangle \in \bigwedge_{i<n} (\varphi \circ \bigwedge_{j \neq i} \theta_j)$. Then, for $i < n$, there is a b_i with $a \ \varphi \ b_i \ (\bigwedge_{j \neq i} \theta_j) \ c$. Consider the $n + 1$ congruences

$$\begin{aligned} x &\equiv a \pmod{\varphi} \\ x &\equiv c \pmod{\theta_0} \\ &\vdots \\ x &\equiv c \pmod{\theta_{n-1}}. \end{aligned}$$

Note $x = c$ is a solution to all but the first of these and $x = b_i$ is a solution to all but $x \equiv c \pmod{\theta_i}$. By (b) there is an element b such that $x = b$ solves all of them. So $a \ \varphi \ b \ (\bigwedge_{i<n} \theta_i) \ c$, showing $\langle a, c \rangle$ is in the left side of $(**)$. The opposite inclusion for $(**)$ is obvious.

Next we show that (iii) $\Rightarrow$ (i). Let $\mathbf{A} = \mathbf{F}_{\mathcal{V}}(x_0, x_1, \ldots, x_n)$. For $i = 0, \ldots, n-1$ let θ_i be the congruence on $\mathbf{A}$ generated by identifying x_j and x_k for all j, k such that $i \notin \{j, k\}$, and let φ be the congruence identifying $x_0, \ldots, x_{n-1}$. Then, since $x_0 \ \varphi \ x_i \ \bigwedge_{j \neq i} \theta_j \ x_n$ for $i < n$, we have $\langle x_0, x_n \rangle \in \varphi \circ \bigwedge_{j \neq i} \theta_j$, for $i < n$. So by (c) $\langle x_0, x_n \rangle \in \varphi \circ \bigwedge_{i<n} \theta_i$. Hence there is a term $t(x_0, \ldots, x_n)$ (as an element of $\mathbf{F}_{\mathcal{V}}(x_0, x_1, \ldots, x_n)$) such that $x_0 \ \varphi \ t \ (\bigwedge_{i<n} \theta_i) \ x_n$. So $x_0 \ \varphi \ t(x_0, \ldots, x_n) \ \varphi \ t(x_0, \ldots, x_0, x_n)$. Since φ is trivial on the subalgebra generated by x_0 and x_n, Theorem 6.9 implies $x_0 = t(x_0, \ldots, x_0, x_n)$ and hence $\mathcal{V}$ satisfies $t(x, \ldots, x, y) \approx x$. Since for each $i < n$ we have $x_n \ \theta_i \ t(x_0, \ldots, x_n) \ \theta_i \ t(x_n, \ldots, x_n, x_i, x_n, \ldots, x_n)$, we can prove that the other equations of (6.10.4) also hold in $\mathcal{V}$.

For the implication (i) $\Rightarrow$ (iv), let $\mathbf{A} \in \mathcal{V}$ and let f be a finite partial operation which preserves all subuniverses of $\mathbf{A}^n$. Let m be the arity of f and let r be the size of the domain of f. First consider the case $r = n$. Let M be an $m \times r$ matrix $(b_{i,j})$ whose columns are the elements of the domain of f and let $\mathbf{S}$ be the subalgebra of $\mathbf{A}^n$ generated by the rows of M. Now applying f to the rows gives

$$\begin{aligned} f(\langle b_{0,0}, \ldots, b_{0,n-1} \rangle, \ldots, \langle b_{m-1,0}, \ldots, b_{m-1,n-1} \rangle) \\ = \langle f(b_{0,0}, \ldots, b_{m-1,0}), \ldots, f(b_{0,n-1}, \ldots, b_{m-1,n-1} \rangle. \end{aligned} \tag{6.10.5}$$

Since f preserves subalgebras of $\mathbf{A}^n$, the right side of the above equation lies in $\mathbf{S}$. Consequently there is a term $s(x_0,\ldots,x_{m-1})$ witnessing this. So s interpolates f. This argument can by used for the case $r < n$ by simply repeating some of the columns.

Now suppose $r = n+1$. Let M be the $m\times r$ matrix as above. Let $M^{(j)}$ be the matrix obtained from M be deleting the jth column. By the previous case applied to $M^{(j)}$, we get a term s_j agreeing with f at all points in its domain except the jth column. Now $t(s_0(x_0,\ldots,x_{m-1}),\ldots,s_n(x_0,\ldots,x_{m-1}))$ interpolates f, as the reader can show. Since an $n+1$-ary near unanimity term is a k-ary near unanimity term for each $k \geq n+1$, the above argument can be repeated for larger r.

Finally to see (iv)$\Rightarrow$(i), we apply (iv) with $\mathbf{A} = \mathbf{F}_{\mathcal{V}}(x,y)$. Let f be the $n+1$-ary partial operation whose domain consists of the $n+1$ elements of $\mathbf{A}^{n+1}$ $\langle x,\ldots,x,y,x,\ldots x\rangle$ with y in the ith place, $i = 0,\ldots,n$, and the value of f at each of these is x. Let B be a subuniverse of $\mathbf{A}^n$ and let $\langle b_{i,0},\ldots,b_{i,n-1}\rangle \in B$, $i = 0,\ldots n$ and, as above, let M be the $n+1\times n$ matrix with these as rows. By assumption, if

$$
\begin{aligned}
\text{(6.10.6)}\qquad & f(\langle b_{0,0},\ldots,b_{0,n-1}\rangle,\ldots,\langle b_{n,0},\ldots,b_{n,n-1}\rangle) \\
&\quad = \langle f(b_{0,0},\ldots,b_{n,0}),\ldots,f(b_{0,n-1},\ldots,b_{n,n-1})
\end{aligned}
$$

is defined, then its value is in B. For the above to be defined we need that each column is in the domain of f. So M has only x's and y's and each column has exactly one y. So there are exactly n y's in M. But there are $n+1$ rows, so there must be a row of only x's, showing $\langle x,\ldots,x\rangle$, which is the value of f applied to the rows, is in B. So by (d) f can be extended to a term operation. Now by Lemma 6.124, $\mathcal{V}$ has a near unanimity term of arity $n+1$. ∎

COROLLARY 6.127 (Kirby A. Baker and Alden F. Pixley 1975). *Let* $\mathbf{A}$ *be a finite algebra that has an* $n+1$*-ary near unanimity term. Then, an operation* f *is a term operation if and only if every subalgebra of* $\mathbf{A}^n$ *is preserved by* f.

Proof. Since A is finite, we can regard f as a finite partial function. So this corollary follows from the theorem. ∎

There are several other conditions equivalent to the existence of a near unanimity term, as well as several consequences; see §3,2 of (Kalle Kaarli and Alden F. Pixley, 2001).

Exercises 6.128

1. Show that the method of proof of Theorem 6.120, if applied to appropriate congruences α, β, and γ on $\mathbf{F}_{\mathcal{V}}(x,y,z,w)$, yields precisely the Hagemann formulas occurring in the proof of Theorem 6.119.

2. (A. F. Bravtsev, B. Csákány) Let A be an eight element cube with one edge direction stipulated as "vertical," and let $\mathbf{G}$ be the eight element group consisting of all rigid motions of A which map each vertical face into a vertical face. ($\mathbf{G}$ is obviously isomorphic to the dihedral group $\mathbf{D}_4$.) Define $\mathbf{A}$ to be the algebra with universe A and with each $g \in G$ acting as a unary operation,

i.e., a **G**-set, as defined in Chapter 3. Prove that **Con A** is not modular, but that every subalgebra of A has regular congruences.

3. (B. Csákány) If **A** is an algebra with regular but not modular congruences, then **A** has at least eight elements.

4. Show that P_* → NL does not hold for individual algebras.

5. We showed the 3-element 0, 1-lattice on $\{0, a, 1\}$ with a unary operation $f(0) = 1$, $f(1) = 0$, and $f(a) = a$ satisfies P_* but does not have a compatible linear order, witnessing NL $\nrightarrow$ P_*. On the other hand, Theorem 6.68 proves that it does have a compatible order that is not an antichain. Find that order.

6. Give a *local* proof of the varietal implication R → NL. This is of interest, since we still do not know whether there exists a local proof of R → P_*.

7. (J. Hagemann (1973b), Corollary 4) Show that a variety $\mathcal{V}$ is k-permutable for some k if and only if every compatible quasiorder on any member of $\mathcal{V}$ is a congruence. Hint: look at the proof of Theorem 6.68.

8. Let **A** be the three element algebra given in the proof of Theorem 6.125. As was shown, **A** generates a congruence distributive variety that does not have a near unanimity term. Show that **A** does not have a cube term. Thus D does not imply CT and therefore M does not imply CT.

9. The variety of groups is in K_2 but not in K_3.

10. The variety of groups of exponent p (a prime) is in K_p but not in K_{p+1}.
Hint: Let $\mathcal{V}$ be a group of exponent p, which of course we can viewed as a vector space over the prime field of order p. If $\mathcal{V}$ has finite dimension n, then $|\mathcal{V}| = p^n$ and proper subalgebras (subspaces) have at most p^{n-1} elements. So the size of the set union of p subspaces is $\leq p \cdot p^{n-1} = p^n$. But for equality to hold the subspaces would have to be disjoint, which is not possible. If $\mathcal{V}$ is infinite and is the set union of p subspaces, we can extend each of these subspaces to a hyperspace (a subspace of co-dimension 1). The intersection of these hyperspaces has finite co-dimension. We can mod out by this intersection and reduce to the finite dimensional case.

On the other hand, if $\mathcal{V}$ has dimension 2, then $|\mathcal{V}| = p^2$ and $\mathcal{V}$ has $p + 1$ one-dimensional subspaces. Counting the elements in the union gives $(p + 1)(p - 1) + 1 = p^2$ elements. So this variety is not in K_{p+1}.

11. The variety of lattices is in D but not in K_2. Thus, in fact, we see from the previous two exercises that none of our varietal properties implies K_2, except of course K_n for $n \geq 2$.

12. (S. Bulman-Fleming and W. Taylor, 1976) $\mathcal{V} \in \mathrm{K}_n$ if and only if the $\mathcal{V}$-free algebra on n generators is not the union of n proper subuniverses.

13. (S. Bulman-Fleming and W. Taylor, 1976) See Exercise 2 on page 18 for the definition of a quasiprimal algebra. If $\mathcal{V} = \boldsymbol{HSP}\mathbf{A}$ for $\mathbf{A}$ a finite quasiprimal algebra, then the relation $\mathcal{V} \in \mathrm{K}_n$ depends on both n and on the subalgebra structure of $\mathbf{A}$. For instance, if $|A| = n + 1$ and the proper subuniverses of $\mathbf{A}$ are precisely the singletons, then $\mathcal{V} \in \mathrm{K}_n - \mathrm{K}_{n+1}$.

14. If $\mathbf{A}$ is primal (quasiprimal with all elements as constants), then $\mathcal{V} = \boldsymbol{HSP}\mathbf{A} \in \mathrm{K}_{m-1} - \mathrm{K}_m$ for $m = \binom{k}{2}$, where $k = |A|$. Consequently, if $\mathbf{A}$ is *any* finite algebra with $|A| = k$ then $\mathcal{V} \notin \mathrm{K}_m$.

15. If $\mathbf{F}_{\mathcal{V}}(2)$ is finite and $\mathcal{V}$ is nontrivial, then $\mathcal{V} \notin \mathrm{K}_n$ for $n = |\mathbf{F}_{\mathcal{V}}(2)|$.

16. Prove that the 12 element subset of $\mathbf{A}_0 \times \mathbf{A}_1$ given in Figure 6.20 is in fact a subuniverse.

17. For the Schmidt variety defined in the proof of Theorem 6.123, exhibit terms f_i of the sort required for condition (iv) of Theorem 6.48.

18. We have seen $\mathrm{P}_* \to \mathrm{NL}$ and have given an example showing the opposite implication is false. Show that for an idempotent variety the two conditions are equivalent. Hint: Use Theorem 6.68.

19. Cube terms are defined in Example 6.66. Show that a near unanimity term is a cube term. Give an example of a variety having a cube term but which is not congruence distributive and hence does not have a near unanimity term.

20. Let $\mathbf{B}$ be the reduct of the two element lattice to the ternary operation $h(a, b, c) = a \wedge (b \vee c)$. Prove that $d_0(x, y, z) = x$, $d_4(x, y, z) = z$ and

$$d_1(x, y, z) = h(x, y, z) \quad d_2(x, y, z) = h(x, z, z) \quad d_3(x, y, z) = h(z, y, x)$$

are Jónsson terms for $\mathbf{B}$, and there is no shorter sequence; see (Kirby A. Baker, 1979). Prove also that $\mathbf{B}$ does not have a near unanimity term.

6.11. Hamiltonian, Semidegenerate, and Abelian Varieties

Abelian varieties were introduced in §4.13 and will be covered in more detail in Chapter 11 of Volume III. This section studies certain concepts which are closely related. The properties considered here have a somewhat different flavor from those that have come before. They are not defined by Mal'tsev conditions and, as we shall see, in conjunction with modularity, they define a very special class of varieties, namely the class of affine varieties. An algebra $\mathbf{A}$ is **affine** if, for some ring $\mathbf{R}$, it is polynomially equivalent to an $\mathbf{R}$-module with universe A. A variety is affine if all its members are.

Recall that an algebra $\mathbf{A}$ is **Abelian** if $Z(\mathbf{A}) = 1_{\mathbf{A}}$, where $Z(\mathbf{A})$, often denoted $\zeta_{\mathbf{A}}$, is the center of $\mathbf{A}$ as defined in §4.13. That definition says that $\mathbf{A}$ is Abelian if and only if it satisfies the implication

$$t(x,\overline{u}) \approx t(x,\overline{v}) \to t(y,\overline{u}) \approx t(y,\overline{v}) \tag{6.11.1}$$

for each term t in the signature of **A**. A variety $\mathcal{V}$ is Abelian if all of its algebras are Abelian, i.e., if each implication (6.11.1) follows from the identities of $\mathcal{V}$. The prototypical examples of Abelian varieties are module varieties and unary varieties. The reader can easily check (6.11.1) in these cases. In particular an affine variety is Abelian.

Such a collection of implications is very different from a Mal'tsev condition. Somewhat closer in spirit to the rest of this chapter is L. Klukovits's closely related notion of a Hamiltonian variety. An algebra **A** is called **Hamiltonian** if every subuniverse of **A** is a block of some congruence on **A**, and a variety $\mathcal{V}$ is called Hamiltonian if each algebra in $\mathcal{V}$ is Hamiltonian. This nomenclature honors W. R. Hamilton (1805-1865), who invented quaternions. The closely allied eight element quaternion group is Hamiltonian, i.e., each of its subgroups is normal. The following characterization of Hamiltonian varieties, which is due to L. Klukovits (1975), is not generally regarded as a true Mal'tsev condition, since it involves the universally quantified function variable f. For a discussion of these more general sorts of conditions, see §10.10 of Volume III.

THEOREM 6.129. *A variety $\mathcal{V}$ is Hamiltonian if and only if for every $\mathcal{V}$-term f there exists a $\mathcal{V}$-term h, such that the following equation is an identity of $\mathcal{V}$:*

$$f(x_0, x_2, \dots, x_n) \approx h(x_0, x_1, f(x_1, \dots, x_n)).$$

Proof. First, suppose that $\mathcal{V}$ is Hamiltonian, and consider the free algebra $\mathbf{F}_{\mathcal{V}}(x_0, x_1, \dots, x_n)$. The subuniverse U generated by x_0, x_1, and $f(x_1, \dots, x_n)$ must be a θ-block for some congruence θ. Since $x_0 \mathrel{\theta} x_1$, we have

$$f(x_0, x_2, \dots, x_n) \mathrel{\theta} f(x_1, x_2, \dots, x_n),$$

i.e., $f(x_0, x_2, \dots, x_n) \in U$. This means that

$$f(x_0, x_2, \dots, x_n) = h(x_0, x_1, f(x_1, \dots, x_n))$$

for some ternary term h. The desired equation now follows.

Conversely, let us suppose that the appropriate terms h exist for every f, and then prove that $\mathcal{V}$ is Hamiltonian. Let **B** be a subalgebra of an algebra $\mathbf{A} \in \mathcal{V}$, and consider the congruence θ generated on **A** by $Z = B^2$. Obviously B is contained in a single θ-block; to complete the proof, it will suffice to show that this θ-block contains nothing outside B. In other words, given $a \mathrel{\theta} b$ with $b \in B$, we must prove that $a \in B$. Now according to Theorem 6.18, we have

$$\begin{aligned} b &= f_1(b_1, b_1', \overline{e}) \\ f_1(b_1', b_1, \overline{e}) &= f_2(b_2, b_2', \overline{e}) \\ &\vdots \\ f_m(b_m', b_m, \overline{e}) &= a \end{aligned}$$

for some $\mathcal{V}$-terms f_i, and some $b_i, b_i' \in B$.

Now our hypothesis (on the existence of h) can be applied both to the term $f_1(x_1, \ldots, x_n)$ and to the term $f_1(x_2, x_1, x_3, \ldots, x_n)$ yielding the identities

$$f_1(u, x_2, \ldots, x_n) \approx h_1(u, x_1, f_1(x_1, \ldots, x_n))$$
$$f_1(x_1, w, x_3, \ldots, x_n) \approx k_1(w, x_2, f_1(x_1, \ldots, x_n))$$

for appropriate $\mathcal{V}$-terms h_1 and k_1. Thus we have

$$\begin{aligned} f_1(b_1', b_1, \overline{e}) &= k_1(b_1, b_1', h_1(b_1', b_1, f_1(b_1, b_1', \overline{e}))) \\ &= k_1(b_1, b_1', h_1(b_1', b_1, b)), \end{aligned}$$

and thus $f_1(b_1', b_1, \overline{e}) \in B$. Continuing in like manner, we inductively prove that $f_i(b_i', b_i, \overline{e}) \in B$ and thus $a \in B$. ■

THEOREM 6.130. *Every Hamiltonian variety is Abelian.*

Proof. In order to establish the implication (6.11.1) for a Hamiltonian variety, we take for each term t, another term h such that

$$t(y, \overline{u}) \approx h(y, x, t(x, \overline{u})).$$

The implication (6.11.1) is now evident. ■

This theorem has a local version. See Exercise 5 for the proof.

THEOREM 6.131. *If $\mathbf{A}^2$ is Hamiltonian then $\mathbf{A}$ is Abelian.* ■

An example showing the converse of Theorem 6.130 fails is given in (Emil Kiss and Matthew Valeriote, 1991). In fact their example is a *strongly Abelian* variety which is not Hamiltonian. (Strongly Abelian is defined in Exercise 3.)

The next corollary shows that every Abelian algebra in a modular variety is Hamiltonian. The assumption that $\mathcal{V}$ is congruence modular can be weakened to $\mathcal{V}$ having a weak difference term using Theorem 11.59 in Chapter 11.

COROLLARY 6.132. *Let $\mathcal{V}$ be congruence modular. Then $\mathcal{V}$ is Abelian if and only if it is Hamiltonian.*

Proof. If $\mathcal{V}$ is Hamiltonian then it is Abelian by Theorem 6.130. On the other hand if $\mathcal{V}$ is modular and Abelian then it is affine by Theorem 6.38, and from this it easily follows that $\mathcal{V}$ is Hamiltonian. ■

We end this section with a sort of "anti-Hamiltonian" property of varieties. A variety $\mathcal{V}$ is called **semidegenerate** if no algebra in $\mathcal{V}$ has a one-element subuniverse, other than, of course, the one-element algebra itself. (Semidegenerate algebras will play a role again in Volume III. See Exercise 15 on page 155 and Exercise 2 on page 257.)

THEOREM 6.133 (Béla Csákány 1976; J. Kollár 1979). *For a variety $\mathcal{V}$, the following conditions are equivalent:*

(i) *$\mathcal{V}$ is semidegenerate;*

(ii) *if* $\mathbf{A} \in \mathcal{V}$ *and* B *is a subuniverse of* $\mathbf{A}$ *with* $B \neq A$, *then* B *is not a block of any congruence on* $\mathbf{A}$;

(iii) *for each* $\mathbf{A} \in \mathcal{V}$, $1_{\mathbf{A}}$ *is a compact element of* **Con A**;

(iv) *for each* $\mathbf{A} \in \mathcal{V}$, $1_{\mathbf{A}}$ *is countably compact in* **Con A**;

(v) *for some* n, *there exist unary* $\mathcal{V}$-*terms* $f_1, \ldots, f_n$ *and* $g_1, \ldots, g_n$ *and ternary* $\mathcal{V}$-*terms* $p_1, \ldots, p_n$ *such that the following are identities of* $\mathcal{V}$.

$$\begin{aligned} x &\approx p_1(x, y, f_1(x)) \\ p_1(x, y, g_1(x)) &\approx p_2(x, y, f_2(x)) \\ &\vdots \\ p_n(x, y, g_n(x)) &\approx y \end{aligned}$$

Proof. We leave the easy equivalence (i) and (ii) to the reader. The rest of the proof will proceed (iii) $\to$ (iv) $\to$ (i) $\to$ (v) $\to$ (iii). The first of these implications is obvious. For (iv) $\to$ (i) let us assume that $\mathcal{V}$ is not semidegenerate, i.e., that some nontrivial $\mathbf{A} \in \mathcal{V}$ has a one-element subuniverse $\{e\}$. To show that (iv) fails for $\mathcal{V}$, we will show that $1_{\mathbf{B}}$ is not countably compact in **Con B**, where **B** is the "weak direct power" consisting of all sequences in A^{ω} which are equal to e in all but finitely many places. For each $i \in \omega$, we define $a\ \theta_i\ b$ if $a_j = b_j$ for all $j \geq i$. It is easy to see that each $\theta_i < 1$ but $\bigvee \theta_i = 1$, contradicting countable compactness.

For (i) $\to$ (v), we assume that $\mathcal{V}$ is semidegenerate, and consider the free algebra $\mathbf{F}_{\mathcal{V}}(x, y)$. Let **B** be the subalgebra generated by x, and θ the congruence generated by $Z = B^2$. Since $\mathbf{B}/\theta$ is a one-element subalgebra of $\mathbf{F}_{\mathcal{V}}(x, y)/\theta$, this latter algebra must itself have only one element, and thus $\langle x, y \rangle \in \theta$. According to Theorem 6.18, we have the following equations holding in $\mathbf{F}_{\mathcal{V}}(x, y)$, for some terms q_i, h_i, and k_i:

$$\begin{aligned} x &= q_1(x, y, h_1(x), k_1(x)) \\ q_1(x, y, k_1(x), h_1(x)) &= q_2(x, y, h_2(x), k_2(x)) \\ q_2(x, y, k_2(x), h_2(x)) &= q_3(x, y, h_3(x), k_3(x)) \\ &\vdots \\ q_m(x, y, k_m(x), h_m(x)) &= y. \end{aligned}$$

We now obtain the Mal'tsev condition of (v) if we define $n = 2m$, and let

$$\begin{aligned} p_{2i-1}(x, y, u) &= q_i(x, y, u, k_i(x)) \\ p_{2i}(x, y, w) &= q_i(x, y, k_i(x), w) \\ f_{2i-1} &= g_{2i} = h_i \\ g_{2i-1} &= f_{2i} = k_i \end{aligned}$$

for $i = 1, \ldots, m$. We leave the details to the reader.

Finally, for (v) $\to$ (iii), we will use the equations (v) to prove that, on any algebra $\mathbf{A} \in \mathcal{V}$, $1_{\mathbf{A}}$ is compact, i.e., finitely generated. We will in fact show that, for any $a \in A$, the congruence θ generated by the n pairs $\langle f_i(a), g_i(a) \rangle$, with f_i and g_i as in (v), is $1_{\mathbf{A}}$.

Thus we have, for arbitrary $b, c \in A$:

$$p_1(a, b, f_1(a)) = a = p_1(a, c, f_1(a))$$
$$p_1(a, b, g_1(a)) \mathrel{\theta} p_1(a, c, g_1(a))$$

By the equations (v), we have

$$p_2(a, b, f_2(a)) \mathrel{\theta} p_2(a, c, f_2(a)),$$

and hence,

$$p_2(a, b, g_2(a)) \mathrel{\theta} p_2(a, c, g_2(a)).$$

Continuing in a like manner, we finally arrive at

$$p_n(a, b, g_n(a)) \mathrel{\theta} p_n(a, c, g_n(a));$$

in other words, $\langle b, c\rangle \in \theta$. Thus $\theta = 1_{\mathbf{A}}$ and the proof is complete. ■

From condition (ii) of this theorem it is evident that the Hamiltonian property is completely incompatible with semidegeneracy, in that the only variety having both properties is the trivial variety defined by $x \approx y$. In the next theorem we prove something a bit stronger. For a related fact concerning congruence distributivity, see Exercise 22.

THEOREM 6.134. *No nontrivial Abelian variety is semidegenerate.*

Proof. By Exercise 1, a variety $\mathcal{V}$ is Abelian if and only if, for each $\mathbf{A} \in \mathcal{V}$, the diagonal subuniverse, $\{\langle a, a\rangle : a \in A\}$, of $\mathbf{A} \times \mathbf{A}$ is the block of some congruence of $\mathbf{A} \times \mathbf{A}$. This, together with the equivalence of (i) and (ii) of Theorem 6.133, prove the result.

We can also prove the theorem by using an argument almost exactly like the one which ended our last proof. Suppose $\mathcal{V}$ is Abelian and semidegenerate, we will prove $\mathcal{V}$ trivial. For b, c any elements of any algebra $\mathbf{A} \in \mathcal{V}$, we need to prove $b = c$. Taking any $a \in A$, and taking the terms p_i, f_i, and g_i as supplied by condition (v) of the previous theorem, we have

$$p_1(a, b, f_1(a)) = a = p_1(a, c, f_1(a)).$$

Since $\mathcal{V}$ is Abelian, we have

$$p_1(a, b, g_1(a)) = p_1(a, c, g_1(a)),$$

and hence, by the equations for p_i, f_i, and g_i, we have

$$p_2(a, b, f_2(a)) = p_2(a, c, f_2(a)).$$

Continuing in like manner, we finally obtain $b = c$. ■

One may easily observe that the variety of bounded lattices (i.e., lattices with 0 and 1 explicitly in the signature) is semidegenerate since in such a lattice with more than one element, $0 \neq 1$, and hence all subalgebras have at least two elements. A similar argument applies to rings with unit. A few more examples appear in the exercises. We conclude the main body of this chapter with Magari's Theorem. Our argument for (iv) $\rightarrow$ (i) in Theorem 6.133 comes from (J. Kollár, 1979); his argument turned out to repeat an important step from (Roberto Magari, 1969); thus we now present Magari's Theorem as a corollary to Theorem 6.133.

COROLLARY 6.135 (Roberto Magari 1969). *Every nontrivial variety contains a simple algebra.*

Proof. Let $\mathcal{V}$ be a nontrivial variety. If $\mathcal{V}$ is semidegenerate, then $\mathcal{V}$ contains a nontrivial algebra $\mathbf{A}$ such that $1_{\mathbf{A}}$ is compact. The Kuratowski-Zorn Lemma yields a maximal proper congruence θ. Clearly $\mathbf{A}/\theta$ is simple.

Otherwise $\mathcal{V}$ is not semidegenerate, and so has an algebra $\mathbf{A}$ with both a one-element subuniverse $\{e\}$ and a further element $a \neq e$. Let $\mathbf{B} = \operatorname{Sg}^{\mathbf{A}}(a, e)$. Then $\mathbf{A}/\operatorname{Cg}^{\mathbf{B}}(a, e)$ has only one element and so $\operatorname{Cg}^{\mathbf{B}}(a, e) = 1_{\mathbf{B}}$, which shows that $1_{\mathbf{B}}$ is compact, and thus as before, $\mathbf{B}$ has a simple quotient. ■

Exercises 6.136

1. Show that an algebra $\mathbf{A}$ is Abelian if and only if the diagonal subalgebra, $\{\langle a, a\rangle : a \in A\}$, of $\mathbf{A} \times \mathbf{A}$ is a block, i.e., congruence class, of some congruence of $\mathbf{A} \times \mathbf{A}$.

2. The Hamiltonian property of varieties is not definable by a Mal'tsev condition. Likewise the Abelian property.

3. A variety $\mathcal{V}$ is called **strongly Abelian** if and only if $\mathcal{V}$ satisfies the following implication for every $\mathcal{V}$-term t:

$$t(x, \overline{u}) \approx t(y, \overline{v}) \rightarrow t(x, \overline{w}) \approx t(y, \overline{w}).$$

Prove that if $\mathcal{V}$ is strongly Abelian, then $\mathcal{V}$ is Abelian, and provide an example showing that the converse is false.

4. Give an example of a Hamiltonian variety which is not congruence permutable.

5. Prove Theorem 6.131. That is, show that if $\mathbf{H}$ is an algebra and $\mathbf{H}^2$ is Hamiltonian then $\mathbf{H}$ is Abelian. Hint: Since $\mathbf{H}^2$ is Hamiltonian the diagonal subalgebra of $\mathbf{H}^2$, $\{\langle x, x\rangle : x \in H\}$, is the block of some congruence Δ on $\mathbf{H}^2$. Then use this congruence to show $\mathbf{H}$ is Abelian.

* 6. (Walter Taylor, 1982) Every k-permutable Abelian variety is affine. (This result is proved in §11.5; see Theorems 11.30 and 11.57 in Volume III.)

7. An algebra $\mathbf{A}$ is affine if and only if $\boldsymbol{HSPA}$ is congruence permutable and $\{\langle a, a\rangle : a \in A\}$ is a block of a congruence on $\mathbf{A} \times \mathbf{A}$.

Call a class of algebras κ-**categorical**, or **categorical in power** κ, if it contains, up to isomorphism, only one algebra with cardinality κ. κ-categorical varieties for κ infinite are characterized by Steven Givant (1979) and E. A. Palyutin (1975). Their results imply that if a variety $\mathcal{V}$ is κ-categorical for sufficiently large κ, then either $\mathcal{V}$ is affine or $\mathcal{V}$ is strongly Abelian (as defined in Exercise 3 above). The following seven exercises provide a description of some κ-categorical affine varieties, and along the way show some interesting examples of reducts of module varieties.

8. Show that if $\mathcal{V}$ is κ-categorical and affine then it is a reduct of a module variety.

9. Let $\mathbf{L}$ be a left ideal of a ring $\mathbf{R}$ with unit. The variety $\mathcal{V}_{\mathbf{R},\mathbf{L}}$ has ternary operations F_r for each $r \in R$, and unary operations G_α for each $\alpha \in L$. It is defined by the laws

$$\begin{aligned}
F_r(x,x,z) &\approx z \\
F_0(x,y,z) &\approx z \\
F_1(x,z,z) &\approx x \\
F_1(x,z,y) &\approx F_1(y,z,x) \\
F_{rs}(x,z,z) &\approx F_r(F_s(x,z,z),z,z) \\
F_{r+s}(x,y,z) &\approx F_1(F_r(x,y,z),z,F_s(x,y,z)) \\
F_1(x,y,F_1(z,v,w)) &\approx F_1(F_1(x,y,z),v,w) \\
F_r(x,y,z) &\approx F_1(F_1(F_r(x,u,u),u,F_{-r}(y,u,u)),u,z) \\
F_r(F_1(x,u,y),u,u) &\approx F_1(F_r(x,u,u),u,F_r(y,u,u)) \\
F_{1+\alpha}(x,y,z) &\approx F_1(G_\alpha(x),G_\alpha(y),z) \\
G_\alpha(G_\beta(x)) &\approx G_{\alpha+\beta+\alpha\beta}(x)
\end{aligned}$$

for all r, $\mathbf{R}$ and all α, $\beta \in L$. Prove that every algebra $\mathbf{A} \in \mathcal{V}_{\mathbf{R},\mathbf{L}}$ can be embedded in an algebra $\langle M, F_r, G_\alpha\rangle$ where $\mathbf{M}$ is an $\mathbf{R}$-module and, on $\mathbf{M}$, we define

$$\begin{aligned}
F_r(x,y,z) &= rx - ry + z \\
G_\alpha(x) &= (1+\alpha)x.
\end{aligned}$$

Conversely each such $\langle M, F_r, G_\alpha\rangle$ is in $\mathcal{V}_{\mathbf{R},\mathbf{L}}$. (Hint: if $\mathbf{A} \in \mathcal{V}_{\mathbf{R},\mathbf{L}}$, consider the free $\mathbf{R}$-module $\mathbf{F}$ generated by the set A and consider the smallest submodule $\mathbf{M}$ such that the map $a \mapsto a$ is a homomorphism $\mathbf{A} \to \mathbf{F}/\mathbf{M}$ for the operations F_r and G_α.) Moreover, in some cases, e.g., $\mathbf{R} = \mathbb{Z}$ and $\mathbf{L} = 2\mathbb{Z}$, there exists such an $\mathbf{A}$ which is not isomorphic to any $\langle M, F_r, G_\alpha\rangle$. (For further related material, see Exercise 10.63.19 on page 182 of Volume III.)

10. Conclude from the previous exercise that the correspondence

$$\begin{aligned}
F_r(x,y,z) &\mapsto rx - ry + z \\
G_\alpha(x) &\mapsto (1+\alpha)x
\end{aligned}$$

establishes an equivalence of $\mathcal{V}_{\mathbf{R},\mathbf{L}}$ with the reduct of the variety of all $\mathbf{R}$-modules to the linear expressions $\sum r_i x_i$ with $\sum (r_i - 1) \in L$.

11. Now assume that the ideal L has a finite orthogonal **R**-basis of idempotents $e_1, \ldots, e_k$. (This means that

$$e_i e_j = \begin{cases} e_i & \text{if } i = j \\ 0 & \text{otherwise,} \end{cases}$$

and that $\mathbf{L} = \mathbf{R}e_1 + \cdots + \mathbf{R}e_k$.) Prove that for every $\mathbf{A} \in \mathcal{V}_{\mathbf{R},\mathbf{L}}$ there exists $a \in A$ with $G_\alpha(a) = a$ for all $\alpha \in L$. Conclude that every $\mathbf{A} \in \mathcal{V}_{\mathbf{R},\mathbf{L}}$ is isomorphic to $\langle M, F_r, G_\alpha \rangle$ (defined in Exercise 9) for some **R**-module **M**.

12. We now let **R** be the ring of $n \times n$ matrices over a skew field **k**, and **L** its left ideal consisting of all matrices whose first m columns are zero. Prove that $\mathcal{V}_{\mathbf{R},\mathbf{L}}$ is categorical in power κ unless $\kappa = |k| \geq \aleph_0$. Prove, moreover, that if **k** is finite, then m, n and $|k|$ can be recovered from the equivalence type (see §4.12) of $\mathcal{V}_{\mathbf{R},\mathbf{L}}$. Specifically, this information can be recovered from the sequence $\langle \omega_i \rangle$ where ω_i is the size of the i-generated free algebra in $\mathcal{V}_{\mathbf{R},\mathbf{L}}$. (By the result of Steven Givant (1979) and E. A. Palyutin (1975), these varieties (with **k** finite) are the only affine varieties which are categorical in all powers.)

13. (I. Š. Aliev, 1966; John T. Baldwin and A. H. Lachlan, 1973) Let the variety $\mathcal{V}$ be defined by the following axioms in one binary operation:

$$xx \approx x \qquad\qquad xy \approx yx$$

$$x(yx) \approx y \approx (xy)x \qquad (ux)(yv) \approx (uy)(xv)$$

Prove that $\mathcal{V}$ is equivalent to $\mathcal{V}_{\mathbf{R},\mathbf{L}}$ for some **R** and **L**. [For a different approach to this variety, see Exercise 3.12.10 on page 127 of Volume I.]

14. Prove that if $\mathcal{V} = \mathcal{V}_{\mathbf{R},\mathbf{L}}$ for some **R**, **L**, then the varietal power $\mathcal{V}^{[k]}$ is equivalent to $\mathcal{V}_{\mathbf{R}',\mathbf{L}'}$ for some $\mathbf{R}'$ and $\mathbf{L}'$. Give an explicit description of $\mathbf{R}'$ and $\mathbf{L}'$. (One example of the varietal power $\mathcal{V}^{[k]}$ was discussed in Exercises 4.38.18 and 19 of Volume I. For the precise definition of $\mathcal{V}^{[k]}$ see §10.6 in Volume III.) See Corollary 10.97 in volume III.

15. For each varietal property P considered in this chapter, we may ask whether either of the following implications holds, where $\mathcal{V}^{[k]}$ denotes the varietal k^{th} power; see §10.6.

$$\mathrm{P}(\mathcal{V}) \to \mathrm{P}(\mathcal{V}^{[k]})$$

$$\mathrm{P}(\mathcal{V}^{[k]}) \to \mathrm{P}(\mathcal{V})$$

Investigate to what extent these hold. (For some information, see pages 22–24 of (O. C. Garcia and W. Taylor, 1984). See also e.g. Corollary 10.97, page 203, Volume III.)

16. If **A** is a finite quasi-primal algebra with no one-element subalgebras, then $V(\mathbf{A})$ is semidegenerate. Write out the terms for condition (v) of Theorem 6.133 in the special case that $\mathbf{A} = \langle A, T, f \rangle$, where T is the ternary discriminator and $f : A \to A$ has no fixed point.

17. If $\mathcal{V}$ is semidegenerate and congruence permutable (such as, e.g., the variety of Exercise 16), then (v) of Theorem 6.133 is satisfied with $n = 1$.

18. (Robert W. Quackenbush, 1982) If $m(x, y, z)$ is a majority term for a variety $\mathcal{V}$ and $a_1, \dots, a_k$ are constants obeying the equation

$$m(\cdots m(m(x, a_1, y), a_2, y), \dots, a_k, y) \approx y,$$

then $\mathcal{V}$ is semidegenerate. (Hint: get terms for condition (v) of Theorem 6.133.) Incidentally, these equations hold on any algebra with exactly k elements having a ternary majority term, if we treat each element as a constant.

19. (Robert W. Quackenbush, 1982) If condition (v) of Theorem 6.133 holds for $\mathcal{V}$, then for all $\mathbf{A} \in \mathcal{V}$, $1_{\mathbf{A}} \in \operatorname{Con} A$ is the join of n principal congruences. Prove that this condition fails for $\mathcal{V} = \boldsymbol{HSP}\mathbf{A}$, with $\mathbf{A} = \langle A, m, a_1, \dots, a_k \rangle$ where $A = \{a_1, \dots, a_k\}$, $k \geq 2n + 2$, m is a majority function, and if $x, y, z \in A$ are distinct, then $m(x, y, z) = a_1$. (Hint: first establish that **Con A** is the same as the congruence lattice of the meet semilattice with universe A diagrammed in the figure. Thus, by Exercise 17, some semidegenerate varieties are not permutable. Moreover, it is now evident that semidegeneracy is not definable by a strong Mal'tsev condition.

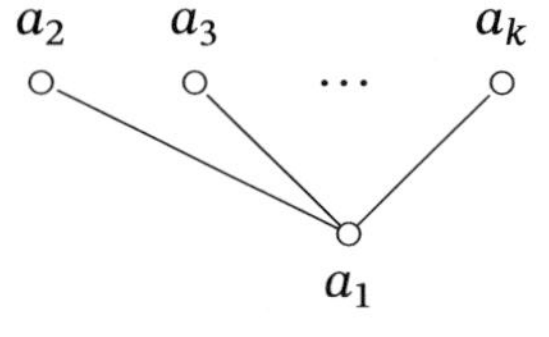

20. (Robert W. Quackenbush, 1982) Write a Mal'tsev condition for the following property of a variety $\mathcal{V}$: every compact congruence on every algebra in $\mathcal{V}$ is the join of n principal congruences (or fewer). Prove that E. T. Schmidt's variety of k-Boolean algebras given in Example 6.14 satisfies this condition for $n = k$ but not for $n = k - 1$.

21. (Béla Csákány, 1964) Consider the following property of a variety $\mathcal{V}$: if $\mathbf{A} \in \mathcal{V}$, $B_1, \dots, B_n$ are subuniverses of $\mathbf{A}$, and F is an n-ary operation of $\mathbf{A}$, then

$$\{F(b_1, \dots, b_n) : b_1 \in B_1, \dots, b_n \in B_n\}$$

is also a subuniverse. Prove that $\mathcal{V}$ satisfies this condition if and only if for every n and for all n-ary operations F and G of $\mathcal{V}$, there exists n-ary terms $\sigma_1, \dots, \sigma_n$ such that $\mathcal{V}$ satisfies the following equations.

$$F(G(x_{11}, \dots, x_{1n}), \dots, G(x_{n1}, \dots, x_{nn})) \\ \approx G(\sigma_1(x_{11}, \dots, x_{1n}), \dots, \sigma_n(x_{n1}, \dots, x_{nn}))$$

22. No nontrivial Abelian variety is congruence distributive. This is an easy corollary of Theorem 6.38 but here we suggest trying a direct proof: apply the Abelian implication (6.11.1) to the Jónsson identities.

23. Let $\mathcal{V}$ be a variety with a constant 0. Show that $\mathcal{V}$ is semidegenerate if and only if for some $n \geq 1$ there are binary terms $t_1, \ldots, t_n$ and constants $a_1, \ldots, a_n$ and $b_1, \ldots, b_n$ such that $\mathcal{V}$ satisfies

$$\begin{aligned} 0 &\approx t_1(x, a_1) \\ t_1(x, b_1) &\approx t_2(x, a_2) \\ &\vdots \\ t_n(x, b_n) &\approx x. \end{aligned}$$

24. Let $\mathbf{A} = \langle \{0, 1\}, m, ' \rangle$ be the algebra where m is the majority function and $'$ is complementation. Show that the variety generated by $\mathbf{A}$ is semidegenerate but has no constant and no unary term which satisfies $t(x) \approx t(y)$.

CHAPTER 7

Equational Logic

7.1. The Set Up

Formal systems of mathematical logic are provided with a means of expression and a means of proof. Equational logic is perhaps the simplest example that is still able to comprehend a considerable portion of mathematics that arises in practice. From one perspective, equational logic is a fragment of elementary (or first-order) logic. In this fragment the only formulas are equations between terms—this logic is provided with neither connectives nor quantifiers (apart from implicit universal quantifiers). In comparison with elementary logic, equational logic has a meager means of expression. As a consequence, many of the powerful methods of model theory that have been developed for elementary logic seem to have limited applicability in equational logic. On the other hand, since the truth of equations is preserved under the formation of homomorphic images, subalgebras, and direct products, the methods of algebra can be brought into play.

The Syntax of Equational Logic

A **signature** is a function which gives natural numbers as outputs. The inputs of a signature are called **operation symbols**. The outputs of a signature are called **ranks**. **Constant symbols** are those operation symbols of rank 0. In any particular investigation, we are free to choose a convenient signature. For example, for the equational logic of groups we might choose a signature that provides one two-place operation symbol (to denote the group product), a one-place operation symbol (to denote the formation of inverses), and one constant symbol (to denote the identity element). We have used the word "signature" here for the concept we called "similarity type" in Volume I. Since Volume I was originally published the word "type" has taken on a completely different meaning, so we have chosen to replace "similarity type" with "signature", a word used in this context by A. I. Mal'tsev.

Fix some signature. An **algebra** $\mathbf{A} = \langle A, F\rangle$ of the given signature is a system made up of a nonempty set A and a system F of operations of finite rank appropriate for the signature. That is, F is a function whose domain is the same as the domain of the signature and $F(Q)$ is an operation on A of the rank assigned by the signature to operation symbol Q. Ordinarily, we dispense with F and use $Q^{\mathbf{A}}$ to denote $F(Q)$. The set A is called the **universe of discourse** of the algebra $\mathbf{A}$. But we will be less formal and refer to it more simply as the **universe** of $\mathbf{A}$. The operations $Q^{\mathbf{A}}$, where Q is an operation symbol, are the **basic** or **fundamental** operations of $\mathbf{A}$.

EXAMPLE 7.1. Suppose our operation symbols are $+, \times, 0, 1$ and our algebra is $\mathbf{M} = \langle M_2(\mathbb{R}), F\rangle$ where $M_2(\mathbb{R})$ is the set of 2×2 matrices over the field of real numbers. Then

$F(+)$ is

$$+^{\mathbf{M}} : M_2(\mathbb{R}) \times M_2(\mathbb{R}) \longrightarrow M_2(\mathbb{R})$$

where $+^{\mathbf{M}}$ is matrix addition. Likewise we take $F(\times) = \times^{\mathbf{M}}$ to be matrix multiplication, $F(1)$ to be the identity matrix, and $F(0)$ is our zero matrix.

We generally write algebras like $\mathbf{A} = \langle A, +, \cdot, \times, \ldots \rangle$ rather than $\langle A, F \rangle$ where F is appropriately defined.

Returning to the syntactical arrangements, the only other essential part of our syntax is a countably infinite list of distinct symbols for **variables**: $v_0, v_1, v_2, \ldots$. For convenience, we also supply ourselves with the symbol $\approx$ for equality.

Terms of a given signature are built up from the variables and the constant symbols with the help of the operation symbols of positive rank. We give a precise definition.

DEFINITION 7.2. The set of **terms** is the smallest set T of finite sequences of operation symbols and variables satisfying the following constraints:

- Every variable belongs to T;
- If Q is an operation symbol of rank r and $t_0, \ldots, t_{r-1}$ belong to T, then $Qt_0 t_1 \cdots t_{r-1}$ also belongs to T.

More informally, every variable is a term and $Qt_0 \cdots t_{r-1}$ is a term, whenever $t_0, \ldots, t_{r-1}$ are terms and Q is an operation symbol with rank r.

According to this definition, operation symbols are to be followed by the terms they combine. This is sensible, since we have allowed operation symbols to have any finite rank. However, it varies from the customary practice of writing terms like this $x \cdot (y + z)$ rather than like this $\cdot x + yz$. This last string of symbols looks odd indeed, but it is nevertheless what is called for by our official definition. There are a couple of virtues that the official definition has. In the first place, we do not have to deal with parentheses—simplifying our syntax. In the second place, while the customary practice works well for two-place operation symbols, there seems to be no natural way to extend it to operation symbols of, let us say, rank seven. This system of notation was promoted by the Polish logician Jan Łukasiewicz in the 1920's and is sometimes called Łukasiewicz or Polish notation. More simply it is called prefix notation.

Terms have a key property that we will employ without much reference: A string of symbols that is a term can be parsed in only one way. That is, if Q is an operation symbol of rank r and $t_0, \ldots, t_{r-1}$ and $s_0, \ldots, s_{r-1}$ are terms such that the string $Qt_0 t_1 \cdots t_{r-1}$ is the same as the string $Qs_0 s_1 \cdots s_{r-1}$, then it must be that $t_0 = s_0, \ldots, t_{r-1} = s_{r-1}$. This may seem so obvious as to require no proof. Nevertheless it is the subject of the next set of exercises and the proof outlined there has some subtlety.

It is helpful to depict terms as (ordered, rooted, labeled) trees. With this scheme each term has a top node: the first symbol in the term. Beneath the top node, should it be an operation symbol of rank r, there will be r nodes arranged from left to right. These nodes will be the top nodes of trees depicting subterms of the original term. The next display gives the idea.

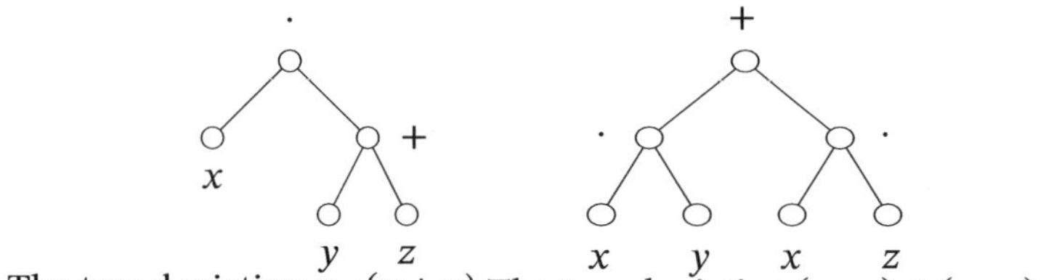

The tree depicting $x \cdot (y + z)$ The tree depicting $(x \cdot y) + (x \cdot z)$

In §6.1 we discussed how to convert such tree depictions of terms into computationally more convenient forms. See Figures 6.2, 6.3, and 6.4.

It is sometimes convenient to change the set of variables in the definition of terms of signature σ, either by taking only finitely many variables or by taking a set of size κ for some infinite cardinal κ. When X is such an adjusted set of variables, we use $T_\sigma(X)$ to denote the resulting set of terms. By imposing operations on this set we arrive at the **term algebra** of signature σ. Here is how the basic operations of the term algebra are defined:

$$Q^{\mathbf{T}_\sigma(X)}(t_0, t_1, \dots, t_{r-1}) := Qt_0t_1 \cdots t_{r-1}$$

for any operation symbol Q and any $t_0, \dots, t_{r-1} \in T_\sigma(X)$, where r is the rank of Q. See pages 229–230 in Volume I for more information.

An **equation** is an ordered pair of terms. We denote these ordered pairs as $s \approx t$, rather than (s, t). So even though $\approx$ has been made officially part of our syntax, it just denotes ordered pair and could be omitted.

The Semantics of Equational Logic

Let $\mathbf{A}$ be an algebra and let t be a term of the same signature. Then $t^{\mathbf{A}}$ will be a certain function from A^ω into A defined, for all $\mathbf{a} = \langle a_0, a_1, \dots \rangle \in A^\omega$, as follows:

- $v_i^{\mathbf{A}}(\mathbf{a}) = a_i$;
- Suppose $Qt_0 \cdots t_{r-1}$ is given, where Q is an operation symbol of rank r and $t_0, \dots, t_{r-1}$ are terms. Then

$$(Qt_0 \cdots t_{r-1})^{\mathbf{A}}(\mathbf{a}) = Q^{\mathbf{A}}\left(t_0^{\mathbf{A}}(\mathbf{a}), t_1^{\mathbf{A}}(\mathbf{a}), \dots, t_{r-1}^{\mathbf{A}}(\mathbf{a})\right)$$

The functions described above are called the **term functions** of $\mathbf{A}$. Of course, even though we have made term functions to have rank ω, really each one depends only on finitely many of its denumerably many inputs. This definition of term functions (or term operations) differs from that adopted in Volume I page 232, which is also described in this Volume on page 7. Roughly speaking, these two approaches to the notion of term operation are interchangeable: the notion adopted here considers a term operation on an algebra to be a specific operation of rank ω, whereas the approach used earlier considers a schema of infinitely many operations of finite rank.

EXAMPLE 7.3. Let's look at an example. Consider the familiar algebra (it is the ring of integers)

$$\mathbf{Z} = \langle \mathbb{Z}, +^{\mathbf{Z}}, \cdot^{\mathbf{Z}}, -^{\mathbf{Z}}, 0, 1 \rangle.$$

Then

$$((v_0 + v_1) + v_2)^{\mathbf{Z}}(3, 7, 5, 4, 4, 4, \dots)$$

is really

$$\left(v_0^{\mathbf{Z}}(3, 7, 5, 4, 4, 4, \dots) +^{\mathbf{Z}} v_1^{\mathbf{Z}}(3, 7, 5, 4, 4, 4, \dots)\right) +^{\mathbf{Z}} v_2^{\mathbf{Z}}(3, 7, 5, 4, 4, 4, \dots)$$

which is really

$$(3 +^{\mathbf{Z}} 7) +^{\mathbf{Z}} 5$$

and of course this is 15.

At last, here is the crucial element of our semantical arrangements—what it means for an equation $s \approx t$ to be **true** in the algebra $\mathbf{A}$. Just as the operation symbols can be regarded as names for the basic operations of $\mathbf{A}$, so the terms can be seen as names for the term functions of $\mathbf{A}$. We say that $s \approx t$ is true in $\mathbf{A}$ if and only if s and t name the same term function of $\mathbf{A}$—that is, if and only if $s^{\mathbf{A}} = t^{\mathbf{A}}$. Here are alternative ways to express this:

- $\mathbf{A} \models s \approx t$;
- $\mathbf{A}$ is a model of $s \approx t$;
- $s \approx t$ is true in $\mathbf{A}$;
- $\mathbf{A} \models \forall \mathbf{y}[s \approx t]$, where $\mathbf{y}$ is any finite string of variables that includes all the variables occurring in the terms s and t.

The last alternative listed above reflects that equational logic is a fragment of elementary logic.

The truth of equations in algebras imposes a two-place relation between the class of all algebras of our fixed signature and the set of all equations of the same signature. We use $\models$ ("models" or "satisfies") to denote this relation. Like any two-place relation, $\models$ gives rise to a Galois connection. This Galois connection is crucial to equational logic. The situation is displayed in Figure 7.1.

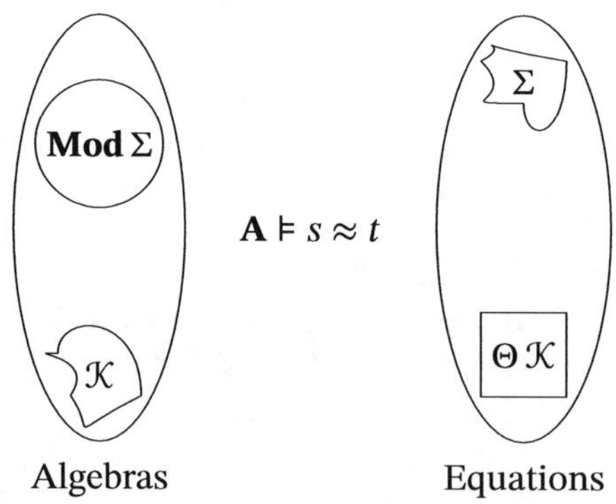

FIGURE 7.1. The Galois Connection Underlying Equational Logic

The polarities of this connection are

$$\mathbf{Mod}\,\Sigma = \{\mathbf{A} \mid \mathbf{A} \models s \approx t \text{ for all } s \approx t \in \Sigma\}, \quad \text{where } \Sigma \text{ is any set of equations.}$$

and

$$\Theta\,\mathcal{K} = \{s \approx t \mid \mathbf{A} \models s \approx t \text{ for all } \mathbf{A} \in \mathcal{K}\}, \quad \text{where } \mathcal{K} \text{ is any class of algebras.}$$

$\mathbf{Mod}\,\Sigma$ is called the **equational class** or the **variety based** on Σ. It is the class of all models of Σ. The set $\Theta\,\mathcal{K}$ is called the **equational theory** of $\mathcal{K}$. It is the set of all equations that are true in every algebra belonging to $\mathcal{K}$.

We described Galois connections in general in § 2.2 (see Volume I, page 51). The basic properties of Galois connections were gathered in Theorem 2.21. For the convenience of our readers, we rephrase Theorem 2.21 for this particular Galois connection.

THEOREM 7.4 (Theorem 2.21 rephrased). *Fix a signature σ and let $\mathcal{A}$ be the class of all algebras of signature σ and E be the set of all equations of signature σ. Then* **Mod** *maps subclasses of $\mathcal{A}$ to subsets of E and Θ maps subsets of E to subclasses of $\mathcal{A}$ so that all of the following hold:*

(i) $\mathcal{K} \subseteq \mathbf{Mod}\,\Theta\,\mathcal{K}$ *for all* $\mathcal{K} \subseteq \mathcal{A}$ *and* $\Sigma \subseteq \Theta\,\mathbf{Mod}\,\Sigma$ *for all* $\Sigma \subseteq E$.

(ii) *If* $\mathcal{J} \subseteq \mathcal{K} \subseteq \mathcal{A}$, *then* $\Theta\,\mathcal{K} \subseteq \Theta\,\mathcal{J}$.

(iii) *If* $\Gamma \subseteq \Sigma \subseteq E$, *then* $\mathbf{Mod}\,\Sigma \subseteq \mathbf{Mod}\,\Gamma$.

(iv) $\Theta\,\mathbf{Mod}\,\Theta\,\mathcal{K} = \Theta\,\mathcal{K}$ *for all* $\mathcal{K} \subseteq \mathcal{A}$ *and* $\mathbf{Mod}\,\Theta\,\mathbf{Mod}\,\Sigma = \mathbf{Mod}\,\Sigma$ *for all* $\Sigma \subseteq E$.

(v) $\mathbf{Mod}\,\Theta$ *is a closure operator on $\mathcal{A}$ whose closed subclasses are exactly the polars of subsets of E, that is the varieties of signature σ.*

(vi) $\Theta\,\mathbf{Mod}$ *is a closure operator on E whose closed subsets are exactly the polars of subclasses of $\mathcal{A}$, that is the equational theories of signature σ.*

(vii) *The lattice of closed subclasses of $\mathcal{A}$ (that is the lattice of varieties of signature σ) is isomorphic to the dual of the lattice of closed subsets of E (that is the lattice of equational theories of signature σ) via the map induced by Θ. Moreover,* **Mod** *induces the inverse isomorphism.* ∎

Actually, the proof of Theorem 2.21 was left as an exercise for our readers. For those who have not done this exercise in Volume I, we provide it in the next exercise set. (See Exercise 6.) In (vii) above, the notion of lattice has been stretched a bit since the universe of a lattice should be a set, whereas here the members of this universe are proper classes. It is (vii) that allows us to escape set-theoretic difficulties by considering the lattice of equational theories.

The first task in the further development of equational logic and the theory of varieties is to provide descriptions of the two closure operators $\mathbf{Mod}\,\Theta$ and $\Theta\,\mathbf{Mod}$.

The closure operator $\mathbf{Mod}\,\Theta$ on the algebra side of the Galois connection takes, as input, the class $\mathcal{K}$ of algebras, and returns as output a class, perhaps somewhat larger, of algebras. Obtaining this output, in the way described, requires a detour through our syntactical arrangements with the help of the semantical notion of truth. What we desire here is a description of this closure operator that is entirely algebraic and avoids this detour. We undertook this task in §4.11 of Volume I, where we find the following theorem:

The HSP Theorem (4.131).
For any class $\mathcal{K}$ of algebras of the same signature, ***HSP***$(\mathcal{K}) = \mathbf{Mod}\,\Theta(\mathcal{K})$. *Thus $\mathcal{K}$ is a variety iff $\mathcal{K}$ is an equational class.*

So the closure operator $\mathbf{Mod}\,\Theta$ is fully described in terms of homomorphisms, subalgebras, and direct products. We take up the description of $\Theta\,\mathbf{Mod}$ in the next section.

Exercises 7.5

1. (After Karl Menger) Let X be a set of variables and L be a set of operation symbols. Define a function λ from the set of finite nonempty sequences of

elements of $X \cup L$ into the integers as follows:

$$\lambda(w) = \begin{cases} -1 & \text{if } w \in X, \\ r-1 & \text{if } w \text{ is an operation symbol of rank } r, \\ \sum_{i<n} \lambda(u_i) & \text{if } w = u_0 u_1 \cdots u_{n-1} \text{ where } u_i \in X \cup L \text{ and } n > 1. \end{cases}$$

Prove that w is a term if and only if $\lambda(w) = -1$ and $\lambda(v) \geq 0$ for every nonempty proper initial segment v of w.

2. Let $w = u_0 u_1 \cdots u_{n-1}$, where $u_i \in X \cup L$ for all $i < n$. Prove that if $\lambda(w) = -1$, then there is a unique cyclic variant $\widehat{w} = u_i u_{i+1} \cdots u_{n-1} u_0 \cdots u_{i-1}$ of w that is a term.

3. Prove that if w is a term and w' is a proper initial segment of w, then w' is not a term.

4. Let $\mathbf{T}$ be the term algebra of L over X. Prove
If Q and P are operation symbols, and $P^{\mathbf{T}}(p_0, p_1, \dots, p_{n-1}) = Q_1^{\mathbf{T}}(q_0, q_1, \dots, q_{m-1})$, then $P = Q$, $n = m$, and $p_i = q_i$ for all $i < n$.

5. Let $\mathbf{A}$ be any algebra and Γ be any collection of fully invariant congruence relations on $\mathbf{A}$. Prove that the join $\bigvee \Gamma$ of the set Γ in $\operatorname{Con} \mathbf{A}$ is again a fully invariant congruence relation on $\mathbf{A}$.

6. This problem, and those that remain in this set of exercises, give an orderly way to prove Theorem 2.21 from Volume I.

Let A and B be two classes and let R be a binary relation with $R \subseteq A \times B$. For $X \subseteq A$ and $Y \subseteq B$ put

$$X^{\rightarrow} = \{b \mid x \mathrel{R} b \text{ for all } x \in X\}$$
$$Y^{\leftarrow} = \{a \mid a \mathrel{R} y \text{ for all } y \in Y\}$$

Prove that if $W \subseteq X \subseteq A$, then $X^{\rightarrow} \subseteq W^{\rightarrow}$. (Likewise if $V \subseteq Y \subseteq B$, then $Y^{\leftarrow} \subseteq V^{\leftarrow}$.)

7. Prove that if $X \subseteq A$, then $X \subseteq X^{\rightarrow\leftarrow}$. (Likewise if $Y \subseteq B$, then $Y \subseteq Y^{\leftarrow\rightarrow}$.)

8. Prove that $X^{\rightarrow\leftarrow\rightarrow} = X^{\rightarrow}$ for all $X \subseteq A$ (and likewise $Y^{\leftarrow\rightarrow\leftarrow} = Y^{\leftarrow}$ for all $Y \subseteq B$).

9. Prove that the collection of subclasses of A of the form $Y^{\leftarrow}$ is closed under the formation of arbitrary intersections. (As is the collection of subclasses of B of the form $X^{\rightarrow}$.) We call classes of the form $Y^{\leftarrow}$ and the form $X^{\rightarrow}$ closed.

10. Let $A = B = \{q \mid 0 < q < 1 \text{ and } q \text{ is rational}\}$. Let R be the usual ordering $\leq$ on this set. Identify the systems of closed sets. How are they ordered with respect to inclusion? This is a recasting of the ideas of Richard Dedekind for grasping the continuity of the unit interval. Dedekind was so struck by his idea that decades later he recalled the date: 24 November 1858, when he was teaching calculus and trying to get it right. See Richard Dedekind (1872), (1888), (1963), (1968).

7.2. The Description of Θ Mod Σ: The Completeness Theorem

The HSP Theorem gives us an algebraic description of the closure operator **Mod** Θ on the algebraic side of the Galois connection established by truth between algebras and equations of a given signature. Our task now is to provide a description of the closure operator Θ **Mod** on the equational side of this Galois connection. Notice that

$$s \approx t \in \Theta\,\mathbf{Mod}\,\Sigma$$

relates the equation $s \approx t$ with the set Σ of equations. These are syntactical objects—made up of strings of symbols. But the closure operator is semantical, bringing in algebras and the truth of equations in algebras. What we seek here is a strictly syntactical description of the closure operator.

It helps to recast matters. Observe

$$s \approx t \in \Theta\,\mathbf{Mod}\,\Sigma \quad \text{if and only if} \quad \mathbf{A} \models s \approx t \text{ for all } \mathbf{A} \text{ such that } \mathbf{A} \models \Sigma.$$

When the condition on the right above is satisfied we say that $s \approx t$ is a **logical consequence** of Σ. We extend the meaning of $\models$ and denote the relation of logical consequence by $\Sigma \models s \approx t$. It is also convenient to abbreviate the compound operator Θ **Mod** and write instead Θ. We say that $\Theta\,\Sigma$ is the equational theory **based on** Σ and that Σ is a **basis** or **base** for this equational theory. Another way to say this is that we have widened the domain of Θ to include sets of equations as well as classes of algebras.

So what we desire is a syntactical description of logical consequence. That is, we would like to devise a means of proof for equational logic that leads from a set Σ of equational axioms by strictly syntactical means to each logical consequence $s \approx t$.

The chief attributes we require of our formal system of inference are:

Soundness: : If the equation $s \approx t$ can be inferred from the set Σ of equations, then $s \approx t \in \Theta\,\mathbf{Mod}\,\Sigma$.

Adequacy: : If $s \approx t \in \Theta\,\mathbf{Mod}\,\Sigma$, then $s \approx t$ can be inferred from Σ.

Effectiveness: : In so far as it is reasonable (for computable signatures), there is an algorithm for the recognition of inferences.

This last stipulation may fail, even for finite signatures. The trouble is that the domain of a signature is a set of operation symbols and these symbols are without restriction. Hence, there is actually a proper class of operation symbols; but only countably many symbols can be suitable as inputs for computational processes.

There is another desire we have for our system of equational inference. We would like our system of inference to be as primitive as possible so that we can readily establish facts about inferences themselves. This desire conflicts with another desire we might have: to make the inferences themselves easy. We favor the first, since our goal is to prove things about the system of inference. Since our system of inference is, in essence, a detailed description of the closure operator Θ **Mod**, this will enable us to obtain far-reaching results about equational theories and varieties. While we will occasionally use our system of inference to actually deduce equations, this will not be its principal purpose. Instead, apart from simplicity, our system of inference makes it easy to demonstrate properties of inferences themselves, and to draw conclusions about the limitations for what is provable in equational logic.

Deductions actually put forward in mathematical practice rarely have a completely formal, syntactical character. In fact, one often encounters a mixture of syntactical and algebraic methods inside such proofs. Below we give an informal, if purely syntactical, example of an equational deduction.

Any ring satisfying the equation $x^3 \approx x$ is commutative. The following derivation of this fact illustrates some of the rules of inference that are commonly used in reasoning about equations. Some steps invoking obvious uses of the axioms of ring theory are omitted. Expressions like $2yxy$ are shorthand for $yxy + yxy$—which itself abbreviates the more elaborate $(yx)y + (yx)y$. In fact, any ring satisfying an equation of the form $x^n \approx x$, where $n > 1$, is a subdirect product of finite fields and hence commutative. This is a celebrated result of N. Jacobson (1945). But every proof of this more general result we know is not purely syntactical, but rather uses a combination of syntactical and algebraic methods. Indeed, that mixture of the syntactic with the algebraic is typical of the derivations of equations in practice. We gave an account of one such proof of Jacobson's result in §4.5 of Volume I (see Corollary 4 on page 177).

(1)	$x^3 - x \approx 0$	by adding $-x$ to both sides of $x^3 \approx x$
(2)	$(x+y)^3 - (x+y) \approx 0$	by substituting $(x+y)$, a term, for the variable x in (1)
(3)	$(x-y)^3 - (x-y) \approx 0$	by substituting $(x-y)$ for x in (1)
(4)	$(x+y)^3 + (x-y)^3 - 2x \approx 0$	by adding (2) and (3)
(5)	$2(x^3 - x + y^2x + yxy + xy^2) \approx 0$	by standard ring theory from (4)
(6)	$2(y^2x + yxy + xy^2) \approx 0$	by replacing $x^3 - x$ by 0 in (5) using (1)
(7)	$2(y^3x - xy^3) \approx 0$	by substituting $(yx - xy)$ for x in (6) and standard ring theory
(8)	$2(yx - xy) \approx 0$	by replacing y^3 with y twice in (7)
(9)	$(x^2 - x)^3 - (x^2 - x) \approx 0$	by substituting $(x^2 - x)$ for x in (1)
(10)	$x^6 - 3x^5 + 3x^4 - x^3 - x^2 + x \approx 0$	by standard ring theory from (9)
(11)	$x^2 - 3x + 3x^2 - x - x^2 + x \approx 0$	by replacing x^3 by x six times in (10)
(12)	$3(x^2 - x) \approx 0$	by standard ring theory from (11)

(13)	$3[(x+y)^2 - (x+y)] \approx 0$	by substituting $(x+y)$ for x in (12)
(14)	$3(x^2 - x + xy + yx + y^2 - y) \approx 0$	by ring theory from (13)
(15)	$3(xy + yx) \approx 0$	by replacing $3(x^2 - x)$ and $3(y^2 - y)$ by 0 in (14)
(16)	$xy + 5yx \approx 0$	by adding (8) to (15)
(17)	$xy + 6yx \approx yx$	by adding yx to both sides of (16)
(18)	$6x^3 \approx 0$	by substituting x for y in (6)
(19)	$6x \approx 0$	by replacing x^3 by x in (18)
(20)	$xy \approx yx$	by replacing $6yx$ by 0 in (17) using (19)

The kind of inference laid out above amounts to a list of equations, each of which is justified by certain axioms—here the axioms of ring theory and the equation $x^3 \approx x$—or by equations earlier in the list. Each justification is made according to some rule of inference: substituting terms for variables, replacing "equals by equals", adding "equals to equals", and so on. The system of inference we are about to introduce is somewhat different. In fact, our inferences turn out to be sequences of terms rather than sequences of equations. Both the proof above and those appropriate to our system are syntactic in character—they are concerned with the formal manipulation of terms and equations considered as strings of symbols—but they embody the semantical notion of logical consequence.

Our notion of equational deduction relies on two syntactical concepts: that of one term being a subterm of another and that of one term being a substitution instance of another. Indeed, these two concepts can be found in the deduction illustrated above. While the first concept might seem straightforward (s is a subterm of t if s is a substring of t that is also a term) and the second unusual in that it refers to an essentially algebraic notion (substitutions are just endomorphisms of the term algebra), here we provide these two concepts with more traditional definitions that appeal to the recursive structure of terms and without mention of endomorphisms of the term algebra.

Fix a signature. For any term t of the signature the set $S(t)$ of **subterms** of t is defined by the following recursion:

(1) $S(x) = \{x\}$ for every variable x.
(2) $S(Qt_0 \cdots t_{r-1}) = \{Qt_0 \cdots t_{r-1}\} \cup \bigcup_{i<r} S(t_i)$ for every operation symbol Q and every sequence $t_0, \ldots, t_{r-1}$ of terms where r is the rank of Q.

We say that a term s is a **subterm** of t provided $s \in S(t)$. This definition is reconciled with the one indicated a paragraph above in the next exercise set.

Now for the notion of substitution instance. Let α be any assignment of terms to the variables. That is, α is a function with domain the set of variables and all of whose values are terms. The substitution operator su_α associated with α is the function from terms to terms defined via the following recursion.

(1) $\mathrm{su}_\alpha(v_i) = \alpha(v_i)$ for each natural number i.
(2) $\mathrm{su}_\alpha(Qt_0 \cdots t_{r-1}) = Q\,\mathrm{su}_\alpha(t_0) \cdots \mathrm{su}_\alpha(t_{r-1})$ for every operation symbol Q and every sequence $t_0, \ldots, t_{r-1}$ of terms where r is the rank of Q.

It is not hard to see that su_α is, in fact, the (necessarily unique) endomorphism of the term algebra that extends the map α. Given a term t and an assignment α of terms to the variables, we say that $\mathrm{su}_\alpha(t)$ is a **substitution instance** of t. If $\alpha(v_i) = u_i$ for every natural number i, we use $t(u_0, u_1, \ldots)$ to denote $\mathrm{su}_\alpha(t)$. Likewise, for any equation $s \approx t$, the equation $\mathrm{su}_\alpha(s) \approx \mathrm{su}_\alpha(t)$ is a **substitution instance** of $s \approx t$. As above we also denote this by $s(u_0, u_1, \ldots) \approx t(u_0, u_1, \ldots)$.

Every equational theory is a congruence relation on the term algebra. More is true. We say that a congruence θ on an algebra $\mathbf{A}$ is *fully invariant* provided for all $a, b \in A$ and every endomorphism f of $\mathbf{A}$, if $a \mathrel{\theta} b$ then, $f(a) \mathrel{\theta} f(b)$. That is, θ is a congruence of the algebra obtained as an expansion of $\mathbf{A}$ by adding all the endomorphisms as one-place basic operations.

THEOREM 7.6 (Theorem on Fully Invariant Congruences of Term Algebras (Garrett Birkhoff, 1935)). *Fix a signature. The equational theories of the signature are exactly the fully invariant congruence relations on the term algebra.*

Proof. We have already seen, in §4.11 of Volume I, that equational theories are intersections of congruences of the term algebra. So they are congruence relations themselves. First, we see that they are fully invariant.

So consider the equational theory $\Theta\,\mathbf{A}$. Suppose that $s \approx t \in \Theta\,\mathbf{A}$ and that f is an endomorphism of the term algebra $\mathbf{T}$. We want to establish that $f(s) \approx f(t) \in \Theta\,\mathbf{A}$ or, what is the same, that $(f(s), f(t)) \in \ker h$ for every homomorphism $h : \mathbf{T} \to \mathbf{A}$. Now $h \circ f$ is a homomorphism from $\mathbf{T}$ into $\mathbf{A}$. Since $s \approx t \in \Theta\,\mathbf{A}$, we know that $(s, t) \in \ker(h \circ f)$, But this is the same as $(f(s), f(t)) \in \ker h$, our desired conclusion.

So each equational theory is, indeed, a fully invariant congruence relation on the term algebra.

We also desire the converse. So let θ be a fully invariant congruence on $\mathbf{T}$. Let $\mathbf{A}$ be $\mathbf{T}/\theta$. We contend that $\Theta\,\mathbf{A}$ and θ are identical. First, suppose that $(s, t) \in \theta$. Let $a_0, a_1, \ldots$ be any elements of A. Pick terms $p_0, p_1, \ldots$ so that $a_i = p_i/\theta$ for each $i \in \omega$. Let f be the endomorphism of $\mathbf{T}$ such that $f(v_i) = p_i$ for all $i \in \omega$. Recalling that $s = s^{\mathbf{T}}(x_0, x_1, \ldots)$ and $t = t^{\mathbf{T}}(x_0, x_1, \ldots)$, observe

$$\begin{aligned}(s, t) \in \theta &\implies (f(s), f(t)) \in \theta \\ &\implies s^{\mathbf{A}}(a_0, a_1, \ldots) = t^{\mathbf{A}}(a_0, a_1, \ldots)\end{aligned}$$

But the ω-tuple $\langle a_0, a_1, \ldots\rangle$ was arbitrary. So $s^{\mathbf{A}} = t^{\mathbf{A}}$. But this is just $s \approx t \in \Theta\,\mathbf{A}$. Our conclusion: $(s, t) \in \theta \implies s \approx t \in \Theta\,\mathbf{A}$.

We also need the reverse implication. To this end, suppose $s \approx t \in \Theta\,\mathbf{A}$. For each $i \in \omega$ let a_i be v_i/θ. Observe

$$\begin{aligned}s \approx t \in \Theta\,\mathbf{A} &\implies s^{\mathbf{A}}(a_0, a_1, \ldots) = t^{\mathbf{A}}(a_0, a_1, \ldots) \\ &\implies s^{\mathbf{T}/\theta}(v_0/\theta, v_1/\theta, \ldots) = t^{\mathbf{T}/\theta}(v_0/\theta, v_1/\theta, \ldots) \\ &\implies s^{\mathbf{T}}(v_0, v_1, \ldots)/\theta = t^{\mathbf{T}}(v_0, v_1, \ldots)/\theta \\ &\implies s/\theta = t/\theta \\ &\implies (s, t) \in \theta.\end{aligned}$$

In this way, we see that each fully invariant congruence of the term algebra $\mathbf{T}$ is indeed an equational theory. ■

A Syntactic Characterization of Θ Mod Σ: The Completeness Theorem for Equational Logic

Now we can reframe our task. Given a set Σ of equations, we see that Θ **Mod** Σ is the smallest equational theory that includes the set Σ. Regarding Σ as a set of ordered pairs of terms, we see that Θ **Mod** Σ is the smallest fully invariant congruence relation on the term algebra that includes the set Σ. That is, Θ **Mod** Σ is the fully invariant congruence relation on **T** that is generated by the set Σ. So our task is to give a description of how the fully invariant congruence relations on the term algebra are generated from a given set Σ.

Given an equation $p \approx q$ and terms r and r' we will say that r and r' are **equivalent in one step using** $p \approx q$ provided there is an endomorphism f of the term algebra and strings u and v, possibly empty, so that

$$\{r, r'\} = \{uf(p)v, uf(q)v\}.$$

We denote this relation by $r \overset{p\approx q}{\longleftrightarrow} r'$. To say this another way, r' is obtained from r by replacing a substitution instance of one side of $p \approx q$ by the same substitution applied to the other side. There is a directional variant of this notion that is useful. We say that r **rewrites in one step to** r' **using** $p \approx q$ provided there is an endomorphism f of the term algebra and strings u and v, possibly empty, so that

$$r \text{ is } uf(p)v \text{ and } r' \text{ is } uf(q)v.$$

We denote this by $r \overset{p\approx q}{\longrightarrow} r'$.

We will say that the equation $s \approx t$ is **deducible** from the set Σ of equations provided there is some finite sequence $r_0, r_1, \dots, r_n$ of terms so that

- s is r_0 and r_n is t, and
- r_i and r_{i+1} are equivalent in one step using some equation from Σ for each $i < n$.

We call the sequence $r_0, \dots, r_n$ a **deduction** of $s \approx t$ from Σ. Notice that deductions with $n = 0$ are permitted. This means $s \approx s$, where s is any term, is deducible from every set of equations, even the empty set. We can display such deductions as

$$s \overset{e_0}{\longleftrightarrow} r_1 \overset{e_1}{\longleftrightarrow} r_2 \overset{e_2}{\longleftrightarrow} \cdots \overset{e_{n-1}}{\longleftrightarrow} t$$

where each e_i is an equation belonging to Σ.

We use $\Sigma \vdash s \approx t$ to denote that there is a deduction of $s \approx t$ from Σ.

THEOREM 7.7 (The Completeness Theorem for Equational Logic). *Let Σ be a set of equations and let $s \approx t$ be an equation. $\Sigma \models s \approx t$ if and only if $\Sigma \vdash s \approx t$.*

Proof. Let $\theta = \{(s,t) \mid \Sigma \vdash s \approx t\}$. We need to show that θ is the smallest fully invariant congruence on the term algebra **T** that includes the set Σ. It is easy to see that θ is an equivalence relation on the set T of terms and that $\Sigma \subseteq \theta$.

Our first goal is to show that $\theta \subseteq \varphi$ for every fully invariant congruence relation φ on **T** that includes Σ. We do this by induction on the length of deductions.

Base Step

In this case, $\Sigma \vdash s \approx t$ is witnessed by a deduction with $n = 0$. This means that s and t are the same. Evidently, (s, s) belongs to every congruence on **T**.

Inductive Step

Here we assume that $(r, r') \in \varphi$ for all fully invariant φ as long as $\Sigma \vdash r \approx r'$ is witnessed by a deduction of length no more than n. Let $\Sigma \vdash s \approx t$ be witnessed by the deduction of length $n+1$ below:

$$s = r_0 \overset{e_0}{\longleftrightarrow} r_1 \overset{e_1}{\longleftrightarrow} r_2 \cdots r_n \overset{e_n}{\longleftrightarrow} r_{n+1} = t.$$

So we see that (s, r_n) belongs to every fully invariant congruence that includes Σ and that $r_n \overset{p\approx q}{\longleftrightarrow} t$, where $p \approx q \in \Sigma$. Let φ be a fully invariant congruence that includes Σ. Let f be an endomorphism of $\mathbf{T}$ and u and v be strings of symbols so that $\{r_n, t\} = \{uf(p)v, uf(q)v\}$. It is harmless to suppose that $r_n = uf(p)v$ and $t = uf(q)v$. We know that $(f(p), f(q)) \in \varphi$ since φ is a fully invariant congruence that includes Σ. Now let y be a variable that occurs in neither r_n nor in t. Let $t^* = uyv$. Observe that t^* is the term obtained from t by replacing the designated occurrence of term $f(q)$ by the new variable y. We contend that $f(p) \equiv f(q) \mod \varphi$ entails that $t^{*\mathbf{T}}(\ldots, f(p), \ldots) \equiv t^{*\mathbf{T}}(\ldots, f(q), \ldots) \mod \varphi$. In fact, this is true about arbitrary algebras and arbitrary congruences on them. A routine induction on the complexity of the term t^* does the job.

So our induction is complete and we know that $\theta \subseteq \varphi$ whenever φ is a fully invariant congruence of $\mathbf{T}$ that includes Σ.

So it only remains for us to show that θ itself is a fully invariant congruence of $\mathbf{T}$. We have already observed that θ is an equivalence relation on the set of terms. Let Q be an operation symbol. To avoid a morass of indices, we show the case when the rank of Q is 2. Suppose that $\Sigma \vdash s_0 \approx t_0$ and $\Sigma \vdash s_1 \approx t_1$. To see that θ is a congruence, we must have $\Sigma \vdash Qs_0s_1 \approx Qt_0t_1$. So let

$$s_0 \overset{e_0}{\longleftrightarrow} \cdots \overset{e_{n-1}}{\longleftrightarrow} t_0$$

and

$$s_1 \overset{g_0}{\longleftrightarrow} \cdots \overset{g_{m-1}}{\longleftrightarrow} t_1$$

be deductions from Σ of $s_0 \approx t_0$ and $s_1 \approx t_1$ respectively. We can piece these two deductions together:

$$Qs_0s_1 \overset{e_0}{\longleftrightarrow} \cdots \overset{e_{n-1}}{\longleftrightarrow} Qt_0s_1 \overset{g_0}{\longleftrightarrow} \cdots \overset{g_{m-1}}{\longleftrightarrow} Qt_0t_1.$$

So we find $\Sigma \vdash Qs_0s_1 \approx Qt_0t_1$, as desired. The case of operation symbols of arbitrary rank holds no mysteries. Our conclusion so far is that θ is a congruence relation of $\mathbf{T}$.

Finally, to see that θ is fully invariant, suppose that $\Sigma \vdash s \approx t$ and the f is an endomorphism of $\mathbf{T}$. We need to see that $\Sigma \vdash f(s) \approx f(t)$. The idea is straightforward. Let

$$s \overset{e_0}{\longleftrightarrow} r_1 \overset{e_1}{\longleftrightarrow} r_2 \overset{e_2}{\longleftrightarrow} \cdots \overset{e_{n-1}}{\longleftrightarrow} t$$

be a deduction of $s \approx t$ from Σ. We claim that

$$f(s) \overset{e_0}{\longleftrightarrow} f(r_1) \overset{e_1}{\longleftrightarrow} f(r_2) \overset{e_2}{\longleftrightarrow} \cdots \overset{e_{n-1}}{\longleftrightarrow} f(t)$$

is also a deduction of $f(s) \approx f(t)$ from Σ. To establish this, we need only consider an arbitrary step in the deduction. So let us suppose that $r \overset{p\approx q}{\longleftrightarrow} r'$. It is harmless to suppose that $r = ug(p)v$ and $r' = ug(q)v$ where g is an endomorphism of $\mathbf{T}$ and u and v are certain strings of symbols. Then $f(r) = \hat{u}f(g(p))\hat{v}$ and $f(r') = \hat{u}f(g(q))\hat{v}$. Here

$\hat{u}$ is obtained from u by replacing each variable v_i by the term $f(v_i)$. The string $\hat{v}$ is obtained from v in the same way. But notice that $f \circ g$ is itself an endomorphism of $\mathbf{T}$. This means $f(r) \overset{p\approx q}{\longleftrightarrow} f(r')$, as desired. It follows that θ is a fully invariant congruence of $\mathbf{T}$. Since $\Sigma \subseteq \theta \subseteq \varphi$ for each fully invariant congruence φ that includes Σ, we find that θ is the least fully invariant congruence of $\mathbf{T}$ that includes Σ. In this way, our theorem is established. ■

We had three aspirations for our system of equational inference:

- Soundness
- Adequacy
- Effectiveness

The left-to-right implication in the Completeness Theorem asserts the adequacy of our system of inference, while the right-to left implication asserts the soundness of our system. The effectiveness for our system would require an algorithm that would be able to recognize which finite sequences of terms are actually deductions. This depends on the signature as well as the set Σ upon which the deductions are based. The operation symbols provided by the signature must be amenable as inputs to an algorithm (this is impossible if the signature is uncountable, or even if it has only one symbol of overwhelming set-theoretic intricacy). On the other hand, many signatures that arise in mathematical practice are even finite and can be taken, without distortion, to be amenable as computational inputs. We must also have the means to decide which equations belong to Σ. But given these caveats, our system of equational inference is as effective as we can hope it to be.

Garrett Birkhoff proved the Theorem on Fully Invariant Congruences of Term Algebras in 1935 and drew from it a completeness theorem for equational logic. Loosely speaking, in Birkhoff's framework a deduction is a sequence of equations, rather than a sequence of terms. His rules of inference reflect the definition of fully invariant congruence relation on the term algebra. Birkhoff's system of equational inference can be found in the next set of exercises. There you can also find a somewhat different system of equational inference put forward by Alfred Tarski. The system we have given, part of the folklore, was inspired by Mal'tsev's description of how to generate congruence relations in arbitrary algebras (see Theorem 6.17 on page 24 and also Theorem 4.19, page 155 in Volume I). It is convenient for giving proofs by induction on the length of deductions.

It is frequently possible to prove theorems concerning deducibility by induction on the length of deductions. We say that a set Γ of terms is **closed with respect to deductions based on** Σ iff $t \in \Gamma$ whenever $s \in \Gamma$ and $\Sigma \vdash s \approx t$.

THEOREM 7.8 (The Principle of Induction on Deductions). *Let Γ be any set of terms and Σ be any set of equations. If*

$$s \overset{e}{\longleftrightarrow} t \text{ implies } t \in \Gamma, \text{ whenever } s \in \Gamma \text{ and } e \in \Sigma$$

then Γ is closed with respect to deductions based on Σ. ■

Now that we have a formal system of inference in hand, we invite the reader to write out a derivation of $xy \approx yx$ from $x^3 \approx x$ and the axioms of ring theory. The deduction provided at the beginning of this section should be of help. (We should also warn the reader that a fully detailed derivation of this within our formal system is fairly long.)

The notion of replacement arises in the next corollary. Let s and t be terms and let Σ be a set of equations. We say that t is obtained from s by **replacement on the basis of** Σ provided there are strings u and v, possibly empty, so that $s = upv$ and $t = uqv$ for some equation $q \approx p \in \Sigma$. Replacement of equals by equals is a familiar step and it occurred in the sample derivation at the beginning of this section.

COROLLARY 7.9. *Let T be a set of equations. The following statements are equivalent:*

(i) *T is an equational theory.*

(ii) (a) *$s \approx s \in T$ for all terms s, and*

(b) *$s \approx t \in T$ whenever $s \approx r \in T$ and $r \overset{e}{\longleftrightarrow} t$ for some term r and some equation $e \in T$.*

(iii) (a) *$s \approx s \in T$ for all terms s, and*

(b) *T is closed under substitution, and*

(c) *$s \approx t \in T$ whenever t is obtained by replacement from s on the basis of some equation in T, and*

(d) *$s \approx t \in T$ whenever, $s \approx r$ and $r \approx t$ belong to T, for some term r.*

Proof. The conditions in (ii) assert little more than that T is closed under deductions of lengths zero, one, and two. We argue by induction on the length of deductions that T is closed under all derivations. Suppose inductively that

$$s \longleftrightarrow p_1 \longleftrightarrow p_2 \longleftrightarrow \cdots \longleftrightarrow p_{n-1} \longleftrightarrow t.$$

The inductive hypothesis gives $s \approx p_{n-1} \in T$. Set $r = p_{n-1}$ and let $e \in T$ be an equation such that $r \overset{e}{\longleftrightarrow} t$. By (b) $s \approx t \in T$. So (ii) implies (i). That (i) implies (ii) is immediate from the Completeness Theorem for Equational Logic.

(iii) is an easy consequence of (i) and the Completeness Theorem for Equational Logic. Suppose that (iii) holds and that $s \approx r \in T$ and that $r \overset{e}{\longleftrightarrow} t$, where $e \in T$. By (iii–b) and (iii–c), $r \approx t \in T$. By (iii-d), $s \approx t \in T$. This means that (iii) implies (ii). ■

Just as knowing that $\mathcal{K}$ is a variety iff $\mathcal{K} = \boldsymbol{HSP}\mathcal{K}$ gives us a way to check, in some instances, whether a given class of algebras is a variety, the corollary above often allows us to determine whether a given set of equations is an equational theory. See, for example, Exercises 12, 16, and 17 on pages 167 *ff.* below.

The notions of subterm and substitution instance underlie our system of inference. Let s and t be terms. We put

$$s \triangleleft t \text{ provided some substitution instance of } s \text{ is a subterm of } t.$$

This binary relation on the set of terms is easily seen to be reflexive and transitive, but not antisymmetric. Put

$$s \bowtie t \text{ provided } s \triangleleft t \text{ and } t \triangleleft s.$$

Then $\bowtie$ is an equivalence relation. It is easy to see that $s \bowtie t$ means that t results from s by a one-to-one renaming of variables (and s results from t by reversing the renaming).

In this event, we say that s and t are **literally similar**. Modulo the equivalence relation of literal similarity, $\lhd$ induces a partial order on (equivalence classes of) terms.

The Logic of Basic Equations

As we will see in the following sections, our system of equational inference is powerful enough to allow us the scope to encode the action of Turing machines, leading to undecidability results in equational logic. On the other hand, it is possible to restrict this notion of inference to capture a still interesting fragment of equational logic that entails computationally decidable reasoning.

We will say that a term t is **basic** provided there is at most one occurrence of an operation symbol of positive rank in t. An equation $s \approx t$ is **basic** provided both s and t are basic terms. See §6.7 for a further introduction to this topic. Elsewhere in the literature basic terms and equations are sometimes referred to as *simple* or as *linear*. Many of the Mal'tsev conditions investigated in Chapter 6 were framed using basic equations.

Restricting our attention to basic equations, we present here a system of inference and prove a completeness theorem for it. We will say that the equation $s \approx t$ is **basically deducible** from the set Σ of basic equations provided there is some finite sequence $r_0, r_1, \ldots, r_n$ of basic terms so that

- s is r_0 and r_n is t, and
- r_i and r_{i+1} are equivalent in one step using some equation from Σ for each $i < n$.

We call the sequence $r_0, \ldots, r_n$ a **basic deduction** of $s \approx t$ from Σ. Notice that basic deductions with $n = 0$ are permitted. This means $s \approx s$, where s is any basic term, is deducible from every set of basic equations, even the empty set.

We use $\Sigma \underset{B}{\vdash} s \approx t$ to denote that there is a basic deduction of $s \approx t$ from Σ. We will also say, in this case, that $s \approx t$ is a **basic consequence** of Σ.

The stipulation that each term in a deduction is a basic term places restrictions on the kind of substitutions that can be used at each step. It was an important property of our full system of equational inference that each equational theory was closed under every substitution. This is no longer true for the set of basic consequences of a set of basic equations. An example illustrates this point. Consider a signature that provides an operation symbol P of rank 2 and an operation symbol Q of rank 1. Let $\Sigma = \{Pxy \approx x, Pxy \approx y\}$. Then $\Sigma \underset{B}{\vdash} x \approx y$ but $\Sigma \underset{B}{\nvdash} Qx \approx Qy$, even though $Qx \approx Qy$ is a substitution instance of $x \approx y$. On the other hand, if α assigns to each variable either a variable or a constant symbol, then the set of basic consequences of a set of basic equations will be closed under the substitution su_α, as can be seen by induction on the length of basic deductions. So in our example we have $\Sigma \underset{B}{\vdash} z \approx w$ for any variables z and w. It is also useful to observe that if Σ is a set of basic equations and $\Sigma \underset{B}{\vdash} s \approx t$, then there is a basic deduction of $s \approx t$ from Σ in which every variable that occurs in the basic deduction also occurs in either s or in t.

Ordinary deductions also have the following property: If

$$s \longleftrightarrow r_1 \longleftrightarrow r_2 \longleftrightarrow \cdots \longleftrightarrow r_{n-1} \longleftrightarrow t$$

and $p = AsB$, where p is a term and A and B are possibly empty strings, then

$$p \longleftrightarrow Ar_1B \longleftrightarrow Ar_2B \cdots \longleftrightarrow Ar_{n-1}B \longleftrightarrow AtB$$

is also a deduction. This property ensures that once $s \approx t$ has been deduced, then it can be used in further deductions. Basic deductions do not enjoy this property, even if s and t are both constant symbols. Here is an example to illustrate this point. Consider a signature that provides a two-place operation symbol Q, a one-place operation symbol P and two constant symbols c and d. Let $\Sigma = \{c \approx Qcd, Qcd \approx d\}$. Then there is a basic deduction of $c \approx d$ from Σ, but no basic deduction of $Pc \approx Pd$.

These two ways in which basic deductions differ from ordinary deductions complicate the proof of our completeness theorem for basic equations. To address the first difference mentioned above, our completeness theorem for basic equations will only apply to those sets Σ of basic equations such that $\Sigma \underset{B}{\nvdash} x \approx y$, where x and y are distinct variables. To address the second difference, we enhance the set Σ of basic equations to $\Sigma \cup \Delta$, where Δ is defined with the help of the following recursion:

$$\begin{aligned} \Delta_0 &:= \emptyset \\ \Delta_{n+1} &:= \{c \approx d \mid \Sigma \cup \Delta_n \underset{B}{\vdash} c \approx d, \text{ where } c \text{ and } d \text{ are constant symbols}\}, \\ &\quad \text{for all natural numbers } n. \end{aligned}$$

Put $\Delta := \bigcup_{n \in \omega} \Delta_n$.

THEOREM 7.10 (The Completeness Theorem for Basic Equations). *Let Σ be a set of basic equations and let $s \approx t$ be a basic equation. Suppose that $\Sigma \underset{B}{\nvdash} v_0 \approx v_1$. Then*

$$\Sigma \cup \Delta \underset{B}{\vdash} s \approx t \textit{ if and only if } \Sigma \models s \approx t.$$

Proof. Since every basic deduction is a deduction, the left-to-right implication in the theorem is secure.

So suppose that $\Sigma \cup \Delta \underset{B}{\nvdash} s \approx t$. We will construct a model of Σ in which $s \approx t$ fails. Let T_B be the set of basic terms and, for any $r, r' \in T_B$, let

$$r \sim r' \text{ if and only if } \Sigma \cup \Delta \underset{B}{\vdash} r \approx r'.$$

Then $\sim$ is an equivalence relation such that s and t lie in different equivalence classes while $u \sim v$ for every $u \approx v \in \Sigma$. Since $\Sigma \underset{B}{\nvdash} v_0 \approx v_1$, we see also that $\Sigma \cup \Delta \underset{B}{\nvdash} x \approx y$, for any distinct variables x and y. This means that the $\sim$-equivalence classes separate variables; that is, no $\sim$-equivalence class contains more than one variable.

From each $\sim$-equivalence class, pick a representative element using the following priorities: If a variable belongs to the class, pick it, if no variable belongs to the class but a constant symbol belongs, pick a constant symbol arbitrarily, and otherwise make an arbitrary choice from the equivalence class. Notice that no $\sim$-equivalence class can contain both a variable x and a constant symbol c, since in that case $\Sigma \cup \Delta \underset{B}{\vdash} x \approx c, y \approx c$ for some variable y different from x and this would support the basic deduction

$$x \longleftrightarrow \cdots \longleftrightarrow c \longleftrightarrow \cdots \longleftrightarrow y.$$

Let T^* be the collection of representative elements. Let $A = T^* \cup \{\infty\}$, where ∞ is a new element. For each operation symbol Q of rank n, for any $t_0, \dots, t_{n-1} \in A$, let z be a variable not occurring in any of the t_i's. For each $i < n$, let t_i^* be t_i if t_i is a variable or a constant symbol and let it be z otherwise. Put

$$Q^{\mathbf{A}}(t_0, \dots, t_{n-1}) := \begin{cases} r & \text{if } Qt_0^* \cdots t_{n-1}^* \sim p \text{ for some basic term } p \text{ in which } z \\ & \text{does not occur and each variable in } p \text{ also occurs} \\ & \text{among the } t_i^*\text{'s; and } r \text{ represents } Qt_0^* \cdots t_{n-1}^*; \\ z & \text{if } Qt_0^* \cdots t_{n-1}^* \sim z; \\ \infty & \text{otherwise} \end{cases}$$

Observe that in the second case, z is the representative of $Qt_0^* \cdots t_{n-1}^*$, so this definition can be cast in a more uniform manner. However, we show that the first two cases are distinct. Suppose otherwise. Then there is a basic deduction

$$z \longleftrightarrow \cdots \longleftrightarrow Qt_0^* \cdots t_{n-1}^* \longleftrightarrow \cdots \leftrightarrow p,$$

where z does not occur in p. Let y be a variable different from z that does not occur in p. Let α be the assignment of terms to variables that assigns y to z and fixes every other variable. By applying su_α to each term in the basic deduction displayed above, we obtain a basic deduction of $y \approx p$. Then reversing this basic deduction and appending it to the one displayed above we obtain a basic deduction of $z \approx y$, contradicting our hypothesis.

Let $\mathbf{A}$ denote the algebra obtained by defining the fundamental operations in this way.

It remains to prove that $\mathbf{A} \models \Sigma$ and that $s \approx t$ fails in $\mathbf{A}$.

To see that $s \approx t$ fails in $\mathbf{A}$, let $\bar{a} := \langle v_0, v_1, \dots \rangle \in A^\omega$. We contend that $s^{\mathbf{A}}(\bar{a}) = r$, where r represents the basic term s and that $t^{\mathbf{A}}(\bar{a}) = r'$, where r' represents t. To verify this for s first notice that if s is the variable v_k, then $s^{\mathbf{A}}(\bar{a}) = v_k$, which represents s. Otherwise $s = Qs_0 \cdots s_{n-1}$ where Q is an operation symbol, n is the rank of Q, and each term s_i is either a variable or a constant symbol. Hence, each $s_i^* = s_i$. Letting $p = s$ we see, by the first case in the definition of $Q^{\mathbf{A}}$, that $s^{\mathbf{A}}(\bar{a}) = r$, as desired. But since s and t lie in different $\sim$-equivalence classes, we have that $r \neq r'$. So $s^{\mathbf{A}}(\bar{a}) \neq t^{\mathbf{A}}(\bar{a})$ and therefore $s \approx t$ fails in $\mathbf{A}$.

To see that $\mathbf{A} \models \Sigma$, let $p \approx q \in \Sigma$ and let $\bar{a}$ be any assignment of members of A to the variables. The members of A are (representatives of) basic terms as well as the default element ∞. Our objective is to show that $p^{\mathbf{A}}(\bar{a}) = q^{\mathbf{A}}(\bar{a})$. We leave to the reader the cases where one or both of p and q are variables and consider the case when $p = Pp_0 \cdots p_{n-1}$ and $q = Qq_0 \cdots q_{m-1}$, where P and Q are operation symbols of ranks n and m, respectively, and the p_i's and q_j's are either variables or constant symbols. Now

$$p^{\mathbf{A}}(\bar{a}) = P^{\mathbf{A}}\left(p_0^{\mathbf{A}}(\bar{a}), \dots, p_{n-1}^{\mathbf{A}}(\bar{a})\right).$$

In case p_i is the variable v_k, we have $p_i^{\mathbf{A}}(\bar{a}) = a_k$ and if p_i is a constant symbol, we have $p_i^{\mathbf{A}}(\bar{a}) = p_i^{\mathbf{A}}$. Similarly,

$$q^{\mathbf{A}}(\bar{a}) = Q^{\mathbf{A}}\left(q_0^{\mathbf{A}}(\bar{a}), \dots, q_{m-1}^{\mathbf{A}}(\bar{a})\right),$$

in case q_j is the variable v_ℓ, we have $q_j^{\mathbf{A}}(\bar{a}) = a_\ell$ and if q_j is a constant symbol, then $q_j^{\mathbf{A}}(\bar{a}) = q_j^{\mathbf{A}}$. Let z be a variable not occurring in

$$p_0^{\mathbf{A}}(\bar{a}), \ldots, p_{n-1}^{\mathbf{A}}(\bar{a}), \ldots, q_0^{\mathbf{A}}(\bar{a}), \ldots, q_{m-1}^{\mathbf{A}}(\bar{a}).$$

Using this z, let $\bar{a^*} = \langle a_0^*, a_1^*, \ldots \rangle$. The values of

$$P^{\mathbf{A}}\left(p_0^{\mathbf{A}}(\bar{a}), \ldots, p_{n-1}^{\mathbf{A}}(\bar{a})\right) \text{ and } Q^{\mathbf{A}}\left(q_0^{\mathbf{A}}(\bar{a}), \ldots, q_{m-1}^{\mathbf{A}}(\bar{a})\right)$$

have been defined with the help of the terms

$$Pp_0^{\mathbf{A}}(\bar{a^*}) \cdots p_{n-1}^{\mathbf{A}}(\bar{a^*}) \text{ and } Qq_0^{\mathbf{A}}(\bar{a^*}) \cdots q_{m-1}^{\mathbf{A}}(\bar{a^*}).$$

Contention $Pp_0^{\mathbf{A}}(\bar{a^*}) \cdots p_{n-1}^{\mathbf{A}}(\bar{a^*}) \sim Qq_0^{\mathbf{A}}(\bar{a^*}) \cdots q_{m-1}^{\mathbf{A}}(\bar{a^*})$.

Proof. Consider first that in case p_i is a constant symbol then $p_i^{\mathbf{A}}(\bar{a^*})$ is the chosen representative of p_i and similarly for constant symbols among the q_j's. In these cases, we know that $p_i \approx p_i^{\mathbf{A}}(\bar{a^*}) \in \Delta$. For each $i < n$ let

$$\hat{p}_i := \begin{cases} p_i & \text{If } p_i \text{ is a variable} \\ p_i^{\mathbf{A}}(\bar{a^*}) & \text{If } p_i \text{ is a constant symbol.} \end{cases}$$

Likewise for each $j < m$ let

$$\hat{q}_j := \begin{cases} q_j & \text{If } q_j \text{ is a variable} \\ q_j^{\mathbf{A}}(\bar{a^*}) & \text{If } q_j \text{ is a constant symbol.} \end{cases}$$

Then there is a basic deduction

$$P\hat{p}_0 \cdots \hat{p}_{n-1} \leftrightarrow \cdots \leftrightarrow Pp_0 \cdots p_{n-1} \leftrightarrow Qq_0 \cdots q_{m-1} \leftrightarrow \cdots \leftrightarrow Q\hat{q}_0 \cdots \hat{q}_{m-1}.$$

The deductions at either end rely on Δ, while the step in the middle is supported by $Pp_0 \cdots p_{n-1} \approx Qq_0 \cdots q_{m-1} \in \Sigma$. Now apply to each term in the basic deduction displayed above the substitution that sends $p_i \mapsto p_i^{\mathbf{A}}(\bar{a^*})$ and $q_j \mapsto q_j^{\mathbf{A}}(\bar{a^*})$ whenever p_i and q_j are variables. In this way, we obtain a basic deduction that witnesses the contention. ■

To complete the proof of the Completeness Theorem for Basic equations it is enough, in view of the Contention, to consult the definitions of $P^{\mathbf{A}}$ and $Q^{\mathbf{A}}$. ■

REMARK 7.11. The Completeness Theorem for Basic Equations has two features that separate it from the Completeness Theorem for Equational Logic: the insertion of the set Δ and the hypothesis that $\Sigma \underset{B}{\nvdash} x \approx y$. This raises two questions: How difficult is it to determine for a set Σ of basic equations whether $\Sigma \underset{B}{\nvdash} x \approx y$? How might the set Δ be constructed syntactically? To address these questions, first observe that if there is a basic deduction of the basic equation $s \approx t$, then there is a basic deduction in which any variable to appear already occurs in either s or t. Then, provided the signature is finite there are only finitely many basic terms in which only variables occurring in s or t appear. This means there are only a finite number of sequences of basic deductions that can be candidates for a basic deduction of $s \approx t$. Provided Σ is finite, a brute-force examination of these will either yield such a deduction or else give the conclusion that no basic deduction exists. This means, in particular, that there is an effective procedure for determining whether $\Sigma \underset{B}{\nvdash} x \approx y$.

Next recall the set Δ is built up in stages by recursion. In case the signature is finite, Δ will be finite and, if Σ is also finite, then this construction of Δ will bc an algorithmic process.

It follows that the set of basic consequences of a finite set Σ of basic equations (subject to the constraint that $\Sigma \not\vdash_B x \approx y$) will be decidable by an algorithm, in contrast to the situation in equational logic in general, see §7.8.

Irredundant Bases of Equational Theories

We see that Θ **Mod** is a closure operator on the set of all equations of a given signature. Closure operators were considered in Volume I §2.2 and §4.6. There we said that a closure operator C on a set A is of **rank** $k+1$, where k is a natural number, if and only if for every $X \subseteq A$, if $C(Z) \subseteq X$ for every $Z \subseteq X$ with $|Z| \leq k$, then $C(X) = X$.

Given a closure operator C on a set A and $X \subseteq A$, call a set $B \subseteq X$ an **irredundant basis** provided $C(B) = X$ and $C(B') \neq C(B)$ for every proper subset B' of B. Given a closed set $X \subseteq A$, we put

$$\nabla X := \{n \mid n \text{ is a natural number and } X \text{ has an irredundant base of cardinality } n\}.$$

We say that a set K of natural numbers is k-**convex**, where k is a natural number, if and only if for every $a, b \in K$ such that $a + k < b$, there is $c \in K$ so that $a < c < b$. Thus K is 1-convex provided K is an interval in the set of natural numbers—that is K is convex in the customary sense.

In Volume I, we gave a proof the following theorem.

Tarski's Interpolation Theorem (4.59).
If C is a closure operator of rank $k+2$ on a set A, where k is a natural number, then ∇X is k-convex for every closed set $X \subseteq A$.

It follows immediately from Corollary 7.9, that the closure operator Θ **Mod** has rank 3. Therefore, for any equational theory T, we conclude that ∇T is a convex set of natural numbers. The convex sets of natural numbers have one of the following three forms:

$$\emptyset, \quad I[n,m], \quad I[n,\infty)$$

where n is any natural number, and in the second case, where $n \leq m$. If the equational theory T has no finite basis, then $\nabla T = \emptyset$. On the other hand, if T has a finite basis, then ∇T has one of the other two forms and the natural number n (and m in the second case) are invariants of T.

Remarks Concerning the History of These Results

Reasoning with equations is, of course, very old. Here is an extract from Book I of Euclid's Elements:

COMMON NOTIONS

1. Things which are equal to the same thing are equal to each other.
2. If equals are added to equals the wholes are equal.
3. If equals are subtracted from equals the remainders are equal.
4. Things which coincide with one another are equal to one another.
5. The whole is greater than the parts.

It seems reasonable to construe common notion 4, in part, as a statement of reflexivity and common notion 1 as "$x \equiv y$ and $z \equiv y$ implies $z \equiv x$". In this we see that Euclid has already the notion of an equivalence relation. While common notions 2 and 3 assert that this equivalence relation respects the given operations. It is not too much of a stretch to think that Euclid has hold of the notion of congruence relation. This is not quite enough for equational reasoning. What is missing is some statement like

When equals are substituted for equals the results are equal.

The 19th century, beginning with Gauss's *Disquisitions*, saw an explosion of detailed equational reasoning. A number of varieties were given equational bases. Still, mathematical logic had not developed to the point where the question of a completeness theorem for any formal system of reasoning arose. It was 1930 when Kurt Gödel, in his doctoral thesis proved a completeness theorem for the predicate logic of countable languages.

The remarkable paper (Garrett Birkhoff, 1935) gave us the HSP Theorem, the first completeness theorem for equational logic, the general notion of a free algebra, and the insight that the equational theories coincide with the fully invariant congruence relations on the algebra of terms. This one paper earns Garrett Birkhoff the role of founder of the theory of varieties and of equational logic.

The formal system of equational reasoning we have framed in this section is a part of the folklore. It combines Birkhoff's insight about fully invariant congruence relations with Anatoly Mal'tsev's (1954) description of how to generate congruences.

The first completeness theorem of basic equations is due to David Kelly (1972), where he devised a system of formal inference for basic equations similar to the system invented by Birkhoff (see Exercise 1 below). David Kelly's system of inference for basic equations is described in Exercise 4 in the set of exercises below. While the notion of basic deduction propounded above resembles the notion of deduction in general, the proof we have given for the Completeness Theorem for Basic Equations has been adapted from the proof given by Keith A. Kearnes, Emil W. Kiss, and Ágnes Szendrei (2016) for David Kelly's system of inference.

Exercises 7.12

1. (Birkhoff's System of Inference, 1935)
Fix a signature. In this system there are three kinds of rules of inference:

The Substitution Rule : From $\{e\}$, it is permitted to infer any substitution instance of the equation e.

Equivalence Rules :

From the any set of equations, it is permitted to infer $s \approx s$, for any term s.

From $\{s \approx t\}$ it is permitted to infer $t \approx s$.

From $\{s \approx t, t \approx u\}$ it is permitted to infer $s \approx u$.

Rules for Operating on Equations : For each operation symbol Q, it is permitted to infer the equation

$$Qp_0p_1\cdots p_{r-1} \approx Qq_0q_1\cdots q_{r-1}$$

from $\{p_i \approx q_i : i < r\}$, where r is the rank of Q.

Take $\Sigma \vdash_B e$ to mean that there is a finite sequence $e_0, e_1, \ldots, e_n$ of equations such that e is just e_n and each member of the sequence either belongs to Σ or is obtainable from some subset of its predecessors by one of the rules of inference above. Prove that

$$\Sigma \vdash_B e \text{ iff } \Sigma \vdash e$$

for all sets Σ of equations and all equations e.

2. (Tarski's System of Inference, 1968)
In this system there are three rules of inference:
The Substitution Rule : From $\{e\}$ it is permitted to infer any substitution instance of the equation e.
The Tautology Rule : From the any set of equations it is permitted to infer any tautology, i.e. any equation of the form $s \approx s$.
The Replacement Rule : From $\{s \approx t, e\}$ it is permitted to infer $t \approx u$ where u is obtained from s by replacement on the basis of the equation e.
Take $\Sigma \vdash_T e$ to mean that there is a finite sequence of equations that ends with e such that each member of the sequence either belongs to Σ or can be inferred from some subset of its predecessors in the sequence by means of the rules given above. Prove that

$$\Sigma \vdash_T e \text{ iff } \Sigma \vdash e$$

for all sets Σ of equations and every equation e.

Birkhoff's system of inference has a rule for each operation symbol and those rules require as many equations as the rank of the operation symbol. The system of Tarski, on the other hand, has three rules of inference and these rules require 0, 1, or 2 equations as inputs. So Tarski's system gives us a rank-3 closure operator.

3. Devise formal inferences of $xy \approx yx$ from $x^3 \approx x$ and the usual axioms of ring theory, using the formal system of inference presented in this section and each of the systems described in the two preceding exercises.

4. (Kelly's system of Inference for Basic Equations 1972) Fix a signature. In this system there are three kinds of rules of inference:
The Substitution Rule: From $\{e\}$, it is permitted to infer any basic substitution instance of the equation e.
Equivalence Rules:
- From any set of basic equations, it is permitted to infer $s \approx s$, for any basic term s.
- From any basic equation $\{s \approx t\}$ it is permitted to infer $t \approx s$.
- From any basic equations $\{s \approx t, t \approx u\}$ it is permitted to infer $s \approx u$.

Rule for Constant Symbols: For each basic term t and any constant symbols c and d, it is permitted to infer from $c \approx d$ the equation $t \approx t'$, where t' is obtained from t by replacing one occurrence of c by d.
Take $\Sigma \vdash_K e$ to mean that there is a finite sequence $e_0, e_1, \ldots, e_n$ of basic equations such that e is just e_n and each member of the sequence either belongs to Σ or is obtainable from some subset of its predecessors by one of the rules

of inference above. Prove that

$$\Sigma \vdash_K e \text{ iff } \Sigma \vdash e$$

for all sets Σ of of basic equations such that $\Sigma \nvdash_K x \approx y$, where x and y are distinct variables, and all basic equations e.

5. Let $p, u_0, u_1, \ldots$ be terms of the same signature and let $q = p(u_0, u_1, \ldots)$ be the result of substituting u_i for v_i in p for all natural numbers i. Let **A** be any algebra of the same signature. Prove

$$q^{\mathbf{A}} = p^{\mathbf{A}}(u_0^{\mathbf{A}}, u_1^{\mathbf{A}}, \ldots).$$

***6.** From the usual axioms of the theory of commutative rings with unit, supplemented by $x^6 \approx x$, find a derivation of $x + x \approx 0$.

***7.** (T. O. Hand 1976)
From the usual axioms of the theory of commutative rings with unit, supplemented by $x^{48} \approx x$, find a derivation of $x^2 \approx x$.

***8.** (F. W. Levi, 1944; W. Stormquist, 1974)
Let Γ be the usual set of axioms for group theory. Prove that

$$\Gamma \cup \{x^p y^p \approx (xy)^p, x^q y^q \approx (xy)^q\} \vdash xy \approx yx \text{ iff } 2 = \gcd(p^2 - p, q^2 - q).$$

9. The next few exercises deal with quandles. **Quandles** are algebras in a signature that provides just two operation symbols, both of them binary. We take these operation symbols to be $\triangleright$ and $\blacktriangleright$. The following equations define the variety of all quandles:

$$x \triangleright x \approx x$$
$$(x \triangleright y) \blacktriangleright y \approx x$$
$$(x \blacktriangleright y) \triangleright y \approx x$$
$$(x \triangleright y) \triangleright z \approx (x \triangleright z) \triangleright (y \triangleright z)$$

Prove that all the equations below hold in all quandles.

$$x \blacktriangleright x \approx x$$
$$(x \triangleright y) \blacktriangleright z \approx (x \blacktriangleright z) \triangleright (y \blacktriangleright z)$$
$$(x \blacktriangleright y) \triangleright z \approx (x \triangleright z) \blacktriangleright (y \triangleright z)$$
$$(x \blacktriangleright y) \blacktriangleright z \approx (x \blacktriangleright z) \blacktriangleright (y \blacktriangleright z)$$

It follows that the more symmetrical set of equations is also a base for the variety of all quandles:

$$x \triangleright x \approx x \qquad x \blacktriangleright x \approx x$$
$$(x \triangleright y) \blacktriangleright y \approx x \qquad (x \blacktriangleright y) \triangleright y \approx x$$
$$(x \triangleright y) \triangleright z \approx (x \triangleright z) \triangleright (y \triangleright z) \qquad (x \blacktriangleright y) \blacktriangleright z \approx (x \blacktriangleright z) \blacktriangleright (y \blacktriangleright z)$$

It should be noted that certain finitely presented quandles are invariants associated with knots embeddable in 3-space, see (David Joyce 1979 and 1982).

10. Prove that all the equations below hold in all quandles.

$$x \triangleright (y \triangleright z) \approx ((x \blacktriangleright z) \triangleright y) \triangleright z \qquad x \blacktriangleright (y \blacktriangleright z) \approx ((x \triangleright z) \blacktriangleright y) \blacktriangleright z$$
$$x \triangleright (y \blacktriangleright z) \approx ((x \triangleright z) \blacktriangleright y) \blacktriangleright z \qquad x \blacktriangleright (y \triangleright z) \approx ((x \blacktriangleright z) \blacktriangleright y) \triangleright z$$

Conclude that every quandle term is equivalent, with respect to the equational theory of quandles, to a term that is associated to the left.

11. (David Joyce, 1979, 1982) Let **G** be any group. We define $x \triangleright^{\mathbf{G}} y := yxy^{-1}$ and $x \blacktriangleright^{\mathbf{G}} y := y^{-1}xy$. Prove that $\langle G, \triangleright^{\mathbf{G}}, \blacktriangleright^{\mathbf{G}}\rangle$ is a quandle. We denote it by Conj(**G**). Now prove that if X is any set and **Q** is the free quandle generated by X and **G** is the free group generated by X, then **Q** is embeddable in Conj(**G**). Conclude that any quandle equation that is true in Conj(**G**) for every group **G**, must be true in all quandles. That is, the class of quandles of the form Conj(**G**), where **G** is a group, generates the variety of all quandles.

12. Fix a signature. An equation $s \approx t$ is said to be *regular* provided every variable that occurs in s also occurs in t and conversely that every variable that occurs in t also occurs in s. Prove that each of the following sets of sentences is an equational theory.
(1) $\{s \approx t \mid s \approx t$ is a regular equation $\}$.
(2) $\{s \approx s \mid s$ is a term $\} \cup \{s \approx t \mid$ neither s nor t is a variable$\}$

13. Fix a signature σ with no constant symbols. Denote by $\mathbf{2}_\sigma$ the algebra of signature σ with universe $2 = \{0, 1\}$ such that

$$F(\bar{a}) = \begin{cases} 1 & \text{if every entry in } \bar{a} \text{ is } 1 \\ 0 & \text{otherwise} \end{cases}$$

for each fundamental operation F. Prove that $s \approx t$ is regular iff $\mathbf{2}_\sigma \models s \approx t$, for all equations $s \approx t$ of signature σ.

***14.** (Ewa Graczyńska, 1983)
Let T be any equational theory, containing nonregular equations, in a signature without constant symbols. Let Reg T be the equational theory consisting of all regular equations belonging to T. Prove that Reg $T \cup \{e\} \vdash$ T, for any nonregular equation $e \in$T.

***15.** (J. Płonka, 1967)
Let σ be a signature with no constant symbols. Let $\mathbf{S} = \langle S, \vee \rangle$ be a semilattice and denote by $\leq$ the join semilattice order on **S**. Let $\langle \mathbf{A}_i : i \in S\rangle$ be a system of algebras such that A_i and A_j are disjoint whenever i and j are distinct elements of S. Finally, let $H = \langle h_{ij} : i, j \in S$ and $i \leq j\rangle$ be a system of homomorphisms such that

$$h_{ij} : \mathbf{A}_i \to \mathbf{A}_j \text{ for } i \leq j \text{ in } \mathbf{S}$$

$$h_{ii} \text{ is the identity map on } \mathbf{A}_i \text{ for all } i \in S \text{, and}$$

$$h_{ik} = h_{jk} \circ h_{ij} \text{ for } i \leq j \leq k \text{ in } \mathbf{S}$$

The **Płonka sum** of the system $\langle \mathbf{A}_i : i \in S\rangle$ with respect to the semilattice $\mathbf{S}$ and the system H of homomorphisms is the algebra with universe $A = \bigcup_{i\in S} A_i$ such that for any operation symbol Q and any $a_0, \dots, a_{r-1} \in A$, where r is the rank of Q, we have

$$Q^{\mathbf{A}}(a_0, \dots, a_{r-1}) = Q^{\mathbf{A}_k}(h_{i_0 k}(a_0), \dots, h_{i_{r-1}k}(a_{r-1}))$$

where $a_0 \in A_{i_0}, \dots, a_{r-1} \in A_{i_{r-1}}$ and $k = i_0 \vee \cdots \vee i_{r-1}$. For any class $\mathcal{K}$ of algebras, we say that $\mathcal{K}$ is **closed with respect to Płonka sums** provided every algebra isomorphic to a Płonka sum of a system of algebras from $\mathcal{K}$ belongs, itself, to $\mathcal{K}$. Prove that $\mathcal{V}$ is closed with respect to Płonka sums iff $\mathcal{V}$ can be axiomatized by regular equations, for any variety $\mathcal{V}$.

16. An equation $s \approx t$ is **balanced** iff whenever u is a variable or an operation symbol with rank less than two, then $|s|_u = |t|_u$, where $|t|_u$ is the number of occurrences of the symbol u in t. Prove that in any signature the set of balanced equations is an equational theory.

17. Let u be any term and set $\Delta^u = \{s \approx t : s = t \text{ or both } u \triangleleft s \text{ and } u \triangleleft t\}$. Prove that Δ^u is an equational theory for every term u.

18. Let $\Sigma = \{x(yz) \approx (xy)z\}$. Since it is irrelevant with respect to Σ how a term is associated, in this exercise we suppress parentheses. For any natural number n, let e_n denote the following equation:

$$v_0 v_1 \cdots v_{n-1} v_n v_n v_{n-1} \cdots v_1 v_0 \approx v_n v_{n-1} \cdots v_1 v_0 v_0 v_1 \cdots v_{n-1} v_n$$

Prove that $\Sigma \cup \{e_i : i < n\} \nvdash e_n$, for every natural number n.

19. Write down a detailed definition of the notion "the tree T **depicts** the term t."

20. Prove that a sequence s of symbols is a subterm of the term t according to the recursive definition given at the end of this section iff s is a term and $t = AsB$ for some possibly empty strings A and B of symbols. [The first few exercises in Exercises 7.5 may prove useful.]

21. Are there substitutions f and g such that $f(x + (x + x)) = g((x + x) + x)$?

22. Let F and G be unary operation symbols and let $s = F^2G^2FGx$ and $t = F^2G^3FGx$. Let u be a subterm of s that is different from x. Is there a substitution instance of u that is also a substitution instance of t—possibly by means of a different substitution?

***23.** Call a signature **large** provided it has an operation symbol of rank at least two or at least two unary operation symbols. Prove that there are 2^ω equational theories for any countable large signature.

24. Invent an equational theory without an irredundant basis.

25. For any natural numbers n and m with $n \leq m$, invent an equational theory T so that $\nabla T = [n, m]$.

26. For each natural number n invent an equational theory T so that $\nabla T = [n, \infty)$.

7.3. Equational Theories that are not Finitely Axiomatizable

One question that immediately presents itself about any given equational theory is whether that theory is finitely based. As the subjects are usually presented, rings, groups, lattices, and Boolean algebras, among others, are defined via finite sets of equations. This can be a bit misleading. For example, groups first arose in a concrete setting: they were sets of permutations endowed with the operations of composition of permutations, formation of inverse permutations, and the identity permutation. Abstract groups are those algebras that are isomorphic to concrete groups. In this light, the familiar theorem usually called the Cayley Representation Theorem (see §3.5 in Volume I for a different take on Cayley's Theorem) should really be called the Cayley Finite Basis Theorem. It gives a finite list of equations to axiomatize the class of all (abstract) groups. On the other hand, the equational axioms for rings emerged as a finite list of properties common to a diverse assortment of algebras.

One might ask whether a given variety or even a given algebra is finitely based. Even restricted to finite algebras of finite signature, this question has turned out to be subtle. Our current concern will be to give examples of finite algebras that are not finitely based and devise general means to construct such examples. For the most part, these algebras have a pathological character.

It has turned out that almost all the finite algebras that emerged in the 19th century are finitely based. This applies to each finite group, each finite ring, and each finite lattice—although these conclusions are by no means obvious. It is obvious that the equational theories of one-element algebras are all based on the single equation $x \approx y$. As shown in Chapter 9 in Volume III, it is even true that every 2-element algebra of finite signature is finitely based: see Corollary 9.48 on page 40 in Volume III. Indeed, in an asymptotic sense made precise in Chapter 9, almost every finite algebra is finitely based. As a consequence, most nonfinitely based finite algebras seem pathological.

The Birkhoff Basis

An algebra $\mathbf{A}$ is **locally finite** provided every finitely generated subalgebra of $\mathbf{A}$ is finite. A class $\mathcal{K}$ of algebras is locally finite when each algebra belonging to $\mathcal{K}$ is locally finite.

LEMMA 7.13. *Every variety generated by a finite algebra is locally finite.*

Proof. Let $\mathbf{A}$ be a finite algebra and let $\mathcal{V} = \boldsymbol{HSP}\mathbf{A}$. Let $\mathbf{C}$ be an algebra in $\mathcal{V}$ that is generated by the finite set X. Recall from the proof of the HSP Theorem that we made a subalgebra $\mathbf{B}$ of $\mathbf{A}^{A^C}$ that was generated by the projection functions ρ_c for each $c \in C$. Then we formed the homomorphism $h : \mathbf{B} \twoheadrightarrow \mathbf{C}$. We make here a small change.

Instead, let $\mathbf{B}$ be the subalgebra of $\mathbf{A}^{A^X}$ generated by the projections ρ_x for each $x \in X$. Then $\mathbf{B}$ will be finite since A^{A^X} is finite. So $\mathbf{C}$ must be finite as well. ■

A small change in the argument above yields the next Lemma.

LEMMA 7.14. *Let $\mathcal{V}$ be a locally finite variety and let n be a natural number. Then $\Theta\ \mathcal{V}$ is an equivalence relation of the set of terms on $v_0, \ldots, v_{n-1}$ that has only finitely many equivalence classes.*

Proof. Let $\mathbf{T}_n$ be the term algebra on n generators $\{v_0, \ldots, v_{n-1}\}$ for the signature of $\mathcal{V}$. The quotient of $\mathbf{T}_n$ by $\Theta\ \mathcal{V}$ is an algebra $\mathbf{C} \in \mathcal{V}$ that has $[v_0], \ldots, [v_{n-1}]$ as a finite set of generators. By local finiteness, $\mathbf{C}$ is finite. The congruence classes of $\Theta\ \mathcal{V}$ are in one-one correspondence with C, hence form a finite family. ■

Let $\mathcal{V}$ be a variety and let n be a natural number. By $\mathcal{V}^{(n)}$ we mean the class of algebras, of the same signature as $\mathcal{V}$, that are models of all the equations true in $\mathcal{V}$ that have no more than n distinct variables. So $\mathcal{V}^{(n)}$ is a variety and $\mathcal{V} \subseteq \mathcal{V}^{(n)}$. It is easy to check that

$$\mathcal{V} \subseteq \cdots \subseteq \mathcal{V}^{(n+1)} \subseteq \mathcal{V}^{(n)} \subseteq \cdots \subseteq \mathcal{V}^{(1)} \subseteq \mathcal{V}^{(0)} \qquad \text{and} \qquad \mathcal{V} = \bigcap_{n \in \omega} \mathcal{V}^{(n)}.$$

The sequence $\langle \mathcal{V}^{(n)} \mid n \in \omega \rangle$ is called the **descending varietal chain of $\mathcal{V}$**.

LEMMA 7.15. *Let $\mathcal{V}$ be a variety and n be a natural number. Then for all algebras $\mathbf{B}$, we have $\mathbf{B} \in \mathcal{V}^{(n)}$ if and only if every subalgebra of $\mathbf{B}$ with n or fewer generators belongs to $\mathcal{V}$.*

Proof. First suppose that $\mathbf{B} \in \mathcal{V}^{(n)}$ and $\mathbf{C}$ is a subalgebra of $\mathbf{B}$ generated by $\{b_0, \ldots, b_{n-1}\}$. To see that $\mathbf{C} \in \mathcal{V}$ let $s \approx t$ be any equation true in $\mathcal{V}$. Pick $c_0, c_1, c_2, \cdots \in C$. For each natural number k pick a term $p_k(x_0, \ldots, x_{n-1})$ so that $c_k = p^{\mathbf{B}}(b_0, \ldots, b_{n-1})$. Then the equation

$$s(p_0, p_1, p_2, \ldots) \approx t(p_0, p_1, p_2, \ldots)$$

is a logical consequence of $s \approx t$ and so is true in $\mathcal{V}$. But the displayed equation has only variables from the set $\{x_0, x_1, \ldots, x_{n-1}\}$. So the displayed equation is true in $\mathbf{B}$, since $\mathbf{B} \in \mathcal{V}^{(n)}$. So

$$s^{\mathbf{C}}(p_0^{\mathbf{C}}(b_0, b_1, \ldots, b_{n-1}), \ldots) = t^{\mathbf{C}}(p_0^{\mathbf{C}}(b_0, b_1, \ldots, b_{n-1}), \ldots)$$
$$s^{\mathbf{C}}(c_0, c_1, \ldots) = t^{\mathbf{C}}(c_0, c_1, \ldots).$$

Since $c_0, c_1, \ldots$ were arbitrary elements of C, we see that $s \approx t$ holds true in $\mathbf{C}$. This entails that $\mathbf{C} \in \mathcal{V}$.

For the converse, suppose $\mathbf{B} \notin \mathcal{V}^{(n)}$. Then there is some equation $s \approx t$, in which at most n distinct variables occur, that is true in $\mathcal{V}$ but fails in $\mathbf{B}$. So pick $b_0, b_1, \ldots, b_{n-1} \in B$ so that $s^{\mathbf{B}}(b_0, b_1, \ldots, b_{n-1}) \neq t^{\mathbf{B}}(b_0, b_1, \ldots, b_{n-1})$. Let $\mathbf{C}$ be the subalgebra of $\mathbf{B}$ generated by $\{b_0, b_1, \ldots, b_{n-1}\}$. Evidently, the equation $s \approx t$ fails in $\mathbf{C}$. This means that $\mathbf{B}$ has a subalgebra, generated by at most n elements, that does not belong to $\mathcal{V}$. ■

THEOREM 7.16 (Birkhoff's Finite Basis Theorem (Garrett Birkhoff, 1935)). *Let $\mathcal{V}$ be a locally finite variety of finite signature and let n be a natural number. Then $\mathcal{V}^{(n)}$ is finitely based.*

Proof. We can assume that $\mathcal{V}$ is not the trivial variety, since otherwise $\mathcal{V}^{(n)}$ is also trivial (and hence finitely based) for all $n \geq 2$. We leave it in the hands of our readers to devise finite bases for the varieties $\mathcal{V}^{(1)}$ and $\mathcal{V}^{(0)}$ in case $\mathcal{V}$ is the trivial variety. (It might help to read through the rest of this proof.)

Let $\mathbf{T}_n$ be the subalgebra of the term algebra consisting of those terms in which only variables from $\{v_0, v_1, \ldots, v_{n-1}\}$ occur. Let $\theta = \{(p,q) \mid p,q \in T_n \text{ and } \mathcal{V} \models p \approx q\}$. It is easy to check that θ is a congruence of $\mathbf{T}_n$ and that $\mathbf{T}_n/\theta$ is the $\mathcal{V}$-freely generated algebra on n free generators. (It is even easier to see that $\mathbf{T}_n/\theta \in \mathcal{V}$, and this is all we need.)

Since $\mathcal{V}$ is locally finite this means that $\mathbf{T}_n/\theta$ is finite. In particular, θ partitions T_n into finitely many blocks. From each block pick a representative element that is as short as possible (and is a variable if the block contains a variable). Let Σ be the set of all equations of the form

$$Qt_0t_1 \cdots t_{r-1} \approx s$$

where Q is an operation symbol, r is the rank of Q, the terms $t_0, t_1, \ldots,$ and t_{r-1} are representative terms, and the term s is the representative of the θ-class to which $Qt_0t_1 \cdots t_{r-1}$ belongs. Because there are only finitely many operation symbols and only finitely many representative terms, the set Σ of equations is finite. It is easy to check that each equation in Σ is true in $\mathcal{V}$. We can conclude that Σ is a base for $\mathcal{V}^{(n)}$ if we can prove that every equation using only variables from $\{v_0, \ldots, v_{n-1}\}$ that is true in $\mathcal{V}$ is derivable from Σ. This will follow from the next claim.

CLAIM 1.
Let $p \in T_n$. Then $\Sigma \vdash p \approx q$, where q is the representative of the θ-class to which p belongs.

Proof. We induct on the complexity of p.

The base step of the induction splits into two cases: when p is a variable and when p is a constant symbol. In the first alternative p is already a representative term. In the second alternative, $p \approx q$ has been included in Σ.

For the inductive step, we have $p = Qp_0p_1 \cdots p_{r-1}$ for some operation symbol Q of positive rank r. By the induction hypothesis we know that

$$\Sigma \vdash p_k \approx q_k$$

for each $k < r$, where q_k is the θ-representative of p_k. So

$$\Sigma \vdash Qp_0p_1 \cdots p_{r-1} \approx Qq_0q_1 \cdots q_{r-1}.$$

But $Qq_0q_1 \cdots q_{r-1} \approx q$ is an equation in Σ. So putting things together, we get

$$\Sigma \vdash p \approx q.$$

This finishes the claim. ■

To conclude the proof of Birkhoff's Finite Basis Theorem, notice that if $s \approx t$ is an equation with variables only from $\{v_0, \ldots, v_{n-1}\}$ that is true in $\mathcal{V}$, then the θ-representative of s is the same as the θ-representative of t. Say it is q. Then $\Sigma \vdash s \approx q, t \approx q$. So $\Sigma \vdash s \approx t$, as desired. ■

For a locally finite variety $\mathcal{V}$ of finite signature, the equational basis constructed in the proof above is referred to as the **Birkhoff basis of** $\mathcal{V}^{(n)}$. This theorem can be found in (Garrett Birkhoff, 1935).

COROLLARY 7.17. *Let $\mathcal{V}$ be a locally finite variety of finite signature. Then $\mathcal{V}$ is finitely based if and only if $\mathcal{V} = \mathcal{V}^{(n)}$ for some natural number n.* ■

As a consequence, if $\mathcal{V}$ is a locally finite variety of finite signature and the only basic operations of $\mathcal{V}$ have rank less than 2, then every equation true in $\mathcal{V}$ can have at most two variables. Hence, $\mathcal{V} = \mathcal{V}^{(2)}$ and so $\mathcal{V}$ is finitely based.

The variety generated by a finite algebra is always locally finite, but the variety generated by a locally finite algebra need not be locally finite. The construction of an example is in the next exercise set. We call an algebra **A uniformly locally finite** provided there is a function $b : \omega \to \omega$ on the natural numbers so that every n-generated subalgebra of **A** has cardinality less than $b(n)$ for every natural number n.

Fact 1. *The variety generated by a uniformly locally finite algebra must be locally finite.*

Lyndon's Example of a Nonfinitely Based Finite Algebra

In 1954 Roger Lyndon published an example of an algebra with seven elements that is not finitely based. His example has two basic operations: a two-place operation and a distinguished element. That he used a constant symbol to denote one element of his algebra was convenient but not essential to his argument. So Lyndon's seven-element algebra really needed only one basic operation and that one operation is a two-place operation. Some forty years after Lyndon's result, Zoltán Székely observed that Lyndon's algebra (as well as several others that arose in the intervening decades) is associated with a finite automaton. Székely's observation allows us to present Lyndon's example via the diagram in Figure 7.2.

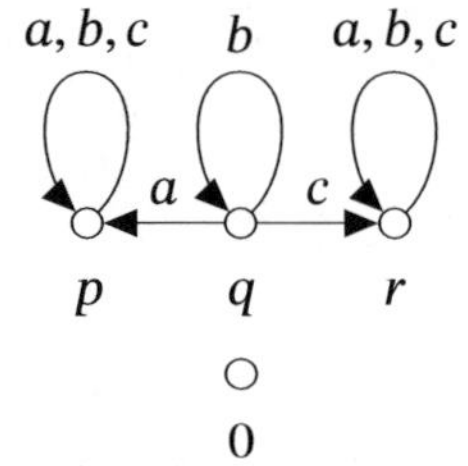

FIGURE 7.2. Lyndon's Algebra **L**

The elements of the algebra **L** fall into three disjoint sets: the 3-element set $\{p, q, r\}$, which we call the set of states, the 3-element set $\{a, b, c\}$, which we call the set of letters, and the 1-element set $\{0\}$. We call 0 the default element. The binary operation, which we denote by juxtaposition, is defined by following the arrows in the diagram. That is

$$\begin{array}{ccc} ap = p & bp = p & cp = p \\ aq = p & bq = q & cq = r \\ ar = r & br = r & cr = r \end{array}$$

with all other products resulting in the default element 0. In this way each letter can be thought to act on the states to produce a state. For instance, the letter a acts on the state q to produce the state p. The reader familiar with the theory of finite automata will notice that any designation of accepting states, found in the definition of a finite automaton, is missing here. In Lyndon's algebra every letter labels one edge directed out of any state, just as in the theory of finite automata. For automatic algebras more generally, we permit states p and letters a so that $ap = 0$. That is, a does not label an outward edge from p to any state—this is one of the attributes of nondeterministic finite automata, however we do not permit more than one outward arrow from any state to be labeled with the same letter, another attribute of nondeterministic finite automata.

Returning to Lyndon's algebra, the equivalence relation θ on L that makes 0 and r equivalent, but isolates all the other elements is easily seen to be a congruence of $\mathbf{L}$. The quotient algebra $\mathbf{L}/\theta$ is isomorphic to $\mathbf{L}^*$, which is displayed in Figure 7.3. Evidently, $\mathbf{L}^*$ is in the variety generated by $\mathbf{L}$. Edmond W. H. Lee (2009) noticed $\mathbf{L}^*$ and proved that $\mathbf{L}$ and $\mathbf{L}^*$ generate the same variety. He also pointed out that $\mathbf{L}^*$ is subdirectly irreducible (our readers can check that $\mathbf{L}^*$ is $\langle p, 0\rangle$-irreducible—see Definition 4.41 in Volume I—or, as we shall say, $\langle p, 0\rangle$ is a **critical pair**, meaning here $p \neq 0$ and that $\langle p, 0\rangle$ belongs to every nontrivial congruence of $\mathbf{L}^*$).

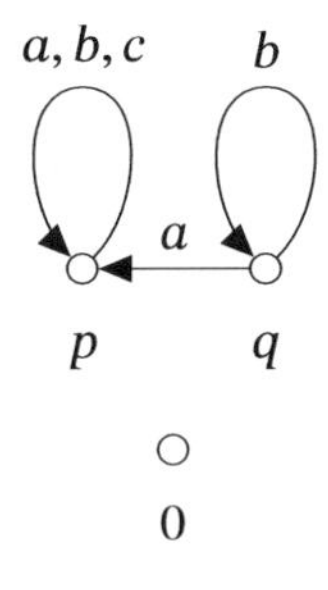

FIGURE 7.3. Lee's Algebra $\mathbf{L}^*$

THEOREM 7.18 (Lyndon's Nonfinite Basis Theorem (Roger C. Lyndon, 1954)). *The algebra* $\mathbf{L}$ *is not finitely based.*

Proof. Let $\mathcal{V}$ denote the variety generated by $\mathbf{L}$. Our plan is to invoke Corollary 7.17. To do this, for each (large enough) natural number n we have to devise an algebra $\mathbf{C}_n$ and an equation ϵ_n satisfying the following constraints:

(a) $\mathbf{C}_n \in \mathcal{V}^{(n)}$,
(b) ϵ_n holds in $\mathcal{V}$, and
(c) ϵ_n fails in $\mathbf{C}_n$.

The last two constraints are equivalent to the constraint that $\mathbf{C}_n \notin \mathcal{V}$. Taken together, all three constraints are equivalent to the contention that $\mathcal{V} \neq \mathcal{V}^{(n)}$ for all n.

So fix a natural number n. We first build the algebra $\mathbf{A}_n$, which will be a subalgebra of the n-fold direct power of $\mathbf{L}^*$. So $\mathbf{A}_n \in \mathcal{V}$, since we know that $\mathbf{L}^*$ is a homomorphic image of $\mathbf{L}$. So the elements of A_n will be n-tuples of elements of L^*. To avoid the accumulation of notation, we drop the commas and parentheses typical of n-tuples and instead write things like $b \quad b \quad b \quad a \quad c \quad c$, which would be a 6-tuple.

Here are some n-tuples that will belong to A_n:

$$\begin{aligned}
\beta_n &= q \quad q \quad q \cdots q \quad q \quad q \\
\alpha_{n-1} &= b \quad b \quad b \cdots b \quad b \quad a \\
\alpha_{n-2} &= b \quad b \quad b \cdots b \quad a \quad c \\
\vdots \quad & \quad \vdots \\
\alpha_1 &= b \quad a \quad c \cdots c \quad c \quad c \\
\alpha_0 &= a \quad c \quad c \cdots c \quad c \quad c
\end{aligned}$$

We take $\mathbf{A}_n$ to be the subalgebra of $(\mathbf{L}^*)^n$ generated by the $n+1$ elements that were just listed. Of course, A_n has more than just these $n+1$ elements. For example, we see that

$$\alpha_{n-1}\beta_n = q \quad q \quad q \cdots q \quad q \quad p.$$

We will call this new element β_{n-1}. More generally,

$$\alpha_k\beta_{k+1} = \beta_k.$$

This makes, for example, $\beta_1 = q \quad p \quad p \cdots p$ and $\beta_0 = p \quad p \quad p \quad p \cdots p \quad p \quad p$.

Here is another product:

$$\alpha_3\beta_2 = \beta_2$$

and in general

$$\alpha_k\beta_j = \beta_j$$

whenever $j \leq k$. But every other kind of product of these n-tuple produces an n-tuple with a 0 in at least one position. Let θ_n be the equivalence relation that collapses all the n-tuples that have at least one entry 0, but that isolates all the other elements of A_n. It is routine to check that θ_n is a congruence on $\mathbf{A}_n$. Let $\mathbf{B}_n = \mathbf{A}_n/\theta_n$. It is convenient to let $b_k = \beta_k/\theta_n = \{\beta_k\}$ for $k \leq n$ and let $a_k = \{\alpha_k\}$ for $k < n$. We accept the ambiguity of letting 0 denote the one large congruence class. So $\mathbf{B}_n$ belongs to $\mathcal{V}$, it has $2n+2$ elements, and just like $\mathbf{L}$ and $\mathbf{L}^*$ it can be displayed in a diagram, as follows.

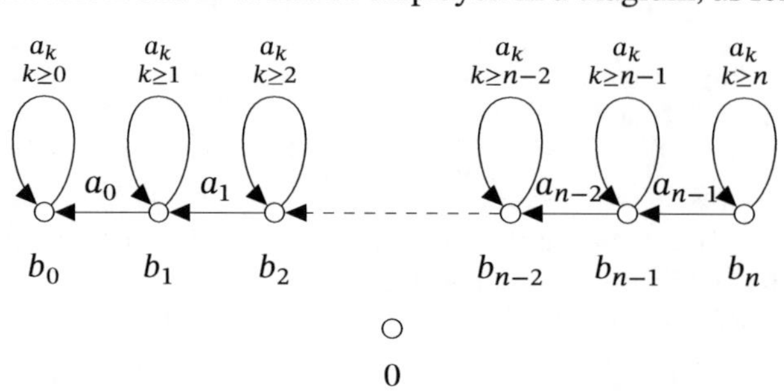

The Algebra $\mathbf{B}_n$

Notice that the rightmost loop actually has no label since there are no a_k's when $k \geq n$. The loop was included in the picture for the sake of uniformity.

In $\mathbf{B}_n$ we have $b_n b_0 = 0$. We obtain the desired algebra $\mathbf{C}_n$ by changing this one product—make $b_n b_0 = b_0$ in $\mathbf{C}_n$, but leave everything else unchanged. Here is the diagram for $\mathbf{C}_n$.

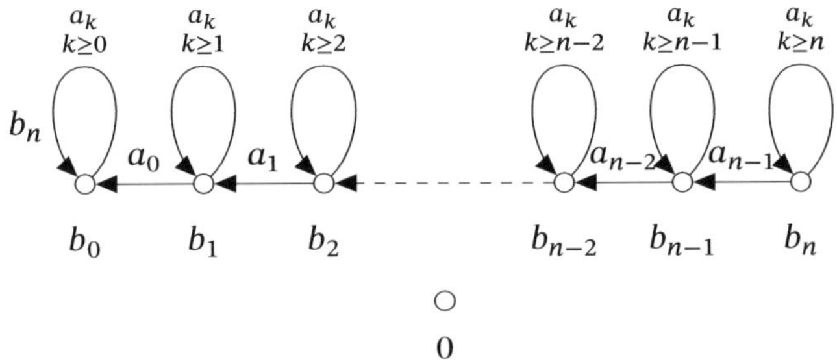

The Algebra $\mathbf{C}_n$

The algebras $\mathbf{B}_n$ and $\mathbf{C}_n$ have the same elements and the operations work almost always the same way. If we were displaying these algebras via multiplication tables, our two tables would differ at only one square of the table: in $\mathbf{B}_n$ the entry would be 0 while in $\mathbf{C}_n$ the entry would be b_0.

This small change is enough to keep $\mathbf{C}_n$ out of $\mathcal{V}$. Indeed, it is easy to see that the equation

$$y(x_0(x_1(x_2 \cdots (x_{n-1}y)\cdots))) \approx yy$$

holds in $\mathbf{L}$ but fails in $\mathbf{C}_n$. (Notice yy always evaluates to 0 in both $\mathbf{L}$ and $\mathbf{C}_n$.) So we take the equation displayed above to be ε_n.

It only remains to show that $\mathbf{C}_n \in \mathcal{V}^{(n)}$. To achieve this we have to show that every subalgebra of $\mathbf{C}_n$ with n or fewer generators belongs to $\mathcal{V}$. Let $\mathbf{D}$ be such a subalgebra. Notice that the $n+1$ elements

$$a_0, a_1, \ldots, a_{n-1}, \text{ and } b_n$$

are not outputs of the basic operation of $\mathbf{C}_n$ and moreover, removing any one of these elements results in a subalgebra of $\mathbf{C}_n$. At least one of these elements cannot be in $\mathbf{D}$. It follows that $\mathbf{D}$ is itself a subalgebra of the subalgebra of $\mathbf{C}_n$ formed by removing that one element. So it suffices to suppose that $\mathbf{D}$ itself arises by removing from $\mathbf{C}_n$ one of the elements listed above.

Suppose first that $\mathbf{D}$ results from removing b_n. Then the diagram below displays $\mathbf{D}$.

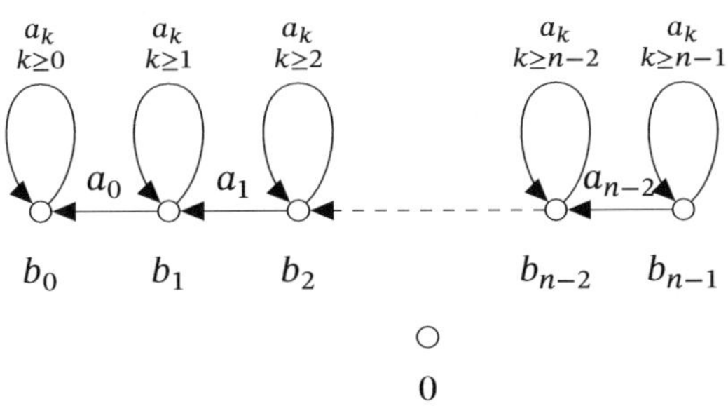

The Algebra $\mathbf{D} = \mathbf{C}_n$ with b_n removed.

We see, in this case, that $\mathbf{D}$ is a subalgebra of $\mathbf{B}_n$, so $\mathbf{D} \in \mathcal{V}$ as desired.

The last case we have to consider is that $\mathbf{D}$ arises by removing a_ℓ from $\mathbf{C}_n$. Then the diagram below displays $\mathbf{D}$.

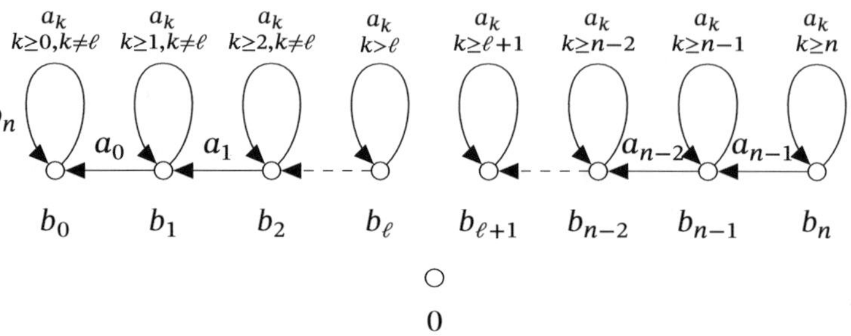

The Algebra $\mathbf{D} = \mathbf{C}_n$ with a_ℓ removed.

This diagram allows us to see two subalgebras of $\mathbf{D}$, namely $\mathbf{H}$, which has universe

$$\{0\} \cup \{b_{\ell+1}, \ldots, b_n\} \cup \{a_k \mid \ell < k < n\},$$

and $\mathbf{K}$, which has universe

$$\{0\} \cup \{b_0, \ldots, b_\ell\} \cup \{a_k \mid \ell \neq k < n\} \cup \{b_n\},$$

where b_n is a label of the leftmost loop, but not the rightmost node.

Evidently, $\mathbf{H}$ is isomorphic to $\mathbf{B}_{n-(\ell+1)}$. This means $\mathbf{H} \in \mathcal{V}$. Now consider the algebra $\mathbf{K}$. Notice that b_n labels an edge but it does not label a node in our diagram. As we want the conclusion that $\mathbf{K} \in \mathcal{V}$, we need to modify our construction of $\mathbf{B}_n$ to add an extra label. Here is how. Let $\mathbf{A}'_n$ be the subalgebra of $(\mathbf{L}^*)^n$ generated by $A_n \cup \{\gamma\}$ where γ is the n-tuple $c \quad c \quad c \cdots c \quad c$. Then let $\mathbf{B}'_n$ be the quotient algebra formed from $\mathbf{A}'_n$ by the congruence that lumps together all n-tuples having at least one entry 0 and isolates all other elements. Let $g = \{\gamma\}$ be the congruence class of γ. The algebra $\mathbf{B}'_n$ has the following diagram.

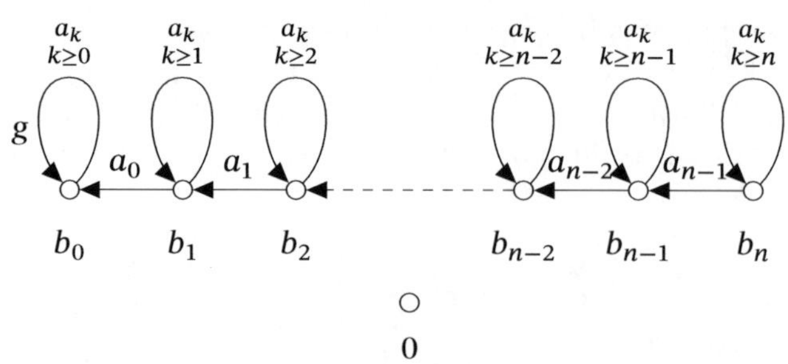

The Algebra $\mathbf{B}'_n$

So $\mathbf{B}'_n \in \mathcal{V}$ and evidently $\mathbf{K}$ is isomorphic to a subalgebra of $\mathbf{B}'_n$. This means that $\mathbf{K} \in \mathcal{V}$. Further, $\mathbf{K} \times \mathbf{H} \in \mathcal{V}$. But we wanted $\mathbf{D} \in \mathcal{V}$. So the last part of our proof will demonstrate that $\mathbf{D}$ is embeddable into $\mathbf{K} \times \mathbf{H}$. What we need is two homomorphisms $\phi : \mathbf{D} \to \mathbf{K}$ and $\psi : \mathbf{D} \to \mathbf{H}$ so that the system $\langle \phi, \psi \rangle$ separates the points of D. Here are the maps:

$$\phi(b_k) := \begin{cases} b_k & \text{if } k \le \ell \\ b_n & \text{if } k = n \\ 0 & \text{Otherwise.} \end{cases} \qquad \psi(b_k) := \begin{cases} b_k & \text{if } k > \ell \\ 0 & \text{Otherwise.} \end{cases}$$

$$\phi(a_k) := a_k \qquad \psi(a_k) := \begin{cases} a_k & \text{If } k > \ell \\ 0 & \text{Otherwise.} \end{cases}$$

$$\phi(0) := 0 \qquad \psi(0) := 0$$

The work of showing that these two maps are homomorphisms is left to our readers. To see the separation of points, observe that the kernel of ϕ places all the a_k's, all the b_k's for $k \le \ell$, and b_n into singleton congruence classes. On the other hand, $\{0\} \cup \{b_k \mid \ell < k < n\}$ is a single congruence class of the kernel of ϕ. But the kernel of ψ separates all the elements in this single big congruence class of the kernel of ϕ. So the system $\langle \phi, \psi \rangle$ separates the points of D. Consequently, $\mathbf{D}$ is embeddable into $\mathbf{K} \times \mathbf{H}$ and so $\mathbf{D} \in \mathcal{V}$, as desired.

This completes the proof of Lyndon's Nonfinite Basis Theorem. ■

Lyndon's original proof, see (Roger C. Lyndon, 1954), is purely syntactical and only a few pages long. The proof above, which may provide a method to show that other finite algebras are not finitely based, is from (Marcel Jackson and George F. McNulty, 2011).

Inherently Nonfinitely Based Equational Theories

An algebra **A** is **inherently nonfinitely based** provided:

(1) **A** has only finitely many basic operations;
(2) **A** belongs to some locally finite variety; and
(3) **A** belongs to no locally finite variety that is finitely based.

Similarly, we say that a locally finite variety $\mathcal{V}$ of finite signature is **inherently nonfinitely based** provided it is not included in any finitely based locally finite variety.

Varieties that are inherently nonfinitely based have the nonfinite basis pathology in a contagious way: if $\mathcal{V}$ is inherently nonfinitely based and $\mathcal{W}$ is a locally finite variety with $\mathcal{V} \subseteq \mathcal{W}$, then $\mathcal{W}$ is also inherently nonfinitely based. Similarly if **A** is inherently nonfinitely based and $\mathcal{W}$ is a locally finite variety with $\mathbf{A} \in \mathcal{W}$, then $\mathcal{W}$ is inherently nonfinitely based. Finally, if **A** is inherently nonfinitely based and $\boldsymbol{HSP}\,\mathbf{B}$ is locally finite (e.g. **B** is finite) and $\mathbf{A} \in \boldsymbol{HSP}\,\mathbf{B}$, then **B** is also inherently nonfinitely based.

From Birkhoff's Finite Basis Theorem we deduce

THEOREM 7.19. *Let $\mathcal{V}$ be a locally finite variety with a finite signature. Then the following conditions are equivalent:*

(i) *$\mathcal{V}$ is inherently nonfinitely based;*
(ii) *The variety $\mathcal{V}^{(n)}$ is not locally finite for any natural number n;*
(iii) *For arbitrarily large natural numbers N, there exists a non-locally finite algebra $\mathbf{B}_N$ whose N-generated subalgebras belong to $\mathcal{V}$.* ■

Thus to show that a locally finite variety $\mathcal{V}$ of finite signature is inherently nonfinitely based, it is enough to construct a family of algebras $\mathbf{B}_n$ (for each $n \in \omega$) so that each $\mathbf{B}_n$ fails to be locally finite, but $\mathbf{B}_n \in \mathcal{V}^{(n)}$.

We say that a variety $\mathcal{W}$ fails to be locally finite **in the finite sense** if there is a natural number p so that $\mathcal{W}$ contains arbitrarily large finite p-generated algebras. For $\mathcal{W}$ to fail to be locally finite, all it needs is an infinite, finitely generated algebra. However, for $\mathcal{W}$ to fail to be locally finite in the finite sense it must have arbitrarily large (but finite) p-generated algebras. A locally finite variety $\mathcal{V}$ is **inherently nonfinitely based in the finite sense** if whenever $\mathcal{V} \subseteq \mathcal{U}$ where $\mathcal{U}$ is a finitely based variety, then there is a natural number p so that $\mathcal{U}$ has arbitrarily large finite p-generated algebras. Equivalently: A locally finite variety $\mathcal{V}$ of finite signature is **inherently nonfinitely based in the finite sense** if and only if $\mathcal{V}^{(n)}$ fails to be locally finite in the finite sense for all natural numbers n.

Combining our ideas thus far, for finite algebras, it is easy to see that

$$\begin{gathered}\text{Inherently nonfinitely based in the finite sense}\\ \Downarrow \\ \text{Inherently nonfinitely based}\\ \Downarrow \\ \text{Nonfinitely based}\end{gathered}$$

In the opposite direction, it is known that there exist finite algebras that are nonfinitely based but that fail to be inherently nonfinitely based. Lyndon's seven-element algebra is one such example. Whether or not a variety being inherently nonfinitely based is enough to conclude that it is inherently nonfinitely based in the finite sense is not yet known. This question is related to a problem posed by Samuel Eilenberg and M. P. Schützenberger (1976):

> **The Problem of Eilenberg and Schützenberger:**
> Let $\mathcal{V}$ be a variety of finite signature generated by a finite algebra. Suppose that $\mathcal{W}$ is a finitely based variety such that $\mathcal{V}$ and $\mathcal{W}$ have exactly the same finite members. Must $\mathcal{V}$ be finitely based?

The same problem can be asked if $\mathcal{V}$ is locally finite, and that problem is still open as well. It turns out that any counterexample to the Problem of Eilenberg of Schützenberger must be inherently nonfinitely based, as observed by Robert Cacioppo (1993). However, a finite algebra that is inherently nonfinitely in the finite sense will not be a counterexample. Mark V. Sapir (1988) has shown that there is no counterexample to the Eilenberg-Schützenberger Problem in semigroups.

Next, we present the Shift Automorphism Theorem, which will entail that many finite algebras are inherently nonfinitely based. To formulate this theorem, we need the notion of a shift-automorphism algebra.

Consider an algebra $\mathbf{A}$ with a finite signature. We call an element $0 \in \mathbf{A}$ an **absorbing element** if any operation evaluated at any tuple containing 0 gives the value 0. The other nonabsorbing elements in $\mathbf{A}$ are called **proper elements**. Note that if it exists, the absorbing element 0 of $\mathbf{A}$ is unique as long as $\mathbf{A}$ has a basic operation of rank at least two. Moreover, if σ is any automorphism of $\mathbf{A}$, the absorbing element will be fixed by σ.

We call any tuple that does not contain the absorbing 0 a **proper tuple** or simply **proper**. If F is a basic operation of $\mathbf{A}$ of rank r, we can regard F as a set of $r+1$-tuples, with the first r entries as the inputs and the last entry as the output. The set of all proper tuples belonging to F is called the **proper part** of F.

Two proper elements of **A** are said to be **operationally related** provided they are entries in a tuple of proper elements that belongs to some basic operation of **A**. An entry in such a tuple will be called an **essential element**.

We define a **shift automorphism algebra** as an infinite, locally finite algebra **A** with only finitely many basic operations, with an absorbing element 0, and with an automorphism σ such that

(1) The only finite σ-orbit of **A** is $\{0\}$;
(2) The proper part of F is partitioned by σ into only finitely many orbits, for each fundamental operation F of **A**;
(3) There is a proper element a of A and a nonconstant unary polynomial function f of **A** such that $f(a) = \sigma(a)$.

A **shift automorphism variety** is any variety generated by a shift automorphism algebra. Notice that we have not stipulated that a shift automorphism algebra be uniformly locally finite. Every shift automorphism algebra has a countably infinite subalgebra that is also a shift automorphism algebra. Thus it is harmless to assume that if **A** is some shift automorphism algebra in some variety, then **A** is countably infinite.

Here is a useful property of shift automorphism algebras: The polynomial mentioned in condition (3) must contain an operation symbol of rank at least two. Let **A** be a shift automorphism algebra. The polynomial function $f(x)$, mentioned in condition (3), has an associated term function. Let $t^{\mathbf{A}}(x, y_1, \ldots, y_{k-1})$ be a term function of **A** so that

$$f(x) = t^{\mathbf{A}}(x, c_1, \ldots, c_{k-1})$$

where each $c_i \in A$. Let $t^{\mathbf{A}}(x, \mathbf{y})$ denote this term function. Note that in a shift automorphism algebra **A** every constant symbol must denote the absorbing element 0. In particular, none of these constant symbols can occur in $t^{\mathbf{A}}(x, \mathbf{y})$.

LEMMA 7.20. *The term $t(x, \mathbf{y})$ as mentioned above contains an operation symbol of rank at least two.*

Proof. With the aim of reaching a contradiction, we begin by supposing $t(x, \mathbf{y})$ contains only unary operation symbols. So $t(x, \mathbf{y})$ is $f(x)$, where f is a string of unary operation symbols. Recall that the σ-orbit of a is infinite and also that the algebra **A** is locally finite. Notice that $f(a) = \sigma(a)$ implies

$$\begin{aligned} f(f(a)) &= f(\sigma(a)) \\ &= \sigma(f(a)) \\ &= \sigma(\sigma(a)) \\ &= \sigma^2(a) \end{aligned}$$

Likewise,

$$\begin{aligned} f(f(f(a))) &= f\left(\sigma^2(a)\right) \\ &= \sigma^2(f(a)) \\ &= \sigma^2(\sigma(a)) \\ &= \sigma^3(a) \end{aligned}$$

Indeed, given that $f^m(a) = \sigma^m(a)$,

$$\begin{aligned} f^{m+1}(a) &= f\left(f^m(a)\right) \\ &= f\left(\sigma^m(a)\right) \\ &= \sigma^m\left(f(a)\right) \\ &= \sigma^m\left(\sigma(a)\right) \\ &= \sigma^{m+1}(a) \end{aligned}$$

Thus the countably infinite set $\{\sigma^m(a) : m \in \omega\}$ can be generated by the element a and applying one-place operations repeatedly. This contradicts the assumption that the algebra **A** is locally finite. ■

Almost all of the finite algebras known to be nonfinitely based come out of arguments involving the following Theorem:

THEOREM 7.21 (The Shift Automorphism Theorem). *Every shift automorphism algebra is inherently nonfinitely based in the finite sense. Furthermore, every shift automorphism variety has a countably infinite subdirectly irreducible member.*

Proof. This proof is long and involved, finishing on page 190. It might help, when working through this proof, to carry out each step in a particular example. Displayed below is an automatic algebra with a particularly simple structure that is also a shift automorphism algebra. It is suitable for this purpose.

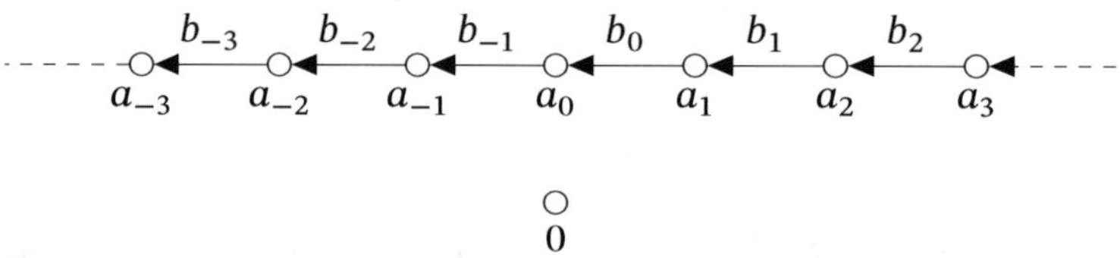

Let **A** be a shift automorphism algebra and let $\mathcal{V}$ denote the shift automorphism variety generated by **A**. We must show three things: that $\mathcal{V}$ has a countably infinite subdirectly irreducible algebra, that $\mathcal{V}$ is locally finite and that $\mathcal{V}^{(n)}$ fails to be locally finite in the finite sense, for all large enough n. Let E denote the set of essential elements of **A**. Let $\mathbf{E}^\sharp$ be the algebra whose universe is $E \cup \{0\}$. We now argue that $\mathbf{E}^\sharp$ is itself a shift automorphism algebra. We begin by showing that σ is an automorphism of $\mathbf{E}^\sharp$.

A set S is **σ-invariant** provided $c \in S$ implies $\sigma(c) \in S$. The automorphism σ will be an automorphism of $\mathbf{E}^\sharp$ provided the set E of essential elements is both σ-invariant and σ^{-1}-invariant.

CLAIM 1.
The automorphism σ when restricted to the algebra $\mathbf{E}^\sharp$ is an automorphism.

Proof. Let $e \in E$ be an essential element of **A**. We aim to show that $\sigma(e)$ is also an essential element of **A**. Since e is essential, it belongs to a tuple of proper entries of a basic operation F of **A**, where F has rank $r+1$. That is,

$$F(a_0, a_1, \ldots, a_r) = a_{r+1}$$

where $e = a_i$ for some $0 \le i \le r+1$ and each a_j is a proper element of $\mathbf{A}$. As σ is an automorphism of $\mathbf{A}$, we see

$$\sigma(F(a_0, a_1, \ldots, a_r)) = \sigma(a_{r+1})$$
$$\Downarrow$$
$$F(\sigma(a_0), \ldots, \sigma(a_r)) = \sigma(a_{r+1})$$

Since each a_j is proper, each $\sigma(a_j)$ is also proper. Hence each $\sigma(a_j)$ is also an essential element whenever a_j is an essential element. As $e = a_i$, we know $\sigma(e) = \sigma(a_i)$ is essential, and thus $\sigma(e) \in E$. Likewise,

$$\sigma^{-1}(F(a_0, a_1, \ldots, a_r)) = \sigma^{-1}(a_{r+1})$$
$$\Downarrow$$
$$F\left(\sigma^{-1}(a_0), \ldots, \sigma^{-1}(a_r)\right) = \sigma^{-1}(a_{r+1})$$

Since each a_j is proper, each $\sigma^{-1}(a_j)$ is also proper. Hence each $\sigma^{-1}(a_j)$ is also an essential element whenever a_j is an essential element. As $e = a_i$, we know $\sigma^{-1}(e) = \sigma^{-1}(a_i)$ is essential, and thus $\sigma^{-1}(e) \in E$. Since the set E is both σ-invariant and σ^{-1}-invariant, the automorphism σ restricted to the algebra $\mathbf{E}^\sharp$ is an automorphism. This completes the proof of Claim 1. ■

Notice that $\mathbf{E}^\sharp$ is partitioned by σ into orbits. Since $\mathbf{A}$ has only finitely many basic operations, by condition (2) of the definition of shift automorphism algebras, the algebra $\mathbf{E}^\sharp$ is partitioned into only finitely many σ-orbits. Notice that $\mathbf{E}^\sharp$ is a countable subalgebra of $\mathbf{A}$ that is itself a shift automorphism algebra. Thus it does no harm to assume that the algebra $\mathbf{A}$ is countable and contains only essential elements together with the default element.

Let m denote the number of σ-orbits into which $A \setminus \{0\}$ is partitioned. Pick representatives $a_0, a_1, \ldots, a_{m-1}$ from these orbits so that $a_0 = a$ where a is the element mentioned in condition (3) of the definition. Arrange the elements of $A \setminus \{0\}$ in a sequence

$$\ldots, a_{-2}, a_{-1}, a_0, a_1, \ldots, a_{m-1}, a_m, \ldots$$

with the order type of the integers so that $\sigma(a_j) = a_{j+m}$. Given this ordering of $A \setminus \{0\}$, it is natural to talk about elements being "to the left of" or "to the right of" other elements in $A \setminus \{0\}$, and the distance between elements in $A \setminus \{0\}$ is defined by the absolute value of the difference of their indices.

By condition (2) of the definition, up to translation by σ, there can only be finitely many pairs of operationally related elements. Let d be the maximum distance between any two operationally related elements of $A \setminus \{0\}$. We require that the choices for representatives for the σ-orbits and their ordering have been chosen in such a way that the parameter d is as small as possible. Notice that the parameters d and m depend only on the algebra $\mathbf{A}$ and its automorphism σ.

We now introduce two new parameters, w_L and w_R. Pick an element a_i from $A \setminus \{0\}$. Let $X = \{a_j : i \le j\}$. Notice that X is the set of elements to the right of a_i together with the element a_i itself. But observe that the subalgebra of $\mathbf{A}$ generated by X may include some elements to the left of a_i. However, any such elements must be generated by the finite set $\{a_j : i \le j < i + d\}$. Since $\mathbf{A}$ is locally finite by assumption, we can conclude that X can generate only finitely many elements to the left of a_i. Let w_i denote the distance between a_i and the element farthest to the left of a_i which is generated by X. Since σ is an automorphism of $\mathbf{A}$, the only numbers that can arise in this way are precisely $w_0, w_1, \ldots, w_{m-1}$. Let w_L be the largest of these numbers. It is meant to denote the farthest distance to the left that we can get from any a_j using generating elements only to the right of a_j. In the same way, let w_R be defined in an analogous manner: it will be the largest member of an m-sequence of numbers.

Let τ be an automorphism of $\mathbf{A}$. We say that a subalgebra $\mathbf{S}$ of $\mathbf{A}$ is τ-**decomposable** provided there is a subalgebra $\mathbf{S}_0$ (called the **core** of the decomposition) such that

$$S = \bigcup_{i \in \mathbb{Z}} \tau^i(S_0)$$

and no element of S_0 is operationally related to any element in any nontrivial τ-translate of S_0. Each τ-translate of $\mathbf{S}_0$ is called a **block** of the decomposition. Note that the τ-decomposition is constructed so that elements from different blocks cannot be operationally related; that is, if F is any basic operation and $s_0, \ldots, s_{u-1} \in S$ such that

$$F(s_0, \ldots, s_{u-1}) \neq 0$$

we are guaranteed that $\{s_0, \ldots, s_{u-1}\} \subset \tau^j(S_0)$ for some unique integer j. Lastly, notice that every τ-decomposable subalgebra of $\mathbf{A}$ is τ-invariant and it is partitioned into τ-orbits.

Some of the subalgebras of $\mathbf{A}$ might turn out to be τ-decomposable while other might not. We need more control of the decomposable subalgebras. Let ℓ be the smallest natural number such that $\ell m > n(d + w_L + w_R)$, where n is an upper bound on the ranks of the basic operations of $\mathbf{A}$. From this point, we reserve τ to denote σ^ℓ. Let $m' = \ell m$. This is the number of τ-orbits into which $A \setminus \{0\}$ is partitioned.

CLAIM 2.
The union of any n τ-orbits of $A \setminus \{0\}$ generates a τ-decomposable subalgebra of $\mathbf{A}$ *whose core is n-generated.*

Proof. Suppose that $Y \subseteq E$ is the union of n τ-orbits. Let $y_0 \in Y$. Scanning the elements of Y to the right we read the sequence

$$y_0, y_1, \ldots, y_{n-1}, y_n = \tau(y_0)$$

The distance between y_0 and y_n is $m' > n(d + w_L + w_R)$. Between adjacent pairs on this sequence there are n gaps. By the Pigeonhole Principle, at least one of these gaps must have length at least $d + w_L + w_R + 1$. So the set Y can be broken up into pieces of size n so that each is the τ translate of one to the left and the gaps that separate these pieces are of length at least $d + w_L + w_R + 1$. Let Y_0 be the piece that contains y_0. Let $\mathbf{S}_0$ denote the subalgebra generated by Y_0. By construction of w_L and w_R, the subalgebra $\mathbf{S}_0$ can extend no farther than w_L points to the left of Y_0 and no farther than w_R points to the right of Y_0. Define $\mathbf{S}_j = \tau^j(\mathbf{S}_0)$ for each integer j. Between any two such translates of $\mathbf{S}_0$, there must be a gap of length at least $d + 1$. So no element of one translate can be operationally related to any other element of another translate. Moreover, the subalgebra generated by Y is

$$\bigcup_{j \in \mathbb{Z}} \tau^j(\mathbf{S}_0)$$

and this subalgebra is τ-decomposable. Claim 2 is proved. ■

The τ-decomposable subalgebra has the structure shown in Figure 7.4. The set of nonzero elements of each block of the decomposition is represented by an ellipse and the element a is the particular element mentioned in condition (3) of the definition of shift automorphism algebra. Condition (3) provides a unary polynomial f and an

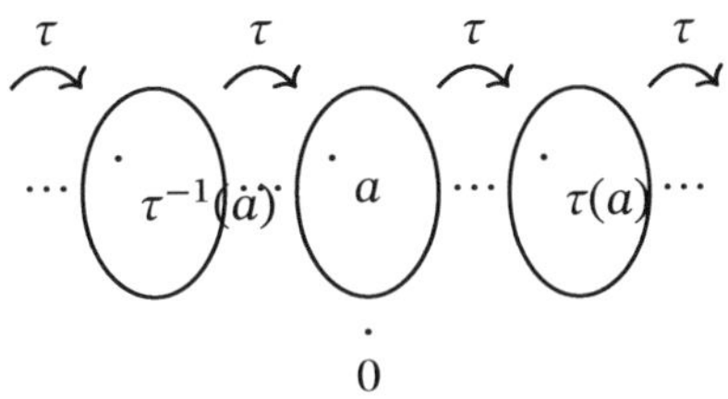

FIGURE 7.4. The τ-Decomposable Subalgebra

essential element a so that $f(a) = \sigma(a)$. We need such a polynomial that will do the same with τ. As we saw above in the proof of Lemma 7.20, let $t(x, y_1, \ldots, y_r)$ be a term in which x occurs and let $c_1, \ldots, c_r$ be essential elements so that $f(x) = t^{\mathbf{A}}(x, c_1, \ldots, c_r)$, which we will denote by $t^{\mathbf{A}}(x, \mathbf{c})$. The polynomial that works with τ and a is $g(x)$ displayed below.

$$g(x) = t^{\mathbf{A}}\left(t^{\mathbf{A}}\left((\cdots t^{\mathbf{A}}(x, \mathbf{c}) \cdots), \sigma^{\ell-2}(c_1), \ldots, \sigma^{\ell-2}(c_r)\right), \sigma^{\ell-1}(c_1), \ldots, \sigma^{\ell-1}(c_r)\right)$$

Notice $g(a) = \tau(a)$. Let $s(x, y_1, \ldots, y_j)$ be a term in which x occurs and let $d_1, \ldots, d_j$ be essential elements so that

$$g(x) = s^{\mathbf{A}}\left(x, d_1, \ldots, d_j\right) \tag{7.3.1}$$

which we will denote by $s^{\mathbf{A}}(x, \mathbf{d})$. By Lemma 7.20, as $t(x, \mathbf{c})$ contains an operation symbol of rank at least two, so must $s(x, \mathbf{d})$. As $g(a) = \tau(a)$, we know

$$\begin{aligned} a &= \tau^{-1}\left(s^{\mathbf{A}}(a, \mathbf{d})\right) \\ &= s^{\mathbf{A}}\left(\tau^{-1}(a), \tau^{-1}(\mathbf{d})\right) \end{aligned}$$

where $\tau^{-1}(\mathbf{d})$ denotes $\tau^{-1}(d_1), \ldots, \tau^{-1}(d_j)$. At some leaf in the evaluation tree displayed in Figure 7.5 whose output is a, we find the element $\tau^{-1}(a)$.

From our argument in Lemma 7.20, we know by starting at $\tau^{-1}(a)$ and following the evaluation tree upward, after only finitely many applications of unary operations, we will come to an operation $Q^{\mathbf{A}}$ of rank at least two, as seen in Figure 7.5. Let H denote this string of unary operation symbols and let $H^{\mathbf{A}}$ be the function obtained by composing the unary operations denoted by the unary operation symbols occurring in the string H.

Thus on the evaluation tree, under the (at least binary) operation $Q^{\mathbf{A}}$ we find at least one node labeled $H^{\mathbf{A}}\left(\tau^{-1}(a)\right)$. Furthermore, since Q is an operation symbol from $\mathbf{A}$, it has rank smaller than n. Since $Q^{\mathbf{A}}$ is not unary, it has at least one other input distinct from $H^{\mathbf{A}}\left(\tau^{-1}(a)\right)$; call these other inputs $d_1, \ldots, d_{k-1}$ where $k \geq 2$ is the rank of Q.

To avoid the accumulation of notation, we will denote $Q^{\mathbf{A}}\left(\tau^{-1}(a), d_1, \ldots, d_{k-1}\right)$ by $Q^{\mathbf{A}}\left(\tau^{-1}(a), \mathbf{d}\right)$. The τ-decomposable subalgebra of $\mathbf{A}$ that is of interest to us is the

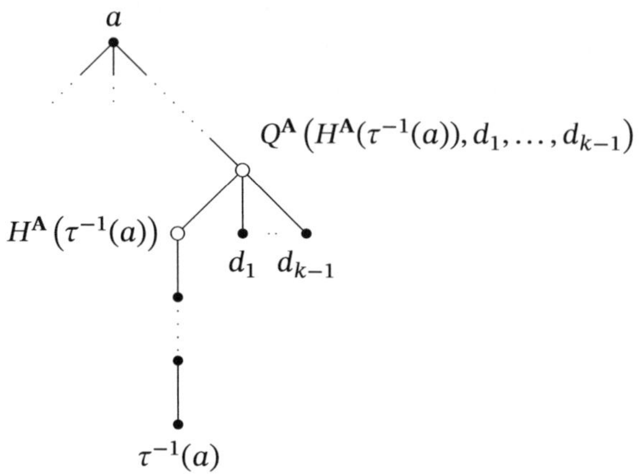

FIGURE 7.5. The Evaluation Tree of the Polynomial $g(x)$

subalgebra generated by the orbits of these leaf elements; namely the orbits of the elements $\tau^{-1}(a), d_1, \ldots, d_{k-1}$. Let **S** be the τ-decomposable subalgebra of **A** generated by the orbits of these elements.

We now proceed to show that the variety $\mathcal{V}$ generated by **A** has a countably infinite subdirectly irreducible member. In order to construct a subdirectly irreducible algebra in $\mathcal{V}$, we begin by making a congruence θ of **A**. The congruence θ is chosen to be a maximal congruence with respect to separating a and 0, where a is the element identified in condition (3). The quotient algebra $\mathbf{A}/\theta$ will be subdirectly irreducible, and provided it is of sufficient size, our construction will be complete. Thus we aim to show that θ has countable many blocks.

We want: Given any natural number j, the element $\sigma^{-j}(a)$ is separated by θ from $\sigma^{-q}(a)$ for any natural number $q \neq j$. The proof of this proceeds by contradiction.

CLAIM 3.
Given any natural number j, if $\tau^{-j}(a)\ \theta\ 0$, then $a\ \theta\ 0$.

Proof of Claim 3. We have seen that a can be written as the output of a term function which contains the input $\tau^{-1}(a)$. In this case, we will say that $\tau^{-1}(a)$ is a **factor** of a or just that $\tau^{-1}(a)$ **divides** a. In general, an element r **divides** an element q if there is an evaluation tree whose output is q and which contains a leaf with the label r. Now,

$$a = s^{\mathbf{A}}\left(\tau^{-1}(a), \tau^{-1}(\mathbf{d})\right)$$
$$\Downarrow$$
$$\tau^{-1}(a) = s^{\mathbf{A}}\left(\tau^{-2}(a), \tau^{-2}(\mathbf{d})\right)$$

and so $\tau^{-2}(a)$ divides $\tau^{-1}(a)$. Combining these, we see

$$a = s^{\mathbf{A}}\left(s^{\mathbf{A}}\left(\tau^{-2}(a), \tau^{-2}(\mathbf{d})\right), \tau^{-1}(\mathbf{d})\right)$$

and so $\tau^{-2}(a)$ divides a. In general, since

$$\tau^{-j}(a) = s^{\mathbf{A}}\left(\tau^{-(j+1)}(a), \tau^{-(j+1)}(\mathbf{d})\right)$$

by back-substituting we see that $\tau^{-j}(a)$ also divides a.

Now suppose that $\tau^{-j}(a)\ \theta\ 0$. Since $\tau^{-j}(a)$ divides a, it is a node on an evaluation tree whose output is a. The top node of any evaluation tree with a node that is θ-related to 0 is itself θ-related to 0. This occurs since 0 is an absorbing element and it is found inside a tuple of an operation symbol of $\mathbf{A}$. Thus if $\tau^{-j}(a)$ is θ-related to 0, it follows that $a\ \theta\ 0$.

FACT 1. *Let j and q be distinct natural numbers. Then*

$$\tau^{-j}(a)\ \theta\ \tau^{-q}(a) \implies \tau^{-(j-1)}(a)\ \theta\ 0$$

Proof. We begin by supposing that $\tau^{-j}(a)$ and $\tau^{-q}(a)$ are related by θ. Since τ is an automorphism, we can apply τ^{-1} to the evaluation tree by applying it to each leaf. This produces an evaluation tree whose output is $\tau^{-1}(a)$ and which has the node

$$Q^{\mathbf{A}}\left(H^{\mathbf{A}}\left(\tau^{-2}(a)\right), \tau^{-1}(\mathbf{d})\right)$$

Repeating this process, the element $\tau^{-(j-1)}(a)$ is the output of an evaluation tree which has the node

$$Q^{\mathbf{A}}\left(H^{\mathbf{A}}\left(\tau^{-j}(a)\right), \tau^{-(j-1)}(\mathbf{d})\right)$$

Since $\tau^{-q}(a)$ and $\tau^{-j}(a)$ are θ-related, we know that

$$H^{\mathbf{A}}\left(\tau^{-j}(a)\right)\ \theta\ H^{\mathbf{A}}\left(\tau^{-q}(a)\right).$$

since H is a composition of unary operation symbols and we have assumed that $\tau^{-j}(a)\ \theta\ \tau^{-q}(a)$. Consider the following:

$$\begin{array}{rcl} H^{\mathbf{A}}(\tau^{-j}(a)) & \theta & H^{\mathbf{A}}(\tau^{-q}(a)) \\ \tau^{-(j-1)}(d_1) & \theta & \tau^{-(j-1)}(d_1) \\ \vdots & & \vdots \\ \tau^{-(j-1)}(d_{k-1}) & \theta & \tau^{-(j-1)}(d_{k-1}) \\ \hline Q^{\mathbf{A}}\left(H^{\mathbf{A}}(\tau^{-j}(a)), \tau^{-(j-1)}(\mathbf{d})\right) & \theta & Q^{\mathbf{A}}\left(H^{\mathbf{A}}(\tau^{-q}(a)), \tau^{-(j-1)}(\mathbf{d})\right) \end{array}$$

The right-side term $Q^{\mathbf{A}}\left(H^{\mathbf{A}}\left(\tau^{-q}(a)\right), \tau^{-(j-1)}(\mathbf{d})\right)$ contains inputs from two different τ-decomposition blocks, so it must evaluate to the default element 0. Meanwhile, the left side is a factor of $\tau^{-(j-1)}(a)$. Since θ is a congruence, this implies $\tau^{-(j-1)}(a)$ is θ-related to 0. This finishes the proof of Claim 1. ■

As a and 0 are not related by θ, we have shown that

$$\left(\tau^{-j}(a), \tau^{-q}(a)\right) \notin \theta$$

for any distinct natural numbers j and q. The quotient algebra $\mathbf{A}/\theta$ will have countably infinitely many elements and it is subdirectly irreducible by the maximality of θ with respect to separating a and 0. Furthermore, this algebra is in $\mathcal{V}$. This concludes the proof of Claim 3. ■

Since the variety $\mathcal{V}$ is not assumed to be generated by a finite algebra, we must show directly that it is locally finite.

CLAIM 4.
The variety $\mathcal{V}$ is locally finite.

Proof. It is enough to find a function $b(n)$ on the positive integers so that every n-generated subalgebra of $\mathbf{A}$ has no more than $b(n)$ elements. Without loss, let n be any natural number larger than the rank of any fundamental operation of $\mathbf{A}$. Let X be any subset of A so that $|X| = n$. Apart from perhaps the default element 0, the only elements that can be generated from X must be essential elements. By insisting that $b(n) > n + 1$, we can suppose that X contains only essential elements. We now pick n τ-orbits in such a way that each element of X lies in one of the selected orbits. Thus the subalgebra generated by X is included in the subalgebra generated by these orbits. Thus the subalgebra generated by X is included in the subalgebra generated by the orbits of the elements in X. Since we have chosen n τ-orbits, the algebra generated by X will be τ-decomposable. Let $\mathbf{S}_0$ be the core of its τ-decomposition. Then the subalgebra generated by X will be contained in the union of the τ-images of the core; that is,

$$\mathrm{Sg}^{\mathbf{A}}(X) \subseteq \bigcup_{i \in \mathbb{Z}} \tau^i(S_0).$$

Since the blocks of the decomposition are operationally unrelated, any finite union of blocks forms a subalgebra. That means that the size of the subalgebra generated by X can be no larger than n times the size of the core S_0. This number still depends on the cardinality of S_0, which is itself an n-generated subalgebra of $\mathbf{A}$. To bound this number, notice that there are only m' many τ-orbits in total. There are $\binom{m'}{n}$ ways to pick n of these orbits. For each such subalgebra we can pick a core algebra for the corresponding τ-decomposable subalgebra. Then n times the size of the largest of these will suffice for $b(n)$. This completes the proof of Claim 4. ■

Given that $\mathcal{V}$ is locally finite, then $\mathcal{V}$ is inherently nonfinitely based in the finite sense provided there is a natural number p so that for all large enough natural numbers n we have that $\mathcal{V}^{(n)}$ has arbitrarily large finite p-generated algebras for some natural number p. Hence our goal is to find a natural number p and finite algebras $\mathbf{B}_{n,k}$ for infinitely many k so that

- Each $\mathbf{B}_{n,k}$ has at least k elements;
- Each $\mathbf{B}_{n,k}$ is generated by a set of p elements; and
- Each $\mathbf{B}_{n,k}$ is in $\mathcal{V}^{(n)}$.

We will now construct the algebras $\mathbf{B}_{n,k}$.

Recall that all nonzero elements of A are assumed to be essential. We consider a partial subalgebra of the $2k$-fold direct power of this algebra. Let e be an essential element and let U be a subset of $\{0, 1, \ldots, 2k-1\}$. By $e|U$ we mean the $2k$-tuple that has e at the jth position for all $j \in U$ and 0 at all other positions.

Let μ be the map from the set $\mathbb{Z}$ of integers to $\{0, 1, \ldots, 2k-1\}$ defined by

$$\mu(z) \equiv z \pmod{2k} \text{ for all } z \in \mathbb{Z}.$$

We say that the subset $U \subseteq \{0, 1, \ldots, 2k-1\}$ is **contiguous modulo** $2k$ provided it is the image under the map μ of an interval in $\mathbb{Z}$. Define G as the set of $2k$-tuples $e|U$ for which

- e is an essential element;
- U is a nonempty contiguous set of integers modulo $2k$; and
- $|U| < k$.

The set G is the universe of a partial subalgebra $\mathbf{G}$. We use $\mathbf{G}^\sharp$ to denote the subalgebra of $\mathbf{A}^{2k}$ obtained by adjoining the constantly 0 tuple to $\mathbf{G}$, where the value 0 to all products undefined in $\mathbf{G}$.

We now provide an example of such an $e|U$. Rather than considering $e|U$ as a usual $2k$-tuple, we will consider it as having a circular form where the positions are as they appear on an analog clock. Let $k = 4$, let $U = \{0, 1\}$, and let e be an essential element of $\mathbf{A}$. Then $e|U$ has the following form:

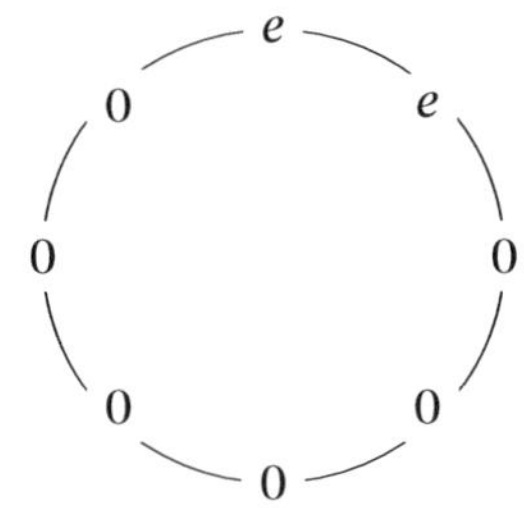

Notice that this depiction gives rise to two automorphisms in a natural way. The automorphism τ of $\mathbf{A}$ extends naturally to a coordinatewise defined automorphism on $\mathbf{G}^\sharp$; namely, we can define τ on $\mathbf{G}^\sharp$ as follows:

$$\tau(e|U) := \tau(e)|U$$

Another automorphism becomes clear using the circular depiction of $e|U$. Let ρ denote the cyclic shift on the coordinate set U. We note that the automorphisms τ and ρ commute. Furthermore, by applying $\tau \circ \rho$ to the particular $e|U$ that appears above, we find

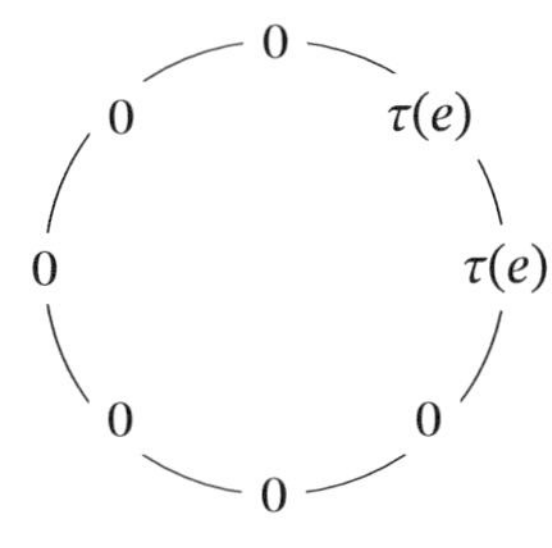

and so $\tau \circ \rho\,(e|U) = \tau(e)|\rho(U)$. Notice that the collection of U's that play a role in G is closed with respect to the cyclic shift.

CLAIM 5.
$\tau \circ \rho$ partitions the partial algebra $\mathbf{G}$ into $(2k^2 - k)m'$ orbits and these orbits are congruence classes of some congruence relation of $\mathbf{G}$.

Proof. As we noted before, there are m' τ-orbits of $A \setminus \{0\}$. The number of $\tau \circ \rho$-orbits of $\mathbf{G}$ will be the product of m' and the number of possible choices for the cyclically contiguous nonempty sets U with cardinality no more than k. To define a particular U, it is enough to describe the cardinality of U and the starting point of the contiguous interval (modulo $2k$). There are $2k$ choices for the starting point of the contiguous interval (modulo $2k$) of U. Given a starting point of this interval, there are $k-1$ choices for its length. Thus there are $2k \cdot (k-1) = 2k^2 - k$ possible sets U. Hence the total number of $\tau \circ \rho$-orbits in $\mathbf{G}$ is $(2k^2 - k) \cdot m'$, as claimed.

Since $\tau \circ \rho$ partitions $\mathbf{G}$, these orbits form the equivalence classes of some equivalence relation. We just need to show that this equivalence relation is a congruence. Suppose Q is a basic operation symbol of rank r and that

$$Q^{\mathbf{G}}(g_0|U_0, \ldots, g_{r-1}|U_{r-1}) = g_r|U_r$$
$$Q^{\mathbf{G}}(h_0|V_0, \ldots, h_{r-1}|V_{r-1}) = h_r|V_r$$
$$(\tau \circ \rho)^{e_j} g_j|U_j = h_j|V_j \text{ for } j < r.$$

We need to verify that $g_r|U_r$ and $h_r|V_r$ lie in the same $\tau \circ \rho$-orbit.

Given that 0 is an absorbing element, the result of $Q^{\mathbf{G}}(g_0|U_0, \ldots, g_{r-1}|U_{r-1})$ must contain 0 at each coordinate in the complement of $U_0 \cap \cdots \cap U_{r-1}$. That is,

$$U_0 \cap U_1 \cap \cdots \cap U_{r-1} = U_r.$$

Furthermore, $Q^{\mathbf{A}}(g_0, \ldots, g_{r-1}) = g_r$. Likewise,

$$V_0 \cap V_1 \cap \cdots \cap V_{r-1} = V_r \text{ and}$$
$$Q^{\mathbf{A}}(h_0, \ldots, h_{r-1}) = h_r.$$

Since $\tau \circ \rho$ acts in a coordinatewise manner, we have

$$\tau^{e_j}(g_j) = h_j \text{ for all } j < r, \text{ and}$$
$$\rho^{e_j}(U_j) = V_j \text{ for all } j < r$$

Provided that $e_0 = e_1 = \cdots = e_{r-1}$, we will have our desired conclusion.

Having chosen n so that it is greater than the rank of any basic operation of $\mathbf{A}$, the elements $g_0, \ldots, g_{r-1}$ cannot be in more than n τ-orbits. Since each g_j is τ-related to each h_j (where $j < r$), the elements $h_0, \ldots, h_{r-1}$ are in those same τ-orbits. All these essential elements must belong to a single block of the τ-decomposable subalgebra. Since $g_0, \ldots, g_{r-1}$ are operationally related, they must belong to a single τ-translate of the core. Likewise, since $h_0, \ldots, h_{r-1}$ are operationally related, they must belong to a single τ-translate of the core. Therefore, there is a single e so that τ^e carries the block containing the g_j's to the block containing the h_j's. Hence $e = e_0 = e_1 = \cdots = e_{r-1}$. ■

We now know that the $\tau \circ \rho$-orbits of $\mathbf{G}$ are really congruence classes of some congruence relation of $\mathbf{G}$. Let γ denote that congruence of $\mathbf{G}$. Define $\mathbf{G}_{n,k}$ as the algebra made by adding the default absorbing element to $\mathbf{G}/\gamma$. Since $\mathbf{G}/\gamma$ has $2k^2m'$ elements by Claim 5, the algebra $\mathbf{G}_{n,k}$ has cardinality $2k^2m' + 1$.

CLAIM 6.
$\mathbf{G}_{n,k} \in \mathcal{V}^{(n)}$.

Proof. Consider n proper elements of $\mathbf{G}/\gamma$. These correspond to n $\tau \circ \rho$-orbits of $\mathbf{G}$. From these, we can obtain n τ-orbits of A since each $\tau \circ \rho$-orbit has an associated τ-orbit. Let $\mathbf{D}$ be the subalgebra generated by the union of these orbits. Since $\mathbf{D}$ is n-generated, it is τ-decomposable by Claim 2. Let $\mathbf{S}$ be the core of this decomposition. Now we can pick elements $g_0, \dots, g_{n-1}$ of S and appropriate nonempty subsets $U_0, \dots, U_{n-1}$ of $\{0, 1, \dots, 2k-1\}$ so that

$$g_0|U_0, \dots, g_{n-1}|U_{n-1}$$

is a system of representatives of our original $\tau \circ \rho$ orbits. Let $\mathbf{H}$ be the subalgebra of $\mathbf{G}^\sharp$ generated by

$$\{g_0|U_0, \dots, g_{n-1}|U_{n-1}\}.$$

Let π be the map from H to $G_{n,k}$ defined so that each proper element of H is assigned the $\tau \circ \rho$-orbit to which it belongs and the absorbing element of $\mathbf{H}$ is assigned the absorbing element of $\mathbf{G}_{n,k}$. We aim to show that π is a homomorphism from $\mathbf{H}$ into $\mathbf{G}_{n,k}$.

Let Q be a basic operation symbol of rank r and let $h_0|V_0, \dots, h_{r-1}|V_{r-1}$ be proper elements of H. Since the operation $Q^{\mathbf{H}}$ is defined coordinatewise, we have

$$\pi\left(Q^{\mathbf{H}}(h_0|V_0, \dots, h_{r-1}|V_{r-1})\right) = \pi\left(Q^{\mathbf{A}}(h_0, \dots, h_{r-1})\,|V_0 \cap \dots \cap V_{r-1}\right). \tag{7.3.2}$$

Since each $h_j|V_j$ is a proper element (where $j < r$), the map π sends each $h_j|V_j$ to $h_j|V_j/\gamma$. Thus

$$Q^{\mathbf{G}_{n,k}}(\pi(h_0|V_0), \dots, \pi(h_{r-1}|V_{r-1})) = Q^{\mathbf{G}_{n,k}}(h_0|V_0/\gamma, \dots, h_{r-1}|V_{r-1}/\gamma). \tag{7.3.3}$$

We aim to show that equation (7.3.2) and equation (7.3.3) are logically equivalent. If the right side of equation (7.3.3) is the default element, the equivalence is clear. If it is not the default element, then there must be representatives $w_0|W_0, \dots, w_{r-1}|W_{r-1}$ of the γ-classes such that

$$Q^{\mathbf{G}}(w_0|W_0, \dots, w_{r-1}|W_{r-1})$$

is defined in $\mathbf{G}$. In this case, the τ-decomposability of $\mathbf{D}$ entails that

$$Q^{\mathbf{G}}(h_0|V_0, \dots, h_{r-1}|V_{r-1})$$

is defined as well. Given that this is defined, all four sides of the above equations are equivalent, and hence π is a homomorphism.

The original n elements of $\mathbf{G}_{n,k}$ all belong to the image of the homomorphism π, and so the subalgebra they generate belongs to $\mathcal{V}$ as $\mathbf{H}$ belongs to $\mathcal{V}$. This completes Claim 6. ■

The algebra $\mathbf{G}_{n,k}$ has more than k elements and is found in $\mathcal{V}^{(n)}$; however, its generating set is not of the right cardinality. We will now define p and the algebra $\mathbf{B}_{n,k}$.

In equation (7.3.1), we constructed a polynomial $g(x)$ so that $g(a) = \tau(a)$ where a is the element mentioned in condition (3). The term $s(x, y_1, \dots, y_j)$ and the essential elements $d_1, \dots, d_j$ of $\mathbf{A}$ so that

$$g(x) = s^{\mathbf{A}}(x, d_1, \dots, d_j)$$

were also specified. Let $p = j + 1$. Notice p is one greater than the number of "coefficients" of $g(x)$. Furthermore, notice that $p = \ell r + 1$, where r is the number of "coefficients" in the original polynomial $f(x)$. Let $Y_i = \{0, \dots, i-1\}$ for each $i < k$. Let Z_i be the preimage of Y_i under the cyclic shift ρ. That is, Z_i is Y_i turned one notch counter-clockwise; explicitly,

$$Z_i = \{2k-1, 0, 1, \dots, i-2\}.$$

Let $\mathbf{B}_{n,k}$ be the subalgebra of $\mathbf{G}_{n,k}$ generated by the following elements:

$$a|Y_k/\gamma, d_1|Y_k/\gamma, \dots, d_{p-1}|Y_k/\gamma$$

where each d_i (for $i < p$) are those found in equation (7.3.1).

As $\mathbf{B}_{n,k}$ is generated by p elements and is found in the variety $\mathcal{V}^{(n)}$, the only thing left to show is that $\mathbf{B}_{n,k}$ has at least k elements. We claim that the following k elements

$$a|Z_1/\gamma, a|Z_2/\gamma, \ldots, a|Z_k/\gamma$$

are distinct and all belong to $\mathbf{B}_{n,k}$.

Notice

$$\begin{aligned} s^{\mathbf{G}}\left(a|Y_k, d_1|Y_k, \ldots, d_{p-1}|Y_k\right) &= s^{\mathbf{A}}\left(a, d_1, \ldots, d_{p-1}\right)|Y_k \cap \cdots \cap Y_k \\ &= s^{\mathbf{A}}\left(a, d_1, \ldots, d_{p-1}\right)|Y_k \\ &= \tau(a)|Y_k. \end{aligned}$$

Now, $\tau(a)|Y_k$ is γ-related to $a|Z_k$ since the preimage of $\tau(a)$ under τ is a and the preimage of Z_k under ρ is Y_k. Using a similar argument, we see

$$\begin{aligned} s^{\mathbf{G}}\left(a|Z_k, d_1|Y_k, \ldots, d_{p-1}|Y_k\right) &= s^{\mathbf{A}}\left(a, d_1, \ldots, d_{p-1}\right)|Z_k \cap Y_k \cap \cdots \cap Y_k \\ &= s^{\mathbf{A}}\left(a, d_1, \ldots, d_{p-1}\right)|Y_{k-1} \\ &= \tau(a)|Y_{k-1} \\ &\gamma\ a|Z_{k-1} \end{aligned}$$

and

$$\begin{aligned} s^{\mathbf{G}}\left(a|Z_{k-1}, d_1|Y_k, \ldots, d_{p-1}|Y_k\right) &= s^{\mathbf{A}}\left(a, d_1, \ldots, d_{p-1}\right)|Z_{k-1} \cap Y_k \cap \cdots \cap Y_k \\ &= s^{\mathbf{A}}\left(a, d_1, \ldots, d_{p-1}\right)|Y_{k-2} \\ &= \tau(a)|Y_{k-2} \\ &\gamma\ a|Z_{k-2}. \end{aligned}$$

We can use this same argument to verify that

$$a|Z_k/\gamma, a|Z_{k-1}/\gamma, \ldots, a|Z_1/\gamma$$

are all distinct elements. Since they all belong to $\mathbf{B}_{n,k}$, it has more than k elements, as desired.

Taking everything into account, we have shown that $\mathcal{V}$ is a locally finite variety, that $\mathcal{V}^{(n)}$ fails to be locally finite in the finite sense, for all large enough values of n (so $\mathcal{V}$ is inherently nonfinitely based in the finite sense), and that $\mathcal{V}$ contains an infinite subdirectly irreducible algebra. This finishes the proof of the Shift Automorphism Theorem. ■

The Shift Automorphism Theorem asserts two conclusions beyond the central conclusion that shift automorphism algebras are inherently nonfinitely based. The first conclusion, that they are inherently nonfinitely based in the finite sense, means that the Shift Automorphism Theorem cannot be used to construct directly a counterexample to the Eilenberg-Schützenberger Conjecture. The second conclusion, that shift automorphism varieties must always have infinite subdirectly irreducible members, means that this theorem cannot be used directly to construct a counterexample to

Jónsson's Speculation

Is every finite algebra of finite signature, which generates a variety with a finite residual bound, actually finitely based?

A variety is said to have residual bound κ provided every subdirectly irreducible algebra in the variety has cardinality less than κ. Bjarni Jónsson offered this speculation in the mid-1970's, noting finite algebras that generate varieties so that the congruence lattice of each algebra in the variety is distributive have finite residual bounds. In 1971, Kirby Baker had proven that such finite algebras of finite signature are finitely based. This significant result ultimately appeared as (Kirby A. Baker, 1977). These results of Jónsson and of Baker will be elaborated in Chapter 8.

The Shift Automorphism Theorem grew out of preliminary work done by V. L. Murskiĭ (1979). The original version of this theorem comes from (Kirby A. Baker, George F. McNulty, and Heinrich Werner, 1989). Later work by Kathryn S. Owens (2010) and George F. McNulty, Zoltán Székely, and Ross Willard (2008) addressed the residual bound and the property of being inherently nonfinitely based in the finite sense, respectively.

Examples of Inherently Nonfinitely Based Finite Algebras

Our first example is like Lyndon's algebra. It is the automatic algebra **R** with the diagram displayed below:

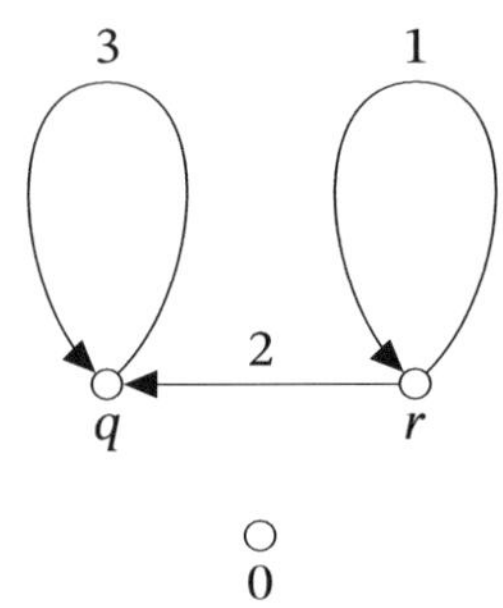

FIGURE 7.6. McKenzie's automatic algebra **R**

This algebra appears in (Ralph McKenzie, 1996b). Of course, **R** is finite, so we cannot apply the Shift Automorphism Theorem to it directly. Instead, we will construct an algebra $\mathbf{A} \in \boldsymbol{HSP}\,\mathbf{R}$ to which the Shift Automorphism Theorem does apply. Here it is.

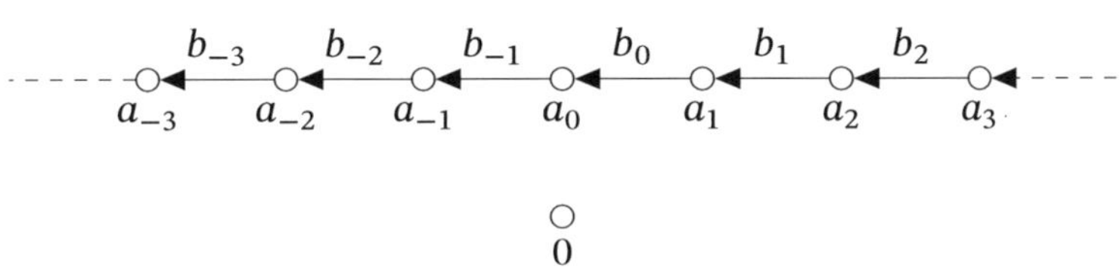

The algebra **A**

Here, we define the product $\cdot$ as follows:

$$b_k \cdot a_{k+1} = a_k$$

and all other products are 0. The automorphism σ that we need works as follows

$$\sigma(a_{k+1}) = a_k \qquad \sigma(b_{k+1}) = b_k \qquad \sigma(0) = 0.$$

So there are two infinite orbits and one finite orbit, namely $\{0\}$. The proper triples have the form (b_k, a_{k+1}, a_k) and σ induces just one orbit on these triples. Finally, $\sigma(a_1) = b_0 \cdot a_1 = f(a_1)$ where $f(x) = b_0 \cdot x$. So $\mathbf{A}$ is a shift automorphism algebra.

Fact 2. $\mathbf{A} \in \boldsymbol{HSP}\,\mathbf{R}$.

Proof. In $\mathbf{R}^{\mathbb{Z}}$, let B be the set of all $\mathbb{Z}$-tuples of the following form: For each $k \in \mathbb{Z}$, define α_k and β_k as follows:

$$\begin{array}{rcccccccc} \alpha_k := & \cdots & r & r & r & q & q & q & q & \cdots \\ \beta_k := & \cdots & 1 & 1 & 1 & 2 & 3 & 3 & 3 & \cdots \end{array}$$

The k is meant to designate the rightmost occurrence of r in α_k, and likewise the rightmost occurrence of 1 in β_k. Notice that $\beta_k \cdot \alpha_{k+1} = \alpha_k$. No matter how we multiply some $\beta_n \cdot \alpha_m$, we'll either get some α_{m-1} or we'll get a string that contains a 0. Now let B consist of all the α_k's and all the β_k's as well as every member of $A^{\mathbb{Z}}$ that has at least one entry that is 0.

The point is that B is a subuniverse of $\mathbf{R}^{\mathbb{Z}}$. Let θ be the equivalence relation on B that lumps together all $\mathbb{Z}$-tuples containing 0 and isolates everything else. That is, θ is going to isolate each α_i by itself and it will isolate each β_j by itself.

We contend that θ is a congruence of the algebra $\mathbf{B}$.

Evidently, θ an equivalence relation. It also respects the operation: just consider

$$\begin{array}{ccc} \delta & \theta & \gamma \\ \mu & \theta & \nu \\ \hline \mu\cdot\delta & \theta & \nu\cdot\gamma \end{array}$$

No matter how we pick δ, γ, μ, and ν in B, this will always work out. This is because our big lump will absorb everything containing 0.

But we have arranged matters so that $\mathbf{A} \cong \mathbf{B}/\theta$. Hence, $\mathbf{A} \in \boldsymbol{HSP}\,\mathbf{R}$, as desired. ∎

It follows that $\mathbf{R}$ is inherently nonfinitely based.

Our second example has only 3 elements. It is made from the graph below.

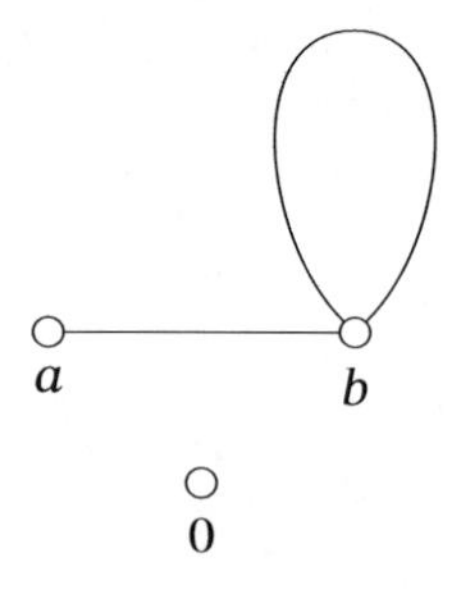

FIGURE 7.7. Murskiĭ's algebra $\mathbf{M}$

This algebra appears in (V. L. Murskiĭ, 1965), where it is shown not to be finitely based. V. L. Murskiĭ (1979) proves that it is inherently nonfinitely based. We reach this conclusion on page 193. In graph algebras like $\mathbf{M}$ the universe consists of the vertices of the graph together with a default element 0 and the basic two place operation works as follows:

$$u \cdot v = \begin{cases} u & \text{if } u \text{ is adjacent to } v \\ 0 & \text{otherwise} \end{cases}$$

Notice that the graph in Figure 7.7 has two vertices, labeled a and b and two edges, one a loop. The edges are not labeled, unlike with automatic algebras. The universe of $\mathbf{M}$ consists of the vertices of the graph and a default element. The operation is defined in terms of adjacency in the graph.

Once more, we construct a shift automorphism algebra $\mathbf{A} \in \boldsymbol{HSP}\,\mathbf{M}$. Here it is

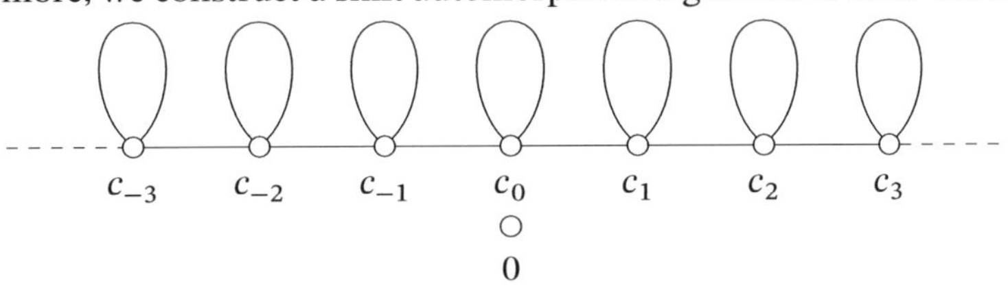

The Shift Automorphism Algebra $\mathbf{A}$

In this algebra, the set $A = \{c_k \mid k \in \mathbb{Z}\} \cup \{0\}$ and the operation is defined as it was in $\mathbf{M}$:

$$u \cdot v = \begin{cases} u & \text{if } u \text{ is adjacent to } v \\ 0 & \text{otherwise} \end{cases}$$

The automorphism σ works so that $\sigma(c_{k+1}) = c_k$ for all $k \in \mathbb{Z}$ and $\sigma(0) = 0$. Consequently, there are just two orbits: $\{c_k \mid k \in \mathbb{Z}\}$ and $\{0\}$. Finally, $\sigma(c_1) = c_0 = c_0 \cdot c_1 = f(c_1)$, where $f(x) = c_0 \cdot x$. So $\mathbf{A}$ is a shift automorphism algebra.

As in the first example, it remains to show that $\mathbf{A} \in \boldsymbol{HSP}\,\mathbf{M}$. We will make short work of this. In $M^{\mathbb{Z}}$, for each integer k, let

$$\gamma_k := \quad \cdots \quad b \quad b \quad b \quad a \quad b \quad a \quad b \quad b \quad a \quad b \quad b \quad b \quad \cdots$$

where the leftmost a occurs at position k. Observe that γ_k has an infinite string of b's running to the left, while to the right the occurrences of a are separated by longer and longer strings of b's. Now note

$$\begin{array}{rcccccccccccccc} \gamma_{k+1} = & \cdots & b & b & b & a & b & a & b & b & a & b & b & b & \cdots \\ \gamma_k = & \cdots & b & b & a & b & a & b & b & a & b & b & b & a & \cdots \\ \hline \gamma_k \cdot \gamma_{k+1} = \gamma_k = & \cdots & b & b & a & b & a & b & b & a & b & b & b & a & \cdots \end{array}$$

and

$$\begin{array}{rcccccccccccccc} \gamma_k = & \cdots & b & b & b & a & b & a & b & b & a & b & b & b & \cdots \\ \gamma_{k+1} = & \cdots & b & b & b & b & a & b & a & b & b & a & b & b & \cdots \\ \hline \gamma_{k+1} \cdot \gamma_k = \gamma_{k+1} = & \cdots & b & b & b & b & a & b & a & b & b & a & b & b & \cdots \end{array}$$

On the other hand, any other product of the γ_k's will produce a $\mathbb{Z}$-tuple with at least one entry that is 0. The rest of the argument proceeds as in the first example.

So Murskiĭ's algebra $\mathbf{M}$ is inherently nonfinitely based.

These two examples explain why we have called *shift* automorphism algebras what we did. The automorphisms given here are indeed shifts.

To construct these four examples, the algebras of Lyndon, Lee, McKenzie and Murskiĭ, we built algebras from finite automata and from finite graphs (and we could have done something similar with directed graphs). Since most of the rich assortment of (finite) algebras that arose in the 19th century turn out to be finitely based, we have used finite mathematical structures that came into their own in the 20th century.

Exercises 7.22

1. Construct a locally finite algebra **A** so that the variety ***HSP*** **A** is not locally finite.

2. Prove that if **A** is uniformly locally finite, then ***HSP*** **A** is locally finite.

3. (Edmond W. H. Lee, 2009) Prove that Lyndon's algebra **L** and Lee's algebra $\mathbf{L}^*$ generate the same variety. [Hint: in the text we saw the Lee's algebra is a quotient of Lyndon's algebra. Discover a second congruence relation of Lyndon's algebra that also produces Lee's algebra. Form the subdirect product.]

***4.** (Tamás Bajusz, George McNulty, and Ágnes Szendrei, 1990) Add a couple of elements to Lyndon's Algebra **L** to obtain a finite algebra **B** so that **L** is a subalgebra of **B** and **B** is finitely based. In this way, establish that even though **L** fails to be finitely based, it also fails to be inherently nonfinitely based.

5. (Robert E. Park, 1976, 1980) Prove that the four-element algebra **P**, with just one basic operation and that a two-place operation whose operation table is displayed below is inherently nonfinitely based

·	0	a	b	c
0	0	0	0	0
a	0	a	b	0
b	0	b	b	c
c	0	0	c	c

6. (Mark V. Sapir, 1988; Robert Cacioppo, 1987, 1993) Prove than any counterexample to the Eilenberg-Schützenberger Conjecture must be inherently nonfinitely based.

7. (Kirby A. Baker, George F. McNulty, and Heinrich Werner, 1989) Prove that the graph algebras associated with the three graphs below are each inherently nonfinitely based.

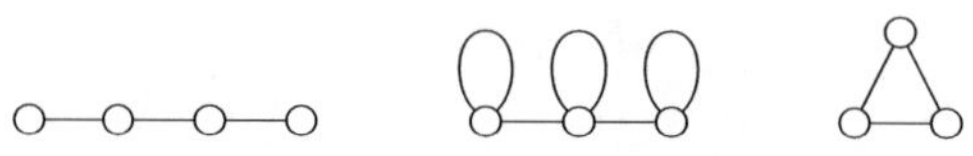

8. Find finite bases for the graph algebras associated to the path of length 2 and to the looped path of length 1 displayed below.

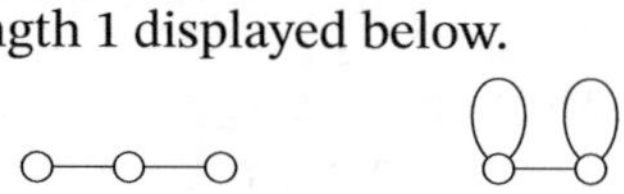

9. (Kirby A. Baker, George F. McNulty, and Heinrich Werner, 1989) Let **F** be the partial lattice displayed below.

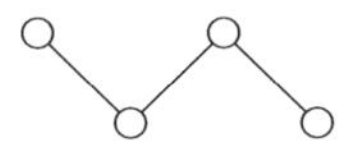

Let $\mathbf{F}^{\#}$ be the algebra obtained from $\mathbf{F}$ by adjoining a new default element 0 and assigning 0 as the value of all operations undefined in $\mathbf{F}$. Prove that $\mathbf{F}^{\#}$ is inherently nonfinitely based.

10. (Kirby A. Baker, George F. McNulty, and Heinrich Werner, 1989) Let $\mathbf{H}$ be the partial algebra on the 3-element set $\{r, s, t\}$ that has one operation (that one being a 2-place operation) described by

$$rr = r \quad ss = s \quad tt = t \quad rs = t.$$

Let $\mathbf{H}^{\#}$ be the algebra obtained from $\mathbf{H}$ by adjoining a new default element 0 and assigning the value 0 to any product left undefined in $\mathbf{H}$. Prove $\mathbf{H}^{\#}$ is inherently nonfinitely based.

* **11.** (Jaroslav Ježek, 1985) Let $\mathbf{G}$ be a finite idempotent algebra with just one fundamental operation, that one being a 2-place operation. Suppose that $\mathbf{G}$ has the further property

$$\mathbf{G} \models xy \approx t \text{ if and only if } xx \approx x \models xy \approx t$$

for every term t. Then $\mathbf{G}$ is inherently nonfinitely based. Prove the corresponding result when the commutative law is added and

$$\mathbf{G} \models xy \approx t \text{ if and only if } xx \approx x, xy \approx yx \models xy \approx t$$

for every term t. The exercise above may prove useful.

Observe that the algebras satisfying Ježek's hypothesis need no absorbing element, but they generate varieties that have shift automorphism algebras.

* **12.** (Jaroslav Ježek, 1985) Invent 3 idempotent 3-element algebras that satisfy the conditions of the problem above, and hence are inherently nonfinitely based.

** **13.** (Peter Perkins, 1984; Kirby A. Baker, George F. McNulty, and Heinrich Werner, 1989). Let $\mathbf{A}$ be a finite algebra with just one fundamental operation, that one a 2-place operation. Suppose further that $\mathbf{A}$ has a zero element 0 (so $0a = a0 = 0$ for all $a \in A$) and a unit element 1 (so that $1a = a1 = a$ for all $a \in A$), and that $\mathbf{A}$ fails to be commutative or associative or to satisfy any equation of the form $t \approx x$ where the term t is not a variable. Prove that $\mathbf{A}$ is inherently nonfinitely based. Construct such an $\mathbf{A}$ with just four elements.

** **14.** (Peter Perkins, 1966, 1969) Let $\mathbf{B}_2^1$ be the semigroup consisting of the following 6 matrices over the 2-element field.

$$\begin{pmatrix}0&0\\0&0\end{pmatrix} \quad \begin{pmatrix}1&0\\0&1\end{pmatrix} \quad \begin{pmatrix}1&0\\0&0\end{pmatrix} \quad \begin{pmatrix}0&1\\0&0\end{pmatrix} \quad \begin{pmatrix}0&0\\1&0\end{pmatrix} \quad \begin{pmatrix}0&0\\0&1\end{pmatrix}$$

where matrix multiplication is the operation. Prove that $\mathbf{B}_2^1$ is not finitely based. The notation $\mathbf{B}_2^1$ follows the practice in the theory of semigroups. This is one of the Brandt semigroups. Actually, this semigroup is inherently nonfinitely based. Mark V. Sapir (1987) proved this as part of his algorithmic characterization of inherently nonfinitely based finite semigroups. This characterization, which relies on methods appropriate to semigroups, is beyond the scope of this work.

7.4. Equational Theories that are Finitely Axiomatizable

Sheila Oates and M. B. Powell (1964) proved that every finite group is finitely based. Almost a decade later Robert L. Kruse (1973) and independently I. V. L'vov (1973) proved that every finite ring is finitely based. These theorems rely on a detailed development of the theories of groups and rings, putting their proofs beyond the scope of this book. A nice exposition of the finite basis result for finite groups, together with the necessary group theoretic background, can be found in (Hanna Neumann, 1967).

Between these two finite basis theorems, Ralph McKenzie (1970) proved that every finite lattice is finitely based. Subsequently, this theorem was extended, first by Kirby A. Baker (1977) to finite algebras that belong to congruence distributive varieties, then by Ralph McKenzie (1987) to finite algebras that belong to congruence modular varieties with a finite residual bound, and then by Ross Willard (2000) to finite algebras that belong to congruence meet-semidistributive varieties with a finite residual bound. And finally to finite algebras generating varieties with a finite residual bound and a difference term by Keith A. Kearnes, Ágnes Szendrei, and Ross Willard (2016, 2022). And there have been alternative proofs to many of these results as well as various extensions to quasivarieties. The proofs of all these generalizations have a heavily algebraic character. In particular, they all depend on an analysis of subdirectly irreducible algebras in some way.

We prove Willard's Finite Basis Theorem later in this section (see page 204). Baker's Finite Basis Theorem is an easy consequence of Willard's Theorem, given some theorems from model theory. We shall provide that argument in §8.4. Given the influential place of Baker's Finite Basis Theorem, we provide two additional proofs in §8.5. McKenzie's 1987 Finite Basis Theorem can be found as Theorem 11.119 on page 367 in Volume III, once we have developed commutator theory. The proof of the finite basis theorem of Kearnes, Szendrei, and Willard, requires tame congruence theory, putting it beyond reach.

On the other hand, the attractive prospect that Baker's Finite Basis Theorem might extend to the statement *Every finite algebra of finite signature that belongs to a congruence modular variety is finitely based*, thus capturing the finite basis results for finite groups and rings, as well as Baker's Theorem, was destroyed by a series of counterexamples, all of them even belonging to congruence permutable varieties. S. V. Polin (1976) produced the first counterexample. Perhaps the most notable of a number of other counterexamples is the finite pointed group of (Roger M. Bryant, 1982) (a pointed group is just a group expanded by an additional constant symbol to the signature to denote a particular element) and the inherently nonfinitely based 5-dimensional vector space (which may be taken over any finite field) expanded by a new pathological two-place operation devised by I. M. Isaev (1989). We provide a new proof of Isaev's Theorem in Volume III; see Theorem 11.117 on page 357 in Volume III.

Given the algebraic character of the extensions of McKenzie's Finite Basis Theorem for Finite Lattices, it is striking, then, in retrospect, that McKenzie's original proof is syntactic, relying on the invention of a particular normal form function for terms. So it seems reasonable to take another look at that original argument with an eye toward seeing if the argument itself can suggest a fresh more syntactic approach to finite basis theorems. What follows is, to all intents and purposes McKenzie's original published argument, although framed a little differently.

Every Finite Lattice is Finitely Based

For every natural number n let $\ell(n) := n^m + 1$, where m is the smallest integer bigger than 1 so that $n < 2^{m+1}$. For example, $\ell(2) = 2^2 + 1 = 5$, $\ell(12) = 12^3 + 1$ and $\ell(120) = 120^6 + 1$. The function $\ell(n)$ is, roughly, $2^{(\log_2(n))^2} = n^{\log_2(n)}$ and it gets above and stays above any polynomial.

THEOREM 7.23 (McKenzie's Finite Basis Theorem For Finite Lattices). *Every finite lattice* $\mathbf{L}$ *of cardinality* n *has an equational basis using no more than* $\ell(n)$ *variables.*

Proof. We let $\mathbf{L}$ be a lattice of cardinality n and we put $\ell := \ell(n)$. We also take the number m described during the definition of ℓ as fixed.

Let $\mathcal{V}$ denote the variety generated by $\mathbf{L}$. Our aim is to show that $\mathcal{V} = \mathcal{V}^{(\ell)}$.

The variety of all lattices has a base using just 3 variables. Since we want all the algebras we have to contend with to be lattices, we insist that n be large enough so that $3 \leq \ell(n)$. This means we want $n > 1$. Of course, we know the 1-element lattice is based on the equation $x \approx y$, which has 2 variables. Since $\ell(1) = 2$, the theorem holds in case $n = 1$.

Let X be any finite set of variables. Below, we will define a (normal form) function η_X on terms whose variables are drawn from X that has the following properties for all terms s and t whose variables lie in X. In what follows, X can always be understood from the context so we use η in place of η_X.

(A) If $s \approx t$ is true in $\mathbf{L}$, then $\eta(s) \approx \eta(t)$ is true in $\mathcal{V}^{(\ell)}$, and
(B) $s \approx \eta(s)$ is true in $\mathcal{V}^{(\ell)}$.

Once such functions η are in hand, the proof of the theorem will be complete. Functions with properties (A) and (B) are an adaptation of the notion of a normal form function that was discussed briefly on page 239 in Volume I.

Let $Y \subseteq \{v_0, v_1, \dots\}$ be a finite set of variables, let A be any finite set, and let $f : Y \to A$. We let μ_f be the substitution determined by

$$\mu_f(x) = \begin{cases} \bigwedge_{f(z)=f(x)} z & \text{if } x \in Y \\ x & \text{if } x \in \{v_0, v_1, \dots\} \setminus Y \end{cases}$$

(The alert reader may have noted that $\bigwedge_{f(z)=f(y)} z$ is not, properly speaking, a term—one must actually order the meetands and install parentheses appropriately—we leave that in the hands of the reader, here and below). We also let μ_f^o be the the substitution $y \mapsto y_f$, where y_f is the variable in Y of least index such that $f(y_f) = f(y)$ and which fixes every other variable.

For each finite $X \subseteq \{v_0, v_1, \dots\}$, the function η_X we want is defined by

$$\eta_X(s) := \bigvee_{f:X\to L} (\mu_f(s)) \text{ for all terms } s.$$

For any terms s and t whose variables all come from Y and for all $f : Y \to A$ all of the following hold.

(i) $\mu_f(s \wedge t) = \mu_f(s) \wedge \mu_f(t)$.
(ii) $\mu_f(s \vee t) = \mu_f(s) \vee \mu_f(t)$.
(iii) $\mu_f(s) = \mu_f(\mu_f^{o}(s))$.

The first two items follow since substitutions are endomorphisms of the algebra of terms, while the third follows directly from the definitions.

The first of the two things we have to verify, namely (A), is in reach. Suppose s and t are terms whose variables are drawn from X and that $s \approx t$ is true in $\mathbf{L}$. We need to prove that $\eta(s) \approx \eta(t)$ holds in every lattice in $\mathcal{V}^{(\ell)}$. Let $f : X \to L$. From $s \approx t$ we derive $\mu_f^{o}(s) \approx \mu_f^{o}(t)$. This is just a substitution instance of $s \approx t$ and it has at most n variables. Since $n \le \ell$, we see that $\mu_f^{o}(s) \approx \mu_f^{o}(t)$ is true in $\mathcal{V}^{(\ell)}$. Any substitution instance of this equation must also be true in $\mathcal{V}^{(\ell)}$. In particular,

$$\mu_f(\mu_f^{o}(s)) \approx \mu_f(\mu_f^{o}(t)) \text{ is true in } \mathcal{V}^{(\ell)}.$$

But we saw above that $\mu_f(\mu_f^{o}(r)) = \mu_f(r)$ for any term r. So we have

$$\mu_f(s) \approx \mu_f(t) \text{ is true in } \mathcal{V}^{(\ell)}.$$

Now form the joins of both sides as f runs through all the functions from X in L, to conclude that the equation below is true in $\mathcal{V}^{(\ell)}$.

$$\eta(s) = \bigvee_{f:X\to L} \mu_f(s) \approx \bigvee_{f:X\to L} \mu_f I(t) = \eta(t)$$

So it only remains to show (B), that is

$$s \approx \eta(s) \text{ is true in } \mathcal{V}^{(\ell)}.$$

Observe that $\mu_f(s) \le s$ holds in every lattice, since the lattice operations are monotone. It follows that $\eta(s) \le s$ in every lattice, as well. So we only need to show that $s \le \eta(s)$ holds in $\mathcal{V}^{(\ell)}$. We prove this by induction on the complexity of s. The base step (when s is just a variable) is trivial. The induction step falls into two parts depending on whether $s = u \vee v$ or $s = u \wedge v$. The case with $\vee$ presents no difficulties. So we are left with establishing

$$(\star) \qquad \begin{aligned} &\text{If } s = u \wedge v \text{ and both } u \le \eta(u) \text{ and } v \le \eta(v) \text{ hold in } \mathcal{V}^{(\ell)},\\ &\qquad\qquad \text{then } s \le \eta(s) \text{ holds in } \mathcal{V}^{(\ell)}.\end{aligned}$$

Before tackling $(\star)$ we develop some further properties of the kind of substitutions defined above.

FACT 1. *Let $f : X \to A$ and $g : X \to B$ so that $f(x) = f(y) \implies g(x) = g(y)$ for all $x, y \in X$. Then $\mu_g(s) \le \mu_f(s)$ holds in all lattices for all terms s with variables drawn from X.*

This is a simple consequence of the monotonicity of the lattice operations.

FACT 2. *Suppose $f : Y \to A$. Let $Y_f := \{y_f \mid y \in Y\}$. Let $g : Y_f \to B$. Let $h : Y \to B$ be defined via $h(y) := g(y_f)$. The equation $\mu_h(t) \approx \mu_f(\mu_g(\mu_f^o(t)))$ is true in all lattices, for any term t with variables from Y.*

Proof. We prove this by induction on the complexity of t.

The base step is that t is a variable, say x. Then

$$\begin{aligned}\mu_f(\mu_g(\mu_f^o(x))) &= \mu_f(\mu_g(x_f)) \\ &= \mu_f\Big(\bigwedge_{g(y_f)=g(x_f)} y_f\Big) \\ &= \bigwedge_{g(y_f)=g(x_f)} \mu_f(y_f) \\ &= \bigwedge_{g(y_f)=g(x_f)} \bigwedge_{f(z)=f(y_f)} z \\ &= \bigwedge_{g(y_f)=g(x_f)} \bigwedge_{z_f=y_f} z \\ &= \bigwedge_{g(y_f)=g(x_f)} y \\ &= \bigwedge_{h(y)=h(x)} y \\ &= \mu_h(x).\end{aligned}$$

The part of lattice theory that enters here is just the commutative and associative laws for $\wedge$.

The inductive step is immediate, since μ_f, μ_g, and μ_f^o are all endomorphisms of the term algebra. ■

Key Lemma.
Let $g : X \to L^m$. Let s be any term with variables drawn from X and let y be a variable not belonging to X. The following inequalities all hold in $\mathcal{V}^{(\ell)}$.

(i) $y \wedge \mu_g(s) \leq \bigvee_{f:X\to L}(y \wedge \mu_f(s))$.
(ii) $\mu_g(s) \leq \eta(s)$.
(iii) $y \wedge \eta(s) \leq \bigvee_{f:X\to L}(y \wedge \mu_f(s))$.

Proof. Let $X' = \{x_g \mid x \in X\}$. Let t be any term with variables drawn from X'. Notice that the inequality

$$y \wedge t \leq \bigvee_{h:X'\to L} (y \wedge \mu_h(t))$$

has no more than $n^m + 1 = \ell$ distinct variables. We will show that this inequality holds in $\mathbf{L}$ and therefore also in $\mathcal{V}^{(\ell)}$.

Let $\bar{a}$ be any X'-tuple of elements of L and let $b \in L$. In the left side of our inequality, plug b in for y and $\bar{a}$ in for the variables in t. The effect of μ_g^o is to identify certain variables.

Now the maps h appearing on the right side are assignments of elements of L to the variables in X'. One of these assignments h is precisely the assignment $\bar{a}$. This means that $h(x) = c \in L$ exactly when the entry on the tuple $\bar{a}$ associated with x is

c. Hence, when $\bigwedge_{h(y)=h(x)} y$ is evaluated under the assignment $\bar{a}$ the result will be $c \wedge c \wedge c \wedge \cdots \wedge c = c$. Consequently, the particular joinand on the right associated with this h, when evaluated at $\bar{a}$ is actually the value of the left side at $\bar{a}$. This verifies the inequality in **L**.

Now let $t = \mu_g^o(s)$. So

$$y \wedge \mu_g^o(s) \leq \bigvee_{h:X' \to L} (y \wedge \mu_h(\mu_g^o(s))$$

holds in $\mathcal{V}^{(\ell)}$. Apply μ_g to both sides:

$$\begin{aligned}
\mu_g(y \wedge \mu_g^o(s)) &\leq \mu_g\left(\bigvee_{h:X' \to L} (y \wedge \mu_h(\mu_g^o(s)))\right) \\
\mu_g(y) \wedge \mu_g(\mu_g^o(s)) &\leq \bigvee_{h:X' \to L} (\mu_g(y) \wedge (\mu_g(\mu_h(\mu_g^o(s)))) \\
y \wedge \mu_g(s) &\leq \bigvee_{h:X' \to L} (y \wedge \mu_g(\mu_h(\mu_g^o(s)))) \\
y \wedge \mu_g(s) &\leq \bigvee_{h:X' \to L} (y \wedge \mu_{h^*}(s))
\end{aligned}$$

where $h^*(w) = h(w_g)$ for all $w \in X$, according to Fact 2. Since

$$\bigvee_{h:X' \to L} (y \wedge \mu_{h^*}(s)) \leq \bigvee_{f:X \to L} (y \wedge \mu_f(s))$$

we are finished with (i).

Part (ii) is an immediate consequence of part (a), obtained by setting y to the join of all the $\mu_f(s)$'s and $\mu_g(s)$.

To establish (iii) we need a bit more groundwork. First consider the following equation:

$$(*) \qquad y \wedge \bigvee_{i<m+1} x_i \approx \bigvee_{f:m \to m+1} \left(y \wedge \bigvee_{j<m} x_{f(j)} \right).$$

This equation has $m+2$ variables. Some simple calculations show that $m+2 \leq \ell$. To see that this equation is true in $\mathcal{V}^{(\ell)}$ we only need to see that it is true in **L**.

Let $A \subseteq L$ have $m+1$ elements. So A has 2^{m+1} subsets. Since $n < 2^{m+1}$ we see that two distinct subsets of A must have the same joins. It follows that A has a proper subset A' so that A and A' have the same join. This means that under any assignment of members of L to the variables in $(*)$ the left side of the equation will be one of the joinands on the right side. On the other hand, each joinand on the right is dominated by the left side. Hence, the equation $(*)$ holds in **L** and hence in $\mathcal{V}^{(\ell)}$. Actually, from $(*)$, with the help of lattice theory, we can deduce

$$(**) \qquad y \wedge \bigvee_{i<k} x_i \approx \bigvee_{f:m \to k} \left(y \wedge \bigvee_{j<m} x_{f(j)} \right).$$

for every $k > m$. So $(**)$ holds in $\mathcal{V}^{(\ell)}$ for every $k > m$.

To finish the proof of this lemma, we need to show that

$$\text{(c)} \qquad y \wedge \bigvee_{f:X\to L} (\mu_f(s)) \leq \bigvee_{f:X\to L} (y \wedge \mu_f(s))$$

holds in $\mathcal{V}^{(\ell)}$. Now the join on the left has $n^{|X|}$ joinands. It is easy to see that $n > m$, so we can use the equation $(**)$ to change (c) to

$$\text{(c')} \qquad \bigvee_{h:m\to K} \left(y \wedge \bigvee_{j<m} \mu_{h(j)}(s) \right) \leq \bigvee_{f:X\to L} (y \wedge \mu_f(s))$$

where K is the set of functions from X into L. This reduces our task to demonstrating that each joinand on the left is dominated by the right side. So let $f_0, \ldots, f_{m-1}$ be functions from X to L. Let $g : X \to L^m$ be the function defined via

$$g(x) := (f_0(x), \ldots, f_{m-1}(x)).$$

Observe that $\mu_{f_i}(s) \leq \mu_g(s)$ holds in all lattices for all $i < m$, by Fact 1. So we get

$$y \wedge \bigvee_{i<m} \mu_{f_i}(s) \leq y \wedge \mu_g(s) \leq \bigvee_{f:X\to L} y \wedge \mu_f(s).$$

The rightmost inequality follows from part (i). This concludes the demonstration of part (iii), and thus that of the Key Lemma. ■

With the help of the Key Lemma, we can complete the inductive step to establish (B) and finish the proof of the Theorem. Recall that we were left with establishing

($\star$) If $s = u \wedge v$ and both $u \leq \eta(u)$ and $v \leq \eta(v)$ hold in $\mathcal{V}^{(\ell)}$,

then $s \leq \eta(s)$ holds in $\mathcal{V}^{(\ell)}$.

Observe

$$\begin{aligned} s = u \wedge v \leq \eta(u) \wedge \eta(v) &\leq \bigvee_{f:X\to L} \eta(u) \wedge \mu_f(v) \\ &\leq \bigvee_{f,h:X\to L} \mu_h(u) \wedge \mu_f(v) \end{aligned}$$

by two applications of part (c) of the Key Lemma. So we need to show that

$$\mu_h(u) \wedge \mu_f(v) \leq \eta(s)$$

holds in $\mathcal{V}^{(\ell)}$ for all $f, h : X \to L$. Let $g : X \to L^m$ be the function defined via

$$g(x) := (h(x), f(x), f(x), \ldots, f(x)).$$

Here is where we need that $2 \leq m$. By Fact 1 we have that

$$\mu_f(w) \leq \mu_g(w) \text{ and } \mu_h(w) \leq \mu_g(w) \text{ for all terms } w \text{ with variables from } X.$$

Finally we see

$$\mu_h(u) \wedge \mu_f(v) \leq \mu_g(u) \wedge \mu_g(v) = \mu_g(u \wedge v) = \mu_g(s) \leq \eta(s),$$

where the rightmost inequality is part (ii) of the Key Lemma. This is what we needed.

This completes the proof of Theorem 7.23. ■

Willard's Finite Basis Theorem

A variety $\mathcal{V}$ is said to be **congruence meet-semidistributive** provided **Con A** is a meet-semidistributive lattice for all $\mathbf{A} \in \mathcal{V}$. Recall that a lattice is meet-semidistributive provided the following implication holds in the lattice:

$$x \wedge y \approx x \wedge z \Rightarrow x \wedge (y \vee z) \approx (x \wedge y) \vee (x \wedge z).$$

This is a weakening of the distributive property.

In §6.4, congruence meet-semi distributivity is discussed in some detail. In particular, Theorem 6.23 has as a consequence that if $\mathcal{V}$ is a variety that has among its operation symbols a two-place operation symbol $\wedge$ so that

$$\mathcal{V} \models x \wedge (y \wedge z) \approx (x \wedge y) \wedge z \quad x \wedge y \approx y \wedge x \quad x \wedge x \approx x;$$

that is, $\wedge$ denotes a semilattice operation on each algebra in $\mathcal{V}$, then $\mathcal{V}$ is congruence meet-semidistributive. Among other things, this means that every variety of lattices, as well as most varieties of algebras arising from logic, are congruence meet-semidistributive varieties.

We use this observation extensively later in this chapter. For now, let us consider **flat graph algebras**. Given a graph **G**, perhaps with loops at some of the vertices, the flat graph algebra $\mathbf{G}_{\mathbf{G}}$ associated with **G** is the algebra with universe $F = V \cup \{0\}$, where V is the set of vertices of **G** and $0 \notin V$, and two fundamental operations $\cdot$ (denoted by juxtaposition) and $\wedge$ defined via

$$ab = \begin{cases} a & \text{if } a \text{ and } b \text{ are adjacent vertices of } \mathbf{G} \\ 0 & \text{otherwise} \end{cases}$$

$$a \wedge b = \begin{cases} a & \text{if } a = b \\ 0 & \text{otherwise} \end{cases}$$

The word "flat" here is appropriate since the order imposed by the semilattice operation makes 0 the least element and puts all the other elements (that is, the vertices of **G**) on one level above 0. At the end of §7.3 we sketched a proof that the variety generated by Murskiĭ's graph algebra contains a shift automorphism algebra. This proof is easily extended to the associated flat graph algebra. So we have an example of a finite algebra of finite signature that generates an inherently nonfinitely based congruence meet-semidistributive variety with an infinite subdirectly irreducible algebra. To obtain a finite basis result for finite algebras of finite signature belonging to a congruence meet-semidistributive variety some additional hypothesis will be needed.

In §6.4 several conditions were established that characterize, for varieties $\mathcal{V}$, congruence meet-semidistributivity. We need one of those conditions, namely the $\diamond$-condition.

The $\diamond$-Condition

> There exists a finite set T of terms such that, for all $\mathbf{A} \in \mathcal{V}$ and all finite sequences $p = r_0, \dots, r_n = q \in A$, if $p \neq q$, there are $i < n$ and $e, f \in A$ with $e \neq f$ so that
>
> $$\{r_i, r_{i+1}\} \looparrowright_1^2 \{e, f\} \text{ and } \{p, q\} \looparrowright_1^2 \{e, f\},$$
>
> with respect to T-enhanced translations.

Here is a diagram of the $\diamond$-Condition.

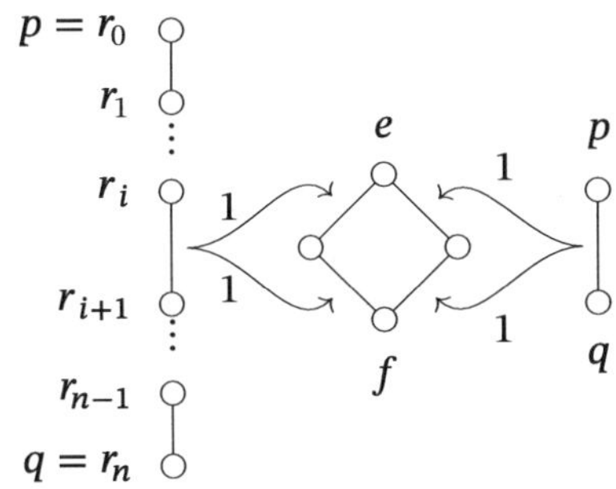

The ◇-Condition

THEOREM 7.24 (Willard's Finite Basis Theorem). *Every congruence meet-semidistributive variety of finite signature that has a finite residual bound is finitely based.*

Our proof of this theorem appears on page 204. The original proof of this theorem (Ross Willard, 2000) followed closely Kirby Baker's proof: see (Kirby A. Baker, 1977) for the congruence distributive case; see also Corollary 8.59 page 357 for a proof of Baker's Theorem based on Willard's Theorem. In fact, we shall first state a somewhat stronger theorem (7.25 below). Our main effort will then be in proving 7.25. A class $\mathcal{K}$ of algebras, all of the same signature, is said to have **bounded critical depth** provided there is a number ℓ such that for every $\mathbf{A} \in \mathcal{K}_{\text{si}}$ and for all $a, b, p, q \in A$ with $a \neq b$ and (p, q) a critical pair (that is, a non-identical pair belonging to the monolith), we have $\{a, b\} \looparrowright^n_\ell \{p, q\}$ for some natural number n.

We say an algebra $\mathbf{A}$ is **finitely subdirectly irreducible** provided $\theta \cap \phi$ is nontrivial whenever θ and ψ are nontrivial congruences of $\mathbf{A}$. Another way to express this is to say that the trivial congruence 0_A is meet irreducible, whereas $\mathbf{A}$ is subdirectly irreducible if and only if 0_A is completely meet irreducible. So every subdirectly irreducible algebra is finitely subdirectly irreducible. Also. it follows that every finite algebra that is finitely subdirectly irreducible is, in fact, subdirectly irreducible. In the exercises there is an example of an infinite finitely subdirectly irreducible algebra that is not subdirectly irreducible. For any class $\mathcal{K}$ of algebras we use $\mathcal{K}_{\text{fsi}}$ to denote the class of finitely subdirectly irreducible algebras that belong to $\mathcal{K}$.

The proof below uses the notions of elementary (i.e. first-order) formula and sentence and invokes the Compactness Theorem (Theorem 8.30). Readers who need a refresher on these matters might look at the first few sections of Chapter 8. For example, we say a class $\mathcal{K}$ of algebras, all of the same signature, is *finitely axiomatizable* provided there is a finite set Σ of elementary sentences so that $\mathcal{K}$ is the class of all algebras in which each sentence belonging to Σ is true. Also, we extend the operator **Mod** to apply to sets of elementary sentences.

THEOREM 7.25 (The ◇ Finite Basis Theorem). *Let $\mathcal{V}$ be a variety of finite signature such that*

(i) *$\mathcal{V}$ has bounded critical depth,*
(ii) *$\mathcal{V}$ is congruence meet-semidistributive,*

(iii) *$\mathcal{V}$ is locally finite, and*
(iv) *$\mathcal{V}_{\text{fsi}}$ is finitely axiomatizable.*

Then $\mathcal{V}$ is finitely based.

This theorem is a variation on the following theorem proved in Chapter 8.

Theorem 8.66 (Jónsson's Finite Basis Theorem (Bjarni Jónsson, 1979)).
Let $\mathcal{V}$ be a variety of finite signature. If $\mathcal{V}$ is congruence distributive and $\mathcal{V}_{\text{fsi}}$ is finitely axiomatizable, then $\mathcal{V}$ is finitely based.

The $\diamond$ Finite Basis Theorem, in addition to replacing congruence distributivity by congruence meet-semidistributivity, needs two additional hypotheses not present in Jónsson's Finite Basis Theorem—namely that $\mathcal{V}$ is locally finite and of bounded critical depth. Nevertheless, the proof here for the $\diamond$ Finite Basis Theorem traces its heritage to Jónsson's proof. Jónsson's proof of the theorem above is based on another of his theorems, namely Theorem 8.54, which is proved in §8.3.

A proof of Willard's Finite Basis Theorem.
Willard's Finite Basis Theorem (7.24) can be deduced from the $\diamond$ Finite Basis Theorem as follows. In fact, congruence meet-semidistributivity plays no role in this deduction, apart from being passed along. So assume the hypotheses of Willard's theorem. Given the finite signature and the finite residual bound, we see that $\mathcal{V}_{\text{si}}$ consists of finitely many finite algebras, up to isomorphism. It follows that $\mathcal{V}$ has bounded critical depth and that $\mathcal{V}_{\text{si}}$ is a finitely axiomatizable elementary class. As Ross Willard (2000) observed, it is part of the folklore that $\mathcal{V}_{\text{si}} = \mathcal{V}_{\text{fsi}}$ in these circumstances. This means that $\mathcal{V}_{\text{fsi}}$ is also finitely axiomatizable. Thus, all the hypotheses of the $\diamond$ Finite Basis Theorem are fulfilled. ■

Proof of the $\diamond$ Finite Basis Theorem. The $\diamond$ Finite Basis Theorem (7.25) will be an immediate consequence of two lemmas that we formulate and prove next.

Let $\mathcal{K}$ be a class of algebras of the same signature. We say $\mathcal{K}$ has the **definite atoms property** provided there is an elementary sentence σ true in $\mathcal{V}$ and there is a finite set T of terms so that

- **Mod** σ has the $\diamond$-condition witnessed by the set T, and
- there is a natural number ℓ such that for all $\mathbf{A} \in \mathbf{Mod}\,\sigma$ and for all $a, b \in A$ and all atoms $\alpha \subseteq \mathrm{Cg}^{\mathbf{A}}(a, b)$ and for all $(p, q) \in \alpha$ such that $p \neq q$, we have
$$\mathbf{A} \models \exists u, v[\neg u \approx v \curlywedge \{a, b\} \looparrowright^1_\ell \circ \looparrowright^2_1 \{u, v\} \curlywedge \{p, q\} \looparrowright^2_1 \{u, v\}]$$
with respect to T-enhanced translations.

See the diagram below for an illustration of the second stipulation in the definite atoms property.

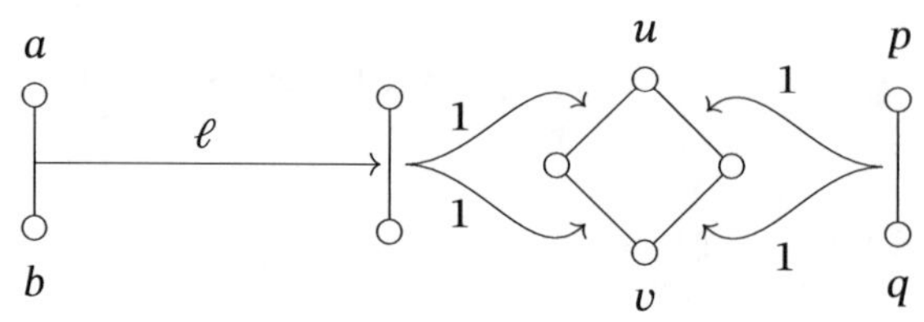

Continuing with the proof of the ◇ Finite Basis Theorem (7.25), we need two lemmas. The first, which traces its ancestry to (Bjarni Jónsson, 1979), yields a proof of Kirby Baker's Finite Basis Theorem, and has a close connection to S. Burris's (1979) proof of Baker's Theorem. We state it here and conclude the proof on page 207. The second is the ◇ Single Sequence Lemma, which will be stated on page 207.

The Definite Atoms Finite Basis Lemma.
Let $\mathcal{V}$ be a variety in a finite signature. If

- *the variety $\mathcal{V}$ has the definite atoms property,*
- *$\mathcal{V}$ is locally finite, and*
- *$\mathcal{V}_{\text{fsi}}$ is finitely axiomatizable,*

then $\mathcal{V}$ is finitely based.

Proof. Let σ be sentence and T be a finite set of terms witnessing the definite atoms property. Let ℓ be a bound guaranteed by the second stipulation of the definite atoms property. For any natural number m let $\pi_m(x_0, y_0, x_1, y_1)$ denote the formula

$$\exists u, v\left[\neg u \approx v \mathbin{\&} \{x_0, y_0\} \looparrowright^1_m \circ \looparrowright^4_2 \{u, v\} \mathbin{\&} \{x_1, y_1\} \looparrowright^1_m \circ \looparrowright^4_2 \{u, v\}\right]$$

where $\looparrowright^n_\ell$ is interpreted in the context of T-enhanced translations. Figure 7.8 illustrates the formula $\pi_m(x_0, y_0, x_1, y_1)$.

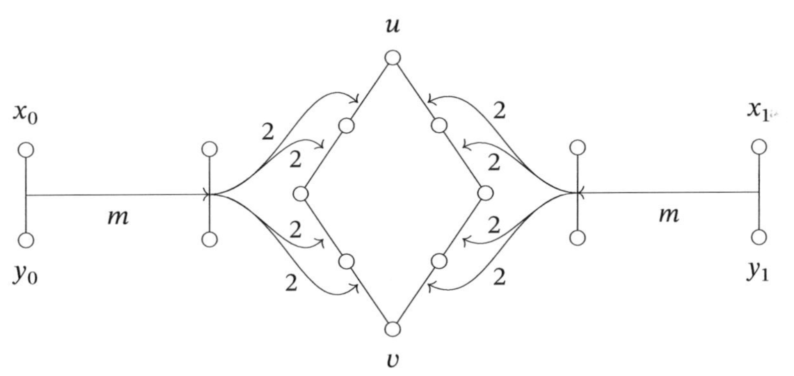

FIGURE 7.8. An illustration of $\pi_m(x_0, y_0, x_1, y_1)$

Contention 1. *If $\mathbf{A} \in \mathcal{V}$ and $a, b, c, d \in A$ so that $\operatorname{Cg}^{\mathbf{A}}(a, b) \cap \operatorname{Cg}^{\mathbf{A}}(c, d)$ is not trivial, then $\mathbf{A} \models \pi_\ell(a, b, c, d)$.*

Proof of contention. Let $\mathbf{A} \in \mathcal{V}$ and let $a, b, c, d \in A$ for which $\operatorname{Cg}^A(a, b) \cap \operatorname{Cg}^A(c, d)$ is not trivial. Now pick $p', q' \in A$ with $p' \neq q'$ so that $(p', q') \in \operatorname{Cg}^A(a, b) \cap \operatorname{Cg}^A(c, d)$. The computations witnessing this can be carried out in a finitely generated subalgebra $\mathbf{B}$ of $\mathbf{A}$. But $\mathcal{V}$ is locally finite, so $\mathbf{B}$ is finite. Observe that $\mathbf{B} \in \mathcal{V}$. Since $\mathbf{B}$ is finite every nontrivial congruence of $\mathbf{B}$ lies above an atom. Now pick p, q with $p \neq q$ and $(p, q) \in \operatorname{Cg}^{\mathbf{B}}(p', q')$ so that $\operatorname{Cg}^{\mathbf{B}}(p, q)$ is an atom. Using the definite atoms property, we pick $p_0, p_1 \in B$ with $p_0 \neq p_1$ and

$$\mathbf{B} \models \{a, b\} \looparrowright^1_\ell \circ \looparrowright^2_1 \{p_0, p_1\} \mathbin{\&} \{p, q\} \looparrowright^2_1 \{p_0, p_1\}.$$

Notice that $(p_0, p_1) \in \mathrm{Cg}^{\mathbf{B}}(c,d)$. Invoking the definite atoms property again, we obtain $r_0, r_1 \in B$ with $r_0 \neq r_1$ so that

$$\mathbf{B} \models \{c,d\} \looparrowright^1_\ell \circ \looparrowright^2_1 \{r_0,r_1\} \wedge \{p_0,p_1\} \looparrowright^2_1 \{r_0,r_1\}.$$

Now $\{a,b\} \looparrowright^1_\ell \circ \looparrowright^2_1 \{p_0,p_1\} \looparrowright^2_1 \{r_0,r_1\}$ yields $\{a,b\} \looparrowright^1_\ell \circ \looparrowright^4_2 \{r_0,r_1\}$. Hence

$$\mathbf{B} \models \{a,b\} \looparrowright^1_\ell \circ \looparrowright^4_2 \{r_0,r_1\} \wedge \{c,d\} \looparrowright^1_\ell \circ \looparrowright^4_2 \{r_0,r_1\}.$$

That is $\mathbf{B} \models \pi_\ell(a,b,c,d)$. But this is an existential formula. So $\mathbf{A}$ is a model as well, as desired. Contention 1 is now proved. ■

Let ψ be the following sentence

$$\forall x_0, y_0, x_1, y_1 \, [\pi_{\ell+1}(x_0,y_0,x_1,y_1) \implies \pi_\ell(x_0,y_0,x_1,y_1)] \,.$$

It follows from the contention above that $\mathcal{V} \models \psi$. Let $\mathcal{W} = \mathbf{Mod}\{\sigma, \psi\}$. We see $\mathcal{W}$ is a finitely axiomatizable class that includes $\mathcal{V}$. Let ϕ be the sentence

$$\forall x_0, y_0, x_1, y_1 \, [(\neg x_0 \approx y_0 \wedge \neg x_1 \approx y_1) \Leftrightarrow \pi_\ell(x_0,y_0,x_1.y_1)] \,.$$

It is our contention that $\{\sigma, \psi, \phi\}$ axiomatizes $\mathcal{W}_{\mathrm{fsi}}$. Certainly every model of those three sentences belongs to $\mathcal{W}_{\mathrm{fsi}}$. We will concern ourselves with the reverse inclusion. Let $\mathbf{A} \in \mathcal{W}_{\mathrm{fsi}}$. We know $\mathbf{A} \models \{\sigma, \psi\}$. We have to establish $\mathbf{A} \models \phi$. To this end, let $a,b,c,d \in A$. First suppose that $a \neq b$ and $c \neq d$. Since $\mathbf{A}$ is finitely subdirectly irreducible, we can pick $r, s \in A$ with $r \neq s$ and $(r,s) \in \mathrm{Cg}^{\mathbf{A}}(a,b) \cap \mathrm{Cg}^{\mathbf{A}}(c,d)$. By Mal'tsev's description of principal congruences, there are numbers n and m so that

$$\{a,b\} \looparrowright^n_m \{r,s\} \text{ and } \{c,d\} \looparrowright^n_m \{r,s\}.$$

Since $\mathbf{A} \models \sigma$, we can apply the definite atom property (twice as we did in the proof of Contention 1), to obtain

$$\mathbf{A} \models \pi_m(a,b,c,d).$$

This doesn't quite get us $\mathbf{A} \models \phi$ since m might be larger than ℓ. But now we can invoke ψ to decrease m step by step to ℓ. Suppose $m > \ell$. Consider the long horizontal arrow labeled m to the left in Figure 7.8. This stands for a translation of complexity m. This translation is the composite of m basic T-enhanced translations. So reading this finer understanding of the diagram from left toward the middle, taking one basic translation at a time we arrive at a position $\ell + 1$ away from the middle of the diagram. Let a' label the upper node at this point and b' the lower node. Doing the same on the right of Figure 7.8, we have an upper node labeled, say, c' and and a lower node labeled d'. So we see

$$\pi_{\ell+1}(a',b',c',d')$$

holds. Applying ψ, we can change the index $\ell + 1$ to ℓ, finding that

$$\pi_\ell(a',b',c',d')$$

holds, possibly with altered value assigned to the middle of the diagram. Hence, we have $\pi_{m-1}(a,b,c,d)$. Continuing in this way we arrive at

$$\pi_\ell(a,b,c,d).$$

So we have established the left-to-right direction within ϕ. To obtain the other implication, let us suppose $a = b$. Then $\pi_\ell(a,b,c,d)$ cannot hold since we would have to have some $r_0, r_1 \in A$ with $r_0 \neq r_1$ and $(r_0,r_1) \in \mathrm{Cg}^{\mathbf{A}}(a,b)$. The case when $c \neq d$ is similar. This means that $\mathbf{A} \models \phi$ as desired.

Let us gather together what we know. There is a finitely axiomatizable class $\mathcal{W}$ such that $\mathcal{V} \subseteq \mathcal{W}$ and $\mathcal{W}_{\text{fsi}}$ is finitely axiomatizable. We also know that $\mathcal{V}_{\text{fsi}}$ is finitely axiomatizable (it is one of our hypotheses).

Let τ be a sentence that axiomatizes $\mathcal{V}_{\text{fsi}}$. Then we have

$$\mathcal{V} \models \sigma \wedge \psi \wedge (\phi \implies \tau).$$

Invoking the Compactness Theorem for elementary logic (see Corollary 8.31), let Γ be a finite set of equations true in $\mathcal{V}$ so that $\Gamma \models \sigma \wedge \psi \wedge (\phi \implies \tau)$ and put $\mathcal{V}' = \mathbf{Mod}\,\Gamma$. Then $\mathcal{V} \subseteq \mathcal{V}' \subseteq \mathcal{W}$ and $\mathcal{V}_{\text{fsi}} = \mathcal{V}'_{\text{fsi}}$, so $\mathcal{V} = \mathcal{V}'$. Since $\mathcal{V}'$ is finitely based, we conclude that $\mathcal{V}$ is also. This completes the proof of the Definite Atoms Finite Basis Lemma. ■

Continuing with the proof of the $\diamond$ Finite Basis Theorem (7.25), we still need to obtain the definite atoms property. Our second lemma does that. It traces its heritage ultimately back to Kirby Baker's original proof of Baker's Finite Basis Theorem.

The $\diamond$ Single Sequence Lemma.
Every locally finite variety of finite signature with the $\diamond$-condition and with bounded critical depth, has the definite atoms property.

Proof. Let $\mathcal{V}$ be a variety satisfying the hypotheses of the lemma. Then $\mathcal{V}$ is congruence meet-semidistributive. So according to § 6.4, there is a sentence asserting that $\mathcal{V}$ is congruence meet-semidistributive. There is also a finite set T of terms witnessing the $\diamond$-condition for $\mathcal{V}$. Let ℓ witness the bounded critical depth property for $\mathcal{V}$. Our aim is to show that T and ℓ witness the second stipulation in the definite atoms property for $\mathcal{V}$ and Σ. Let $\mathbf{A} \in \mathcal{V}$, $a, b \in A$, $\alpha \subseteq \mathrm{Cg}^{\mathbf{A}}(a, b)$, and $(p, q) \in \alpha$ be as in the second stipulation of the definite atoms property.

Pick a congruence $\theta \in \mathbf{Con}\,A$ that is maximal with respect to separating p and q. Then $\mathbf{A}/\theta$ is subdirectly irreducible and $(p/\theta, q/\theta)$ is a critical pair of $\mathbf{A}/\theta$. Notice that θ must also separate a and b.

Now $(p/\theta, q/\theta) \in \mathrm{Cg}^{\mathbf{A}/\theta}(a/\theta, b/\theta)$. According to the bounded critical depth property, there is finite sequence $r'_0, \dots, r'_k$ in $\mathbf{A}/\theta$ and translations $\lambda'_0, \dots, \lambda'_k$ such that

$$\begin{aligned} p/\theta &= r'_0 \\ \{\lambda'_i(a/\theta), \lambda'_i(b/\theta)\} &= \{r'_i, r'_{i+1}\} \text{ for all } i < k \\ r'_k &= q/\theta \end{aligned}$$

and, moreover, the complexity of the translations λ'_i never exceeds ℓ. We pick representatives of the congruence classes and denote this by removing the primes. We get

$$\begin{aligned} p \;&\theta\; r_0 \\ \{\lambda_i(a), \lambda_i(b)\} \;&\theta\; \{r_i, r_{i+1}\} \text{ for all } i < k \\ r_k \;&\theta\; q. \end{aligned}$$

To clarify the middle line of the above display, for each $i < k$ pick a_i and b_i so that $\{a_i, b_i\} = \{a, b\}$ with $\lambda_i(a_i)\ \theta\ r_i$ and $\lambda_i(b_i)\ \theta\ r_{i+1}$. We obtain

$$\begin{gathered} p\ \theta\ r_0\ \theta\ \lambda_0(a_0) \\ \lambda_1(a_1)\ \theta\ r_1\ \theta\ \lambda_0(b_0) \\ \lambda_1(b_1)\ \theta\ r_2\ \theta\ \lambda_2(a_2) \\ \vdots \\ \lambda_{k-1}(b_{k-1})\ \theta\ r_k\ \theta\ q \end{gathered}$$

where we have made the harmless assumption that k is even. This leads to

$$\begin{gathered} p\ \theta\ \lambda_0(a_0) \\ \lambda_1(a_1)\ \theta\ \lambda_0(b_0) \\ \lambda_1(b_1)\ \theta\ \lambda_2(a_2) \\ \vdots \\ \lambda_{k-1}(b_{k-1})\ \theta\ q. \end{gathered}$$

Consider the *single sequence* made by traversing the display above in a back-and-forth manner:

$$p,\ \lambda_0(a_0),\ \lambda_0(b_0),\ \lambda_1(a_1),\ \lambda_1(b_1),\ \ldots,\ \lambda_{k-1}(a_{k-1}),\ \lambda_{k-1}(b_{k-1}),\ q.$$

According to the $\diamond$-condition, there is a pair (c, d) of adjacent members of this sequence so that $\{c, d\} \not\rightsquigarrow^2_1 \{e, f\}$ and $\{p, q\} \not\rightsquigarrow^2_1 \{e, f\}$. Consider the diagram in Figure 7.9. There the single sequence is displayed vertically on the left. Notice that every other

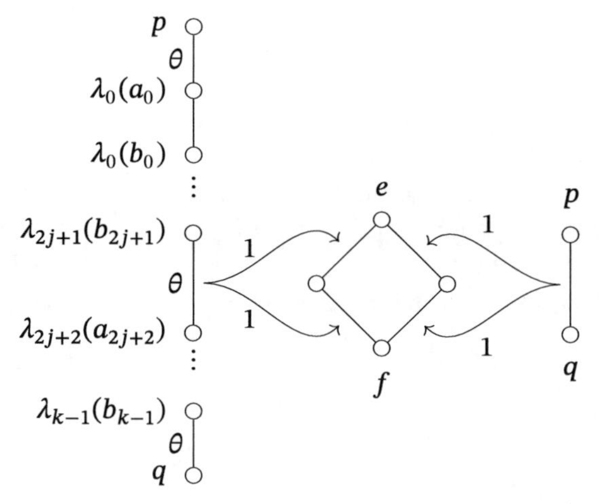

FIGURE 7.9. Proof of the Definite Atoms Property

link indicates a pair in θ. The other links join $\lambda_i(a_i)$ and $\lambda_i(b_i)$, for some i.

First, suppose that the $\diamond$-condition is witnessed on an edge labeled θ, as shown in the diagram. Then we see that $(e, f) \in \mathrm{Cg}^{\mathbf{A}}(p, q)$. But $\mathrm{Cg}^{\mathbf{A}}(p, q)$ is an atom and $e \neq f$. Hence (e, f) generates this atom. But we also see that $(e, f) \in \theta$. Now θ lies above the atom. This is impossible, since θ was chosen to separate p and q. So the $\diamond$-condition must be witnessed by an edge linking $\lambda_i(a_i)$ and $\lambda_i(b_i)$ for some i. Then λ_i witnesses that $\{a, b\} \looparrowright_{\ell}^{1} \{c, d\}$, as desired. This completes the proof of the $\diamond$ Single Sequence Lemma. ■

Now the $\diamond$ Finite Basis Theorem (7.25) follows immediately from the $\diamond$ Single Sequence Lemma and the Definite Atoms Finite Basis Lemma. ■

The Definite Atoms Finite Basis Lemma houses that part of the argument that relies on neither the congruence meet semidistributivity nor the finite bound on the critical depth. The $\diamond$ Single Sequence Lemma is where those properties come into play.

The $\diamond$ Finite Basis Theorem is a variant of a theorem found in (Kirby A. Baker, George F. McNulty, and Ju Wang, 2004). A slightly stronger version of this theorem can be found in (George F. McNulty and Ross Willard, 2018).

There are several corollaries, apart from Willard's Finite Basis Theorem, that result from replacing the hypothesis of bounded critical depth in the $\diamond$ Finite Basis Theorem.

A class $\mathcal{K}$ of algebras, all of the same finite signature, is said to have **term finite principal congruences** provided there is a natural number ℓ so that for all $\mathbf{A} \in \mathcal{K}$ and for all $a, b, c, d \in A$, we have $\langle a, b\rangle \in \mathrm{Cg}^{\mathbf{A}}(c, d)$ if and only if $\{c, d\} \looparrowright_{\ell}^{n} \{a, b\}$ for some natural number n. This notion was introduced by Ju Wang (1988), see also (Ju Wang, 1990), under the name *finite principal length*. Independently, this notion was also introduced in (David M. Clark, Brian A. Davey, Ralph S. Freese, and Marcel Jackson, 2004). Of course, if $\mathcal{V}_{\mathrm{si}}$ has term finite principal congruences then in particular $\mathcal{V}$ has bounded critical depth.

COROLLARY 7.26. *Every locally finite congruence meet-semidistributive variety $\mathcal{V}$ of finite signature such that $\mathcal{V}_{\mathrm{si}}$ has term finite principal congruences is finitely based.* ■

A class $\mathcal{K}$ of algebras, all of the same finite signature, is said to have **bounded critical diameter** provided there is a natural number ℓ so that for every finite $\mathbf{A} \in \mathcal{V}_{\mathrm{si}}$ and all $a, b, c, d \in A$ such that $\langle a, b\rangle$ and $\langle c, d\rangle$ are critical pairs, we have $\{a, b\} \looparrowright_{\ell}^{n} \{c, d\}$.

COROLLARY 7.27. *Let $\mathcal{V}$ be locally finite congruence meet-semidistributive variety of finite signature with bounded critical diameter such that $\mathcal{V}_{\mathrm{fsi}}$ is finitely axiomatizable. Then $\mathcal{V}$ is finitely based.*

The proof of this corollary follows from the proof of the $\diamond$ Finite Basis Theorem. In particular, it is in the proof of the $\diamond$ Single Sequence Lemma where the finite critical bound comes into play. At that point, the bound on the critical diameter will serve.

Exercises 7.28

1. Extend McKenzie's Finite Basis Theorem for Finite Lattices to finite lattices expanded by finitely many monotone operations. (Recall an operation F on a lattice $\mathbf{L}$ is monotone provided if $a_i \leq b_i$ in $\mathbf{L}$ for all $i < r$, where r is the rank of F, then $F(\bar{a}) \leq F(\bar{b})$.)

2. Prove that if $\mathcal{V}$ is a variety so that $\mathcal{V}_{\mathrm{si}}$ is axiomatizable by a set of elementary sentences, then every algebra in $\mathcal{V}_{\mathrm{fsi}}$ is embeddable into some algebra in $\mathcal{V}_{\mathrm{si}}$.

[Hint: This lemma from the folklore is a consequence of the Compactness Theorem. Those unfamiliar with model theory can consult §8.3.]

***3.** (Ralph McKenzie, 1978) Let $\mathcal{V}$ be a locally finite variety of finite signature. Prove that if $\mathcal{V}_{\text{si}}$ is finitely axiomatizable and $\mathcal{V}$ has definable principal congruences, then $\mathcal{V}$ is finitely based. [A class $\mathcal{K}$ of algebras, all of the same signature, is said to have **definable principal congruences** provided there is an elementary formula $\varphi(x, y, z, w)$ such that for every $\mathbf{A} \in \mathcal{V}$ and all $a, b, c, d \in A$ we have $\langle c, d\rangle \in \mathrm{Cg}^{\mathbf{A}}(a, b)$ if and only if $\mathbf{A} \models \varphi(a, b, c, d)$. The formula, in fact, is the one associated with $\{a, b\} \rightsquigarrow^n_\ell \{c, d\}$ for fixed n and ℓ.]

4. (Dejan Delić, 2001) Prove that the flat graph algebras associated with the three graphs below are each inherently nonfinitely based.

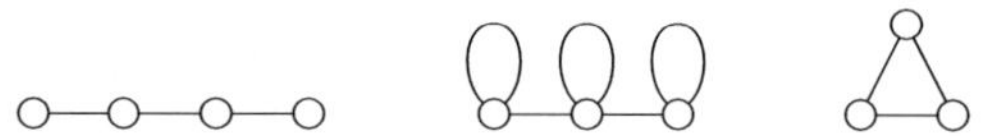

***5.** (Dejan Delić, 2001) Let $\mathbf{F}$ be the flat graph algebra associated with the graph displayed below.

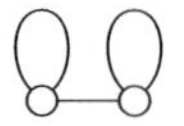

and let $\mathcal{V}$ be the variety generated by $\mathbf{F}$.

Prove that

(i) $\mathcal{V}_{\text{si}}$ consists, up to isomorphism, of those flat graph algebras of complete graphs with a loop at every vertex.
(ii) $\mathcal{V}_{\text{si}}$ is finitely axiomatizable.
(iii) $\mathcal{V}_{\text{si}}$ is closed with respect to the formation of subalgebras.
(iv) $\mathcal{V}_{\text{si}} = \mathcal{V}_{\text{fsi}}$.
(v) Every algebra in $\mathcal{V}_{\text{si}}$ is simple.
(vi) $\mathcal{V}$ has bounded critical depth.
(vii) $\mathcal{V}$ has bounded critical diameter.
(viii) $\mathcal{V}$ is finitely based.

This provides an example where the $\diamond$ Finite Basis Theorem applies but Willard's Finite Basis Theorem does not. We note that Delić's method of proving that $\mathcal{V}$ is finitely based is different than the one we have suggested. Delić makes an appeal to the definable principal congruences by way of Exercise 4.

***6.** (George F. McNulty and Ross Willard, 2018) The definite atoms property, at least in the locally finite case, seems very close to the $\diamond$ condition, and so almost equivalent to congruence meet semidistributivity. Here is an example to show that the two properties are not equivalent,

Let $\mathbf{K}$ be any finite field and let $\mathcal{V}$ be the variety of vector spaces over $\mathbf{K}$. It is easy to check that $\mathcal{V}$ is finitely based, then $\mathcal{V}_{\text{si}} = \mathcal{V}_{\text{fsi}}$ is the class of 1-dimensional vector spaces over $\mathbf{K}$.

Prove

(i) $\mathcal{V}$ has the definite atoms property.
(ii) $\mathcal{V}$ is not congruence meet semidistributive.

7.5. First Interlude: Algebraic Lattices as Congruence Lattices

A deeper understanding of equational logic and, in particular, of the lattice of equational theories is facilitated by the study of how algebraic lattices can be represented as congruence lattices of algebras. Recall from Definition 2.15 of Volume I, that an *algebraic lattice* is a complete lattice in which every element is the join of some set of compact elements. For an algebra **A** it is easy to see that **Con A** is an algebraic lattice; its compact elements are just the finitely generated congruences.

We begin with the following theorem, known as the Grätzer-Schmidt Theorem. It is the converse of the easy remark in the previous paragraph and its proof was a milestone in the field.

THEOREM 7.29 (G. Grätzer and E. T. Schmidt 1963). *For each algebraic lattice* **L** *there is an algebra* **A** *such that* $\mathbf{L} \cong \operatorname{Con} \mathbf{A}$. *Moreover, if* **L** *has only countably many elements, then such an algebra* **A** *can be chosen to be countably infinite with countably infinite list of basic operations, all of rank* 1*; whereas, if* **L** *has* κ *compact elements, where* κ *is an infinite cardinal, then* **A** *can be chosen to have cardinality* κ *and so that the basic operations of* **A** *will be* κ *in number and all of them will have rank* 1.

Proof. Let **L** be an algebraic lattice. The top and bottom elements of **L** will be noted by 1 and 0 respectively. The bottom element 0 is compact. If there are no other compact elements, then **L** is trivial and it is isomorphic to the congruence lattice of any trivial algebra. So we suppose that **L** has a compact element d with $d \neq 0$.

The nonzero compact elements of **L** form a join-semilattice **C**. An *ideal* in this join-semilattice is a subset $I \subseteq C$ with the following properties:

- If $a \in I$ and $b \leq a$ with $b \in C$, then $b \in I$.
- If $a, b \in I$, then $a \vee b \in I$.

Observe that $\emptyset$ is an ideal in this sense. The intersection of any nonempty collection of ideals is again an ideal. It follows that the collection $\mathfrak{I}$ of all ideals of the join-semilattice of nonzero compact elements of **L** is a lattice. In fact, this lattice of ideals is isomorphic to **L**. An isomorphism is given by

$$p \mapsto \{a \mid a \leq p \text{ and } a \neq 0 \text{ is compact}\} \text{ for all } p \in L.$$

In what follows we will construct an algebra **A** so that **Con A** is isomorphic to the lattice of ideals of the join-semilattice of nonzero compact elements of **L**.

The algebra we seek is based on a certain kind of graph. Let **G** be a graph without loops at the vertices and with edges labeled by nonzero compact elements of **L**. Let A be the set of vertices of **G**. We say a function $F : A \to A$ is **stable for G** (or simply **stable**), provided for all $u, v \in A$ that are joined by an edge, either $F(u) = F(v)$ or $F(u)$ and $F(v)$ are joined by an edge and the label on the edge joining u and v is the same as the label on the edge joining $F(u)$ and $F(v)$.

We will say that the pair $\langle \mathbf{G}, \mathcal{F} \rangle$ is **solid** for the algebraic lattice **L**, provided **G** is a simple graph whose edges are labeled by nonzero compact elements of **L**, $\mathcal{F}$ is a set of stable maps on the vertices of **G**, and the following five conditions are satisfied:

(1) Every nonzero compact element of **L** labels some edge of **G**.
(2) For all $u, v \in A$ so that u and v are joined by an edge, if that edge is labeled with c and $c \leq a \vee b$ where a and b are nonzero compact elements of **L**, then

there is a path connecting u and v whose edges are labeled with a and b, and both labels occur on the path.

(3) Let $a_0, a_1, \dots, a_n$ label the edges of a cycle in $\mathbf{G}$. Then $a_0 \leq a_1 \vee \cdots \vee a_n$.

(4) If $u, v, w, z \in A$ and u and v are joined by an edge and w and z are joined by an edge and both of these edges are labeled with the same compact element of $\mathbf{L}$, then there is a stable function $F \in \mathcal{F}$ so that $\{w, z\} = \{F(u), F(v)\}$.

(5) If $u, v \in A$ with $u \neq v$, then there is a path connecting u and v so that each pair of vertices, connected by edges in the path, regarded as ordered pairs of elements of A, belong to $\mathrm{Cg}^{\mathbf{A}_{\mathbf{G}\mathcal{F}}}(u, v)$.

Here $A_{\mathbf{G}\mathcal{F}}$ is the algebra with universe A, i.e. the set of vertices of $\mathbf{G}$, and whose basic operations are the stable functions belonging to $\mathcal{F}$. We have supplied no signature for this algebra; but the congruence lattice would be the same, regardless of the signature. This completes the definition of what it means to say the that the pair $\langle \mathbf{G}, \mathcal{F} \rangle$ is solid for the algebraic lattice $\mathbf{L}$.

Observe that, according to Mal'tsev's description of the generation of principal congruences, property (5) can be rephrased in terms of sequences of elements of A and maps belonging to $\mathcal{F}$.

LEMMA 7.30 (Main Lemma). *Let $\mathbf{G}$ be a graph with edges labeled by nonzero compact elements of $\mathbf{L}$ and let $\mathcal{F}$ be a set of stable maps on the set A of vertices of $\mathbf{G}$. If $\langle \mathbf{G}, \mathcal{F} \rangle$ is solid for $\mathbf{L}$, then $\mathbf{L} \cong \mathbf{Con}\,\mathbf{A}_{\mathbf{G}\mathcal{F}}$.*

Proof. Let $\theta \in \operatorname{Con} \mathbf{A}_{\mathbf{G}\mathcal{F}}$. Then θ is a set of ordered pairs of vertices of $\mathbf{G}$. Define $\mathcal{I}\colon \operatorname{Con} \mathbf{A}_{\mathbf{G}\mathcal{F}} \to \mathfrak{I}$ via

$$\mathcal{I}(\theta) = \{c \mid c \text{ labels an edge joining } u \text{ and } v \text{ for some } (u, v) \in \theta\}.$$

In the other direction define $\Theta\colon \mathfrak{I} \to \operatorname{Con} \mathbf{A}_{\mathbf{G}\mathcal{F}}$ so that for any $I \in \mathfrak{I}$, we take $\Theta(I)$ to be the least equivalence relation on A containing

$$\{(u, v) \mid \text{ there is an edge joining } u \text{ and } v \text{ whose label belongs to } I\}.$$

It is our contention that these two maps are isomorphisms that are inverses of each other.

Our first step is to see that these definitions are sound.

So let $\theta \in \operatorname{Con} \mathbf{A}_{\mathbf{G}\mathcal{F}}$. To see that $\mathcal{I}(\theta)$ is an ideal, it is enough to show that if $a, b \in \mathcal{I}(\theta)$ and $c \leq a \vee b$ and c is compact, then $c \in \mathcal{I}(\theta)$. By property (1), there are $u, v \in A$ such that the edge joining u and v is labeled with c. Then by property (2) there is a path connecting u and v whose edges are labeled with either a and b and such that both labels occur along this path. Since $a, b \in \mathcal{I}(\theta)$, there are $x, y, z, w \in A$ so that the edge joining x and y is labeled a and the edge joining z and w is labeled b and $(x, y), (z, w) \in \theta$. But now by property (4), the pairs of vertices that constitute the path connecting u and v must belong to θ as well, since the stable maps respect congruence relations. This means that $(u, v) \in \theta$ and so the label c of the edge joining u and v belongs to $\mathcal{I}(\theta)$. So $\mathcal{I}(\theta)$ is an ideal, as desired.

Next, let $I \in \mathfrak{I}$. To see that $\Theta(I)$ is a congruence, it is enough to see that $\Theta(I)$ is preserved by each stable map. So let $u, v \in A$ so that there is an edge joining u and v and that edge is labeled with $c \in I$. Let F be any stable map. There are two cases. First, suppose $F(u) = F(v)$. Then $(F(u), F(v)) \in \Theta(I)$. Second, suppose that $F(u) \neq F(v)$. Then $F(u)$ and $F(v)$ must also be joined by an edge with the same label c. So $(F(u), F(v)) \in \Theta(I)$. The other ordered pairs that belong to $\Theta(I)$ arise from taking the transitive closure of the pairs we have already checked. So suppose that u and v are connected by a path whose labels belong to I. Let F be any stable map belonging to $\mathcal{F}$. Apply F to each vertex along the path. The result will be a path (perhaps shorter) connecting $F(u)$ and $F(v)$ and each edge along this path produces an ordered pair belonging to $\mathrm{Cg}^{\mathbf{A}_{\mathbf{G}\mathcal{F}}}(u, v) \subseteq \Theta(I)$. Hence, $\Theta(I)$ is a congruence, as desired.

Since the maps $\mathcal{J}$ and Θ evidently respect set inclusion, it only remains to show that $\mathcal{J} \circ \Theta$ is the identity map on $\mathfrak{I}$ and that $\Theta \circ \mathcal{J}$ is the identity map on $\mathrm{Con}\,\mathbf{A}_{\mathbf{G}\mathcal{F}}$.

For the first task, let $I \in \mathfrak{I}$. We see that $I \subseteq \mathcal{J}(\Theta(I))$ since by property (1) every nonzero compact element labels some edge in $\mathbf{G}$. For the reverse inclusion, suppose $a \in \mathcal{J}(\Theta(I))$. So pick $u, v \in A$ so that a labels the edge joining u and v and so that there is a path connecting u and v all of whose edges are labeled with elements of I. This path, together with the edge joining u and v, is a cycle. Let us say that $a, b_0, \ldots, b_{m-1}$ label the edges of this cycle. So by property (3) we find $a \leq b_0 \vee b_1 \vee \cdots \vee b_{m-1}$, where $b_i \in I$ for all $i < m$. Therefore $a \in I$, as desired.

For the second task, in view of what we have already proven, it is enough to demonstrate that Θ is one-to-one. So let φ and ψ be distinct congruences of $\mathbf{A}_{\mathbf{G}\mathcal{F}}$. It is harmless to suppose that $(u, v) \in \varphi$ but $(u, v) \notin \psi$. By property (5) there is path connecting u and v so that each edge in the path connects an order pair belonging to $\mathrm{Cg}^{\mathbf{A}_{\mathbf{G}\mathcal{F}}}(u, v) \subseteq \varphi$. On the other hand, $(u, v) \notin \psi$, so there must be an edge along the path that connects an ordered pair not belonging to ψ. So the label of that edge belongs to $\mathcal{J}(\varphi)$ but not to $\mathcal{J}(\psi)$. Hence, $\mathcal{J}$ is one-to-one.

Our conclusion is that $\mathbf{Con}\,\mathbf{A}_{\mathbf{G}\mathcal{F}} \cong \mathbf{L}$, completing the proof of the Lemma. ∎

All that is left to do is to construct a graph $\mathbf{G}$ with edges labeled with nonzero compact elements of the algebraic lattice $\mathbf{L}$ and to specify a set $\mathcal{F}$ of stable maps so that properties (1)–(5) in the definition of solid are satisfied.

We will construct the desired graph $\mathbf{G}_\omega$ in stages.

We take $\mathbf{G}_0$ to be the graph with one vertex and no edges.

We take $\mathbf{G}_1$ to be an extension of $\mathbf{G}_0$ by adding one more vertex, connecting it with an edge and labeling that edge with d, our compact element of $\mathbf{L}$ with $d \neq 0$.

Now suppose the labeled graphs $\mathbf{G}_n$ and $\mathbf{G}_{n+1}$ are in hand. We produce $\mathbf{G}_{n+2}$ with the following construction. Let a be a compact element of $\mathbf{L}$ and let u and v be any two distinct vertices joined by an edge labeled a in $\mathbf{G}_{n+1}$ but so that this edge is not in $\mathbf{G}_n$. We define a graph $\mathbf{P}_a(u, v)$ with edges labeled by nonzero compact elements of $\mathbf{L}$ as follows. Start with a single edge connecting vertices u and v and labeled with a. For each pair $\langle b, c\rangle$ of nonzero compact elements such that $a \leq b \vee c$, add new vertices x, y, and z and edges $\{u, x\}, \{x, y\}, \{y, z\}$, and $\{z, v\}$. We label these new edges alternately with b and c, as indicated in the diagram below.

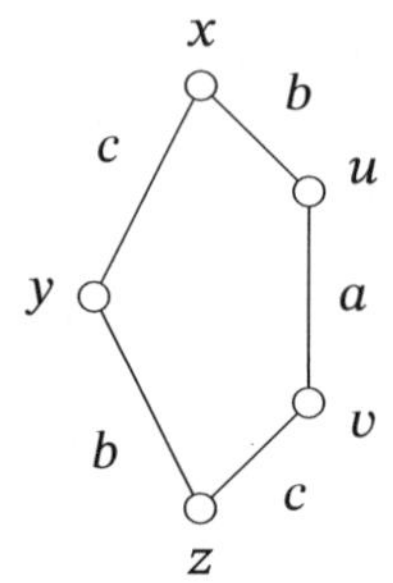

The $\langle b, c\rangle$ flap of $\mathbf{P}_a(u, v)$

The graph $\mathbf{P}_a(u, v)$ has a pentagonal flap for each pair $\langle b, c\rangle$ so that $a \le b \vee c$, so it looks like a rotor around the vertical axis connecting u to v, with the axis labeled a. Observe that none of the new vertices has a loop (so the graph $\mathbf{P}_a(u, v)$ is simple) and every new edge is given exactly one label, which is a nonzero compact element of $\mathbf{L}$. This pentagonal flap witnesses property (2) for the edge joining u and v. The graph $\mathbf{G}_{n+2}$ is obtained from $\mathbf{G}_{n+1}$ by adding such a rotor to each edge of $\mathbf{G}_{n+1}$ that is not an edge of $\mathbf{G}_n$.

This means $\mathbf{G}_2$ is the graph $\mathbf{P}_d(u, v)$. Observe that $d \le d \vee a$ for every nonzero compact element a of $\mathbf{L}$. Consequently, all the nonzero compact elements of $\mathbf{L}$ occur as labels of edges in the graph $\mathbf{G}_2$, ensuring that property (1) will hold. Also, observe that $\mathbf{G}_2$ must have at least 3 new vertices and 4 new edges.

Observe that in this construction each $\mathbf{G}_{n+1}$ is obtained from $\mathbf{G}_n$ by adding new vertices and new edges and labeling the new edges while retaining the labelings of the old edges. Moreover, $\mathbf{G}_{n+1}$ is a simple graph and each edge has exactly one label, which is a nonzero compact element of $\mathbf{L}$. This means that we can let $\mathbf{G}_\omega$ be the labeled graph that is the union of all the $\mathbf{G}_n$ as n runs through the natural numbers.

It is worth noting that the cardinality of $\mathbf{G}_\omega$ will be infinite both because the construction goes through infinitely many stages, but also because the graphs $\mathbf{P}_a(u, v)$ that arise in the construction might also be infinite. Their cardinalities depend on the join-semilattice of nonzero compact elements of $\mathbf{L}$. Of course, if L is countable, $\mathbf{G}_\omega$ will be countably infinite.

On the basis of this construction, we see that $\mathbf{G}_\omega$ satisfies properties (1) and (2) of the definition of solid.

To obtain condition (3) of the definition of solid for $\mathbf{G}_\omega$ we argue by induction that $\mathbf{G}_n$ satisfies condition (3) for all n. As base steps, we observe that both $\mathbf{G}_0$ and $\mathbf{G}_1$ satisfy this condition vacuously. For the inductive step, our hypothesis is that $\mathbf{G}_k$ and $\mathbf{G}_{k+1}$ satisfy the condition. Consider any cycle with edges $e_0, \ldots, e_{m_1}$ labeled $a_0, \ldots, a_{m-1}$ in $\mathbf{G}_{k+2}$. There are two cases.

The first case is that the edge e_0 of the cycle labeled by a_0 belongs to $\mathbf{G}_{k+1}$. Then by replacing every portion of the cycle lying on a flap in $\mathbf{G}_{k+2}$ but not in $\mathbf{G}_{k+1}$ by the axis of that rotor we obtain a new cycle $e_0, f_1, \ldots, f_{\ell-1}$ that lies in $\mathbf{G}_{k+1}$. Let d_j be the label of the edge f_j for each j with $1 \le j < \ell$. Because the join of the labels around a flap dominates the label of its axis, we have, invoking the induction hypothesis as well,

$$a_0 \le \bigvee_{1 \le j < \ell} d_j \le \bigvee_{1 \le i < m} a_i,$$

as desired.

The second case is that the edge e_0 labeled by a_0 does not belong to $\mathbf{G}_{k+1}$. This means that e_0 is a new edge on a rotor added to $\mathbf{G}_{k+1}$ in the construction of $\mathbf{G}_{k+2}$. But observe that on such a flap every label of a new edge appears twice. Therefore for some i with $1 \le i < m$ we must have $a_0 = a_i$. It follows that

$$a_0 \le \bigvee_{1 \le i < m} a_i,$$

as desired. This completes the induction. So each $\mathbf{G}_n$ satisfies condition (3). Since any cycle in $\mathbf{G}_\omega$ must lie in some $\mathbf{G}_n$, we have secured condition (3) of the definition of stable for $\mathbf{G}_\omega$.

Now we must specify an appropriate set $\mathcal{F}$ of stable maps. The next lemma is the key to this task.

LEMMA 7.31.
Let n and m be positive natural numbers. Any stable map from $\mathbf{G}_n$ into $\mathbf{G}_m$ can be extended to a stable map from $\mathbf{G}_\omega$ into $\mathbf{G}_\omega$.

Proof of Lemma 7.31. Let F be a stable map from $\mathbf{G}_n$ into $\mathbf{G}_m$. It is enough to show that F can be extended to a stable map $\hat{F}$ from $\mathbf{G}_{n+1}$ into $\mathbf{G}_{m+1}$. We define $\hat{F}(x)$ for all $x \in G_{n+1}$ as follows.

$$\hat{F}(x) := \begin{cases} F(x) & \text{if } x \in G_n \\ F(u) & \text{if } x \notin G_n, \{u,v\} \text{ is the axis of the flap containing } x \text{ and} \\ & F(u) = F(v) \\ x' & \text{if } x \notin G_n, \{u,v\} \text{ is the axis of the flap containing } x \text{ and} \\ & F(u) \neq F(v), \text{ where } x' \text{ occupies the position corresponding to} \\ & x \text{ in } \mathbf{G}_{m+1} \text{ on the rotor with axis } \{F(u), F(v)\} \end{cases}$$

The situation in the last alternative in the definition of $\hat{F}(x)$ is illustrated in the diagram below, where x can be any one of the vertices w, y or z.

The $\langle b,c\rangle$ flap of $\mathbf{P}_a(u,v)$ in $\mathbf{G}_{n+1}$ The $\langle b,c\rangle$ flap of $\mathbf{P}_a(F(u),F(v))$in $\mathbf{G}_{m+1}$

This definition of $\hat{F}(x)$ is sound since if $F(u) \neq F(v)$ then the edges $\{u,v\}$ and $\{F(u), F(v)\}$ must have the same label because F is stable. Evidently, $\hat{F}$ is stable. This finishes the proof of Lemma 7.31. ■

Now we describe the stable maps on $\mathbf{G}_n$ that we need.

Assuming $2 \leq n$, we will describe four stable maps for each pentagonal flap that was added to $\mathbf{G}_{n-1}$ to form $\mathbf{G}_n$. First, consider the flap in the following diagram.

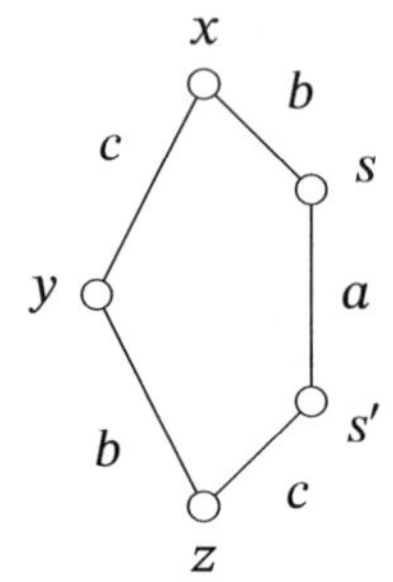

The $\langle b, c\rangle$ flap of $\mathbf{P}_a(s, s')$

The edge $\{s', s\}$ is the axis and it belongs to $\mathbf{G}_{n-1}$. In this case define $F_{x,s}$ for all $w \in G_n$ as follows:

$$F_{x,s}(w) : = \begin{cases} x & \text{if } w \in \{x, y\} \\ s & \text{if } w = z \\ s & \text{otherwise} \end{cases}$$

Notice that $F_{x,s}(y) = x$, $F_{x,s}(x) = x$ and $F_{x,s}(z) = s$, so this function maps one of the edges labeled b to the other edge of flap labeled b. In fact, everything except x and y is sent to s. It is easy to check that $F_{x,s}$ is a stable map from $\mathbf{G}_n$ to $\mathbf{G}_n$. The map in the other direction, $F_{y,z}$ is defined via

$$F_{y,z}(w) : = \begin{cases} y & \text{if } w \in \{x, y\} \\ z & \text{if } w = s \\ z & \text{otherwise} \end{cases}$$

Notice that $F_{y,z}(x) = t$ and $F_{y,z}(s) = z$, so this function maps one of the edges labeled b to the other edge of the flap labeled b. In fact, everything except x is sent to $s = z$. It is easy to check that $F_{y,z}$ is a stable map from $\mathbf{G}_n$ to $\mathbf{G}_n$. The other two maps $F_{z,s'}$ and $F_{x,y}$ are defined in a similar way for the edges of the flap that are labeled with c. We let $F^{\bullet}_{x,s}, F^{\bullet}_{y,z}, F^{\bullet}_{z,s'}$, and $F^{\bullet}_{x,y}$ denote the extensions of these maps to $\mathbf{G}_\omega$ via Lemma 7.31.

LEMMA 7.32.
Let $\{u, v\}$ and $\{w, y\}$ be edges of $\mathbf{G}_\omega$ that have the same label. Then there is a stable map $F : \mathbf{G}_\omega \to \mathbf{G}_\omega$ such that $F(u) = w$ and $F(v) = y$.

Proof of Lemma 7.32. Let n be the least natural number so that $\{u, v\}$ is an edge in $\mathbf{G}_n$ and let m be the least natural number so that $\{w, y\}$ is an edge in $\mathbf{G}_m$. We suppose, without harm, that $v \notin G_{n-1}$. Let $F_{u,v}$ be the stable map from G_n to G_n defined above. Let g be the function with domain $\{u, v\}$ so that $g(u) = w$ and $g(v) = y$. Then $g \circ F_{u,v}$ is a stable map from $\mathbf{G}_n$ into $\mathbf{G}_m$ that maps u to w and v to y. Extend this map to a stable map $F_{u,v,w,y}$ from $\mathbf{G}_\omega$ into $\mathbf{G}_\omega$. ■

This establishes property (4) of the definition of solid.

The set $\mathcal{F}$ of all the stable maps on G_ω includes all those of the forms $F^\bullet_{x,s}, F^\bullet_{y,z}, F^\bullet_{z,s'}$, $F^\bullet_{x,y}$, and $F_{u,v,w,y}$ as described above.

Finally, we have to show that $\mathbf{G}_\omega$ satisfies property (5) in the definition of solid. We prove by induction on n that $\mathbf{G}_n$ has property (5). So let $u, v \in A$ with $u \neq v$.

This means the base step of our induction is $n = 1$. Now since $\mathbf{G}_1$ is just a graph with two vertices that are joined by an edge labeled with d, there is virtually nothing to prove.

For the induction step, we assume that $1 \leq n$ and that $\mathbf{G}_n$ has property (5). Our aim is to show that $\mathbf{G}_{n+1}$ has property (5). That is, we seek a path in G_{n+1} connecting u and v so that each edge, regarded as a set of two ordered pairs of vertices, in the path belongs to the congruence generated by identifying u and v. Finding stable maps that map $\{u, v\}$ onto each of the path will do this. In this case, we will say that the path *stably* connects u and v. So let u and v be distinct vertices in $\mathbf{G}_{n+1}$. If they both belong to $\mathbf{G}_n$, then we can appeal directly to the induction hypothesis. So let us suppose that $u \notin G_n$. This means that u is one of the three new elements on a pentagonal flap of one of the rotors $\mathbf{P}_a(s, s')$ with axis the edge $\{s, s'\}$ of $\mathbf{G}_n$ in the construction of $\mathbf{G}_{n+1}$. Now there are three possibilities for the location of v: that $v \in G_n$, that v belongs to a different pentagonal flap of $\mathbf{P}_b(t, t')$, (note that $\{s, s'\}$ and $\{t, t'\}$ might be the same—that is u and v are on different flaps of the same rotor), or that v belongs to the flap of $\mathbf{P}_a(s, s')$ along with u. Consider the last situation. Should an edge join u and v, there is little to prove. So we consider the case here that no edge joins u and v. The diagram below illustrates this situation.

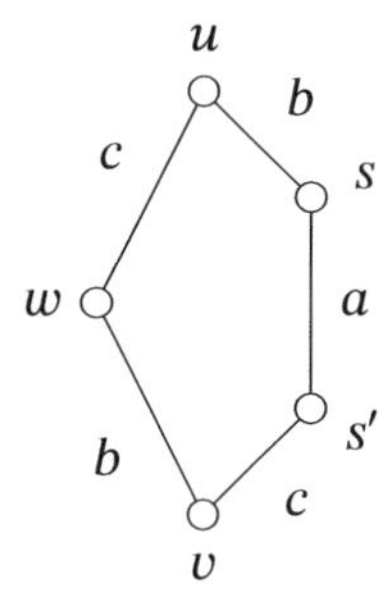

The $\langle b, c\rangle$ flap of $\mathbf{P}_a(s, s')$ in $\mathbf{G}_n$

Observe that $F_{u,w}(u) = u$ and $F_{u,w}(v) = w$ while $F_{v,w}(v) = v$ and $F_{v,w}(u) = w$. So that path uwv stably connects u and v. This settles the case when u and v are not in G_n but belong to the same flap.

So the situation reduces to that in the diagram below where u is one of x, w, or y and v is not among x, w, and y. If $u = x$ then $F_{u,s}(u) = u$ and $F_{u,s}(v) = s$, so the edge $\{u, s\}$ can be used in a path to stably connect u and s. If $u = w$, then the path $u\ x\ s$ can be used, since $F_{u,x}(u) = w$ and $F_{v,x}(s') = x$ while $F_{x,s}(u) = x$ and $F_{x,s}(v) = s$. Finally, if $u = y$, then the edge $\{y, s'\}$ can be used. Let us suppose that u is connected to s by a stable path. Likewise, if $v \notin G_n$, we can suppose that v is connected to t by a stable path. By the induction hypothesis, s and t are stably connected. If $s = t$ we have in hand the desired path. So suppose $s \neq t$. So we see that $(s, t) \in \mathrm{Cg}^{\mathbf{A}_{G_{n+1}}}(u, v)$. Since $s, t \in G_n$ we appeal to the induction hypothesis to obtain a path from s to t so that each edge in the path gives an ordered pair belonging to $\mathrm{Cg}^{\mathbf{A}_{G_{n+1}}}(s, t)$. But since $(s, t) \in \mathrm{Cg}^{\mathbf{A}_{G_{n+1}}}(u, v)$. We see that we have a path connecting u to v in $\mathbf{G}_{n+1}$ with the desired property. Finally, consider the possibility that $v \in G_n$. In this case, s and v are stably connected by the induction hypothesis, as above. Since we have already connected u to s our path is complete. So property (5) holds for $\mathbf{G}_{n+1}$.

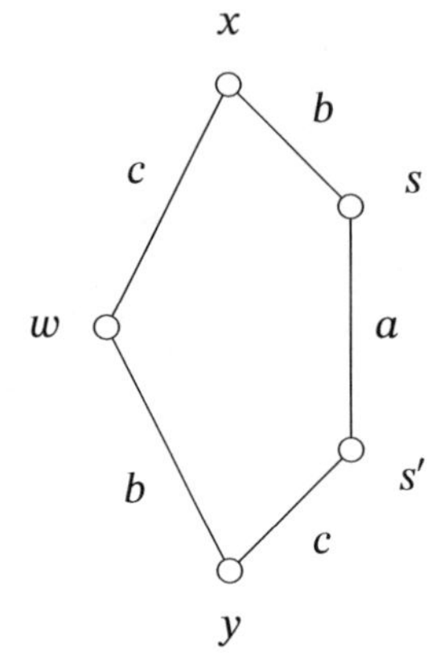

The $\langle b, c\rangle$ flap of $\mathbf{P}_a(s, s')$ in $\mathbf{G}_n$

Consequently, $\mathbf{G}_\omega$ satisfies property (5), as well as all the other properties listed in the definition of solid.

So according to Lemma 7.30, we conclude that $\mathbf{L} \cong \mathbf{Con}\,\mathbf{A}_{G\mathcal{F}}$.

Finally, if $\mathbf{L}$ has κ nonzero compact elements, then each rotor $\mathbf{P}_a$ that comes up in the construction will have κ vertices if κ is infinite and finitely many vertices if κ is finite. So $\mathbf{A}$ will have cardinality which is the least infinite cardinal at least as large as κ, and the list of basic operations of $\mathbf{A}$ will consist of operations of rank 1. We have taken all the stable maps on $\mathbf{G}_\omega$ that belong to $\mathcal{F}$ has the basic operations. Now $\mathcal{F}$ must have 4 maps for each flap in the construction of $\mathbf{G}_\omega$ as well as one map for each choice of $u, v, w, y \in A$ so that the edges $\{u, v\}$ and $\{w, y\}$ have the same labels. So there are altogether at least κ stable maps in $\mathcal{F}$.

So the Grätzer-Schmidt Theorem is established. ■

As G. Grätzer and E. T. Schmidt (1963) proved this theorem, the construction was more elaborate and required the use of partial algebras. The proof we have given follows the method in (Pavel Pudlák, 1976).

Even when the construction in the proof above is started with a finite nontrivial lattice $\mathbf{L}$, the resulting algebra $\mathbf{A}$ will be countably infinite, although each $\mathbf{G}_n$ will be finite. Of the several known proofs of the Grätzer-Schmidt Theorem, all have this feature that the resulting algebra $\mathbf{A}$ will be countably infinite when $\mathbf{L}$ is a finite lattice. One of the longest standing—and apparently very difficult—open problems concerning congruence lattices is the following:

THE FINITE LATTICE CONGRUENCE REPRESENTATION PROBLEM

Is every finite lattice isomorphic to the lattice of congruences of a finite algebra?

According to Theorem 4.62 in Volume 1, every lattice is isomorphic to a sublattice of the lattice of equivalence relations on some set X. This is a classical result of P. Whitman (1946). In the proof given in Volume I, if the lattice is finite the set X will be countably infinite. However, a difficult proof of Pavel Pudlák and J. Tůma (1980) asserts that X can be taken to be finite, whenever the lattice is finite.

Another difficult and surprising result is the proof by P. P. Pálfy and P. Pudlák (1980) that the Finite Lattice Congruence Representation Problem is equivalent to the following problem:

THE FINITE LATTICE SUBGROUP REPRESENTATION PROBLEM
Is every finite lattice isomorphic to an interval in the subgroup lattice of some finite group?

We note that it is the *problems* that are logically equivalent. So if there is a finite lattice **L** that is a counterexample to one of these problems, then there must be a finite lattice **L**′ that is a counterexample to the other problem, but these finite lattices might well be different.

With this proof of the Grätzer-Schmidt Theorem, it is also the case that the signature of the representing algebras will be infinite, although each basic operation will have rank 1. This raises the question of whether such a representation might be achieved using an algebra of finite signature. This prospect must be rejected.

THEOREM 7.33 (Ralph Freese, William A. Lampe, and Walter Taylor 1979). *If* **V** *is an infinite dimensional vector space over an uncountable field* **F**, *then the congruence lattice of* **V** *cannot be represented as a congruence lattice without using at least* $|F|$ *basic operations of rank at least* 1.

Before taking up the proof of this theorem we establish three lemmas. We need the following property about congruences on **V**:

(∗) For every compact ψ, there exist compact θ and φ such that

$$\theta \vee \varphi \geq \psi \text{ and } \theta \wedge \psi = \varphi \wedge \psi = 0.$$

LEMMA 7.34 (Due to W. Lampe). *If* **Con A** *satisfies* (∗), *then*

$$\text{(T)} \qquad \mathbf{A} \models \exists \mathbf{x}\,(t(\mathbf{x},\mathbf{p}) \approx t(\mathbf{x},\mathbf{q})) \Rightarrow \forall \mathbf{x}\,(t(\mathbf{x},\mathbf{p}) \approx t(\mathbf{x},\mathbf{q})),$$

where $\mathbf{x} = \langle x_0, \ldots, x_{n-1}\rangle$ *is an* n*-tuple of variables for some natural number* n, **p** *and* **q** *are* m*-tuples of elements of* A *for some natural number* m, *and* t *is any term on* $n+m$ *variables.*

The condition **(T)** is equivalent to

$$\text{(T}'\text{)} \qquad \mathbf{A} \models \forall \mathbf{x}\forall \mathbf{y}\,[t(\mathbf{x},\mathbf{p}) \approx t(\mathbf{x},\mathbf{q}) \Rightarrow t(\mathbf{y},\mathbf{p}) \approx t(\mathbf{y},\mathbf{q})],$$

which is a quasi-equation. Quasi-equations and the classes of algebras they axiomatize are discussed beginning on page 402.

Proof. Suppose that $t^{\mathbf{A}}(\mathbf{a},\mathbf{p}) = t^{\mathbf{A}}(\mathbf{a},\mathbf{q})$, where $\mathbf{a} \in A^n$. Let $\mathbf{b} \in A^n$. We need $t^{\mathbf{A}}(\mathbf{b},\mathbf{p}) = t^{\mathbf{A}}(\mathbf{b},\mathbf{q})$. To see this, we apply (∗) to the compact congruence

$$\psi := \mathrm{Cg}^{\mathbf{A}}(\mathbf{a},\mathbf{b}) \vee \mathrm{Cg}^{\mathbf{A}}(\mathbf{p},\mathbf{q})$$

Here $\mathrm{Cg}^{\mathbf{A}}(\mathbf{a},\mathbf{b}) = \mathrm{Cg}^{\mathbf{A}}\{\langle a_i, b_i\rangle \mid i < n\}$. $\mathrm{Cg}^{\mathbf{A}}(\mathbf{p},\mathbf{q})$ has a like meaning. According to $(*)$ we have congruences θ and φ so that

$$\mathrm{Cg}^{\mathbf{A}}(\mathbf{a},\mathbf{b}) \subseteq \psi \subseteq \theta \vee \varphi,$$

and so,

$$\mathbf{a} = \mathbf{s}_0 \;\theta\; \mathbf{s}_1 \;\varphi\; \mathbf{s}_2 \;\theta \cdots \mathbf{s}_k = \mathbf{b}$$

for appropriate n-tuples $\mathbf{s}_0, \dots, \mathbf{s}_k \in A^n$. Since $\mathbf{s}_0 \;\theta\; \mathbf{s}_1$, we have

$$t^{\mathbf{A}}(\mathbf{s}_1,\mathbf{p}) \;\theta\; t^{\mathbf{A}}(\mathbf{s}_0,\mathbf{p}) = t^{\mathbf{A}}(\mathbf{s}_0,\mathbf{q}) \;\theta\; t^{\mathbf{A}}(\mathbf{s}_1,\mathbf{q}).$$

Since $\mathbf{p} \;\psi\; \mathbf{q}$, we also have $t^{\mathbf{A}}(\mathbf{s}_1,\mathbf{p}) \;\psi\; t^{\mathbf{A}}(\mathbf{s}_1,\mathbf{q})$. Finally, since $\theta \cap \psi = 0_A$, we have $t^{\mathbf{A}}(\mathbf{s}_1,\mathbf{p}) = t^{\mathbf{A}}(\mathbf{s}_1,\mathbf{q})$. A similar argument—with φ in place of θ—yields $t^{\mathbf{A}}(\mathbf{s}_2,\mathbf{p}) = t^{\mathbf{A}}(\mathbf{s}_2,\mathbf{q})$. Proceeding by induction, we obtain $t^{\mathbf{A}}(\mathbf{b},\mathbf{p}) = t^{\mathbf{A}}(\mathbf{s}_k,\mathbf{p}) = t^{\mathbf{A}}(\mathbf{s}_k,\mathbf{q}) = t^{\mathbf{A}}(\mathbf{b},\mathbf{q})$. ■

Embedded in this proof is an important tool of the general theory of algebraic systems, namely the term condition, which is described and exploited in Chapter 11 in Volume III. See, particularly, Equation (11.2.2) on page 275 of that volume.

LEMMA 7.35. *Let* $\mathbf{A}$ *be an algebra such that* $\mathbf{Con\ A}$ *obeys* $(*)$ *and let* κ *be the cardinality of the set of basic operations of* $\mathbf{A}$ *of rank at least* 1. *Then every congruence class of every compact congruence of* $\mathbf{A}$ *has cardinality no more that* $\kappa + \aleph_0$.

Proof of Lemma 7.35. Let S be a congruence class of a compact congruence θ. Thus $\theta = \mathrm{Cg}^{\mathbf{A}}(\mathbf{a},\mathbf{b})$, where $\mathbf{a},\mathbf{b} \in A^n$ for some natural number n. Fixing $c \in S$ and considering arbitrary $d \in S$, we have $\langle c, d\rangle \in \mathrm{Cg}^{\mathbf{A}}(\mathbf{a},\mathbf{b})$. Therefore, for some terms $t_0, \dots, t_{n-1}$ and some $\mathbf{p}_0, \dots, \mathbf{p}_{n-1} \in A^m$, where m is a natural number, we have (according to Theorem 6.18)

$$\begin{gathered} c = t_0^{\mathbf{A}}(\mathbf{a},\mathbf{b},\mathbf{p}_0) \\ t_0^{\mathbf{A}}(\mathbf{b},\mathbf{a},\mathbf{p}_0) = t_1^{\mathbf{A}}(\mathbf{a},\mathbf{b},\mathbf{p}_1) \\ \vdots \\ t_{n-2}^{\mathbf{A}}(\mathbf{b},\mathbf{a},\mathbf{p}_{n-2}) = t_{n-1}^{\mathbf{A}}(\mathbf{a},\mathbf{b},\mathbf{p}_{n-1}) = d \end{gathered}$$

To see our upper bound on $|S|$, it is clearly enough—by counting the number of finite sequences of terms—to see that d depends only on the the sequence $\langle t_0, \dots, t_{n-1}\rangle$ and *not* on $\langle \mathbf{p}_0, \dots, \mathbf{p}_{n-1}\rangle$. For this purpose suppose $\langle \mathbf{p}'_0, \dots, \mathbf{p}'_{n-1}\rangle$ is an n-tuple so that

$$\begin{gathered} c = t_0^{\mathbf{A}}(\mathbf{a},\mathbf{b},\mathbf{p}'_0) \\ t_0^{\mathbf{A}}(\mathbf{b},\mathbf{a},\mathbf{p}'_0) = t_1^{\mathbf{A}}(\mathbf{a},\mathbf{b},\mathbf{p}'_1) \\ \vdots \\ t_{n-2}^{\mathbf{A}}(\mathbf{b},\mathbf{a},\mathbf{p}'_{n-2}) = t_{n-1}^{\mathbf{A}}(\mathbf{a},\mathbf{b},\mathbf{p}'_{n-1}) = d' \end{gathered}$$

From $t_0^{\mathbf{A}}(\mathbf{a},\mathbf{b},\mathbf{p}_0) = c = t_0^{\mathbf{A}}(\mathbf{a},\mathbf{b},\mathbf{p}'_0)$ we obtain, with the help of Lemma 7.34,

$$t_1^{\mathbf{A}}(\mathbf{a},\mathbf{b},\mathbf{p}_1) = t_0^{\mathbf{A}}(\mathbf{b},\mathbf{a},\mathbf{p}_0) = t_0^{\mathbf{A}}(\mathbf{b},\mathbf{a},\mathbf{p}'_0) = t_1^{\mathbf{A}}(\mathbf{a},\mathbf{b},\mathbf{p}'_1).$$

Proceeding by induction, we obtain

$$d = t_{n-1}^{\mathbf{A}}(\mathbf{a},\mathbf{b},\mathbf{p}_{n-1}) = t_{n-1}^{\mathbf{A}}(\mathbf{a},\mathbf{b},\mathbf{p}'_{n-1}) = d'.$$

This completes the proof of the lemma. ■

LEMMA 7.36. *Let* $\mathbf{A}$ *be an algebra, let* $\psi \in \operatorname{Con}\mathbf{A}$, *let* $X \subseteq \operatorname{Con}\mathbf{A}$ *with* X *infinite, and assume that all distinct* $\theta, \varphi \in X$ *satisfy* $\theta \cap \varphi = 0_A$ *and* $\theta \vee \varphi = \psi$. *Then each nontrivial congruence class of* ψ *has cardinality at least* $|X|$.

Proof of Lemma 7.36. Partition X into 2-element subsets. Since X is infinite, this partition has cardinality $|X|$. Now take $\langle c, e\rangle \in \psi$ with $c \neq e$. For block $\{\theta, \varphi\}$ of the partition of X we see $\theta \vee \varphi = \psi$. So we have

$$(\dagger) \qquad c \; \rho_0 \; d \; \rho_1 \; \cdots \; \rho_k \; e,$$

where $\rho_j \in \{\theta, \varphi\}$ and d and the rest of the intermediate elements belong to A. Of course, we may assume $d \neq c$. But if $c \; \rho' \; d$ were to hold where ρ' belongs to a different block of our partition of X into 2-element sets, then we would have $c = d$ since $\rho_0 \cap \rho' = 0_A$. This means that if $(\dagger)$ were undertaken with a different block of the partition into 2-element sets, the resulting d's would all be distinct. Hence, each congruence class of ψ must have cardinality at least $|X|$, completing the proof of the lemma. ■

With these three lemmas in hand, we can prove the theorem.

Proof of Theorem 7.33. Let $\mathbf{A}$ be an algebra such that $\mathbf{Con\,V} \cong \mathbf{Con\,A}$.

First we observe that $\mathbf{Con\,V}$ obeys $(*)$. Since $\theta \mapsto 0/\theta$ is an isomorphism between $\mathbf{Con\,V}$ and $\operatorname{Sub}\mathbf{V}$, we can deal with subspaces of $\mathbf{V}$ in place of congruences. So let $\mathbf{W}$ be a finite dimensional subspace, let us say of dimension n, of $\mathbf{V}$. Let $\{w_0, \dots, w_{n-1}\}$ be a basis for $\mathbf{W}$. Since $\mathbf{V}$ is infinite dimensional, we can pick n more elements $u_0, \dots,$ $u_{n-1} \in V \setminus W$ so that $\{w_0, \dots, w_{n-1}, u_0, \dots,\}$ is linearly independent. Let $\mathbf{U}$ be the span of $\{u_0, \dots, u_{n-1}\}$. Finally, let $z_i = v_i - u_i$ for $i < n$. Let $\mathbf{Z}$ be the span of $\{z_i \mid i < n\}$. Then

$$\mathbf{W} \subseteq \mathbf{U} \vee \mathbf{Z} \text{ and } U \cap W = Z \cap W = \{0\}.$$

So $\mathbf{Con\,V}$ obeys $(*)$. So $\mathbf{Con\,A}$ must also obey $(*)$.

Now observe that for $\mathbf{Con\ V}$ there is $\mathbf{W}$, a finite dimensional subspace of $\mathbf{V}$ of dimension at least 2 and an infinite set X of subspaces that fulfill the hypotheses of Lemma 7.36 with $|X| = |F|$. To see this, for each ordinal $\alpha < |F|$ pick distinct one-dimensional subspaces $\mathbf{U}_\alpha$ of $\mathbf{W}$. So $\mathbf{Con\ A}$ must also fulfill the same hypotheses. In particular, there is a compact congruence on $\mathbf{A}$ with a congruence class of cardinality at least $|F|$.

On the other hand, by Lemma 7.35, the congruence class produced in the last paragraph has cardinality no more that $\kappa + \aleph_0$, where κ is the cardinality of the set of basic operations of $\mathbf{A}$ of rank at least 1. So we have

$$|F| \leq \kappa + \aleph_0 = \kappa.$$

Since the field $\mathbf{F}$ is uncountable, we conclude that $\mathbf{A}$ must have at least $|F|$ basic operations of rank at least 1. This establishes the theorem. ■

Theorem 7.33 raises the problem of finding sufficient conditions on an algebraic lattice under which the lattice admits a representation as the congruence lattice of an algebra of finite signature. Even if the algebraic lattice admits a representation as a congruence lattice of an algebra of finite signature, there is the question of which finite signatures might serve. The following is a notable result. Its method, visible in the first paragraph of the proof, involves replacing $\mathbf{A}$ by a subtler algebra with universe A^k.

This construction, and this method, will be explored further in Theorem 10.110 and its corollaries, on pages 207*ff* of Volume III.

THEOREM 7.37 (Ralph McKenzie 1984). *For every algebra* **A** *of finite signature, there is an algebra* **B** *in a signature providing only one 2-place operation symbol and one 1-place operation symbol so that* **Con A** $\cong$ **Con B**.

Proof. It is harmless to suppose that all the basic operations of **A** are of the same rank, say $r > 1$. We let $\mathbf{A} = \langle A, Q_0, \dots, Q_{n-1}\rangle$. Let $k = 1 + \max(n, r)$. Let B be the set of all k-tuples $\mathbf{b} \in A^k$. Define $* : B \times B \to B$ and $F : B \to B$ so that for all $\mathbf{x}, \mathbf{y} \in B$

$$\mathbf{x} * \mathbf{y} := \langle y_{k-1}, x_0, \dots, x_{k-2}\rangle$$
$$F(\mathbf{x}) := \langle Q_0(\mathbf{x}'), Q_1(\mathbf{x}'), \dots, Q_{n-1}(\mathbf{x}'), x_0, \dots, x_0\rangle$$

where $\mathbf{x}'$ is the restriction of $\mathbf{x}$ to its first r entries.

We take $\mathbf{B} := \langle B, *, F\rangle$.

Let θ be a 2-place relation on A. Let

$$\overline{\theta} := \{\langle \mathbf{x}, \mathbf{y}\rangle \mid \mathbf{x}, \mathbf{y} \in B \text{ and } x_j \mathrel{\theta} y_j \text{ for all } j < k\}.$$

It is clear that $\theta \subseteq \varphi$ if and only if $\overline{\theta} \subseteq \overline{\varphi}$. Hence, the map $\theta \mapsto \overline{\theta}$ is one-to-one. Also

$$\overline{\theta \cap \varphi} = \overline{\theta} \cap \overline{\varphi} \qquad \overline{\theta^{\smile}} = \overline{\theta}^{\smile} \qquad \overline{\theta \circ \varphi} = \overline{\theta} \circ \overline{\varphi}.$$

Here $\theta^{\smile} = \{\langle y, x\rangle \mid \langle x, y\rangle \in \theta\}$. This entails that $\overline{\theta}$ is an equivalence relation if θ is an equivalence relation.

Now suppose that $\theta \in \operatorname{Con} \mathbf{A}$, that $\mathbf{x}\overline{\theta}\mathbf{y}$, and that $\mathbf{u} \mathrel{\overline{\theta}} \mathbf{v}$. So $x_j \mathrel{\theta} y_j$ for all $j < k$. Thus,

$$Q_i(x_0, \dots, x_{r-1}) \mathrel{\theta} Q_i(y_0, \dots, y_{r-1}) \text{ for all } i < n.$$

Likewise, $u_j \mathrel{\theta} v_j$ for all $j < k$, and

$$Q_i(u_0, \dots, u_{r-1}) \mathrel{\theta} Q_i(v_0, \dots, v_{r-1}) \text{ for all } i < n.$$

But then

$$\begin{aligned} F(\mathbf{x}) &= \langle Q_0(\mathbf{x}'), Q_1(\mathbf{x}'), \dots, Q_{n-1}(\mathbf{x}'), x_0, \dots, x_0\rangle \\ &\mathrel{\overline{\theta}} \langle Q_0(\mathbf{y}'), Q_1(\mathbf{y}'), \dots, Q_{n-1}(\mathbf{y}'), y_0, \dots, y_0\rangle \\ &\mathrel{\overline{\theta}} F(\mathbf{y}). \end{aligned}$$

So $\overline{\theta}$ respects F. The reader can verify that $\overline{\theta}$ also respects $*$. So if $\theta \in \operatorname{Con} \mathbf{A}$, then $\overline{\theta} \in \operatorname{Con} \mathbf{B}$.

It remains to show that $\overline{}$ maps $\operatorname{Con} \mathbf{A}$ onto $\operatorname{Con} \mathbf{B}$. So let $\psi \in \operatorname{Con} \mathbf{B}$. Put

$$\theta := \{\langle u, v\rangle \mid u, v \in A \text{ and } \langle u, u, u, \dots\rangle \mathrel{\psi} \langle v, v, v, \dots\rangle\}.$$

CLAIM 1.
$\theta \in \operatorname{Con} \mathbf{A}$ *and* $\overline{\theta} = \psi$.

Proof. We define a number of term-operations of **B** to facilitate this proof.

Let $\mathbf{x} \in B$. Put

$$G(\mathbf{x}) := \mathbf{x} * \mathbf{x} = \langle x_{k-1}, x_0, \dots, x_{k-2}\rangle.$$

So G gives a cyclic permutation of the entries of its inputs. In particular, for $j < k$ we have

$$G^{k-j}(\mathbf{x}) = \langle x_j, x_{j+1}, \dots, x_{k-1}, x_0, \dots, x_{j-1}\rangle.$$

Now observe

$$G^{k-j}(\mathbf{x}) * G^{k-j-1}(\mathbf{x}) = \langle x_j, x_j, x_{j+1}, \dots, x_0, \dots, x_{j-2}\rangle.$$

By iterating this construction, we arrive at the term operation

$$H_j(\mathbf{x}) := \langle x_j, x_j, \dots, x_j\rangle.$$

Let $\mathbf{y} \in B$ and $j < k$. We have the left-to-right direction of the following equivalence.

$$(\Diamond) \qquad \mathbf{x}\ \psi\ \mathbf{y} \Longleftrightarrow x_k\ \theta\ y_k \text{ for every natural number } k.,$$

since we can employ each H_j to obtain

$$\mathbf{x}\ \psi\ \mathbf{y} \Rightarrow \langle x_k, x_k, \dots\rangle\ \psi\ \langle y_k, y_k, \dots\rangle \text{ for every natural number } k.$$

For the right-to-left direction of $(\Diamond)$, we define another term operation on $\mathbf{B}$. For $\mathbf{x}, \mathbf{y}, \mathbf{z} \in B$ observe

$$\big(\mathbf{z} * G^{k-1}(\mathbf{y})\big) * G^{k-1}(\mathbf{x}) = \langle x_0, y_0, z_0, z_1, \dots\rangle.$$

By iterating this construction we can obtain a term operation K on $\mathbf{B}$ of rank k. Let ${}_0\mathbf{x}, \dots, {}_{k-1}\mathbf{x} \in B$. Then we will have

$$K({}_0\mathbf{x}, \dots, {}_{k-1}\mathbf{x}) = \langle {}_0x_0, {}_1x_0, \dots, {}_{k-1}x_0\rangle.$$

The term operation K simply builds its output by taking the first entries of each of its inputs. In particular,

$$K\left(H_0(\mathbf{x}), H_1(\mathbf{x}), \dots, H_{k-1}(\mathbf{x})\right) = \mathbf{x}.$$

To establish the right-to-left direction of $(\Diamond)$, suppose $\mathbf{x}, \mathbf{y} \in B$ so that $x_j\ \theta\ y_j$ for all $j < k$. But then

$$\mathbf{x} = K\left(H_0(\mathbf{x}), H_1(\mathbf{x}), \dots, H_{k-1}(\mathbf{x})\right)\ \psi\ K\left(H_0(\mathbf{y}), H_1(\mathbf{y}), \dots, H_{k-1}(\mathbf{y})\right) = \mathbf{y}.$$

So we see that

$$\mathbf{x}\ \psi\ \mathbf{y} \text{ if and only if } x_k\ \theta\ y_k \text{ for every natural number } k.$$

It follows that $\overline{\theta} = \psi$.

It only remains to show that $\theta \in \operatorname{Con}\mathbf{A}$. Certainly, θ is an equivalence relation. To see that θ respects each basic operation Q_j of $\mathbf{A}$, pick $u_0, \dots, u_{r-1}, v_0, \dots, v_{r-1} \in A$ so that $u_i\ \theta\ v_i$ for $i < r$. Put

$$\mathbf{x} := \langle u_0, \dots, u_{r-1}, u_0, \dots u_0\rangle \in B$$
$$\mathbf{y} := \langle v_0, \dots, v_{r-1}, v_0, \dots v_0\rangle \in B.$$

Then

$$
\begin{gathered}
u_i \;\theta\; v_i \text{ for all } i < r \\
\Downarrow \\
\mathbf{x} \;\psi\; \mathbf{y} \\
\Downarrow \\
F(\mathbf{x}) \;\psi\; F(\mathbf{y}) \\
\Downarrow \\
\langle Q_0(\mathbf{x}), Q_1(\mathbf{x}), \dots, Q_{n-1}(\mathbf{x}), u_0, \dots, u_0\rangle \;\psi\; \langle Q_0(\mathbf{y}), Q_1(\mathbf{y}), \dots, Q_{n-1}(\mathbf{y}), v_0, \dots, v_0\rangle \\
\Downarrow \\
Q_j(u_0, \dots, u_{r-1}) \;\theta\; Q_j(v_0, \dots, v_{r-1}) \text{ for all } j < n.
\end{gathered}
$$

Hence $\theta \in \operatorname{Con} \mathbf{A}$. This completes the proof of the claim. ■

So **Con A** $\cong$ **Con B**, and the theorem is established. ■

Which algebraic lattices admit a representation as a congruence lattice of an algebra of finite signature? There are several known conditions an algebraic lattice might satisfy to ensure such a representation. We present here one such condition.

Let **L** be an algebraic lattice and let $p \in L$. We say that p is a **pinch point** of **L** provided

- p is a compact element of L, and
- p is comparable with every compact element of **L**.

As every element of an algebraic lattice is a join of compact elements, the last stipulation in the definition of pinch point is equivalent to the assertion that p is comparable with every element of L. Figure 7.10 illustrates a pinch point.

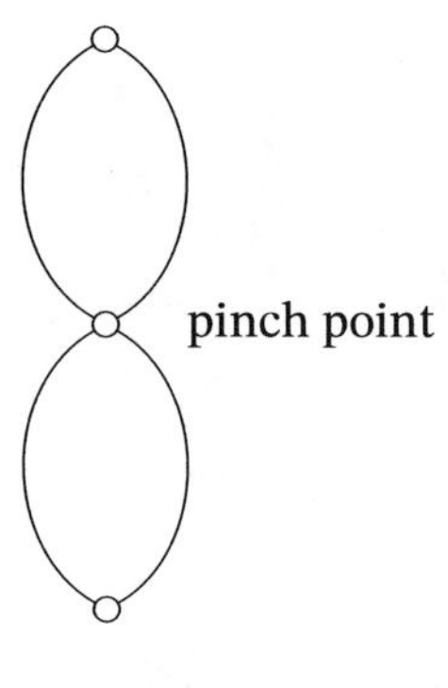

FIGURE 7.10. An algebraic lattice with a pinch point

An algebraic lattice **L**, with top element 1 and bottom element 0, is a **pinched lattice** provided the top element of **L** is the join of some set of pinch points. The most transparent kind of pinched lattice is an algebraic lattice whose top element is compact. As 0 is compact, it is a pinch point. Observe that the set of all pinch points is linearly ordered with least element 0. Moreover, if **L** is a pinched lattice, then there must be a set J of pinch points that witnesses this and that is well-ordered by the lattice ordering of **L**. To demonstrate this, we can start with any set I that witnesses that **L** is a pinched lattice, harmlessly assuming $0 \in I$. If 1 is compact, we can let $J = \{0, 1\}$. If 1 is not compact, then we undertake a recursive procedure on the ordinals to define a sequence $\langle p_0, p_1, \ldots \rangle$ of members of I as follows:

Stage 0: Let $p_0 = 0$.

Successor Stages: Since p_α is compact, we see $p_\alpha \neq 1$. So pick $p_{\alpha+1} \in I$ with $p_\alpha < p_{\alpha+1}$.

Limit Stages: Let λ be a limit ordinal such that p_α has already been selected, for each $\alpha \in \lambda$. If $\bigvee\{p_\alpha \mid \alpha \in \lambda\} = 1$, then the construction halts and we take $J = \{p_\alpha \mid \alpha \in \lambda\}$. Otherwise, pick $p_\lambda \in I$ with $p_\alpha < p_\lambda$ for all $\alpha \in \lambda$.

Recall that a *groupoid* is an algebra that has only one basic operation and the rank of that operation is 2. A *pointed groupoid* is the expansion of a groupoid by the adjunction of a basic operation of rank zero.

We need one last ingredient. We know that the meet in **Con A** is just intersection and that for $\theta, \varphi \in \operatorname{Con} \mathbf{A}$ we have

$$\theta \vee \varphi = (\theta \circ \varphi) \cup (\theta \circ \varphi \circ \theta) \cup (\theta \circ \varphi \circ \theta \circ \varphi) \cup \cdots.$$

This is, in fact a union of an ascending sequence:

$$\theta \circ \varphi \subseteq \theta \circ \varphi \circ \theta \subseteq \theta \circ \varphi \circ \theta \circ \varphi \subseteq \cdots.$$

The lattice **Con A** is said to have **type** 3 **joins** provided

$$\theta \vee \varphi = \theta \circ \varphi \circ \theta \circ \varphi$$

for all $\theta, \varphi \in \operatorname{Con} \mathbf{A}$. Type n joins has a similar definition.

The Pinched Lattice Theorem (William A. Lampe, 1982).
For every pinched lattice **L** *there is a groupoid* **A** *such that* $\mathbf{L} \cong$ **Con** *A and* **Con A** *has type* 3 *joins.*

The involved and lengthy proof of this theorem puts it beyond the reach of our exposition.

7.6. The Lattice of Equational Theories

For a fixed signature, the equational theories are ordered by set-inclusion $\subseteq$. More is true. There is a largest equational theory $\top$: the set of all equations of the signature. The set $\{x \approx y\}$, where x and y are different variables, is a base for $\top$. Equational theories other than $\top$ are called **proper** equational theories. There is a smallest equational theory $\bot$: the set of all equations of the form $s \approx s$, where s is a term. These equations are called **tautologies**. The theory $\bot$ is the **trivial** equational theory and is based on $\{x \approx x\}$. Mostly we will be interested in proper nontrivial equational theories. Moreover every set $\mathcal{F}$ of equational theories has a greatest lower bound or meet, namely $\bigcap \mathcal{F}$, provided $\mathcal{F}$ is not empty and the largest equational theory if $\mathcal{F}$ is empty. In symbols we write

$$\bigwedge \mathcal{F} = \begin{cases} \bigcap \mathcal{F} & \text{if } \mathcal{F} \neq \emptyset \\ \top & \text{otherwise} \end{cases}$$

Also, every set $\mathcal{F}$ has a least upper bound or join, namely $\Theta\,\mathbf{Mod}(\bigcup \mathcal{F})$. In symbols

$$\bigvee \mathcal{F} = \Theta\,\mathbf{Mod} \bigcup \mathcal{F}.$$

Indeed, all these things hold for the closed sets on either side of any Galois connection. Any lattice **L** is referred to as a **complete lattice** provided every subset of L has both a greatest lower bound and a least upper bound with respect to the lattice ordering.

Suppose **L** is a complete lattice. An element $c \in L$ is said to be **compact** provided if $c \leq \bigvee \mathcal{F}$, then $c \leq \bigvee \mathcal{F}^*$ for some finite $\mathcal{F}^* \subseteq \mathcal{F}$. The complete lattice **L** is said to be **algebraic** whenever each element of L is the join of some set of compact elements. It is not hard to see that the compact elements of the lattice of all equational theories are just those equational theories that are finitely based. It is also clear that every equational theory is the join of all finitely based theories it contains. That is, for any fixed signature, the lattice of all equational theories is an algebraic lattice whose top element is compact. For a signature σ we denote the lattice of all equational theories of signature σ by $\mathcal{L}_\sigma$.

The ordering in the lattice of equational theories provides a way to compare the strength of equational theories—the larger the theory the stronger it is. So understanding the structure of the lattice of equational theories of a given signature is one of the routes to a deeper study of equational logic and variety theory. One principal tool in this study is that of ω-universal sets of terms.

κ-universal sets and systems

At a number of points in the rest of this chapter we will need some combinatorial properties of terms. Such properties often facilitate the transfer of results from a convenient signature to almost arbitrary signatures. The transfer of information between different signatures can be seen as part of the algebraic theory of interpretations, to which we now turn.

We will summarize the definition of **interpretation of varieties** from § 4.12 in Volume I, and from §6.1 in this Volume. Our formal basis will be conditions (1)–(4) of Definitions 6.2–6.3 on pages 8–9. Let $\sigma : I \longrightarrow \omega$ and $\tau : J \longrightarrow \omega$ be signatures. An **interpretation of σ in τ**, or a **system of definitions of σ in τ**, is a map D from I to the set T_τ of all τ-terms, obeying conditions (1)-(2) of Definition 6.2. If **A** is an algebra of signature τ, then $\mathbf{A}^D$ denotes the σ-algebra $\mathbf{B} = \langle A, D(Q)^{\mathbf{A}} \rangle_{Q \in I}$. Let us further suppose that $\mathcal{V}$ and $\mathcal{W}$ are varieties of signatures σ and τ, respectively. A system $D : I \longrightarrow T_\tau$ is an **interpretation of $\mathcal{V}$ in $\mathcal{W}$** if $\mathbf{A}^D \in \mathcal{V}$ for each $\mathbf{A} \in \mathcal{W}$.

Each system $D : I \longrightarrow T_\tau$ has a unique extension to a clone morphism $\mathrm{Tr}_D : T_\sigma \longrightarrow T_\tau$. (This means simply that if p is the substitution instance $q(u_0, u_1, \dots)$, then $\mathrm{Tr}_D(p)$ is the substitution instance $\mathrm{Tr}_D(q)(\mathrm{Tr}_D(u_0), \mathrm{Tr}_D(u_1), \dots)$.) This Tr_D is known as the **translation function associated to** D. If Σ is a set of σ-equations, then $\mathrm{Tr}_D\Sigma$ will denote the set of all τ-equations $\mathrm{Tr}_D(p) \approx \mathrm{Tr}_D(q)$ for $(p \approx q) \in \Sigma$.

In Lemma 6.4, page 10, we showed that for all $\mathbf{A} \in \mathcal{W}$ and all $t \in T_\tau$, the realization of $\mathrm{Tr}_D(t)$ in **A** is the same as the realization of t in $\mathbf{A}^D$. We then deduced, in Equation (6.1.4), that $\mathbf{A}^D \models s \approx t$ if and only if $\mathbf{A} \models \mathrm{Tr}_D(s) \approx \mathrm{Tr}_D(t)$. From this we easily saw that if $\Sigma \vdash s \approx t$ in signature σ, then $\mathrm{Tr}_D(\Sigma) \vdash \mathrm{Tr}_D(s) \approx \mathrm{Tr}_D(t)$ in signature τ. However, in general the converse fails without some further restriction on the interpretation $\mathbf{A}^D$. A useful restriction here is that of κ-**universality**, to which we now turn. (See Theorem 7.42.)

DEFINITION 7.38. Term operations on an algebra **A** can be seen as taking ω-tuples, i.e. members of A^ω, as inputs. In this context each term operation depends only on finitely many of the coordinates of its input ω-tuples. More generally, for any $F\colon A^\omega \longrightarrow A$, we define the **support** of F to be the set of coordinates on which F depends. Thus a term operation always has finite support; more generally we say that an arbitrary function $F : A^\omega \longrightarrow A$ is **term-like** iff its support is finite. If the support of F is contained in $\{0, \dots, n-1\}$, then there exists an n-ary operation G such that $F(a_0, a_1, \dots) = G(a_0, a_1, \dots, a_{n-1})$ for all $a_0, a_1, \dots \in A$. More generally, replacing A^ω by A^κ for an infinite cardinal κ, we may speak of **term-like operations on** A^κ.

DEFINITION 7.39. Let κ be an infinite cardinal. A set Δ of terms is said to be κ-**universal** provided for any assignment f of term-like operations on A^κ to the members of Δ such that the support of $f(t)$ is included in the support of t for all $t \in \Delta$, there is an algebra $\mathbf{A}$ in the signature of Δ and with universe κ such that

$$f(t) = t^{\mathbf{A}} \text{ for all } t \in \Delta.$$

Our application of κ-universality will be restricted to ω-universality. See (Walter Taylor, 1981) for some results for other values of κ.

DEFINITION 7.40. For an infinite cardinal κ and a non-nullary signature $\tau : I \longrightarrow \omega$, we will say that a system of definitions $D : I \longrightarrow T_\tau$ is κ-**universal** provided that for each $Q \in I$, exactly $\sigma(Q)$ variables occur in $D(Q)$; and that every σ-algebra $\mathbf{B}$ with universe of power κ is equal to $\mathbf{A}^D$ for some τ-algebra $\mathbf{A}$.

The κ-universality of sets and that of systems are evidently closely related; the details of this connection are recorded in the next easy lemma, whose proof we omit. In the remainder of §7.6 we shall use both forms of universality, while tacitly invoking the lemma as needed. In Exercise 7.48.2 we briefly look at κ-universality for uncountable κ, but for all the rest of this volume we consider only ω-universality.

LEMMA 7.41. *Let $\sigma : I \longrightarrow \omega$ and $\tau : J \longrightarrow \omega$ be non-nullary signatures,*

(i) *Let $D : I \longrightarrow T_\tau$ be an injective system of definitions of σ in τ. Then D is a κ-universal interpretation if and only if $D[I]$ is a κ-universal set of terms.*
(ii) *If $D : I \longrightarrow T_\tau$ is not injective, then D is not a κ-universal interpretation for any κ.*
(iii) *Let S be a subset of T_τ. Then S is a κ-universal set of terms if and only if the embedding $S \hookrightarrow T_\tau$ is a κ-universal interpretation.* ■

THEOREM 7.42 (The ω-Universal Translation Theorem). *Let σ be a countable non-nullary signature and let τ be a countable signature. Let D be an ω-universal interpretation of σ in τ. Then for any set $\Sigma \cup \{s \approx t\}$ of equations of signature σ,*

$$\Sigma \vdash s \approx t \text{ if and only if } \mathrm{Tr}_D(\Sigma) \vdash \mathrm{Tr}_D(s \approx t).$$

Proof. First, let us assume that $\Sigma \vdash s \approx t$ and deduce that $\mathrm{Tr}_D(\Sigma) \vdash \mathrm{Tr}_D(s \approx t)$. So suppose that $\mathbf{A} \models \mathrm{Tr}_D(\Sigma)$. Then $\mathbf{A}^D \models \Sigma$. So also $\mathbf{A}^D \models s \approx t$. But then $\mathbf{A} \models \mathrm{Tr}_D(s \approx t)$ as desired.

Now let us assume that $\Sigma \nvdash s \approx t$ and deduce that $\mathrm{Tr}_D(\Sigma) \nvdash \mathrm{Tr}_D(s \approx t)$. Let $\mathbf{B} = \langle \omega, F_Q \rangle_{Q \in I}$ be a model of Σ in which $s \approx t$ fails. Let $\mathbf{A}$ be an algebra of signature

τ with universe ω such that $\mathbf{B} = \mathbf{A}^D$. Such an algebra exists since D is ω-universal. Consequently, $\mathbf{A}$ is a model of $\mathrm{Tr}_D(\Sigma)$ in which $\mathrm{Tr}_D(s \approx t)$ fails, as desired. ■

Of course the usefulness of this theorem depends on the availability of an ω-universal system D of definitions for σ in τ.

DEFINITION 7.43. A set Δ of terms is **nonoverlapping** provided

- No variable belongs to Δ; and
- For all $d, d' \in \Delta$ and each p a subterm of d that is not a variable and $u_0, u_1, \ldots$ and $w_0, w_1, \ldots$ terms, if

$$p(u_0, u_1, \ldots) = d'(w_0, w_1, \ldots)$$

then $p = d = d'$ and $u_i = w_i$ for all i occurring in d.

THEOREM 7.44. *Every interpretation D of one countable signature σ in another τ that is one-to-one and yields a non-overlapping set of terms $\{D(Q) \mid Q \in I\}$ is ω-universal.*

Proof. Let Δ be a set of nonoverlapping terms associated with D. In place of the set ω of natural numbers, we use the countably infinite set T of all terms of the signature τ. So T is the universe of the term algebra $\mathbf{T}$. Let F assign to each $Q \in I$ a finitary function F_Q on T so that D and F agree according to rank. We want:

> There is an algebra $\mathbf{A}$ of signature τ with universe the set T of terms so that $F_Q = D(Q)^{\mathbf{A}}$ for all $Q \in I$.

We let the universe of $\mathbf{A}$ be $A = T$, the set of terms. Let P be an operation symbol of τ and let r be the rank of P. Then for sequences $\mathbf{t} = \langle t_0, \ldots, t_{r-1}\rangle$ of terms, we put

$$P^{\mathbf{A}}(\mathbf{t}) := \begin{cases} F_Q(\mathbf{u}) & \text{if } P\mathbf{t} = D(Q)(u_0, u_1, \ldots) \text{ for some } Q \in I \text{ and some } \mathbf{u} \in T^\omega \\ P\mathbf{t} & \text{otherwise} \end{cases}$$

Is this definition sound? It looks like there are only two cases, but there really are more: there is a case for each $Q \in I$ and I could be large. In other words, we need to make sure that our cases do not conflict.

Suppose $Q, Q' \in I$ and $\mathbf{u}, \mathbf{w} \in T^\omega$ so that

$$D(Q)(\mathbf{u}) = D(Q')(\mathbf{w})$$

Because Δ is nonoverlapping, we get $D(Q') = D(Q)$ and $u_i = w_i$ if x_i occurs in $D(Q)$. So $F_Q(\mathbf{u}) = F_{Q'}(\mathbf{w})$.

Our contention:

> If $Q \in I$ and s is a proper subterm of $D(Q)$, then $s^{\mathbf{A}}(u_0, u_1, \ldots) = s(u_0, u_1, \ldots) = s^{\mathbf{T}}(u_0, u_1, \ldots)$ for all $\mathbf{u} \in T^\omega$.

We will prove this by induction on the complexity of s. The base step presents no difficulty.

For the inductive step, say $s = Ps_0 \cdots s_{r-1}$, where P is an operation symbol of τ and r is its rank. Then

$$\begin{aligned} s^{\mathbf{A}}(\mathbf{u}) &= P^{\mathbf{A}}\left(s_0^{\mathbf{A}}(\mathbf{u}), \ldots, s_{r-1}^{\mathbf{A}}(\mathbf{u})\right) \\ &= P^{\mathbf{A}}\left(s_0(\mathbf{u}), \ldots, s_{r-1}(\mathbf{u})\right) \text{ by the inductive hypothesis} \\ &= Ps_0(\mathbf{u}) \cdots s_{r-1}(\mathbf{u}) = s(\mathbf{u}) \end{aligned}$$

So $P^{\mathbf{A}}(s_0(\mathbf{u}), \ldots, s_{r-1}(\mathbf{u})) = s(\mathbf{u}) = s^{\mathbf{A}}(\mathbf{u})$.

Now we claim that $F_Q = D(Q)^{\mathbf{A}}$ for all $Q \in I$. To prove this, let $Q \in I$ and $\mathbf{u} \in T^{\omega}$, and suppose $D(Q)$ is $Pt_0 \cdots t_{r-1}$. Then,

$$\begin{aligned} D(Q)^{\mathbf{A}}(\mathbf{u}) &= P^{\mathbf{A}}\left(t_0^{\mathbf{A}}(\mathbf{u}), \ldots, t_{r-1}^{\mathbf{A}}(\mathbf{u})\right) \\ &= P^{\mathbf{A}}\left(t_0(\mathbf{u}), \ldots, t_{r-1}(\mathbf{u})\right) \\ &= F_Q(\mathbf{u}) \end{aligned}$$

where the last equality follows by the definition of $P^{\mathbf{A}}$ since

$$Pt_0(\mathbf{u}) \cdots t_{r-1}(\mathbf{u}) = D(Q)(\mathbf{u}).$$

■

The notion of an ω-universal set of terms was introduced by Jan Mycielski, see (John R. Isbell, 1966). The concept of a nonoverlapping system of terms originated with Ralph McKenzie in the 1960's. "Nonoverlapping" was used in several papers by Don Pigozzi in the 1970's. But actually, these ideas trace back to (Stefan Banach, 1935) and (W. Sierpiński, 1935) and were used later by Jerzy Łoś (1950).

EXAMPLE 7.45. Let B and D be two binary operation symbols. Consider the term $DDxyBxy$ as d. It looks like

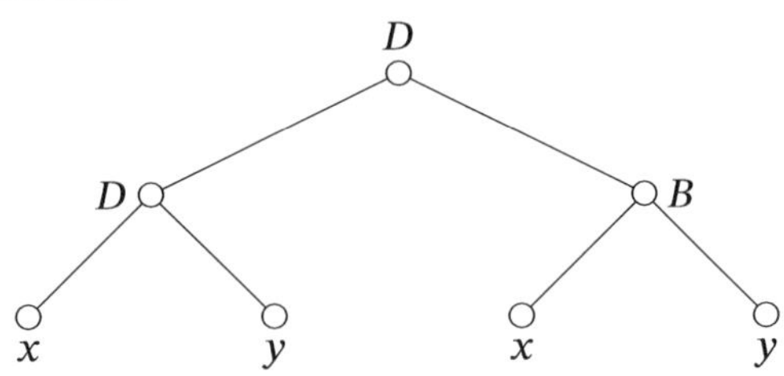

Notice that Dxy is a proper subterm of d. But

$$x \mapsto Dxy \text{ and } y \mapsto Bxy$$

which sends Dxy to d.

Let σ be the signature that provides just one operation symbol Q, that one being binary. Let τ be the signature providing just the binary operation symbols B and D. Let E be the interpretation of σ in τ such that $E(Q) = d$. We claim that E is ω-universal even though $\{d\}$ fails to be nonoverlapping.

Let $F : \omega \times \omega \to \omega$. Define $\mathbf{A}$ so that $A = \omega$ and

$$B^{\mathbf{A}}(a, b) = F(a, b) \text{ and } D^{\mathbf{A}}(a, b) = b$$

Notice that

$$\begin{aligned} d^{\mathbf{A}}(a, b) &= D^{\mathbf{A}}\left(D^{\mathbf{A}}(a, b), B^{\mathbf{A}}(a, b)\right) \\ &= D^{\mathbf{A}}(b, F(a, b)) \\ &= F(a, b) \end{aligned}$$

So $d^{\mathbf{A}} = F$.

EXAMPLE 7.46. Let Q be a binary operation symbol. Let

$$d_n = QxQ^n x^{n+1} = QxQQ \cdots Qxxx \cdots x$$

We claim that $\{d_n : n \in \omega\}$ is nonoverlapping and hence is ω-universal. So let s be a subterm of d_n with $s \neq x$. Suppose u and w are terms and $m \in \omega$ and $s(u) = d_m(w)$. To have nonoverlapping, we need to check that $s = d_n = d_m$ and $u = w$. Here is a diagram of d_3.

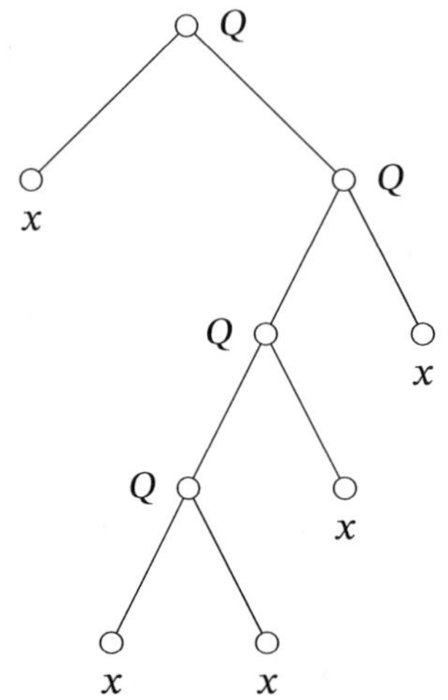

Notice that each d_n is heavily weighted to the right but each proper nonvariable subterm is heavily weighted to the left, with the exception of Qxx, evenly weighted. This weighting disparity persists when an arbitrary term is substituted for x. It is easy to adapt this example to the case where the operation symbol Q has any rank greater than 1.

EXAMPLE 7.47. Let F and G be unary operation symbols. Let

$$d_n = FGF^{n+1}G^2x.$$

We leave it to the reader to show that $\{d_n \mid n \in \omega\}$ is nonoverlapping and hence is ω-universal.

Exercises 7.48

1. An infinite cardinal κ is called **singular** if there is a family $\mathcal{F}$ such that $|\mathcal{F}| < \kappa$, such that $|X| < \kappa$ for each $X \in \mathcal{F}$, and such that $\kappa = \bigcup \mathcal{F}$. If no such $\mathcal{F}$ exists, then infinite cardinal κ is called **regular**. (For finite n, $\aleph_n$ is regular; $\aleph_\omega$ is singular.) Verify that regularity of κ is equivalent to the following condition. Let E, S, W be operations on κ satisfying $E(Sx, Wx) \approx x$. Then there exist a constant (nullary operation) a, and unary operations R and T, either satisfying $R(S(T(x))) \approx x$ or satisfying $R(E(a, T(x))) \approx x$.

*2. (Walter Taylor, 1981) This exercise illustrates that κ-universality can depend on κ. We consider a signature $\sigma : \{0, 1, \dots, 8\} \longrightarrow \omega$, with values (arities) $\langle 0, 0, 2, 1, 1, 4, 4, 1, 1\rangle$, and a signature τ with six nullary operations $\{0, 1, a, b, c, d\}$, four unaries $\{R, S, T, W\}$, one binary $\{E\}$, two quaternaries $\{H, J\}$, and one heptary $\{K\}$. This exercise concerns the system of definitions $D : \{0, \dots, 8\} \longrightarrow T_\tau$

that is defined by the following list of nine τ-terms:

$$\begin{gathered} 0,\ 1,\ Exy,\ Sx,\ Wx,\ Hxyzw,\ Jxyzw, \\ K(x, 1, Jaacd, J01cd, Haacd, Habcc, a) \\ K(x, H(x, RSTx, x, RE(Sa, Tx)), c, d, 0, 0, E(Sa, Wa)). \end{gathered}$$

In other words, $D(0) = 0$, $D(1) = 1$, $D(2) = Exy$, and so on. Prove that D is κ-universal if and only if κ is a regular cardinal.

(Hint. Suppose that κ is regular. Given nine operations on κ corresponding to our nine terms, the first seven operations specify the τ-operations $0, 1, E, S, W, H, J$. We need to provide further operations a, b, c, d, R, T, K so that the two K-terms represent the last two of the given unary functions. Consider the equations

$$Jaacd \approx c, \quad J01cd \approx d, \quad Haacd \approx 0, \quad Habcc \approx 0, \quad a \approx E(Sa, Wa). \tag{7.6.1}$$

If any of these failed, even once, then we would be done (regardless of regularity of κ), since the last two terms would specify K on disjoint subsets of κ^7. Thus we shall assume that the equations (7.6.1) hold identically in all the variables a, b, c, d. One easily checks that, in a universe of more than one element, (7.6.1) entails $0 \neq 1$.

We now take b, c, d to be arbitrary elements (nullary operations) of κ. By Exercise 1, we have nullary a, and unary R and T, satisfying $R(S(T(x))) \approx x$ or satisfying $R(E(S(a), T(x))) \approx x$. In the former case we obviously have $H(x, RSTx, x, RE(Sa, Tx)) \approx H(x, x, x, RE(Sa, Tx)) \approx 0$, and so the two K-terms apply to two disjoint subsets of κ^7, and we finish as before. The latter case is similar. The opposite direction uses the same ideas.)

3. Suppose that $\mathcal{V}$ and $\mathcal{W}$ are term-equivalent varieties. Prove that the lattice of subvarieties of $\mathcal{V}$ is isomorphic to the lattice of subvarieties of $\mathcal{W}$.

Maximal and Minimal Equational Theories

Once a signature is given, the lattice of equational theories will always have a top element. An equational theory T is said to be **maximal** or **equationally complete** provided T is a proper theory and if T' is an equational theory such that $T \subseteq T'$, then either $T = T'$ or $T' = \mathsf{T}$. That is, there is no equational theory properly between T and T.

THEOREM 7.49. *Every proper equational theory is contained in a maximal equational theory.*

Proof. This is a more or less standard Zorn's Lemma argument. Let T be a proper equational theory and let

$$\mathfrak{F} = \{S : T \subseteq S \text{ and } S \text{ is a a proper equational theory}\}.$$

Suppose $\mathfrak{C} \subseteq \mathfrak{F}$ is linearly ordered by $\subseteq$. We claim $\bigcup \mathfrak{C} \in \mathfrak{F}$. Certainly, $T \subseteq \bigcup \mathfrak{C}$ since $T \subseteq S$ for all $S \in \mathfrak{C}$.

Notice that $\bigcup \mathfrak{C}$ is an equational theory since it's closed under logical consequence. To see this, suppose $\bigcup \mathfrak{C} \models s \approx t$. Then there is a finite $\Delta \subseteq \bigcup \mathfrak{C}$ so that $\Delta \models s \approx t$. Pick $S_0, S_1, \ldots, S_n \in \mathfrak{C}$ so that $\Delta \subseteq S_0 \cup \cdots \cup S_n$. Without loss of generality, since $\mathfrak{C}$ is a chain, we have $S_0 \subseteq S_1 \subseteq \cdots \subseteq S_n$. So $S_n \models s \approx t$, so $s \approx t \in S_n$ since S_n is an equational theory.

Lastly, $\bigcup \mathfrak{C}$ is proper. To see this, suppose to the contrary that $\mathfrak{C} \models x \approx y$. This would imply that $\Delta \models x \approx y$ for some finite $\Delta \subseteq \bigcup \mathfrak{C}$, which would imply that $S \models x \approx y$ for some $S \in \mathfrak{C}$. Thus the family $\mathfrak{F}$ has the property that is assumed for Zorn's Lemma: each of its chains has an upper bound. Thus by Zorn's Lemma, $\mathfrak{F}$ has a maximal element. ■

On the other hand, apart from certain meager signatures, the lattice of equational theories has no minimal nontrivial elements. In the language of lattice theory we would say that the lattice of equational theories has no atoms.

THEOREM 7.50. *If the signature provides an operation symbol of positive rank, then any nontrivial equational theory has a proper nontrivial subtheory.*

Proof. Let $s \approx t \in T$ with $s \neq t$. It is harmless to suppose that $\{s \approx t\}$ is a basis for T. Let Q be an operation symbol of positive rank. Pick m larger than the length of s and t. Let $T' = \Theta\{Q^m ss \cdots s \approx Q^m tt \cdots t\}$. So $T' \subseteq T$ since $Q^m ss \cdots s \approx Q^m tt \cdots t$ is a consequence of $s \approx t$. T' is not trivial, since $s \neq t$. Finally, $s \approx t \notin T'$, because the terms s and t are too short for the equations in the basis of T' to come into play in any deduction. So $T' \neq T$. That is, T' is a proper nontrivial subtheory of T. ■

For signatures that do not provide operation symbols of positive rank, it is possible to describe the lattice of equational theories. If fewer than 2 constant symbols are provided, then there are only two equational theories: $\top$ and $\bot$. If at least 2 constant symbols are available, then the lattice of equational theories is isomorphic to the lattice of equivalence relations on the set of constant symbols with a new top element adjoined.

THEOREM 7.51. *Suppose the signature provides at least two operation symbols of rank 1 or at least one operation symbol of rank at least 2. Then there are at least 2^ω equationally complete theories.*

Proof.
First consider the case that the signature provides a countably infinite list $H_0, H_1, \ldots$ of one-place operation symbols.

Let X be a set of natural numbers. Define

$$\Sigma_X = \{H_n x \approx x : n \in X\} \cup \{H_n x \approx H_n y : n \notin X\}$$

where x and y are distinct variables. That first set says that H_n denotes the identity function when $n \in X$ and the second set says that H_n denotes a constant function when $n \notin X$.

Let T_X be the equational theory based on Σ_X. It is apparent that T_X has a countably infinite model and that $T_X \cup T_Y$ has only trivial models when $X \neq Y$, since only on sets of size 1 can the identity function be constant.

Now we know that there are infinite nonoverlapping sets of terms in two unary operation symbols and also in any signature that supplies an operation symbol of rank at least two. So in either of these cases, we can devise an ω-universal system of definitions of the H_n's and invoke the ω-Universal Translation Theorem to complete our proof, noting that we have already seen that proper equational theories can be extended to maximal equational theories. ∎

This theorem was first proved for signatures that provide a 2-place operation symbol by Jan Kalicki (1955) and later extended by A. D. Bol'bot (1970) and Jaroslav Ježek (1970).

Intervals in the Lattice of Equational Theories

The intervals in the lattice $\mathcal{L}_\sigma$ hold a special interest—particularly those intervals of the form

$$\{S \mid S \in \mathcal{L}_\sigma \text{ and } T \subseteq S \subseteq \top\}$$

for a specified equational theory T. These are the principal filters in the lattice $\mathcal{L}_\sigma$. For instance, if T is taken to be the equational theory of groups, then the equational theories in this filter will be exactly the equational theories of groups, that is equational theories that include the group axioms. One of the oldest open problems in equational logic is to characterize those lattices isomorphic to some principal filter in the lattice of equational theories, see (Garrett Birkhoff, 1946) and (A. I. Mal'tsev, 1968). For certain choices of T some detailed information has been obtained—this applies, for example, when T is the equational theory of groups or when T is the equational theory of lattices—but in general, little is known. When T is the equational theory of a finite lattice (or, indeed, of a finite algebra belonging to a congruence distributive variety), then the variety based on T has, up to isomorphism, only finitely many subdirectly irreducible algebras—as follows from Jónsson's Lemma (Theorem 8.57 on page 355 in §8.4). Consequently the principal filter of all equational theories that include T is a finite distributive lattice. For a number of finite lattices, this filter has been completely determined. Another classical case in which the filter has been completely described is when T is the equational theory of bands—a **band** is an idempotent semigroup. So the equational theory of bands is based on $\{x(yz) \approx (xy)z, xx \approx x\}$. To give some idea of the complexity of results of this kind, we display the infinite principal filter above the equational theory of bands in Figure 7.11.

The lattice of equational theories of bands was first described in three independent papers: (A. P. Birjukov, 1970; Charles Fennemore, 1970; J. A. Gerhard, 1970). J. A. Gerhard and Mario Petrich (1989) provide perhaps the best exposition of this result. The proofs are so long and involved that we cannot include them in our account. Nevertheless, it is useful to remark that each of the countable infinity of these equational theories is based on adding some single equation to the two equations that form a base for the theory of bands—in fact the discovery and systematic analysis of this countable infinity of equations lies at the heart of the description of this lattice. The detail of the top 18 elements of the lattice of equational theories of bands is displayed on the right in Figure 7.11 and is labeled with the additional equations. The filled node in this detail is the equational theory of rectangular bands—that is the equational theory based on $\{x(yz) \approx (xy)z, xx \approx x, x(yz) \approx xz\}$. Rectangular bands were introduced on page 115 in Volume I. Trevor Evans (1981) gives a nontechnical account of what was

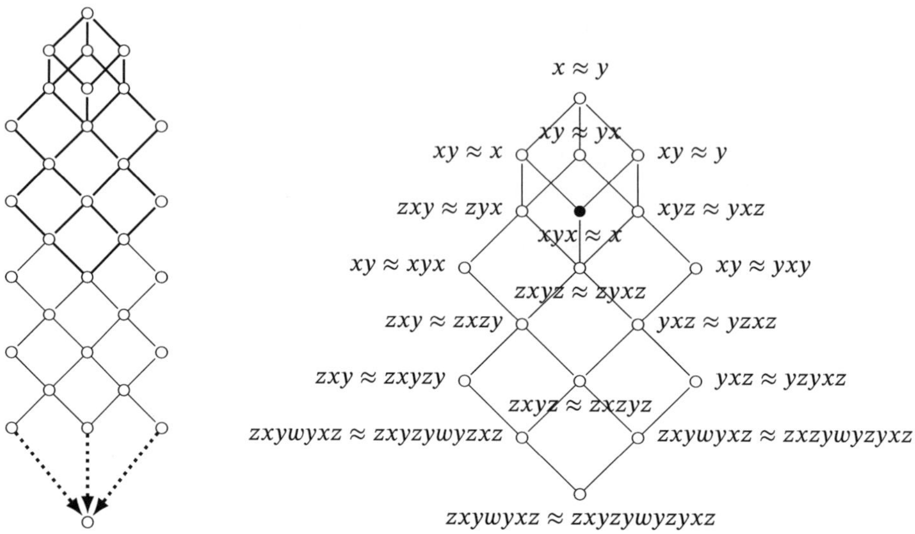

FIGURE 7.11. The lattice of equational theories of bands, with a detail of the top

known about the lattice of equational theories of semigroups in 1980. A more recent and much fuller account is provided by L. N. Shevrin, B. M. Vernikov, and M. V. Volkov (2009). A very recent detailed exposition of the lattice of equational theories of monoids can be found in (Sergey Gusev, Edmond Lee, and Boris Vernikov, 2020). Even in these special cases, there are only a handful of equational theories where the principal filter above the equational theory can be described completely.

While our information about principal filters in the lattice of equational theories is fragmentary and specialized, there are a few general results concerning intervals in lattices of equational theories. We provide four theorems of this general kind (Theorems 7.52–7.55, whose proofs will occupy us for the next several pages).

THEOREM 7.52 (Ježek's Interval Embedding Theorem, (Jaroslav Ježek, 1976)). *Let σ be any signature that provides at least two one-place operation symbols or at least one operation symbol of rank exceeding* 1. *Then any algebraic lattice with only countably many compact elements is isomorphic to some interval in the lattice $\mathcal{L}_\sigma$.*

Proof. According to the remarks following the proof of the Grätzer-Schmidt Theorem the algebraic lattices with only countably many compact elements are, up to isomorphism, exactly the congruence lattices of countably infinite algebras with countably many basic operations, all of rank 1. So our objective is to represent such congruence lattices as intervals in the lattice of equational theories of signature σ. Let $\mathbf{A} = \langle \omega, H_0, H_1, H_2, \dots \rangle$ be such an algebra—recalling $\omega = \{0, 1, 2, \dots\}$ is the set of natural numbers. It is harmless to stipulate, in addition, that H_{2n}, where n is a natural number, is the identity function.

There are two cases to consider.

First suppose that the signature provides an operation symbol Q of rank at least 2. Let r be the rank of Q.

We need a notion of protected terms. In §7.1 we explained how terms could be displayed as certain finite labeled rooted ordered trees with a root node at the top and the other nodes hanging on branches beneath it. With this in mind, we call the nodes that are not leaves of such a tree *operational* nodes. A finite rooted ordered tree is *slim* and *r-ary* provided

(1) each operational node has at most one child node that is operational and that node must be either the leftmost child or the rightmost child;
(2) the lowest operational node must be the leftmost child of its parent node;
(3) there must be an operational node that is the rightmost child of its parent node;
(4) each operational node has exactly r children.

Stipulation (1) ensures that the operational nodes form a path in the tree. The lowest node on this path must have a parent node, according to stipulation (2). So there must be at least two operational nodes. According to stipulation (3) there is a node that is a rightmost child of its parent node. This entails that there must be at least three operational nodes. The path that connects the operational nodes breaks into segments that consist of nodes that are, one after the other, either all leftmost children or all rightmost children. So the tree has a zigzag appearance. The diagrams below illustrate slim 3-ary trees.

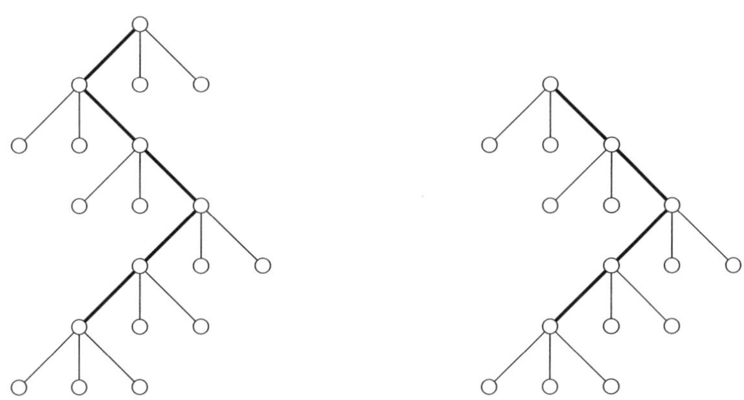

The *protected* terms arise from labeling any slim tree as follows:

- Each operational node is labeled with the operation symbol Q;
- Each leaf is labeled with some variable;
- The leftmost and rightmost child of the lowest operational node are labeled with the same variable.

Here is a diagram depicting two protected terms.

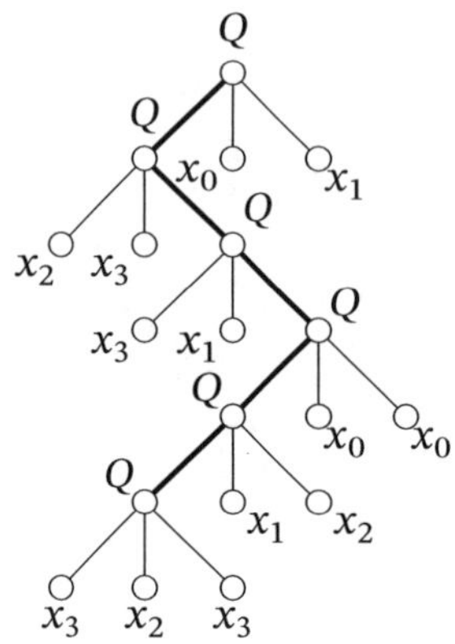

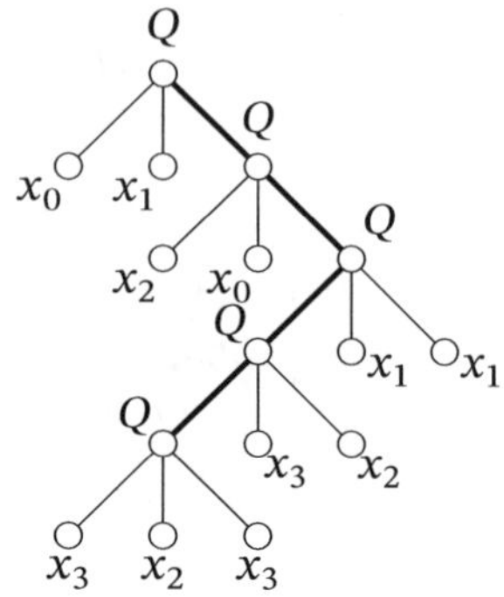

The term p
$QQx_2x_3Qx_3x_1QQQx_3x_2x_3x_1x_2x_0x_0x_0x_1$

The term q
$Qx_0x_1Qx_2x_0QQQx_3x_2x_3x_2x_1x_1$

A protected term can be decomposed using the zigzag structure of the associated slim tree. Here is how to do this for the terms depicted above. Let y be a variable that does not occur in the term. For the term p let s be $QQQx_3x_2x_3x_1x_2x_0x_0$, let $t^r(y)$ be $Qx_2x_3Qx_3x_1y$, and let t^ℓ be Qyx_0x_1. Then $p = t^\ell(t^r(s))$. To decompose the term q, let s be $QQQx_3x_2x_3x_3x_2x_1x_1$ and let $t^\ell(y)$ be $Qx_0x_1Qx_2x_0y$. Then q is $t^\ell(s)$. Terms of the form $Q^k\mathbf{x}$ where k is a positive natural number and $\mathbf{x}$ is a string of arbitrary variables of the appropriate length are not protected. Both of the s's mentioned above are of this kind. We enhance such terms to $Q\mathbf{y}Q^k\mathbf{x}$ where $\mathbf{y}$ is an arbitrary string of variables (with the string having length $r-1$) and again $\mathbf{x}$ is any string of variables of the appropriate length. Such an enhanced term *is* protected. We call protected terms of this kind *leftward* terms.

Finally, for each leftward segment of a slim tree, we associate an *index* that is one less than the number of edges in the segment. The term p has two leftward segments, one associated with s, which has index 1, and one associated with t^ℓ, which has index 0. The term q has only the leftward segment associated with s and it has index 1.

There are actually two kinds of protected terms: ones like p that begin with a leftward segment and ones like q that begin with a rightward segment. We will say that two protected terms have the *same parity* provided that both begin with leftward segments or they both begin with rightward segments. With each protected term that begins with a leftward segment we associate the sequence of its indices, omitting the first. With each protected term that begins with a rightward segment we associate its sequence of indices (keeping them all). For each protected term u let

$$\rho(u) := H_{i_{n-1}}(H_{i_{n-2}}(\cdots H_{i_1}(i_0)\cdots)),$$

where $\langle i_{n-1}, \dots, i_0\rangle$ is the sequence of indices associated with u. (Here i_0 is the index associated be s.)

Let $V = \{p \mid p \text{ is a term and no substitution instance of } p \text{ is protected}\}$. Any term that involves an operation symbol other than Q belongs to V and no protected term belongs to V.

With this preparation, we can now describe the isomorphism we need between the given **Con A** and a certain interval in the lattice of equational theories. Our isomorphism is denoted T. For every $\varphi \in \operatorname{Con}\mathbf{A}$, we define $\mathrm{T}(\varphi)$ to be the set of equations that is the union of the following three sets:

$$\{p \approx p \mid p \text{ is a term}\}$$
$$\{p \approx q \mid p, q \in V\}$$
$$\{p \approx q \mid \rho(p)\ \varphi\ \rho(q) \text{ where } p \text{ and } q \text{ are protected terms of the same parity and in which the same variables occur, and if } p \text{ and } q \text{ begin with a leftward segment, then these segments have the same index}\}$$

We shall denote $\mathrm{T}(\varphi)$ by T_φ. As we shall see presently T_φ is an equational theory.

We define $T^* := \mathrm{T}_{1_\mathbf{A}}$ and $T_* := \mathrm{T}_{0_\mathbf{A}}$. For every equational theory T with $T_* \subseteq T \subseteq T^*$, we define Φ_T via

$$\langle i, j\rangle \in \Phi_T \Leftrightarrow s \approx s' \in T \text{ where } x \text{ is any variable and } s \text{ and } s' \text{ are the leftward terms in the single variable } x, \text{ and where } \rho(s) = i \text{ and } \rho(s') = j.$$

Our aim is to show that **Con A** is isomorphic to the interval $I[T_*, T^*]$ in the lattice of equational theories via the isomorphism T and its inverse Φ (meaning the map $T \mapsto \Phi_T$). We do this by establishing the six contentions below.

Contention 1. *For each $\varphi \in \operatorname{Con}\mathbf{A}$, we have T_φ is an equational theory.*

To see that T_φ is closed with respect to deductions, it is enough to show this for deductions consisting of one step. So suppose v and v' are terms and $v \xrightarrow{p\approx q} v'$ where $p \approx q \in \mathrm{T}_\varphi$. Now T_φ is the union of three sets. If $p \approx q$ belongs to the first set, then $p = q$ and $v = v'$, so $v \approx v'$ also belongs to the first subset of T_φ. If $p \approx q$ belongs to the second subset, then $p, q \in V$. It follows that $v, v' \in V$. So $v \approx v'$ belongs to the second subset of T_φ. Finally, if $p \approx q$ belongs to the third subset, then both p and q are protected and of the same parity, $\rho(p)\ \varphi\ \rho(q)$, and there is an assignment $\mathbf{w}$ of terms to variables so that $p(\mathbf{w})$ is a subterm of v and v' is obtained from v by replacing $p(\mathbf{w})$ by $q(\mathbf{w})$. Observe first that both v and v' will belong to V, (and, hence, that $v \approx v' \in \mathrm{T}_\varphi$ as desired) unless v and v' are subterms of some protected terms. We lose no generality by assuming this. Since p and q are protected, so must v and v' be protected. It follows that $\mathbf{w}$ must be an assignment of variables to variables. This means that $\rho(p(\mathbf{w})) = \rho(p)$ and $\rho(q(\mathbf{w})) = \rho(q)$. Now there is an operational node in v that is the top node of $p(\mathbf{w})$. If that node is the rightmost child of its parent node, then the leftward segments of v above that node will also be the leftward segments above that node in v'. It follows, in this case, that $\rho(v)\ \varphi\ \rho(v')$ and so that $v \approx v'$ belongs to the third subset of T_φ. In the remaining case, when the top node of $p(\mathbf{w})$ is the leftmost child of its parent node in v, there is no difficulty if p (and hence q) has rightward parity. On the other hand if p and q have leftward parity then they have the same beginning index, but this index plays no role in calculating $\rho(p)$ and $\rho(q)$. In v and v' these beginning leftward segments of p and q are extended to leftward segments of v and v' of the same index. So again $v \approx v'$ belongs to the third subset of T_φ, and the contention is established.

Contention 2. *Suppose that T is an equational theory with $T_* \subseteq T \subseteq T^*$. Then Φ_T is a congruence of* **A**.

That Φ_T is an equivalence relation on ω is immediate. So let us suppose that $\langle i, j\rangle \in \Phi_T$ and that k is some natural number. We need to see that $\langle H_k(i), H_k(j)\rangle \in \Phi_T$.

Let $s_i := QyQ^{i+1}\mathbf{x}$ where $\mathbf{x}$ is the string of x's of the appropriate length. Likewise, let $s_j := QyQ^{j+1}\mathbf{x}$ where $\mathbf{x}$ is the string of x's of the appropriate length. So $\rho(s_i) = i$ and $\rho(s_j) = j$ and $s_i \approx s_j \in T$. Let $t(y)$ be $Q^{k+2}\mathbf{x}y\mathbf{z}$ where $\mathbf{x}$ is a string of x's of length $r-1$ (recall r is the rank of Q), and $\mathbf{z}$ is a string of z's of the appropriate length. Let $u := Q\mathbf{x}t(s_i)$ and let $u' := Q\mathbf{x}t(s_j)$. Then $u \approx u' \in T$ and $\rho(u) = H_k(i)$ and $\rho(u') = H_k(j)$.

Notice that $\rho(u) = H_k(i) = \rho(s_{H_k(i)})$ and $\rho(u') = H_k(j) = \rho(s_{H_k(j)})$. This means that $u \approx s_{H_k(i)} \in T_*$ and that $u' \approx s_{H_k(j)} \in T_*$. Therefore

$$s_{H_k(i)} \approx s_{H_k(j)} \in T,$$

since $T_* \subseteq T$. This means that $\langle H_k(i), H_k(j)\rangle \in \Phi_T$, establishing the contention.

Contention 3. *Let $\varphi \in \operatorname{Con} \mathbf{A}$. Then $\Phi_{\mathrm{T}_\varphi} = \varphi$.*

Observe

$$\begin{aligned}\langle i, j\rangle \in \Phi_{\mathrm{T}_\varphi} &\Leftrightarrow s_i \approx s_j \in \mathrm{T}_\varphi \text{ where } s_i \text{ and } s_j \text{ are leftward terms with}\\ &\qquad \rho(s_i) = i \text{ and } \rho(s_j) = j\\ &\Leftrightarrow s_i \text{ and } s_j \text{ are leftward terms in which the same variables}\\ &\qquad \text{occur, and } i = \rho(s_i)\ \varphi\ \rho(s_j) = j\\ &\Leftrightarrow \langle i, j\rangle \in \varphi.\end{aligned}$$

This establishes the contention.

Contention 4. *Let T be an equational theory with $T_* \subseteq T \subseteq T^*$. Then $\mathrm{T}_{\Phi_T} = T$.*

Observe

$$\begin{aligned}v \approx v' \in \mathrm{T}_{\Phi_T} &\Leftrightarrow v = v' \text{ or } v, v' \in V \text{ or } v \text{ and } v' \text{ are protected terms of the same}\\ &\qquad \text{parity and in which the same variables occur, and if they begin}\\ &\qquad \text{with a leftward segment, then these segments have the same}\\ &\qquad \text{index and } \rho(v)\ \Phi_T\ \rho(v')\\ &\Leftrightarrow v = v' \text{ or } v, v' \in V \text{ or } v \text{ and } v' \text{ are protected terms of the same}\\ &\qquad \text{parity and in which the same variables occur, and if they begin}\\ &\qquad \text{with a leftward segment, then these segments have the same}\\ &\qquad \text{index and } s \approx s' \in T \text{ whenever } s \text{ and } s' \text{ are leftward terms in}\\ &\qquad \text{the same single variable with } \rho(s) = \rho(v) \text{ and } \rho(s') = \rho(v')\\ &\Leftrightarrow v \approx v' \in T.\end{aligned}$$

For the left-to-right implication notice that we have $s \approx s' \in T$ where s and s' are leftward terms in one variable with $\rho(s) = \rho(v)$ and $\rho(s') = \rho(v')$. It is harmless to suppose that the variable in s occurs also in v. Let $y_0, \ldots, y_{m-1}$ be the remaining variables occurring in v. We enhance s and s' as follows:

$$s^* = Q\mathbf{y_0}Q\mathbf{y_1}\cdots Q\mathbf{y_{m-1}}s$$
$$s'^* = Q\mathbf{y_0}Q\mathbf{y_1}\cdots Q\mathbf{y_{m-1}}s'$$

where $\mathbf{y_i}$ is a string of y_i's of length $r-1$. So $s^* \approx s'^* \in T$, $\rho(s^*) = \rho(s)$, $\rho(s'^*) = \rho(s')$, and the same variables occur in v and s^*. Now if v and v' have rightward parity, then $v \approx s^*, v' \approx s'^* \in T_* \subseteq T$. In this case, it follows that $v \approx v' \in T$, as desired. On the other hand, if v and v' have leftward parity and the leftward segments they begin with have index k, then we modify s^* and s'^* further, letting

$$s^\bullet = Q^{k+1}s^*\mathbf{x}$$
$$s'^\bullet = Q^{k+1}s'^*\mathbf{x},$$

where $\mathbf{x}$ is a string of x's of the appropriate length. It follows that $s^\bullet \approx s'^\bullet \in T$, $\rho(s^\bullet) = \rho(s)$, $\rho(s'^\bullet) = \rho(s')$, the same variables occur in v and $s^\bullet$, and v, $s^\bullet, v'$, and $s'^\bullet$ have leftward parity and the indices of their beginning leftward segments are the same. So $v \approx s^\bullet, v' \approx s'^\bullet \in T_* \subseteq T$. Again it follows that $v \approx v' \in T$, as desired.

For the right-to-left implication, we start with $v \approx v' \in T \subseteq T^*$. It follows that $v = v'$ or $v, v' \in V$ or v and v' are protected terms of the same parity and if they begin with a leftward segment, then these segments have the same index. We only need to consider the last alternative. So let s and s' be the leftward terms in the variable x with $\rho(v) = \rho(s)$ and $\rho(v') = \rho(s')$. We need to show that $s \approx s' \in T$. Let $\mathbf{x}$ be the tuple that assigns the variable x to every variable. Then $v(\mathbf{x}) \approx v'(\mathbf{x}) \in T$, since T is an equational theory. Observe that $\rho(v(\mathbf{x})) = \rho(v)$ and $\rho(v'(\mathbf{x})) = \rho(v')$. In the event that v and v' have rightward parity, we have that $v(\mathbf{x}) \approx s, v'(\mathbf{x}) \approx s' \in T_* \subseteq T$. So $s \approx s' \in T$, as desired. Otherwise, v and v' have leftward parity and the leftward segments they begin with have the same index. Let this index be k. There are two subcases: whether k is even or odd. If k is even we let

$$v^*(\mathbf{x}) = Q\mathbf{w}v(\mathbf{x})$$
$$v'^*(\mathbf{x}) = Q\mathbf{w}v'(\mathbf{x}),$$

where $\mathbf{w}$ is a string of x's of length $r-1$. Alternatively, when k is odd, we let

$$v^*(\mathbf{x}) = Q\mathbf{w}Qv(\mathbf{x})\mathbf{w}$$
$$v'^*(\mathbf{x}) = Q\mathbf{w}Qv'(\mathbf{x})\mathbf{w},$$

where, again, $\mathbf{w}$ is a string of x's of length $r-1$. In either case $v^*(\mathbf{x}) \approx v'^*(\mathbf{x}) \in T$, the terms $v^*(\mathbf{x})$ and $v'^*(\mathbf{x})$ have rightward parity, and their first leftward segment is of even index. But recall that H_j is the identity function whenever j is even. It follows that $\rho(v^*(\mathbf{x})) = \rho(v(\mathbf{x})) = \rho(s)$ and $\rho(v'^*(\mathbf{x})) = \rho(v'(\mathbf{x})) = \rho(s')$. It follows that $s \approx v^*(\mathbf{x}), s' \approx v'^*(\mathbf{x}) \in T_* \subseteq T$. Hence, $s \approx s' \in T$, as desired. So the contention is established.

The next two contentions follow immediately from the definitions.

Contention 5. *Let T_0 and T_1 be equational theories so that $T_* \subseteq T_0 \subseteq T_1 \subseteq T^*$. Then $\Phi_{T_0} \subseteq \Phi_{T_1}$.*

Contention 6. *Let $\varphi, \psi \in \operatorname{Con} \mathbf{A}$ with $\varphi \subseteq \psi$. Then $\mathrm{T}_\varphi \subseteq \mathrm{T}_\psi$.*

As a consequence of these six contentions, we conclude that T is an isomorphism from $\operatorname{Con} \mathbf{A}$ onto the interval $I[T_*, T^*]$ in the lattice of equational theories. We also see that Φ is the isomorphism inverse to T.

The proof of the theorem is complete in the case when the signature provides an operation symbol of rank at least 2. The case when the signature provides two operation symbols of rank 1 can be established by a similar argument. That argument is easier since terms involving only such operation symbols can involve only one variable and the notion of leftward and rightward parity can play no role. We have made this case a problem in the next set of exercises. ■

The theorem just proven asserts, for all but fairly uninteresting signatures, that $\mathcal{L}_\sigma$ is structurally diverse, having intervals that are essentially as general as they can be. One strength of this theorem is that it applies to most interesting (even finite) signatures. A weakness, on the other hand, is that the location of these diverse intervals is difficult to determine, depending on the algebraic lattice to be represented. In particular, they are not principal filters in $\mathcal{L}_\sigma$. The next theorem addresses principal filters of a particular kind, but at the cost of placing restrictions on the signature.

THEOREM 7.53 (Pigozzi's Embedding Theorem).

(i) *Every algebraic lattice with a compact top element and with a unique maximal element is isomorphic to a principal filter in the lattice $\mathcal{L}_\sigma$, for some signature σ.*
(ii) *Every algebraic lattice is isomorphic to an interval $I[T_*, T^*]$, where T^* is the only equationally complete extension of T_*, in the lattice $\mathcal{L}_\sigma$, for some signature σ.*

Proof. The two statements in the formulation of the theorem are equivalent. To see this first suppose (ii) holds. Let $\mathbf{L}$ be an algebraic lattice with a compact top element and with a unique maximal element t. Let b be the bottom element of $\mathbf{L}$. Then the interval $I[b, t]$ is an algebraic lattice isomorphic to $\mathbf{L}$ with the top element removed. According to (b) there are equational theories $T_* \subseteq T^*$ so that T^* is the only equationally complete extension of T_* and $I[b, t] \cong I[T_*, T^*]$. Now the filter above T_* in $\mathcal{L}_\sigma$ is just $I[T_*, T^*]$ with $\top$, the top element of $\mathcal{L}_\sigma$ adjoined. Hence, $\mathbf{L}$ is isomorphic to the filter of equational theories above T_*. Thus (ii) implies (i). We will prove formulation (ii). We leave it to our readers to prove that (i) implies (ii).

According to the Grätzer-Schmidt Theorem, up to isomorphism the algebraic lattices coincide with congruence lattices of algebras all of whose operations are of rank 1. So fix an infinite cardinal κ and an algebra $\mathbf{A} = \langle A, H_\alpha \rangle_{\alpha < \kappa}$ where $|A| = \kappa$ and each H_α has rank 1. Let $\langle a_\alpha \mid \alpha < \kappa \rangle$ be a one-to-one listing of the elements of A.

The signature σ will provide, in addition to unary operation symbols provided by the signature for $\mathbf{A}$, two lists $\langle \mathbf{a}_\alpha \mid \alpha < \kappa \rangle$ and $\langle \mathbf{d}_\alpha \mid \alpha < \kappa \rangle$ of constant symbols, one list $\langle J_\alpha \mid \alpha < \kappa \rangle$ of operation symbols of rank 1, and one list $\langle D_\alpha \mid \alpha < \kappa \rangle$ of operation symbols of rank 3.

We devise two algebras $\mathbf{B}^*$ and $\mathbf{B}_*$ whose equational theories will be, respectively, T^* and T_*.

Let $\Sigma = \{J_\alpha \mid \alpha < \kappa\}$ be an alphabet of size κ. We denote by Σ^* the set of all finite strings of elements of Σ. Let d_0 be some new element. Then the set $\{wd_0 \mid w \in \Sigma^*\}$ has cardinality κ. Observe that d_0 belongs to this set. Let $\langle d_\alpha \mid \alpha < \kappa\rangle$ be a one-to-one listing of this set. Let a_0 be a new element. Then

$$B^* := \{a_0\} \cup \{d_\alpha \mid \alpha < \kappa\}.$$

The basic operations of $\mathbf{B}^*$ are defined as follows:

$$\begin{aligned} \mathbf{a}_\alpha^{\mathbf{B}^*} &:= a_0 \text{ for each } \alpha < \kappa \\ \mathbf{d}_\alpha^{\mathbf{B}^*} &:= d_\alpha \text{ for each } \alpha < \kappa \\ J_\alpha^{\mathbf{B}^*}(a_0) &:= a_0 \text{ for each } \alpha < \kappa \\ J_\alpha^{\mathbf{B}^*}(wd_0) &:= J_\alpha w d_0 \text{ for each } w \in \Sigma^* \text{ and each } \alpha < \kappa \\ D_\alpha^{\mathbf{B}^*}(x,y,z) &:= \begin{cases} y & \text{if } x = d_\alpha \\ z & \text{if } x \neq d_\alpha \end{cases} \quad \text{for each } \alpha < \kappa. \end{aligned}$$

Let T^* be $\Theta\,\mathbf{B}^*$, the equational theory of $\mathbf{B}^*$. It is possible to prove that T^* is equationally complete by showing that $\mathbf{B}^*$ is embeddable into any nontrivial model of T^*. However, the equational completeness of T^* will follow from other parts of this proof.

Let $B_* := A \cup \{d_\alpha \mid \alpha < \kappa\}$. The basic operations of $\mathbf{B}_*$ are defined as follows:

$$\begin{aligned} \mathbf{a}_\alpha^{\mathbf{B}_*} &:= a_\alpha \text{ for each } \alpha < \kappa \\ \mathbf{d}_\alpha^{\mathbf{B}_*} &:= d_\alpha \text{ for each } \alpha < \kappa \\ J_\alpha^{\mathbf{B}_*}(a_\beta) &:= H_\alpha(a_\beta) \text{ for each } \alpha, \beta < \kappa \\ J_\alpha^{\mathbf{B}_*}(wd_0) &:= J_\alpha w d_0 \text{ for each } w \in \Sigma^* \text{ and each } \alpha < \kappa \\ D_\alpha^{\mathbf{B}_*}(x,y,z) &:= \begin{cases} y & \text{if } x = d_\alpha \\ z & \text{if } x \neq d_\alpha \end{cases} \quad \text{for each } \alpha < \kappa. \end{aligned}$$

Let $T_* := \Theta\,\mathbf{B}_*$, the equational theory of $\mathbf{B}_*$. Observe that A is closed with respect to the operations of $\mathbf{B}_*$ of positive rank and that A is a subuniverse of the corresponding reduct of $\mathbf{B}_*$. For each $\varphi \in \operatorname{Con} \mathbf{A}$ observe that $\widehat{\varphi} := \varphi \cup \{\langle d_\alpha, d_\alpha\rangle \mid \alpha < \kappa\rangle\}$ is a congruence of $\mathbf{B}_*$.

Let $G : B_* \to B^*$ be defined by

$$G(d_\alpha) := d_\alpha \quad \text{and} \quad G(a_\alpha) = a_0$$

for each $\alpha < \kappa$.

Contention 1. *The map G is a homomorphism from $\mathbf{B}_*$ onto $\mathbf{B}^*$ and* $\ker G = \widehat{1_A}$.

First observe that $G(a_\alpha^{\mathbf{B}_*}) = a_0 = \mathbf{a}_\alpha^{\mathbf{B}^*}$ and $G(d_\alpha^{\mathbf{B}_*}) = d_\alpha = \mathbf{d}_\alpha^{\mathbf{B}^*}$ for all $\alpha < \kappa$. Next, for all $\alpha, \beta < \kappa$ observe that there is $\gamma < \kappa$ so that $H_\alpha^{\mathbf{B}_*}(a_\beta) = a_\gamma$. Also observe $H_\alpha^{\mathbf{B}_*}(d_\alpha) = H_\alpha^{\mathbf{B}^*}(d_\alpha)$. So G respects all the unary operations. That G respects all the three-place basic operations is immediate from the definitions of those operations. Finally, observe that

$$\begin{aligned}\langle b,c\rangle \in \ker G &\Leftrightarrow G(b) = G(c)\\ &\Leftrightarrow \text{ either } b,c \in A \text{ or } b = c = d_\alpha \text{ for some } \alpha < \kappa\\ &\Leftrightarrow \langle b,c\rangle \in \widehat{1_A}.\end{aligned}$$

As a consequence of this contention, $T_* \subseteq T^*$.

Now suppose that T is an equational theory with a nontrivial model and also that $T_* \subseteq T$. Define $\Phi(T) := \{\langle a_\alpha, a_\beta\rangle \mid T \models \mathbf{a}_\alpha \approx \mathbf{a}_\beta \text{ for } \alpha,\beta < \kappa\}$.

Contention 2. *Let T be an equational theory with a nontrivial model so that $T_* \subseteq T$. Then $\Phi(T) \in \operatorname{Con}\mathbf{A}$ and $\mathbf{B}_*/\widehat{\Phi(T)} \models T$. In consequence, $T \subseteq T^*$ and T^* is the only equationally complete theory extending T_*.*

Checking that $\Phi(T) \in \operatorname{Con}\mathbf{A}$ is routine, but it also follows from the remainder of this argument.

Let $\mathbf{E}$ be a nontrivial model of T. Define $G\colon B_* \to E$ by

$$G(\mathbf{a}_\alpha) := \mathbf{a}_\alpha^{\mathbf{E}} \text{ and } G(d_\alpha) := \mathbf{d}_\alpha^{\mathbf{E}} \text{ for all } \alpha < \kappa.$$

Since each element of B_* is named by a constant symbol, G is a homomorphism from $\mathbf{B}_*$ into $\mathbf{E}$. Let $\mathbf{E}'$ be the subalgebra of $\mathbf{E}$ that is the image of $\mathbf{B}_*$ with respect to G. Then $\mathbf{E}'$ is also a model of T. Now let $e_0, e_1 \in E$ with $e_0 \neq e_1$. Let $\alpha < \kappa$. Since the equation $D_\alpha \mathbf{d}_\alpha yz \approx y \in T_*$ it must be true in $\mathbf{E}$. This also applies to the equation $D_\alpha \mathbf{d}_\beta yz \approx z$, where $\alpha \neq \beta$ and to the equations $D_\alpha \mathbf{a}_\beta yz \approx z$, where α and β need not be different. Now let $c \in B_*$ with $c \neq d_\alpha$ and let $\mathbf{c}$ be the constant symbol that names c. Then observe

$$\begin{aligned} D_\alpha^{\mathbf{E}}(G(d_\alpha), e_0, e_1) &= D_\alpha^{\mathbf{E}}(\mathbf{d}_\alpha^{\mathbf{E}}, e_0, e_1) = e_0\\ D_\alpha^{\mathbf{E}}(G(c), e_0, e_1) &= D_\alpha^{\mathbf{E}}(\mathbf{c}^{\mathbf{E}}, e_0, e_1) = e_1.\end{aligned}$$

Since $e_0 \neq e_1$, it follows that $G(d_\alpha) \neq G(c)$. So G is one-to-one on the d_α's and $G(a_\beta) \neq G(d_\alpha)$ for any $\alpha,\beta < \kappa$. In particular, $\mathbf{E}'$ is not trivial. Let $\varphi = \ker G$. So $\mathbf{B}_*/\varphi \cong \mathbf{E}' \models T$.

Now suppose $\mathbf{a}_\alpha \approx \mathbf{a}_\beta$ does not belong to T. Then T has a model $\mathbf{E}_{\alpha,\beta}$ that is not a model of $\mathbf{a}_\alpha \approx \mathbf{a}_\beta$. Let $G_{\alpha,\beta}\colon B_* \to E_{\alpha,\beta}$ be defined via

$$G_{\alpha,\beta}(a_\gamma) := \mathbf{a}_\gamma^{\mathbf{E}_{\alpha,\beta}} \text{ and } G_{\alpha,\beta}(d_\gamma) := \mathbf{d}_\gamma^{\mathbf{E}_{\alpha,\beta}} \text{ for all } \gamma < \kappa.$$

Again, this map is a homomorphism. Let $\mathbf{E}'_{\alpha,\beta}$ be the image of $\mathbf{B}_*$ with respect to this homomorphism and let $\varphi_{\alpha,\beta}$ be the kernel of $G_{\alpha,\beta}$.

Let $\mathcal{F} = \{\varphi_{\alpha,\beta} \mid \mathbf{a}_\alpha \approx \mathbf{a}_\beta \notin T\} \cup \{\varphi\}$ and let $\theta = \bigcap \mathcal{F}$. We will prove that $\theta = \widehat{\Phi(T)}$. Suppose that $s \approx t$ is any equation and let $\mathbf{c} = \langle c_0, c_1, \dots\rangle$ is any assignment of elements of B_* to the variables. We use $\mathbf{c}/\psi$ to stand for $\langle c_0/\psi, c_1/\psi, \dots\rangle$ where ψ is any congruence on $\mathbf{B}_*$. Then

$$\begin{aligned} s^{\mathbf{B}_*/\theta}(\mathbf{c}/\theta) = t^{\mathbf{B}_*/\theta}(\mathbf{c}/\theta) &\iff s^{\mathbf{B}_*}(\mathbf{c})/\theta = t^{\mathbf{B}_*}(\mathbf{c})/\theta \\ &\iff \langle s^{\mathbf{B}_*}(\mathbf{c}), t^{\mathbf{B}_*}(\mathbf{c})\rangle \in \theta = \bigcap \mathcal{F} \\ &\iff \langle s^{\mathbf{B}_*}(\mathbf{c}), t^{\mathbf{B}_*}(\mathbf{c})\rangle \in \varphi \text{ and } \langle s^{\mathbf{B}_*}(\mathbf{c}), t^{\mathbf{B}_*}(\mathbf{c})\rangle \in \varphi_{\alpha,\beta} \\ &\qquad \text{for all } \alpha,\beta \text{ such that } \mathbf{a}_\alpha \approx \mathbf{a}_\beta \notin T \\ &\iff s^{\mathbf{E}'}(\mathbf{c}/\varphi) = t^{\mathbf{E}'}(\mathbf{c}/\varphi) \text{ and} \\ &\qquad s^{\mathbf{E}'_{\alpha,\beta}}(\mathbf{c}/\varphi_{\alpha,\beta}) = t^{\mathbf{E}'_{\alpha,\beta}}(\mathbf{c}/\varphi_{\alpha,\beta}) \\ &\qquad \text{for all } \alpha,\beta \text{ such that } \mathbf{a}_\alpha \approx \mathbf{a}_\beta \notin T \end{aligned}$$

As a consequence, $\mathbf{B}_*/\theta \models s \approx t$ if and only if $\mathbf{E}' \models s \approx t$ and $\mathbf{E}'_{\alpha,\beta} \models s \approx t$ for all α and β such that $\mathbf{a}_\alpha \approx \mathbf{a}_\beta \notin T$. In particular,

- $\mathbf{B}_*/\theta \models T$;
- $\mathbf{B}_*/\theta \not\models \mathbf{d}_\alpha \approx \mathbf{d}_\beta$ whenever $\alpha \neq \beta$;
- $\mathbf{B}_*/\theta \not\models \mathbf{d}_\alpha \approx \mathbf{a}_\beta$ for all α and β;
- $\mathbf{B}_*/\theta \models \mathbf{a}_\alpha \approx \mathbf{a}_\beta$ if and only if $\mathbf{a}_\alpha \approx \mathbf{a}_\beta \in T$.

For each $\varphi \in \operatorname{Con} \mathbf{A}$ let $\mathrm{T}(\varphi)$ be the equational theory based on

$$T_* \cup \{\mathbf{a}_\alpha \approx \mathbf{a}_\beta \mid a_\alpha \;\varphi\; a_\beta \text{ and } \alpha, \beta < \kappa\}.$$

So T is a map from $\operatorname{Con} \mathbf{A}$ into the interval $I[T_*, T^*]$. The proof of this theorem will be completed by showing that T is an isomorphism from $\operatorname{Con} \mathbf{A}$ onto the interval $I[T_*, T^*]$ and that T^* is the only equational completion of T_*.

Contention 3. $\Phi(\mathrm{T}(\varphi)) = \varphi$ *for all* $\varphi \in \operatorname{Con} \mathbf{A}$.

To see this, observe for all a_α, a_β with $\alpha, \beta < \kappa$

$$\begin{aligned} \langle a_\alpha, a_\beta\rangle \in \Phi(\mathrm{T}(\varphi)) &\iff \mathbf{a}_\alpha \approx \mathbf{a}_\beta \in \mathrm{T}(\varphi) \\ &\iff \langle a_\alpha, a_\beta\rangle \in \varphi. \end{aligned}$$

For the last equivalence, the upward direction is part of the definition of $\mathrm{T}(\varphi)$, while the downward direction follows since $\mathbf{B}_*/\widehat{\varphi} \models \mathrm{T}(\varphi)$.

Contention 4. $\mathrm{T}(\Phi(T)) = T$ *for every equational theory* T *so that* $T_* \subseteq T \subseteq T^*$.

For this contention, observe that $\mathrm{T}(\Phi(T)) \subseteq T$ follows from the definitions. The reverse inclusion reduces to proving that

($\star$) If $s \approx t \in T^*$, then $T_* \cup \{\mathbf{a}_\alpha \approx \mathbf{a}_\beta \mid T_* \cup \{s \approx t\} \models \mathbf{a}_\alpha \approx \mathbf{a}_\beta\} \models s \approx t$.

The following six schemas of equations belonging to T_* prove useful.

For each $\alpha, \beta < \kappa$ there is a $\gamma < \kappa$ so that

$$J_\alpha \mathbf{a}_\beta \approx \mathbf{a}_\gamma \in T_*$$

For each $\alpha, \beta < \kappa$ there is a $\gamma < \kappa$ so that

$$J_\alpha \mathbf{d}_\beta \approx \mathbf{d}_\gamma \in T_*$$
$$D_\alpha \mathbf{d}_\alpha yz \approx y \in T_* \text{ for each } \alpha < \kappa$$
$$D_\alpha \mathbf{d}_\beta yz \approx z \in T_* \text{ for each } \alpha, \beta < \kappa \text{ with } \alpha \neq \beta$$
$$D_\alpha \mathbf{a}_\beta yz \approx z \in T_* \text{ for each } \alpha, \beta < \kappa$$
$$J_\alpha D_\beta xyz \approx D_\beta x J_\alpha y J_\alpha z \in T_* \text{ for each } \alpha, \beta < \kappa\,.$$

Each of these schemas is true in $\mathbf{B}_*$. The last schema provides a distributive property, while the other five schemas support the deletion of operation symbols of positive rank under certain circumstance when the equations are used in the left-to-right sense in equational deductions. So using these equations from T_* every term can be simplified.

The set of *simplified* terms is defined by the following recursion:

(1) each constant symbol is simplified;
(2) each term in the signature of $\mathbf{A}$ in which a variable occurs is simplified;
(3) if p, q, and r are simplified and p is not a constant symbol, then $D_\alpha pqr$ is simplified for each $\alpha < \kappa$.

Observe that the terms of the signature of $\mathbf{A}$ in which a variable occurs are just finite strings of unary operation symbols followed by a variable—in particular, each variable is a simplified term. We call terms of types (1) and (2) *basic terms*. Every simplified term is built up from basic terms with the help of the ternary operation symbols. Indeed, given any simplified term t there is a term $t^\circ(x_0, \dots, x_{m-1})$ and basic terms $p_0, \dots, p_{m-1}$ so that $t = t^\circ(p_0, \dots, p_{m-1})$, where t° involves some of the ternary operation symbols only and each of the variables $x_0, \dots, x_{m-1}$ occurs in t° exactly once.

Induction on the complexity of terms, with the help of the six schemas displayed above, shows that for every term t there is a simplified term $t^\bullet$ so that $T_* \models t \approx t^\bullet$.

Modulo T_* every equation is equivalent to an equation between simplified terms. So it is harmless to suppose that s and t are simplified.

If $s \approx t \in T_*$, our work is done,so we suppose that $s \approx t \notin T_*$. This means that there is at least one assignment $\mathbf{c} = \langle c_0, c_1, \dots \rangle$ of elements of B_* to the variables so that $s^{\mathbf{B}_*}(\mathbf{c}) \neq t^{\mathbf{B}_*}(\mathbf{c})$. Each c_i is named by a constant symbol $\mathbf{c}_i$. So another way to express that inequality is $\mathbf{B}_* \not\models s(\mathbf{c}) \approx t(\mathbf{c})$. On the other hand, since $s \approx t \in T^*$, we know that $\mathbf{B}^* \models s(\mathbf{c}) \approx t(\mathbf{c})$. It follows that $s^{\mathbf{B}_*}(\mathbf{c}), t^{\mathbf{B}_*}(\mathbf{c}) \in A$.

Let s be any simplified term with basic subterms $p_0, \dots, p_{m-1}$ and let $\mathbf{c}$ be an assignment of elements of B_* to the variables. Induction on the complexity of s shows that $s^{\mathbf{B}_*}(\mathbf{c}) = p_i^{\mathbf{B}_*}(\mathbf{c})$, for some $i < m$. We say that the basic term p_i is *exposed* by $\mathbf{c}$. Now let $\mathbf{e}$ be another assignment that differs from $\mathbf{c}$ only at position k. Suppose that $c_k \in A$ while $e_k = d_\beta$ where $|d_\beta| > |d_\alpha|$ for all $\alpha < \kappa$ such that D_α occurs in s. Then $\mathbf{c}$ and $\mathbf{e}$ expose the same basic subterm of s. This can be shown by induction on the complexity of s—here is the inductive step. Suppose $s = D_\alpha s_0 s_1 s_2$ where the simplified terms s_0, s_1, and s_2 have the desired property. Evaluating these three terms at $\mathbf{c}$ and $\mathbf{e}$ exposes the same basic terms. If x_k occurs in the exposed basic term of s_0, then evaluating s_0 at $\mathbf{c}$ gives value in A whereas evaluating s_0 at $\mathbf{e}$ gives a d_γ with $\gamma \neq \alpha$. So in this case s and s_2 have the same exposed basic subterm, On the other hand, if $s_0^{\mathbf{B}^*}$ agrees on $\mathbf{c}$ and $\mathbf{e}$ for the basic exposed term s_0, then the value assigned to s under either assignment is either the value assigned to s_1 or else the value assigned s_2. In either case, the inductive hypothesis shows that the same basic subterm is exposed by these two assignments.

Recall that $\mathbf{c}$ gives different values to s and t in $\mathbf{B}_*$ and that these values belong to A. Let p be the basic subterm of s exposed by $\mathbf{c}$ and let q by the basic subterm of t exposed by $\mathbf{c}$. We assert that p and q are constant symbols naming elements of A. Otherwise we may assume that p is a term in the signature of $\mathbf{A}$ in which the variable x occurs. It follows that $\mathbf{c}$ assigns an element of A to x. Let $\mathbf{e}$ be an assignment that differs from $\mathbf{c}$ only in assigning x the value d_γ where $|d_\gamma| > |d_\alpha|$ for every α so that D_α occurs in either s and t. Then p and q are the basic subterms of s and t exposed by $\mathbf{e}$. Under $\mathbf{e}$ the term p will be assigned some d_δ as a value and this value will also be the one assigned to s. In the event that x does not occur in q, then the value assigned by $\mathbf{e}$ to q (and hence to t) must lie in A. Since $s \approx t$ holds in $\mathbf{B}^*$, this is impossible. So x must also occur in q. But since $\mathbf{c}$ assigns p and q different values, we see that p and q must be different terms. But then $\mathbf{e}$ must also assign them different values of the form d_δ and d_γ. Again since $s \approx t$ holds in $\mathbf{B}^*$, this must be rejected as well. So no variables occur in p and q. So these exposed terms must be constant symbols of the form $\mathbf{a}_\gamma$ and $\mathbf{a}_\delta$. We call $\langle \mathbf{a}_\gamma, \mathbf{a}_\delta \rangle$ the *crucial pair* associated with the assignment $\mathbf{c}$. Observe that $\mathbf{B}_* \models s(\mathbf{c}) \approx \mathbf{a}_\gamma, t(\mathbf{c}) \approx \mathbf{a}_\delta$. This means

$$T_* \cup \{s \approx t\} \vdash \mathbf{a}_\gamma \approx \mathbf{a}_\delta$$

for each pair crucial for some assignment giving different values to s and t.

Now let $t^\flat$ be the term obtained from s by replacing each basic subterm of s that is exposed by some assignment with t. Then given any assignment $\mathbf{e}$ of elements of B_* to the variables, we see that $\mathbf{B}_* \models t(\mathbf{e}) \approx t^\flat(\mathbf{e})$. This means that

$$T_* \vdash t \approx t^\flat$$

.

Let $s^\sharp$ be the term obtained from $t^\flat$ as follows. In the occurrence of t in $t^\flat$ replacing the exposed term $\mathbf{a}_\gamma$ of some crucial pair $\langle \mathbf{a}_\gamma, \mathbf{a}_\delta \rangle$, replace $\mathbf{a}_\gamma$ with $\mathbf{a}_\delta$. Then

$$\{\mathbf{a}_\gamma \approx \mathbf{a}_\delta \mid \langle \mathbf{a}_\gamma, \mathbf{a}_\delta \rangle \text{ is crucial for some assignment }\} \vdash s^\sharp \approx t^\flat.$$

Also, since the basic subterms exposed by an assignment produce the value of the whole term for that assignment, we have

$$T_* \vdash s \approx s^\sharp.$$

The last three displayed expressions support the desired conclusion that

$$T_* \cup \{\mathbf{a}_\alpha \approx \mathbf{a}_\beta \mid \{s \approx t\} \models \mathbf{a}_\alpha \approx \mathbf{a}_\beta\} \vdash s \approx t.$$

In this way, ($\star$) is verified and the contention follows.

Finally, to see that T and Φ are isomorphisms inverting each other between Con $\mathbf{A}$ and $I[T_*, t^*]$ we observe the following contention, which follows immediately from the definitions of T and Φ.

Contention 5.

(i) *If* $T_* \subseteq T_0 \subseteq T_1 \subseteq T^*$, *then* $\Phi(T_0) \subseteq \Phi(T_1)$, *and*

(ii) *If* $\varphi_0 \subseteq \varphi_1$ *with* $\varphi_0, \varphi_1 \in \operatorname{Con} \mathbf{A}$, *then* $\mathrm{T}(\varphi_0) \subseteq \mathrm{T}(\varphi_1)$.

So Con $\mathbf{A} \cong I[T_*, T^*]$. In this way, Pigozzi's Embedding Theorem is established. ■

Don Pigozzi proved Pigozzi's Embedding Theorem in the 1990's and he circulated a preprint with this result. However, Pigozzi's paper was never published, so his theorem appears officially in the literature here for the first time. A key to Pigozzi's proof is the use of the ternary operations D_α, which resemble ternary discriminator functions. Pigozzi's proof has a natural structure, but turns out to also have many technical details. The proof we have given follows the presentation of Pigozzi's Embedding Theorem given by Gabor Tardós in the late 1990's at a seminar in Budapest. This proof of Tardós was also never published. We have followed the method of Tardós because although it has a less straightforward structure, it is also shorter and avoids the more intricate details required by Pigozzi's method.

Pigozzi's Embedding Theorem uses a signature that is infinite. Here is another result that deals with principal filters in lattices of equational theories under the stipulation that the signature is finite. It has been drawn out of the work of Ralph McKenzie (1984).

THEOREM 7.54. *Let τ denote a signature that provides just two operation symbols: a binary operation symbol and a unary operation symbol. Let F be a principal filter in $\mathcal{L}_\sigma$, where σ is a finite signature. Then F is isomorphic to a principal filter in $\mathcal{L}_\tau$.*

A way to paraphrase this theorem is to say that $\mathcal{L}_\tau$ is universal for principal filters in lattices of equational theories of finite signature.

Proof. Let T be the equational theory in $\mathcal{L}_\sigma$ so that F is the filter of all equational theories in $\mathcal{L}_\sigma$ containing T. Let $\mathcal{V} = \mathbf{Mod}\, T$. Then F is dually isomorphic to the lattice of subvarieties of $\mathcal{V}$. According to Corollary 10.111 part (iv) on page 211 in Volume III, we see that F is dually isomorphic to the lattice of subvarieties of $T_k[\mathcal{V}]$, where k is a numerical parameter depending on σ and T_k is the functor provided by (Ralph McKenzie, 1984). So F is isomorphic to the principal filter above $\Theta\, T_k[\mathcal{V}]$. ■

We conclude this section with the fourth theorem promised earlier. It places restrictions on the lattices that can be isomorphic to filters in lattices of equational theories. Let $\mathbf{L}$ be an algebraic lattice. We say that $\mathbf{L}$ satisfies the **Zipper Condition** provided whenever I is a nonempty set, $a, b \in L$ and $a_i \in L$ for all $i \in I$ such that $\bigvee_{i\in I} a_i = \top$ and $a \wedge a_i = b$ for all $i \in I$, then $a = b$. For finite sets I the zipper condition is defined using a first order formula resembling meet-semidistributivity.

$$\bigvee_{i\in I} a_i = \top \mathbin{\&} a \wedge a_i = b \text{ for all } i \in I \implies a \wedge \bigvee_{i\in I} a_i = b.$$

THEOREM 7.55 (Lampe's Zipper Theorem (William A. Lampe, 1986)).
Let σ be any signature. Any principal filter in $\mathcal{L}_\sigma$ satisfies the zipper condition.

The proof of Lampe's Theorem relies on the following observation, see (Ralph McKenzie, 1983).

> Let T be a equational theory. The principal filter of equational theories that contain T is isomorphic to the congruence lattice of an

> algebra with one basic two-place operation and some one-place basic operations, such that the two-place operation has a left unit and a left zero.

Indeed, let A be the set of all terms of the signature of T. The two-place operation $*$ is defined on A as follows: Let $s, t \in A$ and suppose that the variables occurring in s are among $v_0, v_1, \ldots, v_{r-1}$. Then put

$$s * t := s(s, t, v_2, \ldots, v_{r-1}),$$

that is the term s is substituted for v_0 in s and the term t is substituted for the variable v_1 in s. Then the variable v_1 is a left unit of $\langle A, *\rangle$ since $v_1 * t = t$ for all $t \in A$. Also the variable v_0 is a left zero since $v_0 * t = v_0$ for all $s \in A$. Take each endomorphism of the term algebra to be a one-place basic operation. Now let E denote the set of endomorphisms of the term algebra. We know that T can be regarded as a congruence relation of $\langle A, f\rangle_{f\in E}$. Observe that T also respects $*$. By the Correspondence Theorem, the filter of equational theories extending T is isomorphic to the congruence lattice of $\langle A, *, f\rangle_{f\in E}/T$. This establishes McKenzie's observation.

Proof of Lampe's Zipper Theorem. According to the observation, we work with congruences of an algebra $\mathbf{A}$ with a two-place operation $*$ that has a left unit u and a left zero o. So let I be any nonempty set and let $\varphi, \theta, \varphi_i \in \operatorname{Con}\mathbf{A}$ so that $\bigvee_{i\in I}\varphi_i = 1_A$ and so that $\varphi \wedge \varphi_i = \theta$ of all $i \in I$. We have to show that $\varphi = \theta$. Since I is not empty we have $\varphi \wedge \varphi_i = \theta$ for at least one $i \in I$. This means $\theta \subseteq \varphi$. We need only prove $\varphi \subseteq \theta$.

Since $(o, u) \in 1_{\mathbf{A}} = \bigvee_{i\in I}\varphi_i$, there must be some natural number n and some elements $r_0, \ldots, r_n \in A$ so that

$$o = r_0 \qquad r_j \;\varphi_{\hat{j}}\; r_{j+1} \text{ (for some } \hat{j} \in I) \qquad r_n = u.$$

To see that $\varphi \subseteq \theta$, suppose $a \;\varphi\; b$. We claim that $r_j * a \;\theta\; r_j * b$ for all $j \leq n$. We do this with a small induction. When $j = 0$ we have $r_j = o$. But $o * a = o = o * b$. Now suppose we know $r_j * a \;\theta\; r_j * b$. We have to show $r_{j+1} * a \;\theta\; r_{j+1} * b$. Since we also have $r_j \;\varphi_i\; r_{j+1}$, for some $i \in I$, we also know that $r_j * a \;\varphi_i\; r_{j+1} * a$ and $r_j * b \;\varphi_i\; r_{j+1} * b$. But also we have $a \;\varphi\; b$ and hence $r_{j+1} * a \;\varphi\; r_{j+1} * b$. This means $r_{j+1} * a \;\theta\; r_{j+1} * b$, since $\varphi \wedge \varphi_i = \theta$.

Now let $j = n$. Then $r_j = u$. So $a = u * a \;\theta\; u * b = b$. This means $\varphi \subseteq \theta$. The conclusion of Lampe's Zipper Theorem is at hand. ■

The observation of McKenzie that launched the proof of Lampe's Zipper Theorem, has been sharpened by Anvar M. Nurakunov (2008), who observed that principal filters in the lattices of equational theories are isomorphic to the congruence lattices of certain kinds of monoids expanded by two special unary operations.

The diagrams that follow render the first few steps in the proof of Lampe's Zipper Theorem and reveal how the theorem got its name.

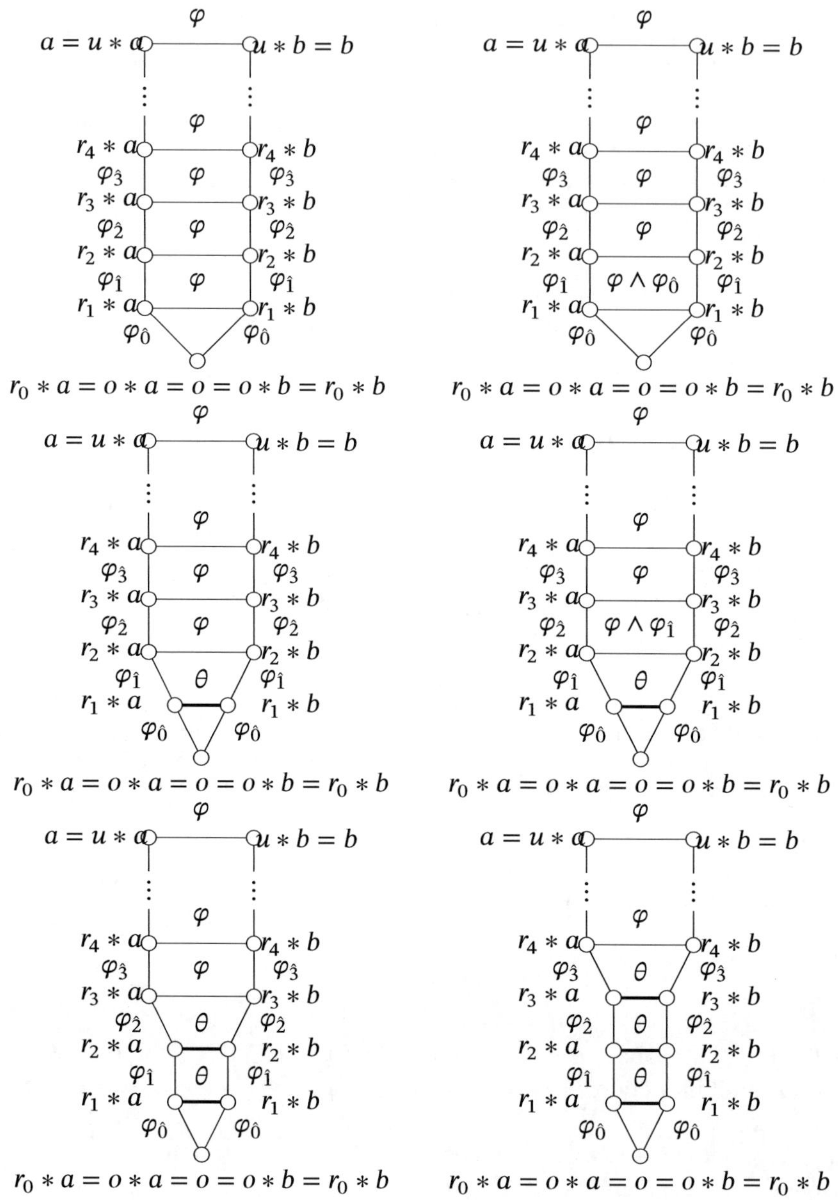

Relying on $\theta = \varphi \wedge \varphi_{\hat{j}}$, each φ on a horizontal rung can be replaced by θ step by step. In this way, the zipper can be closed all the way to the top.

The zipper condition allows us to conclude that certain lattices cannot be isomorphic to principal filters in the lattice of equational theories for any signature. The simplest is the modular lattice $\mathbf{M}_3$:

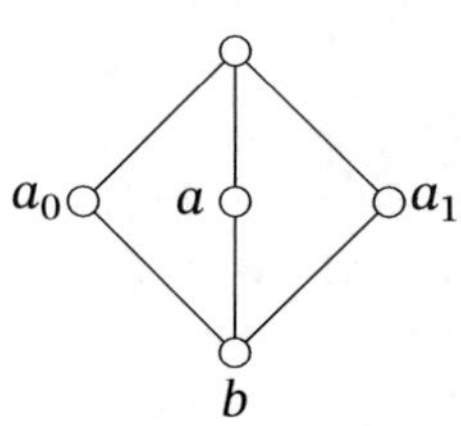

Conditions less restrictive than the zipper condition yet which must still hold in all filters in lattices of equational theories can be found in the literature. For example, (William A. Lampe, 1991). Such conditions are fairly involved.

Some of the most significant general results about lattices of equational theories rely on intricate methods of proof that put them beyond the scope of our exposition. In the late 1960's Ralph McKenzie took up the question of whether the signature σ could be recovered from $\mathcal{L}_\sigma$ considered as an abstract lattice. Of course, this is not really possible since the domain of σ consists of operation symbols and the particulars of their appearance is out of reach. However, Ralph McKenzie (1971) proved that the multiplicity sequence of σ could be recovered. Given a signature σ the **multiplicity sequence** of σ is the sequence of cardinals $\langle \kappa_0, \kappa_1, \dots \rangle$ where for each natural number r, the cardinal κ_r is the number of operation symbols of rank r provided by σ. In the same paper, McKenzie obtained the first results about subsets of $\mathcal{L}_\sigma$ which can be defined, in the language of lattice theory by means of formulas of first-order logic. For example, he showed that for the signature that provides just two operation symbols, both 2-place, the equational theory of lattices is definable by a first order formula. Now sets definable by first order formulas have to be invariant under automorphisms. As a result the equational theory of lattices is fixed by each automorphism of $\mathcal{L}_\sigma$.

In the first half of the 1980's Jaroslav Ježek published his investigation of definability in lattices of equational theories. It spans 100 journal pages; see Jaroslav Ježek (1981a, 1981b, 1982, 1986). He begins with a study of the automorphisms of the lattice $\mathcal{L}_\sigma$. There are obvious automorphisms that have a syntactical basis: two operation symbols of the same rank can be consistently interchanged, transforming one equational theory into another; or for a single operation symbol of rank at least two the order of the inputs can be rearranged. Then, of course, automorphisms of these kinds can be composed to obtain other automorphisms of $\mathcal{L}_\sigma$. These are the *obvious automorphisms.* Below is a short account of some of Ježek's findings:

Ježek's Automorphism Theorem.
Let σ be a signature that provides at least two 1-place operation symbols or at least one operation symbol of rank at least 2. Then $\mathcal{L}_\sigma$ has no nonobvious automorphisms.

Ježek's One Based Definability Theorem.
For any signature σ, the set of one-based equational theories is definable in $\mathcal{L}_\sigma$.

Ježek's Finitely Based Definability Theorem.
For any signature σ the set of finitely based equational theories is a definable subset of $\mathcal{L}_\sigma$.

Ježek's Individual Finitely Based Definability Theorem.
For any signature σ and any finitely based equational theory T of signature σ, the orbit of T under the action of the automorphism group is a definable subset of $\mathcal{L}_\sigma$.

Ježek's Finite Algebra Definability Theorem.
For any signature σ the set of equational theories of finite algebras is a definable subset of $\mathcal{L}_\sigma$.

Ježek's Individual Finite Algebra Definability Theorem.
Let σ be any finite signature and **A** *be any finite algebra of signature σ. Then the orbit of Θ* **A** *under the action of the automorphism group is a definable subset of $\mathcal{L}_\sigma$.*

Jaroslav Ježek and Ralph McKenzie (1993) adapted many of the results above to the lattice of equational theories of semigroups. Once more we can only accommodate here a partial list of their findings.

The Ježek-McKenzie Finite Semigroup Definability Theorem.
The set of equational theories of finite semigroups is definable in the lattice of all equational theories of semigroups.

The Ježek-McKenzie Locally Finite Semigroup Definability Theorem.
The set of locally finite equational theories of semigroups is definable in the lattice of all equational theories of semigroups.

The Ježek-McKenzie Finitely Based Locally Finite Semigroup Definability Theorem.
The set of finitely based locally finite equational theories of semigroups is definable in the lattice of all equational theories of semigroups.

The Ježek-McKenzie Finitely Based Semigroup Definability Theorem.
The set of finitely based equational theories of semigroups is definable in the lattice of all equational theories of semigroups.

Exercises 7.56

1. (Eugene Jacobs and Robert Schwabauer, 1964) Let σ be a signature that provides just one operation symbol and that one has rank 1. Describe the lattice $\mathcal{L}_\sigma$ of all equational theories of signature σ.

2. (Jaroslav Ježek, 1969) Prove that the lattice $\mathcal{L}_\sigma$ of all equational theories of signature σ is countable if and only if σ provides no more than one operation symbol of rank 1 with the remaining operation symbols being finite in number and of rank 0.

***3.** (Jaroslav Ježek, 1969) Let σ be a signature that provides just one operation symbol of rank 1 with the remaining operation symbols having rank 0. Describe the lattice $\mathcal{L}_\sigma$.

4. The set $\{x(yz) \approx (xy)z, xx \approx x, x(yz) \approx xz\}$ is a base for the equational theory of rectangular bands.
   (i) Prove that $\{x(yz) \approx (xy)z, xx \approx x, x(yx) \approx x\}$ is also a base for the equational theory of rectangular bands.
   (ii) Prove that the principal filter in the lattice of equational theories above the theory of rectangular bands has exactly four elements as displayed below.

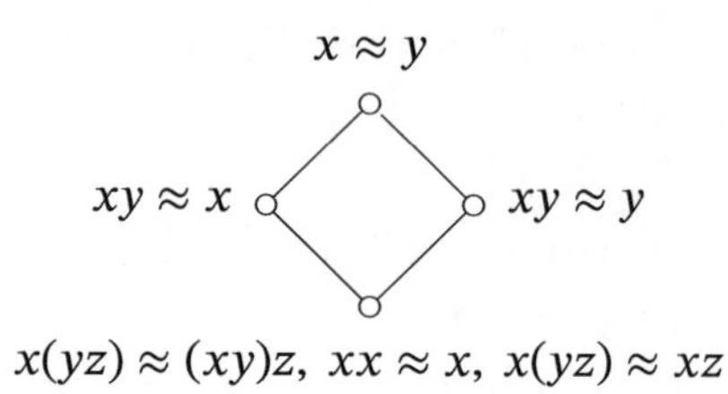

This is near the top of the lattice of equational theories of bands—see Figure 7.11 and the following discussion on page 234.

5. (Dana Scott, 1956) Suppose that **A** is a nontrivial finite algebra of finite signature. Prove that $\Theta\,\mathbf{A}$ has only finitely many maximal (i.e equationally complete) extensions.

6. (Jaroslav Ježek, 1970) Let σ be a signature that provides at least one operation symbol of rank at least 2. Prove that the meet of all the maximal elements of $\mathcal{L}_\sigma$ is the least element $\bot$ of $\mathcal{L}_\sigma$.

7. Let σ be a countable signature that provides at least two operation symbols of rank 1 or at least one operation symbol of rank at least 2. Construct an equational theory that is covered by exactly $\aleph_0$ equational theories in $\mathcal{L}_\sigma$.

8. (Ralph McKenzie, 1971) In a lattice an element a *hereditarily covers* another element b provided $b < a$ and for every element c if $b \wedge c < a \wedge c$, then $a \wedge c$ covers $b \wedge c$. Further, we say that a is *independent* of b provided for every c if $a \leq (c \wedge (a \vee b)) \vee b$ then $a \leq c$. Prove each of the following:

(i) If b hereditarily covers a and $a \wedge c < b \wedge c$, them $b \wedge c$ hereditarily covers $a \wedge c$.

(ii) If b hereditarily covers a and $c \vee d \leq b$ and c is independent of d, then either $c \leq a$ or $d \leq a$.

***9.** (Ralph McKenzie, 1971) Fix a signature σ. Let

$$T_{\text{reg}} = \{s \approx t \mid \text{ every variable occurring in } s \text{ occurs in } t \text{ and conversely}\}$$
$$T_{\text{const}} = \{s \approx t \mid s = t \text{ or operation symbols occur in both } s \text{ and } t\}$$

(i) Prove that T_{reg} and T_{const} are maximal equational theories.

(ii) Suppose, in addition, that σ provides an operation symbol of positive rank or at least two operation symbols of rank 0. Prove that top $\top$ of the lattice $\mathcal{L}_\sigma$ hereditarily covers T_{reg}.

(iii) Suppose σ provides at least two operation symbols of rank 1 or at least one operation symbol of rank at least 2. Prove that $\top$ hereditarily covers both T_{reg} and T_{const}. See Exercise 8. In fact, McKenzie proved that these two equational theories are the only ones hereditarily covered by $\top$.

7.7. Second Interlude: the Rudiments of Computability

Readers familiar with the theory of computable (or recursive) functions should feel free to skip the material below, referring back to it as needed to establish notation. Many substantial accounts of this theory are available in the literature. Among them: (Martin Davis, 1958), (Hartley Rogers, 1987), (Herbert B. Enderton, 2011), (John E. Hopcroft and Jeffrey D. Ullman, 1979), and (Michael Sipser, 2012) What we offer here is a mere sketch of the beginnings of the subject, leaving aside a number of details (mostly uninteresting) as well as a number of philosophical and mathematical issues (mostly interesting) which, properly speaking, belong to such beginnings.

DEFINITION 7.57. A **Turing machine** M is a partial function

$$\text{from } \{0,1,\ldots,m\}\times\{0,1\} \text{ into } \{0,1\}\times\{R,L\}\times\{0,1,\ldots,m\},$$

for some positive natural number m. We will often write $\langle q,s,s',D,q'\rangle \in M$ instead of $M(q,s) = \langle s',D,q'\rangle$. These are called the **instructions** of M. The members of the set $\{0,1,\ldots,m\}$ are called the **states** of M with 0 as the unique halting state and 1 as the initial state, the members of the set $\{0,1\}$ are called the **tape alphabet** of M, and the members of $\{R,L\}$ are called **directions**. To say that 0 is the unique halting state means that the only inputs on which M is not defined are $\langle 0,0\rangle$ and $\langle 0,1\rangle$. An **instantaneous description** or **configuration** for M is a triple $\delta = \langle \alpha,q,z\rangle$, where $\alpha \in \{0,1\}^{\mathbb{Z}}$ such that α takes on the value 1 only finitely many times, q is a state of M, and $z \in \mathbb{Z}$. The two-way infinite string α is called the **tape** of M at δ.

There are many variations of this definition. For instance, some presentations allow any larger, though finite, set to be a tape alphabet. Some presentations have no halting state, while others permit many halting states. The version we have chosen makes it relatively easier to show that certain problems have no computable solutions. But our version makes it relatively difficult to show that even simple functions can be computed by a Turing Machine. This is one reason why most expositions of Turing machines allow larger tape alphabets. The details, as well as the proofs that all these variants lead to the same notion of computability, we leave to the more expansive expositions of computability.

A Turing machine M specifies a transition from one instantaneous description δ to another instantaneous description δ'. If $\delta = \langle \alpha,q,z\rangle$ and $M(q,\alpha(z)) = \langle s',D,q'\rangle$, then $\delta' = \langle \beta,q',z'\rangle$ where β and z' are determined as follows:

$$\beta(r) = \begin{cases} s' & \text{if } r = z \\ \alpha(r) & \text{otherwise} \end{cases}$$

and

$$z' = \begin{cases} z+1 & \text{if } D = R \\ z-1 & \text{if } D = L \end{cases}$$

Under these circumstances, we say M **reaches** δ' from δ **in one step**. We say M **reaches** the description ϵ from δ, if it reaches it after finitely many steps. M is said to **halt at** $\delta = \langle \alpha,q,z\rangle$ if no transition is possible—i.e. if $q = 0$. M **halts from** a description δ if M reaches an instantaneous description from δ at which it halts. Notice that the behavior of a Turing machine M is deterministic in the sense that the transition from one instantaneous description to the next is completely determined by the partial function M.

There are many ways a Turing machine M can be used to specify a function. For example, let F be the partial function from and to instantaneous descriptions for M so that

$$F(\delta) = \begin{cases} \delta' & \text{if } M \text{ reaches } \delta' \text{ from } \delta \text{ and } M \text{ halts at } \delta' \\ \text{undefined} & \text{otherwise.} \end{cases}$$

But, via appropriate codings, M can be viewed as specifying other functions. For example, let $\alpha_{n+1,m}$ be the two-way infinite sequence

$$\ldots 00\underbrace{11\ldots 1}_{n+1}0\underbrace{11\ldots 1}_{m}00\ldots$$

with the leftmost 1 at position 0, and let G be the partial binary operation on the natural numbers such that

$$G(n+1,m) = \begin{cases} k & \text{if, from } \langle \alpha_{n+1,m}, 1, 0\rangle,\ M \text{ halts at a } k\text{-description} \\ \text{undefined} & \text{otherwise.} \end{cases}$$

where, by a k-description we mean any instantaneous description with a tape on which 1 occurs exactly k times. We will regard any function as **Turing-computable**, if, like the function G above, it can be specified by some Turing machine, via an appropriate coding. What amounts to an "appropriate coding" in this context is a subtle point, which, however, has little impact on the issues addressed in this work. Suffice it to say, any such coding must somehow be "computable." For example, later in this chapter we will be interested in computable signatures. This means the mathematical objects that are the operation symbols of the signature must be coded as strings of 0's and 1's. However, we will never descend to this level of detail, always presuming such things have already been arranged.

We subscribe to the

Church-Turing Thesis

> The functions that can be computed via any mechanical or effective process, however conceived, are exactly the Turing-computable functions.

The Church-Turing Thesis was advanced by Alonzo Church (1936) and by A. M. Turing (1936). It is supported by ample evidence and persuasive arguments, some of which can be found in the general references given above. We use this thesis in two ways:

- When it is apparent some function is intuitively computable, we offer no further arguments concerning its computability. In this case, an algorithm by which the function can be computed is usually described.
- When we have demonstrated a function is not Turing-computable, then we leap to the conclusion that it is not computable by any means.

In view of the Church-Turing Thesis, we refer to Turing-computable functions simply as **computable** functions.

A set X is said to be **recursive** or **decidable** iff the characteristic function of X is computable. X is **recursively enumerable** or **semidecidable** or **listable** iff it is the range of a recursive function. Consequently, X is recursively enumerable iff there is a computable function F from the set of natural numbers onto X. Thus, F can be regarded as a means to list the elements of X (allowing repetitions), one after another. Observe that there are only countably many Turing machines. Consequently, only countably many subsets of the set of natural numbers can be decidable. So almost all subsets of the set of natural numbers are undecidable. On the other hand the sets of natural numbers that can be described in some mathematically precise way are countable in number, since such descriptions must be finite strings of symbols chosen from some finite alphabet. We will see shortly that there are also such describable sets that are also undecidable.

Recall that Turing machines are certain kinds of finite sets (of 5-tuples). Turing machines can be appropriately coded as finite strings of 0's and 1's. Similarly, instantaneous descriptions for Turing machines can be coded as finite strings of 0's and 1's. The set of Turing machines is recursive, as is the set of instantaneous descriptions. The following theorem gives an account of the recursive unsolvability of the "halting problem" for Turing machines. It is due to A. M. Turing (1936), but see also (Emil L. Post, 1947). Proofs for this theorem can be found in any of the standard references on the theory of computable functions.

THEOREM 7.58 (The Halting Problem).

(i) *There is a Turing machine M such that the set of instantaneous descriptions from which M halts is not recursive.*

(ii) *Let $\delta_0 = \langle \zeta, 1, 0\rangle$, where ζ is the constantly 0 function on the integers. The set of Turing machines that halt from δ_0 is not recursive.*

(iii) *The set of Turing machines that halt from every instantaneous description is not recursive.* ■

Both the set specified in part (i) and that specified in part (ii) of the theorem above are evidently recursively enumerable. The set specified in part (iii) is the complement of a recursively enumerable set. This theorem can be made more explicit. For example, Turing machines that witness part (i) can be written down explicitly. As to part (ii), there is nothing special about the instantaneous description δ_0 apart from its simplicity.

7.8. Undecidability in Equational Logic

Presentations of algebras and word problems

Given a finite algebra $\mathbf{A}$ of finite signature there is an obvious way to completely describe the algebra: First list all the elements of A and then for each basic operation F of $\mathbf{A}$ list all the equalities of the form $F(a_0, \ldots, a_{r-1}) = b$ that hold in $\mathbf{A}$, where r is the rank of F and $\langle a_0, \ldots, a_{r-1}\rangle$ ranges through all the r-tuples of elements of A. This description of $\mathbf{A}$ is finite. Using such a description it is straightforward to compute the values of the basic operations at any input, by extension, also the values of any term function of $\mathbf{A}$. It follows that the equational theory of any finite algebra of finite signature is decidable.

Here is a way to extend this line of reasoning, at least part way, to certain infinite algebras. Let $\mathcal{V}$ be a variety and let Σ be a basis for $\mathcal{V}$. Let $\mathbf{A} \in \mathcal{V}$ and let G be a finite set of generators for $\mathbf{A}$. Then for every element a of A there is a term p so that $a = p^{\mathbf{A}}(g_0, g_1, \ldots)$ where $g_0, g_1, \ldots \in G$. Actually, there are likely to be many choices of p that will serve this purpose. So if q is another term we would like to know whether $a = q^{\mathbf{A}}(g_0, g_1, \ldots)$ as well. The difficulty is that, unlike in the case when $\mathbf{A}$ is finite, we have no method in hand to evaluate term functions, even in the case $\mathcal{V}$ has finite signature and our basis Σ of $\mathcal{V}$ is finite. To lay a firmer hold on this, let us suppose that $|G| = n$. Let $\mathbf{F}$ be the algebra in $\mathcal{V}$ freely generated by the n-element set $\{x_0, \ldots, x_{n-1}\}$. The map given by $x_i \mapsto g_i$ for all $i < n$ extends uniquely to a homomorphism $h : \mathbf{F} \twoheadrightarrow \mathbf{A}$. Let θ be the kernel of h. So the set of terms p that give term functions evaluating to a

as above is just the appropriate congruence class of θ (modulo, of course, the equational theory of $\mathcal{V}$.). Now let R be a set of ordered pairs of elements of F (we can view these as ordered pairs of terms in the variables $x_0, \ldots, x_{r-1}$) so that $\mathrm{Cg}^{\mathbf{F}}(R) = \theta$. We call the triple $\langle G \mid R \mid \Sigma \rangle$ a **presentation** of $\mathbf{A}$ with respect to $\mathcal{V}$. When both G and R are finite the presentation is said to be **finite** and in this case the algebra $\mathbf{A}$ is said to be **finitely presented for** $\mathcal{V}$. Note an algebra $\mathbf{A}$ is finitely presented for $\mathcal{V}$ if and only if $\mathbf{A} \cong \mathbf{F}/\theta$, where $\mathbf{F}$ is a finitely generated free $\mathcal{V}$-algebra and θ is a compact congruence of $\mathbf{F}$. An early paper concerning presentations of algebras in general is (Trevor Evans, 1951).

Actually, given any set Σ of equations, any set G, and any set R of ordered pairs of elements of the algebra in $\mathbf{Mod}\,\Sigma$ freely generated by G, we can form the presentation $\langle G \mid R \mid \Sigma \rangle$ that will present the quotient algebra of the free algebra modulo the congruence generated by R. In other words, presentations can be formed first and the algebra they present formed later. In $\langle G \mid R \mid \Sigma \rangle$ the set G is referred to as the set of **generators** and R is referred to as the set of **relations** or **relators**.

With this arrangement, the difficulty that arose above becomes the question

$$\text{For any terms } p \text{ and } q \text{ on } n \text{ variables, is } \langle \widehat{p}, \widehat{q} \rangle \in \mathrm{Cg}^{\mathbf{F}}(R)?$$

Above $\widehat{p}$ is obtained from p by substituting g_i for x_i for each $i < n$. Similarly for $\widehat{q}$.

Here is another way to frame this question:

$$\text{For any terms } p \text{ and } q \text{ on } n \text{ variables, is } p^{\mathbf{A}}(g_0, \ldots, g_{m-1}) = q^{\mathbf{A}}(g_0, \ldots, g_{m-1})?$$

The question is known as the **word problem** for the presentation $\langle G \mid R \mid \Sigma \rangle$. This word problem is said to be *solvable* provided there is an algorithm, that is a computational procedure, which upon input of a pair $\langle p, q \rangle$ of terms will determine whether $\langle \widehat{p}, \widehat{q} \rangle \in \mathrm{Cg}^{\mathbf{F}}(R)$. Otherwise, the presentation is said to have an *unsolvable word problem*. A variety $\mathcal{V}$ is said to have an *unsolvable word problem* if there is a finite presentation $\langle G \mid R \mid \Sigma \rangle$ with $\mathbf{Mod}\,\Sigma = \mathcal{V}$ which has an unsolvable word problem.

Here are two theorems that follow virtually from the definitions and whose proofs are left as exercises.

THEOREM 7.59 (W. Dyck 1883). *Let $\mathcal{V}$ be a variety of finite signature based on the finite set Σ of equations and $\langle G \mid R \mid \Sigma \rangle$ be a presentation of the algebra $\mathbf{A}$ with respect to $\mathcal{V}$. Let $\mathbf{B} \in \mathcal{V}$ be generated by a set $G' = \{g'_0, \ldots, g'_{n-1}\}$ with $|G| = |G'|$ and suppose that $p^{\mathbf{B}}(g'_0, \ldots, g'_{n-1}) = q^{\mathbf{B}}(g'_0, \ldots, g'_{n-1})$ for all $\langle p, q \rangle \in R$. Then there is a homomorphism $h \colon \mathbf{A} \twoheadrightarrow \mathbf{B}$. Indeed, $\eta' = h \circ \eta$ where $\eta \colon \mathbf{F} \twoheadrightarrow \mathbf{A}$ is the homomorphism so that $x_i \mapsto g_i$ for all $i < n$ and $\eta' \colon \mathbf{F} \twoheadrightarrow \mathbf{B}$ is the homomorphism so that $x_i \mapsto g'_i$ for all $i < n$.*

Actually, Walther von Dyck proved a result about groups that in later years was rendered as a theorem about group presentations. Particularly, after the work of Emmy Noether, the proof was rendered in terms of homomorphisms. Finally, after Garrett Birkhoff introduced varieties and free algebras in general, it could be seen that von Dyck's proof could be framed in the general manner given above.

THEOREM 7.60. *Let $\mathcal{V}$ be a variety of finite signature based on the finite set Σ of equations and $\langle G \mid R \mid \Sigma \rangle$ be a presentation of the algebra $\mathbf{A}$ with respect to $\mathcal{V}$. Let $\mathbf{B}$ be a finitely presented subalgebra of $\mathbf{A}$. If the presentation of $\mathbf{A}$ has a solvable word problem, then the presentation of $\mathbf{B}$ has a solvable word problem.*

Presentations first arose in the investigations of M. Dehn (1911) into certain infinite groups. Indeed, there is a rich theory of presentations in group theory, called combinatorial group theory, see (Wilhelm Magnus, Abraham Karrass, and Donald Solitar, 2004; Roger C. Lyndon and Paul E. Schupp, 2001; D. L. Johnson, 1997) for example. The first finitely presented algebras with unsolvable word problems were certain finitely presented semigroups constructed, independently, by A. Markov (1947) and Emil L. Post (1947). We give a variant of Post's construction later in this section. The first finitely presented groups with unsolvable word problems were constructed, independently, by William W. Boone (1957) and P. S. Novikov (1955), but see (Ralph McKenzie and Richard J. Thompson, 1973) for a more elementary construction. These results were followed by many other related results concerning decision problems for finitely presented groups and other kinds of algebras. Notable among these is the construction of Angus Macintyre (1973). He showed that if $\mathbf{F}$ is a finitely generated free group and $\mathbf{N}$ is a finitely generated normal subgroup (so $\mathbf{F}/\mathbf{N}$ is a finitely presented group), then there is a division ring $\mathbf{D}$ whose multiplicative group contains $\mathbf{F}$, and which has an element t such that, for $w \in F$, $w \in N$ if and only if $tw = wt$; see Angus Macintyre (1973) and pages 323–325 of P. M. Cohn (1995). Using this fact MacIntyre shows that there is a finitely presented division ring with an unsolvable word problem,. (The word problem for the class of division rings is formulated in terms of universal Horn sentences with a fixed precedent.) It is also a key element in the next theorem. Macintyre's construction relies on the construction of Boone. All of these results about groups and rings make use of methods particular to groups and rings that are beyond the scope of our exposition.

Here we give two ways in which word problems of presentations are connected to equational theories.

First, suppose that in the presentation $\langle G \mid R \mid \Sigma\rangle$ of $\mathbf{A}$ with respect to the variety $\mathcal{V}$, the set R is empty. Then $\mathrm{Cg}^{\mathbf{F}}(R)$ is just the identity relation on $\mathbf{F}$. So $\mathbf{A} \cong \mathbf{F}$. Is this case, the word problem becomes

Is the set of equations true in $\mathcal{V}$ on n or fewer variables decidable?

The most notable result in this direction is the following theorem.

THEOREM.
The set of equations in four or fewer variables true in every modular lattice is undecidable.

The first version of this theorem was proved by Ralph Freese (1980) for equations with five or fewer variables. As we mentioned above, the proof depends on the construction of Macintyre cited above. Christian Herrmann (1983) improved this to four or fewer variables. The result is best possible since in 1899 Richard Dedekind proved that the modular lattice on three free generators, diagrammed in Figure 2.1, page 39 of Volume I, has 28 elements; see Dedekind's collected works (Richard Dedekind, 1968). Herrmann's construction relies on Freese's construction as well as intricate methods from the theory of modular lattices. A full proof of this theorem, depending as it does on the constructions of Macintyre and Boone, is beyond the scope of our exposition.

That the variety of modular lattices has an unsolvable word problem had been shown earlier in (George Hutchinson, 1973, 1977) and (L. Lipshitz, 1974). While this is of course weaker than the above theorem, it does apply to several subvarieties of modular lattices. For example, Hutchinson's results apply to the variety generated by all subgroup lattices of Abelian groups (the congruence variety of Abelian groups as defined in §6.9). This variety has an unsolvable word problem and, by (Ralph Freese, 1994), is not finitely based (see the discussion after Corollary 6.110 on page 109). On the other hand Christian Herrmann and András Huhn (1975) and Christian Herrmann (1973) have shown that the equational theory of this variety of lattices is decidable. George Hutchinson (1995) has an excellent discussion of these and several related results, as well as their proofs.

The second way in which word problems of finite presentations are connected to equational theories results from expanding the signature. If the set G of generators is of size n, the idea is to add n new constant symbols $c_0, \ldots, c_{n-1}$ to the signature and to expand the algebra $\mathbf{A}$ to $\mathbf{A}^\#$ by putting $c_i^{\mathbf{A}^\#} = g_i$ for each $i < n$.

FACT 1. *Let $\mathcal{V}$ be a variety of finite signature based on the finite set Σ of equations. Let $\mathbf{A} \in \mathcal{V}$ be given by the finite presentation $\langle G \mid R \mid \Sigma\rangle$ with $|G| = n$. In the expanded signature that provides new constant symbols to name the elements of G, we let*

$$R^\# := \{s(c_0, \ldots, c_{n-1}) \approx t(c_0, \ldots, c_{n-1}) \mid \langle s, t\rangle \in R\}.$$

Then for any terms p and q on n or fewer variables

$$\langle p, q\rangle \in \operatorname{Cg}^{\mathbf{F}}(R) \text{ if and only if } \Sigma \cup R^\# \models p(c_0, \ldots, c_{n-1}) \approx q(c_0, \ldots, c_{n-1})$$

Recall that an algebra $\mathbf{A}$ is residually finite provided for all $a, a' \in A$ with $a \neq a'$ there is a finite algebra $\mathbf{B}$ and a homomorphism $h: \mathbf{A} \to \mathbf{B}$ so that $h(a) \neq h(a')$. This is the same as asserting that $\mathbf{A}$ is a subdirect product of finite algebras.

THEOREM 7.61. (A. I. Mal'tsev, 1958). *Let $\mathcal{V}$ be a finitely based variety of finite signature and let $\mathbf{A}$ be an algebra finitely presented with respect to $\mathcal{V}$. If $\mathbf{A}$ is residually finite, then it has a solvable word problem*

Proof. Let Σ be a finite basis for $\mathcal{V}$ and let $\langle G \mid R \mid \Sigma\rangle$ is a finite presentation of $\mathbf{A}$. We set up two algorithmic processes. When launched, each of these processes produces, one step at a time, a list of equations in the expanded signature in which no variables occur.

The first process lists those equational consequences of $\Sigma \cup R^\#$ in which no variable occur. The fact that this process is algorithmic (i.e. can be carried out, with suitable encoding, by a Turing machine) follows from the finiteness of the signature by way of the Completeness Theorem for Equational Logic.

The second process lists those equations in which no variables occur that fail in $\mathbf{A}^\#$. Observe that since the signature is finite, we see that up to isomorphism there are only finitely many algebras in the expanded signature of any finite size. So there is an algorithm for listing, as n runs through the natural numbers, all the algebras of the expanded signature that have universes of the form $\{0, \ldots, n\}$. This algorithm produces too many algebras, so as each algebra is produced we test it to see if it is a model of $\Sigma \cup R^\#$, and only let through those algebras that pass this test. This is still not the process we want, but we can use it.

Now suppose that $p^\# \approx q^\#$ is an equation without variables in the expanded signature. Let $p \approx q$ be the equation obtained by replacing each c_i by x_i for $i < n$. Here is how we will decide whether $p^\# \approx q^\#$ is a consequence of $\Sigma \cup R^\#$.

We launch both of our processes, pausing whenever one of the processes lists an output. If the first process has listed $p^\# \approx q^\#$, we stop everything and conclude that $p^\# \approx q^\#$ is a consequence. On the other hand, if the first process has listed some other equation, we continue. If the second process produces a finite algebra $\mathbf{B}^\#$ of the expanded signature, then we check to see whether $p^\# \approx q^\#$ holds in $\mathbf{B}^\#$. If it does not hold, then we stop everything and conclude that $p^\# \approx q^\#$ is not a consequence of $\Sigma \cup R^\#$. Otherwise, we continue.

To see that this is actually a decision procedure we have to show that it halts. It suffices to show that if $p^\# \approx q^\#$ is not a consequence of $\Sigma \cup R^\#$, then $p^\# \approx q^\#$ must fail in some finite model of $\Sigma \cup R^\#$. But we know that if $p^\# \approx q^\#$ is not a consequence then $p^{\mathbf{A}}(g_0, \dots, g_{n-1}) \neq q^{\mathbf{A}}(g_0, \dots, g_{n-1})$. Since $\mathbf{A}$ is residually finite there must be a finite algebra $\mathbf{B}$ and a homomorphism $h : \mathbf{A} \to \mathbf{B}$ so that

$$p^{\mathbf{B}}(h(g_0), \dots, h(g_{n-1})) = h(p^{\mathbf{A}}(g_0, \dots, g_{n-1}) \neq h(q^{\mathbf{A}}(g_0, \dots, g_{n-1}) = q^{\mathbf{B}}(h(g_0), \dots, h(g_{n-1})).$$

Now expand $\mathbf{B}$ to $\mathbf{B}^\#$ by putting $c_i^{\mathbf{B}^\#} = h(g_i)$ for all $i < n$. So $\mathbf{B}^\# \models \Sigma \cup R^\#$ and $p^\# \approx q^\#$ fails in $\mathbf{B}^\#$. So we have the finite algebra that we need. ■

Actually, Mal'tsev proved this in the variety of groups, but his proof uses no group theory.

THEOREM 7.62. (A. V. Kuznetsov, 1956). *Let $\mathcal{V}$ be a finitely based variety of finite signature and let $\mathbf{A}$ be an algebra finitely presented with respect to $\mathcal{V}$. If $\mathbf{A}$ is simple, then it has a solvable word problem.*

Proof. We use the notation established in the proof of Theorem 7.61 and we follow that proof in large, but we have to revise the second process. So suppose $p^\# \approx q^\#$ is not a consequence of $\Sigma \cup R^\#$. It follows that $\langle p, q\rangle \notin \mathrm{Cg}^{\mathbf{F}}(R)$. Since $\mathbf{A}$ is simple, we see that $\mathrm{Cg}^{\mathbf{F}}(R \cup \{\langle p, q\rangle\}) = 1_{\mathbf{F}}$. This means that the only models of $\Sigma \cup R^\# \cup \{p^\# \approx q^\#\}$ that are generated by the constants are 1-element algebras. We can express this with the following finite set Δ of equations:

$$\Delta := \{c_i \approx c_j \mid i, j < n\} \cup \{Q\mathbf{d} \approx c_0 \mid Q \text{ is an operation symbol and } \mathbf{d} \text{ is a string, with length the rank of } Q \text{, of constant symbols}\}$$

The second process that works in this case is to list the consequences of $\Sigma \cup R^\# \cup \{p^\# \approx q^\#\}$ to see if all of the equations in Δ eventually turn up on the list.

So given p and q on n variables, run these listing processes, examining each output as it is listed. If $p^\# \approx q^\#$ turns up on the first list, stop with success. If all the equations in Δ turns up on the second list, stop with failure. Exactly one of these alternatives must happen. ■

Again, this was proven for groups, where the only difference is that Δ in that case has a simpler form.

The last two theorems used the finiteness of R and Σ to conclude that there were algorithmic processes that would list the consequences of various finite sets of equations. In fact, in the proof above, given terms p and q in which no variables occur, we needed algorithms for listing those equational consequences in with no variables occur for the following two sets

$$\Sigma \cup R^{\#} \qquad \text{and} \qquad \Sigma \cup R^{\#} \cup \{p \approx q\}.$$

Actually, such algorithmic listing processes can exist even if R and Σ are infinite. In fact, it is enough that both R and Σ can be listed by algorithmic processes: after listing m equations in R and m equations in Σ, we can then list, say, the first $2m$ equations that are consequences of those $2m$ equations. After that, we can list the first $m+1$ equations in R and the first $m+1$ equations from Σ, and so on. Presentations in which R and Σ are recursive enumerable are said to be **recursive** presentations. Note that we still demand that the generating set G be finite. Then Theorem 7.62 remains true if we replace the finite presentation requirements by only requiring recursive presentations. On the other hand, our proof for Theorem 7.61 falls apart, since the second listing process in that proof requires us to check all the equations in $R^{\#}$ in a finite algebra. Since $R^{\#}$ might be infinite, we are faced with an endless task.

THEOREM 7.63. (Trevor Evans, 1978). *Let $\mathcal{V}$ be a variety of finite signature that provides at least one operation symbol of rank at least* 2, *no operation symbol of rank* 0, *and a recursively enumerable base Σ and let $\mathbf{A}$ be an algebra recursively presented by $\langle G \mid R \mid \Sigma\rangle$. Then $\mathbf{A}$ has a solvable word problem if and only if $\mathbf{A}$ is embeddable into a* 1*-generated simple algebra that is recursively presented with respect to the variety of all algebras of the signature.*

Proof. For the right-to-left direction, suppose that $\mathbf{A}$ is (isomorphic to) a subalgebra of $\mathbf{S}$, where $\mathbf{S}$ is simple and $\langle G' \mid R' \mid \emptyset\rangle$ is a recursive presentation of $\mathbf{S}$. According to the extension of Theorem 7.62, $\langle G' \mid R' \mid \emptyset\rangle$ has a solvable word problem. So by the extension of Theorem 7.60 discussed above, so does the presentation of $\mathbf{A}$.

Now consider the left-to-right direction. We will construct a 1-generated infinite simple algebra that is recursively presented and that has a subalgebra isomorphic to $\mathbf{A}$. Let $E = \{e_0, e_1, \dots\}$ be the set of terms without variables in the signature obtained by expanding the original signature by enough constant symbols to name the generators specified in the presentation of $\mathbf{A}$. Let $H := \{d_0, d_1\} \cup \{h_0, h_1, \dots\}$ with all the elements just listed in H distinct and E and H disjoint. The set of generators of our simple algebra $\mathbf{S}$ is $E \cup H$. For convenience, we suppose Q is a 2-place operation symbol of our signature. We address how to adapt our proof to the case when the signature does not provide a 2-place operation symbol after we complete the construction using Q. The relations of the presentation of $\mathbf{S}$ fall into three groups:

(A) The relations of the presentation of $\mathbf{A}$.

(B) For $i, j \in \omega$ all of the following ordered pairs:

$$\begin{array}{ll}
\langle Q(e,d_0),d_1\rangle \text{ if } e\in E & \langle Q(e,d_1),d_0\rangle \text{ if } e\in E\\
\langle Q(e_i,h_j),d_0\rangle \text{ if } \mathbf{A}^\# \models e_i\approx e_j & \langle Q(e_i,h_j),d_\rangle \text{if } \mathbf{A}^\# \not\models e_i\approx e_j\\
\langle Q(d_0,e),d_0\rangle \text{ if } e\in E & \langle Q(d_1,e),e\rangle \text{ if } e\in E\\
\langle Q(d_0,d_0),e_0\rangle & \\
\langle Q(d_0,d_1),d_0\rangle & \langle Q(d_0,h_i),e_i\rangle\\
\langle Q(d_1,d_0),d_0\rangle & \langle Q(d_1,d_1),h_0\rangle\\
\langle Q(d_1,h_i),h_{i+1}\rangle & \langle Q(h_i,e_j),d_0\rangle\\
\langle Q(h_i,d_0),d_0\rangle & \langle Q(h_i,d_1),d_0\rangle\\
\langle Q(h_i,h_j),d_0\rangle \text{ if } i=j & \langle Q(h_i,h_j),d_1\rangle \text{ if } i\neq j
\end{array}$$

(C) Now let P be any other operation symbol, say of rank n. Let $u_0,\dots,u_{n_1}\in E\cup H$, where at least one u_i is a generator from H. We put the following ordered pairs into the relation presenting **S**:

$$\langle P(u_0,\dots,u_{n-1}),d_0\rangle.$$

An examination of this presentation of **S** leads easily to the conclusion that **S** is generated by $\{d_0\}$ (and in fact the generation process can be restricted to using just the operation $Q^{\mathbf{S}}$).

To see that **S** is recursively presented, observe that those lines of the presentation of **S** are either given explicitly or in terms of how the corresponding operations in $\mathbf{A}^\#$ work or in terms of whether equations without variables hold in $\mathbf{A}^\#$. Since **A** is recursively presented, all the necessary equations to deal with this dependence on $\mathbf{A}^\#$ comprise a recursively enumerable set, since the presentation of **A** has a solvable word problem. Observe that in the presentation of **S**,the generators are the elements of $E\cup H$, rather than the generator d_0. But there is an algorithmic method for assigning each element of $E\cup H$ a term without variables whose only constant symbol is d_0 that denotes the specified element of $E\cup H$, since our presentation of **A** has a solvable word problem. In this way, the presentation of **S** given above can be rendered as a recursively enumerable set of equalities in the generator d_0.

To see that **S** is simple, we first consider what the elements of S must be like. They are all denoted by terms without variables built from the generators in $E\cup H$. We contend that $S=E\cup H$. Here is a short inductive argument based on the complexity terms. The base step is clear. For the inductive step, suppose $Pt_0,\dots,t_{n-1}$ is a term without variables, where the inductive hypothesis is that the elements of S denoted by t_i for each $i<n$ belong to $E\cup H$. If all the elements denoted by the t_i's belong to E, then the element denoted by $Pt_0,\dots,t_{n-1}$ must belong to E since E is closed under all the operations, since it arises from the presentation of **A**. So consider the situation when at least one of the t_i's belongs to H. If P is not the operation symbol Q, then the ordered pair put into the presentation of **S** in Group (C), which becomes an equality in **S** gives us that the element of S denoted by $Pt_0.\dots,P_{n-1}$ is $d_0\in H$. So finally consider the situation when P is the operation symbol Q and at least one of t_0 and t_1 belongs to H. Then examining the order pairs put into the presentation from Group (B), which become equalities is **S**, we can again draw the desired conclusion, completing the induction.

Now to see that **S** is simple, let θ be any nontrivial congruence of **S**. Since S is generated from $\{d_0\}$ with only the help of $Q^{\mathbf{S}}$ we can conclude that θ makes every element of S congruent to every other element, just by showing that $\langle d_0, Q^{\mathbf{S}}(d_0, d_0)\rangle \in \theta$. Since θ is nontrivial, it must collapse two distinct elements of S. Since $S = E \cup H$, there are eight cases to consider. In each case we have to show that θ also collapses d_0 and $Q^{\mathbf{S}}(d_0, d_0)$.

Case: θ collapses e_i and e_j where $i \neq j$.
Observe

$$\begin{aligned} e_i &\equiv e_j \mod \theta \\ Q^{\mathbf{S}}(e_i, h_j) &\equiv Q^{\mathbf{S}}(e_j, h_j) \mod \theta \\ d_0 &\equiv d_1 \mod \theta \\ d_0 = Q^{\mathbf{S}}(d_1, d_0) &\equiv Q^{\mathbf{S}}(d_0, d_0) \mod \theta. \end{aligned}$$

Case: θ collapses e and d_0 where $e \in E$.
Observe

$$\begin{aligned} e &\equiv d_0 \mod \theta \\ d_0 = Q^{\mathbf{S}}(d_0, e) &\equiv Q^{\mathbf{S}}(d_0, d_0) \mod \theta. \end{aligned}$$

Case: θ collapses e and d_1 where $e \in E$.
Observe

$$\begin{aligned} e &\equiv d_1 \mod \theta \\ Q^{\mathbf{S}}(e, e) &\equiv Q^{\mathbf{S}}(e, d_1) \mod \theta \\ Q^{\mathbf{S}}(e, e) &\equiv d_0 \mod \theta. \end{aligned}$$

Since $Q^{\mathbf{S}}(e, e') \in E$, we have reduced to the case we settled just above.

Case: θ collapses e and h_i where $e \in E$ and $i = 0, 1, \ldots$.
Observe

$$\begin{aligned} e &\equiv h_i \mod \theta \\ Q^{\mathbf{S}}(e, e) &\equiv Q^{\mathbf{S}}(h_i, e) \mod \theta \\ Q^{\mathbf{S}}(e, e) &\equiv d_0 \mod \theta \end{aligned}$$

Since $Q^{\mathbf{S}}(e, e) \in E$, we have reduced to a case above we settled above.

Case: θ collapses d_0 and d_1.
Observe

$$\begin{aligned} d_0 &\equiv d_1 \mod \theta \\ d_0 = Q^{\mathbf{S}}(d_0, d_1) &\equiv Q^{\mathbf{S}}(d_0, d_0) \mod \theta \end{aligned}$$

Case: θ collapses d_0 and h_i for some $i = 0, 1, \ldots$.
Observe

$$\begin{aligned} d_0 &\equiv h_i \mod \theta \\ d_0 = Q^{\mathbf{S}}(h_i, d_0) &\equiv Q^{\mathbf{S}}(d_0, d_0) \mod \theta \end{aligned}$$

Case: θ collapses d_1 and h_i for some $i = 0, 1, \ldots$.
Observe

$$d_1 \equiv h_i \mod \theta$$
$$Q^{\mathbf{S}}(d_1, e_0) \equiv Q^{\mathbf{S}}(h_i, e_0) \mod \theta$$
$$Q^{\mathbf{S}}(d_0, d_0) = e_0 \equiv d_0 \mod \theta$$

Case: θ collapses h_i and h_j for some $i, j \in \{0, 1, \ldots\}$ with $i \neq j$.
Observe

$$h_i \equiv h_j \mod \theta$$
$$Q^{\mathbf{S}}(h_i, h_i) \equiv Q^{\mathbf{S}}(h_i, h_j) \mod \theta$$
$$d_0 \equiv d_1 \mod \theta$$

So this case reduces to a case that was settled above.

So we can conclude that **S** is simple.

It remains to argue that **A** is embeddable into **S**. Roughly speaking, the difficulty is the prospect that the equalities stemming from the ordered pair in groups (B) and (C) might support the collapse of elements of E. To deal with this, we use a notion of reduced form. We saw above that any term without variables denotes an element of S and that using the equalities stemming from groups (B) and (C) such a term can be reduced by left-to-right moves to a term in $E \cup H$. On the other hand, the terms in $E \cup H$ admit no such further reductions. To formalize this process we will say a term t without variables *reduces in one step* to a term s provided there is an equality stemming from groups (B) and (C) so that s arises by replacing in t a subterm equal to the left side of the equality by its right side. We say t *reduces to* s provided there is a finite sequence, with the empty sequence allowed, of one-step reductions that leads from t to s. When s admits no one-step-reduction we say it is in *normal forms*. Evidently, all the members of $E \cup H$ are in normal form. So what we showed above is the that every term without variables reduces to a term in normal form.

CLAIM 1.
Every term without variables reduces to a unique term in normal form.

Proof of Claim 1. First suppose that t is a term without variables and that it reduces to both s and s' with $s \neq s'$. We contend that there is a term s'' so that both s and s' reduce to s''. There are three cases to consider: that s reduces to s', that s' reduces to s, and that neither of these hold. The first two cases are easy, so we consider the last alternative. We proceed by induction on the length of t. The term t cannot be just a constant symbol, securing the base step. Let $t = Pu_0, \ldots, u_{n-1}$ where the induction hypothesis holds asserts that our conclusion holds for each u_i. There are several situations to consider:

SITUATION: There are $i, j < n$ with $i \neq j$ so that u_i reduces to u_i' and u_j reduces to u_j' and such that $s = P\ldots, u_i', \ldots$ and $s' = P\ldots, u_j', \ldots$.

In this situation we can take s'' to be the term resulting from making both of these reductions.

SITUATION: There is $i < n$ so that u_i reduces to u_i' and u_i reduces to u_i'' and such that $s = P\ldots, u_i', \ldots$ and $s' = P\ldots, u_i'', \ldots$.

In this situation we can appeal to the induction hypothesis to obtain u_i''' so that u_i' and u_i'' both reduce to u_i'''. Then putting $s'' = P\ldots,u_i''',\ldots$ we see that both s and s' reduce to s''.

SITUATION: There is $i < n$ so that u_i reduces to u_i' and $s = P\ldots,u_i'\ldots$ while t reduces to s' by a sequence of steps at least one of which involves the leading operation symbol.

Examining the equalities stemming from groups (B) and (C), we see that $s'' \in E \cup H$. But $s = P\ldots,u_i',\ldots$ will also reduce to s'. This means that we can take $s'' = s'$.

So our contention holds. But now suppose t reduces to s and s', where both s and s' are in normal form. According to the contention we have just established, there is s'' so that both s and s' reduce to s''. Since neither s nor s' can be reduced, we must have $s = s'' = s'$. The proof of Claim 1 is complete. ■

CLAIM 2.
Let s and t be terms without variables that denote the same element of $\mathbf{S}$ and let s' be the normal form of s and t' be the normal form of t. Then, either

(i) *$s', t' \in E$ and $s' = t'$ in $\mathbf{A}$, or*
(ii) *s' and t' are identical and belong to H.*

Proof of Claim 2. First observe that if s and t are identical, then there is nothing to prove. So we assume they are not identical. According the Fact 1 on page 257, there is an equational deduction of $s \approx t$ from the equations stemming from the presentation of $\mathbf{S}$. Our proof is by induction on the length of such a deduction. The base step of this induction is that the derivation is of length one. That is either s reduces to t in one step (using equations stemming of Group (A) being allowed) or the t reduces to s in one step. It is harmless to suppose that s reduces of t in one step.

THE EQUATION STEMS FROM GROUP (A)
We proceed by induction of the length of s. Let $s = Pu_0,\ldots,u_{n-1}$. If the equation used in the reduction applies to the whole of s, then $t \in E$ and there is nothing more to prove. So consider the situation where the reduction applies to u_j for some $j < n$. Let v_j be the result of applying this reduction to u_j. So $t = P\ldots v_j\ldots$. Now put u_i' to be the normal form of u_i for each $i < n$ and put v_j' be the normal form of v_j. By Claim 1 each of $u_0',\ldots,u_{\{n-1\}}', v_j'$ belongs either to E or to H. By our induction hypothesis, applied to u_j and v_j, we have that either $u_j', v_j' \in B$ and $u_j' = v_j'$ in $\mathbf{A}$ or u_j' and v_j' are identical and belong to H. Considering

$$P\ldots u_j'\ldots \text{ and } P\ldots v_j'\ldots,$$

only at most one further reduction step remains to reach normal form. An examination of these possibilities leads to the conclusion the u' and v' fulfill the claim.

THE EQUATION STEMS FROM GROUP (B) OR GROUP (C)
In this situation, s reduces to t in one step. By Claim 1, s and t must have identical normal forms belonging either to E or to H. So the conclusion of the claim is immediate. ∎

From Claim 2, we see that if $e_i, e_j \in E$, then

$$e_i = e_j \text{ in } \mathbf{S} \text{ if and only if } e_i = e_j \text{ in } \mathbf{A}.$$

It follows that $\mathbf{A}$ is embeddable into $\mathbf{S}$.

Finally, we have to consider the situation where there is no 2-place operation symbol available. If there is an operation symbol O of rank $r > 2$, then we can use the term $Ox, \dots, x, y$ in place of the term Qxy in the Group (B) above. But we need more ordered pairs in relation of the presentation of $\mathbf{S}$.
Group (D)

$$\begin{aligned} &\langle O(u_0, \dots, u_{r-1}), d_0\rangle \text{ if } u_0, \dots, u_{r-2} \text{ are not all equal to the same element} \\ &\qquad\qquad \text{of } H \text{ but at least one belongs to } H \\ &\langle O(u_0, \dots, u_{r-1}), d_0\rangle \text{ if } u_i \in E \text{ for all } i < r-1 \text{ but there are } i, j < r-1 \\ &\qquad\qquad \text{with } i \neq j \text{ so that } u_i \approx u_j \text{ does not hold in } \mathbf{A}^\# \text{ and } h \in H. \end{aligned}$$

It is helpful in the last stage of the proof, that $\mathbf{A}$ is embeddable into $\mathbf{S}$ in this case, to use the following consequences of Group (A): for $e_{i_0}, \dots, e_{i_{r-2}} \in E$ such that $e_{i_0} = e_{i_1} = \dots = e_{i_{r-2}}$ in $\mathbf{A}$:

$$\begin{aligned} &\langle O(e_{i_0}, \dots, e_{i_{r-2}}, d_0), d_1\rangle && \langle O(e_{i_0}, \dots, e_{i_{r-2}}, d_1), d_0\rangle \\ &\langle O(e_{i_0}, \dots, e_{i_{r-2}}, h), d_0\rangle \text{ if } \mathbf{A}^\# \models e_i \approx e_j && \langle O(e_{i_0}, \dots, e_{i_{r-2}}, h), d\rangle \text{if } \mathbf{A}^\# \not\models e_i \approx e_j. \end{aligned}$$

We leave the rest of the details in the hands of the reader. ∎

These theorems allow us to conclude that an assortment of equational theories are decidable.

A finitely based undecidable equational theory

Even restricted to finite signatures, except for the most meager ones, we have already seen that there are uncountably many equational theories. On the other hand, there are only countably many decidable equational theories. So almost all equational theories are undecidable. The theorem on page 256 points out that the variety of modular lattices, which of course is finitely based, has an undecidable equational theory. In this subsection we show that almost all signatures support an undecidable equational theory. To achieve this we begin by associating with each Turing machine M a set of equations. Here we take a version of Turing machines with an arbitrary finite tape alphabet that includes a symbol O for the blank, and uses q_0 as the initial state.

We will make the signature for our set of equations as follows:

- F_a will be a unary operation symbol for each tape symbol a;
- G_q will be a unary operation symbol for each state q;
- For housekeeping purposes, we'll make unary operation symbols H, K, and J.

The set Σ_M of equations that will completely describe our Turing machine is made up as follows. We put into Σ_M all equations of the form

$$F_c G_q F_a x \approx G_r F_c F_b x$$

for each tape symbol c, when $\langle q, a, b, L, r\rangle$ is an instruction of M. Likewise, we include the equation

$$G_q F_a F_c x \approx F_b G_r F_c x$$

when $\langle q, a, b, R, r\rangle$ is an instruction. We include the equation

$$HG_q F_a x \approx HG_r F_0 F_b x$$

whenever (q, a, b, L, r) is an instruction. Likewise, we include the equation

$$G_q F_a Hx \approx F_b G_r F_0 Hx$$

whenever (q, a, b, R, r) is an instruction.

We are not done with Σ_M, but we see that given any word w in the tape alphabet, there is a term t_w that looks, for example, like

$$HG_{q_0} F_a F_b F_a F_a F_b \cdots Hx$$

where q_0 is the initial state and w is $abaab\cdots$ corresponding to the subscripts of the F's. As the computation of M proceeds, the equations we have already put into Σ_M can be used to make an equational derivation. If M halts at some point, it means that we have reached a configuration that contains q, a with no corresponding instruction. In case there is no instruction of the form $q, a, \ldots$, we include in Σ_M the equation

$$G_q F_a x \approx Kx.$$

The last equations to be included in Σ_M are

$$KF_a x \approx Kx, KHx \approx JHx, F_a JHx \approx JHx,$$

for each tape symbol a. The use of these equations is to provide a derivation that ends in $HJHx$ when M halts.

DEFINITION 7.64. A **Post** term looks like

$$HATBHx$$

where A and B are (possibly empty) strings of F_c's for various c's and T is one of J, K, G_q where q is a state.

THEOREM 7.65. *Let M be a Turing Machine and w be a word on the tape alphabet of M. Then M halts on input w if and only if*

$$\Sigma_M \vdash t_w \approx HJHx,$$

where t_w is the term $HG_{q_0} Fw_0 F_{w_1)} \ldots Hx$, each w_i being a letter in the tape alphabet.

Proof.
($\Rightarrow$) Clear from the construction.

($\Leftarrow$) In the forward direction at issue above, the equational derivation emerged from the computation of the Turing machine and the use of each equation in Σ_M was in the left-to-right sense. For this implication, however, we are supposing that $\Sigma_M \vdash t_w \approx$

$HJHx$ and we have to show that M halts on input w. The main difference is that in equational derivations, the equations in Σ_M can also be used in the right-to-left sense.

Let us suppose that

$$(\star) \qquad t_w = t_0, t_1, t_2, \dots, t_n = HJHx$$

is a derivation based on Σ_M. It is easy to see that each term in this derivation is a Post term. There is a leftmost term t_k with $k > 0$ in which J occurs. So the equation $KHx \approx JHx$ must have been applied to t_{k-1} to obtain t_k. In turn, there is a leftmost term t_ℓ with $\ell > 0$ in which K occurs. Some equation of the form $G_qF_ax \approx Kx$, where M has no instruction of the form $(q, a, \dots)$, must have been applied to $t_{\ell-1}$ to obtain t_ℓ. This means we can replace $(\star)$ by

$$(\star\star) \qquad t_w = t_0, t_1, t_2, \dots, t_n = HAGqF_aBHx$$

where the strings A and B might be empty, and where M has no instruction of the form $(q, a, \dots)$. Further, we suppose that the derivation $(\star\star)$ is as short as possible. Now in this derivation if all the equations in Σ_M are applied in the left-to-right sense, then our work is done. So suppose the rightmost step that is done in the right-to-left sense is the one from t_j to t_{j+1}. So a derivation step from t_{j+1} to t_j is a left-to-right step. That is $t_{j+1} = HCG_rF_bDHx$ and M has an instruction of the form $(r, b,)$. One conclusion is that $j + 1 \neq n$, but another is that $t_{j+2} = t_j$, since M has exactly one instruction of the form $(r, b, \dots)$. But then our derivation could be shortened. So all the steps must be in the left-to-right sense. This means that M halts on input w. ■

COROLLARY 7.66. *Let σ be a recursive signature that provides at least two unary operation symbols or some operation symbol of rank at least two. Then there is a finitely based 1-variable equational theory of signature σ that is undecidable.*

This corollary is a consequence of the theorem with the help of the ω-Universal Translation Theorem (7.42).

Theorem 7.65 is a variation of theorems due independently to Emil L. Post (1947) and A. Markov (1947), where finitely presented semigroups with unsolvable word problems were constructed. Indeed, our theorem can be obtained from theirs by appealing to the well-known fact that each semigroup is isomorphic to a semigroup of transformations on some set. Marshall Hall (1949) showed how to obtain the results of Post and Markov using only two generators. However, this only reduces our proof to using five unary operation symbols, since we would still need H, J, and K. The corollary above, in the case of two 1-place operation symbols or one 2-place operation symbol can be found in (A. I. Mal'tsev, 1966a).

Base undecidability: the set up

Suppose someone were to hand you a finite set of equations and ask whether it is a base for the equational theory of groups. You might be able to answer, if the finite set of equations was not too big and the equations in the set were not too involved. On the other hand, you might want to devise a computer program to answer such questions.

An equational theory T of a recursive signature is said to be **base undecidable** provided the set $\{\Delta \mid \Delta \text{ is a finite basis for } T\}$ is not recursive. Might the equational theory of groups be base undecidable? (We return to the consideration of familiar theories on page 273, after completing the proof of The Base Undecidability Theorem (7.69). We look specifically at group theory in Example 7.73 on page 275.)

There is a class of equational theories that are evidently base undecidable: the class of finitely based undecidable equational theories. Just observe that if Σ is a finite base of an undecidable equational theory T then for any equation $s \approx t$ we have

$$\Sigma \cup \{s \approx t\} \text{ is a base for } T \text{ if and only if } s \approx t \in T.$$

To address questions of this kind, we start with a very simple signature. Here we will use σ to denote a fixed signature which provides just two operation symbols D and E, both unary. We reserve Ψ to denote a finite set of equations in the variable x of signature σ that is the base of an undecidable equational theory. Let H and K be two new unary operation symbols.

Now let τ be another signature and let $d = \langle d_D, d_E, d_H, d_K\rangle$ be a system of definitions for σ in τ.

For any equation $s \approx t$ of signature σ we put

$$\Psi(s \approx t) := \Psi \cup \{Hs(Kx) \approx Hs(Ky), Ht(Kx) \approx x\}.$$

LEMMA 7.67. *Let s, t, r, and q be terms of signature σ in the variable x such that $\Psi \nvdash s \approx t$. Then*

$$\Psi(s \approx t) \vdash q \approx r \textit{ if and only if } \Psi \vdash q \approx r.$$

Consequently, if s and t are terms of signature σ in the variable x, then

$$\Psi(s \approx t) \vdash s \approx t \textit{ if and only if } \Psi \vdash s \approx t.$$

Furthermore, there is a countably infinite algebra, denoted $\mathbf{B}^*$, with an element ∞ such that $(Hs(Kx))^{\mathbf{B}^*}$ is the constant function with value ∞, while $(H(t(Kx)))^{\mathbf{B}^*}$ is the identity function. In particular, $\mathbf{B}^* \models \Psi(s \approx t)$.

Proof. The implication from right to left is clear. So suppose that $\Psi \nvdash q \approx r$ and $\Psi \nvdash s \approx t$. We have to show $\Psi(s \approx t) \nvdash q \approx r$. Let $\mathbf{A}$ be a countably infinite model of Ψ in which both $q \approx r$ and $s \approx t$ fail. Pick $a, b \in A$ so that $q^{\mathbf{A}}(a) \neq r^{\mathbf{A}}(a)$ and $s^{\mathbf{A}}(b) = c \neq d = t^{\mathbf{A}}(b)$. Now $\mathbf{A}$ is an algebra of signature σ. We start by improving $\mathbf{A}$ so that $s \approx t$ will have, loosely speaking, infinitely many disjoint failures. As a first step, let $\infty \notin A$ and put $A' = A \cup \{\infty\}$. Make $\mathbf{A}'$ by extending the operations of $\mathbf{A}$ so that $D^{\mathbf{A}'}(\infty) = \infty = E^{\mathbf{A}'}(\infty)$. Since Ψ consists of regular equations we will have $\mathbf{A}' \models \Psi$. Let

$$B = \{\mathbf{e} \mid \mathbf{e} \in A'^{\omega} \text{ and at most one coordinate of } \mathbf{e} \text{ differs from } \infty\}.$$

It is easy to see that B is a subuniverse of $\mathbf{A}'^{\omega}$ and that B is countably infinite. Let $\mathbf{B}$ be the subalgebra of $\mathbf{A}'^{\omega}$ with universe B. So we have $\mathbf{B} \models \Psi$ and $q \approx r$ fails in $\mathbf{B}$.

For $e \in A$ and $i \in \omega$, let $e[i]$ be the ω-tuple with e at the ith position and ∞ at all the other positions. We use $\overline{\infty}$ to denote the ω-tuple all of whose entries are ∞. So we have $s^{\mathbf{B}}(b[i]) = c[i] \neq d[i] = t^{\mathbf{B}}(b[i])$ for all $i \in \omega$.

Now let $\mathbf{B}^*$ be the expansion of $\mathbf{B}$ by taking $K^{\mathbf{B}^*}$ to be a one-to-one function from B onto $\{b[i] \mid i \in \omega\} \cup \{\overline{\infty}\}$ so that $K^{\mathbf{B}^*}(\overline{\infty}) = \overline{\infty}$ and by defining $H^{\mathbf{B}^*}$ as follows:

$$H^{\mathbf{B}^*}(u) = \begin{cases} (K^{\mathbf{B}^*})^{-1}(b[i]) & \text{if } u = d[i] \text{ for some } i \in \omega \\ \overline{\infty} & \text{otherwise} \end{cases}$$

We contend that $\mathbf{B}^* \models \Psi(s \approx t)$. We already have that $\mathbf{B}^* \models \Psi$, so what we need is that $\mathbf{B}^* \models Hs(Kx) \approx Hs(Ky)$ and $\mathbf{B}^* \models Ht(Kx) \approx x$.

Consider the first equation. Because $K^{\mathbf{B}^*}$ produces only the values $b[i]$ for various $i \in \omega$ or $\overline{\infty}$ and because $s^{\mathbf{B}^*}(b[i]) = c[i]$ and $s^{\mathbf{B}^*}(\overline{\infty}) = \overline{\infty}$, we will end up evaluating $H^{\mathbf{B}^*}$ at some $c[i]$ or at $\overline{\infty}$. But since $c[i] \neq d[j]$ for any choices of i and j, we see that $H^{\mathbf{B}^*}(c[i]) = \overline{\infty}$, no matter what value i has. Thus both sides of $Hs(Kx) \approx Hs(Ky)$ evaluate to $\overline{\infty}$ and the equation holds.

Now consider the second equation. Let $u \in B^*$ with $u \neq \overline{\infty}$. Pick $i \in \omega$ so that $K^{\mathbf{B}^*}(u) = b[i]$. Then $t^{\mathbf{B}^*}(b[i]) = d[i]$. Now the definition of $H^{\mathbf{B}^*}$ gives us $H^{\mathbf{B}^*}(d[i]) = u$. When $u = \overline{\infty}$ we see $t^{\mathbf{B}*}(\overline{\infty}) = \overline{\infty}$. The definition of $H^{\mathbf{B}^*}$ gives us $H^{\mathbf{B}^*}(\overline{\infty}) = \overline{\infty} = u$ Putting this together, we have $\mathbf{B}^* \models Ht(Kx) \approx x$, as desired. ■

LEMMA 7.68 (The Base Undecidability Lemma). *Let τ be any recursive signature. Let Γ be any finite set of equations of signature τ. If there are two distinct terms d and $\hat{d}$ in the variable x such that*

- *$\{d, \hat{d}\}$ is ω-universal, and*
- *$\Gamma \vdash d \approx x, \hat{d} \approx x$.*

Then the equational theory based on Γ is base undecidable.

Proof. First put

$$\begin{aligned} d_D &:= d^2(\hat{d}(d(\hat{d}))) \\ d_E &:= d^2(\hat{d}^2(d(\hat{d}))) \\ d_H &:= d^2(\hat{d}^3(d(\hat{d}))) \\ d_K &:= d^2(\hat{d}^4(d(\hat{d}))) \end{aligned}$$

It routine to verify that $\Gamma \vdash d_D \approx x, d_E \approx x, d_H \approx x, d_K \approx x$. That $\{d_D, d_E, d_H, d_K\}$ is ω-universal follows from the fact that $\{d, \hat{d}\}$ is ω-universal and the fact that

$$\{F^2G^{k+1}FGx \mid k \in \omega\}$$

is nonoverlapping and hence ω-universal. To simplify notation we use Tr to denote the translation function with respect to $\{d_D, d_E, d_H, d_K\}$.

Let Ψ be any finite set of equations of a signature σ that provides just two operation symbols D and E, each unary, that is a base for an undecidable 1-variable equational theory. We put

$$\begin{aligned} B(\Gamma, s \approx t) := \operatorname{Tr}(\Psi) \cup \{\operatorname{Tr}(Hs(Kx))(p) \approx \operatorname{Tr}(Hs(Kx))(q) \mid p \approx q \in \Gamma\} \cup \\ \cup \{\operatorname{Tr}(Ht(Kx)) \approx x\} \end{aligned}$$

We reduce the decision problem for the equational theory based on Ψ to the base decidability problem for the equational theory based on Γ. What we need is to show, for each equation $s \approx t$ of signature σ, where x is the only variable to occur in $s \approx t$ and neither s nor t is itself a variable,

$\Psi \vdash s \approx t$ if and only if $B(\Gamma, s \approx t)$ and Γ are bases for the same equational theory.

By the previous lemma and the ω-Universal Translation Lemma, we have

$$\begin{array}{ccc} \Psi \vdash s \approx t & \Longleftrightarrow & \Psi(s \approx t) \vdash s \approx t \\ \Updownarrow & & \Updownarrow \\ \mathrm{Tr}(\Psi) \vdash \mathrm{Tr}(s) \approx \mathrm{Tr}(t) & \Longleftrightarrow & \mathrm{Tr}(\Psi(s \approx t)) \vdash \mathrm{Tr}(s) \approx \mathrm{Tr}(t) \end{array}$$

First, let us suppose that $\Psi \vdash s \approx t$. Then $\mathrm{Tr}(\Psi) \vdash \mathrm{Tr}(s) \approx \mathrm{Tr}(t)$. Consequently, $\mathrm{Tr}(\Psi) \vdash \mathrm{Tr}(Hs(Kx)) \approx \mathrm{Tr}(Ht(Kx))$. From the definition of $B(\Gamma, s \approx t)$ we find that

$$B(\Gamma, s \approx t) \vdash \Gamma.$$

On the other hand, since $\Gamma \vdash d_Q \approx x$ for all $Q \in \{D, E, H, K\}$, we have that $\Gamma \vdash B(\Gamma, s \approx t)$. So we conclude that Γ and $B(\Gamma, s \approx t)$ are bases for the same equational theory.

Now let us suppose that $\Psi \nvdash s \approx t$. So $\mathrm{Tr}(\Psi(s \approx t)) \nvdash \mathrm{Tr}(s) \approx \mathrm{Tr}(t)$. All the equations in $B(\Gamma, s \approx t)$ are substitution instances of equations in $\mathrm{Tr}(\Psi(s \approx t))$, so we have that $\mathrm{Tr}(\Psi(s \approx t)) \vdash B(\Gamma, s \approx t)$. So $B(\Gamma, s \approx t) \nvdash \mathrm{Tr}(s) \approx \mathrm{Tr}(t)$. But we know that

$$\Gamma \vdash \mathrm{Tr}(s) \approx \mathrm{Tr}(t) \approx x.$$

This means that $B(\Gamma, s \approx t)$ and Γ cannot be bases for the same equational theory. ■

The Base Undecidability Theorem

THEOREM 7.69 (The Base Undecidability Theorem (George F. McNulty 1972, 1976a; V. L. Murskiĭ 1971)). *Let T be a finitely based equational theory in a recursive signature such that there is a term t so that $t \approx x \in T$ and either two different unary operation symbols occur in t or some operation symbol of rank at least 2 occurs in t. Then T is base undecidable.*

Proof. We need to construct two distinct terms d and $\hat{d}$ so that $\{d, \hat{d}\}$ is ω-universal and $t \approx x \vdash d \approx x, \hat{d} \approx x$.

We first note that we may assume that x occurs in t. Otherwise $t \approx x \vdash x \approx y$. In this case every equation belongs to T so we may select a suitable replacement for t. We may also suppose that x is the only variable to occur in t, since it does no harm to substitute x for every variable in the equation $t \approx x$.

We consider three cases depending on the term t.

CASE: t **has only unary operation symbols.**
The term t is $F^{n+1}GMF^k x$ where F and G are unary operation symbols, M is a (possibly empty) string of unary operation symbols not ending in F, and n and k are natural numbers.

CLAIM 1.
$F^{n+1}GMF^k x \approx x \vdash F^{n+k+1}GMx \approx x.$

Proof. Suppose $k > 0$, as there is nothing to prove otherwise. In any model of the equation $F^{n+1}GMF^k x \approx x$ the function denoted by F must be bijective and so it is invertible. Hence F^k denotes an invertible function, as well. Thus, $F^{n+k+1}GMx \approx x$ must also be true in the model. ■

This claim allows us to assume, without loss of generality, that $k = 0$ and t is $F^{n+1}GMx$ where M is a (possibly empty) string of unary operation symbols not ending in F. Let $p-1$ be the number of occurrences of F in M. Then define

$$d := (F^{n+1})^{3p}(GM)^{3p}x$$
$$\hat{d} := (F^{n+1})^{2p}(GM)^{2p}(F^{n+1})^{p}(GM)^{p}x$$

It is evident that $t \approx x \vdash d \approx x, \hat{d} \approx x$. To see that d and $\hat{d}$ are nonoverlapping, observe that d begins with a string of $3p(n+1)$ occurrences of F, but after this initial string no strings of consecutive F's can be longer than $p-1$. Likewise $\hat{d}$ begins with a string of $2p(n+1)$ occurrences of F, but after this there is one string of $p(n+1)$ consecutive F's. Elsewhere in $\hat{d}$ no string of consecutive F's is longer than $p-1$. Thus no proper subterm q of d such that q is not x can have a substitution instance $q(u)$ which is of the form $d(w)$ or $\hat{d}(w)$. And likewise for any proper subterm of $\hat{d}$. It is also plain that $d(u)$ and $\hat{d}(w)$ are never the same, regardless of how u and w are chosen. So d and $\hat{d}$ are nonoverlapping and therefore also ω-universal.

CASE: ***t* begins with an operation symbol of rank at least 2.**
We suppose that t is $Qt_0t_1\cdots t_{r-1}$ where r is the rank of Q. So $r \geq 2$. Without loss of generality we will suppose that x occurs in t_0.

Our plan is to construct two terms from t which will be nonoverlapping or at least ω-universal. We will need some easily established symbol counting principles. We are able to limit our attention to those terms in which no variable other than x occurs. For a term q we use $|q|_x$ to denote the number of times the variable x occurs in q and $|q|_c$ to denote the number of occurrences of constant symbols in q. We use $q(w)$ to denote the result of substituting the term w for the variable x in q. Last q^n is defined recursively by $q^0 = x$ and $q^{n+1} = q(q^n)$. The following are easily established:

$$|q(w)|_x = |q|_x|w|_x$$
$$|q^n|_x = |q|_x^n$$
$$|q(w)|_c = |q|_c + |q|_x|w|_c$$
$$|q^n|_c = \begin{cases} |q|_c & \text{if } |q|_x = 0 \\ n|q|_c & \text{if } |q|_x = 1 \\ \frac{|q|_x^n - 1}{|q|_x - 1}|q|_c & \text{if } |q|_x > 1. \end{cases}$$

Let $\mathbf{n}$ be any r-tuple of natural numbers. We say that d is the **associate** of $Qt_0t_1\cdots t_{r-1}$ **corresponding to n** provided d is $Qt^{n_0}(t_0)t^{n_1}(t_1)\cdots t^{n_{r-1}}(t_{r-1})$. We see that $t \approx x \vdash d \approx x$ for every associate d of t. Our plan is to produce two nonoverlapping associates of t by choosing the r-tuples of natural numbers with care.

Let $\mathbf{n}'$ be any r-tuple of natural numbers and let $\hat{d}$ be the associate of t corresponding to $\mathbf{n}'$. Suppose q is a proper subterm of $\hat{d}$ and q is not a variable. So q is a subterm of $t^{n'_i}(t_i)$ for some $i < r$. It follows that q is a substitution instance of a subterm of t that is not a variable. Now let d be the associate of t corresponding to the r-tuple $\mathbf{n}$. Below we will have to exclude the possibility that there are terms w and u so that $q(w) = d(u)$. We can achieve this by instead letting q range over the (finite) collection of subterms of t which are not variables. It is also clear that since d is an associate of t we need only consider those subterms q of t where $q = Qq_0q_1 \cdots q_{r-1}$.

We consider two subcases.

SUBCASE: NO CONSTANT SYMBOLS OCCUR IN t.
We need to find two r-tuples $\mathbf{n}$ and $\mathbf{n}'$ such that the corresponding associates d and $\hat{d}$ of t are nonoverlapping. There are two situations which must be rejected:

(a) q is a proper subterm of $\hat{d}$ which is not a variable and $q(w) = d(u)$ for some terms w and u.

(b) $\hat{d}(u) = d(w)$ for some terms w and u.

In either of these situations, if constant symbols occurred in w or u we could change all the constant symbols to the variable x and obtain a situation without constants. So we assume no constants occur. Moreover, in situation (a) we can instead consider that q is a subterm of t of the form $Qq_0 \cdots q_{r-1}$.

Consider situation (a). For each $i < r$ we have

$$\begin{aligned}
|q_i(w)|_x &= |t^{n_i}(t_i(u))|_x \\
|q_i|_x|w|_x &= |t|_x^{n_i}|t_i|_x|u|_x \\
\frac{|q_0|_x|w|_x}{|q_1|_x|w|_x} &= \frac{|t|_x^{n_0}|t_0|_x|u|_x}{|t|_x^{n_1}|t_1|_x|u|_x} \\
\frac{|q_0|_x}{|q_1|_x} &= \frac{|t|_x^{n_0}|t_0|_x}{|t|_x^{n_1}|t_1|_x} \\
\frac{|q_0|_x|t_1|_x}{|q_1|_x|t_0|_x}|t|_x^{n_1} &= |t|_x^{n_0}
\end{aligned}$$

In this subcase $|t|_x \geq 2$, so let m be a natural number large enough so that

$$\frac{|q_0|_x|t_1|_x}{|q_1|_x|t_0|_x}|t|_x^2 < |t_x|^m$$

for all subterms q_0 and q_1 of t. Here are the desired r-tuples:

$$\begin{aligned}
\mathbf{n} &= \langle m+1, 1, 1, \ldots, 1\rangle \\
\mathbf{n}' &= \langle m, 2, 1, \ldots, 1\rangle
\end{aligned}$$

With these choices, the situation (a) is rejected.

Now consider situation (b). With our $\mathbf{n}$ and $\mathbf{n}'$ where $\hat{d}(u) = d(w)$ we would have

$$\begin{aligned}
t^m(t_0(u)) &= t^{m+1}(t_0(w)) \\
t^2(t_1(u)) &= t^1(t_1(w))
\end{aligned}$$

Since $t^{m+1}(t_0)$ is longer than $t^m(t_0)$, the first equation entails that u is longer than w. However the second equation entails that u is shorter than w, which is impossible since $|t|_x \geq 2$. So the associates $\hat{d}$ and d of t are indeed nonoverlapping, as desired.

SUBCASE: CONSTANT SYMBOLS OCCUR IN t.
As in the last subcase, we need to find two r-tuples $\mathbf{n}$ and $\mathbf{n}'$ such that the corresponding associates d and $\hat{d}$ of t are nonoverlapping. Again, there are two situations which must be rejected:

(a) q is a proper subterm of $\hat{d}$ which is not a variable and $q(w) = d(u)$ for some terms w and u.

(b) $\hat{d}(u) = d(w)$ for some terms w and u.

Moreover, in situation (a) we can instead consider that q is a subterm of t of the form $Qq_0 \cdots q_{r-1}$. Recall that x occurs in t and we have assumed (without loss of generality) that x occurs in t_0.

Since x occurs in t we know that $|t^k(t_0)|_c$ and $|t^k(t_1)|_c$ are strictly increasing functions of k. Pick m so large that

$$|q|_c < |t^m(t_0)|_c \text{ and } |q|_c < |t^m(t_1)|_c \text{ for all subterms } q \text{ of } t.$$

Next pick ℓ so large that

$$|t^\ell(t_0)|_c > \frac{|q_0|_x}{|q_1|_x}|t^{m+1}(t_1)|_c + |q_0|_c$$
$$|t^\ell(t_0)|_x > \frac{|q_0|_x}{|q_1|_x}|t_x^{m+1}|t_1|_x$$

for all subterms q_0 and q_1 of t such that $|q_1|_x \neq 0$.

We take

$$\mathbf{n} := \langle \ell + 1, m, 1, \ldots, 1\rangle$$
$$\mathbf{n}' := \langle \ell, m + 1, 1, \ldots, 1\rangle$$

and let d and $\hat{d}$ be the associates of t corresponding to $\mathbf{n}$ and $\mathbf{n}'$.

Consider situation (a). We have

$$|q_0(w)|_c = |t^{\ell+1}(t_0(u))|_c$$
$$|q_1(w)|_c = |t^m(t_1(u))|_c$$
$$|q_0|_c + |q_0|_x|w|_c = |t^{\ell+1}(t_0)|_c + |t^{\ell+1}(t_0)|_x|u|_c$$
$$|q_1|_c + |q_1|_x|w|_c = |t^m(t_1)|_c + |t^m(t_1)|_x|u|_c$$

In the event that $|q_1|_x = 0$ we get

$$|q_1|_c = |t^m(t_1)|_c + |t^m(t_1)|_x|u|_c$$

But this violates the choice of m. So $|q_1|_x \neq 0$, allowing us to solve the next to the last displayed equation for $|w|_c$. After some manipulation, we obtain

$$\frac{|q_0|_x}{|q_1|_x}|t^m(t_0)|_c + |q_0|_c + \frac{|q_0|_x}{|q_1|_x}|t^m(t_1)|_x|u|_c = |t^{\ell+1}(t_0)|_c + |t^{\ell+1}(t_0)|_x|u|_c + \frac{|q_0|_x}{|q_1|_x}|q_1|_c.$$

But this violates the choice of ℓ. A similar analysis works with $\hat{d}$ in place of d. So the situation (a) is rejected.

Situation (b) is rejected in this subcase the same way it was rejected in the first subcase. So we see that the associates d and $\hat{d}$ of t are nonoverlapping.

CASE: ***t* begins with a unary operation symbol and has operation symbols of rank at least 2.**

In this case, it turns out not to be possible always to obtain two associates of t which are nonoverlapping. But we can still find two associates of t which are ω-universal.

For any term q let q^∂ be the term obtained from q by deleting all the unary operation symbols. Now t^∂ is a term that falls into the last case. Let $\mathbf{n}$ and $\mathbf{n}'$ be two distinct r-tuples whose corresponding associates are nonoverlapping. Let d and $\hat{d}$ be the associates of t corresponding to $\mathbf{n}$ and $\mathbf{n}'$ respectively. Then d^∂ and $\hat{d}^\partial$ are the corresponding associates of t^∂. To see that d and $\hat{d}$ are ω-universal, let $f : \omega \to \omega$ and $g : \omega \to \omega$ be any functions on ω. Since d^∂ and $\hat{d}^\partial$ are nonoverlapping, they are ω-universal. Let $\mathbf{A}$ be an algebra with universe ω and a basic operation for each operation symbol in t^∂ so that $(d^\partial)^{\mathbf{A}} = f$ and $(\hat{d}^\partial)^{\mathbf{A}} = g$. Expand $\mathbf{A}$ to $\mathbf{B}$ by interpreting all the unary operation symbols as the identity function on ω. Then $d^{\mathbf{B}} = f$ and $\hat{d}^{\mathbf{B}} = g$, establishing that d and $\hat{d}$ are ω-universal associates of t. ■

Peter Perkins (1966,1967) proved that the equational theory based on $\{x \approx y\}$, namely the theory $\top$ at the top of lattice $\mathcal{L}_\sigma$ is base undecidable, if σ provides at least two operation symbols of rank 1 or at least one operation symbols of rank at least 2.

As a consequence of the Base Undecidability Theorem, most finitely based equational theories encountered in mathematical practice are base undecidable. The method used to prove the Base Undecidability Theorem can be adapted to prove that various properties of finite sets of equations are undecidable.

Let σ and τ be signatures and let Δ be a set of equations of signature σ and let Γ be a set of equations of signature τ. We say Δ is a **definitional reduct** of Γ provided there is an interpretation L of σ in τ so that for every infinite $\mathbf{A} \models \Delta$ there is an algebra $\mathbf{B} \models \Gamma$ with the same universe as $\mathbf{A}$ so that for every operation symbol Q of signature σ

$$(\mathrm{Tr}_L(Q))^{\mathbf{B}} = (Qv_0v_1 \cdots v_{r-1})^{\mathbf{A}}$$

where r is the rank of Q.

THEOREM 7.70. (George F. McNulty, 1976b) *Let τ be a computable signature that provides at least one operation symbol of rank at least 2. Let Δ be a finite set of equations in some signature without operation symbols of rank 0. Suppose Δ has a nontrivial model. Let $\mathcal{P}$ be a collection of finite sets of equations of signature τ such that*

(i) $\{v_0 \approx v_1\} \in \mathcal{P}$,

(ii) *If $\Gamma \in \mathcal{P}$ and Σ is a finite set of equations so that* $\mathbf{Mod}\,\Gamma = \mathbf{Mod}\,\Sigma$, *then $\Sigma \in \mathcal{P}$, and*

(iii) *if Γ is a finite set of equations of signature τ and Δ is a definitional reduct of Γ, then $\Gamma \notin \mathcal{P}$.*

Then $\mathcal{P}$ is not decidable.

Proof. The proof of this theorem is an adaptation of the proof of the Base Undecidability Theorem and the Base Undecidability Lemma. We use here the notation from those proofs. Let σ be the signature whose operation symbols are those occurring in Δ together with the unary operation symbols D, E, H, and K that do not occur in Δ. According Theorem 7.44 and the examples that follow it, there is an interpretation L of σ in τ that is ω-universal.

For each equation $s \approx t$ in the operation symbols D and E and the variable v_0 let

$$\Omega(s \approx t) := B(v_0 \approx v_1, s \approx t) \cup \mathrm{Tr}_L(\Delta).$$

If $\Psi \vdash s \approx t$, then $\Omega(s \approx t) \vdash v_0 \approx v_1$ and so $\Omega(s \approx t) \in \mathcal{P}$ by conditions (i) and (ii).

Now suppose $\Psi \nvdash s \approx t$. Then $\Psi \cup \Delta \nvdash s \approx t$, since any infinite model of Ψ can be expanded to a model of $\Psi \cup \Delta$.

Let $\mathbf{A}$ be an infinite model of Δ. Expand it to $\mathbf{B}$, a model of $\Psi \cup \Delta$ in which $s \approx t$ fails. Then $\mathrm{Tr}_L(\Psi(s \approx t) \cup \Delta) \nvDash \mathrm{Tr}_L(s \approx t)$. So there is an algebra $\mathbf{C} \models \mathrm{Tr}_L(\Psi(s \approx t) \cup \Delta)$ in which $\mathrm{Tr}_L(s \approx t)$ fails and which has the same universe as $\mathbf{A}$. Notice that all the equations in $\Omega(s \approx t)$ are substitution instances of equations in $\mathrm{Tr}_L(\Psi(s \approx t) \cup \Delta)$ and therefore we have $\mathrm{Tr}_L(\Psi(s \approx t) \cup \Delta) \vdash \Omega(s \approx t)$. So $\mathbf{C} \models \Omega(s \approx t)$. Now $\mathbf{C}$ has the same universe as $\mathbf{A}$ and it witnesses that Δ is a definitional reduct of $\Omega(s \approx t)$. So $\Omega(s \approx t) \notin \mathcal{P}$ by condition (iii).

So $\mathcal{P}$ is undecidable. ■

THEOREM 7.71. (George F. McNulty, 1976b) *Let τ be a computable signature that provides at least one operation symbol of rank at least 2. Let $\mathcal{P}$ be a collection of finite sets of equations of signature τ such that*

(i) *$\mathcal{P}$ is not empty,*

(ii) *If $\Gamma \in \mathcal{P}$ and Σ is a finite set of equations so that $\mathbf{Mod}\,\Gamma = \mathbf{Mod}\,\Sigma$, then $\Sigma \in \mathcal{P}$, and*

(iii) *For each $\Gamma \in \mathcal{P}$ there is a term t in which both v_0 and v_1 occur such that $\Gamma \vdash t \approx v_0$.*

Then $\mathcal{P}$ is not decidable.

Proof. This is again an adaption of earlier proofs, so we retain the notation used before. Let σ be the signature that provides just four operation symbols D, E, H, and K, all unary. Invoking condition (iii), let $\Gamma \in \mathcal{P}$ and p be a term in which both v_0 and v_1 occur so that $\Gamma \vdash p \approx v_0$. As in the proofs of the Base Undecidability Theorem and the Base Undecidability Lemma, there are four distinct terms d_D, d_E, d_H, and d_K that are ω-universal and so that $d_D^\partial, d_E^\partial, d_H^\partial$, and d_K^∂ are nonoverlapping and so that

$$\Gamma \vdash \{d_D \approx v_0, d_E \approx v_0, d_H \approx v_0, d_K \approx v_0\}.$$

We use two ω-universal interpretations of σ in τ. The first, L sends $Q \mapsto d_Q$ for all $Q \in \{D, E, H, K\}$, while the second, L^∂, sends $Q \mapsto d_Q^\partial$ for all $Q \in \{D, E, H, K\}$. Then the set $B(\Gamma, s \approx t)$ uses L while $B^\partial(\Gamma, s \approx t)$ uses L^∂.

Let $s \approx t$ be any equation in which no operation symbols other than D and E occur and in which the only variable to occur is v_0. As in the proof of the Base Undecidability Lemma, if $\Psi \vdash s \approx t$, then $\mathbf{Mod}\, B(\Gamma, s \approx t) = \mathbf{Mod}\,\Gamma$ and so $B(\Gamma, s \approx t) \in \mathcal{P}$, according to condition (ii).

On the other hand, suppose $\Psi \nvdash s \approx t$. As in the proof of Lemma 7.67 there is a countably infinite algebra $\mathbf{B}^*$ that is a model of $\Psi(s \approx t)$ with an element $\overline{\infty}$ so that

$$D^{\mathbf{B}^*}(\overline{\infty}) = \overline{\infty} \quad E^{\mathbf{B}^*}(\overline{\infty}) = \overline{\infty} \quad H^{\mathbf{B}^*}(\overline{\infty}) = \overline{\infty} \quad K^{\mathbf{B}^*}(\overline{\infty}) = \overline{\infty}.$$

and $s \approx t$ fails in $\mathbf{B}^*$.

Now we adapt the proof of Theorem 7.44 according to which a system of nonoverlapping terms is ω-universal. We take the countably infinite universe of $\mathbf{B}^*$ to be the set $T \cup \{\infty^*\}$ where T in the set of all terms of a signature devised for the operation symbols occurring in the terms d_Q^∂ where $Q \in \{D, E, H, K\}$ and $\infty^* \notin T$. The algebra

C will have universe $T \cup \{\infty^*\}$ and operations defined as follows: For each operation symbol P define

$$P^{\mathbf{C}}(b_0,\dots,b_{r-1}) = \begin{cases} \infty^* & \text{if } b_i = \infty^* \text{ for some } i < r \\ Q^{\mathbf{B}^*}(\mathbf{u}) & \text{if } b_i \in T \text{ for all } i < r \text{ and } Pb_0 \cdots = d_Q^{\partial}(\mathbf{u}) \\ & \text{for some } Q \in \{D,E,H,K\} \text{ and some } \mathbf{u} \in T^{\omega} \\ Pb_0 \cdots b_{r-1} & \text{otherwise} \end{cases}$$

where r is the rank of P. That this definition is sound can be established as in the proof of Theorem 7.44. Since $\mathbf{B}^* \models \Psi$ and $\mathbf{B}^* \not\models s \approx t$, we see that $\mathbf{C} \models L^{\partial}(\Psi)$ and $\mathbf{C} \not\models L^{\partial}(s \approx t)$, because v_0 is the only variable occurring in these equations. It was shown in the proof of Lemma 7.67 that $(Hs(Kx))^{\mathbf{B}^*}$ is the constant function with value $\overline{\infty}$, while $(Ht(Kx))^{\mathbf{B}^*}$ is the identity function. Hence, $(Hs(Kx))^{\mathbf{C}}$ is the constant function with value ∞^*, while $(Ht(Kx))^{\mathbf{C}}$ is the identity function. It follows that $\mathbf{C} \models L^{\partial}(\Psi(s \approx t))$.

Now the signature of **C** might not be τ. In the first place, none of the basic operations of **C** is unary. In the second place, there might be operation symbols provided by τ of other ranks that have no correlated basic operation in **C**. We expand **C** to **C**$'$ of signature τ by letting all the unary operation symbols denote the identity function on C and each of the other operation symbols that have no denotation in **C** we let denote the constantly ∞^* function of the proper rank.

Now observe that $\mathbf{C}' \models L(\Psi(s \approx t))$. Also, notice that each equation in $B(s \approx t, \Gamma)$ is a substitution instance of some equation in $L(\Psi(s \approx t))$, so $L(\Psi(s \approx t)) \vdash B(\Gamma, s \approx t)$. This means $\mathbf{C}' \models B(\Gamma, s \approx t)$. Now let r be any term of signature τ in which both v_0 and v_1 occur. Let c be an arbitrary element of C. Then $r^{\mathbf{C}'}(c, \infty^*) = \infty^*$. This means that $B(\Gamma, s \approx t) \nvdash r \approx v_0$. Therefore $B(\Gamma, s \approx t)$ violates condition (iii) and so is not in $\mathcal{P}$.

So $\mathcal{P}$ is undecidable. ■

EXAMPLE 7.72. Let σ be a finite signature that provides at least one operation symbol of rank at least 2. Let $\mathcal{P}$ be the collection of finite sets Γ of equations of signature σ such that if $\mathbf{A} \models \Gamma$ and $\mathbf{A}$ is infinite, then **Con A** is not congruence modular. Then $\mathcal{P}$ is undecidable.

This is a consequence of Theorem 7.70: let Δ be any finite set (M_n) in Theorem 6.32.

EXAMPLE 7.73. Let σ be the signature of group theory: it provides one binary operation symbol and one unary operation symbol. Let $\mathcal{P}$ be the collection of finite bases of nontrivial finite groups. Then $\mathcal{P}$ is undecidable.

This is a consequence of Theorem 7.71. Recall that each finite group is finitely based according to the Oates-Powell Theorem, so $\mathcal{P}$ is not empty.

EXAMPLE 7.74. Let σ be a signature that provides at least one operation symbol of rank at least 2. Let $\mathcal{P}$ be the set of all finite sets of equations of signature σ that are bases of nontrivial varieties with Hobby-McKenzie terms. Then $\mathcal{P}$ is undecidable.

This is a consequence of Theorem 7.71 with the help of Lemma 6.116.

Many other properties of finite sets of equations can be shown to be undecidable by these methods or ones related to them, some with more effort than others. See, for example, (Peter Perkins, 1967; V. L. Murskiĭ, 1971; Douglas D. Smith, 1972; Don Pigozzi, 1976; Cornelia Kalfa, 1986). However, Ralph McKenzie (1975) used completely different methods to prove that the collection of finite sets of equations with a nontrivial finite model is undecidable.

Exercises 7.75

1. Prove that if T is a finitely based equational theory of finite signature and the variety of all models of T is generated by its finite members, then T is decidable. Many equational theories satisfy these criteria. Show this is true for the equational theory of groups and the equational theory of lattices. This is an observation of Trevor Evans.

2. Let $\sigma = \langle \cdot, 1, ^{-1} \rangle$ be our usual signature for group theory, and Γ a finite basis (in this signature) for group theory. Let $\mathbf{A}$ be the group presented by $\langle G \mid R \mid \Gamma \rangle$, and $\mathbf{B}$ the group presented by $\langle H \mid S \mid \Gamma \rangle$, where for convenience we take the generating set G to be disjoint from the generating set H. The exercise here is to find a set Q of relators (equations) such that $\langle G \cup H \mid Q \mid \Gamma \rangle$ is a presentation of $\mathbf{A} \times \mathbf{B}$. If $G \cup H \cup R \cup S$ is finite, then your Q should also be finite. (This situation is atypical; often the product of finitely presented algebras fails to be finitely presented. See (Peter Mayr and Nik Ruškuc, 2018), and references cited there, for a discussion of the general question.)

3. Prove von Dyck's Theorem. It is Theorem 7.59.

4. Prove Theorem 7.60.

5. (Emil L. Post, 1947; A. Markov, 1947) Prove that there is a finitely presented semigroup with an unsolvable word problem. [Hint: Adapt the proof of Theorem 7.65.]

6. Prove that the theory of semigroups is base decidable.

7. Let σ be a finite signature that provides at least one operation symbol of rank at least 2. Prove that the collection of all finite sets of equations of signature σ that are bases of congruence modular varieties is not decidable. [Hint: Apply Theorem 7.71.]

8. Let σ be a finite signature that provides at least one operation symbol of rank at least 2. Prove that the collection of all finite sets of equations of signature σ that are bases of congruence distributive varieties is not decidable.

9. Let σ be a finite signature that provides at least one operation symbol of rank at least 2. Prove that the collection of all finite sets of equations of signature σ that are bases of finite algebras is undecidable.

10. Let σ be a finite signature that provides at least one operation symbol of rank at least 2. Prove that the collection of all finite sets of equations of signature σ whose models are all residually finite is not decidable.

11. Let σ be a finite signature that provides at least one operation symbol of rank at least 2. Let γ be a nontrivial pure lattice congruence identity. Prove the collection of finite bases of varieties of signature σ satisfying γ is not decidable. [Hint: see Theorem 6.113]

7.9. Third Interlude: Residual Bounds

Let $\mathcal{K}$ be a class of algebras of the same signature. The **residual bound** of $\mathcal{K}$ is the least cardinal κ so that every subdirectly irreducible algebra in $\mathcal{K}$ has fewer than κ elements. We also refer to this cardinal as the **residual character** of $\mathcal{K}$. If no such κ exists, then we say $\mathcal{K}$ is **residually large** and we put ∞ as the residual bound. Most varieties are residually large.

We say $\mathcal{K}$ is **residually finite** if the residual bound κ is either finite or equals ω. That is, $\mathcal{K}$ is residually finite provided its residual bound is countable. We say $\mathcal{K}$ is **residually very finite** if the residual bound κ is finite. We say $\mathcal{K}$ is **residually small** if its residual bound is κ for some cardinal κ.

The next two theorems are offered here for information.

THEOREM 8.45 (Walter Taylor, 1972).
A variety $\mathbf{V}$ is residually small if and only if $\mathbf{V}$ has residual bound of $\leq (2^\kappa)^+$, where κ is the cardinality of the set of terms of the signature.

This theorem will be proved, with the help of the Compactness Theorem, in Chapter 8.

A variety with residual bound $\aleph_0$ has arbitrarily large finite subdirectly irreducible algebras but no infinite subdirectly irreducible algebra. A variety with residual bound $\aleph_1$ must have a countably infinite subdirectly irreducible algebra but no uncountable subdirectly irreducible algebra.

THEOREM 7.76 (McKenzie and Shelah, 1970). *Every variety of countable signature that has uncountable subdirectly irreducible algebras must have a subdirectly irreducible algebra of cardinality at least $2^{\aleph_0}$.*

The proof of this theorem requires a more thorough development of model theory than we can provide in Chapter 8.

Notice the following:

I. The variety of one-element algebras has no subdirectly irreducible algebra. This variety will have residual bound 0, since 0 is certainly larger than the cardinality of any subdirectly irreducible algebra in the variety.
II. The cardinal 1 cannot be a residual bound of a variety, since this entails that the variety has a subdirectly irreducible algebra of cardinality 0. Algebras cannot have cardinality 0.
III. The cardinal 2 cannot be the residual bound of a variety because there are no 1-element subdirectly irreducible algebras.

IV. There are varieties of countable signature with residual bounds of

$$0, 3, 4, 5, \ldots, \aleph_0, \aleph_1, (2^{\aleph_0})^+,$$

as well as varieties that are residually large.

In connection with IV above, it should be noted that varieties exemplifying many of the possible residual bounds have been known for a long time. The variety of distributive lattices and the variety of Boolean algebras, have residual bound 3, whereas the variety generated by the 8-element quaternion group is residually large. In fact, most instances can be realized by varieties generated by small finite algebras of finite signature, as indicated in the exercises. In §7.10, we will provide an eight-element algebra of finite signature with residual bound $\aleph_1$. Exercise 7.79.2 provides an example of a variety with residual bound $(2^{\aleph_0})^+$, in the signature that provides just two operation symbols, both unary. We also note that the (modular) lattices $\mathbf{M}_n$ where $3 \leq n$

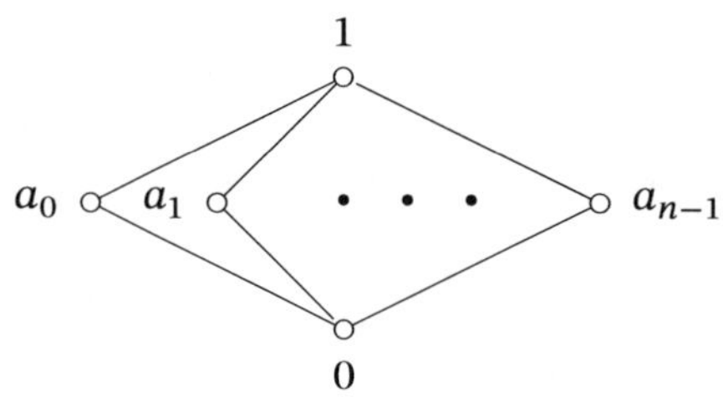

FIGURE 7.12. The Lattice $\mathbf{M}_n$

are simple—by Dedekind's Transposition Principle (Theorem 2.27 in Volume I). So they are subdirectly irreducible. By Corollary 8.58 to Jónsson's Lemma, we see that $\boldsymbol{HSP}\,\mathbf{M}_n$ has residual bound $n + 3$. This means that only the bounds 4 and 5 remain to be represented.

Concerning the residual bound $\aleph_0$, there is a simple but informative theorem that leads to one of the long-standing open problems about residual bounds.

THEOREM 7.77 (The Theorem of W. Dziobiak (1981)). *Let $\mathcal{V}$ be a locally finite variety, let $\mathbf{S} \in \mathcal{V}$ be subdirectly irreducible, and let $\mathbf{B}$ be a finite subalgebra of $\mathbf{S}$. Then $\mathbf{B}$ is embeddable into a finite subdirectly irreducible algebra in $\mathcal{V}$.*

Proof. Let (a, b) be a critical pair for $\mathbf{S}$. Since $\mathcal{V}$ is locally finite, at the cost of enlarging B if necessary, we can suppose that $a, b \in B$. Now for each two element subset $\{c, d\} \subseteq B$ we choose a finite $F_{c,d} \subseteq S$ which consists of all the new constants used along some Mal'tsev chain witnessing that $(a, b) \in \mathrm{Cg}^{\mathbf{S}}(c, d)$. Let $\mathbf{C}$ be the subalgebra of $\mathbf{S}$ generated by $B \cup \bigcup_{c \neq d} F_{c,d}$. This subalgebra is finite since the variety is locally finite. Now let θ be a congruence of $\mathbf{C}$ maximal with respect to not collapsing a and b. Hence θ is a meet irreducible congruence, so $\mathbf{C}/\theta$ is subdirectly irreducible. But, by construction, θ cannot collapse any two distinct elements of B. Therefore the quotient map embeds $\mathbf{B}$ into the finite subdirectly irreducible algebra $\mathbf{C}/\theta$. ■

COROLLARY 7.78 (Quackenbush's Theorem (1971)). *Every locally finite variety with an infinite subdirectly irreducible algebra must have arbitrarily large finite subdirectly irreducible algebras.* ■

Quackenbush actually deduced this corollary from the theorem below, found in §8.3 where its proof is an application of the Compactness Theorem.

THEOREM 8.43 (Alden F. Pixley, 1970).
Let $\mathcal{V}$ be a locally finite variety. Let $\mathcal{K}$ be a finite collection of finite algebras from $\mathcal{V}$. Let **B** *be an algebra of the same signature as $\mathcal{V}$. If every finitely generated subalgebra of* **B** *can be subdirectly represented via algebras in $\mathcal{K}$, then* **B** *can be subdirectly represented by algebras in $\mathcal{K}$.*

In 1971, when Robert Quackenbush published the result in the Corollary above, he asked whether the converse was true. That is

THE CONJECTURE OF QUACKENBUSH

Must every finitely generated variety (of finite signature), that has arbitrarily large finite subdirectly irreducible algebras, have an infinite subdirectly irreducible algebra?

Ralph McKenzie (1996b) published a counterexample: a finite algebra of infinite signature that generates a variety with residual bound $\aleph_0$. The problem remains open for finite signatures.

Keith A. Kearnes and Ross Willard (1999) proved that if $\mathcal{V}$ is a congruence meet-semidistributive variety of finite signature that has arbitrarily large finite subdirectly irreducible algebras, then $\mathcal{V}$ must have an infinite subdirectly irreducible algebra.

Ralph Freese and Ralph McKenzie (1981) proved that if a congruence modular variety is residually small and generated by a finite algebra, then the variety must have a finite residual bound. We prove this in §11.7.

So any finite algebra of finite signature that provides a counterexample to the Quackenbush Conjecture must generate a variety that is neither congruence modular nor congruence meet-semidistributive.

The Variety Generated by McKenzie's Algebra R is Residually Large

From Section 7.3, recall McKenzie's algebra **R**, diagrammed in Figure 7.13.

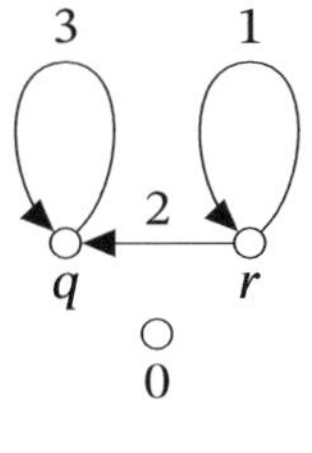

FIGURE 7.13. McKenzie's Automatic Algebra **R**

This is an automatic algebra. It has just one operation and that operation is a two-place operation, which we denote by juxtaposition. The elements of this algebra fall into the set $\{1, 2, 3\}$ of letters and the disjoint set $\{q, r\}$ of states, with an additional default element 0. The operation is defined so that

$$1r = r \qquad 2r = q \qquad 3q = q$$

with all other products resulting in 0. We proved that the variety generated by **R** contains a shift automorphism algebra. As a consequence, we know that ***HSP*** **R** has an infinite subdirectly irreducible algebra, according to the Shift Automorphism Theorem. However it is easy to construct subdirectly irreducible algebras in this variety that have any cardinality greater than 1.

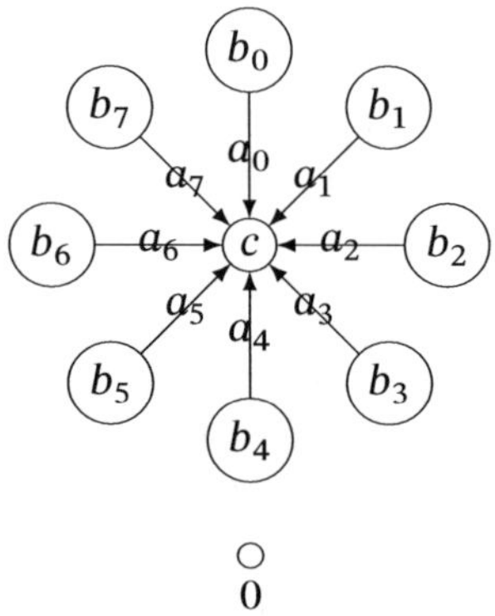

FIGURE 7.14. The Subdirectly Irreducible Algebra $\mathbf{S}_8$

Figure 7.14 gives a diagram of a subdirectly irreducible algebra with 18 elements in this variety and it is plainly possible to replace 8 with any cardinal greater than 1 and also to vary and elaborate this diagram to obtain quite complicated subdirectly irreducible algebras of arbitrary infinite cardinalities. The node in the middle of the diagram is c. Moreover, the operation is defined by following the arrows, just as it was for **R**.

Fact 0. $\mathbf{S}_8$ *is subdirectly irreducible.*

Proof. We argue that $(c, 0)$ is a critical pair. What we must show is that if $u \neq v$ in S_8, then any congruence θ that collapses u and v must also collapse c and 0. Notice that $c = a_i b_i$ for all i and, moreover, that this product is the only nonzero product that involves either b_i or a_i. Without loss of generality, we can suppose that $u \neq 0$. There are three cases: $u = a_i$ for some i, $u = b_i$, for some i, and $u = c$. We consider just the first case and the last case, leaving the middle case in the hands of our readers.

Case: $u = a_i$

We know that $c = a_i b_i$ and $0 = v b_i$, since $v \neq a_i = u$. Thus $c = a_i b_i \ \theta \ v b_i = 0$, as desired.

Case: $u = c$

If $v = 0$ there is nothing to prove. So consider that $c = u \neq v \neq 0$. So $v = a_i$ for some i or $v = b_i$ for some i. We only deal with first alternative. So we have $c = a_i b_i = v b_i \ \theta \ u b_i = c b_i = 0$, as desired. ■

Fact 1. $\mathbf{S}_8 \in \boldsymbol{HSP}\,\mathbf{R}$.

Proof. We define ω-tuples α_i, β_i, and γ, for $i < 8$, like we did in Section 7.3. Were we to replace 8 by another cardinal κ, this construction could be modified (we would have to replace ω by a suitable ordered set). Actually, in the present case, we could replace ω by 8.

$$\begin{array}{rcccccccccc} \alpha_i := & q & q & q & q & r & q & q & q & \cdots \\ \beta_i := & 3 & 3 & 3 & 3 & 2 & 3 & 3 & 3 & \cdots \\ \gamma := & q & q & q & q & q & q & q & q & \cdots \end{array}$$

where the r and the 2 occur at the ith position. As in Section 7.3, we consider the subalgebra $\mathbf{B}$ of $\mathbf{R}^\omega$ generated by the tuples displayed above. The equivalence relation θ that lumps together all tuples in B that contain a 0 and isolates everything else is a congruence. Moreover, $\mathbf{S}_8 \cong \mathbf{B}/\theta$. So $\mathbf{S}_8 \in \boldsymbol{HSP}(\mathbf{R})$. ■

So $\boldsymbol{HSP}\,\mathbf{R}$ is residually large.

Exercises 7.79

1. (Walter Taylor, 1972) Prove that any variety of unary algebras is residually small.

2. (Walter Taylor, 1972) Let $\mathbf{A} = \langle 2^\omega, f, g\rangle$ where

$$f(\langle a_0, a_1, \ldots\rangle) = \langle a_1, a_2, \ldots\rangle$$
$$g(\langle a_0, a_1, \ldots\rangle) = \langle a_0, a_0, \ldots\rangle$$

for all $\langle a_0, a_1, \ldots\rangle \in 2^\omega$. Prove that $\mathbf{A}$ is subdirectly irreducible. Hence, by the exercise above, $\boldsymbol{HSP}\,\mathbf{A}$ has residual bound $(2^\omega)^+$.

7.10. A Finite Algebra of Residual Character $\aleph_1$

During the course of developing material in this section we will deal with algebras that have many basic operations. Among them will always be three denoted by $\cdot$, $\wedge$, and 0. The operation symbol $\cdot$ has rank 2 and will usually denote an automatic algebra operation. The operation symbol $\wedge$ will always be a meet-semilattice operation and 0 will always denote the least element with respect to $\leq$, the underlying semilattice order. Meet semilattices of height one are referred to as *flat semilattices*. 0 is an *absorbing element* for the product $\cdot$ in the sense that $0 \cdot x \approx x \cdot 0 \approx 0$ always holds. The product also satisfies $(x \cdot y) \cdot z \approx 0 \approx x \cdot x$. Therefore, only right associated products can produce results other than 0. For the time being we assume that the remaining operations are term operations built up from $\wedge$, $\cdot$, and 0.

Let $Q_{\mathbb{Z}} = \{0\} \cup \{a_p : p \in \mathbb{Z}\} \cup \{b_p : p \in \mathbb{Z}\}$, where all the a_p's and b_q's are distinct and different from 0. The algebra $\langle Q_{\mathbb{Z}}, \wedge, 0, \ldots\rangle$ is a height 1 semilattice with least element 0. Let $\mathbf{Q}_{\mathbb{Z}} := \langle Q_{\mathbb{Z}}, \cdot, \wedge, 0\rangle$. The product in $\mathbf{Q}_{\mathbb{Z}}$ is defined so that $a_p \cdot b_{p+1} = b_p$ for all $p \in \mathbb{Z}$, with all other products 0. Figure 7.15 gives a diagram of $\mathbf{Q}_{\mathbb{Z}}$.

The operation $\cdot$ could be referred to as an *automatic algebra operation* and the algebra $\langle Q_{\mathbb{Z}}, \cdot, 0\rangle$ as an *automatic algebra*. Any directed labeled graph gives rise to an *automatic*

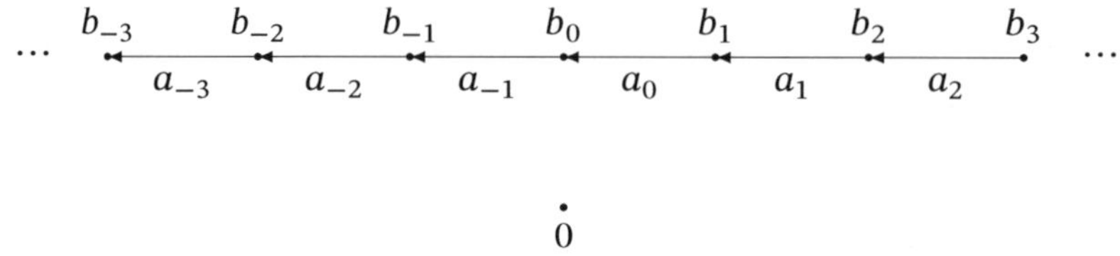

FIGURE 7.15. The algebra $\mathbf{Q}_{\mathbb{Z}}$

algebra provided no two edges directed away from the same vertex have the same label. Specifically, for such a directed graph, $\cdot$ is an *automatic algebra operation on A* provided the elements of A fall into three disjoint sets—vertex elements, edge labels, and a default element 0—and $b = a \cdot d$ holds when a labels an edge directed from vertex d to vertex b, with all other $\cdot$ products result in the default element 0. The algebra $\mathbf{Q}_{\mathbb{Z}}$ is called a *flat automatic algebra*, since the ordering arising from the semilattice operation $\wedge$ is of height 1. In the next section more complicated ternary operations rooted in digraphs with doubly labeled edges will be used to encode Turing machines and their computations.

$\mathbf{Q}_\omega$ denotes the subalgebra of $\mathbf{Q}_{\mathbb{Z}}$ with universe $\{0\} \cup \{a_p : p \in \omega\} \cup \{b_p : p \in \omega\}$. Likewise, for each natural number n, $\mathbf{Q}_n$ denotes the subalgebra with universe $\{0\} \cup \{a_p : 0 \leq p < n\} \cup \{b_p : 0 \leq p \leq n\}$. This algebra has $2n + 2$ elements.

THEOREM 7.80 (Ralph McKenzie 1996b). *Let $\mathbf{Q}_{\mathbb{Z}}$ be an algebra as described above (in particular with the given basic operations $0, \wedge$, and $\cdot$ and with all other basic operations being term operations built from these). The following hold:*

(i) *$\mathbf{Q}_\omega$ is subdirectly irreducible, as is each $\mathbf{Q}_n$;*
(ii) *$\mathbf{Q}_{\mathbb{Z}}$ generates a locally finite variety;*
(iii) *$\mathbf{Q}_{\mathbb{Z}}$ is inherently nonfinitely based.*

Proof. Indeed, for (i) we contend that $(0, b_0)$ belongs to every nontrivial congruence. To see this, let θ be a nontrivial congruence. First, suppose that $b_0 \ \theta \ c$ with $b_0 \neq c$. Then $b_0 = b_0 \wedge b_0 \ \theta \ b_0 \wedge c = 0$. Next, suppose that $0 < p$ and that θ collapses b_p to c where $b_p \neq c$. We obtain $b_{p-1} = a_{p-1} \cdot b_p \ \theta \ a_{p-1} \cdot c = 0$. So, inductively we have $b_0 \ \theta \ 0$. Finally, suppose that θ collapses a_p to c where $a_p \neq c$. Then we obtain $b_p = a_p \cdot b_{p+1} \ \theta \ c \cdot b_{p+1} = 0$, and so also $b_0 \ \theta \ 0$.

For (ii), note that no a_p results from the product operation and that b_p can only result from the product $a_p \cdot b_{p+1}$. So if S is any subset of $\mathbf{Q}_{\mathbb{Z}}$, then $S \cup \{0\} \cup \{b_p : a_p \in S\}$ is a subuniverse of $\mathbf{Q}_{\mathbb{Z}}$. Thus the subuniverse of $\mathbf{Q}_{\mathbb{Z}}$ generated by a set of n elements will have no more than $2n + 1$ elements, and usually a lot less. Hence $\mathbf{Q}_{\mathbb{Z}}$ generates a locally finite variety.

To establish (iii), just observe that the obvious map is a shift automorphism, so that the Shift Automorphism Theorem applies. ■

Our objective is to devise a finite algebra which generates a variety in which, apart from a few small algebras, the only subdirectly irreducible algebras are the $\mathbf{Q}_n$'s and

$\mathbf{Q}_\omega$. As we saw at the end of §7.9, the algebra **R** will not do, but we can adapt it to our purpose.

Finite subdirectly irreducibles generated by finite flat algebras

In this subsection we will suppose that **A** is a finite *flat* algebra (that is, an algebra with basic operations $\wedge$ and 0 so that the algebra has the structure of a meet-semilattice of height one with least element 0 with perhaps other basic operations) and that **S** is a finite subdirectly irreducible algebra in the variety generated by **A**.

Now according to Birkhoff's HSP Theorem, **S** will always arise as a quotient of some **B**, which is in turn a subalgebra of $\mathbf{A}^T$ for some T. Since **S** is subdirectly irreducible, we know that there is a strictly meet irreducible $\theta \in \operatorname{Con}\mathbf{B}$ such that $\mathbf{S} \cong \mathbf{B}/\theta$. The restriction of strict meet irreducibility means that there is a congruence ρ of **B** so that any congruence properly above θ must include ρ. In view of the Correspondence Theorem, the filter of congruences that include θ is isomorphic to **Con A** and ρ is matched with the monolith of **S** under this isomorphism. It is more convenient to work with **B** than with **S**. Since **S** is finite, we can choose T to be finite. Indeed, in this subsection we assume the following:

- $\mathbf{B} \subseteq \mathbf{A}^T$
- $\theta \in \operatorname{Con}\mathbf{B}$
- θ is strictly meet-irreducible in **Con B**.
- $\mathbf{S} \cong \mathbf{B}/\theta$
- T is as small as possible for representing **S** in this way.

In particular this last condition entails that if $t \in T$, then there must be $u, v \in B$ so that $(u, v) \notin \theta$ but $u(s) = v(s)$ for all $s \in T - \{t\}$. Our effort at understanding the finite subdirectly irreducible **S** is largely focused on θ.

First, we locate an element b_0 in B whose image in **S** will be part of a critical pair. Since **B** has a semilattice operation, there are elements $u, v \in B$ with $u < v$ and (u, v) critical over θ, that is $(u, v) \in \rho$. Using the finiteness of B pick p to be minimal, with respect to the semilattice order, among all those $v \in B$ such that (u, v) is critical over θ for some $u < v$.

Fact 0. *If $w < p$, then $(w, p) \notin \theta$.*

Proof. Suppose $w < p$ but $w \;\theta\; p$. Pick $u < p$ with (u, p) critical over θ. Then $w = p \wedge w \;\rho\; u \wedge w$. But this means that either $(w, u \wedge w) \in \theta$ or that $(w, u \wedge w)$ is critical over θ. So by the minimality of p, we have $u \wedge w \;\theta\; w$. But then $u = u \wedge p \;\theta\; u \wedge w \;\theta\; w \;\theta\; p$, contradicting $(u, p) \notin \theta$. ∎

Now for each $t \in T$ pick $(x, y) \in B^2 - \theta$ so that $x(t) \neq y(t)$ but $x(s) = y(s)$ for all $s \in T - \{t\}$. Pick $u < p$ so that (u, p) is critical over θ. This means that $(u, p) \in \theta \vee \operatorname{Cg}^{\mathbf{B}}(x, y)$. Then according to Mal'tsev's Congruence Generation Theorem there is a finite sequence $e_0, e_1, \ldots, e_n$ of elements of B, finite sequence of translations $\lambda_0, \ldots, \lambda_{n-1}$ of **B**, and a finite sequence of two-element subsets $\{z_0, w_0\}, \ldots, \{z_{n-1}, w_{n-1}\}$ each belonging to $\theta \cup \{(x, y)\}$ such that

$$u = e_0 \quad \{e_i, e_{i+1}\} = \{\lambda_i(z_i), \lambda_i(w_i)\} \text{ for all } i < n \quad e_n = p.$$

But now, meeting every element in the sequence of the sequence of e_i's with p, we have

$$\begin{gathered} u = u \wedge p = e_0 \wedge p \\ \{e_i \wedge p, e_{i+1} \wedge p\} = \{\lambda_i(z_i) \wedge p, \lambda_i(w_i) \wedge p\} \text{ for all } i < n \\ e_n \wedge p = p \wedge p = p \end{gathered}$$

Since $u < p$, there is some $i < n$ so that $p \in \{\lambda_i(z_i) \wedge p, \lambda_i(w_i) \wedge p\}$ where $\lambda_i(z_i) \wedge p \neq \lambda_i(w_i) \wedge p$. Let χ_t denote the element of $\{\lambda_i(z_i) \wedge p, \lambda_i(w_i) \wedge p\}$ which is different from p. Evidently $\chi_t < p$. By Fact 0 we see that $(\chi_t, p) \notin \theta$. Hence, $(z_i, w_i) = (x, y)$ and $\{p, \chi_t\} = \{\lambda_i(x) \wedge p, \lambda_i(y) \wedge p\}$. From this construction we obtain:

- $\chi_t(s) = p(s)$ for all $s, t \in T$ with $s \neq t$.
- $\chi_t(t) < p(t)$ for all $t \in T$.
- $\chi_t(t) = 0$ and $0 < p(t)$ for all $t \in T$.

The last item listed above is a consequence of the flatness of $\mathbf{A}$. Thus, χ_t agrees with p at all coordinates with the exception of t, where χ_t is 0 while p is not 0. So χ_t is uniquely determined by p and t (and is independent of the choices of x, y, and λ_i made above). We will eventually see—once enough is specified about $\mathbf{A}$—that p is also uniquely determined.

Fix $t_0 \in T$ so that $u \leq \chi_{t_0}$ for some $u < p$ for which (u, p) is critical over θ. Let $q = \chi_{t_0}$.

Fact 1. *p is a maximal element of $\mathbf{A}^T$. $\chi_t \in B$ and p covers χ_t in $\mathbf{A}^T$ for all $t \in T$. (q, p) is critical over θ. Finally, if $u \in A^T$ and $u < p$, then $u \in B$.*

Proof. Essentially, Fact 1 gathers the conclusions we drew above. To see that (q, p) is critical, notice $(q, p) \notin \theta$ according to Fact 0. Let $u \leq q < p$ with (u, p) critical over θ. Then we have $p \rho u = q \wedge u \;\rho\; q \wedge p = q$. The elements of A^T less than or equal to p form a Boolean algebra in which every element is a meet of the coatoms χ_t. ■

Fact 2. *If $p \;\theta\; x$, then $p = x$.*

Proof. Suppose $p \;\theta\; x$. Meeting both sides with p we also get $p \;\theta\; p \wedge x$. From Fact 0, we conclude that $p \not> p \wedge x$. Thus $p \leq x$. But since p is a maximal element, we arrive at $p = x$. ■

Fact 3. *$x \;\theta\; y$ if and only if $\mu(x) = p \Leftrightarrow \mu(y) = p$ for all translations μ of $\mathbf{B}$.*

Proof. In the forward direction the result follows from Fact 2.

Now for the converse direction, suppose $(x, y) \notin \theta$. By Fact 1, we know $(q, p) \in \theta \vee \mathrm{Cg}^{\mathbf{B}}(x, y)$. Now repeating the analysis that led to the χ_t's we obtain, in fact, a translation $\mu = \lambda \wedge p$ so that $\mu(x) \neq \mu(y)$ but $p \in \{\mu(x), \mu(y)\}$. ■

Fact 4. *If $x < p$, then $(x, x \wedge q) \in \theta$.*

Proof. (p', p) is critical over θ by Fact 1, so $x = x \wedge p \ \varphi\ x \wedge q$, for all $\varphi > \theta$. Hence, either $(x, x \wedge q) \in \theta$ or $(x, x \wedge q)$ is critical over θ. Since $p > x \geq x \wedge q$, it follows from the minimality of p that $x\ \theta\ x \wedge q$. ■

Suppose that x, y, and $z \in B$. Then $(x \wedge y)$ and $(x \wedge z)$ also belong to B and the element x is a common upper bound. Recalling that **B** has the structure of a finite $\wedge$-semilattice, it follows that $(x \wedge y)$ and $(x \wedge z)$ must have a least upper bound—we denote this least upper bound by $(x \wedge y) \vee (x \wedge z)$.

Fact 5. $\mathbf{S} \in \boldsymbol{HS}\mathbf{A}$ *or the ternary operation* $F(x, y, z) = (x \wedge y) \vee (x \wedge z)$ *is not a polynomial of* **B**.

Proof. Suppose $\mathbf{S} \notin \boldsymbol{HS}\mathbf{A}$. Then T has at least two elements. Let $t_1 \in T$ with $t_0 \neq t_1$. Let $q' = \chi_{t_1}$. Since $q' < p$ we have by Fact 4 that $q'\ \theta\ q' \wedge q$. But then, were $(x \wedge y) \vee (x \wedge z)$ a polynomial of **B**, we would have $p = (p \wedge q) \vee (p \wedge q')\ \theta\ (p \wedge q) \vee (p \wedge q \wedge q') = q$. Since $(p, q) \notin \theta$, we conclude that $(x \wedge y) \vee (x \wedge z)$ is not a polynomial. ■

Fact 5 reveals that our investigation of (finite) subdirectly irreducible algebras can be split in two. Since **A** is finite, a complete description of the subdirectly irreducible algebras in HS**A** can be devised given a description of **A**. We only note the obvious upper bound on their cardinality. Most of our effort will concern the alternative case when $(x \wedge y) \vee (x \wedge z)$ is not a polynomial of **B**. It is the subdirectly irreducible algebras arising from these algebras that we want to show must be isomorphic to our $\mathbf{Q}_n$'s.

Here is a lemma that simply gathers together the most salient of the facts just listed.

LEMMA 7.81. *Suppose that* **A** *is a finite flat algebra and that* **S** *is a finite subdirectly irreducible algebra in* $\boldsymbol{HSP}\,\mathbf{A}$. *Choose* T, **B**, *and* $\theta \in \operatorname{Con} \mathbf{B}$ *so that*

- **B** *is a subalgebra of* $\mathbf{A}^T$,
- θ *is (strictly) meet irreducible in* Con **B**.
- $\mathbf{S} \cong \mathbf{B}/\theta$, *and*
- T *is as small as possible subject to the conditions above.*

Then there are elements $p, p' \in B$ *such that*

(i) (p', p) *is critical over* θ,
(ii) $p/\theta = \{p\}$,
(iii) p *is a maximal element of* A^T *(so* $p(s) > 0$ *for all* $s \in T$*), and*
(iv) *for all* $x, y \in B$, $x\ \theta\ y$ *if and only if* $\mu(x) = p \Leftrightarrow \mu(y) = p$ *for all translations* μ *of* **B**.

Moreover, $\mathbf{S} \in \boldsymbol{HS}\mathbf{A}$ *or* $(x \wedge y) \vee (x \wedge z)$ *is not a polynomial of* **B**. ■

The Eight Element Algebra A

McKenzie's six-element algebra **R** which was constructed at the end of §7.9 generates a variety with a lot of finite subdirectly irreducible algebras in addition to the $\mathbf{Q}_n$'s. We must modify this algebra to eliminate subdirectly irreducible algebras like $\mathbf{S}_8$, whose diagrams are not (finite) directed paths. Evidently, for our subdirectly irreducible algebras, we need a kind of unique factorization property:

$$a \cdot b = c \cdot d \neq 0 \Rightarrow a = c \text{ and } b = d.$$

To accomplish this we are going to add some new basic operations and some new elements to the algebra **R**, but we need to take some care since we want $\mathbf{Q}_{\mathbb{Z}}$ to remain essentially unchanged and still to belong to the variety generated by the finite algebra we are trying to devise.

To obtain the unique factorization property we introduce the new basic 4-place operation U^0:

$$U^0(x,y,z,w) = \begin{cases} xy & \text{if } xy = zw \neq 0 \text{ and } x = z \text{ and } y = w, \\ \overline{xy} & \text{if } xy = zw \neq 0 \text{ and either } x \neq z \text{ or } y \neq w, \\ 0 & \text{otherwise.} \end{cases}$$

At the moment, we should understand that the first case corresponds to the situation when the unique factorization property prevails, the second case corresponds to the failure of the unique factorization property, and the remaining case is just a default. For the moment, $\overline{xy}$ is simply a reminder that the output in this case should depend on xy but differ from xy. Our hope is to obtain the unique factorization property by forcing the first case to happen. In essence, this means preventing the second case. For this purpose we introduce a new basic 5-place operation S_2:

$$S_2(u,v,x,y,z) = \begin{cases} (x \wedge y) \vee (x \wedge z) & \text{if } u = \bar{v}, \\ 0 & \text{otherwise.} \end{cases}$$

Recall the algebra **B**. In **B** we know from Fact 5 that $(x \wedge y) \vee (x \wedge z)$ cannot be a polynomial. So S_2 is designed to prevent B from having elements u and v so that $u = \bar{v}$. This, in turn, will prevent the second case in the definition of U^0 from arising.

To give more sense to this, notice that in the six element algebra **R**, a product xy could have only q, r, or 0 as a value. So we introduce two elements $\bar{q}$ and $\bar{r}$ in addition to the six with which we have been dealing. Further, we stipulate that $\bar{u} = q$ if $u = \bar{q}$ and likewise $\bar{u} = r$ if $u = \bar{r}$. In this way, both U^0 and S_2 have unambiguous definitions, once the product and meet have been extended to operations on the new set with eight elements.

These two additional operations and two additional elements are not quite enough. We need the operation below.

$$S_1(u,v,x,y,z) = \begin{cases} (x \wedge y) \vee (x \wedge z) & \text{if } u \in \{1,3\}, \\ 0 & \text{otherwise.} \end{cases}$$

The role of S_1, as we will see, is to ensure that our finite subdirectly irreducible algebra **S** has another property that each $\mathbf{Q}_n$ has—namely, that the labels of the edges are not repeated. Last, here are the operations J and J' which are 3-place operations:

$$J(x,y,z) = \begin{cases} x & \text{if } x = y \neq 0, \\ x \wedge z & \text{if } x = \bar{y}, \\ 0 & \text{otherwise.} \end{cases} \qquad J'(x,y,z) = \begin{cases} x \wedge z & \text{if } x = y \neq 0, \\ x & \text{if } x = \bar{y}, \\ 0 & \text{otherwise.} \end{cases}$$

The role of these operations is less forthright. Since we are really working inside a subalgebra of a direct power, we have to contend with coordinatewise properties. The role of these last two operations is to ensure that we fall into the "good" case at every coordinate.

We are led to a flat algebra **A** with eight elements and eight basic operations.

The universe is $A = \{0\} \cup \{1,2,3\} \cup \{q,\bar{q},r,\bar{r}\}$. Set $U = \{1,2,3\}$ and $W = \{q,\bar{q},r,\bar{r}\}$. We regard $\bar{\ }$ as an involution on W. The basic operations of **A** are denoted by

$$0, \wedge, \cdot, J, J', U^0, S_1, \text{ and } S_2.$$

$\langle A, \wedge, 0\rangle$ is a flat semilattice with least element 0. The operation $\cdot$ is defined to give the default value 0 except when

$$\begin{array}{ll} 1\cdot r = r & 1\cdot \bar{r} = \bar{r} \\ 2\cdot r = q & 2\cdot \bar{r} = \bar{q} \\ 3\cdot q = q & 3\cdot \bar{q} = \bar{q} \end{array}$$

This is an automatic operation. Ordinarily, we represent the product $\cdot$ simply by juxtaposition. The diagram of the automatic algebra **A** is given in Figure 7.16.

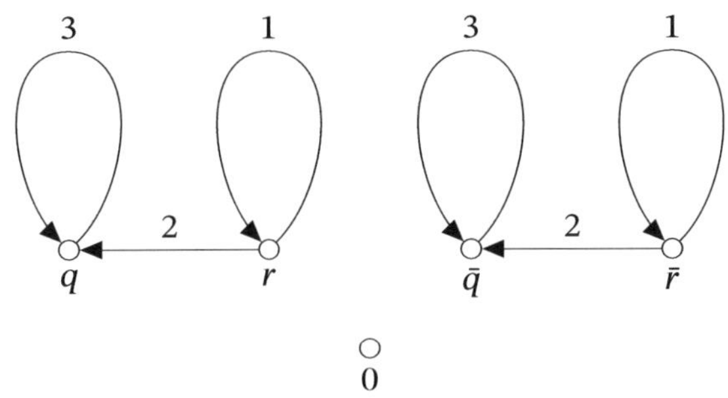

FIGURE 7.16. The eight element flat algebra **A**

The following fact is evident from the definition of the product.

Fact 6. *If λ is a basic translation on* **A** *associated with the product $\cdot$, and $\lambda(a) = \lambda(b) \neq 0$, then $a = b$. The same is true for every translation built using only the product.*

Properties of B based on the eight element algebra A

With the description of our eight element algebra **A** in hand, we continue to develop facts about **B** and its congruence θ. Denote by B_1 the set consisting of p and all its factors with respect to the product $\cdot$. That is

$$B_1 = \{u : \lambda(u) = p \text{ for some nonconstant translation } \lambda \text{ of } \mathbf{B} \text{ built only from the product}\}$$

So $u \in B_1$ if and only if $u = p$ or $u = c_i$ for some factorization $p = c_0c_1\cdots c_m$ (where this latter product is associated to the right).

Let B_0 denote the set of those tuples in B which contain at least one 0. Evidently, $B_0 \subseteq B - B_1$. It is also clear that if $\mathbf{S} \notin HS\mathbf{A}$, then the ranges of the operations S_1 and S_2 are contained in B_0 and hence in $B - B_1$.

The basic operation J of **A** is **monotone** in the sense that if $a \le a', b \le b'$ and $c \le c'$ where all these elements belong to A, then $J(a,b,c) \le J(a',b',c')$. Similarly, all the basic operations of **A** are monotone.

Fact 7. *Let f be a monotone unary polynomial of* $\mathbf{B}$. *If $x < p$ and $f(x) = p$, then $f(q) = p$.*

Proof. By Fact 4 we have $x \; \theta \; x \wedge q$. This entails $p = f(x) \; \theta \; f(x \wedge q)$. So by Fact 2 we get $p = f(x \wedge q)$. But then $p \leq f(q)$ by the monotonicity of f. Thus $p = f(q)$ by the maximality of p. ■

PROVISO: *FACTS 8–17 BELOW ARE ESTABLISHED UNDER THE ASSUMPTION THAT THE RANGES OF S_1 AND S_2 ARE CONTAINED IN $B - B_1$.*

Fact 8. *If $u \in B_1$ and $v \in B$ so that for all $s \in T$ either $u(s) = v(s)$ or $u(s) = \overline{v(s)} \in W$, then $u = v$.*

Proof. First suppose $u = p$.

Let $Y = \{s : p(s) = \overline{v(s)}\}$.

CLAIM: Y is empty.

Proof of the Claim: Since the range of the operation S_2 is disjoint from B_1, it follows that $T \neq Y$. Pick $t' \in T - Y$ and let $q' = \chi_{t'}$. So for each $s \in T$ we have

$$J(p(s), v(s), q'(s)) = \begin{cases} p(s) & \text{if } s \notin Y, \\ p(s) \wedge q'(s) & \text{if } s \in Y. \end{cases}$$

But this entails $J(p, v, q') = p$, since $q'(s) = p(s)$ for all $s \in Y$ because $t' \notin Y$. Therefore, by Fact 7 and the monotonicity of J, we have $J(p, v, q) = p$. But then the definition of J gives us $q(s) = p(s)$ for all $s \in Y$. Since $q(t_0) = 0$, it follows that $t_0 \notin Y$.

Now observe that $J'(p(t_0), v(t_0), q(t_0)) = p(t_0) \wedge 0 = 0$. Hence, $J'(p, v, q) \neq p$. So by Fact 7 and the monotonicity of J', we conclude that $J'(p, v, \chi_t) \neq p$ for all $t \in T$. But for all $s, t \in T$

$$J'(p(s), v(s), \chi_t(s)) = \begin{cases} p(s) \wedge \chi_t(s) & \text{if } s \notin Y, \\ p(s) & \text{if } s \in Y. \end{cases}$$

It follows that $t \notin Y$ for all $t \in Y$. This means Y is empty. So the Claim is established.

Since Y is empty, we also know that $p(s) = v(s)$ for all $s \in T$. Hence $u = v$ as desired.

Now suppose $u \in B_1 - \{p\}$. There are two kinds of elements in $B_1 - \{p\}$: those in U^T and those in W^T. Clearly, we can restrict our attention to the case when $u \in W^T$. Let λ be a translation built from the product such that $\lambda(u) = p$. Set $p' = \lambda(v)$. Since the product respects bars on elements, we see that for each $s \in T$, either $p(s) = p'(s)$ or $p(s) = \overline{p'(s)}$. So by the claim just established, we have $\lambda(u) = p = p' = \lambda(v)$. But then $u = v$ by Fact 6. Fact 8 is now proved. ■

Our basic strategy calls for θ to isolate the members of B_1 and to lump all the elements of $B - B_1$ together. To see that this really does happen, in view of Fact 3 we need the following.

Fact 9. *If $u \in B$ and $\lambda(u) \in B_1$ for some nonconstant translation λ, then $u \in B_1$.*

Proof. The proof is by induction on the complexity of λ. The initial step of the induction is obvious, since the identity function is the only simplest nonconstant translation. The inductive step breaks down into seven cases, one for each basic operation of positive rank.

CASE $\wedge$: $\lambda(x) = \mu(x) \wedge r$, where $r \in B$.

We have $\lambda(u) \leq \mu(u)$. But every element of B_1 is maximal with respect to the semilattice order. So $\lambda(u) = \mu(u) \in B_1$. Now μ must be nonconstant. Invoking the induction hypothesis, we get $u \in B_1$.

CASE $\cdot$: $\lambda(x) = \mu(x)r$ or $\lambda(x) = r\mu(x)$.

Under the first alternative we have $\mu(u)r = \lambda(u) \in B_1$. So $\mu(u), r \in B_1$. Since μ must be nonconstant, we can invoke the induction hypothesis to conclude that $u \in B_1$. The other alternative is similar.

CASE J: $\lambda(x) = J(\mu(x), r, s)$ or $\lambda(x) = J(r, \mu(x), s)$ or $\lambda(x) = J(r, s, \mu(x))$.

Consider the first alternative. We have $\lambda(u) = J(\mu(u), r, s) \leq \mu(u)$. By the maximality of $\lambda(u)$ we get

$$\lambda(u) = J(\mu(u), r, s) = \mu(u) \in B_1.$$

Now μ cannot be constant. Hence we can invoke the inductive hypothesis to conclude that $u \in B_1$. The second alternative is similar, except that Fact 8 comes into play. Under the last alternative, since $r \geq J(r, s, \mu(u)) = \lambda(u)$ is maximal, we see that r and s fulfill the hypotheses of Fact 8. Consequently, $r = s \in B_1$. But then, $\lambda(x) = J(r, s, \mu(x)) = r$ according to the definition of J. This means the third alternative is impossible, since $\lambda(x)$ is not constant.

CASE J': $\lambda(x) = J'(\mu(x), \nu(x), \rho(x))$.

This case is easier than the last one and is omitted.

CASES S_1 AND S_2: Too easy.

CASE U^0: $\lambda(x) = U^0(\mu(x), s, r', s')$ or $\lambda(x) = U^0(r, \mu(x), r', s')$ or $\lambda(x) = U^0(r, s, \mu(x), s')$ or $\lambda(x) = U^0(r, s, r', \mu(x))$.

Consider the first alternative. We have $\lambda(u) = U^0(\mu(u), s, r', s') \in B_1$. Evidently, $\lambda(u)$ and $\mu(u)s$ satisfy the hypotheses of Fact 8. So $\lambda(u) = \mu(u)s$. Since $\lambda(u) \in B_1$, we know that $\mu(u) \in B_1$ by the definition of B_1. Now μ is nonconstant. So $u \in B_1$ by the inductive hypothesis. The second alternative is similar.

Consider the third alternative. We have $\lambda(u) = U^0(r, s, \mu(u), s')$. Evidently, $\lambda(u)$ and rs satisfy the hypotheses of Fact 8. So $\lambda(u) = rs$. Then by the definition of U^0, we have $\lambda(u) = rs = \mu(u)s'$. But then $\mu(u) \in B_1$ and the induction hypotheses applies to yield $u \in B_1$. The fourth alternative is similar. This finishes the proof of Fact 9. ■

Fact 10. *$u/\theta = \{u\}$ for each $u \in B_1$ and $0/\theta = B - B_1$.*

Proof. Suppose $u \in B_1$ and that $u \;\theta\; v$. Let $\lambda(u) = p$ for some translation λ built just using $\cdot$. It follows that $\lambda(v) = p$ by Fact 3. By Fact 6, we conclude that $u = v$.

Fact 9 says that $B - B_1$ is closed with respect to nonconstant translations. Since $p \in B_1$, we have that $\lambda(u) \neq p$ for all $u \in B - B_1$ and all nonconstant translations λ. Hence, by Fact 3, $B - B_1$ is collapsed by θ. But, as we just saw, B_1 is the union of (singleton) θ-classes. Hence $B - B_1$ is a θ-class. Clearly, $0 \in B - B_1$. ■

To establish that $\mathbf{S} \cong \mathbf{Q}_n$ for some natural number n we need to analyze each of our basic operations. We deal with the product first.

Here is the unique factorization property for the product that we require.

Fact 11. *If $ab = cd \in B_1$, then $a = c$ and $b = d$.*

Proof. Let $u = ab$ and $v = U^0(a,b,c,d)$. From the definition of the operation U^0, we see that u and v satisfy the hypotheses of Fact 8. Hence, $ab = U^0(a,b,c,d)$. But then the definition of U^0 gives $a = c$ and $b = d$. ■

In $\mathbf{Q}_{\mathbb{Z}}$ none of the labels of the edges were repeated. We need this property as well. It is the reason why we introduced the operation S_1. The relevant fact is next.

Fact 12. *No factorization of p has repeated factors.*

Proof. It is clear that if $d_0 d_1 \cdots d_{m-1} e = p$ then $e \in W^T$ while $d_0, \ldots, d_{m-1} \in U^T$. Suppose that $d_i = d_j$ with $i < j$. Since the range of the operation S_1 is disjoint from B_1, we conclude that B contains no elements from $\{1,3\}^T$. So pick $s \in T$ so that $d_i(s) = d_j(s) = 2$. Now we see

$$p(s) = d_0(s) \cdots d_{i-1}(s) 2 d_{i+1}(s) \cdots d_{j-1}(s) 2 d_{j+1}(s) \cdots d_{m-1}(s) e(s)$$

So $p(s) = 0$, violating the maximality of p. ■

We are now in a position to describe B_1 more explicitly. Consider the following factorization of p:

$$\begin{aligned} p &= b_0 \\ &= a_0 b_1 \\ &= a_0 a_1 b_2 \\ &\vdots \\ &= a_0 a_1 \cdots a_{n-1} b_n \\ &\vdots \end{aligned}$$

Evidently, $a_i \in U^T$ for all i and according to Fact 12 all the a_i's are distinct. But B_1 is finite, so we suppose without loss of generality that b_n cannot be factored. But the unique factorization property, Fact 11, entails that the factorization of p displayed above is the only way p can be factored. Consequently,

$$B_1 = \{a_0, a_1, \ldots, a_{n-1}\} \cup \{b_0, b_1, \ldots, b_n\}$$

It is also evident that $b_i \in W^T$ for all i. Were $b_i = b_j$ for some $i \neq j$, it would be easy to construct a factorization of p with repeated factors, in violation of Fact 12. This means that B_1 has $2n + 1$ elements, and that $b_i = a_i b_{i+1}$ for each $i < n$. That all the other products of elements chosen from B_1 will belong to B_0, follows easily from the unique factorization property Fact 11. Consequently, at least with respect to the product operation, $\mathbf{S}$ and $\mathbf{Q}_n$ are isomorphic.

Now consider the operation $\wedge$. Since $\wedge$ is obviously a semilattice operation on $\mathbf{S}$, what we need is that $\langle B/\theta, \wedge \rangle$ is flat.

Fact 13. *If $x, y \in B_1$ and $x \neq y$, then $x \wedge y \in B - B_1$.*

Proof. Since $x \neq y$ there is $t \in T$ with $x(t) \neq y(t)$. But then $((x \wedge y)(t) = 0$. Therefore $x \wedge y \in B - B_1$. ■

Finally, we need to know that the remaining basic operations on **S** can be construed as term operations built up from $\cdot, \wedge$, and 0 in a manner dependent only on the hypotheses that **S** is a finite subdirectly irreducible algebra in $HSP\mathbf{A}$ and that $\mathbf{S} \notin HS\mathbf{A}$. That is the content of the next sequence of facts.

Fact 14. *$U^0(x,y,z,w)\ \theta\ (xy) \wedge (zw)$ for all $x,y,z,w \in B$.*

Proof. We must show that either $U^0(x,y,z,w)$ and $(xy) \wedge (zw)$ both belong to $B - B_1$ or else $U^0(x,y,z,w) = (xy) \wedge (zw) \in B_1$. Since $B - B_1$ is a θ-class, we see that Fact 8 forces $U^0(x,y,z,w) \in B - B_1$ except in the case that $xy = zw \in B_1$. In that case,

$$U^0(x,y,z,w) = xy = zw = (xy) \wedge (zw) \in B_1.$$

But also, $(xy) \wedge (zw) \in B - B_1$ except in the case that $xy = zw \in B_1$. In that case as well, $U^0(x,y,z,w) = xy = (xy) \wedge (zw) \in B_1$. Therefore, $U^0(x,y,z,w)\ \theta\ (xy) \wedge (zw)$. ■

Fact 15. *$J(x,y,z)\ \theta\ x \wedge y$ for all $x,y,z \in B$.*

Proof. Again, we must show that either $J(x,y,z)$ and $x \wedge y$ both belong to $B - B_1$ or else $J(x,y,z) = x \wedge y \in B_1$. Now again using that $B - B_1$ is a θ-class and Fact 8, $J(x,y,z) \in B - B_1$, except in the case that $x = y \in B_1$. In that case, $J(x,y,z) = x = y = x \wedge y \in B_1$. But also, $x \wedge y \in B - B_1$, except in the case that $x = y \in B_1$. In that case, $x \wedge y = x = J(x,y,z) \in B_1$. Therefore, $J(x,y,z)\ \theta\ x \wedge y$. ■

Fact 16. *$J'(x,y,z)\ \theta\ x \wedge y \wedge z$ for all $x,y,z \in B$.*

Proof. This is too easy. ■

Fact 17. *$S_1(u,v,x,y,z)\ \theta\ 0\ \theta\ S_2(u,v,x,y,z)$ for all $u,v,x,y,z \in B$.* ■

For each natural number n, we take $\mathbf{Q}_n$ to be an algebra on $2n+2$ elements with the basic operations $\cdot, \wedge$, and 0, and the remaining basic operations determined by the stipulation that the following equations are true in $\mathbf{Q}_n$:

$$U^0(x,y,z,w) \approx (xy) \wedge (zw)$$

$$J(x,y,z) \approx x \wedge y \qquad S_1(u,v,x,y,z) \approx 0$$

$$J'(x,y,z) \approx x \wedge y \wedge z \qquad S_2(u,v,x,y,z) \approx 0$$

Thus we arrive at the desired conclusion.

LEMMA 7.82. *Let **S** be a finite subdirectly irreducible algebra in **HSP A**. Either $\mathbf{S} \in$ **HS A** or else there is a natural number n such that $\mathbf{S} \cong \mathbf{Q}_n$.* ■

What we haven't done yet is prove that any of these expanded $\mathbf{Q}_n$'s belong to the variety generated by our 8-element algebra **A**.

A is Inherently Nonfinitely Based and Has Residual Character ω_1

The algebra $\mathbf{Q}_{\mathbb{Z}}$ and its subalgebras $\mathbf{Q}_\omega$, and $\mathbf{Q}_n$ for each $n \in \omega$, were introduced at the beginning of §7.10. The operations 0, $\wedge$, and $\cdot$ were examined in detail, but the only stipulation about any remaining operations was that they must be defined as term operations of these first three. Five more operation symbols were introduced: U^0, J, J', S_1, and S_2. In the algebras $\mathbf{Q}_{\mathbb{Z}}, \mathbf{Q}_\omega$, and $\mathbf{Q}_n$ these five further basic operations are defined so that the following equations are true:

$$U^0(x,y,z,w) \approx (xy) \wedge (zw)$$

$$J(x,y,z) \approx x \wedge y \qquad S_1(u,v,x,y,z) \approx 0$$
$$J'(x,y,z) \approx x \wedge y \wedge z \qquad S_2(u,v,x,y,z) \approx 0$$

The whole earlier discussion culminating in Theorem 7.80 of these algebras goes through in this expanded setting, with the exception of the last phase: showing they belong to ***HSP*** **A**. The five new operations were not defined on the six element algebra **R** in §7.9. We now want to replace that algebra with the eight element algebra **A** introduced above. What we need is the following theorem.

THEOREM 7.83. $\mathbf{Q}_{\mathbb{Z}}$ *belongs to the variety generated by* **A**.

Proof. We retrace the proof of Fact 2. First, for each $p \in \mathbb{Z}$ we designate elements α_p and β_p of $A^{\mathbb{Z}}$ as before:

$$\begin{array}{lcccccccc} \alpha_p := & \cdots & 1 & 1 & 1 & 2 & 3 & 3 & 3 & \cdots \\ \beta_p := & \cdots & r & r & r & q & q & q & q & \cdots \end{array}$$

where the first change in α_p and the only change in β_p is taking place at the pth position. Next we let $B_1 = \{\alpha_p : p \in \mathbb{Z}\} \cup \{\beta_p : p \in \mathbb{Z}\}$ and we let **B** be the subalgebra of $\mathbf{A}^{\mathbb{Z}}$ generated by B_1. Let B_0 be the set of all elements of B in which 0 occurs. Now let Φ be the map defined from B to $\mathbf{Q}_{\mathbb{Z}}$ via

$$\Phi(x) = \begin{cases} a_p & \text{if } x = \alpha_p \text{ for some } p \in \mathbb{Z}, \\ b_p & \text{if } x = \beta_p \text{ for some } p \in \mathbb{Z}, \\ 0 & \text{otherwise} \end{cases}$$

We contend that $B_0 \cup B_1$ is a subuniverse of **B** (and so $B = B_0 \cup B_1$) and also that Φ is a homomorphism from **B** onto $\mathbf{Q}_{\mathbb{Z}}$. Checking either of these contentions can be done by examining the behavior of each basic operation case by case. We will do this simultaneously.

CASE 0. Plainly, $0 \in B_0$ and $\Phi(0) = 0$. So this case is secure.

CASE $\wedge$. Suppose that $u, v \in B_0 \cup B_1$. Then either $u = v$ and $u \wedge v = u \in B_0 \cup B_1$ or else $u \neq v$ and $u \wedge v \in B_0$. Hence, $B_0 \cup B_1$ is closed under $\wedge$. But also, $\Phi(u \wedge v) = \Phi(u) \wedge \Phi(v)$.

CASE $\cdot$. Suppose that $u, v \in B_0 \cup B_1$. Then either $uv \in B_0$ or for some p we have $u = \alpha_p, v = \beta_{p+1}$ and $uv = \beta_p \in B_1$. It follows that $B_0 \cup B_1$ is closed under $\cdot$ and that Φ preserves $\cdot$.

To handle the remaining cases, the following property of $x, y \in B_0 \cup B_1$ proves useful:

$(\star)$ If for each $s \in \mathbb{Z}$, either $x(s) = y(s) \neq 0$ or $x(s) = \overline{y(s)}$, then $x = y$.

Since any $x, y \in B_0 \cup B_1$ for which the hypothesis of $(\star)$ holds must both belong to B_1, and since bars occur in no member of B_1, $(\star)$ is true.
CASE J. For $J(x, y, z) \notin B_0$, observe that the inputs x and y must satisfy the hypothesis of $(\star)$. Hence, either $J(x, y, z) \in B_0$ or $x = y \in B_1$ and $J(x, y, z) = x \in B_1$. So $B_0 \cup B_1$ is closed under J and Φ preserves J.
CASE J'. This case is very similar to the last case.
CASE U^0. Let $x, y, z, w \in B_0 \cup B_1$. Let $u = U^0(x, y, z, w)$ and $v = xy$. Then $U^0(x, y, z, w) \in B_0$ unless u and v fulfill the hypothesis of $\star$. In that case, we must have $U^0(x, y, z, w) = u = v \in B_1$. Consequently, $B_0 \cup B_1$ is closed under U^0 and Φ preserves U^0.
CASE S_1. Let $u, v, x, y, z \in B_0 \cup B_1$. Then $S_1(u, v, x, y, z) \in B_0$ unless $u \in \{1, 3\}^{\mathbb{Z}}$. But $\{1, 3\}^{\mathbb{Z}}$ and $B_0 \cup B_1$ are disjoint. Consequently, $B_0 \cup B_1$ is closed under S_1 and Φ preserves S_1.
CASE S_2. Let $u, v, x, y, z \in B_0 \cup B_1$. Then $S_2(u, v, x, y, z) \in B_0$ unless $u(s) = \overline{v(s)}$ for all $s \in \mathbb{Z}$. But no element of B_1 has a bar at any of its entries. Consequently, $B_0 \cup B_1$ is closed under S_2 and Φ preserves S_2. ■

At this point we know that the eight-element algebra $\mathbf{A}$, which has eight basic operations, is inherently nonfinitely based, that the finite subdirectly irreducible algebras in the variety generated by $\mathbf{A}$ are the subdirectly irreducible algebras in $\boldsymbol{HS}\mathbf{A}$ and the algebras $\mathbf{Q}_n$ for each $n \in \omega$, and that $\mathbf{Q}_\omega$ is a countably infinite subdirectly irreducible member of the variety.

We will demonstrate that our variety has no other infinite subdirectly irreducible algebras.

Let $\mathbf{S}$ be any infinite subdirectly irreducible algebra in the variety generated by $\mathbf{A}$. According to the Theorem of Dziobiak (7.77) any finite subalgebra of $\mathbf{S}$ can be embedded into arbitrarily large finite subdirectly irreducible algebras in the variety generated by $\mathbf{A}$, i.e. into $\mathbf{Q}_n$ for a sequence of n's approaching infinity. This means that every finitely generated (= finite) subalgebra of $\mathbf{S}$ is embeddable into $\mathbf{Q}_\omega$. Consequently, every universal sentence true in $\mathbf{Q}_\omega$ must be true in $\mathbf{S}$.

Here are some interesting properties of $\mathbf{Q}_\omega$ which can be expressed with universal sentences:

- Any equation true in $\mathbf{Q}_{\mathbb{Z}}$. For example: $U^0(x, y, z, w) \approx (xy) \wedge (zw)$.
- The height is no bigger than 1: $x \not\approx y \to x \wedge y \approx 0$.
- $xy \approx zw \not\approx 0 \to (x \approx z \;\&\; y \approx w)$.
- $xy \not\approx 0 \not\approx xz \to y \approx z$.
- $xy \not\approx 0 \not\approx zy \to x \approx z$.
- $xy \not\approx 0 \to zx \approx 0 \approx yw$.

Consequently, in $\mathbf{S}$, the operations U^0, J, J', S_1, and S_2 are term functions (using the same terms as in $\mathbf{Q}_\omega$) in $0, \wedge$, and $\cdot$. We ignore them from now on. With respect to $\wedge$ and 0, $\mathbf{S}$ is a height 1 meet-semilattice with least element 0. So the balance of our analysis depends primarily on the product $\cdot$. Since $(xy)z \approx 0$ is true in $\mathbf{Q}_\omega$, we see that in **S**, just as in $\mathbf{Q}_\omega$, only right-associated products can differ from 0. The last four properties itemized above put further and severe restrictions on the product in **S**.

We make $S - \{0\}$ into a labeled directed graph as follows. We take as the vertex set those elements which are right factors in nonzero products, outputs, or do not occur in nonzero products. We take as the set of labels those elements which are left factors in nonzero products. Our itemized properties entail that the set of vertices and the set of labels are disjoint. We put an edge from b to c and label it with a provided $ab = c$ in **S**. Our itemized assertions ensure that a vertex can have outdegree at most 1, indegree at most 1, and that every edge has a uniquely determined label which occurs as a label of exactly one edge in the whole graph.

Let C be a connected component of our graph. Let θ_C be the equivalence relation that collapses all the vertices and labels in C to 0, but which isolates every other point. θ_C is a congruence of **S**. Since **S** is subdirectly irreducible, it follows that our graph has only one component. This already implies that **S** is countably infinite. But more is true. There are only three possible countable connected graphs of this kind: the one associated with $\mathbb{Z}$ (and then we would have $\mathbf{S} \cong \mathbf{Q}_\mathbb{Z}$), the one associated with ω (and then we would have $\mathbf{S} \cong \mathbf{Q}_\omega$), and the one associated with the set of nonnegative integers (and then **S** would be isomorphic to an algebra we might as well call $\mathbf{Q}_{-\omega}$). But neither $\mathbf{Q}_\mathbb{Z}$ nor $\mathbf{Q}_{-\omega}$ is subdirectly irreducible. So **S** must be isomorphic to $\mathbf{Q}_\omega$.

We summarize the results in the following theorem.

THEOREM 7.84 (Ralph McKenzie 1996b). *The eight element algebra* **A**, *which has finitely many basic operations, is inherently nonfinitely based. The subdirectly irreducible algebras in the variety generated by* **A** *are, up to isomorphism, exactly the subdirectly irreducible homomorphic images of subalgebras of* **A**, *the algebra* $\mathbf{Q}_\omega$, *and the algebra* $\mathbf{Q}_n$ *for each* $n \in \omega$. ■

This theorem settles in the negative some outstanding problems.

The R-S Conjecture: *Every finitely generated residually small variety is residually very finite.* Here R stands for "residually" and S stands for "small". (David Hobby and Ralph McKenzie, 1988)

The Broader Finite Basis Speculation: *Every finitely generated residually small variety of finite signature is finitely based.*

Theorem 7.84 is a counterexample to both of these. However, the two problems below are closely related and still open.

The Quackenbush Conjecture: *Every finitely generated residually finite variety of finite signature is residually very finite.* (Robert W. Quackenbush, 1971)

The Jónsson/Park Speculation: *Every finitely generated residually very finite variety of finite signature is finitely based.* (Robert E. Park, 1976)

7.11. Undecidable Properties of Finite Algebras

While he was a post-doc at Berkeley, Roger Lyndon, in 1951, proved that every 2-element algebra of finite signature is finitely based (see Corollary 9.48 on page 40 in Volume III) and in 1954 he devised a 7-element algebra of finite signature that is not finitely based (see Theorem 7.18). This led Alfred Tarski to pose the following problem

TARSKI'S FINITE BASIS PROBLEM

Is there an algorithm that determines, upon input of a finite algebra of finite signature, whether the algebra is finitely based?

This problem, which attracted considerable effort over the course of decades, was settled in the negative by Ralph McKenzie in 1996. The remainder of this chapter provides an exposition of this result.

How $\mathbf{A}(M)$ encodes the computations of M

In this subsection we describe, in part, McKenzie's machine algebras and show how they capture the computations of Turing machines. Turing machines are finite objects, but the computations that they produce can be endless. So it is reasonable to expect to use a finite algebra to convey the information of any particular Turing machine. However, finite algebras are too small to hold arbitrary computations. The algebra $\mathbf{Q}_{\mathbb{Z}}$ described in §7.10, however, suggests a way to grapple with arbitrary computations. The idea is to designate certain elements of the machine algebra as configurations of a Turing machine and draw labeled directed edges between configurations to represent the transitions of the machine computation. Then we try to realize these directed edges by new operations applied to certain elements. Next we try to find a finite algebra so that the whole thing is happening coordinatewise inside a big direct power. Finally, we will have to add further operations to control all the finite subdirectly irreducible algebras.

To each Turing machine M we will associate a finite algebra $\mathbf{A}(M)$. The construction of $\mathbf{A}(M)$ begins with the eight-element algebra $\mathbf{A}$ that was introduced in Figure 7.16 on page 287. Its universe $A = \{0, 1, 2, 3, q, r, \overline{q}, \overline{r}\}$ will be enlarged to accommodate some elements to represent certain configurations of M and its signature will be expanded by adding finitely many operations capable of emulating transitions between configurations. We will also need some operations to keep control of the finite subdirectly irreducible algebras.

The analysis of computation itself will go on in $\mathbf{A}(M)^X$ for some large set X [think of $X = \mathbb{Z}$].

Recalling §7.7, we conceive of a Turing machine M as having finitely many internal states $0, 1, \ldots, m$. The machine is always launched in state 1 and we take 0 to be the unique halting state. The Turing machine M has a tape alphabet consisting of the symbols 0 and 1. The Turing machine itself is a finite collection of 5-tuples each of the form:

$$[i, \gamma, \delta, D, j]$$

This 5-tuple is the instruction: "If you are in state i and you are examining a tape square containing the symbol γ, then write the symbol δ on that square, move one square in the direction D (D must be either L for left or R for right), and pass into internal state j". We insist that no 5-tuple begin with 0 and that otherwise the machine must have exactly one instruction which begins $[i, \gamma, \ldots]$ for each state i other than the halting state 0 and each tape symbol γ.

We say Q is a *configuration* for a Turing machine M provided $Q = \langle t, n, i\rangle$ where $t \in \{0, 1\}^{\mathbb{Z}}$, $n \in \mathbb{Z}$, and i is one of the states of M. The idea is that at some stage of a computation, the tape of the machine looks like t, the machine is focused on square n and is itself in state i.

To obtain the desired undecidability results our plan is to use the following version of the Halting Problem:

> The set of Turing machines that halt when started on the blank tape (that is the tape with 0 written on each square) is undecidable.

A significant problem we have to resolve comes from the fact that machine computations, at any given stage, happen at a particular location on the tape, and that these locations are arranged in a sequence with only the adjacent locations available for the next step in the computation. Thus some elements of our "computation algebra" which are used to label those directed edges must also fall into a sequence of "tape locations". To make short work of this point we take the elements a_p of $\mathbf{Q}_{\mathbb{Z}}$ (see Figure 7.15 on page 282) as a model of how elements fall into sequence. Looking at what we had to have in A during the proof of Theorem 7.83 to get these a_p's we recall:

$$\begin{array}{rcccccccccc}
\alpha_p\colon & \cdots & 1 & 1 & 1 & 2 & 3 & 3 & 3 & \cdots \\
\alpha_{p+1}\colon & \cdots & 1 & 1 & 1 & 1 & 2 & 3 & 3 & \cdots \\
\alpha_{p+2}\colon & \cdots & 1 & 1 & 1 & 1 & 1 & 2 & 3 & \cdots
\end{array}$$

So in all our machine algebras we want a subset $U = \{1, 2, 3\}$ making elements like the ones above available in direct powers. To impose the precedence above in the direct power, we impose $3 \prec 3 \prec 2 \prec 1 \prec 1$ on U. We also use $\prec$ to denote the coordinatewise relation in any direct power of a machine algebra. Suppose $\mathbf{B} = \mathbf{A}(M)^X$. A subset $F \subseteq B$ is *sequentiable* provided

- $F \subseteq U^X$,
- 2 occurs at least once in f, for each $f \in F$, and
- $\prec$ gives F a structure isomorphic to some convex substructure of the ordered set of integers with the successor relation.

Since 2 may occur at several places in such an f, sequentiable sets can be more complex than $\{\alpha_p : p \in \mathbb{Z}\}$. For a fixed sequentiable set F the index set X falls into natural pieces that help us see the structure. Look at the following display of the four element sequentiable set $F = \{f_0, f_1, f_2, f_3\}$.

$$\begin{array}{rccccccccccccc}
f_0\colon & 1 & 1 & 2 & 3 & 3 & 3 & 2 & 3 & 3 & 3 & 1 & 2 & 1 \\
f_1\colon & 1 & 1 & 1 & 2 & 3 & 3 & 1 & 3 & 3 & 3 & 1 & 1 & 1 \\
f_2\colon & 1 & 1 & 1 & 1 & 3 & 3 & 1 & 2 & 3 & 2 & 1 & 1 & 1 \\
f_3\colon & 1 & 1 & 1 & 1 & 3 & 2 & 1 & 1 & 3 & 1 & 1 & 1 & 1
\end{array}$$

Examining the 13 columns, we see that several are exactly the same. In this example the set X has 13 elements and some unspecified arrangement of these thirteen elements underlies the display above. But the particular arrangement of X is immaterial from the point of view of the algebra $\mathbf{A}(M)^X$. Thus we are free to rearrange X to make the precedence on F more transparent. Below is the result of such a rearrangement:

$$\begin{array}{rccccccccccccc}
f_0\colon & 1 & 1 & 1 & 1 & 2 & 2 & 2 & 3 & 3 & 3 & 3 & 3 & 3 \\
f_1\colon & 1 & 1 & 1 & 1 & 1 & 1 & 1 & 2 & 3 & 3 & 3 & 3 & 3 \\
f_2\colon & 1 & 1 & 1 & 1 & 1 & 1 & 1 & 1 & 2 & 2 & 3 & 3 & 3 \\
f_3\colon & 1 & 1 & 1 & 1 & 1 & 1 & 1 & 1 & 1 & 1 & 2 & 3 & 3
\end{array}$$

We have put all the columns consisting entirely of 1's to the left. Next we put all the columns beginning with 2 in position 0, then all columns with 2 in position 1, and so on. At the right we have placed all columns consisting entirely of 3's. Doing this, we see that there are only $6 = 4 + 2$ different kinds of columns possible:

$$\begin{matrix} 1 & 2 & 3 & 3 & 3 & 3 \\ 1 & 1 & 2 & 3 & 3 & 3 \\ 1 & 1 & 1 & 2 & 3 & 3 \\ 1 & 1 & 1 & 1 & 2 & 3 \end{matrix}$$

This means our sequentiable set F partitions the index set X into 6 blocks. The blocks can be labeled X_L for the set of all indices of columns that are constantly 1, X_R for the set of all indices of columns that are constantly 3, and X_n for the set of all indices where the necessarily unique 2 occurs at the nth position.

To simplify the presentation a bit and make the pictures understandable, once a sequentiable set F has been specified, we will assume that X is arranged in such a line so that the set X_L is an initial (or left) segment, X_R is a final segment (or right) segment, and each X_n is placed at the obvious position on the line. Since at its biggest, F can be indexed only by $\mathbb{Z}$, we can accommodate such a line like picture if we are willing to place X_L at $-\infty$ and X_R at $+\infty$.

Now let F be the four element sequentiable set above but with the columns collapsed to 6 and arranged as in the last display, and let $Q = \langle t, 2, i\rangle$ be a configuration. We code a portion of Q by

$$\begin{matrix} \beta\colon & r^{0}_{i,t(2)} & r^{t(0)}_{i,t(2)} & r^{t(1)}_{i,t(2)} & H^{t(2)}_{i} & q^{t(3)}_{i,t(2)} & q^{0}_{i,t(2)} \\ \text{block:} & X_L & X_0 & X_1 & X_2 & X_3 & X_R \end{matrix}$$

This gives a real forest of superscripts and subscripts and the truth is that we will need a few more to get to full generality. However, we can decode it a bit. The r's mean "left of the reading head". The q's mean "to the right of the reading head". H locates where the machine reading head is. The index i specifies the state of the machine. The subscript $t(2)$ tells what symbol is written on the tape square scanned by the reading head. Finally, the indices $t(j)$ tell us what is printed on the corresponding square of the tape, unless it is too far off to the left (in X_L) or too far off to the right (in X_R), in which case we have used 0 as a default value (other choices would be okay). So reading across the superscripts is like reading across the tape. In this way, each component of β carries a lot of information about the configuration.

Now X in this example had 13 elements rather than 6, so the β above is too short. However, by duplicating the entries in β the correct number of times (e.g. the first entry $q^{0}_{i,t(2)}$ should occur 4 times while the last entry $r^{0}_{i,t(2)}$ should occur twice) we would get a β of the correct length. That $|X| = 13$ is immaterial. But our particular sequentiable set had only four elements, it was indexed with the convex set $\{0, 1, 2, 3\}$, and we took $n = 2$ in our configuration. To get the general case, let I be any convex subset of $\mathbb{Z}$ and suppose that F is a sequentiable set indexed by I. Let $n \in I$ and let $Q = \langle t, n, i\rangle$ be a configuration. Then we use the β below as a code for Q and we say that *β codes Q over*

F indexed by I.

$$\beta(x) = \begin{cases} r^{0}_{i,t(n)} & \text{if } x \in X_L. \\ r^{t(j)}_{i,t(n)} & \text{if } x \in X_j \text{ and } j < n \text{ and } j \in I. \\ H^{t(n)}_{i} & \text{if } x \in X_j \text{ and } j = n \in I. \\ q^{t(j)}_{i,t(n)} & \text{if } x \in X_j \text{ and } n < j \in I. \\ q^{0}_{i,t(n)} & \text{if } x \in X_R. \end{cases}$$

Capturing the transitions between configurations

To get a grip on how to handle the transition between configurations let $\mathbf{B} = \mathbf{A}(M)^{\mathbb{Z}}$ and let $F = \{\alpha_p : p \in \mathbb{Z}\}$. Then F is a sequentiable set indexed by $\mathbb{Z}$, and the partition imposed on $\mathbb{Z}$ by F consists of singleton sets $\{p\}$. Let $Q = \langle t, n, i\rangle$ be a configuration of M, let $t(n) = \gamma$, and suppose that $[i, \gamma, \delta, L, j]$ is an instruction in M. It also proves convenient to let $t(n-1) = \varepsilon$. Then $M(Q) = \langle s, n-1, j\rangle$ is the configuration following Q in the computation of M, where

$$s(k) = \begin{cases} \delta & \text{if } k = n, \\ t(k) & \text{otherwise.} \end{cases}$$

The configuration Q is coded over F by

$$\beta = \cdots \quad r^{t(n-3)}_{i,\gamma} \quad r^{t(n-2)}_{i,\gamma} \quad r^{\varepsilon}_{i,\gamma} \quad H^{\gamma}_{i} \quad q^{t(n+1)}_{i,\gamma} \quad q^{t(n+2)}_{i,\gamma} \quad q^{t(n+3)}_{i,\gamma} \quad \cdots$$

whereas the configuration $M(Q)$ is coded over F by

$$M(\beta) = \cdots \quad r^{t(n-3)}_{j,\varepsilon} \quad r^{t(n-2)}_{j,\varepsilon} \quad H^{\varepsilon}_{j} \quad q^{\delta}_{j,\varepsilon} \quad q^{t(n+1)}_{j,\varepsilon} \quad q^{t(n+2)}_{j,\varepsilon} \quad q^{t(n+3)}_{j,\varepsilon} \quad \cdots$$

$M(\beta)$ differs from β in several ways. First, the two positions indexed by $n-1$ and n undergo a change of character from r to H and from H to q. Second, the remaining changes amount to changing γ to ε and i to j in various subscripts and superscripts. The idea is to effect this transition with a new operation for the machine instruction $[i, \gamma, \delta, L, j]$. Changes of the first kind have to do with two tape locations. Our new operation must combine the two location elements, α_{n-1} and α_n, with the configuration element β to produce the new configuration element $M(\beta)$—our "instruction" operation should be ternary. To see what is needed to accomplish this, look at

$$\begin{array}{rcccccccc} \alpha_{n-1} = \cdots & 1 & 1 & 2 & 3 & 3 & 3 & 3 & \cdots \\ \alpha_n = \cdots & 1 & 1 & 1 & 2 & 3 & 3 & 3 & \cdots \\ \beta = \cdots & r^{t(n-3)}_{i,\gamma} & r^{t(n-2)}_{i,\gamma} & r^{\varepsilon}_{i,\gamma} & H^{\gamma}_{i} & q^{t(n+1)}_{i,\gamma} & q^{t(n+2)}_{i,\gamma} & q^{t(n+3)}_{i,\gamma} & \cdots \\ \hline M(\beta) = \cdots & r^{t(n-3)}_{j,\varepsilon} & r^{t(n-2)}_{j,\varepsilon} & H^{\varepsilon}_{j} & q^{\delta}_{j,\varepsilon} & q^{t(n+1)}_{j,\varepsilon} & q^{t(n+2)}_{j,\varepsilon} & q^{t(n+3)}_{j,\varepsilon} & \cdots \end{array}$$

The instruction $[i, \gamma, \delta, L, j]$ makes no reference to ε (the symbol written on square $n-1$ of the tape). Since our operation must act coordinatewise, we will build ε into the operation itself. So to each machine instruction we will associate two ternary operations, one for each of the two possible values of ε. Since the machine instructions for a fixed Turing machine M are determined by their first two components we will denote the operations corresponding to the machine instruction above by $F_{i\gamma\varepsilon}$. What must happen in $\mathbf{A}(M)$ to accomplish the transition above is

$$
\begin{aligned}
F_{i\gamma\varepsilon}(1,1,r^{\nu}_{i,\gamma}) &= r^{\nu}_{j,\varepsilon}\\
F_{i\gamma\varepsilon}(3,3,q^{\nu}_{i,\gamma}) &= q^{\nu}_{j,\varepsilon}\\
F_{i\gamma\varepsilon}(2,1,r^{\varepsilon}_{i,\gamma}) &= H^{\varepsilon}_{j}\\
F_{i\gamma\varepsilon}(3,2,H^{\gamma}_{i}) &= q^{\delta}_{j,\varepsilon}
\end{aligned}
$$

We would like to declare that in $\mathbf{A}(M)$ the operation $F_{i\gamma\varepsilon}$ results in the default value 0 except in the cases above. Ultimately, this won't do since we will find it necessary to introduce barred versions of all those q's, r's, and H's with all the attached subscripts and superscripts in order to control the finite subdirectly irreducible algebras. So we will have to revisit the definition of $F_{i\gamma\varepsilon}$. For the present, it is no great distortion to think that all the other values are 0.

A similar analysis of right-moving instructions leads the ternary operations $F_{i\gamma\varepsilon}$ being defined (with caveats about barred elements) in $\mathbf{A}(M)$ via

$$
\begin{aligned}
F_{i\gamma\varepsilon}(1,1,r^{\nu}_{i,\gamma}) &= r^{\nu}_{j,\varepsilon}\\
F_{i\gamma\varepsilon}(3,3,q^{\nu}_{i,\gamma}) &= q^{\nu}_{j,\varepsilon}\\
F_{i\gamma\varepsilon}(2,1,H^{\gamma}_{i}) &= r^{\delta}_{j,\varepsilon}\\
F_{i\gamma\varepsilon}(3,2,q^{\varepsilon}_{i,\gamma}) &= H^{\varepsilon}_{j}
\end{aligned}
$$

With this definition, in $\mathbf{A}(M)^{\mathbb{Z}}$

$$F_{i\gamma\varepsilon}(\alpha_n,\alpha_{n+1},\beta) = M(\beta)$$

provided β is as above, ε is the symbol on tape square $n+1$, and $[i,\gamma,\delta,R,j]$ is an instruction of M. For a given Turing machine M, the definition of $F_{i\gamma\varepsilon}$ is unambiguous, since whether $F_{i\gamma\varepsilon}$ should be left or right moving can be determined from M, i, and γ.

These operations can be envisioned as in something like the way automatic algebra operations were, where, however, the edges representing a particular operation now have two labels. See Figure 7.17

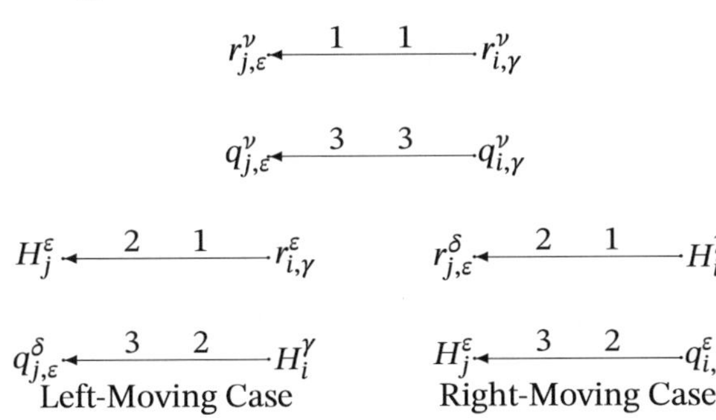

FIGURE 7.17. Machine operations

Here is a useful fact, apparent Figure 7.17.

Fact 0. *If λ is a basic translation on $\mathbf{A}(M)$ associated with one of the operations $F_{i\gamma\varepsilon}$, and $\lambda(a) = \lambda(b) \neq 0$, then $a = b$. The same is true for every translation built only using the basic operations $F_{i\gamma\varepsilon}$, various choices of i, γ, and ε allowed.* ■

Now suppose β codes a configuration Q over a sequentiable F indexed by I. Then we use $M(\beta)$ to be the code of $M(Q)$ over F with index set I. On the basis of these definitions, we obtain the following very useful conclusion.

The Key Coding Lemma: *Let M be a Turing machine, and let X be a set. Let F be a sequentiable set for $\mathbf{A}(M)^X$ and let i be a nonhalting state of M. Finally, let $\gamma, \varepsilon \in \{0, 1\}$ and let f, g, and β be any elements of $A(M)^X$.*

Then $F_{i\gamma\varepsilon}(f, g, \beta) = M(\beta)$ if

- *β codes a configuration Q over F,*
- *i and γ are the first two components of the M instruction determined by Q,*
- *$f, g \in F$ with $f \prec g$ and these two elements refer to the two adjacent tape squares involved in the motion called for in the instruction,*
- *ε is the symbol in the square to which the reading head is being moved, and*
- *$M(\beta)$ codes the configuration $M(Q)$ over F;*

Otherwise 0 *occurs in $F_{i\gamma\varepsilon}(f, g, \beta)$.* ■

$\mathbf{A}(M)$ and what happens if M doesn't halt

The basic plan is to do for $\mathbf{A}(M)$ what we did for $\mathbf{A}$ in §7.10. We were able to prove for $\mathbf{A}$ three crucial things:

(1) $\mathbf{Q}_{\mathbb{Z}}$ is in the variety generated by $\mathbf{A}$ (and hence that variety is inherently nonfinitely based and had a countably infinite subdirectly irreducible member).

(2) Any finite subdirectly irreducible in the variety, except possibly a few very small ones, has a very well determined structure (in fact they are all embeddable into $\mathbf{Q}_{\mathbb{Z}}$).

(3) There are no other infinite subdirectly irreducible algebras in the variety.

It is the second point that compelled us to adjoin additional elements and operations to our original 6-element algebra. Having done that, we had to revisit the first point to assure ourselves that the new elements and operations were innocuous. The third point depends on the first two and Dziobiak's Theorem.

Proceeding along the same lines with $\mathbf{A}(M)$ we are able to do the following:

(1) $\mathbf{Q}_{\mathbb{Z}}$ is in the variety generated by $\mathbf{A}(M)$, provided M does not halt.

(2) In the event that M halts, the cardinality of any finite subdirectly irreducible can be bounded by a function of the size of M and the number of tape squares it visits before halting.

(3) In the event that M halts, the variety generated by $\mathbf{A}(M)$ has no infinite subdirectly irreducible algebras.

(4) In the event that M halts, the variety generated by $\mathbf{A}(M)$ is finitely based.

In the second point, at the cost of adding more elements and more operations to our 8-element algebra $\mathbf{A}$, we can ensure that any sequentiable set arising in the construction of a finite subdirectly irreducible cannot be large enough to accommodate the full halting computation. (The idea is that being able to reach a "halting configuration" would force the forbidden $(x \wedge y) \vee (x \wedge z)$ to be a polynomial.) Then we need to argue that bounding the size of sequentiable sets entails a bound on the subdirectly irreducible algebra itself. In the first point, after making an inessential modification to $\mathbf{Q}_{\mathbb{Z}}$ to make it into an algebra of the correct signature, it is the inaccessibility of the codes of halting configurations that ensures that the extra operations we had to add to accomplish the second point are innocuous. The third point is an immediate consequence of Quackenbush's Theorem (Corollary 7.78). The fourth point is a consequence of Theorem 7.24.

The algebra $\mathbf{A}(M)$

Let M be a Turing machine with states $0, 1, \ldots, m$. The universe of the algebra $\mathbf{A}(M)$ is easiest to describe in pieces. For each of the $4m + 4$ choices of $i = 0, 1, \ldots, m$ and $\gamma, \delta \in \{0, 1\}$, we need four distinct elements denoted by $q^{\delta}_{i,\gamma}$, $\overline{q^{\delta}_{i,\gamma}}$, $r^{\delta}_{i,\gamma}$, and $\overline{r^{\delta}_{i,\gamma}}$. For each of the $2m + 2$ choices of $i = 0, 1, \ldots, m$ and $\gamma \in \{0, 1\}$, we need two elements denoted by H^{γ}_i and $\overline{H^{\gamma}_i}$. The unbarred versions were needed to code configurations. The barred versions help us control the finite subdirectly irreducible algebras. Let V be the set comprised of all $20m + 20$ of these elements. We also let V_i denote the set of 20 elements of V whose first lower index is i. In particular, V_0 contains all the elements used in coding halting configurations. The universe of $\mathbf{A}(M)$ is just

$$A(M) = \{0\} \cup U \cup W \cup V$$

where $U = \{1, 2, 3\}$ and $W = \{q, \bar{q}, r, \bar{r}\}$. Thus the size of $\mathbf{A}(M)$ is $20m + 28$ where m is the number of nonhalting states of M.

The old algebra $\mathbf{A}$ will be a subreduct of $\mathbf{A}(M)$. Indeed, we insist that $\wedge$ make $\mathbf{A}(M)$ into a height 1 meet-semilattice with least element 0, and that any product involving a new element results in 0. The definitions of the remaining old operations are changed little or not at all. Here are the J's:

$$J(x, y, z) = \begin{cases} x & \text{if } x = y \neq 0 \\ x \wedge z & \text{if } x = \bar{y} \in V \cup W \\ 0 & \text{otherwise.} \end{cases}$$

$$J'(x, y, z) = \begin{cases} x \wedge z & \text{if } x = y \neq 0 \\ x & \text{if } x = \bar{y} \in V \cup W \\ 0 & \text{otherwise.} \end{cases}$$

Along with the old S's we insert one more:

$$S_0(u, v, x, y, z) = \begin{cases} (x \wedge y) \vee (x \wedge z) & \text{if } u \in V_0, \\ 0 & \text{otherwise.} \end{cases}$$

$$S_1(u, v, x, y, z) = \begin{cases} (x \wedge y) \vee (x \wedge z) & \text{if } u \in \{1, 3\}, \\ 0 & \text{otherwise.} \end{cases}$$

$$S_2(u, v, x, y, z) = \begin{cases} (x \wedge y) \vee (x \wedge z) & \text{if } u = \bar{v} \in V \cup W, \\ 0 & \text{otherwise.} \end{cases}$$

Along with the old U^0 we insert two new operations $U^1_{i\gamma\varepsilon}$ and $U^2_{i\gamma\varepsilon}$ for each of the $4m$ choices of i, γ, and ε, where i is a nohalting state:

$$U^0(x,y,z,w) = \begin{cases} xy & \text{if } xy = zw \neq 0 \text{ and } x = z \text{ and } y = w \\ \overline{xy} & \text{if } xy = zw \neq 0 \text{ and } x \neq z \text{ or } y \neq w \\ 0 & \text{otherwise.} \end{cases}$$

$$U^1_{i\gamma\varepsilon}(x,y,z,w) = \begin{cases} F_{i\gamma\varepsilon}(x,y,w) & \text{if } x \prec z \text{ and } F_{i\gamma\varepsilon}(x,y,w) \neq 0 \text{ and } y = z \\ \overline{F_{i\gamma\varepsilon}(x,y,w)} & \text{if } x \prec z \text{ and } F_{i\gamma\varepsilon}(x,y,w) \neq 0 \text{ and } y \neq z \\ 0 & \text{otherwise.} \end{cases}$$

$$U^2_{i\gamma\varepsilon}(x,y,z,w) = \begin{cases} F_{i\gamma\varepsilon}(y,z,w) & \text{if } x \prec z \text{ and } F_{i\gamma\varepsilon}(y,z,w) \neq 0 \text{ and } x = y \\ \overline{F_{i\gamma\varepsilon}(y,z,w)} & \text{if } x \prec z \text{ and } F_{i\gamma\varepsilon}(y,z,w) \neq 0 \text{ and } x \neq y \\ 0 & \text{otherwise.} \end{cases}$$

Finally, we need the $4m$ ternary operations $F_{i\gamma\varepsilon}$ introduced above (but extended to accommodate the barred elements of V) and one further unary operation which serves to set up initial configurations:

$$I(x) = \begin{cases} q^0_{1,0} & \text{if } x = 1, \\ H^0_1 & \text{if } x = 2, \\ r^0_{1,0} & \text{if } x = 3, \\ 0 & \text{otherwise.} \end{cases}$$

Notice that for outputs other than 0, the operation I is one-to-one. In this way, the next fact is an extension of Fact 0

Fact 1. *If λ is any translation of $\mathbf{A}(M)$ built only from the basic operations I and $F_{i\gamma\varepsilon}$, various choices of i, γ, and ε allowed, and $\lambda(a) = \lambda(b) \neq 0$, then $a = b$.*

While all this is relatively intricate, the F's and the I plainly help us emulate the computations of the Turing machine. The role of the S's is to prevent certain kinds of elements from getting into the picture during the construction of finite subdirectly irreducible algebras. U^0 was crucial to get a kind of unique decomposition result for $\cdot$ in the finite subdirectly irreducible algebras. The U^1 and U^2 operations play a similar role in connection with the F operations.

What Happens If M Does Not Halt on the Blank Tape

Now we expand $\mathbf{Q}_{\mathbb{Z}}$ to the signature appropriate to M by insisting that all the following equations hold in the expansion:

$$
\begin{aligned}
U^0(x,y,z,w) &\approx (xy)\wedge(zw) & S_0(u,v,x,y,z) &\approx 0\\
J(x,y,z) &\approx x\wedge y & S_1(u,v,x,y,z) &\approx 0\\
J'(x,y,z) &\approx x\wedge y\wedge z & S_2(u,v,x,y,z) &\approx 0\\
F_{i\gamma\varepsilon}(x,y,w) &\approx 0 & I(x) &\approx 0\\
U^1_{i\gamma\varepsilon}(x,y,z,w) &\approx 0 & U^2_{i\gamma\varepsilon}(x,y,z,w) &\approx 0
\end{aligned}
$$

for all choices of i,γ and ε.

This sort of inessential expansion leaves its key properties intact: any locally finite variety to which (this expanded) $\mathbf{Q}_{\mathbb{Z}}$ belongs will be inherently nonfinitely based, and $\mathbf{Q}_{\mathbb{Z}}$ has a countably infinite subalgebra $\mathbf{Q}_\omega$ which is subdirectly irreducible.

THEOREM 7.85. *If M does not halt, then $\mathbf{Q}_{\mathbb{Z}}$ belongs to the variety generated by $\mathbf{A}(M)$. In particular, if M does not halt, then $\mathbf{A}(M)$ is inherently nonfinitely based and the variety it generates has a countably infinite subdirectly irreducible algebra.*

Proof. We follow the pattern set in the proof of Theorem 7.83. For each $p\in\mathbb{Z}$ we take $\alpha_p,\beta_p\in A(M)^{\mathbb{Z}}$ to be the same elements we used before:

$$
\begin{array}{lcccccccccc}
\alpha_p{:=} & \cdots & 1 & 1 & 1 & 2 & 3 & 3 & 3 & \cdots\\
\beta_p{:=} & \cdots & r & r & r & q & q & q & q & \cdots
\end{array}
$$

where the change in β_p is taking place at the pth position. Next we let $B_1=\{\alpha_p : p\in\mathbb{Z}\}\cup\{\beta_p : p\in\mathbb{Z}\}$ and we take $\mathbf{B}$ to be the subalgebra of $\mathbf{A}(M)^{\mathbb{Z}}$ generated by B_1. Let B_0 denote the subset of B consisting of all those $\mathbb{Z}$-tuples in B which contain at least one 0. The set $\{\alpha_p : p\in\mathbb{Z}\}$ is sequentiable and consists of all the tuples in B belonging to $U^{\mathbb{Z}}$, since none of the operations of $\mathbf{A}(M)$ ever produces an element of U. Now for every $p\in\mathbb{Z}$, let

$$
\begin{array}{lcccccccccc}
I(\alpha_p){:=} & \cdots & r^0_{1,0} & r^0_{1,0} & r^0_{1,0} & H^0_1 & q^0_{1,0} & q^0_{1,0} & q^0_{1,0} & \cdots
\end{array}
$$

which gives the code of a configuration (the all-0 tape with the machine in state 1 reading square p). The $F_{i\gamma\varepsilon}$'s may now be applied, step by step, to produce the codes of further configurations reached as the computation of M proceeds. Plainly, all these codes of configurations belong to B. Let C denote the set of all these configuration codes. We will prove that $C\cup B_0\cup B_1$ is a subuniverse of $\mathbf{A}(M)^{\mathbb{Z}}$, and therefore $B=C\cup B_0\cup B_1$.

Now let Φ be the map defined from B to $\mathbf{Q}_{\mathbb{Z}}$ via

$$\Phi(x) = \begin{cases} a_p & \text{if } x = \alpha_p \text{ for some } p \in \mathbb{Z}, \\ b_p & \text{if } x = \beta_p \text{ for some } p \in \mathbb{Z}, \\ 0 & \text{otherwise} \end{cases}$$

We contend that Φ is a homomorphism from $\mathbf{B}$ onto $\mathbf{Q}_{\mathbb{Z}}$. To verify this, as well as that $C \cup B_0 \cup B_1$ is a subuniverse, requires us to examine the behavior of each of our operations on $C \cup B_0 \cup B_1$. For each operation in turn, we show that this set is closed and that Φ preserves the operation.

CASE 0: Evidently $0 = \ldots, 0, 0, 0, 0, \cdots \in B_0$ and so $\Phi(0) = 0$.

CASE $\wedge$: Evidently, $u \wedge v = u$ if $u = v$ and $u \wedge v \in B_0$ if $u \neq v$, for all $u, v \in C \cup B_0 \cup B_1$. Hence, our set is closed under $\wedge$ and $\Phi(u \wedge v) = \Phi(u) \wedge \Phi(v)$.

CASE $\cdot$: Clearly, $\alpha_p \cdot \beta_{p+1} = \beta_p$ for all $p \in \mathbb{Z}$, with all other $\cdot$-products resulting in elements of B_0. So our set is closed under $\cdot$ and Φ preserves $\cdot$.

CASE $F_{i\gamma\varepsilon}$: According to the Key Coding Lemma, the results of applying $F_{i\gamma\varepsilon}$ to members of $C \cup B_0 \cup B_1$ lie in $C \cup B_0$. Hence, $C \cup B_0 \cup B_1$ is closed under this operation and Φ preserves the operation.

CASE I: Applied to elements of $C \cup B_0 \cup B_1$, I produces only elements of $C \cup B_0$. Hence, $C \cup B_0 \cup B_1$ is closed with respect to I, and Φ preserves I.

Observe that no barred elements occur in any of the members of $C \cup B_1$. It follows that

($\star$) if $u, v \in C \cup B_0 \cup B_1$ with $u(p) = v(p) \neq 0$ or $u(p) = \overline{v(p)} \in V \cup W$ for all $p \in \mathbb{Z}$, then $u = v$.

CASE J: Evidently, $J(x, y, z) \in B_0$ if $x \in B_0$ or $y \in B_0$ or $x \neq y$, according to ($\star$). Otherwise, $J(x, y, z) = x$. This entails that $C \cup B_0 \cup B_1$ is closed under J and Φ preserves J.

CASE J': Likewise, $J'(x, y, z) \in B_0$ if $x \in B_0$ or $y \in B_0$ or $x \neq y$, according to ($\star$). Otherwise, $J'(x, y, z) = x \wedge z$. This entails that $C \cup B_0 \cup B_1$ is closed under J and Φ preserves J.

CASE U^0: If $xy \in B_0$ or $zw \in B_0$ or $xy \neq zw$, then we have $U^0(x, y, z, w) \in B_0$ and $(xy) \wedge (zw) \in B_0$, for any elements $x, y, z, w \in C \cup B_0 \cup B_1$. On the other hand, if $xy = zw \notin B_0$ it must be that $x = z = \alpha_p$ and $y = w = \beta_{p+1}$ for some $p \in \mathbb{Z}$. In that case, $U^0(x, y, z, w) = xy = zx = (xy) \wedge (zw)$. Thus, $C \cup B_0 \cup B_1$ is closed under U^0 and Φ preserves U^0.

Observe that for $u, v \in C \cup B_0 \cup B_1$, we have $u \prec v$ only when $u = \alpha_p$ and $v = \alpha_{p+1}$ for some $p \in \mathbb{Z}$. In particular,

($*$) With respect to $\prec$, every element of $C \cup B_0 \cup B_1$ has at most one predecessor and at most one successor.

CASE $U^1_{i\gamma\varepsilon}$: In case $F_{i\gamma\varepsilon}(x,y,w) \in B_0$ or $x \nprec z$, we have $U^1_{i\gamma\varepsilon}(x,y,z,w) \in B_0$. In the alternative case, it follows from the definition of $F_{i\gamma\varepsilon}$ that $x \prec y$. In view of $(*)$ it must be that $y = z$. So $U^1_{i\gamma\varepsilon}(x,y,z,w) = F_{i\gamma\varepsilon}(x,y,w) \in C$. Therefore, the application of $U^1_{i\gamma\varepsilon}$ always results in an element of $C \cup B_0$. Consequently, $C \cup B_0 \cup B_1$ is closed with respect to $U^1_{i\gamma\varepsilon}$ and Φ preserves this operation.

CASE $U^2_{i\gamma\varepsilon}$: This case is like the one above, but it exploits the uniqueness of predecessors instead of successors.

CASE S_0: Since M does not halt, the set $V_0^{\mathbb{Z}}$ is disjoint from $C \cup B_0 \cup B_1$. It follows that the application of S_0 always results in an element of B_0. Thus S_0 is preserved by Φ and $C \cup B_0 \cup B_1$ is closed with respect to S_0. **It should be noted that this is the sole place in the argument where the fact that M does not halt comes into play.**

CASE S_1: The set $\{1,3\}^{\mathbb{Z}}$ is disjoint from $C \cup B_0 \cup B_1$. It follows that the application of S_1 always results in an element of B_0. Thus S_1 is preserved by Φ and $C \cup B_0 \cup B_1$ is closed with respect to S_1.

CASE S_2: It follows from $(\star)$ that the application of S_2 always results in an element of B_0. Thus S_2 is preserved by Φ and $C \cup B_0 \cup B_1$ is closed with respect to S_2.

So $\mathbf{Q}_{\mathbb{Z}}$ belongs to the variety generated by $\mathbf{A}(M)$. ■

When M halts: finite subdirectly irreducible algebras of sequentiable type

Throughout this subsection we assume that M is a Turing machine that eventually halts when started on the all-0 tape. We denote by $\pi(M)$ the number of squares examined by M in the course of its computation. Thus $\pi(M)$ is the length of the stretch of tape which comes into use for this computation. Our ambition is to describe all the finite subdirectly irreducible algebras in the variety generated by $\mathbf{A}(M)$, or at any rate to bound their size. From the facts developed above, we already have a lot of information at our disposal. Once again we take $\mathbf{S}$ to be a finite subdirectly irreducible algebra in the variety and we fix a finite set T, $\mathbf{B}$, and θ, so that

- $\mathbf{B} \subseteq \mathbf{A}(M)^T$
- θ is strictly meet-irreducible in Con $\mathbf{B}$.
- $\mathbf{S}$ is isomorphic to $\mathbf{B}/\theta$.
- T is as small as possible for representing $\mathbf{S}$ in this way.
- $|T| > 1$ (i.e. $\mathbf{S} \notin HS\mathbf{A}(M)$).

Among other things, we know that $(x \wedge y) \vee (x \wedge z)$ is not a polynomial of $\mathbf{B}$ (Fact 5 in §7.10). We also have an element $p \in B$ so that (p,q) is critical over θ. In the previous section the analysis revealed that all the elements of S, except 0, arose from a unique longest factorization of p using the product $\cdot$. We want, loosely speaking, to do the same thing now; but the machine operations I and $F_{i\gamma\varepsilon}$ have to be considered along with $\cdot$. We will change the definition of B_1. Thus, the facts that grew out of our analysis of the old version of B_1 must be re-examined. Also, Fact 9 was proved using an analysis by cases, with one case for each basic operation. Now we have more operations. Finally, we have modified all the old operations by extending their domains, (in the case of J, J', and S_2, we have done this by treating the new elements in V like the elements in W). However, in all its essential features the old analysis can be carried forward.

We take B_0 to be the collection of all elements of B which contain at least one 0. In **B** the ranges of S_0, S_1, and S_2 lie entirely in B_0. Moreover, V_0^T and $\{1,3\}^T$ are disjoint from B and there are no elements $u, v \in B$ so that $u = \bar{v} \in (V \cup W)^T$. This is just a direct consequence of Fact 5, page 285 in §7.10.

Fact 2. *Every sequentiable subset of B has fewer than $\pi(M)$ members.*

Proof. By the Key Coding Lemma any large enough sequentiable set would allow us, using I and the $F_{i\gamma\varepsilon}$'s, to emulate in **B** the entire halting computation of M, producing an element of V_0^T in B. ■

Next we restate a part of Fact 8 of §7.10 in our expanded setting. The only difference is the insertion of V in the statement and the proof.

Fact 3. *If $v \in B$ and $p(s) = v(s)$ or $p(s) = \overline{v(s)} \in V \cup W$ for all $s \in T$, then $p = v$.* ■

The next fact splits our analysis into two cases.

Fact 4. *Either $p \in V^T$ or $p \in W^T$.*

Proof. First notice that there must be a nonconstant translation f and $u \in B$ with $f(u) = p$ but $u \neq p$. Otherwise, it follows from Fact 3 from §7.10 that $B - \{p\}$ is a θ-class. This means that our subdirectly irreducible algebra **S** has only two elements, and indeed is isomorphic to a subalgebra of $\mathbf{A}(M)$. This contradicts our assumption that T has at least two elements.

Let λ be a nonconstant translation of least complexity so that for some $u \in B$ with $u \neq p$ we have $\lambda(u) = p$. Also fix such a u. Now the rest of the argument falls into cases according to the leading operation symbol of λ.
CASE $\wedge$: $\lambda(x) = \mu(x) \wedge r$. Then $p = \mu(u) \wedge r$ Since p is maximal, we conclude that $p = \mu(u)$. This leads to a violation of the minimality of λ.
CASE $\cdot$: The range of λ is included in $B_0 \cup W^T$. This means $p \in W^T$.
CASE I: The range of λ is included in $B_0 \cup V^T$. This means $p \in V^T$.
CASES $F_{i\gamma\varepsilon}$: The range of λ is included in $B_0 \cup V^T$. So $p \in V^T$.
CASES S_i: Impossible: the range of each S_i is included in B_0.
CASES $U^0, U^j_{i\gamma\varepsilon}$: These cases put $p \in W^T$ (for U^0) or $p \in V^T$ (for $U^j_{i\gamma\varepsilon}$'s).
CASE J: $\lambda(x) = J(\mu(x), r, s)$, or $\lambda(x) = J(r, \mu(x), s)$, or $\lambda(x) = J(r, s, \mu(x))$. Under the first alternative, $p = \lambda(u) = J(\mu(u), r, s) \leq \mu(u)$. Then $p = \mu(u) = r$ by Fact 3 and the maximality of p. This violates the minimality of λ. The same reasoning applies to the second alternative. So consider the last alternative. Then $p = J(r, s, \mu(u)) \leq r$. Then $p = r$, and so Fact 3 implies that $p = r = s$. But this means that $\lambda(x) = J(p, p, \mu(x)) = p$, and so λ is constant. This case is impossible.
CASE J': This is like the last case, but easier. ■

S is of *sequentiable* type if $p \in W^T$ and of *machine* type otherwise.

Fact 5. *Finite subdirectly irreducibles of sequentiable type have fewer than $2\pi(M)$ members.*

Proof. We can just follow the old analysis for **A**, paying a modest amount of attention to the additional operations, and observing that a sequentiable set arises in a natural way.

Now $p \in W^T$. Let B_1 be the set of all factors of p with respect to $\cdot$. Now *all* our previously established facts hold, as is evident in all cases except for Fact 9. This fact asserts that, *if $u \in B$ and $\lambda(u) \in B_1$ for some nonconstant translation λ, then $u \in B_1$.* The proof of Fact 9 relied on a case-by-case analysis according to the leading operation symbol. To get a proof for Fact 9 in our expanded signature, we have to consider the operations $I, F_{i\gamma\varepsilon}, U^1_{i\gamma\varepsilon}, U^2_{i\gamma\varepsilon}$, and S_0. (Actually, there are also minor changes in the definitions of J, J', and S_2, which merit a small amount of attention not provided here.) All these cases are trivial because $\lambda(u) \notin B_1$ for any u if the leading operation is any of these, since $B_1 \subseteq U^T \cup W^T$.

As in our analysis for **A**, we have $B_1 = \{a_0, a_1, \dots, a_{n-1}\} \cup \{b_0, b_1, \dots, b_n\}$ where $b_k = a_k b_{k+1}$ for all $k < n$ and $b_0 = p$. Also $B - B_1$ is the θ-class of 0, B_1 splits into singletons modulo θ, and $a_k \in U^T$ and $b_k \in W^T$ for all k. It remains to see that $\{a_k : k < n\}$ is a sequentiable set. Since $\pi(M) - 1$ bounds the size of sequentiable sets, we would be finished. We need $a_k \prec a_{k+1}$ for all $k < n-1$. Let $t \in T$, and suppose first that $a_{k+1}(t) = 1$. Then $b_k(t) \in \{r, \overline{r}\}$, so $a_k(t) \in \{1, 2\}$. Hence $a_k(t) \prec a_{k+1}(t)$. Next, suppose that $a_{k+1}(t) = 2$. Then $b_k(t) \in \{q, \overline{q}\}$, so $a_k(t) = 3$. Hence, $a_k(t) \prec a_{k+1}(t)$. Finally, suppose $a_{k+1}(t) = 3$. Then $b_k(t) \in \{q, \overline{q}\}$, so $a_k(t) = 3 \prec 3 = a_{k+1}(t)$. thus, $a_k \prec a_{k+1}$ and $\{a_k : k < n\}$ is sequentiable. ■

When M halts: finite subdirectly irreducible algebras of machine type

We now consider the case when the finite subdirectly irreducible algebra **S** is of machine type. So we have $p \in V^T$. In this case, we let B_1 be the smallest subset of B which includes p and which is closed under the inverses of all the machine operations I and $F_{i\gamma\varepsilon}$. Hence,

$$B_1 = \{u : \lambda(u) = p \text{ for some nonconstant translation } \lambda \text{ of } \mathbf{A}(M) \text{ built only from the machine operations}\}$$

It is easy to see that since $p \in V^T$, then $B_1 \subseteq U^T \cup V^T$. It also follows that if λ is a translation built up from the machine operations, and $\lambda(u) = p$, then all the elements of $A(M)$ used in the formation of λ also belong to B_1.

Since we have now substantially altered the definition of B_1, we will need to reexamine Facts 8 and 9 (pages 288–288 in §7.10). Here is the new version of Fact 8. It is an immediate consequence of Fact 3 (page 306) and Fact 1 (page 302).

Fact 6. *If $u \in B_1$ and $v \in B$ so that for all $s \in T$ either $u(s) = v(s)$ or $u(s) = \overline{v(s)} \in V \cup W$, then $u = v$.*

Here is the new version of Fact 9. The statement has not changed, but the proof is different, accommodating the change in the definition of B_1.

Fact 7. *If $u \in B$ and $\lambda(u) \in B_1$ for some nonconstant translation λ, then $u \in B_1$.*

Proof. The proof is by induction on the complexity of λ. The initial step of the induction is obvious, since the identity function is the only simplest nonconstant translation.

For the inductive step we take $\lambda(x) = \nu(\mu(x))$, where $\nu(x)$ is a basic translation and $\mu(x)$ is a translation with smaller complexity than λ. The work breaks down into cases according to the basic operation associated with ν.

CASE $\wedge$: $\lambda(x) = \mu(x) \wedge r$. But every element of B_1 is maximal with respect to the semilattice order. So $\lambda(u) = \mu(u) \in B_1$. Invoking the induction hypothesis for $\mu(x)$, we get $u \in B_1$.

CASE $\cdot$: This cannot happen since then the range λ would be included in $B_0 \cup W^T$, which is disjoint from B_1.

CASES $F_{i\gamma\varepsilon}$: Since $(F_{i\gamma\varepsilon}(\mu(u)) = \lambda(u) \in B_1$, it follows from the definition of B_1, that $\mu(u) \in B_1$. Now the induction hypothesis applies.

CASE I: $\lambda(x) = I(\mu(x))$. By the definition of B_1, $\mu(u) \in B_1$. So the induction hypothesis applied.

CASE J: $\nu(x) = J(v, y, z)$, where x is one of v, y, and z, while the remaining two are coefficients. First, suppose x is either v or y. From Fact 6 and the maximality of the members of B_1 it follows that $\mu(u) = \lambda(u) \in B_1$. So the induction hypothesis applies. Now suppose x is z and so v and y are coefficients. In this case, it follows from Fact 6 that $v = y = \lambda(u) \in B_1$. But this means that $\nu(x) = v$ and so λ is constant. That cannot happen.

CASE J': This case is easier than the last one and its discussion is omitted.

CASES S_0, S_1 AND S_2: Too easy—the range of λ would be included in B_0.

CASE U^0: This cannot happen since the range of λ would be included in $B_0 \cup W^T$, which is disjoint from B_1.

CASES $U^j_{i\gamma\varepsilon}$: $\nu(x) = U^j_{i\gamma\varepsilon}(v, y, z, w)$, where exactly one of v, y, z, and w is x and the remaining ones are coefficients, which we will regard as constant functions.

The other cases being similar, we suppose that $j = 1$. Evidently, $\lambda(u)$ and $F_{i\gamma\varepsilon}(v(u), y(u), w(u))$ satisfy the hypotheses of Fact 6. So

$$\lambda(u) = F_{i\gamma\varepsilon}(v(u), y(u), w(u)) = F_{i\gamma\varepsilon}(v(u), z(u), w(u))$$

(since also $y(u) = z(u)$) follows from the definition of $U^1_{i\gamma\varepsilon}$. So $v(u), y(u), z(u), w(u) \in B_1$, by the definition of B_1. So $\mu(u) \in B_1$ and the induction hypothesis applies. ■

Here is the new version of Fact 10 from §7.10. Again, the statement is the same, but B_1 has a new meaning. The proof is like that for Fact 10, but it uses Fact 7 in place of Fact 9 and Fact 1 in place of Fact 6.

Fact 8. *$u/\theta = \{u\}$ for each $u \in B_1$ and $0/\theta = B - B_1$.* ■

Thus to bound the cardinality of $\mathbf{S}$ we need to bound $|B_1|$. This is our next task. However, here we can remark that in fact a complete analysis of finite subdirectly irreducible algebras of machine type, as well as those of sequentiable type, is at hand. This further analysis would describe the behavior of all the operations. We will not pursue this more detailed analysis, except to point out that all these subdirectly irreducible algebras are flat.

We can suppose that no component of $p \in V^T$ is a barred element. (The basic reason is that the operations $F_{i\gamma\varepsilon}$ do not alter whether a symbol is barred. Hence the distribution of bars in any member of $B_1 \cap V^T$ is the same as the distribution of bars in

p.) Now $B_1 \subseteq U^T \cup V^T$. Let $\Omega = B_1 \cap V^T$ and $\Sigma = B_1 \cap U^T$. Look first in more detail at Ω. We define Ω_n by the following recursion.

$$\begin{aligned}\Omega_0 &= \{p\}\\ \Omega_{n+1} &= \Omega_n \cup \{u \in B_1 \,:\, F_{i\gamma\varepsilon}(f,g,u) \in \Omega_n \text{ for some } f,g \in B \text{ and some } i,\gamma,\varepsilon\}\end{aligned}$$

Evidently, $\Omega = \bigcup_n \Omega_n$. We will say that $f \in U^T$ **matches** $v \in V^T$ provided for all $t \in T$

$$\begin{aligned}f(t) = 1 &\Leftrightarrow v(t) \text{ is a } r^{\nu}_{i\gamma}\\ f(t) = 2 &\Leftrightarrow v(t) \text{ is an } H^{\gamma}_{i}\\ f(t) = 3 &\Leftrightarrow v(t) \text{ is a } q^{\nu}_{i\gamma}\end{aligned}$$

Observe that every $v \in V^T$ matches exactly one $f \in U^T$. For each natural number n, we let $\Sigma_n = \{f \in \Sigma \,:\, f \text{ matches } v \text{ for some } v \in \Omega_n\}$. By referring to the definition of $F_{i\gamma\varepsilon}$, we have that the elements of the two element set $\{f,g\}$ match the elements of the two element set $\{u,v\}$ whenever $F_{i\gamma\varepsilon}(f,g,u) = v \in \Omega$ (the order in which this matching occurs depends on whether the underlying Turing machine instruction is right-moving or left-moving). It follows that $\Sigma = \bigcup_n \Sigma_n$.

Fact 9. *Σ is a sequentiable set.*

Proof. We argue by induction that Σ_n is sequentiable.

INITIAL STEP: Observe that Σ_0 has only one element. (Σ_0 cannot be empty, since then our subdirectly irreducible **S** would be in ***HS*** $\mathbf{A}(M)$.) Since $\Sigma_0 \subseteq B_1 \cap U^T$ and B is disjoint for $\{1,3\}^T$, we see that its element has to have H in at least one place. Thus, Σ_0 is a sequentiable set.

INDUCTIVE STEP: Suppose $h \in \Sigma_{n+1} - \Sigma_n$. Pick $u \in \Omega_{n+1} - \Omega_n$ so that h matches u. Further, pick $F_{i\gamma\varepsilon}, f, g$, and v so that $F_{i\gamma\varepsilon}(f,g,u) = v \in \Omega_n$. It does no harm to suppose that we have a left-moving operation. So g matches u and f matches v. It follows that $h = g$, that $f \in \Sigma_n$, and that $f \prec g$. By the inductive hypothesis, we have that Σ_n is sequentiable. Let us display Σ_n as

$$f_a \prec f_{a+1} \prec \cdots \prec f_b$$

In the event that $f = f_b$ we have $\Omega_n \cup \{h\}$ sequentiable as desired. On the other hand, if $f = f_c$ for some $c < b$, then, in view of Fact 6, we know $U^1_{i\gamma\varepsilon}(f,h,f_{c+1},u) = F_{i\gamma\varepsilon}(f,h,u)$. So we would be able to conclude that $h = f_{c+1} \in \Sigma_n$, contrary to our choice of h.

Right-moving operations are handled in a way similar to what we just did for left-moving operations, but using $U^2_{i\gamma\varepsilon}$. ■

Fact 10. *Σ has fewer than $\pi(M)$ elements.* ■

To obtain a bound on the cardinality of Ω we must recall that the sequentiable set Σ partitions T into $T_L, T_a, \dots, T_b, T_R$ where $\Sigma = \{f_a, \dots, f_b\}$.

Fact 11. *$u|_{T_c}$ is constant for each $u \in \Omega$ and each $c \in \{a, \dots, b\}$.*

Proof. The proof is accomplished in stages, each stage showing that more elements of Ω are constant on more T_c's until everything is accomplished. This proof needs some preliminary observations.

Suppose that $u \in \Omega_{n+1} - \Omega_n$ with $F_{i\gamma\varepsilon}(f_c, f_{c+1}, u) = v \in \Omega_n$. In this case we will say that u, c and $c+1$ become *active* at stage $n+1$. (We regard p as the only element active at stage 0 and no member of $c \in \{a, \dots, b\}$ as active at stage 0.) The definition of $F_{i\gamma\varepsilon}$ entails that $u|_{T_c}, u|_{T_{c+1}}, v|_{T_c}$ and $v|_{T_{c+1}}$ are all constant. Moreover, for all d, $u|_{T_d}$ is constant if and only if $v|_{T_d}$ is constant. In checking this, it helps to notice that the relevant subscripts and superscripts can all be determined from $F_{i\gamma\varepsilon}$ and the related Turing machine instruction $[i, \gamma, \delta, M, j]$.

Now we argue by induction on n, that every member of Ω_n is constant on T_c for all c that have become active by stage n and that, for all d and all $v, v' \in \Omega_n$, $v|_{T_d}$ is constant if and only if $v'|_{T_d}$ is constant.

The initial step of the induction holds vacuously.

For the inductive step, suppose $u, u' \in \Omega_{n+1} - \Omega_n$ with

$$F_{i\gamma\varepsilon}(f_c, f_{c+1}, u) = v \in \Omega_n \qquad \text{and} \qquad F_{i'\gamma'\varepsilon'}(f_{c'}, f_{c'+1}, u') = v' \in \Omega_n$$

Now our preliminary observations give the conclusions that u and u' are constant on all the d's active by stage n as well as for $c, c', c+1$, and $c'+1$, some of which may have become active for stage $n+1$. Moreover, we also conclude that, for all d, u is constant on T_d if and only if v is constant on T_d if and only if v' is constant of T_d if and only if u' is constant on T_d. In this way, the inductive step is complete. ■

Now we just count things to obtain:

Fact 12. *Ω has no more than $2^s ms$ elements where $s = |\Sigma|$ and m is the number of non-halting states of M.*

Proof. For each $u \in \Omega$ there are no more than s possibilities for $c \in \{a, \dots, b\}$ so that $u(t) = H_i^\gamma$, for some i and some γ and all $t \in T_c$. Having fixed one of these possibilities there are m choices for i and two choices for γ. Now for d with $a \le d < c$ we must have a ν so that $u(t) = r_{i\gamma}^\nu$ for all $t \in T_d$. Thus for each such d there are no more than two possibilities for ν. Likewise, if $c < d \le b$, then there is some ν so that $u(t) = q_{i\gamma}^\nu$ for all $t \in T_d$. Again, for each such d there are no more than two possibilities for ν. Thus far we have bounded the number of possibilities for u by $2^s ms$, as desired—but we still have to examine what $u(t)$ is like when $t \in T_L \cup T_R$. Suppose $t \in T_L$. Then $f_c(t) = 1$ for all $c \in \{a, \dots, b\}$. From the definition of the operations $F_{i\gamma\varepsilon}$, it follows that $u(t) = r_{i\gamma}^\nu$, where ν is determined by $p(t) = r_{i'\gamma'}^\nu$, and i and γ are the same subscripts that occur throughout u. So u is determined on T_L by our previous choices and by the structure of p. Likewise, u is determined on T_R. So the desired bound is established. ■

THEOREM 7.86. *If M halts, then the cardinality of any subdirectly irreducible member of the variety generated by* $\mathbf{A}(M)$ *is no greater than the maximum of* 2π, $2^{(\pi-1)}m(\pi-1)+\pi$ *and* $20m+28$, *where* π *is the number of tape squares used by M in its halting computation and m is the number of nonhalting states of M; moreover, every subdirectly irreducible algebra in the variety is flat.* ∎

The $20m+28$ that occurs above is just the cardinality of $\mathbf{A}(M)$. It bounds the cardinalities of the subdirectly irreducibles that belong to $HSA(M)$. The 2π bounds the cardinalities of the subdirectly irreducible algebras of sequentiable type. The $2^{(\pi-1)}m(\pi-1)+\pi$ bounds the cardinalities of the subdirectly irreducible algebras of machine type.

It is clear that much more was accomplished than just establishing the bound on subdirectly irreducible algebras given above. Our analysis is very close to a complete description (given a description of the behavior of M) of all the subdirectly irreducible algebras, even in the case that M does not halt. The only way in which the hypothesis that M does not halt entered into consideration of the finite subdirectly irreducible algebras was in bounding their size. Further analysis of their structure, similar to what was done in Theorem 7.84 could be carried out.

Finally, we have in hand all the pieces of McKenzie's undecidability results about finite algebras:

THEOREM 7.87 (Ralph McKenzie 1996a). *The set of finite algebras of finite signature that generate residually very finite varieties is not recursive.* ∎

According to Theorem 7.85, if M does not halt, then $\mathbf{A}(M)$ is inherently nonfinitely based. On the other hand, if M halts, then the variety generated by $\mathbf{A}(M)$ has a finite residual bound. Since $\mathbf{A}(M)$ has a semilattice operation it also generates a variety that is congruence meet-semidistributive. So by Willard's Finite Basis Theorem (Theorem 7.24), if M halts, then $\mathbf{A}(M)$ is finitely based. In this way, we obtain McKenzie's resolution of Tarski's Finite Basis Problem.

THEOREM 7.88 (Ralph McKenzie 1996c). *The set of finite algebras of finite signature that are finitely based is not recursive.* ∎

McKenzie's proof of Theorem 7.88 was substantially different.

Observe that the signature of $\mathbf{A}(M)$ depends on the Turing machine M. While this signature is always finite and the ranks of the operations are no more than 5, the number of basic operations cannot be given a finite bound, since Turing machines can have any finite number of instructions. This unsatisfactory situation was addressed earlier by Ralph McKenzie (1984).

THEOREM 7.89 (Ralph McKenzie 1996c). *Let* τ *be a computable signature that provides just two operation symbols: a binary operation symbol and a unary operation symbol. The set of finite algebras of signature* τ *that are finitely based is not recursive.*

Proof. Here is the required reduction. Given a finite algebra **A** (acceptable as input to a computational process) of finite signature:

(1) Recover the signature of **A**. Call it σ.
(2) In view of Corollary 10.111 part (i) on page 211 of Volume III, calculate the numerical parameter k.
(3) Apply McKenzie's functor T_k from Corollary 10.111 to **A** to obtain a finite algebra of signature τ.

According the Corollary 10.111, **A** is finitely based if and only if $T_k[\mathbf{A}]$ is finitely based. ■

Let ρ be a signature that provides only one operation symbol, that one being binary. Ralph McKenzie (1984) by way of a difficult argument showed that ρ can replace τ in Theorem 7.89.

CHAPTER 8

Rudiments of Model Theory

Model theory, a branch of mathematical logic with points of contact all across mathematics, is a rich, highly developed, and active area of mathematics. Indeed, historically the focus of these volumes grew up alongside model theory into the 1970's. Thereafter, while problems based on model-theoretic ideas have continued to challenge general algebra, the methods of these two fields became quite distinct. Many people have made contributions to the development of both model theory and to the general theory of algebraic systems. This chapter is an introduction to model theory that emphasizes those results most frequently applied in the general theory of algebras. The serious student of our field is well advised to acquire an understanding of the deeper aspects of model theory from one of the excellent books which are available: see, for example, (C. C. Chang and H. J. Keisler, 1990; Wilfrid Hodges, 1993; David Marker, 2002). The next section is a slow-paced presentation of the beginnings of elementary mathematical logic and elementary model theory. Readers familiar with this material might skip this section, referring back to it as needed for notation and the details of definitions.

There is a particular algebraic construction, that of reduced products, that plays a prominent role in this chapter. Suppose $\langle \mathbf{A}_i \mid i \in I \rangle$ is a system of algebras all of the same signature. A **reduced product** of this system of algebras will be a special kind of homomorphic image of $\prod_{i \in I} \mathbf{A}_i$. It turns out in many cases this reduced product will have many properties in common with the factors $\mathbf{A}_i$. To specify how this works, we provide a brief overview of the formalism of elementary logic (also known as first order logic).

8.1. The Formalism of Elementary Logic

$\langle \mathbb{R}, +, \cdot, \leq \rangle$ and $\langle \omega, +, \cdot, \leq \rangle$ are among the most familiar mathematical structures. Almost at the outset in developing properties of these structures, one encounters sentences like

"For all numbers x, y, and z, if $x + y \leq x + z$, then $y \leq z$."

This sentence might be rendered in symbols as follow:

$$\forall x \forall y \forall z [x + y \leq x + z \Rightarrow y \leq z].$$

We see here two departures from the prevailing context of our exposition. First, structures like $\langle \omega, +, \cdot, \leq \rangle$, while they are very much like algebras, fail to be algebras. Not only does this structure have a universe and some fundamental operations, but it has a fundamental (binary) relation as well. Second, the sentence symbolized above, for all its simplicity, is not an equation. Model theory provides the means to address both of these departures. The central issue in elementary model theory is the connection between such mathematical structures on the one hand and the formulas and sentences of formal languages, like the one displayed above, on the other. While our focus here is openly algebraic, motivations based on model theoretic considerations are unmistakable, and the methods of model theory play an essential role. Indeed, Birkhoff's HSP Theorem is one of the earliest results in model theory. However, in most contexts, equations offer only meager means of expression, and even the simple cancellation law cited above cannot be formulated by means of equations alone. The real power of model theory emerges once our means of expression are sufficiently rich.

DEFINITION 8.1. An **elementary language** is a system comprising a set of finitary operation and relation symbols each with a finite rank, a countably infinite sequence $v_0, v_1, v_2, \ldots$ of distinct variables, and the following symbols which are called **logical symbols**: $\approx$, $\wedge$, $\vee$, $\neg$, $\Rightarrow$, $\Leftrightarrow$, $\exists$, $\forall$,), and (. These symbols are always taken to be distinct.

Two elementary languages can differ only in their operation and relation symbols. Consequently, we may refer to a particular language by specifying its set of operation and relation symbols. In practice, we take an elementary language $\mathfrak{L}$ to be the set consisting of all the operation and relation symbols. Each of the operation symbols and each of the relation symbols has a **rank** which must be a natural number in the case of operation symbols and a positive natural number in the case of relation symbols. We will use operation and relation symbols to name the fundamental operations and relations of given mathematical structures. The variables are intended to range over the elements of a given mathematical structure. This restriction on the variables accounts for our use of the word "elementary". While it is customary in the model theory literature to speak of languages, we see that they enhance the notion of signatures by also allowing relation symbols that have been assigned ranks. So in what follows we will sometimes use the words "language" and "signature" interchangeably.

The logical symbols have the following meanings:

$\approx$ "equality"

The **connectives**:

$\wedge$	"and"	$\vee$	"or"
$\Rightarrow$	"implies"	$\Leftrightarrow$	"if and only if"
$\neg$	"not"		

The **quantifiers**:

$\exists$ "there exists" $\qquad$ $\forall$ "for all"

The parentheses) and (are used for punctuation. We note that $\wedge$ and $\vee$ are not the symbols used in model theory, those being $\land$ and $\lor$. We use these variations because we don't want to overload our symbols for lattice operations. While this set of logical symbols provides a natural way to express many notions, at several places below it proves convenient to rely on a smaller set of quantifiers and connectives:

$$\approx, \wedge, \neg, \text{ and, } \exists.$$

The remaining connectives and quantifiers can then be regarded as abbreviations. For example, $\neg\exists x\neg(x \le x+1)$ expresses the same notion as $\forall x(x \le x+1)$.

DEFINITION 8.2. Let $\mathfrak{L}$ be an elementary language. An **$\mathfrak{L}$-structure** is a system $\mathbf{A} = \langle A, F\rangle$ where A is a nonempty set, referred to as the **universe** of $\mathbf{A}$, and F is a function with domain $\mathfrak{L}$ such that $F(Q)$ is an operation on A with the same rank as Q, for each operation symbol Q of $\mathfrak{L}$, and $F(R)$ is a relation on A of the same rank as R, for each relation symbol R of $\mathfrak{L}$; $F(Q)$ is also denoted by $Q^{\mathbf{A}}$ and referred to as a **fundamental** or **basic** operation of $\mathbf{A}$, while $F(R)$ is also denoted by $R^{\mathbf{A}}$ and referred to as a **fundamental** or **basic** relation of $\mathbf{A}$.

An $\mathfrak{L}$-structure is just a nonempty set equipped with a fundamental operation of the right rank for each operation symbol of $\mathfrak{L}$ and a fundamental relation of the right rank for each relation symbol of $\mathfrak{L}$. If $\mathfrak{L}$ has no relation symbols, then $\mathfrak{L}$-structures are just algebras. If $\mathfrak{L}$ has no operation symbols, then each $\mathfrak{L}$-structure is said to be a **relational structure**—ordered sets and graphs (with vertices and edges) are examples of relational structures. A **structure** is an $\mathfrak{L}$-structure for some $\mathfrak{L}$. We frequently display structures as sets equipped with a sequence of fundamental operations and relations. Thus $\langle \mathbb{R}, +, -, \cdot, 0, 1, \le\rangle$ denotes the ordered field of real numbers. Two structures are said to be **similar** provided they are both $\mathfrak{L}$-structures for the same language $\mathfrak{L}$.

Let $\mathfrak{L}$ and $\mathfrak{L}'$ be languages such that $\mathfrak{L} \subseteq \mathfrak{L}'$. Let $\mathbf{A}$ be an $\mathbf{L}$-structure and let $\mathbf{A}'$ be an $\mathfrak{L}'$-structure. We say that $\mathbf{A}'$ is an **expansion** of $\mathbf{A}$ and that $\mathbf{A}$ is a **reduct** of $\mathbf{A}'$ provided

$$\begin{aligned} &A = A', \\ &Q^{\mathbf{A}} = Q^{\mathbf{A}'} \text{ for every operation symbol } Q \in \mathfrak{L} \text{ and} \\ &R^{\mathbf{A}} = R^{\mathbf{A}'} \text{ for every relation symbol } R \in \mathfrak{L} \end{aligned}$$

That is, $\mathbf{A}'$ is an expansion of $\mathbf{A}$ if and only if $\mathbf{A}$ and $\mathbf{A}'$ have the same universe and the fundamental operation and relation symbols of $\mathfrak{L}$ denote the same fundamental operations and relations in both $\mathbf{A}$ and $\mathbf{A}'$. So we could obtain $\mathbf{A}$ from $\mathbf{A}'$ by ignoring some of the fundamental operations and relations. See page 8 where a notion like this was introduced. In this chapter, reduct will usually mean the one associated with the identity interpretation of a signature to itself, in the sense of the first pages in this volume.

Let $\mathfrak{L}$ be an elementary language and let $\mathbf{A}$ and $\mathbf{B}$ be $\mathfrak{L}$-structures. Several notions of what it might mean for $\mathbf{B}$ to be a substructure of $\mathbf{A}$ or for a map from A to B to be a homomorphism suggest themselves. We settle on the following alternatives. We say that $\mathbf{B}$ is a **substructure** of $\mathbf{A}$, and we write $\mathbf{B} \subseteq \mathbf{A}$, provided

(1) $B \subseteq A$,
(2) B is closed with respect to $Q^{\mathbf{A}}$ and $Q^{\mathbf{B}}$ is the restriction to B of $Q^{\mathbf{A}}$, for every fundamental operation symbol Q, and
(3) $R^{\mathbf{B}} = R^{\mathbf{A}} \cap B^r$, where r is the rank of R and R is any basic relation symbol.

The notion of a **subuniverse** remains as it was for algebras: a subset of the universe which is closed with respect to all the basic operations–the basic relations play no role.

Now let $h : A \to B$. We say that h is a **homomorphism** from $\mathbf{A}$ into $\mathbf{B}$, and we write $h : \mathbf{A} \to \mathbf{B}$, provided

(1) $h(Q^{\mathbf{A}}(a_0, a_1, \dots, a_{r-1})) = Q^{\mathbf{B}}(h(a_0), h(a_1), \dots, h(a_{r-1}))$ whenever Q is a basic operation symbol, r is its rank, and $a_0, \dots, a_{r-1} \in A$, and
(2) if $R^{\mathbf{A}}(a_0, a_1, \dots, a_{r-1})$, then $R^{\mathbf{B}}(h(a_0), h(a_1), \dots, h(a_{r-1}))$, whenever R is a basic relation symbol, r is its rank, and $a_0, \dots, a_{r-1} \in A$.

Among the notions which might be reasonably called "homomorphism", the one defined above is weak. This can be an irritation since it produces effects that run counter to those holding for homomorphisms between algebras. We saw some of these effects when we considered isotone maps for lattice-ordered sets in Chapter 7. In particular, it is possible for h to be a homomorphism from a structure $\mathbf{A}$ onto several nonisomorphic structures. In the context of arbitrary structures, we cannot expect the close connection between homomorphic images and quotient structures that prevails among algebras according to the Homomorphism Theorem. One must be careful not to leap to the conclusion that $h(\mathbf{A})$ is a substructure of $\mathbf{B}$—while the operations work in the expected way, when a basic relation of $\mathbf{B}$ is restricted to $h(A)$ the result might be larger then the h-image of the corresponding relation of $\mathbf{A}$. Perhaps more vexing, there can be one-to-one homomorphisms from one structure onto another which should not be called isomorphisms. In fact, we insist that an **isomorphism** from $\mathbf{A}$ onto $\mathbf{B}$ is a one-to-one homomorphism h from $\mathbf{A}$ onto $\mathbf{B}$ such that h^{-1} is a homomorphism from $\mathbf{B}$ onto $\mathbf{A}$.

Congruence relations for structures have the same definition as for algebras. In particular, the fundamental relations play no role and, in the absence of any fundamental operations (i.e. for relational structures) every equivalence relation is a congruence relation. Let $\mathbf{A}$ be any structure and let θ be any congruence on $\mathbf{A}$. We form the **quotient structure** $\mathbf{A}/\theta$ by handling the universe and the basic operations exactly as for algebras, and then defining the basic relations of $\mathbf{A}/\theta$ as follows. Let R be a basic relation symbol and let r be its rank. Define $R^{\mathbf{A}/\theta}$ on A/θ so that for all $a_0, a_1, \dots, a_{r-1} \in A$,we have $R^{\mathbf{A}/\theta}(a_0/\theta, a_1/\theta, \dots, a_{r-1}/\theta)$ if and only if $R^{\mathbf{A}}(b_0, b_1, \dots, b_{r-1})$ for some $b_0, b_1, \dots, b_{r-1}$ such that $b_0 \; \theta \; a_0, b_1 \; \theta \; a_1, \dots, b_{r-1} \; \theta \; a_{r-1}$. Thus, in the quotient structure only those basic relations are made to hold which are necessary in order that the quotient map $a \mapsto a/\theta$ be a homomorphism.

Turning to direct products, let I be any set and let $\mathbf{A}_i$ be an $\mathfrak{L}$-structure for each $i \in I$. The **direct product** $\mathbf{A} = \prod_{i \in I} \mathbf{A}_i$ is defined as for algebras, with the basic relations handled "coordinatewise". That is, if R is a basic relation symbol and r is its rank, and $b_0, b_1, \dots, b_{r-1} \in A$, then

$$R^{\mathbf{A}}(b_0, b_1, \dots, b_{r-1}) \text{ if and only if } R^{\mathbf{A}_i}(b_{0,i}, b_{1,i}, \dots, b_{r-1,i}) \text{ for all } i \in I.$$

With this definition of direct product, the projection functions are homomorphisms. To keep the notation simple, we write $\prod_I \mathbf{A}_i$ if it is clear that the index i ranges over the index set I.

Just as operation symbols name the fundamental operations of a structure, so relation symbols name the fundamental relations. We have seen how the variables and operations symbols can be compounded to form terms and how these terms name the term functions of a given structure. With the essential help of the connectives and quantifiers, we will now see how to make compound expressions called formulas which will serve as names for certain finitary relations on a given structure.

Let $\mathfrak{L}$ be an elementary language. An **$\mathfrak{L}$-expression** is just a finite string of $\mathfrak{L}$-symbols. The simplest $\mathfrak{L}$-formulas are called **atomic formulas** and they consist of all $\mathfrak{L}$-expressions of the following forms:

(1) $s \approx t$, where s and t are any $\mathfrak{L}$-terms, and
(2) $Rt_0 \dots t_{r-1}$, where R is any relation symbol of $\mathfrak{L}$, r is the rank of R, and $t_0, \dots, t_{r-1}$ are any $\mathfrak{L}$-terms.

The $\mathfrak{L}$-formulas are just those expressions which can be built from the atomic formulas with the help of the quantifiers and connectives.

DEFINITION 8.3. Take $\mathfrak{L}$ to be an elementary language. The set of **$\mathfrak{L}$-formulas** is the smallest set F of $\mathfrak{L}$-expressions such that

(1) Every atomic $\mathfrak{L}$-formula belongs to F,
(2) If $\varphi, \psi \in F$, then $\neg\varphi$, $(\varphi \wedge \psi)$, $(\varphi \vee \psi)$, $(\varphi \Rightarrow \psi)$, and $(\varphi \Leftrightarrow \psi) \in F$, and
(3) If $\varphi \in F$ and x is any variable, then $\exists x\varphi, \forall x\varphi \in F$.

This definition of the notion of formulas, like the definition of the notion of terms given as Definition 4.114 in Volume 1, is recursive: more complicated formulas are built from less complicated formulas. Induction is therefore a powerful gambit for proving assertions about formulas. Ultimately, such inductions rest on the fact that formulas are uniquely readable in the sense that there is only one way to decompose a given formula along the lines of the definition above. We dispense with the precise formulation and proof of this unique readability of formulas, but the interested reader can consult Lemma 4.115 in Volume 1 for the corresponding result for terms or see the first four exercises in Exercise Set 7.5.

Our definition of the set of $\mathfrak{L}$-formulas leads to a rather awkward notation for formulas. We follow the customary practice and regard formulas like

$$x \leq y \wedge y \leq z \Rightarrow x \leq z$$

as perfectly acceptable, even though our official definition would yield

$$((\leq xy \wedge \leq yz) \Rightarrow \leq xz)$$

We omit parentheses when the meaning is clear, we use a mixture of parentheses and brackets to enhance readability, and we insert binary relation symbols between two terms rather than in front of them.

In displaying long formulas, we sometimes write

$$\bigwedge_{i \in I} \varphi_i$$

in place of $\varphi_0 \wedge \varphi_1 \wedge \cdots \wedge \varphi_i \wedge \cdots$. We use

$$\bigvee_{i \in I} \varphi_i$$

to replace $\varphi_0 \vee \varphi_1 \vee \cdots$.

Let $\mathfrak{L}$ be the language appropriate to the ordered field of real numbers. Consider the following formulas:

$$x^2 + y^2 \leq 4$$
$$x^2 + y^2 \leq 4 \wedge \exists x[x^2 \approx y]$$
$$\forall x \exists y[x^2 + y^2 \leq 4]$$

Here 4 is an abbreviation of $1+1+1+1$ and x^2 abbreviates $x \cdot x$. For the ordered field $\langle \mathbb{R}, +, -, \cdot, 0, 1, \leq \rangle$ of real numbers, the first formula defines the closed disc of radius 2, the second formula defines the upper half of the disc of radius 2, and the third formula is a statement which is false in the ordered field of real numbers. The reason for the distinction between the meaning of the third formula and the other two stems from the fact that all the occurrences of variables in the last formula are bound under the control of the quantifiers, while the first two formulas have occurrences of variables free of such control. The bound variables are very much like the dummy variable u used in the definite integral $\int_a^b f(u)\,du$. In the first of the formulas displayed above all occurrences of variables are free. In the second formula all occurrences of y are free, as is the leftmost occurrence of x, but the remaining occurrences of x are bound. The last formula has no free occurrences of variables. We formalize this concept.

DEFINITION 8.4. Let $\mathfrak{L}$ be an elementary language. Let Va denote the function which assigns to each term φ the set $\mathrm{Va}(\varphi)$ of all variables that occur in φ. Fv is the function with domain the set of all $\mathfrak{L}$-formulas specified by

$$\mathrm{Fv}(\varphi) = \begin{cases} \mathrm{Va}(\varphi) & \text{if } \varphi \text{ is atomic} \\ \mathrm{Fv}(\theta) \cup \mathrm{Fv}(\psi) & \text{if } \varphi \text{ is } (\theta \wedge \psi), (\theta \vee \psi), (\theta \Rightarrow \psi), \text{ or } (\theta \Leftrightarrow \psi) \\ \mathrm{Fv}(\psi) & \text{if } \varphi \text{ is } \neg\psi \\ \mathrm{Fv}(\psi) \setminus \{x\} & \text{if } \varphi \text{ is } \exists x\psi \text{ or } \forall x\psi \end{cases}$$

$\mathrm{Fv}(\varphi)$ is called the **set of free variables** of φ. A formula without any free variables is called a **sentence**.

Let φ be a formula. We write $\varphi(y_0, y_1, \dots, y_{n-1})$ to mean that all the free variables of φ occur among $y_0, y_1, \dots, y_{n-1}$. More formally, this means $\mathrm{Fv}(\varphi) \subseteq \{y_0, y_1, \dots, y_{n-1}\}$. Notice that we do not insist on equality, but only on set-inclusion. If n is unimportant, we write $\varphi(\mathbf{y})$ in place of $\varphi(y_0, \dots, y_{n-1})$. If it is unimportant to distinguish the variables, we may write

$$\exists \mathbf{y}\varphi(\mathbf{x}, \mathbf{y}) \text{ for } \exists y_0 \exists y_1 \cdots \exists y_{n-1}\varphi(\mathbf{x}, y_0, y_1, \dots, y_{n-1})$$

The universal quantifier $\forall$ is treated in the same manner.

Now we can specify how formulas define relations on structures and what it means for an elementary sentence to be true in a structure. Because our definition goes by recursion on the complexity of formulas, it is inconvenient to set a bound on the number of variables which occur freely in the formulas. To accommodate this, we employ countably infinite sequences of elements. Also, to simplify this definition and subsequent work, we limit our connectives to $\neg$ and $\wedge$, and we take $\exists$ as our only quantifier. This represents no loss of generality since the remaining symbols can be regarded as abbreviating various schemas built with the help of only these three symbols.

DEFINITION 8.5. Let $\mathbf{A}$ be an $\mathfrak{L}$-structure. For every $\mathfrak{L}$-formula φ, we take $\varphi^{\mathbf{A}}$ to be a certain ω-ary relation on A by specifying whether $\varphi^{\mathbf{A}}(\mathbf{a})$ [that is, whether $\mathbf{a} \in \varphi^{\mathbf{A}}$] for each $\mathbf{a} \in A^{\omega}$ as follows:

(1) $(s \approx t)^{\mathbf{A}}(\mathbf{a})$ if and only if $s^{\mathbf{A}}(\mathbf{a}) = t^{\mathbf{A}}(\mathbf{a})$
(2) $(Rt_0t_1 \cdots t_{r-1})^{\mathbf{A}}(\mathbf{a})$ if and only if $R^{\mathbf{A}}(t_0^{\mathbf{A}}(\mathbf{a}), \dots, t_{r-1}^{\mathbf{A}}(\mathbf{a}))$
(3) $(\neg\psi)^{\mathbf{A}}(\mathbf{a})$ if and only if $\psi^{\mathbf{A}}(\mathbf{a})$ fails.
(4) $(\theta \wedge \psi)^{\mathbf{A}}(\mathbf{a})$ if and only if $\theta^{\mathbf{A}}(\mathbf{a})$ and $\psi^{\mathbf{A}}(\mathbf{a})$.
(5) $(\exists v_i \psi)^{\mathbf{A}}(\mathbf{a})$ if and only if $\psi^{\mathbf{A}}(\mathbf{b})$ for some $\mathbf{b} \in A^{\omega}$ such that $a_j = b_j$ for all $j \neq i$.

We use

$$\mathbf{A} \models \varphi(\mathbf{a}), \quad \varphi^{\mathbf{A}}(\mathbf{a}), \quad \text{and } \mathbf{a} \in \varphi^{\mathbf{A}}$$

interchangeably. We read all three of these as "the assignment **a satisfies** φ in **A**."

This definition appears rather involved, but it holds no surprises. What we have done is give to the connectives and quantifiers the meanings we intended for them all along. In this definition, even though $\mathbf{a}$ is an infinite sequence of elements, it is plain that whether $\mathbf{a}$ satisfies φ hinges only on (the indices of) the free variables of φ, a finite set. The sets

$$\{\mathbf{a} \mid \mathbf{a} \in A^{\omega} \text{ and } \varphi^{\mathbf{A}}(\mathbf{a})\}$$

are infinitary relations which depend essentially on only finitely many coordinates. We regard $\varphi^{\mathbf{A}}$ as a finitary relation on A, relegating a completely correct formulation of this to the exercises. Those finitary relations on A which are of the form $\varphi^{\mathbf{A}}$ for some formula φ are said to be **definable** in **A**.

Finally, we say that φ is **true** in **A**, and alternatively that **A** is a **model** of φ if and only if $\mathbf{A} \models \varphi(\mathbf{a})$ for all $\mathbf{a} \in A^{\omega}$. If φ is an equation and **A** is an algebra, then this definition agrees with the definition provided in Chapter 4 of Volume I. We use

$$\mathbf{A} \models \varphi$$

to denote that φ is true in **A**. Observe that for any formula $\varphi(x_0, \dots, x_{n-1})$ we have

$$\mathbf{A} \models \varphi(x_0, \dots, x_{n-1}) \text{ if and only if } \mathbf{A} \models \forall x_0 \cdots \forall x_{n-1}\varphi(x_0, \dots, x_{n-1})$$

Now $\forall x_0 \cdots \forall x_{n-1}\varphi(x_0, \dots, x_{n-1})$ is a sentence. Thus, to know all the formulas true in **A** it suffices to know all the sentences true in **A**. So for the most part, we regard $\models$ as a binary relation between $\mathfrak{L}$-structures and $\mathfrak{L}$-sentences. This relation establishes a Galois connection in a manner similar to the connection between varieties and equational theories described in Chapter 4 of Volume 1. The polarities of the present Galois connection are denoted as follows:

$$\mathbf{Mod}\,\Sigma = \{\mathbf{A} \mid \mathbf{A} \models \varphi \text{ for all } \varphi \in \Sigma\}$$

where Σ is any set of $\mathfrak{L}$-sentences, and

$$\mathbf{Th}\,\mathcal{K} = \{\varphi \mid \mathbf{A} \models \varphi \text{ for all } \mathbf{A} \in \mathcal{K}\}$$

where $\mathcal{K}$ is any class of $\mathfrak{L}$-structures. Classes of the form $\mathbf{Mod}\,\Sigma$ are called **elementary classes** and sets of the form $\mathbf{Th}\,\mathcal{K}$ are called **elementary theories**. An elementary class $\mathcal{K}$ is said to be **finitely axiomatizable** provided $\mathcal{K} = \mathbf{Mod}\,\Sigma$ for some finite set Σ of sentences.

Two sets Σ and Γ of $\mathfrak{L}$-sentences are said to be **logically equivalent** provided $\mathbf{Mod}\,\Sigma = \mathbf{Mod}\,\Gamma$; that is, they have exactly the same models. We say that the sentence θ is a **logical consequence** of the set Σ of $\mathfrak{L}$-sentences, and we write $\Sigma \models \theta$, if and only if every model of Σ is a model of θ. An elementary theory T is said be be **finitely axiomatizable** provided T is logically equivalent to some finite set Σ of elementary sentences.

Two $\mathfrak{L}$-structures $\mathbf{A}$ and $\mathbf{B}$ are **elementarily**, rendered in symbols as $\mathbf{A} \equiv \mathbf{B}$, if and only if $\mathbf{Th}\,\mathbf{A} = \mathbf{Th}\,\mathbf{B}$. Elementary equivalence is an equivalence relation on the class of $\mathfrak{L}$-structures which is coarser than the relation of isomorphism. [That is, if $\mathbf{A} \cong \mathbf{B}$, then $\mathbf{A} \equiv \mathbf{B}$.] In the next section we will see that every infinite structure in a countable language is elementarily equivalent to structures of arbitrarily large infinite cardinality, so that $\equiv$ is indeed much coarser than $\cong$ among infinite structures. For finite structures these relations are the same.

THEOREM 8.6. *Let $\mathfrak{L}$ be an elementary language and let $\mathbf{A}$ and $\mathbf{B}$ be $\mathfrak{L}$-structures. If $\mathbf{A} \equiv \mathbf{B}$ and $\mathbf{A}$ is finite, then $\mathbf{A} \cong \mathbf{B}$.*

Proof. Using just variables and the logical symbols, for each natural number n it is possible to write down a sentence which expresses "there are at least n distinct elements". For example, here is such a sentence expressing "there are at least three elements":

$$\exists x_0 \exists x_1 \exists x_2 [\neg(x_0 \approx x_1) \wedge \neg(x_0 \approx x_2) \wedge \neg(x_1 \approx x_2)]$$

Call this sentence δ_3. The reader can devise an analogous sentence δ_n for each natural number n. The sentence $\neg\delta_{n+1}$ express "there are no more than n elements" and the sentence $\delta_n \wedge \neg\delta_{n+1}$ expresses "there are exactly n elements." It now follows that $\mathbf{B}$ has the same (finite) cardinality as $\mathbf{A}$.

For the sake of contradiction, assume that $\mathbf{A}$ and $\mathbf{B}$ are not isomorphic. There are only finitely many one-to-one maps from A onto B. For each such map pick an operation or relation symbol which the map does not respect. Let $\mathfrak{L}'$ be the finite subset of $\mathfrak{L}$ comprised of these symbols, and let $\mathbf{A}'$ and $\mathbf{B}'$ be the reducts of $\mathbf{A}$ and $\mathbf{B}$ to $\mathfrak{L}'$. Then $\mathbf{A}' \equiv \mathbf{B}'$ and $\mathbf{A}'$ is not isomorphic with $\mathbf{B}'$. Let $\mathbf{a} = \langle a_0, \dots, a_{n-1}\rangle$ be a list of all the distinct elements of A. Let

$$\begin{aligned}\Delta = &\{\varphi(\mathbf{x}) \mid \varphi \text{ is an atomic } \mathfrak{L}'\text{-formula and } \mathbf{A} \models \varphi(\mathbf{a})\}\\ &\cup \{\neg\varphi(\mathbf{x}) \mid \varphi \text{ is an atomic } \mathfrak{L}'\text{-formula and } \mathbf{A} \models \neg\varphi(\mathbf{a})\}\end{aligned}$$

Let σ be the conjunction of all the formulas in Δ [that is, the formula resulting from combining all the formulas in Δ by $\wedge$]. Let ψ be the formula $\sigma \wedge \neg\delta_{n+1}$. Then certainly $\mathbf{A}' \models \exists x_0 \cdots \exists x_{n-1}\psi$. Hence, $\mathbf{B}' \models \exists x_0 \cdots \exists x_{n-1}\psi$. This means we can pick $b_0, \dots, b_{n-1} \in B$ so that

$$\mathbf{B}' \models \psi(b_0, \dots, b_{n-1}).$$

ψ carries all the information necessary to demonstrate that the map $a_i \mapsto b_i$ for all $i < n$ is an isomorphism from $\mathbf{A}'$ onto $\mathbf{B}'$. This is a contradiction. ■

It is interesting to note that the formula ψ in the proof just given is, in some sense, a complete description of $\langle a_0, \dots, a_{n-1}\rangle$ in $\mathbf{A}'$. In fact, taking $\pi = \exists x_0 \cdots \exists x_{n-1}\psi$, we have that if $\mathbf{C} \models \pi$, then $\mathbf{C} \cong \mathbf{A}'$, for all $\mathfrak{L}'$-structures $\mathbf{C}$. Thus, we have the following corollary.

COROLLARY 8.7. *If $\mathfrak{L}$ is a finite elementary language and* **A** *is a finite $\mathfrak{L}$-structure, then there is a sentence π such that* $\mathbf{A} \cong \mathbf{B}$ *if and only if* $\mathbf{B} \models \pi$. ■

COROLLARY 8.8. *If $\mathfrak{L}$ is any elementary language and* **A** *is a finite $\mathfrak{L}$-structure, then* **Mod Th A**, *the smallest elementary class containing* **A**, *consists of precisely the isomorphic images of* **A**. ■

These two corollaries point out the sharp distinction between varieties and equational theories on the one hand, and elementary classes and elementary theories on the other.

The classification of structures into elementary classes is much finer than the classification of algebras into varieties. For example, the class of all fields is an elementary class, as can be seen from the customary axioms of field theory, but it is not a variety. By adding the sentence $1 + 1 \approx 0$ to the field axioms we obtain the class of all fields of characteristic 2. Similar sentences can be devised for every prime p. By adding instead the negations of these sentences, we arrive at the class of all fields of characteristic 0. Even the class of algebraically closed fields is an elementary class. However, there are many significant classes which fail to be elementary. The class of Archimedean ordered fields is of this kind.

One way to measure the price that must be paid for this finer classification scheme is to observe that elementary classes in general are not closed under any of the operators $\boldsymbol{H}$, $\boldsymbol{S}$, or $\boldsymbol{P}$. Consequently, the elaboration of elementary model theory relies on the rich means of expression provided by elementary languages more than it relies on the manipulations of homomorphisms, substructures, congruences, or direct products. Nevertheless, these notions still retain a position of importance in model theory generally. Of course, they are never far from the focus of our attention.

To a large extent, our choices for how to extend the algebraic notions of substructure, homomorphism, and direct product to $\mathfrak{L}$-structures was motivated by the desire to have atomic formulas in general behave like equations. For example, whenever $\mathbf{B} \subseteq \mathbf{A}$, an easy induction on the complexity of terms reveals that for any atomic formula $\varphi(\mathbf{x})$ and any tuple $\mathbf{b}$ from B,

$$\mathbf{B} \models \varphi(\mathbf{b}) \text{ if and only if } \mathbf{A} \models \varphi(\mathbf{b})$$

Likewise, whenever h is a homomorphism from **B** into **A**, we have

$$\mathbf{B} \models \varphi(\mathbf{b}) \text{ implies } \mathbf{A} \models \varphi(h(\mathbf{b})).$$

By the same sort of reasoning, if $\mathbf{A} = \prod\langle \mathbf{A}_i \mid i \in I\rangle$, then the family $\{p_i \mid i \in I\}$ of projection functions separates points in the strong sense that if $\varphi(\mathbf{x})$ is any atomic formula [not just $v_0 \approx v_1$] and $\mathbf{a}$ is a tuple from A such that $\mathbf{A} \not\models \varphi(\mathbf{a})$, then $\mathbf{A}_i \not\models \varphi(p_i(\mathbf{a}))$, for some $i \in I$. Some of the key notions of model theory arise from the effort to extend these conditions from atomic formulas to all formulas.

DEFINITION 8.9. Let **A** and **B** be $\mathfrak{L}$-structures and let h be a function from A into B. We say that h is an **elementary embedding** if and only if $\mathbf{A} \models \varphi(\mathbf{a})$ implies $\mathbf{B} \models \varphi(h(\mathbf{a}))$ for all formulas φ and all tuples $\mathbf{a}$ from A.

The key notion of elementary embedding is due to Alfred Tarski and familiar to the participants in Tarski's seminar at the University of Warsaw in the later 1920's.

That h is an elementary embedding of $\mathbf{A}$ into $\mathbf{B}$ means that any property of any finite sequence of elements of A which can be expressed by $\mathfrak{L}$-formulas must also be a property of the sequence of elements of B corresponding by the map h. As the properties expressed by atomic formulas are respected, we see that elementary embeddings are homomorphisms. To say that $\neg(x \approx y)$ is respected by h is easily the same as saying that h is one-to-one. More generally, since h respects the negations of all atomic formulas, we see that h is an isomorphism from $\mathbf{A}$ onto a substructure of $\mathbf{B}$. So the use of "embedding" rather than "morphism" is justified. But elementary embeddings are much stronger than ordinary embeddings.

EXAMPLE 8.10. The only elementary embedding of $\langle \omega, < \rangle$ into itself is the identity map.

The essential reason for this is that using $<$ and the logical symbols we can compose a formula $\varphi_n(x)$ which expresses "there are exactly n elements less than x", Since $\langle \omega, < \rangle \models \varphi_n(n)$, we see that $\langle \omega, < \rangle \models \varphi_n(h(n))$ for all $n \in \omega$ and for every elementary embedding h. Thus $h(n) = n$ for all $n \in \omega$ and for all elementary embeddings h. On the other hand, $\langle \omega, < \rangle$ has plenty of embedding into itself. The successor function is an embedding.

The structure $\mathbf{A}$ is said to be an **elementary substructure** of $\mathbf{B}$ (and $\mathbf{B}$ is called an **elementary extension** of $\mathbf{A}$) provided $A \subseteq B$ and the inclusion map in an elementary embedding of $\mathbf{A}$ into $\mathbf{B}$. We use $\mathbf{A} \prec \mathbf{B}$ to denote that $\mathbf{A}$ is an elementary substructure of $\mathbf{B}$. The same concept arises when the notion of substructure is modified by insisting that all formulas be respected rather than just atomic formulas. The structure $\langle \omega, < \rangle$ has no proper elementary substructures. On the other hand, $\langle \omega - \{0\}, < \rangle$ is isomorphic with $\langle \omega, < \rangle$. So we have an example of two structures $\mathbf{A}$ and $\mathbf{B}$ such that $\mathbf{A} \subseteq \mathbf{B}$ and $\mathbf{A} \equiv \mathbf{B}$ (even $\mathbf{A} \cong \mathbf{B}$), but $\mathbf{A}$ is not an elementary substructure of $\mathbf{B}$. Here are some easy facts about $\prec$.

THEOREM 8.11. *Let $\mathfrak{L}$ be an elementary language and let $\mathbf{A}$, $\mathbf{B}$, and $\mathbf{C}$ be $\mathfrak{L}$-structures.*

(i) *If $\mathbf{A} \prec \mathbf{B}$ and $\mathbf{B} \prec \mathbf{C}$, then $\mathbf{A} \prec \mathbf{C}$.*
(ii) *If $\mathbf{A} \prec \mathbf{B}$, then $\mathbf{A} \subseteq \mathbf{B}$ and $\mathbf{A} \equiv \mathbf{B}$.*
(iii) *If $\mathbf{A} \prec \mathbf{C}$ and $\mathbf{B} \prec \mathbf{C}$ and $\mathbf{A} \subseteq \mathbf{B}$, then $\mathbf{A} \prec \mathbf{B}$.*

The next theorem, sometimes called the Tarski-Vaught Elementary Extension Test, supplies a useful test for elementary extensions.

THEOREM 8.12 (Tarski and Vaught (1958)). *Let $\mathfrak{L}$ be an elementary language and let $\mathbf{A}$ and $\mathbf{B}$ be $\mathfrak{L}$-structures. $\mathbf{A} \prec \mathbf{B}$ if and only if both conditions below hold:*

(i) *$\mathbf{A} \subseteq \mathbf{B}$, and*
(ii) *For all formulas $\varphi(x, y_0, y_1, \dots)$ and all $a_0, a_1, \dots \in A$, if $\mathbf{B} \models \exists y \varphi(y, a_0, a_1 \dots)$, then $\mathbf{A} \models \varphi(a, a_0 \dots)$ for some $a \in A$.*

Proof. We deal only with the more difficult direction, so we assume (i) and (ii) and prove $\mathbf{A} \prec \mathbf{B}$. We will prove by induction on the complexity of ψ that for all formulas ψ and all $\mathbf{a} \in A^\omega$ that $\mathbf{B} \models \psi(\mathbf{a})$ if and only if $\mathbf{A} \models \psi(\mathbf{a})$. If ψ is atomic, this is just the content of (i). In the event that ψ is not atomic there are three cases.

Case $\neg$: ψ is $\neg\varphi$ for some formula φ.

Just observe

$$\begin{aligned} \mathbf{B} \models \psi(\mathbf{a}) & \text{ if and only if } \mathbf{B} \not\models \varphi(\mathbf{a}) \\ & \text{ if and only if } \mathbf{A} \not\models \varphi(\mathbf{a}) \\ & \text{ if and only if } \mathbf{A} \models \psi(\mathbf{a}). \end{aligned}$$

Case $\wedge$: ψ is $\varphi \wedge \theta$ for some formulas φ and θ.

Just observe

$$\begin{aligned} \mathbf{B} \models \psi(\mathbf{a}) & \text{ if and only if } \mathbf{B} \models \varphi(\mathbf{a}) \text{ and } \mathbf{B} \models \theta(\mathbf{a}) \\ & \text{ if and only if } \mathbf{A} \models \varphi(\mathbf{a}) \text{ and } \mathbf{A} \models \theta(\mathbf{a}) \\ & \text{ if and only if } \mathbf{A} \models \psi(\mathbf{a}). \end{aligned}$$

Case $\exists$: ψ is $\exists y \varphi(y, x_0, x_1, \dots)$.

If $\mathbf{B} \models \psi(\mathbf{a})$, then $\mathbf{A} \models \psi(\mathbf{a})$ according to condition (ii). For the converse, suppose that $\mathbf{A} \models \psi(\mathbf{a})$. Then $\mathbf{A} \models \varphi(c, \mathbf{a})$ for some $c \in A$. The induction hypothesis yields $\mathbf{B} \models \varphi(c, \mathbf{a})$. Hence $\mathbf{B} \models \psi(\mathbf{a})$. ■

COROLLARY 8.13. *Let* $\mathbf{A}$ *and* $\mathbf{B}$ *be* $\mathfrak{L}$*-structures such that* $\mathbf{A} \subseteq \mathbf{B}$. *Suppose that for any finite sequence* $a_0, \dots, a_{n-1}$ *of elements of* A *and for any* $b \in B$, *there is an automorphism of* $\mathbf{B}$ *leaving each of* $a_0, \dots, a_{n-1}$ *fixed but mapping* b *into* A. *Then* $\mathbf{A} \prec \mathbf{B}$. ■

With the help of this corollary it is easy to see that the ordered set of rationals interior to the unit interval is an elementary substructure of $\langle \mathbb{Q}, < \rangle$, as is the ordered set of all dyadic rationals (those of the form $m/2^n$).

Here is our first application to the theory of varieties.

THEOREM 8.14 (due to Robert L. Vaught). *Let* $\mathcal{V}$ *be any variety and let* $X \subseteq Y$ *with* X *infinite. Let* $\mathbf{B}$ *be the algebra free in* $\mathcal{V}$ *over* Y *and let* $\mathbf{A}$ *be the subalgebra of* $\mathbf{B}$ *generated by* X. *Then* $\mathbf{A} \prec \mathbf{B}$.

Proof. We invoke Corollary 8.13. Let $a_0, \dots, a_{n-1} \in A$ and let $b \in B$. Let X' be a large enough finite subset of X so that $a_0, \dots, a_{n-1}$ belong to the subuniverse of $\mathbf{A}$ generated by X'. Let Y' be a large enough finite subset of $Y - X'$ so that b is in the subuniverse generated by $X' \cup Y'$. Let X'' be any subset of $X - X'$ such that $|X''| = |Y'|$. Let h be any permutation of Y that maps Y' onto X'' and leaves every element of $Y - (Y' \cup X'')$ fixed. Since $\mathbf{B}$ is free in $\mathcal{V}$ over Y, we know that h extends to an automorphism of $\mathbf{B}$. Under this automorphism X' is fixed, and so each of $a_0, \dots, a_{n-1}$ is also fixed. On the other hand, $X' \cup Y'$ gets mapped into X. So the image of b under this automorphism lies in A. Thus, $\mathbf{A} \prec \mathbf{B}$. ■

COROLLARY 8.15. *Let* $\mathcal{V}$ *be any nontrivial variety. If* κ *and* λ *are any two infinite cardinals, then* $\mathbf{F}_{\mathcal{V}}(\kappa) \equiv \mathbf{F}_{\mathcal{V}}(\lambda)$. ■

The next theorem is a descendant of one published by Leopold Löwenheim in 1914—possibly the oldest theorem which belongs openly to model theory. It reveals a key property of elementary model theory. Löwenheim's original result was, roughly, "Any elementary sentence which has a model must have a countable model." Thoralf Skolem, in a series of papers stretching from 1920 to 1928, revisited Löwenheim's work, eventually showing it works for any sets of sentences in an arbitrary language (see below for the correct cardinality stipulation). Tarski's name gets attached because of the elementary substructure relation.

THEOREM 8.16 (The Downward Löwenheim-Skolem-Tarski Theorem). *Let $\mathfrak{L}$ be an elementary language and let κ be the cardinality of the set of all $\mathfrak{L}$-formulas. Let* $\mathbf{A}$ *be an $\mathfrak{L}$-structure and let $X \subseteq A$ with $|X| \geq \kappa$. Then* $\mathbf{A}$ *has an elementary substructure* $\mathbf{B}$ *such that $X \subseteq B$ and $|B| = |X|$.*

Proof. Let $B_0 = X$. For each formula $\varphi(y, \mathbf{x})$ and for all appropriate tuples $\mathbf{b}$ from B_0 such that $\mathbf{A} \models \exists y \varphi(y, \mathbf{b})$, select $a \in A$ such that $\mathbf{A} \models \varphi(a, \mathbf{b})$. Let C_0 be the set of all elements a selected as above as $\varphi(y, \mathbf{x})$ ranges through the formulas and $\mathbf{b}$ ranges through all appropriate tuples from B_0. Let $B_1 = B_0 \cup C_0$. Since $\kappa \leq |B_0|$ we see that $|B_1| = |B_0| = |X|$. Now we repeat this process on B_1 to obtain C_1 and then B_2. Continuing in this way through countably many steps, we obtain a tower of sets

$$X = B_0 \subseteq B_1 \subseteq B_2 \subseteq \cdots$$

Let $B = \bigcup_{k \in \omega} B_k$. Clearly, $X \subseteq B$ and $|B| = |X|$. To see that B is a subuniverse of $\mathbf{A}$, let Q be any operation symbol and pick $b_0, \dots, b_{r-1} \in B$, where r is the rank of Q. Pick m so large that $b_0, \dots, b_{r-1} \in B_m$. Now

$$\mathbf{A} \models \exists y [Q b_0 \cdots b_{r-1} \approx y].$$

By the construction, $Q^{\mathbf{A}}(b_0, \dots, b_{r-1}) \in B_{m+1} \subseteq B$. So B is a subuniverse of $\mathbf{A}$. That $\mathbf{B}$ satisfies the second condition of the Tarski-Vaught Elementary Extension Test is built into our construction of B. Thus $\mathbf{B} \prec \mathbf{A}$. ■

At this point, the reader without previous exposure to elementary languages should try to express some common mathematical notions in the elementary setting. There are really two sorts of exercises of this kind:

- The discovery of elementary sentences which axiomatize familiar classes of structures.
- Within a given structure, the discovery of elementary formulas which define familiar sets, relations, or operations.

This section concludes with a set of exercises, among which may be found a number of problems of both kinds. After trying a few of these exercises the reader should have the impression that quite a few common mathematical notions can find elementary expression. On the other hand, the Downward Löwenheim-Skolem-Tarski Theorem informs us that no set of elementary sentences can capture $\langle \mathbb{R}, +, -, \cdot, 0, 1, \leq \rangle$ up to isomorphism—the familiar axiomatizations of Dedekind to the contrary. (It is Dedekind's axiom of **completeness** that cannot be captured by elementary axioms.) The theorems proven in the next section draw much sharper limitations on the classes of structures which can be axiomatized by elementary sentences. Roughly, a class of structures whose ordinary definition in mathematical practice refers only to properties of elements (taken finitely many at a time) of the structures will be an elementary class. Those classes whose ordinary descriptions refer to sets of elements, to sequences of elements, or to external objects like natural numbers are usually not elementary. Of course, just because the customary description of a familiar class of structures appears to be nonelementary, it does not follow that no elementary description exists. The usual way to specify the class of algebraically closed fields involves quantifying over all polynomials. Nevertheless, this class *is* elementary.

The determination of which sets, relations, and operations are definable by elementary formulas in a given structure is usually very difficult. The following results, listed here without proof, offer some insight into the subtlety of definability questions:

A. The set of natural numbers is definable in the ring of integers.
B. The set of integers is definable in the field of rational numbers.
C. The binary operation $x \uparrow y = x^y$ of exponentiation is definable in $\langle \omega, +, \cdot \rangle$.
D. Addition is not definable in $\langle \omega, \cdot \rangle$ nor is multiplication definable in $\langle \omega, + \rangle$, but both addition and multiplication are definable in $\langle \omega, \uparrow \rangle$.
E. The sets definable in the field of real numbers are exactly the unions of finitely many intervals with algebraic endpoints. In particular, the set of rationals is not definable in the field of real numbers.
F. The set of compact elements is definable in the lattice of all equational theories of a given fixed signature.

Perhaps (A) is immediate, but only for those who know Lagrange's theorem to the effect that every natural number is the sum of four squares. (C) and (D) require imagination and persistence. The remaining items are genuinely difficult.

We close this section with a discussion of logically equivalent formulas, followed by an exercise set intended to show that formulas can be given several different kinds of normal forms. These exercises may be particularly useful to those with a small background in mathematical logic. Let $\varphi(x_0, \dots, x_{n-1})$ and $\psi(x_0, \dots, x_{n-1})$ be formulas, of the language $\mathfrak{L}$, all of whose free variables belong to the set $\{x_0, \dots, x_{n-1}\}$. The formulas $\varphi(x_0, \dots, x_{n-1})$ and $\psi(x_0, \dots x_{n-1})$ are said to be **logically equivalent** provided for every $\mathfrak{L}$-structure $\mathbf{A}$ and all $a_0, \dots, a_{n-1} \in A$

$$\mathbf{A} \models \varphi(a_0 \dots, a_{n-1}) \text{ if and only if } \mathbf{A} \models \psi(a_0, \dots, a_{n-1}).$$

Another way to say this is that for all $\mathcal{L}$-structures $\mathbf{A}$

$$\mathbf{A} \models \forall x_0, \dots, x_{n-1}[\varphi(x_0, \dots, x_{n-1}) \Leftrightarrow \psi(x_0, \dots, x_{n-1})].$$

Yet a third way is to assert that in any $\mathfrak{L}$-structure the two formulas define the same n-ary relation.

Exercises 8.17

1. Let φ be an $\mathfrak{L}$-formula and $\mathbf{A}$ be an $\mathfrak{L}$-structure. Let $\mathbf{a}, \mathbf{b} \in A^\omega$ such that $a_i = b_i$ if the variable v_i occurs freely in φ. Prove that $\mathbf{a}$ satisfies φ is $\mathbf{A}$ if and only if $\mathbf{b}$ satisfies φ in $\mathbf{A}$.

2. Devise a definition of satisfaction that uses finite tuples of elements of a structure in place of ω-tuples and prove that the same notions of of truth and logical consequence result from your definition.

3. Let φ be a formula and z be a variable that does not occur in φ. Let x be a variable that occurs freely in φ. Let ψ be the formula obtained for φ by replacing each free occurrence of x by z. Prove that φ and ψ are logically equivalent.

4. Let φ and ψ be $\mathfrak{L}$-formulas. In each part below prove that the formulas displayed are logically equivalent.
(i) $\neg\forall x\varphi$ and $\exists x\neg\varphi$.
(ii) $\neg\exists x\varphi$ and $\forall x\neg\varphi$.
(iii) $\varphi \wedge \forall x\psi$ and $\forall x[\varphi \wedge \psi]$, provided x does not occur free in φ.
(iv) $\varphi \wedge \exists x\psi$ and $\exists x[\varphi \wedge \psi]$, provided x does not occur free in φ.
(v) $\varphi \vee \forall x\psi$ and $\forall x[\varphi \vee \psi]$, provided x does not occur free in φ.
(vi) $\varphi \vee \exists x\psi$ and $\exists x[\varphi \vee \psi]$, provided x does not occur free in φ.
Now prove every formula is logically equivalent to one that has a string of quantifiers at the left end followed by a formula free of quantifiers. (Formulas of the latter kind are said to be in **prenex form**.)

5. Let φ, ψ, and θ be $\mathfrak{L}$-formulas. In each part below prove that the formulas displayed are logically equivalent.
(i) $\neg\neg\varphi$ and φ.
(ii) $\neg(\varphi \wedge \psi)$ and $\neg\varphi \vee \neg\psi$.
(iii) $\neg(\varphi \vee \psi)$ and $\neg\varphi \wedge \neg\psi$.
(iv) $\varphi \wedge (\psi \vee \theta)$ and $(\varphi \wedge \psi) \vee (\varphi \wedge \theta)$.
(v) $\varphi \vee (\psi \wedge \theta)$ and $(\varphi \vee \psi) \wedge (\varphi \vee \theta)$.
Now prove that every formula without quantifiers is logically equivalent to a conjunction of some collection of formulas each of which is a disjunction of a set of atomic or negated atomic formulas. (Formulas of the latter kind are said to be in **conjunctive-disjunctive normal form**. There is also a disjunctive-conjunctive form.)

6. Find an elementary formula $\pi(x)$ in the language of rings with unit that defines the set of prime numbers in the ring $\mathbb{Z}$ of integers.

7. Let Σ be a set of $\mathfrak{L}$ sentences and let θ and σ be $\mathfrak{L}$-sentences. Prove

If $\Sigma \cup \{\theta\} \models \sigma$, then $\Sigma \models \theta \Rightarrow \sigma$.

Prove the converse as well. This statement concerns the notion of logical consequence; we will call it the Deduction Theorem for Logical Consequence. Had we developed a system of syntactical inference complete for elementary logic, as we did for equational logic, and use $\vdash$ for "supports a deduction of", then the statement above could be rendered as

If $\Sigma \cup \{\theta\} \vdash \sigma$, then $\Sigma \vdash \theta \Rightarrow \sigma$.

In this form, it is called the Deduction Theorem for elementary logic.

8. Let **A** be an $\mathfrak{L}$-structure. Prove that every definable relation of **A** is invariant under every automorphism of **A**.

9. Let $\mathbf{A} = \langle \mathbb{N}, \cdot, 1\rangle$, the set of natural numbers endowed with multiplication and the constant 1. Prove that neither ordinary addition nor the usual ordering $<$ is definable in $\mathbf{A}$.

8.2. Ultraproducts and the Compactness Theorem

Let $\mathfrak{L}$ be any elementary language. General considerations concerning Galois connections lead us to conclude that **Mod Th** is a closure operator on the class of all $\mathfrak{L}$-structures and that **Mod Th** $\mathcal{K}$ is the smallest elementary class including $\mathcal{K}$, for all classes $\mathcal{K}$ of $\mathfrak{L}$-structures. In the equational context, Birkhoff's HSP Theorem (Theorem 4.131 in Volume 1) plays a central role because, in part, it provides an analysis of the corresponding closure operator **Mod** Θ. The attempt to provide a similar analysis of **Mod Th** runs into difficulty early. An easy but crucial observation leading to the HSP Theorem is that the truth of equations is preserved under the formation of homomorphic images, subalgebras, and direct products. The corresponding statements are all false when arbitrary elementary sentences are considered in place of equations. Indeed, the sentence asserting "There are exactly two elements.", which can be formulated in any elementary language, is not preserved under any of those algebraic constructions, apart from isomorphism, which we have so far considered. To make any progress in analyzing **Mod Th** we need to consider new algebraic constructions. The ultraproduct construction described in this section will serve our purpose.

Ultraproducts arise as special quotients of direct products. Let $\mathbf{A} = \prod_I \mathbf{A}_i$, where $\mathbf{A}_i$ is an $\mathfrak{L}$-structure for each $i \in I$. The members of A are I-tuples. The intuition in forming the quotient structure is to identify those I-tuples which are substantially the same—that is, the indices for which the identified tuples agree form a large subset of I. To frame this in a mathematically acceptable way, we have to come to grips with the notion of "largeness". Consider two examples where I is the unit interval. The subsets of I which have Lebesgue measure 1 certainly seem to be sets which occupy almost all of I. An even more restrictive notion is to take as "large" those subsets of I which are cofinite. Both of these notions of "large" possess the following properties:

(1) I is large.
(2) If X is large and $X \subseteq Y \subseteq I$, then Y is large.
(3) If X and Y are large, then $X \cap Y$ is large.

Property (3) is a reasonable attribute of largeness. If Y is large it ought to occupy all but a small portion of I, and hence it overlaps all but a small portion of X. Since X occupies all but a small portion of I, we see that $X \cap Y$ occupies all but two small portions of I. For any set I, these three properties characterize those families of subsets of I which are filters in the lattice of all subsets of I. More specifically, a collection F of subsets of I is said to be a **filter** on I provided

(1) $I \in F$,
(2) if $X \in F$ and $X \subseteq Y \subseteq I$, then $Y \in F$, and
(3) if X and Y belong to F, then $X \cap Y \in F$.

Each filter on I poses a concept of largeness. So there is no single notion of largeness, but rather one associated with each of the plethora of filters on I.

There are two ways in which a filter can be trivial. First, it could consist of all subsets of I. Such indiscriminate filters are said to be **improper**. So $\emptyset \notin F$ characterizes a filter F as a **proper** filter. Second, there could be a single subset X of I so that F is the collection of all subsets of I that contain X. Such filters are called **principal** filters.

Let I be any infinite set. The collection of cofinite subsets of I is a filter on I which is both proper and not principal. It is called the **Fréchet filter** on I. On the other hand, it is easy to see that any filter which has a finite member must be principal. In particular, every filter on a finite set must be principal.

Before taking up the task of building filters of more diverse kinds, we present the reduced product construction. Let I be any set, let F be a filter on I, and let $\mathbf{A}_i$ be an $\mathfrak{L}$-structure for each $i \in I$. Let $\mathbf{A} = \prod_I \mathbf{A}_i$. Consider a and a' from A. So

$$a = \langle a_i : i \in I\rangle \text{ and } a' = \langle a'_i : i \in I\rangle.$$

Our idea is to regard a and a' as congruent modulo F provided

$$\{i : a_i = a'_i \text{ and } i \in I\} \in F.$$

Another way to phrase this condition is

$$\{i : \mathbf{A}_i \models p_i(a) \approx p_i(a') \text{ and } i \in I\} \in F$$

which emphasizes that the formula $a \approx a'$ is true in a "large" set of of the factor structures. It is fruitful to introduce some notation, since we will meet such phrases frequently. Let $\varphi(\mathbf{x})$ be any formula and let $\mathbf{a}$ be an appropriate tuple from A. Thus, if $\mathbf{x} = \langle x_0, \ldots, x_{n-1}\rangle$, then $\mathbf{a}$ will be a certain n-tuple, say $\mathbf{a} = \langle a_0, \ldots, a_{n-1}\rangle$. Recall that each a_k is, in turn, an I-tuple. For each $i \in I$ we use p_i to denote the ith projection function. Finally, for each $i \in I$ we abuse our notation by setting

$$p_i(\mathbf{a}) = \langle p_i(a_0), \ldots, p_i(a_{n-1})\rangle$$

This allows us to define

$$[\![\varphi(\mathbf{a})]\!] = \{i : \mathbf{A}_i \models \varphi(p_i(\mathbf{a})) \text{ and } i \in I\}$$

Thus, $[\![\varphi(\mathbf{a})]\!]$ is the set of coordinates on which $\varphi(\mathbf{x})$ is satisfied by the projection of the sequence $\mathbf{a}$ into the factor structure.

Finally, we define a binary relation $\sim_F$ on A, called **congruence modulo** F, so that for all $a, a' \in A$

$$a \sim_F a' \text{ if and only if } [\![a \approx a']\!] \in F.$$

THEOREM 8.18. *Let I be any set, let F be any filter on I, and let $\mathbf{A}_i$ be an $\mathfrak{L}$-structure for each $i \in I$. Then $\sim_F$ is a congruence relation on $\prod_I \mathbf{A}_i$.*

Proof. Let $\mathbf{A} = \prod_I \mathbf{A}_i$. For each $a \in A$, we see that $[\![a \approx a]\!] = I$. Thus, the reflexivity of $\sim_F$ follows from property (i) of filters. The symmetry of $\sim_F$ follows from the observation that $[\![a \approx a']\!] = [\![a' \approx a]\!]$. To obtain transitivity, suppose $[\![a \approx a']\!] \in F$ and $[\![a' \approx a'']\!] \in F$. By property (iii) of filters

$$[\![a \approx a']\!] \cap [\![a' \approx a'']\!] \in F.$$

But it is easy to see that $[\![a \approx a']\!] \cap [\![a' \approx a'']\!] \subseteq [\![a \approx a'']\!]$. So, property (ii) of filters yields the transitivity of $\sim_F$. Therefore, $\sim_F$ is an equivalence relation on A.

Finally, let Q be any operation symbol. Let r denote the rank of Q. Let $a_0, \dots, a_{r-1}$ and $a'_0, \dots, a'_{r-1}$ be members of A so that $a_k \sim_F a'_k$ for all $k < r$. Using property (iii) of filters repeatedly, we see

$$[\![a_0 \approx a'_0]\!] \cap \cdots \cap [\![a_{r-1} \approx a'_{r-1}]\!] \in F.$$

But this intersection is included in $[\![Qa_0 \cdots a_{r-1} \approx Qa'_0 \cdots a'_{r-1}]\!]$, allowing us to conclude that

$$[\![Qa_0 \cdots a_{r-1} \approx Qa'_0 \cdots a'_{r-1}]\!] \in F$$

by property (ii) of filters. Therefore $\sim_F$ is a congruence relation on $\mathbf{A}$. ■

To prevent the buildup of notation in dealing with reduced products, we adopt the following conventions:

- For $a \in A$, we use a/F in place of $a/\sim_F$ to denote the congruence class of a modulo $\sim_F$.
- For $\mathbf{a} = \langle a_0, \dots, a_{r-1} \rangle$ with $a_k \in A$ for each $k < r$, we use $\mathbf{a}/F$ in place of $\langle a_0/F, \dots, a_{r-1}/F \rangle$.
- We use A/F to denote the partition of A into congruence classes modulo $\sim_F$.

DEFINITION 8.19. Let $\mathfrak{L}$ be an elementary language, let I be any set, let F be a filter on I, and let $\mathbf{A}_i$ be an $\mathfrak{L}$-structure for each $i \in I$. The **reduced product of** $\langle \mathbf{A}_i : i \in I \rangle$ **modulo** F is the $\mathfrak{L}$-structure $\mathbf{B}$ such that

(1) the universe of $\mathbf{B}$ is $\prod_I A_i/F$,
(2) $Q^{\mathbf{B}}(a_0/F, \dots, a_{r-1}/F) = Q^{\mathbf{A}}(a_0, \dots, a_{r-1})/F$, where Q is any operation symbol of $\mathfrak{L}$, r is the rank of Q, $\mathbf{A} = \prod_I \mathbf{A}_i$, and $a_0, \dots, a_{r-1} \in A$, and
(3) $R^{\mathbf{B}}(a_0/F, \dots, a_{r-1}/F)$ if and only if $[\![Ra_0 \cdots a_{r-1}]\!] \in F$, where R is any relation symbol of $\mathfrak{L}$, r is the rank of R, and $a_0, \dots, a_{r-1} \in \prod_I \mathbf{A}_i$.

The reduced product is denoted by $\prod_I \mathbf{A}_i/F$.

Theorem 8.18 guarantees that condition (2) is sound: the particular choice of representatives $a_0, \dots, a_{r-1}$ from the congruence classes is immaterial. That condition (3) is also sound is left as an exercise. Observe that the ordinary direct product is a special sort of reduced product which is obtained by setting F to the principal filter $\{I\}$. The following theorem is little more than a reformulation of the definition of reduced product.

THEOREM 8.20. *Let $\prod_I \mathbf{A}_i/F$ be a reduced product and let $\varphi(\mathbf{x})$ be an atomic formula. For any sequence $\mathbf{a}$ from $\prod_I A_i$,*

$$\prod_I \mathbf{A}_i/F \models \varphi(\mathbf{a}/F) \textit{ if and only if } [\![\varphi(\mathbf{a})]\!] \in F.$$

Recall that we are trying to find an algebraic construction which will preserve the truth of elementary sentences. The theorem above is a quite provocative step. Reduced products would work if only this theorem were true for all formulas, not just atomic formulas. However, this is too much to hope for. To see why, let $I = \omega$, and for each $i \in I$, let $A_i = \{0, 1\}$. Take $F = \{\omega\}$ and let the language be empty. The resulting reduced product is the same as the direct power of countably many copies of $\{0, 1\}$. It has continuum many elements. On the other hand, the sentence asserting that there are exactly two elements is true in each $\mathbf{A}_i$.

Were we to attempt to prove a stronger version of Theorem 8.20, the strategy at hand would be induction on formulas. We would use Theorem 8.20 as the initial step of the induction. The next step would take up negation, and would look something like:

$$\begin{aligned}\prod_I \mathbf{A}_i/F \models \neg\varphi(\mathbf{a}/F) &\text{ if and only if } \prod_I \mathbf{A}_i/F \not\models \varphi(\mathbf{a}/F)\\ &\text{ if and only if } [\![\varphi(\mathbf{a})]\!] \notin F\\ &\text{ if and only if } [\![\neg\varphi(\mathbf{a})]\!] \in F\end{aligned}$$

The first equivalence would hold by the definition of satisfaction. The second equivalence would hold by appealing to the induction hypothesis. But the last equivalence might fail. It is easy to see that

$$[\![\neg\varphi(\mathbf{a})]\!] = I - [\![\varphi(\mathbf{a})]\!],$$

and so if F is proper we can have

$$[\![\neg\varphi(\mathbf{a})]\!] \in F \text{ implies } [\![\varphi(\mathbf{a})]\!] \notin F.$$

It is the converse of this implication that need not be true, since it can happen that neither $[\![\varphi(\mathbf{a})]\!]$ nor its complement belongs to F. So we are led to the next definition.

DEFINITION 8.21. Let I be any set. U is an **ultrafilter** on I if and only if U is a proper filter on I such that either $X \in U$ or $(I - X) \in U$, for every $X \subseteq I$. A reduced product modulo an ultrafilter is called an **ultraproduct**.

It is easy to see that an ultrafilter U is principal if and only if there is $i \in I$ so that $X \in U$ if and only if $i \in X \subseteq I$. Conversely, for each $i \in I$, the collection of all subsets of I which contain i is easily seen to be an ultrafilter. An ultraproduct built using a principal ultrafilter will be isomorphic with the factor structure designated by the element i. Leaving aside for the moment the question of whether any nonprincipal ultrafilter exist, we can now present a key result.

THEOREM 8.22 (The Fundamental Theorem for Ultraproducts (Jerzy Łoś, 1955b)). *Let $\prod_I \mathbf{A}_i/U$ be an ultraproduct and let $\varphi(\mathbf{x})$ be any formula. For any sequence $\mathbf{a}$ from $\prod_I A_i$ we have*

$$\prod_I \mathbf{A}_i/U \models \varphi(\mathbf{a}/U) \textit{ if and only if } [\![\varphi(\mathbf{a})]\!] \in U.$$

Proof. We proceed by induction on the complexity of $\varphi(\mathbf{x})$. In view of Theorem 8.20 and the remarks following it, only the cases where φ is a conjunction or an existentialization remain.

Case $\wedge$: $\varphi(\mathbf{x})$ is $\psi(\mathbf{x}) \wedge \theta(\mathbf{x})$.

This case is easily handled by properties (ii) and (iii) of filters, since $[\![\psi(\mathbf{a}) \wedge \theta(\mathbf{a})]\!] = [\![\psi(\mathbf{a})]\!] \cap [\![\theta(\mathbf{a})]\!]$.

Case $\exists$: $\varphi(\mathbf{x})$ is $\exists y \psi(y, \mathbf{x})$

The crucial observation is $[\![\exists y\psi(y, \mathbf{a})]\!] = [\![\psi(b, \mathbf{a})]\!]$ for some $b \in \prod_I A_i$. Indeed, for each $i \in [\![\exists y\psi(y, \mathbf{a})]\!]$, pick $b_i \in A_i$ so that $\mathbf{A}_i \models \psi(b_i, \mathbf{a})$, and for all other $i \in I$ pick b_i arbitrarily from A_i. Setting $b = \langle b_i : i \in I \rangle$, we have

$$[\![\exists y\psi(y, \mathbf{a})]\!] = [\![\psi(b, \mathbf{a})]\!].$$

Now just observe:

$$\begin{aligned}\prod_I \mathbf{A}_i/U \models \exists y\psi(y, \mathbf{a}/U) &\text{ if and only if } \prod_I \mathbf{A}_i/U \models \psi(b/U, \mathbf{a}/U) \\ &\qquad\qquad \text{for some } b \in \prod_I A_i \\ &\text{ if and only if } [\![\psi(b, \mathbf{a})]\!] \in U \text{ for some } b \in \prod_I A_i \\ &\text{ if and only if } [\![\exists y\psi(y, \mathbf{a})]\!] \in U\end{aligned}$$

where the last equivalence follows in one direction from property (ii) of filters, and the other direction is the observation above. ■

COROLLARY 8.23. *Let $\prod_I \mathbf{A}_i/U$ be an ultraproduct and let θ be a sentence. Then $\prod_I \mathbf{A}_i/U \models \theta$ if and only if $[\![\theta]\!] \in U$.* ■

So a sentence is true in an ultraproduct if and only if it is true for a large set of the factors (using the ultrafilter to determine largeness). As a consequence, every elementary class is closed with respect to the formation of ultraproducts.

Suppose that $\mathbf{A}_i = \mathbf{B}$ for all $i \in I$ and that F is a filter on I. We denote $\prod_I \mathbf{A}_i/F$ by $\mathbf{B}^I/F$ and call it a **reduced power** (or **ultrapower** if F is an ultrafilter) of $\mathbf{B}$. In the event that $\mathbf{C} \cong \mathbf{B}^I/U$, where U is an ultrafilter on I, we call $\mathbf{B}$ an **ultraroot** of $\mathbf{C}$. Clearly, $\mathbf{B}^I/U \models \theta$ if and only if $[\![\theta]\!] \in U$ if and only if $\mathbf{B} \models \theta$, for all sentences θ. This means that every structure is elementarily equivalent to each of its ultrapowers.

COROLLARY 8.24. *Let $\mathfrak{L}$ be an elementary language and let $\mathcal{K}$ be any class of $\mathfrak{L}$-structures. If $\mathcal{K}$ is an elementary class, then $\mathcal{K}$ is closed under isomorphism, the formation of ultraproducts, and the formation of ultraroots.* ■

The notion of ultraproduct was formulated in (Jerzy Łoś, 1955b), although earlier special cases of this construction can be found in (Thoralf Skolem, 1934) and (Edwin Hewitt, 1948). After Łos's paper and (C. C. Chang and Anne C. Morel, 1958), Tarski extracted the notion reduced product and the essential properties of reduced products were developed by Frayne, Morel, Scott, and Tarski (see (T. Frayne, A. C. Morel, and D. S. Scott, 1962, 1964), which circulated in the late 1950's, although its publication was delayed.) A. I. Mal'tsev (1966b), based on the work of Frayne, Morel, and Scott, characterized quasivarieties using reduced products, see Theorem 8.108.

So we have in hand several algebraic constructions preserving elementary sentences. The corollary above is like the easy direction of Birkhoff's HSP Theorem. It turns out that the converse of this corollary is also true, although much more difficult to prove, providing us with an analog of the HSP Theorem.

Every structure $\mathbf{A}$ can be embedded into any of its nontrivial direct powers $\mathbf{A}^I$ via the diagonal map $a \mapsto \langle a : i \in I \rangle$. The composition of the diagonal map with the quotient map results in an elementary embedding of a structure into any of its ultrapowers.

DEFINITION 8.25. Let **A** be any structure, I be any nonempty set, and F be any ultrafilter on I. The **canonical embedding** of A into A^I/F is the function $\delta : A \to A^I/F$ such that $\delta(a) = \langle a : i \in I\rangle/F$ for all $a \in A$.

COROLLARY 8.26. *If* $\mathbf{A}^I/U$ *is an ultrapower of* **A**, *then the canonical embedding is an elementary embedding.* ■

Now we have at hand a means to construct elementary extensions of structures. To take advantage of it we have to be able to construct ultrafilters which are not principal.

Let I be a set. A collection C of subsets of I has the **finite intersection property** provided the intersection of any finite nonempty subcollection of C is nonempty. Every proper filter has the finite intersection property. Moreover, if C has the finite intersection property, then C can be enlarged to a proper filter by first forming C', the collection of all sets arising as intersections of finitely many sets from C, and then letting F consist of those sets X such that $Y \subseteq X$ for some $Y \in C'$. The collection F is clearly the smallest filter on I which includes C. We say that C **generates** F.

THEOREM 8.27. *Let I be any nonempty set, and let U be a proper filter on I. The following are equivalent:*

(i) *U is an ultrafilter on I.*
(ii) *For all $X, Y \subseteq I$, if $X \cup Y \in U$, then $X \in U$ or $Y \in U$.*
(iii) *For every $X \in U$ and every partition of X into finitely many blocks, exactly one of the blocks belongs to U.*
(iv) *For every proper filter F on I, if $U \subseteq F$, then $U = F$.*

Proof. It is very easy to establish the equivalence of the first three conditions, so we will only argue that conditions (i) and (iv) are equivalent.

(i) ⇒ (iv)

Suppose U is an ultrafilter and that $U \subseteq F$ where F is a filter and $F \neq U$. Pick $X \in F - U$. Since U is an ultrafilter we have $(I - X) \in U \subseteq F$. Since F is a filter, we conclude that $\emptyset = X \cap (I - X) \in F$. Hence, F is not proper.

(iv) ⇒ (i)

Suppose U is a maximal proper filter. Pick $X \subseteq I$ with $X \notin U$. Since U is maximal, we have that $U \cup \{X\}$ must not generate a proper filter. Hence $U \cup \{X\}$ cannot have the finite intersection property. However, U is closed under finite nonempty intersections, so there must be $Y \in U$ with Y and X disjoint. This entails that $Y \subseteq (I - X) \in U$. Therefore, U is an ultrafilter. ■

Sometimes ultrafilters are called **prime filters** (and then ultraproducts are called **prime products**) because condition (ii) in the theorem above asserts that the notion of ultrafilter is dual to the notion of prime ideal in the distributive lattice of all subsets of I. This allows us to recite the Prime Ideal Theorem for Distributive Lattices (Theorem 2.60 in Volume I):

THEOREM 8.28 (The Ultrafilter Theorem (Alfred Tarski, 1930; M. H. Stone, 1936)). *Let I be any set.*

(i) *If N is any ideal in the lattice of subsets of I and F is any filter on I such that N and F are disjoint, then there is an ultrafilter U on I disjoint from N such that $F \subseteq U$.*

(ii) *Every proper filter on I is the intersection of all the ultrafilters which include it.* ■

Letting N be the collection whose only element is the empty set, we see that the first part of the Ultrafilter Theorem tells us that every proper filter can be extended to an ultrafilter. Of course, this also means that any collection of subsets of I with the finite intersection property can be extended to an ultrafilter. In particular, if I is infinite then the Fréchet filter on I can be extended to an ultrafilter. Thus there really are nonprincipal ultrafilters.

We have framed the theorem above over the collection of all subsets of I, since that is the context we find immediately useful. However, this theorem could be framed for an arbitrary Boolean algebra, in place of the Boolean algebra of all subsets of I. Then the theorem would read:

The Ultrafilter Theorem (Alternate Version).
Let $\mathbf{B}$ be any Boolean algebra.

(i) *If N is any ideal of $\mathbf{B}$ and F is any filter of $\mathbf{B}$ such that N and F are disjoint, then there is an ultrafilter U of $\mathbf{B}$ disjoint from N such that $F \subseteq U$.*

(ii) *Every proper filter of $\mathbf{B}$ is the intersection of all the ultrafilters which include it.*

■

Here is another useful way to obtain ultrafilters.

THEOREM 8.29. (Bjarni Jónsson, 1967) *Let I be any nonempty set, and let $\mathcal{M}$ be a collection of nonempty subsets of I. If*

(i) *$I \in \mathcal{M}$,*

(ii) *For all X and Y, if $X \in \mathcal{M}$ and $X \subseteq Y \subseteq I$, then $Y \in \mathcal{M}$, and*

(iii) *For all X and Y, if $X \cup Y \in \mathcal{M}$, then $X \in \mathcal{M}$ or $Y \in \mathcal{M}$,*

then there is an ultrafilter U on I such that $U \subseteq \mathcal{M}$.

Proof. For the purposes of this proof we will call a collection $\mathcal{M}$ of nonempty subsets of I *prime* if $\mathcal{M}$ fulfills the three conditions set out in the theorem. Prime collections are also known as grills. We need to show that every prime collection includes an ultrafilter.

First observe that a minimal prime collection is an ultrafilter: Suppose $\mathcal{M}$ is minimal, and $X, Y \in \mathcal{M}$. We need to prove that $X \cap Y \in \mathcal{M}$. Let

$$\mathcal{N} = \{Z : Z \subseteq I \text{ and } X \cap Z \in \mathcal{M}\}.$$

It is easy to check that $\mathcal{N}$ is a prime collection. [For example, let $Z \cup Z' \in \mathcal{N}$. Then $(X \cap Z) \cup (X \cap Z') = X \cap (Z \cup Z') \in \mathcal{M}$. So $X \cap Z \in \mathcal{M}$ or $X \cap Z' \in \mathcal{M}$. Thus $Z \in \mathcal{N}$ or $Z' \in \mathcal{N}$.] It is also clear that $\mathcal{N} \subseteq \mathcal{M}$. Since $\mathcal{M}$ is minimal, we have $\mathcal{N} = \mathcal{M}$. But then $Y \in \mathcal{N}$. Hence $X \cap Y \in \mathcal{M}$ and $\mathcal{M}$ is an ultrafilter.

It remains to argue that every prime collection includes a minimal prime collection. So let $\mathcal{M}$ be any prime collection and let

$$\mathcal{K} = \{\mathcal{N} : \mathcal{N} \subseteq \mathcal{M} \text{ and } \mathcal{N} \text{ is a prime collection}\}.$$

By the Hausdorff Maximality Principle (see page 7 of Volume I) there is a maximal chain $\mathcal{C}$ in $\mathcal{K}$. Let $U = \bigcap \mathcal{C}$. It is a straightforward matter of checking (i)–(iii) to see that U is a prime collection. Since $\mathcal{C}$ is a maximal chain, U must be a minimal prime collection. Thus U is an ultrafilter. ■

We are now in a position to prove one of the most fundamental and powerful theorems of model theory. One formulation of this theorem, the Compactness Theorem, reads "The lattice of elementary theories is algebraic", but such a formulation disguises the real power of the theorem, which resides largely in its usefulness in the construction of models of sets of elementary sentences. This theorem, in the case of countable languages, is an easy corollary of Gödel's Completeness Theorem (Kurt Gödel, 1930), but its significance and extension to arbitrary language came six years later at the hands of A. I. Mal'tsev.

THEOREM 8.30 (The Compactness Theorem (A. I. Mal'tsev, 1936)). *Let Σ be a set of elementary sentences. If every finite subset of Σ has a model, then Σ has a model.*

Proof. We make the harmless assumption that Σ is closed under conjunction—that is, that $\varphi \wedge \psi \in \Sigma$ whenever both φ and ψ belong to Σ. Thus we can replace the hypothesis that every finite subset of Σ has a model by the simpler condition that every sentence in Σ has a model. The idea of the proof is to form an ultraproduct of these models to obtain a model of Σ. So for each $\varphi \in \Sigma$, pick $\mathbf{A}_\varphi$ so that

$$\mathbf{A}_\varphi \models \varphi.$$

Our desire is to find an ultrafilter U on Σ so that

$$\prod_\Sigma \mathbf{A}_\varphi / U \models \Sigma.$$

We see by the Fundamental Theorem for Ultraproducts that this is equivalent to

$$[\![\,\theta\,]\!] \in U \text{ for all } \theta \in \Sigma.$$

So we only need an ultrafilter extending $\{[\![\,\theta\,]\!] : \theta \in \Sigma\}$. Therefore, it suffices to show that this collection has the finite intersection property. To this end let $\theta_0, \dots, \theta_{n-1} \in \Sigma$. Since

$$[\![\,\theta_i\,]\!] = \{\varphi : \varphi \in \Sigma \text{ and } \mathbf{A}_\varphi \models \theta_i\},$$

we see that

$$[\![\,\theta_0\,]\!] \cap \cdots \cap [\![\,\theta_{n-1}\,]\!] = \{\varphi : \varphi \in \Sigma \text{ and } \mathbf{A}_\varphi \models \theta_0 \wedge \cdots \wedge \theta_{n-1}\}.$$

But Σ is closed under conjunction, so $(\theta_0 \wedge \cdots \wedge \theta_{n-1}) \in \Sigma$ and we have

$$[\![\,\theta_0\,]\!] \cap \cdots \cap [\![\,\theta_{n-1}\,]\!] = [\![\,\theta_0 \wedge \cdots \wedge \theta_{n-1}\,]\!].$$

By observing that $\theta \in [\![\,\theta\,]\!]$ for all $\theta \in \Sigma$, we conclude that the finite intersection property holds. ■

The proof we have given was suggested by Alfred Tarski, who found a version for the case of Horn sentences; Thomas Frayne carried Tarski's suggestion to completion and the resulting proof, as above, can be found in (T. Frayne, A. C. Morel, and D. S. Scott, 1962). In most model theory texts proofs of the Compactness Theorem usually rely on the method of Leon Henkin (1948), see also (Leon Henkin, 1949). It is interesting to note that the initial result for countable languages of Gödel, the full version of the theorem due to Mal'tsev, and the approach, not given here, of Henkin, were all in their doctoral dissertations, while the work of Frayne, Morel, and Scott was launched while they were graduate students. A. I. Mal'tsev (1941) gave some of the earliest applications of the Compactness Theorem (they were in group theory).

Some immediate corollaries of the Compactness Theorem are close enough to it that they could justly be called by the same name. We close this section with three such corollaries.

COROLLARY 8.31. *Let $\Sigma \cup \{\varphi\}$ be any set of elementary sentences. If $\Sigma \models \varphi$, then there is some finite subset $\Sigma' \subseteq \Sigma$ such that $\Sigma' \models \varphi$.* ■

COROLLARY 8.32. *Let $\mathfrak{L}$ be an elementary language. The lattice of all elementary theories of $\mathfrak{L}$ is algebraic.* ■

COROLLARY 8.33. *Let Σ be any set of elementary sentences. If Σ is logically equivalent to some finite set of sentences, then Σ is logically equivalent to some finite subset of itself.* ■

It follows from this last corollary that if a variety can be axiomatized by a finite set of elementary sentences, then it can also be axiomatized by a finite set of equations.

Exercises 8.34

1. Let $\mathfrak{L}$ be the language for group theory with operation symbols $\cdot, ^{-1}$, and 1. Let T be a set of $\mathfrak{L}$-sentences which includes all the group axioms (so every model of T will be a group). Suppose that for each n, there is a model of T which has no elements, other than 1, of order smaller than n. Prove that there is a model of T such that 1 is the only element of finite order.

2. Suppose that $\mathbf{G}$ is a group which has elements of arbitrarily large finite order. Prove that $\mathbf{G}$ is elementarily equivalent to a group with an element of infinite order. [Hint: Try expanding the language by a new constant symbol.]

3. Let $\langle \mathbb{N}, +, \cdot, 0, 1, \leq \rangle$ be the familiar structure consisting of the natural numbers equipped with addition, multiplication, the two distinguished elements 0 and 1, and the usual order relation. Let T be the set of all the sentences true in $\langle \mathbb{N}, +, \cdot, 0, 1, \leq \rangle$. Prove T has a model $\mathbf{M}$ with an element ω so that all the following are true in $\mathbf{M}$:

$$0 \leq \omega, \ 1 \leq \omega, \ 2 \leq \omega, \ldots.$$

4. Let $\mathfrak{L}$ be the language of rings. Find a set Σ of $\mathcal{L}$-sentences such that $\operatorname{Mod}\Sigma$ is the class of algebraically closed fields. Then prove that there is no finite set of $\mathfrak{L}$-sentences which will serve the same purpose.

5. Let $\mathfrak{L}$ be the language of ordered sets. Prove that there is no set Σ of $\mathfrak{L}$-sentences such that $\operatorname{Mod}\Sigma$ is the class of all well-ordered sets.

8.3. Applications of the Compactness Theorem

Like the ultraproduct construction, the Compactness Theorem can be used to build elementary extensions of structures. Suppose we want to construct $\mathbf{A}$ such that $\mathbf{B} \subseteq \mathbf{A}$ for some given $\mathbf{B}$. If $\mathbf{B}$ is finite and the language is finite, then we saw how to formulate a single elementary sentence that describes $\mathbf{B}$ to within isomorphism. This sentence asserts the existence of $|B|$ elements which satisfy various atomic and negated atomic formulas, and then concludes by asserting that no further elements exist. If the last phrase is omitted, what remains would be a sentence such that any model of it would have a substructure isomorphic to $\mathbf{B}$. If $\mathbf{B}$ is infinite there is more difficulty, since we cannot assert the existence of infinitely many elements in a single sentence, except in very restricted cases. So we cannot talk about how the individual elements of $\mathbf{B}$ are interrelated. Fortunately, a simple device circumvents this difficulty: we add a name for each element of $\mathbf{B}$ to the language.

DEFINITION 8.35. Let $\mathfrak{L}$ be an elementary language, and let $\mathbf{A}$ be an $\mathfrak{L}$-structure. The **diagram language** of $\mathbf{A}$ is $\mathfrak{L} \cup \{c_a : a \in A\}$ where the c_a's are new distinct constant symbols. The structure $\mathbf{A}_A$ is the expansion of $\mathbf{A}$ to the diagram language such that c_a is interpreted in $\mathbf{A}_A$ as a, for each $a \in A$. The **diagram** of $\mathbf{A}$ is the set of all atomic and negated atomic sentences of the diagram language true in $\mathbf{A}_A$ and is denoted by $\Delta_{\mathbf{A}}$. The **atomic diagram** of $\mathbf{A}$ is the set of all atomic sentences in the diagram language that are true in $\mathbf{A}_A$. The **elementary diagram** of $\mathbf{A}$ is $\operatorname{Th}\mathbf{A}_A$ and it is denoted by $D_{\mathbf{A}}$.

The next corollary can be proven by checking the definitions involved.

COROLLARY 8.36. *Let* $\mathbf{A}$ *and* $\mathbf{B}$ *be* $\mathfrak{L}$*-structures.*

(i) $\mathbf{B}$ *is isomorphic to a substructure of* $\mathbf{A}$ *if and only if* $\mathbf{A}$ *is the reduct to* $\mathfrak{L}$ *of a model of* $\Delta_{\mathbf{B}}$.
(ii) $\mathbf{B}$ *can be elementarily embedded into* $\mathbf{A}$ *if and only if* $\mathbf{A}$ *is the reduct to* $\mathfrak{L}$ *of a model of* $D_{\mathbf{B}}$.

■

Complete and Model Complete Elementary Theories

An elementary theory T is said to be **complete** provided T has a model and any two models of T are elementarily equivalent. Another way to express this is to assert that for every sentence φ exactly one of φ and $\neg\varphi$ belongs to T. Apart from the trivial cases (all models of T have exactly one element), no quasivariety (and hence no variety) can have a complete elementary theory. Nevertheless, an understanding of complete theories informs the investigation of varieties and quasivarieties. For example, below we will show that the elementary theory of infinite atomic Boolean algebras is complete. In §8.8 we will use this result to obtain a theorem about direct products that has straightforward applications to varieties and quasivarieties.

An elementary theory T is said to be **model complete** provided that $\mathbf{A} \preccurlyeq \mathbf{B}$ whenever $\mathbf{A}$ and $\mathbf{B}$ are models of T and $\mathbf{A}$ is a substructure of $\mathbf{B}$. The notion of model completeness is due to Abraham Robinson (1956a).

THEOREM 8.37. *Let T be an elementary $\mathfrak{L}$-theory that has a model. The following conditions are equivalent:*

(i) *T is model complete.*
(ii) *For every model $\mathbf{A}$ of T, the elementary theory axiomatized by $T \cup \Delta_{\mathbf{A}}$ is complete.*
(iii) *Suppose $\mathbf{A}$ and $\mathbf{B}$ are models of T such that $\mathbf{A}$ is a substructure of $\mathbf{B}$. If $\varphi(\mathbf{x})$ is a universal formula, and $\mathbf{a}$ is an ω-tuple of elements from A so that $\mathbf{A} \models \varphi(\mathbf{a})$, then $\mathbf{B} \models \varphi(\mathbf{a})$.*
(iv) *For any $\mathfrak{L}$-formula φ, there is a universal $\mathfrak{L}$-formula ψ with the same free variables as φ such that $T \models \varphi \Leftrightarrow \psi$.*

Proof.

(i)⇒(ii)
Let $\mathbf{A} \models T$ and let $\mathbf{B} \models T \cup \Delta_{\mathbf{A}}$. Then the set of elements of $\mathbf{B}$ named by the new constant symbols in the diagram language of $\mathbf{A}$ form a substructure $\mathbf{B}^*$ of $\mathbf{B}$ that is isomorphic to $\mathbf{A}_A$. So the reduct of $\mathbf{B}^*$ to the language of T is a substructure of the reduct of $\mathbf{B}$ to the language of T. Because T is model complete, we see that the reduct of $\mathbf{B}^*$ to the language of T is an elementary substructure of the reduct of $\mathbf{B}$ to the language of T. But then $\mathbf{B}^*$ is also an elementary substructure of $\mathbf{B}$. Hence, $\mathbf{A}_A$ is elementarily equivalent to $\mathbf{B}$. So the elementary theory axiomatized by $T \cup \Delta_{\mathbf{A}}$ is complete.

(ii)⇒(iii)
Let $\mathbf{A}$ and $\mathbf{B}$ be models of T so that $\mathbf{A}$ is a substructure of $\mathbf{B}$ and let $\varphi(\mathbf{x})$ be a universal formula and $\mathbf{a}$ be a tuple of elements of A so that $\mathbf{A} \models \varphi(\mathbf{a})$. Because $\mathbf{A}$ is a substructure of $\mathbf{B}$ we see that $\mathbf{B}_A \models T \cup \Delta_{\mathbf{A}}$. This means that $\mathbf{A}_A$ and $\mathbf{B}_A$ are elementarily equivalent. Now we can regard $\varphi(\mathbf{a})$ as a sentence in the diagram language of $\mathbf{A}$. This sentence, must also be true in $\mathbf{B}_A$. So we can conclude that $\mathbf{B} \models \varphi(\mathbf{a})$, as desired.

(iii)⇒(iv)
First we show that if $\theta(u, x_0, \dots, x_{n-1})$ is a quantifier-free formula then there is a universal formula $\psi(x_0, \dots, x_{n-1})$ so that

$$T \models \exists \mathbf{u} \theta(\mathbf{u}, x_0, \dots, x_{n-1}) \Leftrightarrow \psi(x_0, \dots, x_{n-1}). \tag{8.3.1}$$

Let $\mathfrak{L}'$ be the expansion of $\mathfrak{L}$ by adjoining n new distinct constant symbols $c_0, \dots, c_{n-1}$. Let

$$\Gamma = \{\eta \mid \eta \text{ is a universal } \mathfrak{L}'\text{-sentence and } T \models \exists \mathbf{u} \theta(\mathbf{u}, c_0, \dots, c_{n-1}) \Rightarrow \eta\}.$$

CLAIM 2.
The set $T \cup \Gamma \cup \{\neg \exists \mathbf{u} \theta(\mathbf{u}, c_0, \dots, c_{n-1})\}$ has no model.

Proof of Claim 2. Suppose not. Let $\mathbf{A}' = \langle \mathbf{A}, a_0, \ldots, a_{n-1} \rangle$ be a model of

$$T \cup \Gamma \cup \{\neg \exists \mathbf{u} \theta(\mathbf{u}, c_0, \ldots, c_{n-1})\}.$$

Let $\mathcal{L}''$ be the diagram language of $\mathbf{A}'$. Let $\mathbf{B}''$ be a model of $T \cup \Delta_{\mathbf{A}'}$. To simplify notation we regard $\mathbf{A}'$ as a substructure of the reduct $\mathbf{B}'$ of $\mathbf{B}''$ to the language of $\mathbf{A}'$. Now $\mathbf{A} \models \neg \exists \mathbf{u} \theta(\mathbf{u}, a_0, \ldots, a_{n-1})$. This is the same as $\mathbf{A} \models \forall \mathbf{u} \neg \theta(\mathbf{u}, a_0, \ldots, a_{n-1})$. By condition (iii) we have $\mathbf{B} \models \forall \mathbf{u} \neg \theta(\mathbf{u}, a_0, \ldots, a_{n-1})$. It follows that

$$T \cup \Delta_{\mathbf{A}'} \models \neg \exists \mathbf{u} \theta(\mathbf{u}, c_0, \ldots, c_{n-1}).$$

By the Compactness Theorem there is a finite subset Δ' of $\Delta_{\mathbf{A}'}$ so that

$$T \cup \Delta' \models \neg \exists \mathbf{u} \theta(\mathbf{u}, c_0, \ldots, c_{n-1}).$$

Let $\delta(c_0, \ldots, c_{n-1}, d_0, \ldots, d_{m-1})$ be the conjunction of the sentences belonging to Δ'. Here $d_0, \ldots, d_{m-1}$ are the elements named by the constant symbols added the $\mathfrak{L}'$ to obtain $\mathfrak{L}''$. After applying the Deduction Theorem for Logical Consequence (see Exercise 7.7), we arrive at

$$T \models \delta(c_0, \ldots, c_{n-1}, d_0, \ldots, d_{m-1}) \Rightarrow \neg \exists \mathbf{u} \theta(\mathbf{u}, c_0, \ldots, c_{n-1}).$$

Now T places no constraints on the elements named by the new constant symbols. So,

$$T \models \forall y_0, \ldots, y_{m-1} [\delta(c_0, \ldots, c_{n-1}, y_0, \ldots, y_{m-1}) \Rightarrow \neg \exists \mathbf{u} \theta(\mathbf{u}, c_0, \ldots, c_{n-1})].$$

Replacing one formula by a formula logically equivalent to it, we obtain

$$T \models \big(\exists y_0, \ldots, y_{m-1} \delta(c_0, \ldots, c_{n-1}, y_0, \ldots, y_{m-1})\big) \Rightarrow \neg \exists \mathbf{u} \theta(\mathbf{u}, c_0, \ldots, c_{n-1}).$$

Again using logical equivalence we have

$$T \models \exists \mathbf{u} \theta(\mathbf{u}, c_0, \ldots, c_{n_1}) \Rightarrow \forall y_0, \ldots, y_{m-1} \neg \delta(c_0, \ldots, c_{n-1}, y_0, \ldots, y_{m-1}).$$

This means that $\forall y_0, \ldots, y_{m-1} \neg \delta(c_0, \ldots, c_{n-1}, y_0, \ldots, y_{m-1}) \in \Gamma$. Since $\mathbf{A}' \models \Gamma$ we see that

$$\mathbf{A}' \models \forall y_0, \ldots, y_{m-1} \neg \delta(c_0, \ldots, c_{n-1}, y_0, \ldots, y_{m-1}).$$

But $\delta(c_0, \ldots, c_{n-1}, d_0, \ldots, d_{m-1})$ belongs to the diagram of $\mathbf{A}'$, and as a consequence we know that $\mathbf{A}' \models \exists y_0, \ldots, y_{m-1} \delta(c_0, \ldots, c_{n-1})$. So we have arrived at a contradiction. ■

By the Compactness Theorem, there is a finite subset $\{\eta_0, \ldots, \eta_{m-1}\}$ of Γ so that $T \cup \{\eta_0, \ldots, \eta_{m-1}\} \cup \{\neg \exists u \theta(u, c_0, \ldots, c_{n-1})\}$ has no model. By the Deduction Theorem for Logical Consequence (see Exercise 8.17.7)

$$T \models [(\eta_0 \wedge \cdots \wedge \eta_{m-1}) \Rightarrow \exists \mathbf{u} \theta(\mathbf{u}, c_0, \ldots, c_{n-1})].$$

Since the conjunction of universal formulas is logically equivalent to a universal formula and since $\eta_j \in \Gamma$ for all $j < m$, there is a universal sentence $\psi(c_0, \ldots, c_{n-1}) \in \Gamma$ logically equivalent to $\eta_0 \wedge \cdots \wedge \eta_{m-1}$. So

$$T \models \psi(c_0, \ldots, c_{n-1}) \Rightarrow \exists \mathbf{u} \theta(\mathbf{u}, c_0, \ldots, c_{n-1}).$$

But since $\psi(c_0, \ldots, c_{n-1}) \in \Gamma$ we also have the converse implication. This means

$$T \models \psi(c_0, \ldots, c_{n-1}) \Leftrightarrow \exists \mathbf{u} \theta(\mathbf{u}, c_0, \ldots, c_{n-1}).$$

But notice that the new constant symbols do not occur in T. So T places no constraints on the elements denoted by these constant symbols in any model of T. We conclude that

$$T \models \psi(x_0, \ldots, x_{n-1}) \Leftrightarrow \exists \mathbf{u} \theta(\mathbf{u}, x_0, \ldots, x_{n-1}).$$

So the assertion (8.3.1) has been established and every existential formula is equivalent, modulo T, to a universal formula.

We finish this part of the argument by induction on the complexity of φ. Taking $\vee$ and $\neg$ as our connectives and $\forall$ as our quantifier, we would have to consider the base case (φ is atomic), and one inductive step for each connective and one step for the quantifier. All of these are straightforward except the step for negation. Here it is.

Step: $\varphi(\mathbf{x})$ is $\neg\theta(\mathbf{x})$ where $\theta(\mathbf{x})$ is a universal formula
Then $\varphi(\mathbf{x})$ is logically equivalent to an existential formula. But every existential formula is equivalent, modulo T, to a universal formula. Thus there is a universal formula $\psi(\mathbf{x})$ so that $T \models \varphi(\mathbf{x}) \Leftrightarrow \psi(\mathbf{x})$, confirming (iv).

(iv)$\Rightarrow$(i)
Let $\mathbf{A}$ and $\mathbf{B}$ be models of T such that $\mathbf{A}$ is a substructure of $\mathbf{B}$. Let $\varphi(\mathbf{x})$ be a formula and $\mathbf{a}$ be a tuple of elements of A so that $\mathbf{B} \models \varphi(\mathbf{a})$. We need to show that $\mathbf{A} \models \varphi(\mathbf{a})$. Let $\psi(\mathbf{x})$ be a universal formula such that $T \models \varphi \Leftrightarrow \psi$. It follows that $\mathbf{B} \models \psi(\mathbf{a})$. Since ψ is universal, we obtain that $\mathbf{A} \models \psi(\mathbf{a})$. But then $\mathbf{A} \models \varphi(\mathbf{a})$ as desired, since $\mathbf{A} \models T$. ■

Recall that a is an **atom** in a Boolean algebra provided a covers 0. A Boolean algebra $\mathbf{B}$ is **atomic** provided for all $b \in B$ with $0 < b$ there is an atom a so that $a \leq b$. These notions can be expressed by elementary formulas in the language of Boolean algebras. Evidently, the class of infinite atomic Boolean algebras is an elementary class. Below we will show that the elementary theory of infinite atomic Boolean algebras is complete. It is an exercise to show that, on the other hand this theory is not model complete.

Let Σ_{At} axiomatize the elementary theory of infinite atomic Boolean algebras. Expand the language of Boolean algebras by adjoining a new basic one-place relation symbol A and let $\Sigma^*_{\mathrm{At}} = \Sigma_{\mathrm{At}} \cup \{\forall x[Ax \Leftrightarrow$ "x is an atom" $]\}$. We will call the elementary theory axiomatized by this set of sentence the **enhanced elementary theory of infinite atomic Boolean algebras.**.

THEOREM 8.38. *The enhanced elementary theory of infinite atomic Boolean algebras is both model complete and complete.*

Proof. Let T be the enhanced elementary theory of infinite atomic Boolean algebras. We will appeal to condition (iii) of Theorem 8.37. So suppose that $\mathbf{B}_0$ and $\mathbf{B}_1$ are infinite atomic Boolean algebras and that $\langle \mathbf{B}_0, A_0\rangle$ and $\langle \mathbf{B}_1, A_1\rangle$ are the corresponding expansions. Suppose also that $\langle \mathbf{B}_0, A_0\rangle$ is a substructure of $\langle \mathbf{B}_1, A_1\rangle$. Now let $\varphi(\mathbf{x})$ be a universal formula and $\mathbf{a}$ is an ω-tuple of elements of B_0 so that $\langle \mathbf{B}_1, A_1\rangle \not\models \varphi(\mathbf{a})$. We want to show that $\langle \mathbf{B}_0, A_0\rangle \not\models \varphi(\mathbf{a})$. Let $\varphi(\mathbf{x})$ be $\forall y_0, \dots, y_{m-1}\psi(\mathbf{x}, \mathbf{y})$, where $\psi(\mathbf{x}, \mathbf{y})$ is quantifier free. So pick an m-tuple $\langle c_0, \dots, c_{m-1}\rangle$ of elements of B_1 so that $\langle \mathbf{B}_1, A_1\rangle \not\models \psi(\mathbf{a}, \mathbf{c})$. Let $\mathbf{B}^*_1$ be that substructure of $\langle \mathbf{B}_1, A_1\rangle$ generated by $\{a_0, \dots, a_{n-1}\} \cup \{c_0, \dots, c_{m-1}\}$. Observe that since ψ is quantifier free, we have $\mathbf{B}^*_1 \models \neg\psi(\mathbf{a}, \mathbf{c})$. Since the variety of all Boolean algebras is generated by the 2-element Boolean algebra, it follows that the variety of Boolean algebras is locally finite by Theorem 4.49 in on page 222 in Volume I. Hence, $\mathbf{B}^*_1$ is finite. So it is an atomic Boolean algebra. Now consider any one-to-one map f that maps the atoms of $\mathbf{B}^*_1$ to atoms in $\langle \mathbf{B}_0, A_0\rangle$, which is an expansion of an infinite atomic Boolean algebra, and which is the identity function on the atoms of

$\mathbf{B}_1^*$ that belong to B_0. It is easy to verify that f can be extended to an embedding of $\mathbf{B}_1^*$ into $\langle \mathbf{B}_0, A_0 \rangle$ that is the identity function on those elements of $\mathbf{B}_1^*$ that belong to B_0. This means that $\langle \mathbf{B}_0, A_0 \rangle$ has a substructure in which $\neg\psi(\mathbf{a}, f(\mathbf{c}))$ holds. Consequently, $\neg\psi(\mathbf{a}, f(\mathbf{c}))$ holds in $\langle \mathbf{B}_0, A_0 \rangle$. We conclude that $\langle \mathbf{B}_0, A_0 \rangle \not\models \varphi(\mathbf{a})$, verifying condition (iii) of Theorem 8.37. So we find that T is model complete.

By condition (iv) of Theorem 8.37 every formula is equivalent, modulo T, to a universal formula. So to demonstrate the completeness of T, it suffices to show that every universal sentence true in one model of T is true in every model of T. So let φ be $\forall x_0, \ldots, x_{n-1}\psi(x_0, \ldots, x_{n-1})$, where ψ is quantifier free. Let $\langle \mathbf{B}_0, A_0 \rangle$ and $\langle \mathbf{B}_1, A_1 \rangle$ be models of T and suppose $\langle \mathbf{B}_0, A_0 \rangle \models \neg\varphi$. Pick $c_0, \ldots, c_{n-1} \in B_0$ so that $\langle \mathbf{B}_0, A_0 \rangle \models \neg\psi(c_0, \ldots, c_{n-1})$. Let $\mathbf{B}_0^*$ be the substructure of $\langle \mathbf{B}_0, A_0 \rangle$ generated $\{c_0, \ldots, c_{n-1}\}$. As above, $\mathbf{B}_0^*$ is embeddable into $\langle \mathbf{B}_1, A_1 \rangle$. And once more we can conclude that $\langle \mathbf{B}_1, A_1 \rangle \models \neg\varphi$. Hence T is complete. ■

THEOREM 8.39. *The elementary theory of infinite atomic Boolean algebras is complete.*

Proof. We have just seen in Theorem 8.38 that the enhanced elementary theory of infinite atomic Boolean algebras is complete. By restricting our attention to sentences in the language of Boolean algebras, it follows at once the elementary theory of infinite atomic Boolean algebras is also complete. ■

Applications of Compactness with Elementary Diagrams

Using the Compactness Theorem and elementary diagrams, it is easy to produce elementary extensions of infinite structures.

THEOREM 8.40 (The Upward Löwenheim-Skolem-Tarski Theorem).
Let $\mathbf{A}$ *be an* $\mathfrak{L}$*-structure which is infinite. Let* κ *be any cardinal at least as large as* $|A|$ *and at least as large as the number of* $\mathfrak{L}$*-formulas.* $\mathbf{A}$ *has an elementary extension of cardinality* κ*.*

Proof. Let $\mathfrak{L}'$ be the language obtained by adjoining κ new distinct constant symbols, d_α for each $\alpha \in \kappa$, to the diagram language of $\mathbf{A}$. Let

$$\Sigma = D_{\mathbf{A}} \cup \{\neg(d_\alpha \approx d_\beta) : \alpha < \beta < \kappa\}.$$

Any model of Σ will be a model of $D_{\mathbf{A}}$, and so its reduct to $\mathfrak{L}$ will be (isomorphic to) an elementary extension of $\mathbf{A}$. On the other hand such a model must give each d_α a distinct interpretation, and so the model must have cardinality at least κ. Finally, an application of the Downward Löwenheim-Skolem-Tarski Theorem produces an elementary extension of $\mathbf{A}$ of the desired cardinality, in view of Theorem 8.11. So it is enough to prove that Σ has a model. But any given finite subset Σ' of Σ involves only finitely many of the d_α's. Hence $\mathbf{A}$ can be easily expanded to a model of Σ'. So by the Compactness Theorem, Σ has a model. ■

Some of the limitations of elementary languages should now be apparent. Such familiar structures as $\langle \omega, +, \cdot, \leq \rangle$ and $\langle \mathbb{R}, +, \cdot, \leq \rangle$ cannot be described to within isomorphism by any sets of elementary sentences. The well-known axiom systems of Peano and Dedekind for these structures make references to arbitrary sets of numbers (in the induction and completeness axioms). We see now, from simple cardinality considerations alone, that these references cannot be eliminated in favor of elementary sentences, however cleverly we may attempt to frame them.

Another limitation on the expressive power of elementary sentences is provided by the next theorem. We give two proofs in order to demonstrate how proofs using the Compactness Theorem and proofs using ultraproducts are related.

THEOREM 8.41. *Let Σ be a set of elementary sentences. If Σ has arbitrarily large finite models, then Σ has an infinite model.*

Proof.

Using Ultraproducts:

Let I be an infinite set of natural numbers such that for every $n \in I$ there is a model $\mathbf{A}_n$ of Σ of cardinality n. Let U be any ultrafilter on I which extends the Fréchet filter on I. Let $\mathbf{B} = \prod_I \mathbf{A}_n/U$. By the Fundamental Theorem of Ultraproducts, $\mathbf{B} \models \Sigma$. To see that $\mathbf{B}$ is infinite, let δ_k be a sentence expressing "there are at least k elements". Plainly, $\{n : \mathbf{A}_n \models \delta_k\}$ is a cofinite subset of I for each natural number k. So $\{n : \mathbf{A}_n \models \delta_k\} \in U$. But this means $\mathbf{B} \models \delta_k$ for all natural numbers k. Hence $\mathbf{B}$ must be infinite.

Using the Compactness Theorem:

Since Σ has arbitrarily large finite models, we see that every finite subset of $\Sigma \cup \{\delta_k : k \in \omega\}$ has a model. So by the Compactness Theorem $\Sigma \cup \{\delta_k : k \in \omega\}$ has a model. Such a model must be infinite. ■

Actually, the ultraproduct proof produces a slightly sharper result: If $\mathcal{K}$ is a class of similar structures such that $\mathcal{K}$ contains arbitrarily large finite structures and $\mathcal{K}$ is closed under ultraproducts, then $\mathcal{K}$ must contain an infinite structure.

THEOREM 8.42. *Every structure $\mathbf{A}$ is isomorphic to a substructure of an ultraproduct of finitely generated substructures of $\mathbf{A}$.*

Proof. Let Σ be the set of all sentences which are conjunctions of finitely many sentences each belonging to $\Delta_{\mathbf{A}}$. Thus Σ and $\Delta_{\mathbf{A}}$ have the same models. Our idea is to follow the proof of the Compactness Theorem to produce a model of Σ which is an ultraproduct. Note that each $\varphi \in \Sigma$ mentions only finitely many of the new constants used to name elements of $\mathbf{A}$. Let $\mathbf{B}_\varphi$ be the substructure of $\mathbf{A}$ generated by these elements and let $\mathbf{B}'_\varphi$ be some expansion of $\mathbf{B}_\varphi$ to the diagram language of $\mathbf{A}$ that interprets the new constants correctly in so far as that is possible. As in the proof of the Compactness Theorem, $\{ [\![\theta]\!] : \theta \in \Sigma\}$ has the finite intersection property. Using the Ultrafilter Theorem, obtain an ultrafilter U on Σ extending this set. Then $\prod_\Sigma \mathbf{B}'_\varphi/U$ is a model of Σ, and hence of $\Delta_{\mathbf{A}}$. But then $\mathbf{A}$ is isomorphic to a substructure of $\prod_\Sigma \mathbf{B}/U$, since this last ultraproduct is just the reduct to $\mathfrak{L}$ of $\prod_\Sigma \mathbf{B}'_\varphi/U$. ■

A sharper variant of this theorem is useful.

THEOREM 8.43 (Alden F. Pixley 1970). *Let $\mathcal{V}$ be a locally finite variety. Let $\mathcal{K}$ be a finite collection of finite algebras from $\mathcal{V}$. Let $\mathbf{B}$ be an algebra of the same signature as*

$\mathcal{V}$. If every finitely generated subalgebra of **B** *can be subdirectly represented via algebras in $\mathcal{K}$, then* **B** *can be subdirectly represented by algebras in $\mathcal{K}$.*

Proof. What is needed is for each $b, b' \in B$ with $b \neq b'$, to find an algebra $\mathbf{C} \in \mathcal{K}$ and a homomorphism $h : \mathbf{B} \to \mathbf{C}$ such that $h(b) \neq h(b')$.

Let $\mathbf{A}_0, \ldots, \mathbf{A}_{n-1}$ be the algebras that belong to $\mathcal{K}$. We assume that the universes $B, A_0, \ldots, A_{n-1}$ are pairwise disjoint. We expand the signature by adjoining new constants to name each of the elements of $B \cup A_0 \cup \cdots \cup A_{n-1}$. In addition we take $B, A_0, \ldots, A_{n-1}$ as one-place relations and we take h to be a new one-place operation symbol. Now let Σ be the union of the elementary diagrams of **B** and the algebras $\mathbf{A}_i$. Let Γ be the collection of sentences that assert that each A_i has at least $|A_i|$ elements. Let Ψ be the set of sentences that assert that h is a homomorphism separating b and b'. Finally let θ be the following sentence

$$\forall x[x \in B \implies h(x) \in A_0] \mathbin{\mathbb{V}} \cdots \mathbin{\mathbb{V}} \forall x[x \in B \implies h(x) \in A_{n-1}].$$

What we need is a model of $\Sigma \cup \Gamma \cup \Psi \cup \{\theta\}$. We will invoke the Compactness Theorem. So let Φ be any finite subset of this set of sentences. We assume b and b' occur in Φ. Only finitely many constant symbols naming elements of B can occur. Let **D** be the subalgebra of **B** generated by these finitely many named elements. By hypothesis, **D** has a subdirect representation over $\mathcal{K}$. This entails that $\mathbf{D} \in \mathcal{V}$. So **D** is finite since $\mathcal{V}$ is locally finite. Now, without loss of generality, we assume that Φ includes the diagram of **D**. Then the subdirect representation of **D** over $\mathcal{K}$ provides us with a particular $\mathbf{A}_i$ and a particular $h : \mathbf{D} \to \mathbf{A}_i$ such that $h(b) \neq h(b')$. Using this h we see that Φ has a model. So by the Compactness Theorem the set $\Sigma \cup \Gamma \cup \Psi \cup \{\theta\}$ has a model. The map named by h in this model will restrict to a homomorphism from **B** into $\mathbf{A}_i$ that separates b and b'. Since we can do this for every pair of distinct elements of B, we conclude that **B** has a subdirect representation over $\mathcal{K}$. ■

A different way to sharpen Theorem 8.42 was shown to us by J. B. Nation. His theorem modifies the result that every algebra can be embedded into an ultraproduct of subalgebras, to obtain a similar result for subdirectly irreducible algebras. We also note Nation's Theorem has the theorem of Robert W. Quackenbush (1971) (Corollary 7.78) as a direct consequence; see Exercise 1.

We need a new notion to formulate this theorem. Let σ be a signature. Expand σ by one 2-place relation symbol R. Suppose that $\langle \mathbf{A}_i \mid i \in I \rangle$ is a system of algebras of signature σ and that U is an ultrafilter on I. It is straightforward to construct a set Σ of sentences in the expanded language that asserts that R denotes a congruence of **A** if and only if $\langle \mathbf{A}, C \rangle \models \Sigma$, for any algebra of signature σ. So if $\theta \in \operatorname{Con} \mathbf{A}_i$ for all $i \in I$, then R will denote a congruence relation of $\prod_i \mathbf{A}_i/U$, by the Fundamental Theorem of Ultraproducts. This congruence is called the **ultraproduct congruence of the system** $\langle \theta_i \mid i \in I \rangle$ of congruences on the factors.

THEOREM 8.44 (J. B. Nation, unpublished). *If* **B** *is subdirectly irreducible then* **B** *is a subalgebra of an ultraproduct of a family* $\mathbf{B}_i$*, $i \in I$, of finitely generated, subdirectly irreducible algebras lying in* ***HS***(**B**)*. Moreover, if β_i is the monolith of* $\mathbf{B}_i$*, and β is the ultraproduct congruence determined by the β_i's, then β restricted to* **B** *is nonzero.*

Proof. Let $\mathrm{Cg}^{\mathbf{B}}(a,b)$ be the monolith of $\mathbf{B}$. Let $\mathcal{S}$ be the set of all finite subsets of B that contain a and b. For $S \in \mathcal{S}$ let $\psi_S \in \mathrm{Con}(\mathbf{Sg}^{\mathbf{B}}(S))$ be a maximal congruence not containing $\langle a,b\rangle$ and let $\mathbf{B}_S = \mathbf{Sg}^{\mathbf{B}}(S)/\psi_S$. Let $\mathcal{F}$ be the filter on $\mathcal{S}$ consisting of all $\mathcal{T} \subseteq \mathcal{S}$ such that there is an $S_0 \in \mathcal{S}$ with $\mathcal{T}_{S_0} := \{S \in \mathcal{S} : S \supseteq S_0\} \subseteq \mathcal{T}$. Let $\mathcal{U}$ be an ultrafilter extending $\mathcal{F}$ and let $\mathbf{D} = \prod_{S\in\mathcal{S}} \mathbf{Sg}^{\mathbf{B}}(S)/\psi_S = \prod_{S\in\mathcal{S}} \mathbf{B}_S$. Let $\varphi : \mathbf{B} \to \mathbf{D}$ be the map sending $x \in B$ to the element whose S coordinate is x/ψ_S if $x \in S$ and is an arbitrary element of $\mathbf{B}_S$ otherwise. Let $\pi : \mathbf{D} \twoheadrightarrow \mathbf{D}/\mathcal{U}$ be the natural projection of $\mathbf{D}$ onto the ultraproduct and also let $\pi_S : \mathbf{D} \twoheadrightarrow \mathbf{B}_S$ be the natural projection to $\mathbf{B}_S$. The map $\pi \circ \varphi : \mathbf{B} \to \mathbf{D}/\mathcal{U}$ is a homomorphism. Indeed, suppose f is an n-ary operation symbol in the signature of $\mathbf{B}$ and $x_1,\dots,x_n \in B$ and let $S_0 = \{a,b,x_1,\dots,x_n,f^{\mathbf{B}}(x_1,\dots,x_n)\}$. The set of coordinates $S \in \mathcal{S}$ where $f^{\mathbf{B}_S}(\pi_S(x_1),\dots,\pi_S(x_n)) = \pi_S(f^{\mathbf{B}}(x_1,\dots,x_n))$ contains $\mathcal{T}_{S_0}$ which is in $\mathcal{F}$ and so is in $\mathcal{U}$. This implies that $\pi\circ\varphi$ is a homomorphism. It is an embedding because $\pi(\varphi(a)) \neq \pi(\varphi(b))$. The last part the theorem is a consequence of the fact that $\langle \pi(\varphi(a)), \pi(\varphi(b))\rangle$ is in the ultraproduct congruence β determined by the monoliths of the factors. ■

This theorem will play a role in the proof of Theorem 11.105 on page 346 in Volume III. That theorem generalizes Jónsson's Lemma (Theorem 8.57 in the next section) to congruence modular varieties subject to certain restrictions.

The next theorem, already mentioned in §7.9, requires a result from combinatorial set theory that extends Ramsey's Theorem.

A THEOREM OF PAUL ERDŐS FROM 1942

Let κ be any infinite cardinal. Let A be any set with $|A| > 2^\kappa$. If the two-element subsets of A are colored with κ colors, then there is an infinite subset $D \subseteq A$ so that it is monochromatic, that is the two-element subsets of D have all been given the same color.

THEOREM 8.45 (Walter Taylor 1972). *A variety $\mathcal{V}$ is residually small if and only if $\mathcal{V}$ has residual bound of $\leq (2^\kappa)^+$, where κ is the cardinality of the set of terms of the signature.*

Proof. Only the forward implication needs a proof. We address the contrapositive. So let $\mathbf{A} \in \mathcal{V}$ be subdirectly irreducible and $|A| > 2^\kappa$. Let $\langle c,d\rangle$ be a critical pair for $\mathbf{A}$. Now for any $a,b \in A$ with $a \neq b$, we see that $\langle c,d\rangle \in \mathrm{Cg}^{\mathbf{A}}(a,b)$. So there will be a principal congruence formula $\pi_{a,b}(x,y,z,w)$ so that $\mathbf{A} \models \pi_{a,b}(a,b,c,d)$. There are at most κ principal congruence formulas. So we use $\pi_{a,b}$ to color the two-element set $\{a,b\}$. According to the Theorem of Erdős quoted above, there is an infinite subset $D \subseteq A$ all of whose two-element subsets have been assigned the same principal congruence formula π.

We have to show that $\mathcal{V}$ has arbitrarily large subdirectly irreducible algebras. To this end, let λ be any infinite cardinal. Expand the signature by adjoining λ new constant symbols c_α, one for each $\alpha \in \lambda$. Also, take c and d to be two more constant symbols. Let Σ be an equational base for $\mathcal{V}$. Let Γ be the following set of sentences:

$$\{\neg c_\alpha \approx c_\beta \wedge \pi(c_\alpha, c_\beta, c, d) \mid \alpha,\beta \in \lambda \text{ and } \alpha \neq \beta\} \cup \Sigma \cup \{\neg c \approx d\}.$$

Any model of Γ will have cardinality at least λ. The reduct of such a model to the original signature will belong to $\mathcal{V}$. First, we show that Γ has a model. By the Compactness

Theorem, it is enough to show that every finite subsets of Γ has a model. So let Γ' be a finite subset of Γ. Only finitely many of the constants c_α can occur in Γ. So we can expand **A** by naming some elements of D with these finitely many constant symbols. This will result the desired model of Γ'. So let **B** be a model of Γ. Let θ be a congruence of **B** that is maximal with respect to separating the elements named by c and d. Then $\mathbf{B}/\theta$ will be subdirectly irreducible and θ will also separate the elements named by the other constant symbols, since $\mathbf{B} \models \pi(c_\alpha, c_\beta, c, d)$ for all $\alpha \neq \beta$. So the cardinality of $\mathbf{B}/\theta$ is at least λ. The same conclusions apply to the reduct of $\mathbf{B}/\theta$ to the original signature. So we have found in $\mathcal{V}$ a large subdirectly irreducible algebra, as desired. ∎

The attempt to replace the diagram of **A** by the elementary diagram of **A** in the proof given for Theorem 8.42, leads to the following result.

THEOREM 8.46 (Frayne's Theorem). $\mathbf{A} \equiv \mathbf{B}$ *if and only if* **A** *is isomorphic to an elementary substructure of some ultrapower of* **B**.

Proof. If $\mathbf{A} \prec \mathbf{B}^I/U$, where U is an ultrafilter on I, then $\mathbf{A} \equiv \mathbf{B}$, since the canonical embedding of **B** into $\mathbf{B}^I/U$ is elementary.

For the converse, let $\Sigma = D_{\mathbf{A}}$. Note that Σ is closed under finite conjunctions. Each $\varphi \in \Sigma$ is of the form $\psi(\mathbf{c_a})$, where $\psi(\mathbf{x})$ is a formula in the language of **A** and $\mathbf{A} \models \psi(\mathbf{a})$. Thus, for $\psi(\mathbf{c_a}) \in \Sigma$, we have $\mathbf{A} \models \exists \mathbf{x} \psi(\mathbf{x})$. Since $\mathbf{A} \equiv \mathbf{B}$, we get $\mathbf{B} \models \exists \mathbf{x} \psi(\mathbf{x})$. So there is a tuple $\mathbf{b}$ from B such that $\mathbf{B} \models \psi(\mathbf{b})$. But this is just another way of saying that for each $\varphi \in \Sigma$, we can expand **B** to $\mathbf{B}_\varphi$ so that $\mathbf{B}_\varphi \models \varphi$. Now let

$$[\![\theta]\!] = \{\varphi \mid \varphi \in \Sigma \text{ and } \mathbf{B}_\varphi \models \theta\}.$$

A routine proof shows that $\{[\![\theta]\!] \mid \theta \in \Sigma\}$ has the finite intersection property. Let U be an ultrafilter extending this set. By the Fundamental Theorem of Ultraproducts $\prod_\Sigma \mathbf{B}_\varphi/U \models D_{\mathbf{A}}$. Since $\mathbf{B}^\Sigma/U$ is the reduct of this ultraproduct to the language of **A**, we conclude that **A** is isomorphic to an elementary substructure of $\mathbf{B}^\Sigma/U$. ∎

While Frayne's Theorem is a characterization of elementary equivalence, it is one that still refers to the sentences and formulas of elementary languages. The next theorem provides a characterization that relies only on the notions of isomorphism and ultrapower.

THEOREM 8.47 (The Keisler-Shelah Isomorphism Theorem (H. Jerome Keisler, 1961; Saharon Shelah, 1971)). $\mathbf{A} \equiv \mathbf{B}$ *if and only if* $\mathbf{A}^I/U \cong \mathbf{B}^I/U$ *for some nonempty set* I *and some ultrafilter* U *on* I.

The proof of this theorem is rather intricate, and it makes use of a number of features of the arithmetic of infinite cardinals. We refer the reader to (C. C. Chang and H. J. Keisler, 1990) for a proof.

Let $\mathcal{K}$ be any class of $\mathfrak{L}$-structures. We introduce the following notation to supplement $\boldsymbol{H}$, $\boldsymbol{S}$, and $\boldsymbol{P}$:

$\mathbf{A} \in \boldsymbol{P}_u\mathcal{K}$ provided **A** is isomorphic to an ultraproduct of members of $\mathcal{K}$.

$\mathbf{A} \in \boldsymbol{P}_r\mathcal{K}$ provided **A** is isomorphic to a reduced product of members of $\mathcal{K}$.

$\mathbf{A} \in \sqrt[u]{\mathcal{K}}$ provided **A** is isomorphic to an ultraroot of a member of $\mathcal{K}$.

$\mathbf{A} \in \boldsymbol{S}^{\prec}\mathcal{K}$ provided $\mathbf{A}$ is elementarily embeddable into some member of $\mathcal{K}$.

The next theorem concerns how the operators $\boldsymbol{P_u}$ and $\boldsymbol{P_r}$ are related to the operators $\boldsymbol{H}$, $\boldsymbol{S}$, and $\boldsymbol{P}$. The theorem is stated in a compact form which hides some information. Roughly speaking, all the parts of this theorem are proven by using the natural constructions, and the fact that it is the natural constructions which work is, itself, occasionally useful.

THEOREM 8.48. *Let $\mathcal{K}$ be a class of $\mathfrak{L}$-structures.*

(i) $\mathcal{K} \subseteq \boldsymbol{P_u}\mathcal{K} \subseteq \boldsymbol{P_r}\mathcal{K} \subseteq \boldsymbol{HP}\mathcal{K}$ *and* $\boldsymbol{P}\mathcal{K} \subseteq \boldsymbol{P_r}\mathcal{K}$.

(ii) $\boldsymbol{P_uS}\mathcal{K} \subseteq \boldsymbol{SP_u}\mathcal{K}$ *and* $\boldsymbol{P_rS}\mathcal{K} \subseteq \boldsymbol{SP_r}\mathcal{K}$.

(iii) $\boldsymbol{P_uH}\mathcal{K} \subseteq \boldsymbol{HP_u}\mathcal{K}$ *and* $\boldsymbol{P_rH}\mathcal{K} \subseteq \boldsymbol{HP_r}\mathcal{K}$.

(iv) $\boldsymbol{P_uP_u}\mathcal{K} = \boldsymbol{P_u}\mathcal{K}$ *and* $\boldsymbol{P_rP_r}\mathcal{K} = \boldsymbol{P_r}\mathcal{K}$.

(v) $\boldsymbol{P_r}\mathcal{K} \subseteq \boldsymbol{SPP_u}\mathcal{K}$ *and* $\boldsymbol{P_uP}\mathcal{K} \subseteq \boldsymbol{SPP_u}\mathcal{K}$.

Proof.
(i) All of these inclusions are easy. For instance, $\mathbf{A} \cong \mathbf{A}^I/U$ if U is a principal ultrafilter on I. Hence $\mathcal{K} \subseteq \boldsymbol{P_u}\mathcal{K}$. The direct product $\prod_I \mathbf{A}_i$ is isomorphic with $\prod_I \mathbf{A}_i/F$, if $F = \{I\}$. Hence, $\boldsymbol{P}\mathcal{K} \subseteq \boldsymbol{P_r}\mathcal{K}$.

(ii) Suppose F is any filter on I and $\mathbf{B}_i \subseteq \mathbf{A}_i$ for all $i \in I$. It is straightforward to check that $\prod_I \mathbf{B}_i/F \subseteq \prod_I \mathbf{A}_i/F$.

(iii) Suppose F is any filter on I and $h_i : \mathbf{A}_i \twoheadrightarrow \mathbf{B}_i$ for all $i \in I$. Define $h : \prod_I A_i/F \twoheadrightarrow \prod B_i/F$ so that

$$h(\langle a_i \mid i \in I\rangle/F) = \langle h_i(a_i) \mid i \in I\rangle/F.$$

It is routine to verify that this definition is sound and that h is a homomorphism from $\prod_I \mathbf{A}_i/F$ onto $\prod_I \mathbf{B}_i/F$.

(iv) This one takes more work. Suppose that F is a filter on I and, for each $i \in I$, that G_i is a filter on J_i. For each $i \in I$ and each $j \in J_i$, suppose $\mathbf{A}_{ij} \in \mathcal{K}$. Let $\mathbf{B} = \prod_I \left(\prod_{J_i} \mathbf{A}_{ij}/G_i\right)/F$. So $\mathbf{B}$ is a typical member of $\boldsymbol{P_rP_r}\mathcal{K}$. It does not harm the generality of our conclusion to assume that the J_i's are pairwise disjoint. This allows us to simplify the notation, replacing $\mathbf{A}_{ij}$ by $\mathbf{A}_j$, since j now determines i. Let $J = \bigcup_I J_i$, and define D as follows: for any $X \subseteq J$

$$X \in D \text{ if and only if } \{i \mid X \cap J_i \in G_i\} \in F.$$

Thus, D consists of those sets X which "meet a large set of coordinate-sets in a large set." We will prove that D is a filter on J (even an ultrafilter if F and each G_i are ultrafilters) and then that

$$\prod_J \mathbf{A}_j/D \cong \prod_I \left(\prod_{J_i} \mathbf{A}_{ij}/G_i\right)/F.$$

CLAIM 1.
D is a filter on J and, moreover, if F is an ultrafilter and each G_i is an ultrafilter, then D is an ultrafilter.

Proof of Claim 1. $J \in D$ since $J \cap J_i = J_i \in G_i$ for all $i \in I$. If $X \in D$ and $X \subseteq Y \subseteq J$, then $X \cap J_i \in G_i$ implies $Y \cap J_i \in G_i$, for all $i \in I$. So we obtain $Y \in D$. Now suppose $X, Y \in D$. Then $\{i \mid X \cap J_i \in G_i\} \in F$ and $\{i \mid Y \cap J_i \in G_i\} \in F$. So $\{i \mid X \cap J_i \in G_i \text{ and } Y \cap J_i \in G_i\} \in F$. Consequently, $\{i \mid (X \cap Y) \cap J_i \in G_i\} \in F$. Therefore, $X \cap Y \in D$. We conclude that D is a filter. To finish the Claim, suppose F and all the G_i's are ultrafilters. Then

$$\begin{aligned}(X \cup Y) \in D &\text{ implies } \{i \mid (X \cup Y) \cap J_i \in G_i\} \in F \\ &\text{ implies } \{i \mid (X \cap J_i) \cup (Y \cap J_i) \in G_i\} \in F\end{aligned}$$

But each G_i is an ultrafilter, so either $X \cap J_i \in G_i$ or $Y \cap J_i \in G_i$, for each i in the set above. This means that

$$\{i \mid (X \cap J_i) \cup (Y \cap J_i) \in G_i\} \subseteq \{i \mid X \cap J_i \in G_i\} \cup \{i \mid Y \cap J_i \in G_i\}.$$

Consequently,

$$\begin{aligned}(X \cup Y) \in D &\text{ implies } \{i \mid X \cap J_i \in G_i\} \cup \{i \mid Y \cap J_i \in G_i\} \in F \\ &\text{ implies } \{i \mid X \cap J_i \in G_i\} \in F \text{ or } \{i \mid Y \cap J_i \in G_i\} \in F \\ &\text{ implies } X \in D \text{ or } Y \in D.\end{aligned}$$

Thus, D is an ultrafilter. ■

CLAIM 2.
$\prod_J \mathbf{A}_j/D \cong \prod_I \left(\prod_{J_i} \mathbf{A}_j/G_i\right)/F.$

Proof of Claim 2. The isomorphism is the natural one: for $a \in \prod_J A_j$ define

$$h(a/D) = \langle a^i/G_i \mid i \in I\rangle/F$$

where $a^i = \langle a_j \mid j \in J_i\rangle$ for all $i \in I$. Checking the details involves no new ideas. ■

(v) The second inclusion follows from the first with the help of (i) and (iv). So we prove only $\boldsymbol{P_r}\mathcal{K} \subseteq \boldsymbol{SPP_u}\mathcal{K}$.

Let $\prod_I \mathbf{A}_i/F$ be a reduced product of structures in $\mathcal{K}$. Let

$$J = \{U \mid U \text{ is an ultrafilter of } I \text{ extending } F\}.$$

Let $\mathbf{B}_U = \prod_I \mathbf{A}_i/U$ for each $U \in J$. Both $\sim_F$ and $\sim_U$ are congruences on $\prod_I \mathbf{A}_i$, and $\sim_F$ is included in $\sim_U$ for all $U \in J$. It follows that the map h_U which sends $a/F \mapsto a/U$ for all $a \in \prod_I A_i$, is a homomorphism from $\prod_I \mathbf{A}_i/F$ onto $\mathbf{B}_U$. Now let $\mathbf{B} = \prod_J \mathbf{B}_U$. We can embed $\prod_I \mathbf{A}_i/F$ into $\mathbf{B}$ using the map h which is defined so that

$$h(a/F) = \langle a/U \mid U \in J\rangle$$

for all $a \in \prod_I A_i$. That h is a homomorphism follows directly from the fact that h_U is a homomorphism for each $U \in J$. To see that h is an embedding we must verify the following claim. (Recall that one-to-one homomorphisms need not be embeddings.)

CLAIM 3.
For each atomic formula $\varphi(\mathbf{x})$ and each sequence $\mathbf{a}$ from $\prod_I A_i$, if $\prod_I \mathbf{A}_i/F \models \neg\varphi(\mathbf{a}/F)$, then $\prod_I \mathbf{A}_i/U \models \neg\varphi(\mathbf{a}/U)$, for some $U \in J$.

Proof of Claim 3. Since $\prod_I \mathbf{A}_i/F \models \neg\varphi(\mathbf{a}/F)$, we know that $[\![\varphi(\mathbf{a})]\!] \notin F$. According to the Ultrafilter Theorem, there is an ultrafilter U extending F such that $[\![\varphi(\mathbf{a})]\!] \notin U$. But for this U we have $[\![\neg\varphi(\mathbf{a})]\!] \in U$. Thus, $\prod_I \mathbf{A}_i/U \models \neg\varphi(\mathbf{a}/U)$, by the Fundamental Theorem of Ultraproducts. ■

This completes the proof of the theorem. ■

Further relationships similar to those put forward in the theorem above, but focusing on the ultraroot operator and other constructions can be found among the exercises. An intriguing relationship which is, in essence, the Keisler-Shelah Isomorphism Theorem, is $\boldsymbol{S}^{\prec}\mathcal{K} \subseteq \sqrt[u]{\boldsymbol{P}_u\mathcal{K}}$, for any class $\mathcal{K}$ of $\mathfrak{L}$-structures.

We are now in a position to write down a theorem that does for elementary classes what Birkhoff's HSP Theorem does for varieties.

THEOREM 8.49 (The Elementary Closure Theorem). *Let $\mathcal{K}$ be any class of $\mathfrak{L}$-structures, where $\mathfrak{L}$ is an elementary language.*

(i) $\mathbf{Mod\,Th}\,\mathcal{K} = \{\mathbf{A} : \mathbf{A} \equiv \mathbf{B}$ *for some* $\mathbf{B} \in \boldsymbol{P}_u\mathcal{K}\}$.
(ii) $\mathbf{Mod\,Th}\,\mathcal{K} = \boldsymbol{S}^{\prec}\boldsymbol{P}_u\mathcal{K}$.
(iii) $\mathbf{Mod\,Th}\,\mathcal{K} = \sqrt[u]{\boldsymbol{P}_u\mathcal{K}}$.

Proof.

(i) The inclusion $\mathbf{Mod\,Th}\,\mathcal{K} \supseteq \{\mathbf{A} : \mathbf{A} \equiv \mathbf{B}$ for some $\mathbf{B} \in \boldsymbol{P}_u\mathcal{K}\}$ follows from the Fundamental Theorem for Ultraproducts. For the reverse inclusion suppose $\mathbf{A} \in \mathbf{Mod\,Th}\,\mathcal{K}$. Let θ be a sentence true in $\mathbf{A}$. Hence $\neg\theta \notin \mathbf{Th}\,\mathcal{K}$, since $\mathbf{A} \models \mathbf{Th}\,\mathcal{K}$. So pick $\mathbf{B}_\theta \in \mathcal{K}$ such that $\mathbf{B}_\theta \models \theta$. Now let $I = \mathbf{Th}\,\mathbf{A}$. For each $\theta \in I$ we have $[\![\theta]\!] = \{\varphi \mid \varphi \in I$ and $\mathbf{B}_\varphi \models \theta\}$. As in several earlier proofs the collection

$$\{[\![\theta]\!] \mid \theta \in I\}$$

has the finite intersection property. Let U be an ultrafilter on I extending this collection. That $\mathbf{A} \equiv \prod_I \mathbf{B}_\theta/U$ follows by that Fundamental Theorem for Ultraproducts, since $[\![\theta]\!] \in U$ for all $\theta \in \mathbf{Th}\,\mathbf{A} = I$.

(ii) This follows from (i) by Frayne's Theorem and $\boldsymbol{P}_u\boldsymbol{P}_u\mathcal{K} = \boldsymbol{P}_u\mathcal{K}$.

(iii) this follows from (i) by the Keisler-Shelah Isomorphism Theorem and the equality $\boldsymbol{P}_u\boldsymbol{P}_u\mathcal{K} = \boldsymbol{P}_u\mathcal{K}$. ■

COROLLARY 8.50. *Let $\mathcal{K}$ be any class of $\mathfrak{L}$-structures, where $\mathfrak{L}$ is any elementary language. The following are equivalent.*

(i) *$\mathcal{K}$ is an elementary class.*
(ii) $\mathcal{K} = \boldsymbol{S}^{\prec}\boldsymbol{P}_u\mathcal{K}$.
(iii) $\mathcal{K} = \boldsymbol{S}^{\prec}\boldsymbol{P}_u\mathcal{K}'$, *for some class $\mathcal{K}'$ of $\mathfrak{L}$-structures.*
(iv) $\mathcal{K} = \boldsymbol{P}_u\mathcal{K} = \boldsymbol{S}^{\prec}\mathcal{K}$.
(v) $\mathcal{K} = \boldsymbol{P}_u\mathcal{K}$ *and $\mathcal{K}$ is closed with respect to elementary equivalence.*
(vi) $\mathcal{K} = \sqrt[u]{\boldsymbol{P}_u\mathcal{K}}$.
(vii) $\mathcal{K} = \sqrt[u]{\boldsymbol{P}_u\mathcal{K}'}$ *for some class $\mathcal{K}'$ of $\mathfrak{L}$-structures.*
(viii) $\mathcal{K} = \boldsymbol{P}_u\mathcal{K} = \sqrt[u]{\mathcal{K}}$. ■

COROLLARY 8.51. *Let $\mathcal{K}$ be a finite set of finite $\mathfrak{L}$-structures, and let $\boldsymbol{I}\,\mathcal{K}$ be the class of isomorphic images of members of $\mathcal{K}$. Then $\boldsymbol{I}\,\mathcal{K} = \mathbf{P}_u\mathcal{K}$ and $\boldsymbol{I}\,\mathcal{K}$ is an elementary class.*

Proof. We use the equivalence of (i) and (v) in Corollary 8.50. According to Theorem 8.6, $\boldsymbol{I}\mathcal{K}$ is closed under elementary equivalence. To see that $\boldsymbol{I}\mathcal{K}$ is also closed with respect to the formation of ultraproducts, let I be any nonempty set, let U be any ultrafilter on I, and let $\mathbf{A}_i \in \mathcal{K}$ for each $i \in I$. Now partition I by placing i and j in the same block provided $\mathbf{A}_i \cong \mathbf{A}_j$. Since $\mathcal{K}$ is finite, we have partitioned I into finitely many blocks. Since U is an ultrafilter on I, exactly one of these blocks belongs to U. Let us say it is the block to which j belongs. By the Fundamental Theorem of Ultraproducts, $\prod_I \mathbf{A}_i/U \equiv \mathbf{A}_j$, and so the ultraproduct belongs to $\boldsymbol{I}\mathcal{K}$, since we already know that $\boldsymbol{I}\mathcal{K}$ is closed with respect to elementary equivalence. ∎

Recall that an elementary class $\mathcal{K}$ is said to be **finitely axiomatizable** provided there is a finite set Σ of sentences such that $\mathcal{K} = \mathbf{Mod}\,\Sigma$. Likewise, an elementary theory is said to be *finitely axiomatizable* provided it is logically equivalent to some finite set of sentences. For any class $\mathcal{K}$ of $\mathfrak{L}$-structures, we use $\mathcal{K}^c$ to denote the complement of $\mathcal{K}$. That is

$$\mathcal{K}^c = \{\mathbf{A} : \mathbf{A} \text{ is an } \mathfrak{L}\text{-structure and } \mathbf{A} \notin \mathcal{K}\}.$$

Ultraproducts offer a technique for dealing with finite axiomatizability.

THEOREM 8.52 (The Finite Axiomatizability Theorem). *Let $\mathfrak{L}$ be an elementary language, and let $\mathcal{K}$ be any class of $\mathfrak{L}$-structures. The following are equivalent.*

(i) *$\mathcal{K}$ is a finitely axiomatizable elementary class.*
(ii) *$\mathcal{K}^c$ is a finitely axiomatizable elementary class.*
(iii) *Both $\mathcal{K}$ and $\mathcal{K}^c$ are elementary classes.*
(iv) *Both $\mathcal{K}$ and $\mathcal{K}^c$ are closed under isomorphism and the formation of ultraproducts.*

Proof. (i)⇔(ii)⇒(iii)⇒(iv) are all easy, since any finite set of sentences is equivalent to a single sentence, namely the conjunction of the set. The negation of this single sentence axiomatizes the complement of the original class of models. (iv)⇒(iii) follows immediately from the Keisler-Shelah Isomorphism Theorem. To prove (iii)⇒(i) we invoke the Compactness Theorem. Let Σ and Γ be sets of sentences so that $\mathcal{K} = \mathbf{Mod}\,\Sigma$ and $\mathcal{K}^c = \mathbf{Mod}\,\Gamma$. Then $\Sigma\cup\Gamma$ has no model since $\mathcal{K}\cap\mathcal{K}^c$ is empty. By the Compactness Theorem there is a finite set $\Sigma' \subseteq \Sigma$ and a finite set $\Gamma' \subseteq \Gamma$ so that $\Sigma' \cup \Gamma'$ has no model. We claim that $\mathcal{K} = \mathbf{Mod}\,\Sigma'$. Clearly $\mathcal{K} \subseteq \mathbf{Mod}\,\Sigma'$. For the reverse inclusion, let $\mathbf{A} \in \mathbf{Mod}\,\Sigma'$. Then $\mathbf{A} \notin \mathbf{Mod}\,\Gamma'$ since $\Sigma' \cup \Gamma'$ has no models. But then $\mathbf{A} \notin \mathbf{Mod}\,\Gamma$, and so $\mathbf{A} \notin \mathcal{K}^c$. Therefore $\mathbf{A} \in \mathcal{K}$ as desired. Hence $\mathcal{K}$ is finitely axiomatizable. ∎

The most common use of the Finite Axiomatizability Theorem is to prove that certain classes are *not* finitely axiomatizable. In the typical case, a class $\mathcal{K}$ of structures is given, then certain structures in $\mathcal{K}^c$ are selected and an ultraproduct of these is shown to be in $\mathcal{K}$. As a consequence, $\mathcal{K}$ is not finitely axiomatizable. It is useful to notice that this method invokes (i)⇒(iv), which does not depend on the difficult Keisler-Shelah Isomorphism Theorem.

EXAMPLE 8.53. The class of fields of characteristic zero is not finitely axiomatizable.

Proof. Let I be the set of prime numbers and let $\mathcal{K}$ be the class of fields of characteristic zero. For each $p \in I$, let $\mathbf{F}_p$ be the field with p elements. Let U be any nonprincipal ultrafilter on I. Using the Fundamental Theorem for Ultraproducts, it is easy to see that $\prod_I \mathbf{F}_p/U$ is a field of characteristic zero. Hence $\mathcal{K}^c$ is not closed under ultraproducts. Consequently, $\mathcal{K}$ is not finitely axiomatizable. As an aside, observe that $\mathcal{K}$ is an elementary class, since the customary field axioms are openly elementary and supplementing this (finite) set of axioms with infinitely many sentences of the form $\neg(1 + \cdots + 1 \approx 0)$ produces a set of elementary sentences axiomatizing the class of fields of characteristic zero. ■

For varieties of algebras, it is possible to take advantage of a knowledge of subdirectly irreducible algebras in the variety to frame conditions sufficient for finite axiomatizability. Recall that a variety $\mathcal{V}$ is said to be **finitely based** provided $\mathcal{V}$ is finitely axiomatizable.

THEOREM 8.54 (Bjarni Jónsson 1979). *Let $\mathcal{V}$ be a variety. Let $\mathcal{U}$ be a finitely axiomatizable elementary class with $\mathcal{V} \subseteq \mathcal{U}$. Let $\mathcal{W}$ be a class closed with respect to the formation of ultraproducts. If*

(i) *every subdirectly irreducible algebra in $\mathcal{U}$ belongs to $\mathcal{V} \cup \mathcal{W}$, and*
(ii) *$\mathcal{V} \cap \mathcal{W}$ is finitely axiomatizable,*

then $\mathcal{V}$ is finitely based.

Proof. We will assume that $\mathcal{V}$ is not finitely based and that (i) holds, and prove that $\mathcal{V} \cap \mathcal{W}$ is not finitely axiomatizable.

Since $\mathcal{V}$ is not finitely axiomatizable, there must be an algebra $\mathbf{B} \in \mathcal{V}$ such that $\mathbf{B} = \prod_I \mathbf{A}_i/U$, where U is an ultrafilter on I and $\mathbf{A}_i \notin \mathcal{V}$ for all $i \in I$. Let δ be a sentence which axiomatizes $\mathcal{U}$. Then $\mathbf{B} \models \delta$ and so $[\![\delta]\!] \in U$ by the Fundamental Theorem for Ultraproducts. Now for each $i \in I$ select an algebra $\mathbf{A}'_i$ according to the following stipulations: if $i \in [\![\delta]\!]$, then take $\mathbf{A}'_i = \mathbf{A}_i$, otherwise choose $\mathbf{A}'_i$ arbitrarily from $\mathcal{U} - \mathcal{V}$. Let $\mathbf{B}' = \prod_I \mathbf{A}'_i/U$. The Fundamental Theorem for Ultraproducts easily yields $\mathbf{B}' \equiv \mathbf{B}$. So $\mathbf{B}' \in \mathcal{V}$, and $\mathbf{A}'_i \in \mathcal{U} - \mathcal{V}$ for all $i \in I$. Since $\mathcal{V}$ is a variety, by Birkhoff's Subdirect Representation Theorem (Theorem 4.44 in Volume 1) select, for each $i \in I$, a subdirectly irreducible factor $\mathbf{C}_i$ of $\mathbf{A}'_i$ so that $\mathbf{C}_i \notin \mathcal{V}$. Let $\mathbf{B}'' = \prod_I \mathbf{C}_i/U$. Since $\mathbf{C}_i$ is a homomorphic image of $\mathbf{A}'_i$ for each $i \in I$, we see that $\mathbf{B}''$ is a homomorphic image of $\mathbf{B}'$. Consequently, $\mathbf{B}'' \in \mathcal{V}$. Hence, $\mathbf{B}'' \models \delta$, and the Fundamental Theorem for Ultraproducts tells us that $\{i \mid \mathbf{C}_i \models \delta\} \in U$. As above, pick $\mathbf{D}_j$ so that $\mathbf{D}_j = \mathbf{C}_j$ if $j \in \{i \mid \mathbf{C}_i \models \delta\}$, and otherwise choose $\mathbf{D}_j$ to be any subdirectly irreducible algebra not in $\mathcal{V}$. Finally, let $\mathbf{B}''' = \prod_I \mathbf{D}_i/U$. As before, $\mathbf{B}''' \equiv \mathbf{B}''$ by the Fundamental Theorem for Ultraproducts, and so $\mathbf{B}''' \in \mathcal{V}$. Now $\{i \mid i \in I \text{ and } \mathbf{D}_i \in \mathcal{U}\} \in U$, so $\{i \mid i \in I \text{ and } \mathbf{D}_i \in \mathcal{V} \cup \mathcal{W}\} \in U$, in view of condition (i). However, the $\mathbf{D}_i$'s were all chosen to lie outside of $\mathcal{V}$. Hence, $\{i \mid i \in I \text{ and } \mathbf{D}_i \in \mathcal{W}\} \in U$. Since $\mathcal{W}$ is closed under ultraproducts, we infer $\mathbf{B}''' \in \mathcal{W}$. So $\mathbf{B}''' \in \mathcal{V} \cap \mathcal{W}$. Since $\mathbf{B}'''$ is an ultraproduct of algebras not in $\mathcal{V} \cap \mathcal{W}$ (actually, not even in $\mathcal{V}$), we conclude that $\mathcal{V} \cap \mathcal{W}$ is not finitely axiomatizable, by the Finite Axiomatizability Theorem. ■

A good example of how to use this theorem is given in the proof of Theorem 8.66 on page 362.

The proof above is nonconstructive in the sense that it provides no method to find a finite base for $\mathcal{V}$. With some additional restrictions, however, such a constructive method is available. Suppose, in addition to the stipulations of the theorem, all the following hold:

(a) The sentence θ axiomatizes $\mathcal{U}$.
(b) The sentence φ axiomatizes $\mathcal{W}$.
(c) The sentence ψ axiomatizes $\mathcal{V} \cap \mathcal{W}$
(d) There is an algorithm for enumerating the equations true in $\mathcal{V}$.

Then notice that the sentence $\theta \mathbin{\&} (\varphi \Rightarrow \psi)$ is true in $\mathcal{V}$. By Birkhoff's HSP Theorem, $\mathcal{V}$ is axiomatized by some set of equations. Therefore, in view of the Compactness Theorem, there is a finite set Σ of equations true in $\mathcal{V}$ such that $\Sigma \models \theta \mathbin{\&} (\varphi \Rightarrow \psi)$. According to the well-known Completeness Theorem of elementary logic, given any finite set of sentences, like Σ, the elementary sentences which are its logical consequences can be enumerated algorithmically. Hence, if we first begin enumerating the equations true in $\mathcal{V}$ to obtain finite sets of equations, and then for each finite set spawn a subprocess to enumerate its elementary logical consequences, we must eventually discover a finite set Σ of equations such that $\Sigma \models \theta \mathbin{\&} (\varphi \Rightarrow \psi)$. This Σ must be a base for $\mathcal{V}$, since any subdirectly irreducible model $\mathbf{B}$ of Σ belongs to $\mathcal{U}$ (since $\Sigma \models \theta$). But then $\mathbf{B} \in \mathcal{V}$ or $\mathbf{B} \in \mathcal{W}$. In the latter case, we have $\mathbf{B} \models \varphi$. But since $\mathbf{B} \models \varphi \Rightarrow \psi$, we also conclude that $\mathbf{B} \models \psi$ and hence that $\mathbf{B} \in \mathcal{V} \cap \mathcal{W}$. So in any event, $\mathbf{B} \in \mathcal{V}$. Thus every subdirectly irreducible model of Σ is in $\mathcal{V}$. It follows from Birkhoff's Subdirect Representation Theorem, that Σ is a base for $\mathcal{V}$.

The restrictions (a) and (c) enumerated above do not substantially alter the sense of Theorem 8.54. Restriction (b) does openly limit the range of applicability, but at least it is a reasonable restriction to impose in order to make the theorem constructive. Restriction (d), at first glance, seems to be rather severe. However, in practice it is frequently met. For example, $\mathcal{V}$ may be described by an infinite but recursive list of axioms, or $\mathcal{V}$ may be the variety generated by a given finite set of finite algebras. In all cases of this kind, the restriction (d) will be satisfied.

The method described for obtaining a finite basis for $\mathcal{V}$, under the additional restrictions, is effective. However, it depends on algorithms for listing all the logical consequences of a given finite set of sentences. Such algorithms are notoriously slow and expensive in computational resources. It would be useful to have more efficient algorithms for constructing finite bases of varieties, at least in some special cases.

Let us turn to considering the class of finitely subdirectly *reducible* algebras in a variety, as opposed to those that drew our attention above, which were finitely subdirectly irreducible.

Anvar M. Nurakunov (2017) observed that, for a fixed signature the class of algebras that are finitely subdirectly reducible (as opposed to finitely subdirectly irreducible) is closed under the formation of ultraproducts. He could even do this for algebraic structures. We prove his theorem next.

A homomorphism h from $\mathbf{A}$ onto $\mathbf{B}$ between two algebraic structures of the same signature is **proper** provided for some basic relational symbol R (with $\approx$ treated as a basic relational symbol), say of rank r, there is an r-tuple $\vec{a} \in A^r$ so that $\vec{a} \notin r^{\mathbf{A}}$ but

$h(\vec{a}) \in R^{\mathbf{B}}$. Note that this is simply the statement that h is not an isomorphism between $\mathbf{A}$ and $\mathbf{B}$. That is, an onto homomorphism is proper provided it is not an isomorphism.

Let $\mathbf{A}$ be an algebraic structure, let J be a set and for each $j \in J$ let $h_j : \mathbf{A} \twoheadrightarrow \mathbf{B}_j$ be a homomorphism onto the structure $\mathbf{B}_j$. We say that the system $\langle h_j \mid j \in J\rangle$ of homomorphisms **separates basic relations** provided for each basic relation symbol R (including the equality symbol $\approx$) and each tuple $\vec{a} \in A^r$, where r is the rank of R, such that $\vec{a} \notin R^{\mathbf{A}}$, there is $j \in J$ such that $h_j(\vec{a}) \notin R^{\mathbf{B}_j}$. The system $\langle h_j \mid j \in J\rangle$ is a **subdirect representation** of $\mathbf{A}$ if and only if it separates relations.

The structure $\mathbf{A}$ is **finitely subdirectly reducible** provided there is a finite subdirect representation of $\mathbf{A}$ made up of proper homomorphisms. This is easily seen to be equivalent to the existence of proper homomorphisms $h : \mathbf{A} \twoheadrightarrow \mathbf{B}$ and $g : \mathbf{A} \twoheadrightarrow \mathbf{C}$ that separate relations.

Thus if $\mathbf{A}$ is finitely subdirectly reducible witnessed as above, then $\mathbf{A}$ is isomorphic to a substructure of $\mathbf{B} \times \mathbf{C}$ via an isomorphism so that the projections restricted to the images of $\mathbf{A}$ are proper homomorphisms. The converse is also true.

THEOREM 8.55 (Anvar M. Nurakunov 2017). *Fix a signature that provides only finitely many relation symbols but any number of operation symbols. The class of all finitely subdirectly reducible structures is closed under ultraproducts.*

Proof. Let I be a set and suppose for each $i \in I$ that the algebraic structure $\mathbf{A}_i$ is finitely subdirectly reducible. Let D be ultrafilter on the set I. Our aim is to show that the ultraproduct

$$\prod_I \mathbf{A}_i/D$$

is also finitely subdirectly reducible.

To this end, for each $i \in I$ pick structures $\mathbf{B}_i$ and $\mathbf{C}_i$ and proper homomorphism $h_i : \mathbf{A}_i \twoheadrightarrow B_i$ and $g_i : \mathbf{A}_i \twoheadrightarrow \mathbf{C}_i$ so that $\langle h_i, g_i\rangle$ separates all the basic relations as well as the equality relation.

In the natural way, define

$$h\colon \prod_I \mathbf{A}_i/D \twoheadrightarrow \prod_I \mathbf{B}_i/D$$

and

$$g\colon \prod_I \mathbf{A}_i/D \twoheadrightarrow \prod_I \mathbf{C}_i/D$$

by

$$h(\langle a_i \mid i \in I\rangle/D) = \langle h_i(a_i) \mid i \in I\rangle/D$$
$$g(\langle a_i \mid i \in I\rangle/D) = \langle g_i(a_i) \mid i \in I\rangle/D$$

for all $\langle a_i \mid i \in I\rangle \in \prod_I A_i$.

To ease notation, let $\mathbf{A}^*$ denote $\prod_I \mathbf{A}_i/D$. Similarly, the $\mathbf{B}^*$ and $\mathbf{C}^*$ denote the other two ultraproducts.

It is our contention that $\langle h, g\rangle$ is a subdirect representation of $\mathbf{A}^*$ by proper homomorphisms.

It is first of all necessary to prove that the definitions of h and g are sound and that they are surjective homomorphisms. The proofs are routine.

It is also necessary to prove that h and g are proper and that $\langle h, g\rangle$ separates relations.

To see that h is proper we must find

$$\text{a basic relation symbol } R \text{ and elements } \vec{a}^0/D, \dots, \vec{a}^{(r-1)}/D \text{ of } A^*$$

so that

$$\langle \vec{a}^0/D, \dots, \vec{a}^{(r-1)}/D\rangle \notin R^{\mathbf{A}^*}$$

but

$$\langle h(\vec{a}^0/D), \dots, h(\vec{a}^{(r-1)}/D)\rangle \in R^{\mathbf{B}^*}.$$

For each basic relation symbol T let

$$J_T : \; = \{i \mid i \in I \text{ and } T \text{ witnesses that } h_i \text{ is proper}\}$$

Since h_i is proper for each $i \in I$, we see that I is the union of all the J_T's. But there are only finitely many basic relations and D is an ultrafilter. So pick a basic relation symbol R so that $J_R \in D$. Let r be the rank of R. For each $i \in J_R$ pick $a_i^0, \dots, a_i^{(r-1)} \in A_i$ so that $\langle a_i^0, \dots, a_i^{(r-1)}\rangle \notin R^{\mathbf{A}_i}$ but $\langle h_i(a_i^0), \dots, h_i(a_i^{(r-1)})\rangle \in R^{\mathbf{B}_i}$. For $i \in I$ but $i \notin J_R$ and $k < r$, let a_i^k be some default element. In this way $\vec{a}^0, \dots, \vec{a}^{(r-1)}$ are determined. Since J_R belongs to the ultrafilter, we see that

$$\langle \vec{a}^0/D, \dots, \vec{a}^{(r-1)}/D\rangle \notin R^{\mathbf{A}^*}$$

and

$$\langle h(\vec{a}^0)/D, \dots, h(\vec{a}^{(r-1)})/D\rangle \in R^{\mathbf{B}^*}.$$

In other words, h is proper. Likewise g is also proper.

Last, we need to see that $\langle h, g\rangle$ separates relations. So let R be a basic relation symbol. Let r be the rank of R. Let $\vec{a}^0/D, \dots, \vec{a}^{(r-1)}/D$ be elements of A^* so that

$$\langle \vec{a}^0/D, \dots, \vec{a}^{(r-1)}/D\rangle \notin R^{\mathbf{A}^*}.$$

This entails that

$$Y : \; = \{i \mid i \in I \text{ and } \langle a_i^0, \dots, a_i^{(r-1)}\rangle \notin R^{\mathbf{A}_i}\} \in D.$$

For each $i \in Y$, we know that $\langle h_i, g_i\rangle$ separates relations. This means that either

$$\langle h_i(a_i^0), \dots, h_i(a^{(r-1)})\rangle \notin R^{\mathbf{B}_i},$$

or

$$\langle g_i(a_i^0), \dots, g_i(a^{(r-1)})\rangle \notin R^{\mathbf{C}_i}.$$

So

$$Y = \{i \mid i \in I \text{ and } \langle h_i(a_i^0), \dots, h_i(a^{(r-1)}\rangle \notin R^{\mathbf{B}_i}\} \cup$$
$$\cup \{i \mid i \in I \text{ and } \langle g_i(a_i^0), \dots, g_i(a^{(r-1)}\rangle \notin R^{\mathbf{C}_i}\}$$

Because Y belongs to the ultrafilter, one of the two sets in the union above must also belong to the ultrafilter. Without loss of generality, suppose that

$$\{i \mid i \in I \text{ and } \langle h_i(a_i^0), \dots, h_i(a^{(r-1)}\rangle \notin R^{\mathbf{B}_i}\} \in D.$$

But this means

$$\langle h(\vec{a}^0)/D, \dots, h(\vec{a}^{(r-1)})/D\rangle \notin R^{\mathbf{B}^*}.$$

In this way, we see that $\langle h, g\rangle$ separates relations. ∎

Let $\mathcal{K}$ be a class of structures, all of the same signature, that is closed with respect to the formation of ultraproducts. Theorem 8.55 can be reworked relative to $\mathcal{K}$, with the proviso that the definition of subdirect representations is changed to mean subdirect $\mathcal{K}$-representation where the target structures must also belong to $\mathcal{K}$. When $\mathcal{K}$ is axiomatizable by a set of positive elementary sentences it will be closed with respect to ultraproducts and the formation of homomorphic images so that the subdirect $\mathcal{K}$-representations will just be ordinary subdirect representations. In particular, when $\mathcal{K}$ is a variety of algebras, then Nurakunov's observation relativizes in the most natural way.

In the set of exercises below we gather together some rudimentary properties of varieties $\mathcal{V}$ of finite signature whose class $\mathcal{V}_{\mathrm{si}}$ of subdirectly irreducible members is finitely axiomatizable. This list is assembled from the work of many people, including Bjarni Jónsson, Ralph McKenzie, Kirby Baker, Ju Wang, Anvar Nurakunov, and Ross Willard. Surely a lot of it is folklore.

We use $\mathcal{V}_{\mathrm{fsi}}$ for the class of finitely subdirectly irreducible algebras in $\mathcal{V}$.

Exercises 8.56

1. Use Theorem 8.44 to prove prove Quackenbush's result (Corollary 7.78) that if a locally finite variety has an infinite subdirectly irreducible algebra, then it has arbitrarily large finite subdirectly irreducibles.

2. Prove that every finitely subdirectly irreducible algebra is embeddable into an ultraproduct of its subdirectly irreducible images.

3. Let $\mathcal{V}$ be a variety such that $\mathcal{V}_{\mathrm{si}}$ is axiomatizable by a set of elementary sentences. Prove that every finitely subdirectly irreducible algebra in $\mathcal{V}$ is embeddable into some subdirectly irreducible algebra in $\mathcal{V}$. Thus in a variety $\mathcal{V}$ of finite signature with a finite residual bound b, every finitely subdirectly irreducible algebra must have cardinality bounded by b as well. It follows that every finitely subdirectly irreducible algebra is finite and so it is subdirectly irreducible. Thus, $\mathcal{V}_{\mathrm{fsi}} = \mathcal{V}_{\mathrm{si}}$. In consequence, $\mathcal{V}_{\mathrm{fsi}}$ is also finitely axiomatizable.

4. Let $\mathcal{V}$ be a variety so that $\mathcal{V}_{\mathrm{si}}$ is finitely axiomatizable. Prove that $\mathcal{V}$ is finitely based if and only if there is a finitely based variety $\mathcal{W}$ so that
   - $\mathcal{V} \subseteq \mathcal{W}$, and
   - $\mathcal{W}_{\mathrm{si}}$ is finitely axiomatizable.

5. An algebra **A** is **critical** provided it is finite and not in the variety generated by the class of smaller algebras belonging to $\boldsymbol{HS}\mathbf{A}$. When $\mathcal{K}$ is a class of algebras, we use $\mathcal{K}_{\mathrm{cr}}$ to denote the class of critical algebras belonging to $\mathcal{K}$. Suppose $\mathcal{V}$ and $\mathcal{W}$ are varieties so that $\mathcal{V} \subseteq \mathcal{W}$ where $\mathcal{W}$ is locally finite. Suppose further that both $\mathcal{V}_{\mathrm{cr}}$ and $\mathcal{W}_{\mathrm{cr}}$ are finitely axiomatizable. Prove that if $\mathcal{W}$ is finitely based, then $\mathcal{V}$ is finitely based.

6. Give a proof of the following statement.

Let $\mathcal{V}$ and $\mathcal{W}$ be classes of algebras of the same signature with $\mathcal{V} \subseteq \mathcal{W}$ such that

(i) $\mathcal{V}$ is a variety,

(ii) $\mathcal{W}$ is finitely axiomatizable, and

(iii) Both $\mathcal{V}_{\mathrm{fsi}}$ and $\mathcal{W}_{\mathrm{fsi}}$ are finitely axiomatizable.

Then $\mathcal{V}$ is finitely based.

***7.** For each natural number k let σ_k be the following elementary sentence:

$$\exists z, w[z \not\approx w \mathbin{\bigwedge} \forall x, y(x \not\approx y \Rightarrow (\{x,y\} \looparrowright_k^k \{z,w\}))].$$

(The notation $\{x,y\} \looparrowright_k^k \{z,w\}$) is based on Mal'tsev's Congruence Generation Theorem; see Corollary 6.19 on page 25.) Thus, σ_k asserts, among other things, that there is a critical pair.

Prove the following statement.

Let $\mathcal{V}$ be an elementary class of finite signature. The following conditions are equivalent:

(i) $\mathcal{V}_{\mathrm{si}}$ is an elementary class.

(ii) $\mathcal{V}_{\mathrm{si}}$ is closed with respect to ultraproducts.

(iii) There is a natural number ℓ such that $\mathbf{A} \in \mathcal{V}_{\mathrm{si}}$ if and only if $\mathbf{A} \in \mathcal{V}$ and $\mathbf{A} \models \sigma_\ell$.

(iv) $\mathcal{V}_{\mathrm{si}}$ is finitely axiomatizable relative to $\mathcal{V}$.

A version of this observation holds with $\mathcal{V}_{\mathrm{fsi}}$ in place of $\mathcal{V}_{\mathrm{si}}$ and τ_ℓ in place of σ_ℓ, where τ_ℓ is the following sentence:

$$\forall x, y, z, w\,[(x \not\approx y \mathbin{\bigwedge} z \not\approx w) \Rightarrow$$
$$\exists u, v\,\big(u \not\approx v \mathbin{\bigwedge} (\{x,y\} \looparrowright_\ell^\ell \{u,v\}) \mathbin{\bigwedge} (\{z,w\} \looparrowright_\ell^\ell \{u,v\})\big)\big]$$

8. Prove the statement below.

Let $\mathcal{V}$ be a variety. The following are equivalent:

(i) $\mathcal{V}$ has a finite residual bound;

(ii) there is a positive integer n so that for any algebra $\mathbf{A} \in \mathcal{V}$ and for any $a_0, a_1, \dots, a_n \in A$,

$$\bigcap_{i<j\leq n} \mathrm{Cg}^{\mathbf{A}}(a_i, a_j) = 0_A;$$

(iii) There is a finite upper bound on the cardinalities of the finitely subdirectly irreducible algebras in $\mathcal{V}$.

(iv) there is a positive number n such that for any algebra $\mathbf{A} \in \mathcal{V}$, any elements $a_0, a_1, \dots, a_n \in A$, and for any congruence $\theta \in \mathbf{Con}\,A$,

$$\bigcap_{i<j\leq n} \mathrm{Cg}^{\mathbf{A}}(a_i, a_j) \vee \theta = \theta;$$

(v) there is a positive number n such that for any algebra $\mathbf{F}$ that is $\mathcal{V}$-freely generated by a set $\{x_0, x_1, \dots\}$ with more than n elements, and for any congruence $\theta \in \mathbf{Con}\,F$,

$$\bigcap_{i<j\leq n} \mathrm{Cg}^{\mathbf{F}}(x_i, x_j) \vee \theta = \theta.$$

9. Let p be a term with variables from among $x_0, \ldots, x_{k-1}$. Denote by p^{ij} the term on $k-1$ variables obtained from p by changing variable x_j to variable x_i.

Prove the statement below.

Let $\mathcal{V}$ be a variety residually bounded by n. Then

$$\mathcal{V} \models \forall x_0 x_1 \ldots x_{k-1} \left[\bigwedge_{i<j\leq n} (p^{ij} \approx q^{ij}) \Rightarrow p \approx q \right]$$

for any $k > n$ and any terms p and q with variables among $x_0, \ldots, x_{k-1}$.

10. (Anvar M. Nurakunov, 2017) Let κ be an infinite cardinal. An ultrafilter D on a set I is **κ-complete** provided that the intersection of any nonempty collection of sets in the ultrafilter again belongs to the ultrafilter, if the collection has cardinality less then κ. Prove that Theorem 8.55 remains true for signatures that provide fewer than κ basic relation symbols when the ultraproduct is taken with respect to a κ-complete ultrafilter. It should be noted here that every ultrafilter is ω-complete, and for uncountable κ, existence of κ-complete ultrafilters leads into deep set-theoretical water, like the existence of uncountable measurable cardinals and the failure of $V = L$. See (C. C. Chang and H. J. Keisler, 1990) for more on this point.

11. Prove that the elementary theory of infinite atomic Boolean algebras fails to be model complete.

8.4. Jónsson's Lemma for Congruence Distributive Varieties

THEOREM 8.57 (Jónsson's Lemma (Bjarni Jónsson, 1967)).

(i) *Let $\langle \mathbf{A}_i \mid i \in I \rangle$ be a system of algebras, all of the same signature, let $\mathbf{B}$ be a subalgebra of $\prod_I \mathbf{A}_i$ so that $\mathbf{Con\ B}$ is a distributive lattice, and let $\theta \in \operatorname{Con} \mathbf{B}$ such that $\mathbf{B}/\theta$ is finitely subdirectly irreducible. Then there is an ultrafilter U on I such that the restriction of $\sim_U$ to B is included in θ.*

(ii) *Let $\mathcal{V}$ be a congruence distributive variety and suppose $\mathcal{V} = \mathbf{V}\mathcal{K}$. Then $\mathcal{V}_{\text{fsi}} \subseteq \mathbf{HSP_u}\mathcal{K}$, and therefore $\mathcal{V} = \mathbf{P_s HSP_u}\mathcal{K}$.*

Proof. Statement (i) requires us to find a certain ultrafilter U. Every subset X of I which belongs to this ultrafilter is required to have the following property:

($\star$) For any $b, b' \in B$, if $X \subseteq [\![b \approx b']\!]$, then $b\ \theta\ b'$.

It is just property ($\star$) which will ensure that $\sim_U$ restricted to B is included in θ, as desired.

So for every $X \subseteq I$ define φ_X, a binary relation on B, by

$$b\ \varphi_X\ b' \text{ if and only if } X \subseteq [\![b \approx b']\!].$$

Then φ_X is easily seen to be a congruence relation on $\mathbf{B}$. Notice X satisfies ($\star$) if and only if $\varphi_X \subseteq \theta$.

Now let $\mathcal{M}$ be the collection of all subsets X of I which satisfy ($\star$). What we need is an ultrafilter included in $\mathcal{M}$. Theorem 8.29 supplies the following sufficient conditions for $\mathcal{M}$ to include an ultrafilter:

(a) $I \in \mathcal{M}$.
(b) If $X \subseteq Y \subseteq I$ and $X \in \mathcal{M}$, then $Y \in \mathcal{M}$.
(c) If $X \cup Y \in \mathcal{M}$, then either $X \in \mathcal{M}$ or $Y \in \mathcal{M}$.

Now (a) holds since φ_I is the identity relation, and (b) holds since $\varphi_Y \subseteq \varphi_X$, when $X \subseteq Y$. To deduce (c) suppose $X \cup Y \in \mathcal{M}$. Observe first that $\varphi_{X\cup Y} = \varphi_X \cap \varphi_Y$. Since $X \cup Y \in \mathcal{M}$, we have

$$\begin{aligned}\theta &= \varphi_{X\cup Y} \vee \theta\\ &= (\varphi_X \wedge \varphi_Y) \vee \theta\\ &= (\varphi_X \vee \theta) \wedge (\varphi_Y \vee \theta)\end{aligned}$$

where the last equality holds because **Con B** is distributive. Since $\mathbf{B}/\theta$ is finitely subdirectly irreducible, we know that θ is meet irreducible. Therefore, either $\varphi_X \vee \theta = \theta$ or $\varphi_Y \vee \theta = \theta$. But this entails that either $X \in \mathcal{M}$ or $Y \in \mathcal{M}$. Therefore (c) holds. Now observe that $\varphi_\emptyset = B \times B$. Since $\mathbf{B}/\theta$ is finitely subdirectly irreducible it must have more than one element. Therefore $\theta \neq B \times B$, and so $\emptyset \notin \mathcal{M}$. Consequently, any ultrafilter contained in $\mathcal{M}$ is proper.

Thus, $\mathcal{M}$ includes an ultrafilter U by Theorem 8.29. Evidently, $\sim_U$ is included in θ.

To obtain (ii), suppose that $\mathbf{C}$ is a finitely subdirectly irreducible member of $\mathcal{V}$. By Birkhoff's HSP Theorem $\mathbf{C} \in \boldsymbol{HSP}\mathcal{K}$. So we select a set I, and for each $i \in I$ an algebra $\mathbf{A}_i \in \mathcal{K}$, so that $\prod_I \mathbf{A}_i$ has a subalgebra $\mathbf{B}$, and $\mathbf{B}$ has a congruence θ such that $\mathbf{C} \cong \mathbf{B}/\theta$. By statement (i) let U be an ultrafilter on I so that $\sim_U$ restricted to B is included in θ. Let h be the quotient map from $\prod_I \mathbf{A}_i$ onto $\prod_I \mathbf{A}_i/U$ and let h' be the restriction of h to B. Then the kernel of h' is the restriction of $\sim_U$ to B. Hence, the kernel of h' is included in θ. So by the Second Isomorphism Theorem (Theorem 4.10 in Volume 1), we have

$$\mathbf{C} \cong \mathbf{B}/\theta \cong (\mathbf{B}/\ker h')/(\theta/\ker h')$$

and

$$\mathbf{B}/\ker h' \cong h(\mathbf{B}) \subseteq \prod_I \mathbf{A}_i/U.$$

So $\mathbf{C} \in \boldsymbol{HSP_u}\mathcal{K}$, as desired. That $\mathcal{V} = \boldsymbol{P_s HSP_u}\mathcal{K}$ follows from Birkhoff's Subdirect Representation Theorem. ■

At first glance Jónsson's Lemma—which gives $\mathcal{V} = \boldsymbol{P_s HSP_u}\mathcal{K}$ for congruence distributive varieties $\mathcal{V}$—seems less strong than Birkhoff's HSP Theorem—which gives $\mathcal{V} = \boldsymbol{HSP}\mathcal{K}$ for *all* varieties $\mathcal{V}$. A closer look reveals something different. Because ultraproducts preserve elementary properties, we typically have much better information about the algebras in $\boldsymbol{P_u}\mathcal{K}$ than we do about the algebras in $\boldsymbol{P}\mathcal{K}$. In many cases, this applies to algebras in $\boldsymbol{HSP_u}\mathcal{K}$ as well, and this last class includes of the subdirectly irreducible algebras in $\mathcal{V}$. Another way to compare Jónsson's Lemma with Birkhoff's HSP Theorem is to notice that the HSP Theorem focuses on building algebras free in $\mathcal{V}$ early (in $\boldsymbol{SP}\mathcal{K}$), whereas Jónsson's Lemma is based on getting the subdirectly irreducible algebras early (in $\boldsymbol{HSP_u}\mathcal{K}$). Although as a general matter, the description of the free algebras and the description of the subdirectly irreducible algebras in a variety both pose overwhelming difficulties, there are quite a few varieties where the subdirectly irreducible algebras are more accessible than the free algebras. The varieties of lattices are congruence distributive. Jónsson's Lemma has proven to be a key tool for the investigation of these varieties.

COROLLARY 8.58. *If $\mathcal{K}$ is a finite set of finite algebras, and $\mathbf{V}\mathcal{K}$ is congruence distributive, then all the subdirectly irreducible algebras in $\mathbf{V}\mathcal{K}$ belong to $\mathbf{HS}\mathcal{K}$. Consequently, there is a finite upper bound on the cardinality of the subdirectly irreducible members of $\mathbf{V}\mathcal{K}$ and $\mathbf{V}\mathcal{K} = \mathbf{P}_s\mathbf{HS}\mathcal{K}$.* ■

COROLLARY 8.59 (Baker's Finite Basis Theorem (Kirby A. Baker, 1977)). *Let $\mathbf{A}$ be a finite algebra of finite signature. If $\mathbf{A}$ belongs to a congruence distributive variety, then $\mathbf{A}$ is finitely based.*

This follows from Theorem 7.24 (Willard's Finite Basis Theorem) and the previous corollary.

COROLLARY 8.60. *If $\mathcal{K}$ is a finite set of finite algebras and $\mathbf{V}\mathcal{K}$ is congruence distributive, then $\mathbf{V}\mathcal{K}$ has only finitely many subvarieties.* ■

COROLLARY 8.61. *Let $\mathcal{V}$ be a congruence distributive variety, and suppose that $\mathbf{A}$ and $\mathbf{B}$ are finite subdirectly irreducible members of $\mathcal{V}$. Then $\mathbf{A} \cong \mathbf{B}$ if and only if $\Theta\,\mathbf{A} = \Theta\,\mathbf{B}$.* ■

This last corollary can be viewed as a variation on the theme set out in Theorem 8.6 on page 320: Two finite algebras $\mathbf{A}$ and $\mathbf{B}$ are isomorphic if and only if $\mathbf{Th}\,\mathbf{A} = \mathbf{Th}\,\mathbf{B}$. In the case of finite subdirectly irreducible algebras belonging to a congruence distributive variety, it is enough that their equational theories are the same.

COROLLARY 8.62. *If $\mathcal{V}$ is a congruence distributive variety for a finite language, and $\mathcal{V}$ is generated by a finite set of finite algebras, then the class of subdirectly irreducible members of $\mathcal{V}$ is a finitely axiomatizable elementary class, as is the class of finitely subdirectly irreducible members of $\mathcal{V}$.* ■

COROLLARY 8.63. *Let $\mathcal{V}$ be a congruence distributive variety. The map defined on the set of subvarieties of $\mathcal{V}$ which sends $\mathcal{U} \mapsto \mathcal{U}_{si}$ for all subvarieties $\mathcal{U}$ of $\mathcal{V}$ is an embedding of the lattice of subvarieties of $\mathcal{V}$ into a lattice of subclasses of $\mathcal{V}$, where the lattice operations are intersection and union.* ■

The essence of this last corollary is $(\mathcal{U} \vee \mathcal{W})_{si} = \mathcal{U}_{si} \cup \mathcal{W}_{si}$.

The subvariety lattice of the variety of lattices

In §7.6 we considered the lattice of equational theories, where we had a particular interest in the principal filters in such a lattice. Jónsson's Lemma gives us the means to grasp the subdirectly irreducible algebras in a congruence distributive variety. The publication of Jónsson Lemma launched a detailed investigation of lattices of subvarieties—dually isomorphic to lattices of equational theories—particularly for the variety of lattices. It is immediately evident, as Jónsson pointed out, that a variety generated by a finite lattice can have, up to isomorphism, only finitely many subdirectly irreducible algebras and, according the Birkhoff's Subdirect Representation Theorem, such a variety can have only finitely many subvarieties. So each variety of lattices that is generated by a finite lattice has finite height in the lattice of all varieties of lattices. Subsequently, over the course of almost twenty years, a detailed study of the bottom of the lattice of all lattice varieties was carried out by a large group of researchers. These results were gathered and expanded by Peter Jipsen and Henry Rose (1992). All the subvarieties of finite height discovered during this period were generated by finite lattices, supporting the follow conjecture:

THE FINITE HEIGHT CONJECTURE

> Every variety of lattices that is of finite height in the lattice of all varieties of lattices is generated by a finite lattice.

The Finite Height Conjecture was open for more than 30 years.

EXAMPLE 8.64. (J. B. Nation, 1996)
Let **L** be the lattice displayed in Figure 8.1 and let **F** be the lattice displayed in Figure 8.2. Then **L** is subdirectly irreducible and ***V*** **L** covers ***V*** **F** in the lattice of lattice varieties. Therefore, ***V*** **L** is a variety of finite height that is not generated by a finite lattice. This is a counterexample to the Finite Height Conjecture.

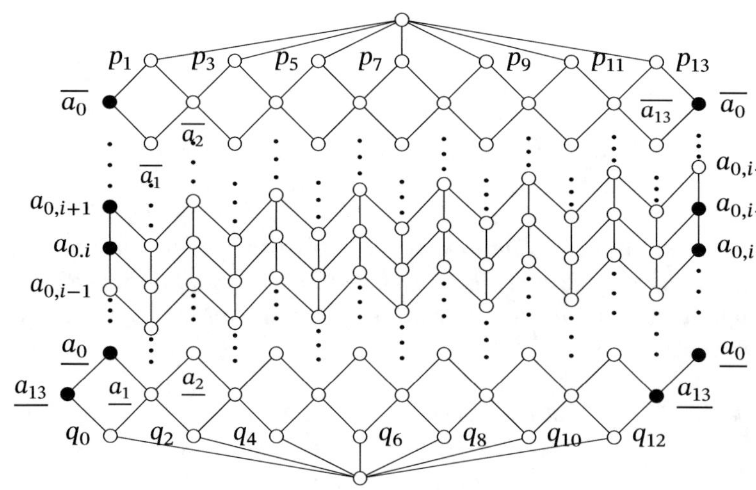

FIGURE 8.1. The lattice **L**

While **L** displayed in Figure 8.1 * looks planar, actually the left edge and the right edge coincide and the filled nodes with the same labels represent one node. So **L** should be understood as drawn on the surface of a cylinder with a cap at the top comprising 21 nodes, a bowl at the bottom comprising 21 nodes, and 14 chains ordered like the integers running vertically up the side of the cylinder. In fact, the cap, the bowl, and

*The authors are grateful to J. B. Nation for sharing with us his tikz code for this display.

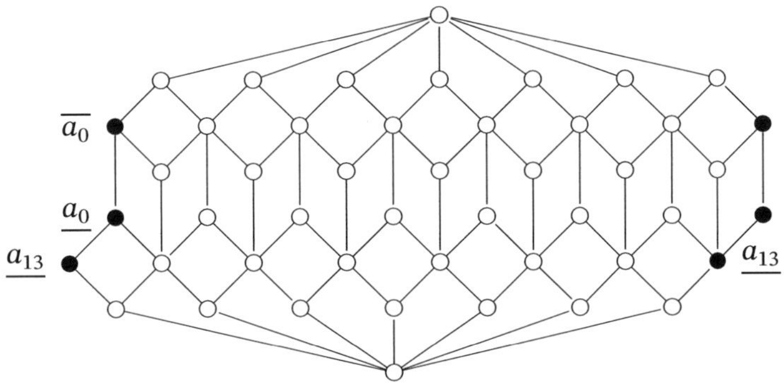

FIGURE 8.2. The lattice **F**

the chains form a partition of **L**. What appears to be three fences in the middle of the diagram is actually one two-way infinite fence wrapped helically around the cylinder. Observe that **F** results from **L** upon removal of the 14 $\mathbb{Z}$-chains. We have not labeled all the nodes, only enough to suggest how the remaining nodes can be labeled.

Proof of Example 8.64. While all the coverings in **F** are displayed and all those in **L** are adequately suggested, one should check that these Hasse diagrams actually are those of lattices. We leave this to the reader, but note that trying to reduce the number of vertical $\mathbb{Z}$-chains can lead to diagrams that do not display lattices.

We note that **F** is a sublattice of **L** so $\boldsymbol{V}\mathbf{F} \leq \boldsymbol{V}\mathbf{L}$. It also easy to see that the 42 elements in the cap and the bowl together with any one element belonging to a $\mathbb{Z}$-chain generate the lattice **L**. So **L** is not locally finite and cannot belong to any finitely generated variety. Thus

$$\boldsymbol{V}\mathbf{F} < \boldsymbol{V}\mathbf{L}.$$

Now for any varieties $\mathcal{U} < \mathcal{W}$ of the same signature there is an equation true in $\mathcal{U}$ that fails in $\mathcal{W}$. This equation, in fact, must fail in some finitely generated algebra belonging to $\mathcal{W}$. Indeed, such a finitely generated algebra can be taken to be subdirectly irreducible. This means that there is a finitely generated subdirectly irreducible lattice **S** such that

$$\mathbf{S} \in \boldsymbol{V}\mathbf{L} \text{ and } \mathbf{S} \notin \boldsymbol{V}\mathbf{F}.$$

CLAIM 1.
Every subdirectly irreducible lattice in $\boldsymbol{V}\,\mathbf{L}$ *belongs to* $\boldsymbol{HS}\,\mathbf{L}$.

Proof. Let **S** be a subdirectly irreducible lattice in $\boldsymbol{V}\ \mathbf{L}$. By Jónsson's Lemma, we see $\mathbf{S} \in \boldsymbol{HSP}_u\ \mathbf{L}$. So we have

$$\mathbf{A} = \prod \mathbf{L}/U$$
$$\mathbf{B} \leq \mathbf{A}$$
$$\mathbf{S} \cong \mathbf{B}/\theta$$

for some ultrafilter U, some lattices **A** and **B**, and some strictly meet irreducible congruence θ on **B**.

By adding new constant symbols to name the 43 elements of the cap and the bowl and adding $A_0, \dots, A_{13}$ as new one-place relation symbols to name the 14 $\mathbb{Z}$-chains in $\mathbf{L}$, we can expand $\mathbf{L}$ to $\mathbf{L}^*$ and obtain

$$\mathbf{A}^* = \prod \mathbf{L}^*/U.$$

In $\mathbf{L}$ we see that the map $\lambda_0(x) := x \wedge \overline{a_1}$ imposes an isomorphism from $\mathbf{A}_0^{\mathbf{L}}$ onto $\mathbf{A}_1^{\mathbf{L}}$. The remaining maps λ_k with $k = 2, \dots, 13$ have similar definitions. Likewise $\lambda_1(x) := x \vee \underline{a_2}$ imposes an isomorphism from $\mathbf{A}_1^{\mathbf{L}}$ onto $\mathbf{A}_2^{\mathbf{L}}$. So all 14 $\mathbb{Z}$-chains in $\mathbf{L}$ are isomorphic to each other via such isomorphisms and their compositions.

Now by the Fundamental Theorem of Ultraproducts, $\mathbf{A}^*$ is partitioned into a cap (isomorphic to the cap of $\mathbf{L}$), a bowl (isomorphic to the bowl of $\mathbf{L}$) and 14 isomorphic discrete chains (in which every element covers another element and every element has a cover). We note that discrete chains are ordinal sums of $\mathbb{Z}$-chains. So there is a linearly ordered set $\mathbf{I}$ (depending on the ultrafilter U) so that each of the 14 chains of $\mathbf{A}^*$ is obtained by replacing each $i \in I$ by a $\mathbb{Z}$-chain, which we will refer to as the ith *component*. For any $i \in I$, consider the ith component of $A_0^{\mathbf{A}^*}$. Applying λ_1 to an element u of that component produces an element in a component of $A_1^{\mathbf{L}^*}$—a component we say is *linked* to the ith component of $A_0^{\mathbf{A}^*}$. Continuing to apply the λ_k, in order, leads to more components linked to the ith of $\mathbf{A}_0^{\mathbf{A}^*}$. Notice that eventually applying λ_{13} produces an element that covers u and so belongs to the ith component of $\mathbf{A}_0^{\mathbf{A}^*}$. In this way, we see that for every $i \in I$ there is a two-way infinite fence helically wrapped through the ith component and those components of the remaining chains linked to it.

Let $i \in I$, let C be the ith component of $\mathbf{A}_0^{\mathbf{A}^*}$, and let ξ_i be the equivalence relation on $\mathbf{A}$ that

- identifies all the points on any of the 14 chains above the component linked to C with the least member of the cap above that chain.
- identifies all the points on any of the 14 chains below the component linked to C with the greatest member of the bowl below that chain.
- Isolates every other element of A.

The reader can check that each ξ_i is a congruence of $\mathbf{A}$ and $\mathbf{A}/\xi_i \cong \mathbf{L}$. Notice

$$\bigcap_{i \in I} \xi_i = 0_A.$$

For each $i \in I$ let ψ_i be the restriction of ξ_i to B. Then

$$\bigcap_{i \in I} \psi_i = 0_B.$$

As θ is strictly meet irreducible in **Con B**, we see that $\theta \geq \psi_i$ for some $i \in I$. So

$$\mathbf{S} \cong \mathbf{B}/\theta \cong (\mathbf{B}/\psi_i)/(\theta/\psi_i) \in \boldsymbol{HS}\,\mathbf{L},$$

as desired. ∎

We are reduced to considering subdirectly irreducible lattices $\mathbf{S} \in \boldsymbol{HS}\,\mathbf{L}$. Let μ be the equivalence relation on L that collapses each of the 14 $\mathbb{Z}$-chains to a point and isolates all the other elements of L. The reader can check that μ is a congruence of $\mathbf{L}$ and, indeed, every nontrivial congruence of $\mathbf{L}$ includes μ. This means that μ is the monolith of $\mathbf{L}$ and $\mathbf{L}$ is subdirectly irreducible.

Now **L** is finitely generated, so every proper sublattice of **L** is contained in a maximal sublattice of **L**. If we can show that all the maximal sublattices of **L** belong to $\boldsymbol{V}$ **F**, then we will be able to conclude that, up to isomorphism, **L** is the only subdirectly irreducible lattice is $\boldsymbol{V}$ **L** $\smallsetminus$ $\boldsymbol{V}$ **F** and, hence, that $\boldsymbol{V}$ **L** covers $\boldsymbol{V}$ **F**, establishing the example.

So all that remains is the following

CLAIM 2.
If **B** *is a proper sublattice of* **L** *so that B includes some element of one of the* 14 $\mathbb{Z}$*-chains, then* **B** *is included in some maximal sublattice* **M** *of* **L** *and* **M** $\in$ $\boldsymbol{V}$ **F**.

Proof of Claim 2. **F** is a maximal sublattice of **L**. There are 14 other maximal sublattices, which we describe below. **L** has 7 coatoms. These coatoms generate **F**. Each coatom is the join of two atoms. Since B has an element on one of the 14 $\mathbb{Z}$-chains it cannot contain all the coatoms, as **B** is a proper sublattice of **L**. Let p be any coatom not in B and let q and q' be the two distinct atoms so that $p = q \vee q'$. So we must have either $B \cap I[q, p] = \emptyset$ or $B \cap I[q', p] = \emptyset$. Thus

$$B \subseteq L \smallsetminus I[q, p] \text{ or } B \subseteq L \smallsetminus I[q', p].$$

So to finish this proof we need to show two things:

- $L \smallsetminus I[q, p]$ and $L \smallsetminus I[q', p]$ are subuniverses of **L**.
- the sublattices with the universes above belong to $\boldsymbol{V}$ **F**.

We leave the first item above in the hands of the reader—Figure 8.3 helps.

In addition to **F**, each of the other 14 maximal sublattices has universe of the form $L \smallsetminus I[q, p]$, where p is a coatom and q is one of the two atoms that join to p. Sets of this kind result from removing two adjacent $\mathbb{Z}$-chains from L as well as the upper and lower bounds of these chains and the atom and the coatom. In view of the automorphisms of **L**, it is harmless to suppose that p_7 is a coatom missing from B and we see $p_7 = q_4 \vee q_{10}$.

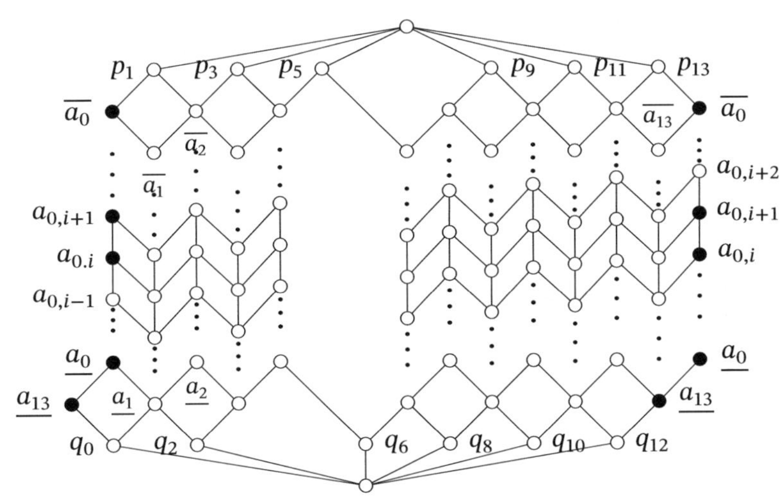

FIGURE 8.3. $L \smallsetminus I[q_4, p_7]$

Removing the 12 remaining $\mathbb{Z}$-chains leaves the sublattice **G** of **F** displayed in Figure 8.4.

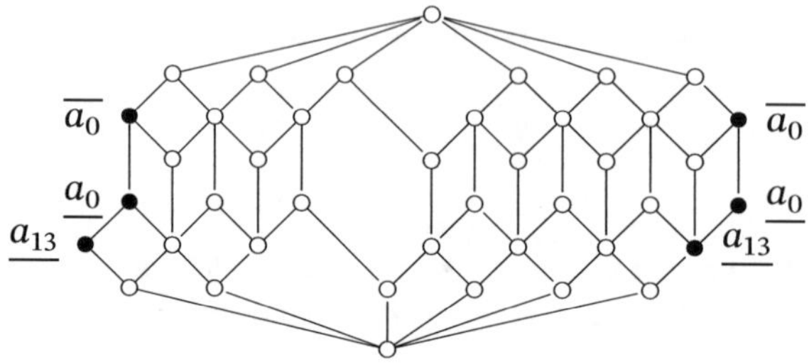

FIGURE 8.4. The sublattice **G** of **F**

To confirm that $\mathbf{L} \setminus \mathbf{I}[q_4, p_7]$ belongs to $\boldsymbol{V}$ **F**, we create a subdirect representation $\langle h_r \mid r \in \mathbb{Z}\rangle$—that is a system of homomorphisms from $\mathbf{L} \setminus \mathbf{I}[q_4, p_7]$ onto **G** that separates points. Let $r \in \mathbb{Z}$. Define $h_r : \mathbf{L} \setminus \mathbf{I}[q_4, p_7] \twoheadrightarrow \mathbf{G}$ via

$$h_r(x) = \begin{cases} \overline{a_i} & \text{If } x = a_{i,t} \text{ and } r \leq t \text{ for some } t \in \mathbb{Z} \text{ and some } i < 14 \text{ with } i \neq 5, 6 \\ \underline{a_i} & \text{If } x = a_{i,t} \text{ and } t < r \text{ for some } t \in \mathbb{Z} \text{ and some } i < 14 \text{ with } i \neq 5, 6 \\ x & \text{otherwise.} \end{cases}$$

It is routine to verify that each h_r is a homomorphism and easy to see that the system $\langle h_r \mid r \in \mathbb{Z}\rangle$ separates points.

The reasoning for $L \setminus I[q_{10}, p_7]$ is similar. ■

With Claim 2 in hand, the example is established. ■

The following corollary is implicit in (J. B. Nation, 1996).

COROLLARY 8.65. *Let* **L** *be the countably infinite lattice in Example* 8.64. *The subdirectly irreducible lattices in* $\boldsymbol{V}$ **L** *are, up to isomorphism, the lattice* **L** *and finitely many finite subdirectly irreducible lattices in* $\boldsymbol{V}$ **F**. *So* $\boldsymbol{V}$ **L** *has residual bound* $\aleph_1$ *and there is a finite upper bound on the cardinalities of the finite subdirectly irreducible lattices in* $\boldsymbol{V}$ **L**. ■

Compare this corollary with Theorem 7.84 of Ralph McKenzie (1996b) on page 294 which provides an example of a finitely generated variety of finite signature with residual bound $\aleph_1$.

Finite axiomatizability of congruence distributive varieties

THEOREM 8.66 (Bjarni Jónsson 1979). *Let* $\mathcal{V}$ *be a congruence distributive variety of finite signature. If* $\mathcal{V}_{\text{fsi}}$ *is finitely axiomatizable, then* $\mathcal{V}$ *is finitely based,*

Baker's Finite Basis Theorem is a corollary of this theorem and Jónsson's Lemma.

The proof of this theorem relies on four lemmas of independent interest. We formulate and prove them before beginning the proof of the theorem. They are all found in (Kirby A. Baker, 1977) and again in (Bjarni Jónsson, 1979). They use the characterization of congruence distributivity given in Theorem 6.1. The complexity of translations in these lemmas is calculated treating the Jónsson terms as if they were basic operations symbols. Basic translations were defined on page 23 of this volume.

LEMMA 8.67. *Let $\mathcal{V}$ be congruence distributive, realizing Δ_m, and $\mathbf{A} \in \mathcal{V}$ with $r_0, \ldots, r_n \in A$ and $r_0 \neq r_n$. Then for some $i < n$ there are $e, f \in A$ with $e \neq f$ and basic translations λ and μ so that*

$$\{r_0, r_n\} \underset{\lambda}{\longrightarrow} \{e, f\} \text{ and } \{r_i, r_{i+1}\} \underset{\mu}{\longrightarrow} \{e, f\}.$$

Proof. Let $r_{i,j} = d_j^{\mathbf{A}}(r_0, r_i, r_n)$ for $i \leq n$ and $j \leq m$. Pick q as small as possible so that the elements $r_{i,q}$ are not all r_0. Note $0 < q$.

In case q is odd, then $r_{0,q} = r_{0,q-1} = r_0$ and we then choose $i < n$ so that $r_{i,q} = r_0 \neq r_{i+1,q}$. In this case take $e = r_0$ and $f = r_{i+1,q}$. Then use $\mu(x) = d_q^{\mathbf{A}}(r_0, r_{i+1}, x)$ and $\lambda(x) = d_q^{\mathbf{A}}(r_0, x, r_n)$.

In case q is even, then $r_{n,q} = r_{n,q-1} = r_0$. So we choose $i < n$ so that $r_{i,q} \neq r_0 = r_{i+1,q}$. In this case take $e = r_{i,q}$ and $f = r_0$. Now use $\lambda(x) = d_q^{\mathbf{A}}(r_0, r_i, x)$ and $\mu(x) = d_q^{\mathbf{A}}(r_0, x, r_n)$. ■

We will refer to the property described in this lemma as the Collapsed $\diamond$-Condition. Here is a diagram of this condition.

The Collapsed $\diamond$-Condition

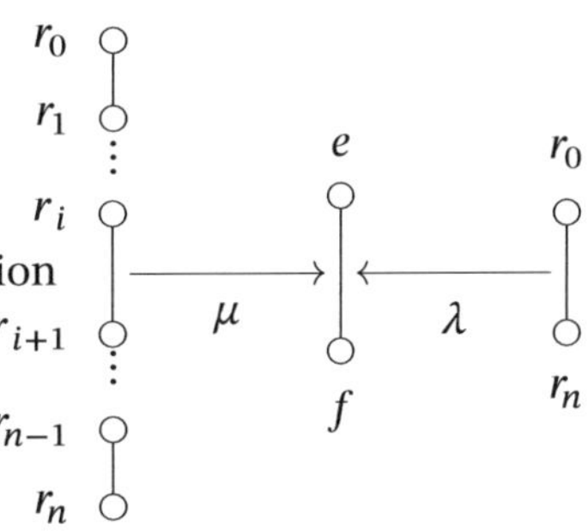

It would be helpful to compare this diagram with the corresponding diagram of the $\diamond$-Condition found on page 28. The symbol $\looparrowright$ was defined on page 25; it provides a detailed picture of congruence generation in an algebra.

LEMMA 8.68. *Let $\mathcal{V}$ be congruence distributive and let $\mathbf{A} \in \mathcal{V}$. If $a, b, a', b' \in A$ with $\mathrm{Cg}^{\mathbf{A}}(a, b) \cap \mathrm{Cg}^{\mathbf{A}}(a', b') \neq 0_A$, there is a natural number k and elements $e, f \in A$ with $e \neq f$ so that*

$$\{a, b\} \looparrowright_k \{e, f\} \text{ and } \{a', b'\} \looparrowright_k \{e, f\}.$$

Proof. Let $(p, q) \in \mathrm{Cg}^{\mathbf{A}}(a, b) \cap \mathrm{Cg}^{\mathbf{A}}(a', b')$ with $p \neq q$. According to Mal'tsev's Principal Congruence Generation Theorem there are natural numbers n and m so that $\{a, b\} \looparrowright_m^n \{p, q\}$. Apply the Collapsed $\diamond$ property to the principal congruence sequence to obtain e_0 and f_0 with $e_0 \neq f_0$ so that $\{a, b\} \looparrowright_{m+1} \{e_0, f_0\}$ and $\{p, q\} \looparrowright_1 \{e_0, f_0\}$. Notice that $(e_0, f_0) \in \mathrm{Cg}^{\mathbf{A}}(a, b) \cap \mathrm{Cg}^{\mathbf{A}}(a', b')$. So again by Mal'tsev's Principal Congruence Generation Theorem there are natural numbers n' and m' so that $\{a', b',\} \looparrowright_{m'}^{n'} \{e_0, f_0\}$.

Apply the Collapsed $\diamond$ Property to the new principal congruence sequence to obtain e and f with $e \neq f$ so that $\{a', b'\} \rightsquigarrow_{m'+1} \{e, f\}$ and $\{e_0, f_0\} \rightsquigarrow_1 \{e, f\}$. But we already know $\{a, b\} \rightsquigarrow_{m+1} \{e_0, f_0\}$, so $\{a, b\} \rightsquigarrow_{m+2} \{e, f\}$. So let k be the larger of $m' + 1$ and $m + 2$. ■

Notice that the value of k in this lemma depends on the choice of a, b, a', and b'. If there is a value of k that works for all such choices of elements, then the least such k is called the *radius* of $\mathbf{A}$. If $\mathcal{K}$ is a class of algebras, then the *radius* of $\mathcal{K}$ is the supremum of the radii of the algebras belonging to $\mathcal{K}$.

LEMMA 8.69. *Let $\mathcal{U}$ be a congruence distributive variety of finite signature and let $\mathcal{W}$ be an elementary subclass of $\mathcal{U}$. Then $\mathcal{W}_{\text{fsi}}$ is elementary if and only if $\mathcal{W}_{\text{fsi}}$ has a finite radius.*

Proof. Because the signature is finite the list of terms needed to witness translations of complexity at most k is finite. So $\{x, y\} \rightsquigarrow_k \{z, w\}$ can be regarded as an elementary formula. By the preceding lemma, an algebra $\mathbf{A} \in \mathcal{U}$ is finitely subdirectly irreducible if and only if it satisfies the (infinite) disjunction of the formulas $x \approx y$, $z \approx w$, and $\{x, y\} \rightsquigarrow_k \{z, w\}$. This means that if $\mathcal{W}_{\text{fsi}}$ is elementary, then it must satisfy the disjunction of finitely many of those formulas. In fact, as k grows larger the formulas $\{x, y\} \rightsquigarrow_k \{z, w\}$ grow weaker. So there is a value of k so that if $\mathcal{W}_{\text{fsi}}$ is elementary, then $\mathcal{W}_{\text{fsi}}$ must satisfy $x \approx y \vee z \approx x \vee \{x, y\} \rightsquigarrow_k \{z.w\}$. That k is clearly the radius of $\mathcal{W}_{\text{fsi}}$.

Conversely, if the radius of $\mathcal{W}_{\text{fsi}}$ is k, then $\mathcal{W}_{\text{fsi}}$ is the class of all algebras in the elementary class $\mathcal{W}$ that satisfy $x \approx y \vee z \approx x \vee \{x, y\} \rightsquigarrow_k \{z.w\}$. Thus $\mathcal{W}_{\text{fsi}}$ is elementary. ■

LEMMA 8.70. *Let $\mathcal{V}$ be a congruence distributive variety. If $\mathcal{V}_{\text{fsi}}$ has finite radius k, then $\mathcal{V}$ has radius no more than $k + 2$.*

Proof. Let $\mathbf{A} \in \mathcal{V}$ and suppose that $a_0, b_0, a_1, b_1 \in A$ so that

$$\mathrm{Cg}^{\mathbf{A}}(a_0, b_0) \cap \mathrm{Cg}^{\mathbf{A}}(a_1, b_1) \neq 0_A.$$

Pick $(c, d) \in \mathrm{Cg}^{\mathbf{A}}(a_0, b_0) \cap \mathrm{Cg}^{\mathbf{A}}(a_1, b_1)$ with $c \neq d$.

By Birkhoff's Subdirect Representation Theorem there is $h : \mathbf{A} \twoheadrightarrow \mathbf{B}$ so that $h(c) \neq h(d)$ and $\mathbf{B}$ is subdirectly irreducible. Since $(c, d) \in \mathrm{Cg}^{\mathbf{A}}(a_0, b_0) \cap \mathrm{Cg}^{\mathbf{A}}(a_1, b_1)$, we see that $h(a_0) \neq h(b_0)$ and $h(a_1) \neq h(b_1)$. Now $\mathbf{B} \in \mathcal{V}_{\text{fsi}}$ since it is subdirectly irreducible. Notice that

$$(h(c), h(d)) \in \mathrm{Cg}^{\mathbf{B}}(h(a_0), h(b_0)) \cap \mathrm{Cg}^{\mathbf{B}}(h(a_1), h(b_1)).$$

So there are $p, q \in B$ with $p \neq q$ so that

$$\{h(a_0), h(b_0)\} \rightsquigarrow_k \{p, q\} \text{ and } \{h(a_1), h(b_1)\} \rightsquigarrow_k \{p, q\}.$$

Since h maps $\mathbf{A}$ onto $\mathbf{B}$, we can pull each of these back to $\mathbf{A}$ by picking pre-images of the coefficients occurring in the translations. In this way we obtain

$$\{a_0, b_0\} \rightsquigarrow_k \{p_0, q_0\} \text{ and } \{a_1, b_1\} \rightsquigarrow_k \{p_1, q_1\},$$

where we know $h(p_0) = h(p_1) = p$ and $h(q_0) = h(q_1) = q$. Now choose j so that the element $p^* = d_j^{\mathbf{A}}(p_0, p_1, q_0)$ and $q^* = d_j^{\mathbf{A}}(p_0, q_1, q_0)$ are distinct. This is possible since in $\mathbf{B}$ we cannot have $d_j^{\mathbf{B}}(p,p,q) = d_j^{\mathbf{B}}(p,q,q)$ for all j, because that would give

$$p = d_0^{\mathbf{B}}(p,p,q) = d_1^{\mathbf{B}}(p,q,q) = d_1^{\mathbf{B}}(p,p,q) = \cdots = d_n^{\mathbf{B}}(p,p,q) = q,$$

which is impossible. So we find $\{a_1,b_1\} \looparrowright_{k+1} \{p^*,q^*\}$.

Taking $\lambda_0(x) = d_j^{\mathbf{A}}(p_0,p_1,x)$, and $\lambda_1(x) = d_j^{\mathbf{A}}(p_0,q_1,x)$, gives us

$$\{p_0,q_0\} \underset{\lambda_0}{\longrightarrow} \{p^*,p_0\} \text{ and } \{p_0,q_0\} \underset{\lambda_1}{\longrightarrow} \{p_0,q^*\}.$$

Consider the sequence $p^*, p_0, q^* \in A$. By the Collapsed $\diamond$-Condition, there are distinct elements e and f and basic translations λ_2 and μ so that either

$$\{p^*,p_0\} \underset{\lambda_2}{\longrightarrow} \{e,f\} \text{ and } \{p^*,q^*\} \underset{\mu}{\longrightarrow} \{e,f\}$$

or

$$\{p_0,q^*\} \underset{\lambda_2}{\longrightarrow} \{e,f\} \text{ and } \{p^*,q^*\} \underset{\mu}{\longrightarrow} \{e,f\}.$$

In either case, we find

$$\{a_0,b_0\} \looparrowright_{k+2} \{e,f\} \text{ and } \{a_1,b_1\} \looparrowright_{k+2} \{e,f\}.$$

Therefore, $\mathcal{V}$ has radius no more than $k+2$, as desired. ■

Proof of Theorem 8.66. We will invoke Theorem 8.54. Since $\mathcal{V}$ is congruence distributive it must realize one of the finite sets Δ_n of equations described in Theorem 6.1 on page 2. Let $\mathcal{U}$ be the variety based on Δ_n. By Lemma 8.69, the radius of $\mathcal{V}_{\text{fsi}}$ is no more than k for some natural number k. So by Lemma 8.70, the radius of $\mathcal{V}$ is no more than $k+2$. Let $\mathcal{Z}$ be the class of all algebras in $\mathcal{U}$ with radius no more than $k+2$. Since $\mathcal{U}$ is finitely axiomatizable and since having radius no more than $k+2$ is expressible by an elementary sentence (because the signature is finite), we see that $\mathcal{Z}$ is finitely axiomatizable. Evidently, the radius of $\mathcal{Z}$ is no more than $k+2$. So by Lemma 8.69, $\mathcal{Z}_{\text{fsi}}$ is finitely axiomatizable. Since $\mathcal{V} \cap \mathcal{Z}_{\text{fsi}} = \mathcal{V}_{\text{fsi}}$ is finitely axiomatizable by hypothesis, we can put $\mathcal{W} = \mathcal{Z}_{\text{fsi}}$ and apply Theorem 8.54 to conclude that $\mathcal{V}$ is finitely based. ■

Some varieties of lattices that are not finitely based

Jónsson's Lemma, the Fundamental Theorem of Ultraproducts, and the Finite Axiomatizability Theorem can be used jointly to show that certain varieties are not finitely based. The general strategy is

(A) use Jónsson's Lemma to obtain a complete description of the subdirectly irreducible algebras in the variety $\mathcal{V}$, and then

(B) find infinitely many subdirectly irreducible algebras in $\mathcal{V}^c$ whose ultraproduct belongs to $\mathcal{V}$.

Of course, (B) and the Fundamental Theorem of Ultraproducts entail that $\mathcal{V}^c$ is not an elementary class. Then, as a consequence of the Finite Axiomatizability Theorem, we have that $\mathcal{V}$ is not finitely based.

EXAMPLE 8.71. (Ralph McKenzie, 1971; Anvar M. Nurakunov, 2017) Let $\mathbf{L}_f$ denote the lattice displayed in Figure 8.5. Let $\mathcal{V}$ be the variety generated by $\mathbf{L}_f$. Then $\mathcal{V}$ is not finitely based. Moreover, $\mathcal{V}_{\text{fsi}}$ is axiomatizable and not finitely axiomatizable. (We will complete a proof of these assertions on page 368.)

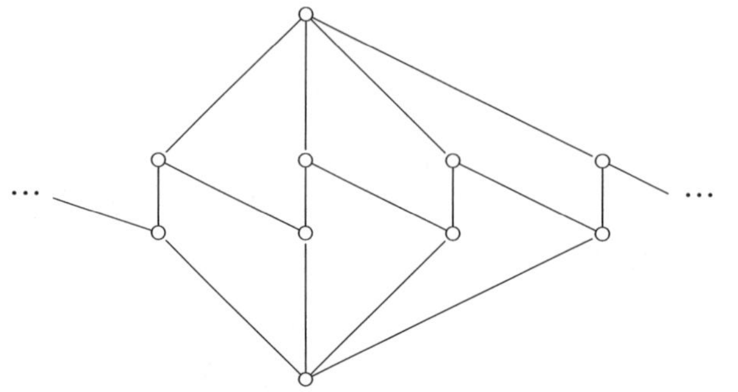

FIGURE 8.5. The lattice $\mathbf{L}_f$

Proof. Now $\mathbf{L}_f$ is a lattice of length 3 consisting of a largest element $\top$, a smallest element $\bot$, a set of atoms, and a set of coatoms. In $\mathbf{L}_f$ every atom is covered by exactly two coatoms and every coatom covers exactly two atoms. These features of $\mathbf{L}_f$ can all be expressed with elementary sentences. Certain finite lattices have these properties as well. The lattice $\mathbf{L}_{C_4}$ is displayed in Figure 8.6. This lattice has four atoms and four coatoms. For $n \geq 4$, the lattice $\mathbf{L}_{C_n}$ resembles $\mathbf{L}_{C_4}$, except it has n atoms and n coatoms. When $n = 2$ the corresponding ordered set fails to be a lattice. For $n = 3$ it is the 8-element Boolean lattice and so is distributive. For each $n \geq 4$, it is easy to invent an elementary sentence that is true in $\mathbf{L}_f$ but fails in $\mathbf{L}_{c_n}$. The ordered set that remains after the largest and smallest elements of $\mathbf{L}_f$ are removed is sometimes called a *two-way infinite fence.* When the largest and smallest elements of $\mathbf{L}_{C_n}$ are removed, the ordered set remaining is called an *n-crown.*

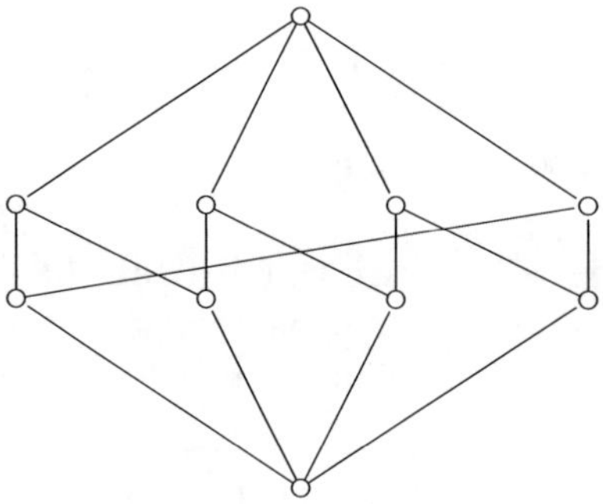

FIGURE 8.6. The lattice $\mathbf{L}_{C_4}$

Jónsson's Lemma tells us that every finitely subdirectly irreducible lattice in $\mathcal{V}$ belongs to $\boldsymbol{HSP_u}\mathbf{L}_f$. Our ambition is to establish for $n \geq 4$

(A) $\mathbf{L}_{C_n} \notin \boldsymbol{HSP_u}\mathbf{L}_f$,
(B) $\mathbf{L}_{C_n}$ is finitely subdirectly irreducible and
(C) a nonprincipal ultraproduct of the $\mathbf{L}_{C_n}$'s belongs to $\mathcal{V}_{\text{fsi}}$.

With these three properties in hand, the Finite Axiomatizability Theorem (Theorem 8.52 on page 348) tells us that each of $\mathcal{V}$ and $\mathcal{V}_{\text{fsi}}$ is not finitely axiomatizable.

To establish (A), suppose, to the contrary, that $\mathbf{L}_{C_n} \in \boldsymbol{HSP_u}\mathbf{L}_f$. Because $\mathbf{L}_{C_n}$ is finite, there must be a finitely generated sublattice $\mathbf{F}$ of an ultrapower of $\mathbf{L}_f$ and a congruence θ on $\mathbf{F}$ so that $\mathbf{L}_{C_n} \cong \mathbf{F}/\theta$. Now by the Fundamental Theorem for Ultraproducts, every elementary sentence true in $\mathbf{L}_f$ must be true in every ultrapower of $\mathbf{L}_f$. Thus an ultrapower of $\mathbf{L}_f$ resembles $\mathbf{L}_f$ itself, except instead of one infinite fence across the middle the ultrapower is provided with a (usually large infinite) pairwise disjoint collection of infinite fences. Since $\mathbf{F}$ is a finitely generated sublattice of such an ultrapower, it is not hard to see that it must be finite, and, moreover, it is isomorphic to a sublattice of $\mathbf{L}_f$ itself. Roughly speaking, since F must have at least $2n+2$ elements (with $n \geq 3$), $\mathbf{F}$ looks like what would result by erasing connected stretches of the fence in $\mathbf{L}_f$ leaving only finitely many connected pieces. Apart from lattices of cardinality 4 or less, the homomorphic images of such lattices are again lattices of the same kind. Hence $\mathbf{L}_{C_n} \notin \boldsymbol{HSP_u}\mathbf{L}_f$.

A condition stronger than (B) holds:

$$\mathbf{L}_{C_n} \text{ is simple, for each } n > 3.$$

To see this, let $a_0, a_1, \ldots, a_{n-1}$ be a list of the atoms, let $c_0, c_1, \ldots, c_{n-1}$, be a listing of the coatoms, and suppose matters are arranged so that $a_i \vee a_{i+1} = c_i$ holds for all i modulo n. Then $c_i \wedge c_{i+1} = a_{i+1}$ will also hold for all i modulo n. To demonstrate that $\mathbf{L}_{C_n}$ is simple, it is only necessary to show that $\mathrm{Cg}^{\mathbf{L}_{C_n}}(a_1, c_1)$ and $\mathrm{Cg}^{\mathbf{L}_{C_n}}(c_0, 1)$ both collapse 0 to 1. Indeed, since $c_0 = c_0 \vee a_1$ and $1 \vee c_1 = 1$, it follows that $\mathrm{Cg}^{\mathbf{L}_{C_n}}(c_0, 1) \subseteq \mathrm{Cg}^{\mathbf{L}_{C_n}}(a_1, c_1)$. So we only need to prove that $\langle 0, 1\rangle \in \mathrm{Cg}^{\mathbf{L}_{C_n}}(c_0, 1)$. The following reasoning establishes this. Let $\theta = \mathrm{Cg}^{\mathbf{L}_{C_n}}(c_0, 1)$. Observe that $c_i \wedge c_j = 0$ unless i and j differ by at most 1, modulo n. So assuming $n \geq 5$, $c_0 \wedge c_2 = c_0 \wedge c_3 = 0$. Hence,

$$0 = c_0 \wedge c_2 \;\theta\; 1 \wedge c_2 = c_2.$$

Similarly, $0 \;\theta\; c_3$, and so $0 = 0 \vee 0 \;\theta\; c_2 \vee c_3 = 1$. The proof that $\mathbf{L}_{C_4}$ is simple is only slightly harder and is left to the reader. And, of course, we can leave $\mathbf{L}_{C_4}$ out of the ultraproduct without changing it.

Finally, to establish (C), let $I = \{3, 4, 5, \ldots\}$ and let U be any nonprincipal ultrafilter on I. Let $\mathbf{L} = \prod_I \mathbf{L}_{C_n}/U$. For each $n \in I$, let σ_n be a sentence which asserts that every atom is in the middle of a fence of length $2n+1$. For instance, σ_3 would assert, "For every atom a_3, there are six additional atoms a_0, a_1, a_2, a_4, a_5, and a_6 and seven distinct coatoms $c_0, c_1, c_2, c_3, c_4, c_5$, and c_6 so that $a_i \vee a_{i+1} = c_i$ for all $i < 7$ and $c_i \wedge c_{i+1} = a_{i+1}$ for all $i < 7$." Since each of these sentences is true in all but finitely many of the lattices $\mathbf{L}_{C_n}$, the Fundamental Theorem of Ultraproducts tells us that each of these sentences is true in $\mathbf{L}$. Consequently, $\mathbf{L}$ might be described as a certain (infinite) antichain of disjoint two-way infinite fences with a top and bottom element adjoined. In other words, it looks like an ultrapower of $\mathbf{L}_f$. We leave it as an exercise to show, in fact, that $\mathbf{L} = \prod_I \mathbf{L}_{C_n}/U$ is embeddable into $\mathbf{L}_f^I/U$. It remains to show that $\mathbf{L} \in \mathcal{V}_{\mathrm{fsi}}$.

CLAIM 1.

Suppose that $\mathbf{J} \in \boldsymbol{SP_u}\mathbf{L}_f$ has at least six atoms and at least six coatoms. Then

(i) *If θ is a congruence of $\mathbf{J}$ and $\theta \neq 1_\mathbf{J}$ and $a \in J \setminus \{\bot, \top\}$, then neither $\langle \bot, a\rangle$ nor $\langle a, \top\rangle$ belongs to θ.*

(ii) *If θ is a congruence of $\mathbf{J}$ and $\theta \neq 1_\mathbf{J}$ and $a, b \in J$ so that $\bot < a < b < \top$ and $\langle a, b\rangle \in \theta$, then both a and b are doubly irreducible.*

(iii) **J** *is finitely subdirectly irreducible if and only if there are no doubly irreducible elements* $a, b, c, d \in J$ *such that* $a < b$ *and* $c < d$.

Proof of Claim 1. We saw above that **J** must consist of a number (perhaps infinite) of copies of $\mathbf{L}_f$ sharing top and bottom elements, but from which some stretches of the fences have been deleted. Now consider any congruence θ of **J**. Since **J** has at least six atoms and at least six coatoms, for each $a \in J$ that is neither the top nor the bottom element, we can pick two more elements b and c so that a, b, and c are distinct and so that these three elements together with the top and bottom elements constitute a copy of $\mathbf{M}_3$. It follows that if $\langle \bot, a\rangle \in \theta$, then $\langle \bot, \top\rangle \in \theta$. So in this case, $\theta = 1_{\mathbf{J}}$. The same applies if $\langle a, \top\rangle \in \theta$.

This means that if $\theta \neq 1_{\mathbf{J}}$, then $\langle \bot, a\rangle \notin \theta$ and also $\langle a, \top\rangle \notin \theta$, for every $a \in J \setminus \{\bot, \top\}$.

Now suppose $\bot < a < b < \top$ and $\langle a, b\rangle \in \theta$. There is no element c so that $a = b \wedge c$ since this would force $\theta = 1_{\mathbf{J}}$. So a is meet irreducible and also join irreducible (since it is an atom). The element b is likewise doubly irreducible.

Now suppose there are four distinct doubly irreducible elements $a, b, c, d \in J$ so that $\bot < a < b < \top$ and $\bot < c < d < \top$. Then the sets $\theta_0 = \{\langle a, b\rangle, \langle b, a\rangle\} \cup \{\langle x, x\rangle \mid x \in J\}$ and $\theta_1 = \{\langle c, d\rangle, \langle d, c\rangle\} \cup \{\langle x, x\rangle \mid x \in J\}$ are congruences of **J**. We see $\theta_0 \cap \theta_1 = 0_{\mathbf{J}}$. So **J** is finitely subdirectly reducible.

For the converse, suppose **J** is finitely subdirectly reducible. So there are two nontrivial congruences θ_0 and θ_1 so that $\theta_0 \cap \theta_1 = 0_{\mathbf{J}}$. These congruences must be proper so there must be distinct doubly irreducible elements $a, b, c, d \in J$ with $\bot < a < b < \top$ and $\bot < c < d < \top$ and $\langle a, b\rangle \in \theta_0$ as well as $\langle c, d\rangle \in \theta_1$. ■

Observe that part (iii) of Claim 1 leads to the conclusion that $\mathcal{V}_{\text{fsi}}$ is axiomatizable.

So let $\mathbf{J} \in \boldsymbol{SP}_u\mathbf{L}_f$ such that **J** has at least six atoms and at least six coatoms. Let θ be any proper congruence of **J**. Then $\mathbf{J}/\theta$ just identifies certain comparable pairs of doubly irreducible elements. This is the same as deleting, say, the lower element of each such pair. In other words, $\boldsymbol{HSP_u}\mathbf{L}_f = \boldsymbol{SP}_u\mathbf{L}_f$. We already know that $\mathbf{L} \in \boldsymbol{SP}_u\mathbf{L}_f$. But also $\mathbf{L} = \prod_I \mathbf{L}_{C_n}/U$. Since $\mathbf{L}_{C_n}$ lacks doubly irreducible elements for all $n > 3$, the same must apply to **L**. So by part (iii) of Claim 1, the lattice $\mathbf{L} \in \mathcal{V}_{\text{fsi}}$. This establishes (C) and so the demonstration of Example 8.71 is complete. ■

It is worth pointing out that $\mathbf{L}_f$ fails to be inherently nonfinitely based, as is shown in (Ralph Freese, George F. McNulty, and J. B. Nation, 2002)

EXAMPLE 8.72. (Kirby A. Baker, 1971) Let $\mathcal{K}$ be the class of all lattices of length at most 4, and let $\mathcal{V}$ be the variety generated by $\mathcal{K}$. Then $\mathcal{V}$ is not finitely based.

The **length** of a lattice is the least upper bound on length of any chain (a set of pairwise comparable elements). The length of a chain of $n + 1$ elements is n, so the length counts the number of links in the chain rather than the number of elements. The length of a lattice is sometimes referred to as the height of the lattice.

Proof. It is clear that $\mathcal{K}$ is closed with respect to $\boldsymbol{H}$, $\boldsymbol{S}$, and $\boldsymbol{P_u}$. So by Jónsson's Lemma, every subdirectly irreducible algebra in $\mathcal{V}$ belongs to $\mathcal{K}$.

For each natural number n, let $\mathbf{L}_n$ be the lattice illustrated in Figure 8.7, where the two nodes labeled with p should be identified and the same with those labeled with q. The lattice $\mathbf{L}_n$ has $4n + 16$ elements. Arranged circularly, this rather Escheresque diagram resembles the lattice $\mathbf{L}_{C_n}$. Once again, we will have established that $\mathcal{V}$ is not finitely based if we prove

(A) $\mathbf{L}_n \notin \mathcal{K}$,
(B) $\mathbf{L}_n$ is subdirectly irreducible, so $\mathbf{L}_n \notin \mathcal{V}$, and
(C) a nonprincipal ultraproduct of the $\mathbf{L}_n$'s belongs to $\mathcal{V}$.

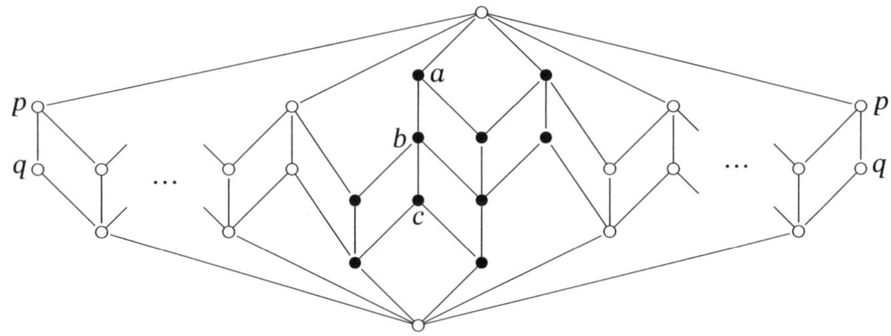

FIGURE 8.7. The lattice $\mathbf{L}_n$; it is circular

Now (A) is obvious, since the length of $\mathbf{L}_n$ is 5 for every natural number n.

To prove (B) notice $\mathbf{L}_n$ is subdirectly irreducible and the congruence μ generated by $\langle a, b\rangle$ is the monolith. Moreover, for $n \geq 0$, $\mathbf{L}_n/\mu \cong \mathbf{L}_{C_{n+3}}$. Since the latter is simple for $n \geq 1$, the congruence lattice of $\mathbf{L}_n$ is a three element chain for $n \geq 1$, and **Con** $\mathbf{L}_0$ is the lattice obtained by adding a new least element to the 8 element Boolean lattice. Providing the details to support this is Exercise 9 below.

To prove (C), let U be any nonprincipal ultrafilter on the set ω of natural numbers. Add three constants a, b, and c to the language of lattices. Let $\mathbf{L}'_n$ be the expansion of $\mathbf{L}_n$ which assigns these new constant symbols as names to the three elements labeled in Figure 8.7. Let $\mathbf{L}' = \prod_\omega \mathbf{L}'_n/U$ and let $\mathbf{L}$ be the reduct of $\mathbf{L}'$ to the language of lattices. Now in the middle of each $\mathbf{L}_n$ there is a unique configuration comprising the 10 black elements which, with the top and bottom elements of the lattice adjoined, includes all the chains of length 5 which occur in $\mathbf{L}_n$. The fact that there is exactly one configuration like this can be expressed by an elementary sentence. By the Fundamental Theorem for Ultraproducts, this sentence is also true for $\mathbf{L}$. It follows that $\mathbf{L}$ can be roughly described as comprising a top element, a bottom element, and an (infinite) antichain of components each of which is a "doubled two-way infinite fence", with the exception of exactly one component which has our 10-configuration inserted into the middle of a "two-way infinite doubled fence". Let φ be the principal congruence relation on $\mathbf{L}$ which collapses the elements denoted in $\mathbf{L}'$ by a and b. Let ψ be the principal congruence relation on $\mathbf{L}$ which collapses the elements denoted in $\mathbf{L}'$ by b and c. It is easy to show that $\varphi \cap \psi = 0_L$. It follows that $\mathbf{L}$ is embeddable into $\mathbf{L}/\varphi \times \mathbf{L}/\psi$. As it is also clear that both $\mathbf{L}/\varphi$ and $\mathbf{L}/\psi$ are lattices of length 4, we conclude that $\mathbf{L} \in \mathcal{V}$. ■

EXAMPLE 8.73. (Kirby A. Baker, 1971; Ralph Freese, 1977) Let $\mathcal{K}$ be the class of all modular lattices whose width does not exceed 5. The variety $\mathcal{M}_5^\infty$ generated by $\mathcal{K}$ is not finitely based.

The **width** of a lattice is the least upper bound of the sizes of all antichains (a set of pairwise incomparable elements).

Proof. Once again, our ambition is to devise lattices $\mathbf{W}_n$, this time modular, and then to establish

(A) $\mathbf{W}_n \notin \mathcal{M}_5^\infty$,
(B) $\mathbf{W}_n$ is finitely subdirectly irreducible and
(C) a nonprincipal ultraproduct of the $\mathbf{W}_n$'s belongs to $\mathcal{M}_5^\infty$.

With these three properties in hand, the Finite Axiomatizability Theorem (Theorem 8.52 on page 348) tells us that $\mathcal{V} = \mathcal{M}_5^\infty$ and $\mathcal{V}_{\text{fsi}}$ are not finitely based.

The lattices $\mathbf{W}_n$, for $2 \leq n$, have height $2n$. They are displayed in Figure 8.8.

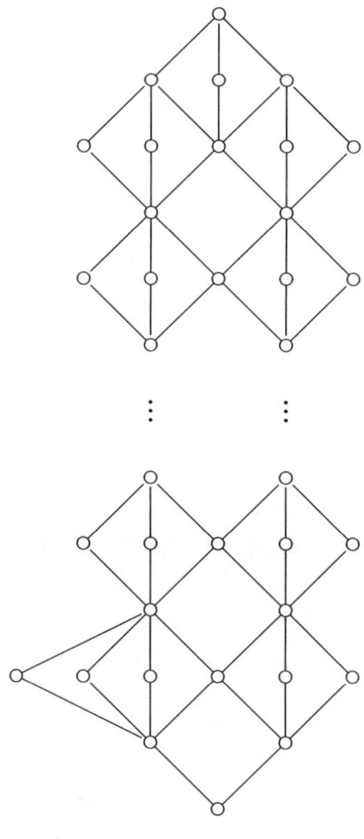

FIGURE 8.8. The lattice $\mathbf{W}_n$

Observe that $\mathbf{W}_n$ is a modular lattice of width 6 and that it is simple (and hence finitely subdirectly irreducible). Dedekind's Transposition Principle (Theorem 2.27 on page 57 of Volume I) is helpful in proving the simplicity. So at least stipulation (B) is secure.

To obtain (A), suppose for the sake of contradiction that it fails. That is, suppose $\mathbf{W}_n \in \mathcal{M}_5^\infty$. By Jónsson's Lemma we must have $\mathbf{W}_n \in \boldsymbol{HSP}_u\mathcal{K}$. The sentence that says "the width is at most 5" is easily formulated as an elementary sentence of a particularly nice form. Even for 5 this sentence is long, so to give the idea, we formulate the sentence that works for 2:

$$\forall x_0, x_1, x_2[(x_0 \approx x_0 \wedge x_1) \mathbin{\mathbb{V}} (x_1 \approx (x_0 \wedge x_1)) \mathbin{\mathbb{V}} (x_0 \approx x_0 \wedge x_2) \mathbin{\mathbb{V}} (x_2 \approx (x_0 \wedge x_2)) \\ \mathbin{\mathbb{V}} \cdots \mathbin{\mathbb{V}} (x_2 \approx x_2 \wedge x_3) \mathbin{\mathbb{V}} (x_3 \approx (x_2 \wedge x_3))].$$

Loosely, the sentence we want says "given any 6 elements the first is less than or equal to the second or the first is less than or equal to the third or $\cdots$ or the sixth is less than or equal to the fifth." The nice thing about this sentence is that it is universal (that is, it begins with string of universal quantifiers followed by a formula without quantifiers) and that the quantifier-free part is a disjunction of equations. Now this elementary sentence is true in $\mathcal{K}$, so it is also true in $\boldsymbol{P_u}\mathcal{K}$. Since it is a universal sentence, it must also be true in $\boldsymbol{SP_u}\mathcal{K}$. Finally, since the quantifier-free part is a disjunction of equations, it must also be true in $\boldsymbol{HSP_u}\mathcal{K}$. But this sentence fails in $\mathbf{W}_n$. This is a contradiction. So (A) must hold.

Finally, to secure (C) we take $\mathbf{W}$ to be any ultraproduct of the $\mathbf{W}_n$'s over the index set $I = \{2, 3, 4, \dots\}$ using a nonprincipal ultrafilter on I. Then $\mathbf{W}$ must be an infinite modular lattice of width at most 6 that has its top 11 elements configured as in $\mathbf{W}_n$ and bottom 11 elements configured as in $\mathbf{W}_n$ (since these are elementary properties holding in each $\mathbf{W}_n$). Moreover, the top of $\mathbf{W}$ must look like the top of $\mathbf{W}_3$ and the bottom of $\mathbf{W}$ must look like the bottom of $\mathbf{W}_3$. These two configurations can be described by elementary sentences. Finally, in $\mathbf{W}$ there is only one antichain of size 6 (the one occurring in the bottom configuration) and this can also be expressed with an elementary sentence. This means that we have a rough description of $\mathbf{W}$: it is the ordered sum consisting of one denumerable sublattice, which we call the bottom piece and denote by $\mathbf{B}$, and above which is some linearly ordered arrangement of denumerable sublattices (all isomorphic to each other), above which is a denumerable sublattice, we call the top piece. These sublattices are displayed in Figure 8.9

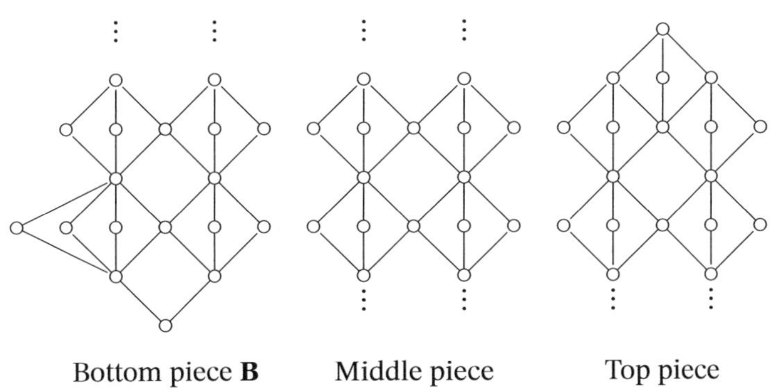

FIGURE 8.9. Certain denumerable sublattices of $\mathbf{W}$

Actually, $\mathbf{W}$ is in the variety generated by the three lattices displayed in Figure 8.9, see Exercise 10. The middle piece and the top piece belong to $\mathcal{M}_5^\infty$ (and, in fact, to $\mathcal{K}$). So we only need to prove that the bottom piece belongs to $\mathcal{M}_5^\infty$. To see this let θ_0 be the congruence on $\mathbf{B}$ that collapses the left atom to the least element and let θ_1 be the congruence collapsing the other atom to the least element. (Notice, that these are proper congruences.) It is straightforward to see that $\theta_0 \cap \theta_1 = 0_B$. This means that the bottom piece is a subdirect product of the two quotient lattices. These lattices are displayed in Figure 8.10. Evidently, they both belong the $\mathcal{K}$, so the bottom piece, and therefore **W**, belongs to $\mathcal{M}_5^\infty$. This establishes property (C) and finishes the demonstration of the example. ■

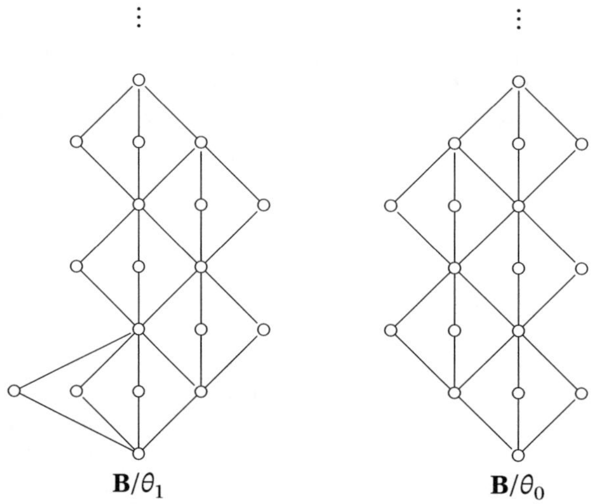

FIGURE 8.10. Subdirect factors of the bottom lattice

This reasoning can be easily adjusted to show that $\mathcal{M}_n^\infty$ is not finitely based, for each $n \geq 5$. The varieties $\mathcal{M}_1^\infty$ and $\mathcal{M}_2^\infty$ are both seen to be the variety of distributive lattices. Bjarni Jónsson (1968) has shown that $\mathcal{M}_3^\infty$ is finitely based and Ralph Freese (1977) has shown that $\mathcal{M}_4^\infty$ is finitely based.

Exercises 8.74

The first seven exercises below are drawn from (Bjarni Jónsson, 1967).

1. Let $\mathcal{V}$ be a congruence distributive variety and suppose that $\mathbf{A}, \mathbf{B} \in \mathcal{V}_{\text{si}}$. Prove that if **A** and **B** are finite and not isomorphic and $|A| \leq |B|$, then there is an equation that holds in **A** but fails in **B**.

2. Let $\mathcal{V}$, $\mathcal{U}$, and $\mathcal{W}$ be varieties so that $\mathcal{W}$ is the least variety that contains both $\mathcal{V}$ and $\mathcal{U}$. Suppose $\mathcal{W}$ is congruence distributive. Prove that every member of $\mathcal{W}$ is a subdirect product of a member of $\mathcal{V}$ and a member of $\mathcal{U}$. Conclude that every subdirectly irreducible member of $\mathcal{W}$ belongs to either $\mathcal{V}$ or to $\mathcal{U}$.

3. Prove that every lattice can be embedded into a subdirectly irreducible lattice.

4. Let $\mathcal{K}$ be the class of all finite lattices. Prove that $\boldsymbol{HSP_u}\mathcal{K}$ is the class of all lattices.

5. Prove that the lattice of equational theories of lattices (§7.6) is distributive.

6. Prove that in the lattice of equational theories of lattices, every equational theory (except the equational theory of all lattices) covers another equational theory of lattices.

7. Prove that the lattice of equational theories of lattices has no atoms,

***8.** (Kirby A. Baker, 1977) Let $\mathcal{K}$ be the class of all lattices of width no more than 4. Show that the variety generated by $\mathcal{K}$ is not finitely based. (As we have pointed out just above, $\mathcal{M}_4^\infty$ *is* finitely based.)

9. Let the lattices $\mathbf{L}_n$ be as described in Example 8.72. Prove that $\mathbf{L}_n$ is subdirectly irreducible and the congruence μ generated by $\langle a, b\rangle$ is the monolith. Moreover, for $n \geq 0$, $\mathbf{L}_n/\mu \cong \mathbf{L}_{C_{n+3}}$. Since the latter is simple for $n \geq 1$, the congruence lattice of $\mathbf{L}_n$ is a three element chain for $n \geq 1$.

10. Let $\mathbf{L}_0$ and $\mathbf{L}_1$ be lattices, which we assume are disjoint. We say that the lattice $\mathbf{L}$ is the *ordered sum* of $\mathbf{L}_0$ and $\mathbf{L}_1$ provided $L = L_0 \cup L_1$ and we define the lattice ordering (and hence the lattice) by letting L_0 and L_1 have their given orderings and insisting that $a < b$ for all $a \in L_0$ and $b \in L_1$. Prove that if at least one of $\mathbf{L}_0$ and $\mathbf{L}_1$ is nontrivial, then their ordered sum is in the variety generated by $\mathbf{L}_0$ and $\mathbf{L}_1$.

8.5. Definable Congruences and Baker's Finite Basis Theorem

In the last section we saw how to deduce Baker's Finite Basis Theorem from Willard's Finite Basis Theorem and Jónsson's Lemma and we even gave a second proof of Baker's Finite Basis Theorem. Because Baker's Finite Basis Theorem was a turning point in the investigation of finite basis properties for varieties, we provide here a third proof.

An elementary formula $\varphi(x, y, z, w)$ is called a **positive existential formula** provided it is an existential formula whose quantifier free part uses only the connectives $\vee\!\!\!\vee$ and $\wedge\!\!\!\wedge$. In particular, neither $\neg$ nor $\Rightarrow$ make an appearance. Positive existential formulas with the following property $(\star)$ are called **principal congruence formulas**:

$$(\star) \qquad \models \forall x, z, w[\varphi(x, x, z, w) \Rightarrow z \approx w].$$

One can show (by induction on the complexity of $\psi(x_0, \dots, x_{m-1})$) that each positive formula $\psi(x_0, \dots, x_{m-1})$ is preserved under the formation of quotient algebras—see Lemma 8.134 on page 435. That is, for all algebras $\mathbf{B}$ and all $\theta \in \operatorname{Con}\mathbf{B}$ and all $b_0, \dots, b_{m-1} \in B$:

$$\text{If } \mathbf{B} \models \psi(b_0, \dots, b_{m-1}), \text{ then } \mathbf{B}/\theta \models \psi(b_0/\theta, \dots, b_{m-1}/\theta).$$

Then taking $\theta = \operatorname{Cg}^{\mathbf{A}}(a, b)$, we see, with the help of $(\star)$, that if φ is a principal congruence formula and $\mathbf{A} \models \varphi(a, b, p, q)$, then $(p, q) \in \operatorname{Cg}^{\mathbf{A}}(a, b)$.

This allows us to recast Mal'tsev's Principal Congruence Generation Theorem (Theorem 6.17 on page 24) a second time.

MAL'TSEV'S PRINCIPAL CONGRUENCE GENERATION THEOREM
Let **A** be an algebra and a, b, p, and q be elements of A. Then $\langle p, q\rangle \in \mathrm{Cg}^{\mathbf{A}}(a, b)$ if and only if there is a principal congruence formula $\varphi(x, y, z, w)$ so that $\mathbf{A} \models \varphi(a, b, p, q)$.

A class $\mathcal{K}$ of algebras, all of the same signature, is said to have **definable principal congruences** provided there is an elementary formula $\varphi(x, y, z, w)$ such that for every $\mathbf{A} \in \mathcal{K}$ and all $a, b, c, d \in A$ we have $\langle c, d\rangle \in \mathrm{Cg}^{\mathbf{A}}(a, b)$ if and only if $\mathbf{A} \models \varphi(a, b, c, d)$. If a variety $\mathcal{V}$ has definable principal congruences, then the defining formula φ can be taken to be a principal congruence formula. The definability of principal congruences is a demanding restriction on a variety—few varieties arising in mathematical practice have definable principal congruences.

A class $\mathcal{K}$ of algebras, all of the same signature, is said to have **definable principal subcongruences** provided there are two principal congruence formulas $\varphi(x, y, z, w)$ and $\psi(x, y, z, w)$ so that for all $\mathbf{A} \in \mathcal{K}$ and all $a, b \in A$ with $a \neq b$, there are $c, d \in A$ with $c \neq d$ so that $\mathbf{A} \models \varphi(a, b, c, d)$ and $\psi^{\mathbf{A}}(c, d, z, w)$ defines $\mathrm{Cg}^{\mathbf{A}}(c, d)$, where $\psi(c, d, z, w)$ is understood as a formula in the diagram language of **A**.

THEOREM 8.75 (Kirby A. Baker and Ju Wang 2002). *Let $\mathcal{V}$ be a locally finite variety of finite signature with definable principal subcongruences. Then $\mathcal{V}$ is finitely based if and only if $\mathcal{V}_{\mathrm{si}}$ is finitely axiomatizable.*

Proof. Let φ and ψ be formulas witnessing that $\mathcal{V}$ has definable principal subcongruences. Let $\pi_{\psi}(z, w)$ be the formula that asserts, in each algebra **A** and for any $a, b \in A$ that the set

$$\{\langle c, d\rangle \mid c, d \in A \text{ and } \mathbf{A} \models \psi(a, b, c, d)\}$$

is a congruence. So $\pi_{\psi}(z, w)$ must assert that the set displayed above is an equivalence relation that respects each basic operation. Such a formula exists since the signature is finite. With the help of $\pi_{\psi}(z, w)$ we can devise a sentence θ that asserts the definability of principal subcongruences witnessed by φ and ψ. Here is one such sentence.

$$\theta \text{ is } \forall x, y\left[\neg x \approx y \implies \exists z, w[\neg z \approx w \wedge \varphi(x, y, z, w) \wedge \pi_{\psi}(z, w)]\right].$$

Now suppose that $\mathcal{V}_{\mathrm{si}}$ is axiomatized by the finite set Δ of sentences. We take γ to be a single sentence axiomatizing $\mathcal{V}_{\mathrm{si}}$.

We see that $\mathcal{V} \models \theta$. Since $\mathcal{V}$ is axiomatized by some set of equations, we conclude from the Compactness Theorem that θ is a logical consequence of some finite set of equations, all true in $\mathcal{V}$. This finite set involves only finitely many variables; say no more than m variables. This means that $\mathcal{V}^{(m)} \models \theta$. So $\mathcal{V}^{(m)}$ has definable principal congruences witnessed by φ and ψ. So in $\mathcal{V}^{(m)}_{\mathrm{si}}$ the sentence γ^* displayed below holds.

$$\exists u, v\left[\neg u \approx v \Rightarrow \forall x, y\left[\neg x \approx y \Rightarrow \exists z, w[\varphi(x, y, z, w) \wedge \psi(z, w, u, v)]\right]\right].$$

Consequently $\mathcal{V} \models \gamma^* \Rightarrow \gamma$. As before, we can find a natural number n so that $\mathcal{V}^{(n)} \models \gamma^* \Rightarrow \gamma$. It is harmless to suppose the $n \geq m$. So $\mathcal{V}$ and $\mathcal{V}^{(n)}$ have the same subdirectly irreducible algebras. So $\mathcal{V} = \mathcal{V}^{(n)}$. But the latter variety, according to Birkhoff, is finitely based.

For the converse, suppose that $\mathcal{V}$ is based on the finite set Σ of equations. Then $\Sigma \cup \{\gamma^*\}$ axiomatizes $\mathcal{V}_{\mathrm{si}}$. ■

Before turning to our second proof of Baker's Finite Basis Theorem, we record here a useful corollary:

COROLLARY 8.76 (Ralph McKenzie 1978). *Any variety of finite signature that has definable principal congruences and a finite residual bound is finitely based.* ■

Actually, the corollary above holds with definable principal subcongruences replacing the hypothesis of definable principal congruences.

THEOREM 8.77 (Kirby A. Baker and Ju Wang 2002). *Let* **A** *be a finite algebra of finite signature and let* $\mathcal{V}$ *be the variety generated by* **A**. *If* $\mathcal{V}$ *is congruence distributive, then* $\mathcal{V}$ *has definable principal subcongruences.*

Proof. According to Jónsson's Lemma, $\mathcal{V}$ has, up to isomorphism only finitely many subdirectly irreducible algebras, all finite. This means the minimum complexities of translations for the algebras in $\mathcal{V}_{\text{si}}$ has an upper bound—see page 23. Since the signature is finite, it follows that a finite list $\pi_0, \ldots, \pi_{m-1}$ of principal congruence formulas suffices to witness all the relations of the form $\langle c, d\rangle \in \mathrm{Cg}^{\mathbf{S}}(a, b)$ where $\mathbf{S} \in \mathcal{V}_{\text{si}}$ and $a, b, c, d \in S$. In fact, the class $\mathcal{V}_{\text{si}}$ of finitely many finite algebras, has definable principal subcongruences in the strong sense that the subcongruence can be taken to be the monolith of the subdirectly irreducible algebra.

Our plan is to lift the definability of principal subcongruences from $\mathcal{V}_{\text{si}}$ to $\mathcal{V}$. It is here that the congruence distributivity of $\mathcal{V}$ plays a crucial role. To begin, let m be the cardinality of the largest algebra in $\mathcal{V}_{\text{si}}$. Let n be the cardinality of the largest algebra in $\mathcal{V}$ with m generators. Let $\langle \mathbf{S}_i \mid i \in I\rangle$ be a list of the elements of $\mathcal{V}_{\text{si}}$.

Now let **B** be a nontrivial algebra in $\mathcal{V}$ and let a and b be any two distinct elements of B. Pick an algebra, as large as possible, $\mathbf{S}_k \in \mathcal{V}_{\text{si}}$, with $k \in I$, and pick a homomorphism $h_k : \mathbf{B} \to \mathbf{S}_k$ so that $h_k(a) \neq h_k(b)$. Indeed, we have a subdirect representation $\langle h_i \mid i \in I\rangle$ of **B** by algebras in $\mathcal{V}_{\text{si}}$. Now $|S_k| \leq n$, so we can select a subset $X \subseteq B$ consisting of one h_k-preimage of each element of S_k, being careful to select the elements a and b. Let **D** be the subalgebra of **B** generated by X. So $|D| \leq m$. Let h'_i be the restriction of h_i to D for each $i \in I$. Then $\ker h'_0$ is a completely meet irreducible congruence of **D**, since h'_0 maps **D** onto $\mathbf{S}_k$. By congruence distributivity, the interval $I[0_D, \ker h'_k]$ is a prime ideal in **Con** D. It follows that the complement of this prime ideal is a filter in **Con** D with least member δ that is join prime. Since every congruence is the join of principal congruences and since this congruence lattice is finite, we conclude that δ is a principal congruence; say $\delta = \mathrm{Cg}^{\mathbf{D}}(c, d)$ with $c \neq d$. Since $\langle a, b\rangle \notin \ker h'$, we see that $\langle c, d\rangle \in \mathrm{Cg}^{\mathbf{D}}(a, b)$. Since the cardinality of D is bounded only in terms of **A**, there is a principal congruence formula $\varphi(x, y, z, w)$ such that for any $\mathbf{B} \in \mathcal{V}$ and any $a, b \in B$ with $a \neq b$ there are $c, d \in B$ with $c \neq d$ so that $\mathbf{B} \models \varphi(a, b, c, d)$.

Now suppose $i \in I$ so that $h_i(c) \neq h_i(d)$. As above, let h'_i be the restriction of h_i to D. Notice that $\ker h'_i$ is not in the filter above δ. This means $\ker h'_i \subseteq \ker h'_k$. Hence, $|h'_i(D)| \geq |h'_0(D)| = |S_k|$. But we also have $|S_i| = |h_i(B)| \geq |h'_i(D)|$. But $|S_k| \geq |S_i|$, by the maximal way $\mathbf{S}_k$ was chosen. So we conclude that h'_i maps D onto S_i.

Now let $\langle r, s\rangle \in \mathrm{Cg}^{\mathbf{B}}(c, d)$. Finally, let **C** be the subalgebra of **B** generated by $X \cup \{r, s\}$. Observe that a finite upper bound of the cardinality of C does not really depend on the particular choices made in assembling $X \cup \{r, s\}$ but only on the number of elements in this set and on the algebra $\mathbf{A}$. Now let h_i^* be the restriction of h_i to C for each $i \in I$. Observe that if h_i^* separates c and d, so $h_i^*(C) = S_i$, since $D \subseteq C$ and $h_i'(D) = S_i$. So $\langle h_i^*(r), h_i^*(s)\rangle \in \mathrm{Cg}^{\mathbf{S}_i}(h_i^*(c), h_i^*(d))$, which entails $\langle r, s\rangle \in \mathrm{Cg}^{\mathbf{C}}(c,d) \vee \ker h_i^*$. On the other hand, if $h_i^*(c) = h_i^*(d)$, then $h_i^*(r) = h_i^*(s)$ since $\langle r, s\rangle \in \mathrm{Cg}^{\mathbf{B}}(c,d)$. In either case, $\langle r, s\rangle \in \mathrm{Cg}^{\mathbf{C}}(c,d) \vee \ker h_i^*$. This means

$$\begin{aligned}\langle r, s\rangle \in \bigcap_{i\in I}\left(\mathrm{Cg}^{\mathbf{C}}(c,d) \vee \ker h_i^*\right) &= \mathrm{Cg}^{\mathbf{C}}(c,d) \vee \bigcap_{i\in I}\ker h_i^* \\ &= \mathrm{Cg}^{\mathbf{C}}(c,d) \vee 0_C = \mathrm{Cg}^{\mathbf{C}}(c,d).\end{aligned}$$

The upshot of this paragraph is that if $\langle r, s\rangle \in \mathrm{Cg}^{\mathbf{B}}(c,d)$, then $\langle r, s\rangle \in \mathrm{Cg}^{\mathbf{C}}(c,d)$ for some subalgebra $\mathbf{C}$ of $\mathbf{B}$ that is bounded in size by a natural number depending only on $\mathbf{A}$. So there is a principal congruence formula $\psi(x,y,z,w)$ such that $\mathbf{B} \models \psi(c,d,r,s)$ whenever $\langle r, s\rangle \in \mathrm{Cg}^{\mathbf{B}}(c,d)$. Hence, $\psi^{\mathbf{B}}(c,d,z,w)$ defines $\mathrm{Cg}^{\mathbf{B}}(c,d)$.

The formulas $\varphi(x,y,z,w)$ and $\psi(x,y,z,w)$ witness that $\mathcal{V}$ has definable principal subcongruences. ■

Baker's Finite Basis Theorem follows from these two theorems and Jónsson's Lemma. It is noteworthy that congruence distributivity enters only in Jónsson's Lemma and Theorem 8.77. In the proof of that theorem it is used three times: once to conclude that a certain ideal in the congruence lattice of an algebra is a prime ideal, second to obtain the complementary filter (see Theorem 2.60 in Volume I), and, finally, in the formulas displayed above. The concept of definable principal subcongruences and the definite atoms property introduced in §6.4 seem to be similar in spirit and Theorem 8.77 plays a role in this finite basis proof that the ◇ Single Sequence Lemma played in §6.4. It is not known whether congruence meet-semidistributive varieties of finite signature that have finite residual bounds also have definable principal subcongruences.

We conclude this section with a theorem that illustrates the utility of the notion of definable (sub)congruences. A variety $\mathcal{V}$ is **directly representable** provided $\mathcal{V}$ is locally finite and there is a finite set $\mathcal{M}$ of finite algebras with $\mathcal{M} \subseteq \mathcal{V}$ such that every finite algebra in $\mathcal{V}$ is isomorphic to a direct product of algebras belonging to $\mathcal{M}$. Directly representable varieties are investigated further in §9.3 and §11.9 in Volume III.

THEOREM 8.78 (Ralph McKenzie 1978). *Every locally finite variety $\mathcal{V}$ that is directly representable has a finite residual bound and definable principal congruences. Moreover, if $\mathcal{V}$ has a finite signature, then $\mathcal{V}$ is finitely based.*

Proof. Let $\mathcal{M}$ be a finite set of finite algebras witnessing that $\mathcal{V}$ is directly representable.

First, let $\mathbf{A} \in \mathcal{V}$ be a finite subdirectly irreducible algebra. By direct representability of $\mathcal{V}$ we see that $\mathbf{A}$ must be isomorphic with a direct product of algebras belong to $\mathcal{M}$. Since $\mathbf{A}$ is subdirectly irreducible, it must be isomorphic to an algebra in $\mathcal{M}$. It follows that $\mathcal{V}_{\mathrm{fin}}$ has a finite residual bound. So by Exercise 8.79.2 below, $\mathcal{V}$ itself must have a finite residual bound.

Now suppose, for the sake of contradiction, that $\mathcal{V}$ does not have definable principal congruences. Using Exercise 8.79.3 below, we can assume that for every principal congruence formula $\varphi(x,y,u,v)$, there is a finite algebra $\mathbf{A} \in \mathcal{V}$ and $a,b,c,d \in A$

such that $(a, b) \in \mathrm{Cg}^{\mathbf{A}}(c, d)$ and $\mathbf{A} \models \neg\varphi(a, b, c, d)$. The set Φ of all congruence formulas is closed under the formation of finite disjunctions. So for any algebra $\mathbf{C}$ and any $a, b, c, d \in C$ we have

$$\begin{gathered}(a, b) \in \mathrm{Cg}^{\mathbf{C}}(c, d) \\ \text{if and only if} \\ \text{for some finite } \Phi' \subseteq \Phi \text{ we have } \mathbf{C} \models \bigvee_{\varphi \in \Phi'} \varphi(a, b, c, d).\end{gathered}$$

This means that there is an infinite sequence $\varphi_0, \varphi_1, \ldots$ of congruence formulas, and a sequence $\mathbf{A}_0.\mathbf{A}_1, \cdots \in \mathcal{V}_{\text{fin}}$, and $a_n, b_n, c_n, d_n \in A_n$ for each natural number n satisfying:

$$\begin{aligned} &\models \forall x, y, u, v[\varphi_n \Rightarrow \varphi_{n+1}] \\ \mathbf{A}_n &\models \varphi_{n+1}(a_n, b_n, c_n, d_n) \\ \mathbf{A}_n &\models \neg\varphi_n(a_n, b_n, c_n, d_n). \end{aligned}$$

By direct representability, we can assume that for each n there is a finite set I_n so that $\mathbf{A}_n = \prod_{i \in I_n} \mathbf{B}_i$, where $\mathbf{B}_i \in \mathcal{M}$ for each $i \in I_n$.

There are only finitely many 5-tuples $\langle a, b, c, d, \mathbf{B}\rangle$ with $a, b, c, d \in B$ and $\mathbf{B} \in \mathcal{M}$. By the Pigeon-Hole-Principle, there exist m and n with $n < m$ so that

$$\{\langle a_n(i), b_n(i), c_n(i), d_n(i), \mathbf{B}_i\rangle \mid i \in I_n\} = \{\langle a_m(j), b_m(j), c_m(j), d_m(j), \mathbf{B}_j\rangle \mid j \in I_m\}.$$

But since φ_m is an existential positive formula and $\mathbf{A}_n \models \varphi_m(a_n, b_n, c_n, d_n)$, the set equality above implies that $\mathbf{A}_m \models \varphi_m(a_n, b_n, c_n, d_n)$, a contradiction. ∎

Exercises 8.79

1. If a variety has definable principal congruences, then some principal congruence formula defines principal congruences over that variety.

2. If $\mathcal{V}$ is a locally finite variety and a congruence formula $\phi(x, y, u, v)$ defines principal congruences over $\mathcal{V}_{\text{fin}}$, then ϕ defines principal congruences over $\mathcal{V}$ itself.

3. If $\mathcal{V}$ is locally finite and $\mathcal{V}_{\text{fin}}$ has a finite residual bound, then $\mathcal{V}$ has the same finite residual bound. [Hint: Quackenbush's Theorem 7.9 is useful.]

8.6. Boolean Powers

The Boolean power construction, described in this section, is special kind of subdirect power. To develop its properties, it is convenient to develop the theory of Boolean algebras further and, in particular, to obtain Stone's Duality Theorem for Boolean algebras[2].

In Volume I we described the class of Boolean algebras by offering a finite equational basis (see page 17 in Volume I). There we conceived Boolean algebras as lattices

[2] After reading an earlier version of this section, Brian Davey suggested a restructuring of the section and supplied us with simpler, more informative proofs for Theorems 8.82, 8.84, and 8.93. The authors welcome his contributions.

expanded by a unary operation of complementation. It proves convenient here to use a signature that, in addition, provides two constant symbols to denote the top and bottom elements of the Boolean algebra. This approach, leads to a variety that is term equivalent to the variety described in Volume I. Thus we take $\mathbf{B} = \langle B, \wedge, \vee, ', 0, 1\rangle$ to be a Boolean algebra provided $\langle B, \wedge, \vee\rangle$ is a distributive lattice and $\mathbf{B}$ satisfies the following equations.

$$\begin{array}{cc}
(x \wedge y)' \approx x' \vee y' & (x \vee y)' \approx x' \wedge y' \\
x \wedge x' \approx 0 & x \vee x' \approx 1 \\
x \wedge 0 \approx 0 & x \vee 0 \approx x \\
x \wedge 1 \approx x & x \vee 1 \approx 1 \\
(x')' \approx x &
\end{array}$$

Notice that here we have used $'$ to denote complementation, rather than $\overline{}$, which was used in Volume I.

We use $\mathbf{2}$ to denote the 2-element Boolean algebra with universe $2 = \{0, 1\}$. We will need two theorems that highlight particular properties of the algebra $\mathbf{2}$.

THEOREM 8.80 (Emil L. Post 1920, 1921). *Every finitary operation on* $\{0, 1\}$ *is a term operation of* $\mathbf{2}$.

Proof. Let f be finitary operation on $\{0, 1\}$ and let r be the rank of f. With every r-tuple $\alpha \in \{0, 1\}^r$ we associate a term s_α with variables among $x_0, \dots, x_{r-1}$ as follows. Put $s_\alpha(x_0, \dots, x_{r-1}) = y_0 \wedge \cdots \wedge y_{r-1}$ where for each $i < r$

$$y_i = \begin{cases} x_i & \text{if the } i^{\text{th}} \text{ entry of } \alpha \text{ is } 1 \\ x_i' & \text{if the } i^{\text{th}} \text{ entry of } \alpha \text{ is } 0. \end{cases}$$

Then for every r-tuple β of 0's and 1's we see that s_α evaluated at β is 1 if $\beta = \alpha$ and otherwise it is 0. Now let F be the set of r-tuples that f assign the value 1. So put $t = \bigvee_{\alpha \in F} s_\alpha(x_0, \dots, x_{r-1})$, recalling that the join of an empty collection is 0. The reader can check that t denotes the operation f in $\mathbf{2}$. ∎

Post framed this as a result about the sentential connectives $\boldsymbol{\wedge}, \boldsymbol{\vee}, \neg, \bot, \top$ acting on the 2-element set $\{\mathrm{F}, \mathrm{T}\}$. Another proof can be based on Theorem 9.57 on page 50 in Volume III.

THEOREM 2.62 (Stone's Representation Theorem, (M. H. Stone, 1936)). *Every Boolean algebra is a subdirect power of* $\mathbf{2}$. ∎

In Volume I we provided two proofs of this theorem. One relies essentially on the Ultrafilter Theorem (see page 88 in Volume I), while the other is a consequence of Birkhoff's Subdirect Representation Theorem (see Example 1 on page 172 of Volume I).

In this way, we saw that every Boolean algebra could be viewed as an algebra whose elements were certain subsets of some large set (the universe) and whose operations were union, intersection, set complementation, with the empty set and the whole universe as the bottom and top elements. But actually in 1936, Marshall Stone proved a much stronger duality theorem, one that plays a key role in this section. Loosely speaking, this duality theorem sets up a correspondence between Boolean algebras and certain topological spaces—called Stone spaces. This correspondence supports the use of topological methods to address algebraic questions and the use of algebraic methods to tackle topological problems.

In fact, Stone's Duality Theorem is the first of many similar theorems in an active and significant branch of the general theory of algebraic systems that is known as the theory of natural dualities. While we can only give a taste of this broad area of research, the field itself is supplied with two excellent expositions: (David M. Clark and Brian A. Davey, 1998) and (Jane Pitkethly and Brian Davey, 2005).

The theory of natural dualities is most conveniently framed via category theory. §3.6 in Volume I supplies the background in category theory we will need here.

In the following pages, we will describe Stone's duality as a way of introducing the concepts and notation underlying all natural dualities. As this description progresses, we will make a number of assertions, whose proofs are left aside, and we simply let the description flow. The proof of the Stone Duality Theorem which follows, will establish those assertions.

By a **Stone space** we mean a topological space X that is compact, Hausdorff, and totally disconnected. This last condition is equivalent to saying that X has a basis of clopen sets. Consider the class $\mathcal{S}$ of all Stone spaces and the class $\mathcal{B}$ of all Boolean algebras. We view each of these classes as a category: the maps of the category $\mathcal{S}$ are just the continuous maps, while the maps of $\mathcal{B}$ are just the homomorphisms. To emphasize this usage, we use **CAT** $\mathcal{B}$ to denote the category of all homomorphisms between Boolean algebras. We use **2** to denote the 2-element Boolean algebra and 2 to denote the topological space on $\{0,1\}$ with the discrete topology. The set $\{0,1\}$ is two-faced: it plays both an algebraic role and a topological role in the duality that we are in the midst of describing. In the language of duality theory the topological space 2 is called the **alter ego** of the algebra **2**.

Given a Boolean algebra **B**, let $\mathcal{D}(\mathbf{B})$ be the set of all homomorphisms from **B** into **2**. Observe that $\mathcal{D}(\mathbf{B}) \subseteq \{0,1\}^B$. The space $\mathcal{D}(\mathbf{B})$ inherits its topology from the product topology on $2^{\,B}$. This $\mathcal{D}(\mathbf{B})$ is the topological space $\mathbf{P}(\mathbf{B})$ that was foretold at the bottom of page 88 in Volume I. It turns out that this inherited topology makes $\mathcal{D}(\mathbf{B})$ into a Stone space. In the other direction, suppose X is a Stone space. Let $\mathbf{E}(\mathrm{X})$ be the set of continuous maps from X into 2. Observe $\mathbf{E}(\mathrm{X}) \subseteq \{0,1\}^{\mathrm{X}}$. It turns out that $\mathbf{E}(\mathrm{X})$ is a subuniverse of $\mathbf{2}^{\mathrm{X}}$. Any homomorphism $h\colon \mathbf{B} \to \mathbf{2}$ is associated with a unique ultrafilter U on the Boolean algebra **B**. Namely, U is the set of elements mapped by h to 1. So in some presentations $\mathcal{D}(\mathbf{B})$ is taken to be the set of ultrafilters on **B**. Likewise, each continuous $f\colon \mathrm{X} \to 2$ partitions X into two complementary clopen sets. So in some presentations of the Stone duality these complementary sets are taken as $\mathbf{E}(\mathrm{X})$.

Given two Boolean algebras $\mathbf{B}_0$ and $\mathbf{B}_1$ and a homomorphism $h\colon \mathbf{B}_0 \to \mathbf{B}_1$, the map $\mathcal{D}(h)\colon \mathcal{D}(\mathbf{B}_1) \to \mathcal{D}(\mathbf{B}_0)$ is defined, for each $f \in \mathcal{D}(\mathbf{B}_1)$, by

$$\mathcal{D}(h)(f) = f \circ h. \tag{8.6.1}$$

It turns out that $\mathcal{D}(h)$ is continuous.

With these definitions it is clear that for each Boolean algebra **B** we have

$$\mathcal{D}(\mathrm{id}_B) = \mathrm{id}_{\mathcal{D}(\mathbf{B})},$$

where id_U stands for the identity morphism on the object U. Also for homomorphisms $h\colon \mathbf{B}_0 \to \mathbf{B}_1$ and $g\colon \mathbf{B}_1 \to \mathbf{B}_2$ on Boolean algebras we have

$$\mathcal{D}(g \circ h) = \mathcal{D}(h) \circ \mathcal{D}(g).$$

Given two Stone spaces X and Y and a continuous $g\colon \mathrm{X} \to \mathrm{Y}$, the map $\mathbf{E}(g)\colon \mathbf{E}(\mathrm{Y}) \to \mathbf{E}(\mathrm{X})$ is defined, for each $h \in \mathbf{E}(\mathrm{Y})$, by

$$\mathbf{E}(g)(h) = h \circ g. \tag{8.6.2}$$

It turns out that $\mathbf{E}(g)$ is a homomorphism.

With these definitions it is clear that for each Stone space X we have

$$\mathbf{E}(\mathrm{id}_{\mathrm{X}}) = \mathrm{id}_{\mathbf{E}(\mathrm{X})}.$$

Also for continuous maps $h\colon \mathrm{X}_0 \to \mathrm{X}_1$ and $g\colon \mathrm{X}_1 \to \mathrm{X}_2$ on Stone spaces we have

$$\mathbf{E}(g \circ h) = \mathbf{E}(h) \circ \mathbf{E}(g).$$

In this way the maps $\mathcal{D}$ and $\mathbf{E}$ are contravariant functors between the category of Boolean algebras and the category of Stone spaces.

Given a Boolean algebra $\mathbf{B}$ there is a natural evaluation map $e_{\mathbf{B}}\colon \mathbf{B} \to \mathbf{E}\mathcal{D}(\mathbf{B})$ defined by $e_{\mathbf{B}}(b)(f) = f(b)$ for all $b \in B$ and all $f \in \mathcal{D}(\mathbf{B})$. Likewise, for each Stone space X there is a natural evaluation map $\varepsilon_{\mathrm{X}}\colon \mathrm{X} \to \mathcal{D}\mathbf{E}(\mathrm{X})$ defined by $\varepsilon_{\mathrm{X}}(x)(f) = f(x)$ for all $x \in \mathrm{X}$ and all $f \in \mathbf{E}(\mathrm{X})$. Following the usage in the theory of natural dualities, we say that $\mathrm{2}$ yields a **natural duality** on the category of Boolean algebras provided the contravariant functors $\mathcal{D}$ and $\mathbf{E}$ behave as described and the maps $e_{\mathbf{B}}$ are isomorphisms. Further, we say that the duality is **full** provided the maps ε_{X} are homeomorphisms as well. We also say that $\mathbf{2}$ and its alter ego $\mathrm{2}$ support a (full) duality between the category of Boolean algebras and the category of Stone spaces.

As preparation for the Stone Duality Theorem, we have the following lemma, interesting in its own right.

LEMMA 8.81. *Let* $\mathbf{A}$ *and* $\mathbf{B}$ *be algebras of the same signature. Let* X *be a topological space and endow* A *and* B *with the discrete topology, resulting in the topological spaces* A *and* B*. Then*

(i) *the set* $\hom(\mathrm{X}, \mathrm{A})$ *of continuous maps from* X *into* A *is a subuniverse of* $\mathbf{A}^{\mathrm{X}}$*, and*

(ii) *the set* $\hom(\mathbf{A}, \mathbf{B})$ *of homomorphisms from* $\mathbf{A}$ *into* $\mathbf{B}$ *is a closed subset of* B^A.

Proof. For part (i), let $f_0, \dots, f_{r-1} \in \hom(\mathrm{X}, \mathrm{A})$ and let Q be an operation symbol and let r be the rank of Q. Note that the function $Q^{\mathbf{A}}(f_0, \dots, f_{r-1})$ can be written as

$$Q^{\mathbf{A}}(f_0, \dots, f_{r-1}) = Q^{\mathbf{A}} \circ (f_0 \sqcap \cdots \sqcap f_{r-1}),$$

where $f_0 \sqcap \cdots \sqcap f_{r-1}\colon \mathrm{X} \to A^r$ is the natural product map given by

$$(f_0 \sqcap \cdots \sqcap f_{r-1})(x) := \langle f_0(x), \dots, f_{r-1}(x)\rangle \text{ for all } x \in \mathrm{X}.$$

As the reader can easily see, $f_0 \sqcap \cdots \sqcap f_{r-1}$ is continuous, and $Q^{\mathbf{A}}\colon A^r \to A$ is continuous as the topology on A is discrete. As the composition of two continuous maps is continuous, it follows at once that $Q(f_0, \dots, f_{r-1})$ is continuous and so belongs to $\hom(\mathrm{X}, \mathrm{A})$. We conclude that $\hom(\mathrm{X}, \mathrm{A})$ is a subuniverse of $\mathbf{A}^{\mathrm{X}}$.

For part (ii), suppose that $f: A \to B$ is not a homomorphism. We must have an open set around f that contains no homomorphisms. Since f is not a homomorphism there is an operation symbol Q, say of rank r, and elements $a_0, \dots, a_{r-1} \in A$ such that

$$f(Q^{\mathbf{A}}(a_0, \dots, a_{r-1})) \neq Q^{\mathbf{B}}(f(a_0), \dots, f(a_{r-1})).$$

Let $V := \{a_0. \dots .a_{r-1}, Q^{\mathbf{A}}(a_0, \dots, a_{r-1})\}$. Let

$$U := \bigcap_{v \in V} \{g \mid g: A \to B \text{ and } f(v) = g(v)\}.$$

Then $f \in U$ and no member of U is a homomorphism. Since V is finite, it only remains to show that $U_v := \{g \mid g: A \to B \text{ and } f(v) = g(v)\}$ is open (for all $v \in V$). Since B has the discrete topology the set $\{f(v)\}$ open. But U_v is the inverse image of that open set with respect to the projection function associated with v. Since projection functions are continuous, we see that U_v is open, as desired. ∎

THEOREM 8.82 (Stone's Duality Theorem (M. H. Stone, 1936, 1937)). *The two element algebra* **2** *and its alter ego* 2 *support a full duality between the category of all Boolean algebras and the category of Stone spaces.*

Proof. Let $\mathcal{S}$ be the class of all Stone spaces.

Let $\mathbf{B} \in \mathcal{B}$ and $\mathrm{X} \in \mathcal{S}$. By Lemma 8.81, we have

- the hom-set $\mathbf{E}(\mathrm{X}) = \hom(\mathrm{X}, 2)$ is a subuniverse of $\mathbf{2}^X$, and
- the hom-set $\mathcal{D}(\mathbf{B}) = \hom(\mathbf{B}, \mathbf{2})$ is a closed subset of $2^{\,B}$.

Since the algebra **2** is finite, we have that $2^{\,B}$ is a Stone space, and, hence, that $\hom(\mathbf{B}, \mathbf{2})$ is also a Stone space. That is, $\mathcal{D}(\mathbf{B}) = \hom(\mathbf{B}, \mathbf{2})$ is the Stone space with topology inherited from 2^B.

Let $\mathbf{B}_0, \mathbf{B}_1 \in \mathcal{B}$ and let $h: \mathbf{B}_0 \to \mathbf{B}_1$. Further, let $\mathrm{X}_0, \mathrm{X}_1 \in \mathcal{S}$ and let $g: \mathrm{X}_0 \to \mathrm{X}_1$. Then Equations 8.6.1 and 8.6.2 define $\mathcal{D}(h)$ and $\mathbf{E}(g)$ respectively. In this way, **E** and $\mathcal{D}$ are seen to be contravariant functors:

$$\mathcal{D}: \mathbf{CAT}\,\mathcal{B} \to \mathbf{CAT}\,\mathcal{S}$$
$$\mathbf{E}: \mathbf{CAT}\,\mathcal{S} \to \mathbf{CAT}\,\mathcal{B}.$$

The proof is completed by the following contentions.

Contentions.

(A) *The map* $e_{\mathbf{B}}: \mathbf{B} \to \mathbf{E}(\mathcal{D}(\mathbf{B}))$ *is an isomorphism, for all* $\mathbf{B} \in \mathcal{B}$.
(B) *The map* $\varepsilon_{\mathrm{X}}: \mathrm{X} \to \mathcal{D}(\mathbf{E}(\mathrm{X}))$ *is a homeomorphism, for all* $\mathrm{X} \in \mathcal{S}$.

To establish (A), let $\mathbf{B} \in \mathcal{B}$. Since $\mathcal{B} = \boldsymbol{P_s}\mathbf{2}$, we conclude that the set $\hom(\mathbf{B}, \mathbf{2})$ separates the points of B. It follows that the homomorphism $e_{\mathbf{B}}$ is one-to-one—that is, it is an embedding.

We now prove that $e_{\mathbf{B}}$ surjective. The proof is a special case of the proof of the First Duality Theorem (David M. Clark and Brian A. Davey, 1998, Theorem 2.2.2). Let $f: \hom(\mathbf{B}, \mathbf{2}) \to 2$ be continuous, i.e., $f \in \mathbf{E}(\mathcal{D}(\mathbf{B}))$. We shall find $a \in B$ such that $e_{\mathbf{B}}(a) = f$, i.e., $u(a) = f(u)$, for all $u \in \hom(\mathbf{B}, \mathbf{2})$. Since $\hom(\mathbf{B}, \mathbf{2})$ is a closed subset of $2^{\,B}$, by Exercise 8.86.1 there is a continuous map $g: 2^B \to 2$ with $g\restriction_{\hom(\mathbf{B},\mathbf{2})} = f$. By Exercise 8.86.2, the map g has a finite support, say $F \subseteq B$. Define $\pi_F: 2^B \to 2^F$ by $\pi_F(u) = u\restriction_F$, for all $u \in 2^B$. Since F is a finite support for g, we have $\ker \pi_F \subseteq \ker g$, and hence there exists a map $h\colon 2^F \to 2$ with $h \circ \pi_F = g$. Let $F = \{a_0, \dots, a_{n-1}\}$. Since each finitary operation on 2 is a term operation of **2**, there exists a term $t(v_0, \dots, v_{n-1})$ such that, for all $v\colon F \to 2$, we have $h(v) = t^{\mathbf{2}}(v(a_0), \dots, v(a_{n-1}))$. Now define $a := t^{\mathbf{2}}(a_0, \dots, a_{n-1})$. Then, for all $u \in \hom(\mathbf{B}, \mathbf{2})$,

$$\begin{aligned} f(u) = g(u) = (h \circ \pi_F)(u) = h(u{\restriction_F}) &= t^{\mathbf{2}}(u{\restriction_F}(a_0), \dots, u{\restriction_F}(a_{n-1})) \\ &= t^{\mathbf{2}}(u(a_0), \dots, u(a_{n-1})) = u(t^{\{}(a_0, \dots, a_{n-1})) = u(a). \end{aligned}$$

Hence $e_{\mathbf{B}}$ is surjective and so is an isomorphism and Contention (A) is established.

To establish Contention (B), let $\mathrm{X} \in \mathcal{S}$. Since 2 is nontrivial and distinct points of X are separated by clopen sets, it follows almost at once that $\hom(\mathrm{X}, 2)$ separates the points of X. Thus ε_{X} is one-to-one. Hence ε_{X} is a continuous one-to-one map from a compact space to a Hausdorff space and is, consequently, a homeomorphism onto its image.

It only remains to prove that ε_{X} maps X onto $\mathcal{D}(\mathbf{E}(\mathrm{X}))$. The proof is a special case of (David M. Clark and Brian A. Davey, 1998, Exercise 9.9 page 290). We require two easily proved facts.

FACT 1. *Let* X *be a Stone space and let* Y *be a proper closed subset of* X*. Then there exist* $\alpha, \beta \in \mathbf{E}(\mathrm{X})$ *with* $\alpha{\restriction_Y} = \beta{\restriction_Y}$ *but* $\alpha \neq \beta$.

Proof of Fact 1. Let $x \in X \setminus Y$. A simple compactness argument using the fact that, for every element y of Y, there exists a clopen subset U of X with $y \in U$ and $x \notin U$, shows that there is a clopen subset V of X with $Y \subseteq V$ and $x \notin V$. Now take $\alpha\colon \mathrm{X} \to 2$ to be the constant function onto 1 and β to be the characteristic function of V. ∎

FACT 2. *Let* **B** *be a Boolean algebra. If* Y *is a closed subset of* $\mathcal{D}(\mathbf{B})$ *that separates the points of* B *(i.e., for all* $b, c \in B$ *with* $b \neq c$*, there exists* $y \in Y$ *with* $y(b) \neq y(c)$*), then* $Y = \mathcal{D}(\mathbf{B})$.

Proof of Fact 2. We prove the contrapositive. Assume that Y is a proper closed subset of $\mathcal{D}(\mathbf{B})$. By Fact 1, there exist $\alpha, \beta \in \mathbf{E}(\mathcal{D}(\mathbf{B}))$ with $\alpha{\restriction_Y} = \beta{\restriction_Y}$ but $\alpha \neq \beta$. By Contention (A), the map $e_{\mathbf{B}}\colon \mathbf{B} \to \mathbf{E}(\mathcal{D}(\mathbf{B}))$ is an isomorphism. Hence there exist $a, b \in B$ with $e_{\mathbf{B}}(a) = \alpha$ and $e_{\mathbf{B}}(b) = \beta$. Of course, $a \neq b$. For all $y \in Y$, we have

$$y(a) = e_{\mathbf{B}}(a)(y) = \alpha(y) = \beta(y) = e_{\mathbf{B}}(b)(y) = y(b),$$

whence Y does not separate the points of **B**. ∎

We can now use Fact 2 to prove that ε_{X} maps X onto $\mathcal{D}(\mathbf{E}(\mathrm{X}))$. By Fact 2 it suffices to prove $\varepsilon_{\mathrm{X}}(X)$ is a closed subset of $\mathcal{D}(\mathbf{E}(\mathrm{X}))$ that separates the points of $\mathbf{E}(\mathrm{X})$. That $\varepsilon_{\mathrm{X}}(\mathrm{X})$ is a closed subset of $\mathcal{D}(\mathbf{E}(\mathrm{X}))$ follows from the the compactness of X, $\mathcal{D}(\mathbf{E}(\mathrm{X}))$ is Hausdorff, and ε_{X} is continuous. (More generally, every continuous map from a compact space to a Hausdorff space maps closed sets to closed sets.) That $\varepsilon_{\mathrm{X}}(X)$ separates the points of $\mathbf{E}(\mathrm{X})$ is almost trivial: let $\alpha, \beta \in \mathbf{E}(\mathrm{X})$ with $\alpha \neq \beta$, then there exists $x \in \mathrm{X}$ with $\alpha(x) \neq \beta(x)$, and hence $\varepsilon_{\mathrm{X}}(x)(\alpha) = \alpha(x) \neq \beta(x) = \varepsilon_{\mathrm{X}}(x)(\beta)$. This establishes Contention (B).

This completes the proof of Stone's Duality Theorem. ∎

DEFINITION 8.83. A contravariant functor $\Gamma\colon \mathbf{C} \to \mathbf{D}$ between two categories $\mathbf{C}$ and $\mathbf{D}$ is called a **dual equivalence functor** if

(1) for all $D \in \mathbf{D}$, there exists $C \in \mathbf{C}$ with $\Gamma(C) \cong D$, and
(2) Γ maps $\hom(C_0, C_1)$ bijectively onto $\hom(\Gamma(C_1), \Gamma(C_0))$, for all $C_0, C_1 \in \mathbf{C}$.

We say that categories $\mathbf{C}$ and $\mathbf{D}$ are **dually equivalent** if there exists a dual equivalence functor $\Gamma\colon \mathbf{C} \to \mathbf{D}$.

THEOREM 8.84. *The functors $\mathcal{D}$ and $\mathbf{E}$ are dual equivalence functors. Hence* $\mathbf{CAT}\,\mathcal{B}$ *and* $\mathbf{CAT}\,\mathcal{S}$ *are dually equivalent categories.*

Proof of Theorem 8.84. We demonstrate that $\mathcal{D}$ is a dual equivalence functor. From this, the result for $\mathbf{E}$ is immediate.

We must prove that

(A) for all $\mathrm{X} \in \mathcal{S}$, there exists $\mathbf{B} \in \mathcal{B}$ with $\mathcal{D}(\mathbf{B}) \cong \mathrm{X}$, and
(B) $\mathcal{D}$ maps $\hom(\mathbf{B}_0, \mathbf{B}_1)$ bijectively onto $\hom(\mathcal{D}(\mathbf{B}_1), \mathcal{D}(\mathbf{B}_0))$, for all $\mathbf{B}_0, \mathbf{B}_1 \in \mathcal{B}$.

Since, by Stone's Duality Theorem, the map $\varepsilon_{\mathrm{X}}\colon \mathrm{X} \to \mathcal{D}(\mathbf{E}(\mathrm{X}))$ is an isomorphism, (A) is easy; simply choose $\mathbf{B} = \mathbf{E}(\mathrm{X})$.

We now prove (B). We will employ Facts 1 and 2 below. (They are properties of all dual equivalences). We leave the verifications of these facts to the reader.

Let $\mathbf{B}_0, \mathbf{B}_1 \in \mathcal{B}$ and let $\mathrm{X}, \mathrm{Y} \in \mathcal{S}$.

FACT 1. *For $u \in \hom(\mathbf{B}_0, \mathbf{B}_1)$ and $\varphi \in \hom(\mathrm{X}, \mathrm{Y})$, we have*

$$\text{(a)}\quad e_{\mathbf{B}_1} \circ u = \mathbf{E}\mathcal{D}(u) \circ e_{\mathbf{B}_0} \quad \text{and} \quad \text{(b)}\quad \varepsilon_{\mathrm{Y}} \circ \varphi = \mathcal{D}\mathbf{E}(\varphi) \circ \varepsilon_{\mathrm{X}},$$

FACT 2. *For $u \in \hom(\mathbf{B}_0, \mathbf{E}(\mathrm{X}))$ and $\varphi \in \hom(\mathrm{X}, \mathcal{D}(\mathbf{B}_0))$, we have*

$$\text{(a)}\quad u = \mathbf{E}(\mathcal{D}(u) \circ \varepsilon_{\mathrm{X}}) \circ e_{\mathbf{B}_0} \quad \text{and} \quad \text{(b)}\quad \varphi = \mathcal{D}(\mathbf{E}(\varphi) \circ e_{\mathbf{B}_0}) \circ \varepsilon_{\mathrm{X}}.$$

See Figure 8.11.

Choose $u, v \in \hom(\mathbf{B}_0, \mathbf{B}_1)$ with $\mathcal{D}(u) = \mathcal{D}(v)$. Applying Fact 1 and the fact that $e_{\mathbf{B}_1}$ is an isomorphism, we obtain

$$u = e_{\mathbf{B}_1}^{-1} \circ \mathbf{E}\mathcal{D}(u) \circ e_{\mathbf{B}_0} = e_{\mathbf{B}_1}^{-1} \circ \mathbf{E}\mathcal{D}(v) \circ e_{\mathbf{B}_0} = v.$$

Hence $\mathcal{D}\colon \hom(\mathbf{B}_0, \mathbf{B}_1) \to \hom(\mathcal{D}(\mathbf{B}_1), \mathcal{D}(\mathbf{B}_0))$ is one-to-one.

We now prove that $\mathcal{D}\colon \hom(\mathbf{B}_0, \mathbf{B}_1) \to \hom(\mathcal{D}(\mathbf{B}_1), \mathcal{D}(\mathbf{B}_0))$ is surjective. Choose $\psi \in \hom(\mathcal{D}(\mathbf{B}_1), \mathcal{D}(\mathbf{B}_0))$. We claim that $\psi = \mathcal{D}(v)$ where $v\colon \mathbf{B}_0 \to \mathbf{B}_1$ is defined by

$$v = e_{\mathbf{B}_1}^{-1} \circ \mathbf{E}(\psi) \circ e_{\mathbf{B}_0}.$$

We first apply Fact 2 with $\mathrm{X} = \mathcal{D}(\mathbf{B}_1)$, with $\varphi = \mathrm{id}_{\mathcal{D}(\mathbf{B}_1)}$ and with $\mathbf{B}_0$ replaced by $\mathbf{B}_1$. This gives us $\mathcal{D}(e_{\mathbf{B}_1}) \circ \varepsilon_{\mathcal{D}(\mathbf{B}_1)} = \mathrm{id}_{\mathcal{D}(\mathbf{B}_1)}$ (since $\mathbf{E}(\mathrm{id}_{\mathcal{D}(\mathbf{B}_1)}) = \mathrm{id}_{\mathbf{E}\mathcal{D}(\mathbf{B}_1)}$), and hence (again using the fact that $e_{\mathbf{B}_1}$ is an isomorphism),

$$\varepsilon_{\mathcal{D}(\mathbf{B}_1)} = \mathcal{D}(e_{\mathbf{B}_1})^{-1}.$$

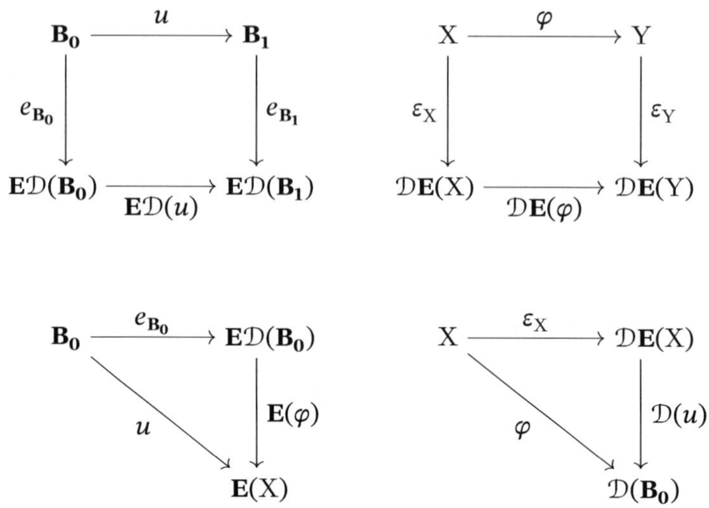

FIGURE 8.11. Two squares and two triangles

Next we apply Fact 2 again, this time with $X = \mathcal{D}(\mathbf{B}_1)$ and $\varphi = \psi$. This yields $\psi = \mathcal{D}(\mathbf{E}(\psi) \circ e_{\mathbf{B}_0}) \circ \varepsilon_{\mathcal{D}(\mathbf{B}_1)}$. Hence we obtain

$$\begin{aligned}\psi &= \mathcal{D}(\mathbf{E}(\psi) \circ e_{\mathbf{B}_0}) \circ \varepsilon_{\mathcal{D}(\mathbf{B}_1)} = \mathcal{D}(\mathbf{E}(\psi) \circ e_{\mathbf{B}_0}) \circ \mathcal{D}(e_{\mathbf{B}_1})^{-1} \\ &= \mathcal{D}(\mathbf{E}(\psi) \circ e_{\mathbf{B}_0}) \circ \mathcal{D}(e_{\mathbf{B}_1}^{-1}) = \mathcal{D}(e_{\mathbf{B}_1}^{-1} \circ \mathbf{E}(\psi) \circ e_{\mathbf{B}_0}) = \mathcal{D}(\upsilon),\end{aligned}$$

as required.

This completes the proof of Theorem 8.84. ■

The proof above is a special case of the proof of Lemma 1.1.2 in (David M. Clark and Brian A. Davey, 1998).

COROLLARY 8.85. *Let* $\mathbf{B}$ *be a Boolean algebra. The lattice of all filters on* $\mathbf{B}$ *is dually isomorphic to the lattice of all closed subsets of* $\mathcal{D}(\mathbf{B})$*. Consequently,* $\mathbf{Con}\ \mathbf{B}$ *is dually isomorphic to the lattice of closed subsets of* $\mathcal{D}(\mathbf{B})$*.*

Proof. Let $R := \{\langle b, f\rangle \mid b \in B, f \in \mathcal{D}(\mathbf{B}), \text{ and } f(b) = 1\}$. The relation R imposes a Galois connection between B and $\mathcal{D}(\mathbf{B})$. It is our contention that the Galois-closed subsets of B are precisely the filters of $\mathbf{B}$ while the Galois-closed subsets of $\mathcal{D}(\mathbf{B})$ are precisely the topologically closed subsets. The first part of this contention follows from the fact that every filter is the intersection of the collection of all ultrafilters that include it (see the alternate form of Theorem 8.28). The second part follows from the observation that $f(b) = 0 \Leftrightarrow f(b') = 1$ and so each topologically closed subset of $\mathcal{D}(\mathbf{B})$ is the intersection of some nonempty collection of sets of the form $\{f \mid f \in \mathcal{D}(\mathbf{B}) \text{ and } f(b) = 1\}$, where $b \in B$. ■

Exercises 8.86

1. Prove that every finite discretely topologized set P is *injective* in the category of Stone spaces. That is, for every Stone space X and every closed subset Y of X,

each continuous map $f : Y \to P$ extends to a continuous map $g : X \to P$ with $g\restriction_Y = f$. [Hint. First assume that $|P| = 2$; this case involves a straightforward double-compactness argument. Then show that every power and every continuous retract of an injective Stone space is again injective. Finally, use the almost trivial fact that every finite discretely topologized space is a continuous retract of a finite power of a two-element discretely topologized space.]

2. Let P be a finite discretely topologized set, let S be a set and let $g : P^S \to P$ be continuous. Prove that there exists a finite subset F of S such that, for all $u, v \in P^S$, we have $u\restriction_F = v\restriction_F \Rightarrow g(u) = g(v)$. We say that F is a *finite support* for g.

Primal algebras

As mentioned in Exercise 11 on page 149 in Volume I, a finite nontrivial algebra **P** is said to be **primal** provided each finitary operation on P is a term operation of **P**. We have seen that the 2-element Boolean algebra **2** is primal. An examination of the proofs of Theorem 8.82 (Stone's Duality Theorem) and Thereon 8.84 shows that the only properties the **2** that came into play were

(A) **2** is a primal algebra;
(B) Every algebra in the variety generated by **2** is a subdirect power of **2**.

This means that if we can show that (A) implies (B), then we can obtain duality and dual equivalence theorems that apply to all primal algebras, not just to **2**.

THEOREM 8.87. *Let* **P** *be a primal algebra and let* $\mathcal{P}$ *be the variety generated by* **P**. *Then* $\mathcal{P} = \boldsymbol{P}_s\mathbf{P}$.

Proof. Since **P** is primal, the ternary majority function on P is denoted by a term. Hence, $\mathcal{P}$ is congruence distributive. By Jónsson's Lemma

$$\mathcal{P} = \boldsymbol{P}_s\boldsymbol{H}\boldsymbol{S}\boldsymbol{P}_u\mathbf{P}.$$

Because P is finite, this reduces to

$$\mathcal{P} = \boldsymbol{P}_s\boldsymbol{H}\boldsymbol{S}\mathbf{P}.$$

But since **P** is primal, the reader can easily check that **P** has no proper subalgebras and no proper homomorphic images. So we obtain

$$\mathcal{P} = \boldsymbol{P}_s\mathbf{P},$$

as desired. ■

As promised, the next theorem follows immediately.

THEOREM 8.88. *Let* **P** *be a primal algebra and let* $\mathcal{P}$ *be the variety generated by* **P**. *Let* P *be the topological space obtained by giving* P *the discrete topology. The algebra* **P** *and its alter ego* P *support a full duality between the category* **CAT** $\mathcal{P}$ *and the category* $\mathcal{S}$ *of Stone spaces. Moreover,* **CAT** $\mathcal{P}$ *and* $\mathcal{S}$ *are dually equivalent.* ■

Let **P** be a primal algebra and $\mathcal{P}$ be the variety it generates. Theorem 8.88 entails a close connection between $\mathcal{B}$ and $\mathcal{P}$. To visualize this connection, we let $\mathcal{D}$ and **E** retain the meanings we gave them in the context of Boolean algebras, but use $\widehat{\mathcal{D}}$ and $\widehat{\mathbf{E}}$ denote the corresponding contravariant functors with respect to $\mathcal{P}$. Then we see

$$\mathbf{CAT}\,\mathcal{B} \underset{\mathcal{D}}{\longrightarrow} \mathcal{S} \underset{\widehat{\mathcal{D}}}{\longleftarrow} \mathbf{CAT}\,\mathcal{P}.$$

Rewriting this, we see

$$\mathbf{CAT}\,\mathcal{B} \underset{\mathcal{D}}{\longrightarrow} \mathcal{S} \underset{\widehat{\mathbf{E}}}{\longrightarrow} \mathbf{CAT}\,\mathcal{P}.$$

Both $\mathcal{D}$ and $\widehat{\mathbf{E}}$ support dual equivalences. Forming the composite functor, we obtain

$$\mathbf{CAT}\,\mathcal{B} \underset{\widehat{\mathbf{E}}\circ\mathcal{D}}{\longrightarrow} \mathbf{CAT}\,\mathcal{P}. \tag{8.6.3}$$

Now consider $\mathbf{B} \in \mathcal{B}$. Let $\mathcal{D}(\mathbf{B}) = \mathrm{X}$. Then we see

$$\widehat{\mathbf{E}}(\mathcal{D}(\mathbf{B})) = \widehat{\mathbf{E}}(\mathrm{X}) = \hom(\mathrm{X}, \mathrm{P}).$$

We know $\hom(\mathrm{X}, \mathrm{P})$ is a subalgebra of $\mathbf{P}^X$.

What happens if we do not stipulate that **P** is primal or even finite?

The Boolean power construction

DEFINITION 8.89. Let **B** be a Boolean algebra and let **A** be an arbitrary algebra. The **Boolean power of A with respect to B** is the subalgebra $\mathbf{A}[\mathbf{B}]$ of the direct power $\mathbf{A}^{\mathcal{D}(\mathbf{B})}$ with universe $A[\mathbf{B}]$, which is the collection of all continuous functions from $\mathcal{D}(\mathbf{B})$ to A, where A is the set A endowed with the discrete topology.

Observe that $A[\mathbf{B}]$ is actually a subuniverse of $\mathbf{A}^{\mathcal{D}(\mathbf{D})}$ by Lemma 8.81.

The Boolean power construction is present in the work of Richard F. Arens and Irving Kaplansky (1948), which focused on rings, and it has forerunners in earlier work. Alfred L. Foster (1953a,b) put this construction into the framework of general algebraic systems (and G. C. Nelson (1984) expanded it to structures more generally). Tah Kai Hu (1969) refined and generalized the 1953 work of Foster. Hu uses "characteristic algebra" to refer to what we have called Boolean powers. S. Burris (1975) named the construction and developed many of its basic properties. Our presentation and notation here differ from those in (S. Burris, 1975). In that paper, two constructions were defined. One he called the Boolean power and denoted there by $\mathbf{A}[\mathbf{B}]$ and the other he called the bounded Boolean power and denoted there by $\mathbf{A}[\mathbf{B}]^*$. What we call the Boolean power, Burris called the bounded Boolean power. To simplify notation, we have dropped the * from the notation. While the first construction found in (S. Burris, 1975) will not be explored here, we suggest *Boolean extension* as a name for it.

If **B** is the one-element Boolean algebra, then $\mathbf{A}[\mathbf{B}]$ is also a one-element algebra. Since $\mathcal{D}(\mathbf{2})$ has only one element, we see that $\mathbf{A}[\mathbf{2}]$ is isomorphic to **A**. More generally, if **B** is nontrivial, then **A** is embeddable into $\mathbf{A}[\mathbf{B}]$ since every constant function from $\mathcal{D}(\mathbf{B})$ into A is continuous. For the same reason, we observe that $\mathbf{A}[\mathbf{B}]$ is always a subdirect power of **A**. This last was observed by Alfred Foster.

THEOREM 8.90. *Let $\mathbf{A}_0$ and $\mathbf{A}_1$ be algebras of the same signature and let $\mathbf{B}_0$ and $\mathbf{B}_1$ be Boolean algebras. Each of the following hold:*

(i) $\mathbf{A}_0[\mathbf{B}_0 \times \mathbf{B}_1] \cong \mathbf{A}_0[\mathbf{B}_0] \times \mathbf{A}_0[\mathbf{B}_1]$;

(ii) $(\mathbf{A}_0 \times \mathbf{A}_1)[\mathbf{B}_0] \cong \mathbf{A}_0[\mathbf{B}_0] \times \mathbf{A}_1[\mathbf{B}_0]$;

(iii) *Define* Γ *by*

$$(\Gamma(h))(\gamma) = \gamma \circ \mathcal{D}(h) \textit{ for each } \gamma \in A_0[\mathbf{B}_0] \textit{ and each } h \in \hom(\mathbf{B}_0, \mathbf{B}_1).$$

Then $\Gamma\colon \hom(\mathbf{B}_0, \mathbf{B}_1) \to \hom(\mathbf{A}_0[\mathbf{B}_0], A_0[\mathbf{B}_1])$. *That is, if* $h \in \hom(\mathbf{B}_0, \mathbf{B}_1)$, *then* $\Gamma(h) \in \hom(\mathbf{A}_0[\mathbf{B}_0], \mathbf{A}_0[\mathbf{B}_1])$. *Moreover, the map* Γ *is one-to-one as long as* A_0 *has at least* 2 *elements.*

Proof. For part (i) it is harmless to suppose that the two Stone spaces $\mathcal{D}(\mathbf{B}_0)$ and $\mathcal{D}(\mathbf{B}_1)$ are disjoint. Then their union is also a Stone space. Let $\mathbf{C}(\mathcal{D}(B_0) \cup \mathcal{D}(\mathbf{B}_1), \mathbf{A}_0)$ denote the corresponding algebra of continuous maps. We saw above that the Stone spaces $\mathcal{D}(\mathbf{B}_0 \times \mathbf{B}_1)$ and $\mathcal{D}(\mathbf{B}_0) \cup \mathcal{D}(\mathbf{B}_1)$ are homeomorphic. Hence $C(\mathcal{D}(\mathbf{B}_0) \cup \mathcal{D}(\mathbf{B}_1), \mathbf{A}_0) \cong \mathbf{A}_0[\mathbf{B}_0 \times \mathbf{B}_1]$. Now for each $f \in C(\mathcal{D}(\mathbf{B}_0) \cup \mathcal{D}(\mathbf{B}_1), \mathbf{A}_0)$ define

$$\alpha(f) := \langle f \restriction \mathcal{D}(\mathbf{B}_0), f \restriction \mathcal{D}(\mathbf{B}_1)\rangle.$$

Then

$$\alpha\colon \mathbf{C}(\mathcal{D}(\mathbf{B}_0) \cup \mathcal{D}(\mathbf{B}_1), \mathbf{A}_0) \rightarrowtail\!\!\!\twoheadrightarrow \mathbf{A}_0[\mathbf{B}_0] \times \mathbf{A}_0[\mathbf{B}_1],$$

as the reader can easily check. Putting things together, we obtain (i).

To prove part (ii), for all $f_0 \in A_0[\mathbf{B}_0]$, $f_1 \in A_1[\mathbf{B}_0]$, and all $x \in \mathcal{D}(\mathbf{B}_0)$, define

$$\beta(\langle f_0, f_1\rangle)(x) := \langle f_0(x), f_1(x)\rangle,$$

giving us a map

$$\beta\colon A_0[\mathbf{B}_0] \times A_1[\mathbf{B}_0] \to (A_0 \times A_1)[\mathbf{B}_0].$$

We contend that β is the desired isomorphism. That β is one-to-one is left in the hands of the reader.

To see that β is onto, let $f \in (A_0 \times A_1)[\mathbf{B}_0]$. Since $\mathcal{D}(\mathbf{B}_0)$ is a Stone space, the range of f is finite—giving rise to a partition of $\mathcal{D}(\mathbf{B}_0)$ into finitely many clopen sets, on each of which f is constant. Let $p_i\colon A_0 \times A_1 \to A_i$ for $i = 0, 1$ denote the projection functions. Then $p_i \circ f \in A_i[\mathbf{B}_0]$, for $i = 0, 1$, since these two composite functions are each constant on every block of the partition of $\mathcal{D}(\mathbf{B}_0)$. Now observe that $\beta(p_0 \circ f, p_1 \circ f) = f$ to conclude that (ii) holds.

Finally, for (iii) first notice that $(\Gamma(h))(\gamma)$ for each $\gamma \in A_0[\mathbf{B}_0]$ is continuous because it is the composition of continuous functions. To see that $\Gamma(h)$ is a homomorphism, consider any operation symbol Q, taking r to be the rank of Q and suppose that $\gamma_0, \dots, \gamma_{r-1} \in A_0[\mathbf{B}_0]$ and that $f \in \mathcal{D}(B_1)$. Observe

$$\begin{aligned}
(\Gamma(h))(Q^{\mathbf{A}_0[\mathbf{B}_0]}(\gamma_0, \dots, \gamma_{r-1}))(f) &= Q^{\mathbf{A}_0[\mathbf{B}_0]}(\gamma_0, \dots, \gamma_{r-1})(f \circ h) \\
&= Q^{\mathbf{A}_0}(\gamma_0(f \circ h), \dots, \gamma_{r-1}(f \circ h)) \\
&= Q^{\mathbf{A}_0}((\Gamma(h))(\gamma_0)(f), \dots, (\Gamma(h))(\gamma_{r-1}(f)) \\
&= Q^{\mathbf{A}_0[B_1]}((\Gamma(h))(\gamma_0), \dots, (\Gamma(h))(\gamma_{r-1}))(f)
\end{aligned}$$

To see that Γ is one-to-one, let $h, g\colon \mathbf{B}_0 \to \mathbf{B}_1$ while $h \neq g$. Pick $b \in B_0$ so that $h(b) \neq g(b)$. So in $\mathbf{B}_1$ at least one of $h(b)$ and $g(b)$ is not 0. Harmlessly, we take $h(b) \neq 0$. An adaptation of Theorem 8.28 ensures that there is an ultrafilter on $\mathbf{B}_1$ that contains $h(b)$ but not $g(b)$. So there is $f\colon \mathbf{B}_1 \to \mathbf{2}$ so that $(f \circ h)(b) = 1$ while $(f \circ g)(b) = 0$. This means $f \circ h \neq f \circ g$. Now both $f \circ h, f \circ g \in \mathcal{D}(\mathbf{B}_0)$. Since $\mathcal{D}(\mathbf{B}_0)$ is a Stone space there is a clopen set V that contains $f \circ h$ and whose complement contains $f \circ g$. Let a_0 and a_1 be distinct elements of A_0. Let $\gamma \in \mathbf{A}_0[\mathbf{B}_0]$ be defined by

$$\gamma(v) = \begin{cases} a_0 & \text{if } v \in V \\ a_1 & \text{otherwise.} \end{cases}$$

Then γ is continuous and so $\gamma \in \mathbf{A}_0[\mathbf{B}_0]$. But observe $((\Gamma(h))(\gamma))(f) = \gamma(f \circ h) = a_0$ while $((\Gamma(g))(\gamma))(f) = \gamma(f \circ g) = a_1 \neq a_0$. So the map Γ is one-to-one, as desired.

This completes the proof of the theorem. ■

We adopt some notation. Let $\mathbf{B}$ be a Boolean algebra and $\mathbf{A}$ be any algebra. Let $c, d \in A^{\mathcal{D}(\mathbf{B})}$. Let N be a clopen subset of $\mathcal{D}(\mathbf{B})$. We will use N° to denote its complement, which is also clopen. Using N we can form a new $\mathcal{D}(\mathbf{B})$-tuple, denoted by $c \sqcap_N d$, as follows

$$(c \sqcap_N d)(f) = \begin{cases} c(f) & \text{if } f \in N \\ d(f) & \text{if } f \in N^\circ \end{cases}$$

for all $f \in \mathcal{D}(\mathbf{B})$.

THEOREM 8.91. *Let* $\mathbf{B}$ *be a Boolean algebra and* $\mathbf{A}$ *be any algebra and suppose that* $S \subseteq A^{\mathcal{D}(\mathbf{B})}$. *Then* $S = A[\mathbf{B}]$ *if the following hold:*

(i) *the constant functions of* $A^{\mathcal{D}(\mathbf{B})}$ *belong to* S, *and*

(ii) $c \sqcap_N d \in S$ *for all* $c, d \in S$ *and all clopen subsets* N *of* $\mathcal{D}(\mathbf{B})$.

Proof. It is useful to notice that $A[\mathbf{B}]$ has the properties (i) and (ii) attributed to S. Leaving (i) to the reader, we check (ii). So let N be a clopen subset of $\mathcal{D}(\mathbf{B})$ and let $c, d \in A[\mathbf{B}]$. Put $e = c \sqcap_N d$. We must show that e is a continuous function from $\mathcal{D}(\mathbf{B})$ into A. It suffices to show that the inverse image of any point in A, with respect to e, is clopen. So let $a \in A$ and observe for all $f \in \mathcal{D}(\mathbf{B})$we have

$$\begin{aligned} f \in e^{-1}(a) &\Leftrightarrow e(f) = a \\ &\Leftrightarrow c(f) = a \text{ if } f \in N \text{ and } d(f) = a \text{ otherwise} \\ &\Leftrightarrow f \in c^{-1}(a) \cap N \text{ or } f \in d^{-1}(a) \cap N^\circ \\ &\Leftrightarrow f \in (c^{-1}(a) \cap N) \cup (d^{-1}(a) \cap N^\circ). \end{aligned}$$

This means that $e^{-1}(a) = (c^{-1}(a) \cap N) \cup (d^{-1}(a) \cap N^\circ)$. Hence $e^{-1}(a)$ is clopen and so e is continuous and thus belongs to $A[\mathbf{B}]$.

Now suppose that S has properties (i) and (ii). Let $c \in A[\mathbf{B}]$. We must prove that $c \in S$. In view of the compactness of $\mathcal{D}(\mathbf{B})$, we observe that the function c can take on only finitely many values. We argue that $c \in S$ by induction on the number of values c has.

Base Step This is the case when c has only one value. This puts $c \in S$ according to property (i).

Induction Step Our induction hypothesis is that all members of $A[\mathbf{B}]$ with just n values belong to S. So suppose c has $n + 1$ values. Let these values be $a_0, \dots, a_n$. Let $N_j = c^{-1}(a_j)$ for each $j \leq n$. Then each N_j is a clopen set. Define $c' \in A[\mathbf{B}]$ by

$$c'(f) = \begin{cases} c(f) & \text{if } f \in \bigcup_{j<n} N_j \\ a_{n-1} & \text{if } f \in N_n \end{cases}$$

for all $f \in \mathcal{D}(\mathbf{B})$. As c' is continuous, we have $c' \in A[\mathbf{B}]$. Now c' has just n values and so belongs to S by the inductive hypothesis. Let $d \in A[\mathbf{B}]$ be the constant function with value a_n. Then $d \in S$ by property (i). Finally, let $N = \bigcup_{j<n} N_j$. By property (ii), we arrive at $c' \sqcap_N d \in S$. But checking the definitions, we see that $c = c' \sqcap_N d \in S$. This completes the inductive step and hence the proof of the theorem. ■

An application of Boolean powers to primal algebras

DEFINITION 8.92. A functor $\Gamma : \mathbf{CAT}\,\mathcal{V} \to \mathbf{CAT}\,\mathcal{W}$ between two varieties $\mathcal{V}$ and $\mathcal{W}$ construed as categories, is called an **equivalence functor** if and only if two conditions below are satisfied:

(1) for all $\mathbf{B} \in \mathcal{W}$ there exists $\mathbf{A} \in \mathcal{V}$ with $\Gamma(\mathbf{A}) \cong \mathbf{B}$, and
(2) Γ maps $\hom(\mathbf{A}_0, \mathbf{A}_1)$ bijectively onto $\hom(\Gamma(\mathbf{A}_0), \Gamma(\mathbf{A}_1))$, for all $\mathbf{A}_0, \mathbf{A}_1 \in \mathcal{V}$.

Varieties $\mathcal{V}$ and $\mathcal{W}$ are called **categorically equivalent** if and only if there exists an equivalence functor between them.

We leave it to the reader to verify that categorical equivalence is, indeed, an equivalence relation; in particular, it is symmetric (but see also, exercises 15 and 16 on page 215 in Volume III).

Observe that if $\Gamma_0 : \mathbf{C}_0 \to \mathbf{C}_1$ and $\Gamma_1 : \mathbf{C}_1 \to \mathbf{C}_2$ are dual equivalence functors, then

$$\Gamma_1 \circ \Gamma_0 : \mathbf{C}_0 \to \mathbf{C}_2$$

is an equivalence functor.

This definition of equivalence functor is a special case of the definition on pages 135–136 in §3.6 of Volume I. It will be recalled in §10.6 in Volume III. The whole of §10.7 in Volume III is devoted to categorical equivalence of varieties. Indeed, another route to the theorem below is laid out in Example 2 on page 220 in Volume III.

Let $\mathbf{P}$ be a primal algebra and let $\mathcal{P}$ be the variety it generates. As a result of Theorems 8.84 and 8.88 we saw in Display 8.6.3 that the functor $\Gamma = \widehat{\mathbf{E}} \circ \mathcal{D}$ is an equivalence functor between the category $\mathbf{CAT}\,\mathcal{B}$ of Boolean algebras and $\mathbf{CAT}\,\mathcal{P}$. So we obtain the following theorem.

THEOREM 8.93. *Let $\mathcal{B}$ be the variety of Boolean algebras. Let $\mathbf{P}$ be a primal algebra and let $\mathcal{P}$ be the variety it generates. Let the functor $\Gamma : \mathbf{CAT}\,\mathcal{B} \to \mathbf{CAT}\,\mathcal{P}$ be defined by*

(i) *$\Gamma(\mathbf{B}) = \mathbf{P}[\mathbf{B}]$ for each object $\mathbf{B} \in \mathcal{B}$;*
(ii) *$\Gamma(h)$ is the map defined by $(\Gamma(h))(\gamma) = \gamma \circ \mathcal{D}(h)$ for all objects $\mathbf{B}_0, \mathbf{B}_1 \in \mathcal{B}$ and all morphisms $h \in \hom(\mathbf{B}_0, \mathbf{B}_1)$.*

Then Γ is an equivalence functor. ■

Parts of Theorem 8.93 were proved by Alfred Foster in two papers, (Alfred L. Foster 1953, 1953b). A proof of a generalization of this theorem was given by Tah Kai Hu (1969).

Theorem 8.93 above has a converse: Every variety categorically equivalent to the variety of Boolean algebras is generated by a primal algebra. So a variety is categorically equivalent to the variety of Boolean algebras if and only if it is generated by a primal algebra. This will be proved as Theorem 9.78 on page 58 in Volume III.

Exercises 8.94

1. Let **P** be a primal algebra. Prove that **P** is simple, has no proper subalgebras, and no automorphisms other than the identity.

2. Let **P** be a primal algebra and let $\mathbf{B}_0$ and $\mathbf{B}_1$ be Boolean algebras. Prove that $\mathbf{P}[\mathbf{B}_0] \cong \mathbf{P}[\mathbf{B}_1]$ if and only if $\mathbf{B}_0 \cong \mathbf{B}_1$.

3. Let **C** be an algebra of signature σ, and let **B** be a Boolean algebra, with Stone space X. Let **D** be the Boolean power $\mathbf{C}[\mathbf{B}]$, with universe the set D of all continuous functions $\mathrm{X} \longrightarrow C$ (as described in Definition 8.89). Here we describe a way of introducing operations on D that commute with all the σ-operations of **D**.

Let n be any positive integer and let $\phi : \mathrm{X} \longrightarrow \mathrm{X} \dot{\cup} \cdots \dot{\cup} \mathrm{X}$ (the disjoint union of n copies of X) be any continuous function. Let $F_\phi : D^n \longrightarrow D$ be defined by the formula $F_\phi(\zeta_0, \ldots, \zeta_{n-1}) = (\zeta_0 \sqcup \cdots \sqcup \zeta_{n-1}) \circ \phi$, where $\zeta_0 \sqcup \cdots \sqcup \zeta_{n-1}$ is the amalgamated function $\zeta : \mathcal{X} \dot{\cup} \cdots \dot{\cup} \mathcal{X} \longrightarrow \mathcal{X}$ that satisfies $\zeta \circ q_i^n = \zeta_i$ for each i. (Here q_i^n is the ith co-projection embedding X into $\mathrm{X} \dot{\cup} \cdots \dot{\cup} \mathrm{X}$.) The definition of "co-product" says this ζ exists uniquely.

Let G be an m-ary σ-operation of D. The exercise is to prove that G and F_ϕ **commute**, which is to say

$$\begin{aligned} &G(F_\phi(z_{0,0}, \ldots, z_{0,n-1}), \ldots, F_\phi(z_{m-1,0}, \ldots, z_{m-1,n-1})) \\ &\qquad = F_\phi(G(z_{0,0}, \ldots, z_{m-1,0}), \ldots, G(z_{0,n-1}, \ldots, z_{m-1,n-1})), \end{aligned}$$

for all values of the variables z_{ij}.

Another Application of Boolean Powers.

For any signature σ we may let $\mathcal{K}_\sigma$ denote the class of all algebras of signature σ modulo the relation of isomorphism. Then $\mathcal{K}_\sigma$ may be considered as a commutative monoid under the operation of forming direct products $\langle \mathbf{A}, \mathbf{B} \rangle \longmapsto \mathbf{A} \times \mathbf{B}$. (Any one-element algebra of $\mathcal{K}_\sigma$ will serve as a unit element.) Various classes of algebras, such as varieties, quasivarieties, and their subclasses of finite algebras, and so on, have been studied as subsemigroups of $\mathcal{K}_\sigma$.

This analysis of $\mathcal{K}_\sigma$ and its submonoids was introduced in §3.3 (Volume I), and continued into Chapter 5, where we focused on unique factorization. In that chapter we examined some (factor-closed) submonoids of $\mathcal{K}_\sigma$, using natural generalizations of the axioms of finite group theory, which indeed turn out to yield unique-factorization theorems. Finite algebras with 0 (defined below) form such a class, as do algebras **A** such that **A** has a one-element subalgebra and such that **Con A** is a modular lattice of finite height. In such a realm of uniquely factorable finite algebras, the corresponding submonoid of $\mathcal{K}_\sigma$ takes the simple form of being isomorphic to $\langle \omega, \cdot \rangle$ (see page 261 of Volume I). But when we look beyond uniquely factorable algebras, there is no limit to the possible finite submonoids of $\mathcal{K}_\sigma$.

Thus in Chapter 5 we found assorted failures of unique factorization, and other non-intuitive factorization possibilities. For instance, we described Bjarni Jónsson's example (1945–57) of infinite torsion-free Abelian groups **G** and **J**, both directly indecomposable, such that $\mathbf{G}\times\mathbf{G}\cong\mathbf{J}\times\mathbf{J}$, but $\mathbf{G}\ncong\mathbf{J}$. (Volume I, Exercise 5.1(6), page 266; see also Exercise 3 here.) And in Exercise 5.1(9) we exhibited a Boolean algebra **B** such that $\mathbf{B}\cong\mathbf{B}\times\mathbf{B}_2\times\mathbf{B}_2$, but $\mathbf{B}\ncong\mathbf{B}\times\mathbf{B}_2$ (with $\mathbf{B}_2$ the two-element Boolean algebra). (This is a variation on an example due to William Hanf (1957). See also Exercise 4.)

Jussi Ketonen (1978) went on to prove that every countable commutative semigroup can be embedded into the semigroup of all isomorphism classes of countable Boolean algebras under $\times$. Bjarni Jónsson (1957) had given a method, based on Boolean powers, for translating such Boolean-algebra results to the field of centerless groups, or more generally, algebras with 0 that have center equal to $\{0\}$. (Definitions will be supplied below.) We now commit the rest of §8.6 to an explanation of Jónsson's method.

An **algebra with zero** is an algebra $\mathbf{A}=\langle A;+,0,\cdots F\cdots\rangle$ obeying $x+0\approx 0+x\approx x$ and all the laws needed to say that $\{0\}$ is a one-element subuniverse (Volume I, §5.4). (These laws will be applied without further mention in the proof of Theorem 8.98.) In keeping with our introductory paragraphs, we will fix a similarity type σ containing a binary symbol $+$ and a nullary symbol 0. Thus all direct-product monoids of algebras with zero will automatically be submonoids of $\mathcal{K}_\sigma$.

A **central subuniverse** of such an **A** (Volume I, §5.5) is a subuniverse C such that

(1) $(\forall b\in C)(\exists c\in C)(b+c=0)$.
(2) $(\forall a_1,a_2\cdots\in A)(\forall c_1,c_2\cdots\in C)$

$$F(a_1+c_1,\,a_2+c_2,\cdots) \;=\; F(a_1,a_2,\cdots)+F(c_1,c_2,\cdots)$$

for every **A**-operation F.

For each algebra **A** with zero, there is a largest central subuniverse C (Volume I, p. 296, Lemma 1). This C is called **the center of A**. If this center is $\{0\}$, then we call **A centerless**. It is not hard to check that this definition of the center accords with the group-theoretic definition. In other words, if **A** happens to be a group (written additively, with 0 as unit element), then the center of **A** is the set of elements that commute (under $+$) with every element of **A** (Volume I, p. 300, Exercise 2).

Now let **A** be an arbitrary algebra of any signature. For $\psi\in\mathbf{Con\,A}$ and $\Gamma\subseteq\mathbf{Con\,A}$, we will write $\psi=\prod\Gamma$ if and only if $\psi=\bigwedge\Gamma$ and

$$(\forall a\in A^\Gamma)(\exists b\in A)(\forall\theta\in\Gamma)\,[\langle a_\theta,b\rangle\in\theta]\,. \tag{8.6.4}$$

The practical meaning of (8.6.4) is that the natural homomorphism from $\mathbf{A}/\psi$ to $\prod_\Gamma A/\theta$ is surjective. This surjectivity makes possible an analysis of factorizations of **A** in terms of the members of **Con A**—see §5.2 of Volume I for more information. In words, we say that ψ is **the product of the congruences** $\theta\in\Gamma$. If the congruences in Γ are named, e.g. if $\Gamma=\{\theta_0,\theta_1,\dots\}$, then we may write $\theta_0\times\theta_1\times\cdots$ for the product congruence. We say that $\theta\in\mathbf{Con\,A}$ is **indecomposable** iff $\theta\neq 1$ and $\theta=\theta_0\times\theta_1$ implies that $\theta_0=1$ or $\theta_1=1$. A congruence θ on **A** is called a **factor congruence of A** if there exists a congruence β such that $0=\alpha\times\beta$. The family of all factor congruences on **A** will be denoted $\mathcal{F}(\mathbf{A})$.

The **strict refinement property** (§5.6) for an algebra **A** asserts that if

$$0 = \prod_{i\in I} \beta_i = \prod_{j\in J} \gamma_j,$$

for congruences β_i, γ_j on **A**, then there exist congruences δ_{ij} ($i \in I, j \in J$) such that, for all i and j,

$$\beta_i = \prod_{j\in J} \delta_{ij} \qquad \text{and} \qquad \gamma_j = \prod_{i\in I} \delta_{ij}.$$

LEMMA 8.95 (Exercise 5.15(8)). *Let* **A** *be an algebra with zero. If* **A** *is a subdirect product of centerless algebras, then* **A** *is centerless.* ∎

LEMMA 8.96 (Volume 1, p. 304, Corollary 2). *A centerless algebra with zero has the strict refinement property.* ∎

LEMMA 8.97 (Theorem 5.17, (i) $\Longrightarrow$ (iii)). *If* **A** *is an algebra that satisfies the strict refinement property, then $\mathcal{F}(\mathbf{A})$ is a Boolean lattice. (I.e., it is a complemented sublattice of* **Con A**, *and the distributivity laws hold on this sublattice.)* ∎

THEOREM 8.98 (Bjarni Jónsson 1957). *Let* **G** *be a non-trivial finite or denumerable indecomposable centerless algebra with zero, and let* **B** *be an infinite Boolean algebra. Then the Boolean power* $\mathbf{G}[\mathbf{B}]$ *is a centerless algebra with zero, and there exists a function F with the following properties:*

(i) *F maps* **B** *isomorphically onto the Boolean algebra* $\mathcal{F}(\mathbf{G}[\mathbf{B}])$.
(ii) *For all $a, b \in$* **B**, *we have* $\mathbf{G}[\mathbf{B}]/F(a) \cong \mathbf{G}[\mathbf{B}]/F(b)$ *if and only if* $\mathbf{I}[0,a] \cong \mathbf{I}[0,b]$ (isomorphism of intervals in **B**).
(iii) $\mathbf{G}[\mathbf{B}]/F(b) \cong \mathbf{G}$, *for every atom b of* **B**.

Remark on intervals in Boolean algebras. For a an element of a Boolean algebra **B**, the interval $\mathbf{I}[0,a]$, is a subuniverse if we disregard the complementation operation. It is still a Boolean lattice, so we may (uniquely) restore complementation; in fact complementation for $\mathbf{I}[0,a]$ has the formula $c(x) = a \wedge (x^-)$. In places such as Condition (ii) just above, it is immaterial whether we understand isomorphism of intervals relative to all three operations or relative to meet and join alone. And in Equation (8.6.6) below, in order to construct $\mathbf{G}[\mathbf{I}[0,b]]$, we must know the Boolean space corresponding to the interval $\mathbf{I}[0,b]$; it is made up of maximal ideals, which can be described without mention of the complement operation. So our various mentions of intervals remain valid without express availability of complementation.

Proof of Theorem 8.98.

Definition of F: The fact that $\mathbf{G}[\mathbf{B}]$ is centerless is evident from Lemma 8.95 and our assumption that **G** is centerless; the fact that $\mathcal{F}(\mathbf{G}[\mathbf{B}])$ is a Boolean algebra is evident from Lemmas 8.95–8.97. We may assume that for some Stone (compact, totally disconnected) topological space $\mathcal{X}$ (the Stone space of **B**), the algebra **B** is the set of continuous maps $\mathcal{X} \to 2$ (equivalently, the set of clopen subsets of $\mathcal{X}$). The Boolean complement b^- of $b \in B$ is its complement in the set $\mathcal{X}$. For b an element of **B** we may consider b

to be a clopen subset of $\mathcal{X}$, and then we may of course consider the interval sublattice $\mathbf{I}[0,b]$ to be the set of continuous functions from b to $\{0,1\}$. In this setting, the Boolean power $\mathbf{G}[\mathbf{B}]$ is the set of continuous maps $\mathcal{X} \longrightarrow \mathbf{G}$. If $c \in B$ and $\mathbf{I}[0,c]$ denotes the corresponding interval sublattice (or principal ideal) then the (smaller) Boolean power $\mathbf{G}[\mathbf{I}[0,c]]$ may be taken as the set of continuous functions from the clopen set c to the algebra $\mathbf{G}$.

Having established these conventions, we may define the function F that is needed for Conditions (i–iii) of the theorem. For b an element of $\mathbf{B}$ (i.e. for b a clopen subset of $\mathcal{X}$), we define F(b) to be the congruence on $\mathbf{G}[\mathbf{B}]$ defined by

$$\langle \phi, \psi \rangle \in F(b) \quad \text{iff} \quad \phi \restriction b^- = \psi \restriction b^-, \tag{8.6.5}$$

for elements $\phi, \psi \in \mathbf{G}[\mathbf{B}]$. Directly from the definition one can see the following isomorphism of algebras with zero:

$$\mathbf{G}[\mathbf{B}]/F(b) \cong \mathbf{G}[\mathbf{I}[0,b]]. \tag{8.6.6}$$

Part (i): **basics plus injectivity.** The reader may check that (8.6.5) defines $F(b)$ as a congruence relation on $\mathbf{G}[\mathbf{B}]$. Moreover $F(b)$ has $F(b^-)$ as a Boolean complement for every $b \in B$, so obviously $F(b)$ is a factor congruence for every b. We leave it to the reader to prove that F is a homomorphism and F is injective.

Part (i): **surjectivity.** To finish the proof of Part (i), we will prove that F is surjective. Suppose η_0 is any factor congruence on $\mathbf{G}[\mathbf{B}]]$; this means that for some η_1 we have $0 = \eta_0 \times \eta_1$ in $\mathbf{Con}\,\mathbf{G}[\mathbf{B}]]$. We must prove that there exists $b \in B$ such that $F(b) = \eta_0$.

We now consider an arbitrary $p \in \mathcal{X}$ (our Stone space). For each p there is a homomorphism of algebras with zero, $\pi_p : \mathbf{G}[\mathbf{B}] \longrightarrow \mathbf{G}$, which picks up function values at p. Namely, $\pi_p(f) := f(p)$. (Effectively, π_p is the pth coordinate projection from the subdirect power $\mathbf{G}[\mathbf{B}]$.) We now define, for each $p \in \mathcal{X}$, two congruences on $\mathbf{G}$:

$$\alpha_0 = \alpha_0(p) = \pi_p[\eta_0 \vee \ker \pi_p]; \quad \alpha_1 = \alpha_1(p) = \pi_p[\eta_1 \vee \ker \pi_p]. \tag{8.6.7}$$

We omit the straightforward arguments that $\alpha_i \in \mathbf{Con}\,\mathbf{G}$ for $i = 0, 1$, that $\alpha_0 \wedge \alpha_1 = \bot$ and that $\alpha_0 \circ \alpha_1 = \top$ in $\mathbf{Con}\,\mathbf{G}$. In other words, α_0 and α_1 are complementary factor congruences on $\mathbf{G}$. Since $\mathbf{G}$ was assumed to be indecomposable, we must have $\alpha_0(p) = \bot$ or $\alpha_1(p) = \bot$.

Now letting p range over the space $\mathcal{X}$, we may define two complementary subsets of $\mathcal{X}$:

$$U_0 \;=\; \{p \in \mathcal{X} \mid \alpha_0(p) = \bot\}; \quad U_1 \;=\; \{p \in \mathcal{X} : \alpha_1(p) = \bot\}.$$

We now set out to prove that U_0 (hence also U_1) is a clopen subset of $\mathcal{X}$.

The algebra $\mathbf{G}$ contains an element $g \neq 0$. Let $\hat{g}$ be the constant function on $\mathcal{X}$ with value g. Similarly $\hat{0}$ is the constant function with value 0. As mentioned above, $\hat{g}$ may be written uniquely as $\hat{g} = g_0 + g_1$, where g_0 and g_1 are continuous functions $\mathcal{X} \longrightarrow G$, and where $\langle g_i, \hat{0} \rangle \in \eta_i$ for $i = 0, 1$. Our claim is now that

$$U_0 \subseteq \{p \in \mathcal{X} \mid g_0(p) = 0\} \quad \text{and} \quad U_1 \subseteq \{p \in \mathcal{X} \mid g_1(p) = 0\}. \tag{8.6.8}$$

These two inclusions have formally identical proofs, so we prove only the first. It follows readily from (8.6.7) that $\alpha_0(p) = \bot$ iff $\eta_0 \subseteq \ker \pi_p$. Therefore our claim reduces to proving that for all p, if $\eta_0 \subseteq \ker \pi_p$ then $g_0(p) = 0$. Suppose that $\eta_0 \subseteq \ker \pi_p$. Then we have $\langle g_0, \hat{0} \rangle \in \ker \pi_p$, and so $g_0(p) = \hat{0}(p) = 0$. This completes the proof of (8.6.8). We now have the two complementary sets, U_0 and U_1, appearing as subsets of the disjoint sets $\{p \in \mathcal{X} \mid g_0(p) = 0\}$ and $\{p \in \mathcal{X} \mid g_1(p) = 0\}$. (Disjoint because g was taken to be non-zero.) The only way this can happen is that (8.6.8) holds with equality in place of each inclusion relation. Thus (by the continuity of g_0) U_0 and U_1 are a complementary pair of clopen subsets of $\mathcal{X}$. In other words, U_0 and U_1 are a complementary pair in our Boolean algebra **B**.

To finish surjectivity, we will prove that $F(U_0) = \eta_0$. We first prove that $\eta_0 \subseteq F(U_0)$ (and of course $\eta_1 \subseteq F(U_1)$). (Recall that F was defined in (8.6.5).) For $\phi, \psi : \mathcal{X} \longrightarrow \mathbf{G}$, let us assume that $\langle \phi, \psi \rangle \in \eta_0$. To check that $\langle \phi, \psi \rangle \in F(U_0)$, we need to verify, for all $p \in U_0$, that $\phi(p) = \psi(p)$, which is to say that $\pi_p(\phi) = \pi_p(\psi)$. This conclusion is immediate from $\eta_0 \subseteq \ker \pi_p$, proved above. We now have $\eta_i \subseteq F(U_i)$ for $i = 0, 1$, plus $\eta_0 \times \eta_1 = F(U_0) \times F(U_1) = 0$. Thus Exercise 1 tells us that $\eta_0 = F(U_0)$, which yields surjectivity of F.

Part (ii): By (8.6.6) it will suffice to prove that $\mathbf{I}[0, a] \cong \mathbf{I}[0, b]$ if and only if $\mathbf{G}[\mathbf{I}[0, a]] \cong \mathbf{G}[\mathbf{I}[0, b]]$. The forward direction is obvious. Conversely, if we are given $\mathbf{G}[\mathbf{I}[0, a]] \cong \mathbf{G}[\mathbf{I}[0, b]]$, then by Part (i), applied to the Boolean algebras $\mathbf{I}[0, a]$ and $\mathbf{I}[0, b]$, we have

$$\mathbf{I}[0, a] \cong \mathcal{F}(\mathbf{G}[\mathbf{I}[0, a]]) \cong \mathcal{F}(\mathbf{G}[\mathbf{I}[0, b]]) \cong \mathbf{I}[0, b].$$

Part (iii): This is a special case of (8.6.6). ■

Application of Theorem 8.98: Tarski's cube problem

One motivator for the study of direct factorizations was "Tarski's cube problem" (Alfred Tarski, 1949), which asked if there is an infinite algebra **A** such that $\mathbf{A}^3 \cong \mathbf{A}$ but $\mathbf{A}^2 \ncong \mathbf{A}$. Actually, Tarski posed this question, among others, in the context of Boolean algebras. Later he promoted this problem in a wider context. A. L. S. Corner (1969) found an Abelian p-group with the property. For Boolean algebras, this question was resolved affirmatively by Jussi Ketonen (1978), with **A** a countable Boolean algebra. In fact he proved, much more generally, that any countable commutative semigroup can be represented isomorphically by (isomorphism classes of) certain denumerable Boolean algebras under the direct product operation. The proof of Ketonen's result goes beyond the scope of this book, and we merely cite it.

At the time of Ketonen's discovery it was of course understood that Jónsson's Theorem 8.98 can be applied to transfer such results from Boolean algebras to another realm, such as group theory. Here we will illustrate this technique by transferring the Boolean-algebra solution of Tarski's cube problem to groups.

First, to apply the Ketonen result to the cube problem, we consider the commutative semigroup $\mathbf{S} = \langle \{a, b\}, \cdot \rangle$, where $b \cdot b = a \cdot a = b$ and $a \cdot b = b \cdot a = a$. (In fact this is the commutative semigroup presented by $\langle a \mid a^3 = a \rangle$. For any further application of this type, such as Exercise 3 below, one might first look at an appropriate presentation.) So, according to Ketonen, we may represent **S** as $\langle \{\mathbf{A}, \mathbf{B}\}, \times \rangle$ by nonisomorphic countable Boolean algebras **A**, **B**. In particular, these two Boolean algebras must satisfy $\mathbf{A}^3 \cong \mathbf{A}$ and $\mathbf{A}^2 \cong \mathbf{B} \ncong \mathbf{A}$.

Let **G** be a group that is centerless and directly indecomposable, such as any simple group, and let **A** and **B** be the already described Boolean algebras. Let **H** be the Boolean power $\mathbf{G}[\mathbf{A}]$. We will prove that the group **H** is a solution to Tarski's cube problem; we begin with $\mathbf{H}^3 \cong \mathbf{H}$. Since $\mathbf{A} \cong \mathbf{A}^3$, there exist congruences α_i $(i = 0,1,2)$ on $\mathbf{A}$ such that $0 = \alpha_0 \times \alpha_1 \times \alpha_2$, and with $\mathbf{A}/\alpha_i \cong \mathbf{A}$ for $i = 0,1,2$. By Exercise 2, there exist $a_0, a_1, a_2 \in A$, obeying

$$a_0 \wedge a_1 = a_0 \wedge a_2 = a_1 \wedge a_2 = 0; \quad a_0 \vee a_1 \vee a_2 = 1. \tag{8.6.9}$$

such that, for each i, the congruence α_i is the same as $\theta(a_i)$ as defined in Exercise 2. Moreover, by Exercise 2, each interval $\mathbf{I}[0, a_i] \subseteq \mathbf{A}$ is isomorphic to $\mathbf{A}/\alpha_i$, which in turn is isomorphic to $\mathbf{A}$.

We now turn to Jónsson's Theorem 8.98. Part (i) of that theorem yields a homomorphism F which provides us with factor congruences $F(a_i)$ $(i = 0,1,2)$ of $\mathbf{H}$. Since F is a homomorphism, Equation (8.6.9) yields $F(a_0)\times F(a_1)\times F(a_2) = 0$ in $\mathbf{Con\,H}$. Now Part (ii) of Jónsson's theorem tells us that $\mathbf{H}/F(a_0) \cong \mathbf{H}/F(a_1) \cong \mathbf{H}/F(a_2) \cong \mathbf{H}/\bot = \mathbf{H}$ (since $\mathbf{I}[0, a_0] \cong \mathbf{I}[0, a_1] \cong \mathbf{I}[0, a_2] \cong \mathbf{I}[0, 1]$). Thus $\mathbf{H} \cong \mathbf{H}^3$.

Now let $b = a_0 \vee a_1$ in $\mathbf{A}$, and let $\beta = \theta(b)$. We have $\mathbf{A}/\beta \cong \mathbf{A}^2 \not\cong \mathbf{A} \cong \mathbf{A}/\bot$. Thus $\mathbf{A}/\beta \not\cong \mathbf{A}/\bot$. Thus by Exercise 2, Part (4), $\mathbf{I}[0, b] \not\cong \mathbf{I}[0, 1]$. Thus by Part (ii) of Jónsson's Theorem, we have $\mathbf{H}/\beta \not\cong \mathbf{H}$. Finally $\mathbf{H}^2 = \mathbf{H}\times\mathbf{H} \cong \mathbf{H}/\alpha_0 \times \mathbf{H}/\alpha_1 \cong \mathbf{H}/\beta \not\cong \mathbf{H}$.

Exercises 8.99

1. Let α, β, γ and δ be equivalence relations on a set A, with $\alpha \subseteq \gamma$ and $\beta \subseteq \delta$. If $\alpha \times \beta = \gamma \times \delta = 0_A$, then $\alpha = \gamma$ and $\beta = \delta$.

2. If $\mathbf{B}$ has a semilattice operation $\wedge$, and if $b \in B$, we can define an equivalence relation $\theta(b)$ on B as follows: $\langle x, y\rangle \in \theta(b)$ iff $x \wedge b = y \wedge b$. We now assume that $\mathbf{B}$ is a Boolean algebra: $\mathbf{B} = \langle B; \wedge, \vee, ' \rangle$. (1) If $b_0, \ldots, b_{k-1}$ are mutually orthogonal ($b_i \wedge b_j = 0$ for $i < j < k$) and if they join to 1, then $\theta(b_0) \times \cdots \times \theta(b_{k-1}) = 0$. (2) Conversely, if $\beta_i \in \mathbf{Con\,B}$ $(0 = i < k)$ and if $\beta_0 \times \cdots \times \beta_{k-1} = 0$, then there exist $b_0, \ldots, b_{k-1} \in B$ such that $\theta(b_i) = \beta_i$ for each i. (3) In fact each b_i' must be the largest element of the class $0/\beta_i$, and so each b_i is uniquely determined by β_i. (4) For each i we have $\mathbf{B}/\beta_i \cong \mathbf{I}[0, b_i]$ (as Boolean lattices). (These facts can be proved directly from the axioms of Boolean algebra. It may seem easier to prove them from Stone's Representation Theorem (8.82, on page 381).)

3. In Volume 1, page 266, we saw Bjarni Jónsson's construction of torsion-free Abelian groups $\mathbf{G}$ and $\mathbf{J}$ such that $\mathbf{G}\times\mathbf{G} \cong \mathbf{J}\times\mathbf{J}$, but $\mathbf{G} \not\cong \mathbf{J}$. Here we ask for a pair of non-Abelian groups exhibiting this pattern, this time formed from Ketonen's theorem and Theorem 8.98.

4. Let $\mathbf{G}$ be a non-trivial finite group that is centerless and directly indecomposable, such as any simple group. Construct an uncountable group $\mathbf{H}$ such that $\mathbf{H}\times\mathbf{G}\times\mathbf{G} \cong \mathbf{H}$, but $\mathbf{H}\times\mathbf{G} \cong \mathbf{H}$. (In fact we have already realized this pattern in groups, but here we have the added feature that $\mathbf{G}$ is finite.) (Hint. In Volume I, Exercise 5.1(9), there is an uncountable Boolean algebra $\mathbf{B}$ such that $\mathbf{B} \cong \mathbf{B}\times\mathbf{B}_2\times\mathbf{B}_2$, but $\mathbf{B} \not\cong \mathbf{B}\times\mathbf{B}_2$ (with $\mathbf{B}_2$ the two-element Boolean algebra).)

8.7. Universal Classes and Quasivarieties

Classes That Are Relativized Reducts of Elementary Classes

Suppose $\mathcal{K}$ is a class of structures of the same signature. Under what circumstances can we be sure that $\mathcal{K}$ is an elementary class? We know, from Theorem 8.49, one answer involving ultraproducts and ultraroots. In contrast with Birkhoff's HSP Theorem, this characterization of elementary classes is difficult to apply. In this section, we consider two special kinds of elementary classes, which like varieties, have more amendable characterizations.

The first step is to introduce a notion wider than elementary class.

Let us look first at an example. A group **G** has a **faithful** n**-dimensional representation** provided **G** can be embedded into the group of $n \times n$ invertible matrices over some field. Let $\mathcal{K}_n$ be the class of all groups with faithful n-dimensional representations. We would like to know whether there is some set Γ of sentences in the signature of group theory that axiomatizes this class $\mathcal{K}_n$. The basic difficulty is that the definition we gave talks about the existence of things like embeddings and some nice vector space over some field ... none of these things are elements of the groups at hand. But we can deal with them if we are willing to contend with a richer signature. Here is what we could use:

- A one-place relation symbol G to name the universe of our group.
- A one-place relation symbol F to name the universe of a field.
- A one-place relation symbol V to name the universe of a vector space.
- Operation symbols to stand for the group operations on G.
- Operation symbols to stand for the ring operations on F.
- Operation symbols to stand for the vector space operations on V.
- A two-place operation symbol $\star$ to stand for the action of the group on the vector space.

Now we would have to write down in detail a whole bunch of elementary sentences to try to capture $\mathcal{K}_n$. We only indicate what has to be said.

- "G is closed under the group operations".
- "Under the group operations, G becomes a group" (This is basically a handful of equations.)
- "F is closed under the ring operations".
- "Under the ring operations, F becomes a field".
- "V is closed under the vector space operations." (Careful about scalar multiplication!)
- "Under the vector space operations, V becomes a vector space over the field."
- "The dimension of the vector space is n".
- "For all $g \in G$ it turns out that $g \star v$ is a function from V to V".
- "For all $g \in G$ the function $g \star v$ is a linear operator on V".
- "The assignment $g \mapsto g \star v$ is a group embedding."

Let Λ be the set of all these elementary sentences.

Now the idea is that $\mathbf{G} \in \mathcal{K}_n$ if and only if there is some model $\mathbf{M}$ of Λ so that **G** is just the "group part" of **M**. This is not enough to say that $\mathcal{K}_n$ is an elementary class. On the other hand, it is closely tied to one. Classes like $\mathcal{K}_n$ that admit such descriptions are known as **relativized reducts** of elementary classes. That is, roughly speaking,

$\mathcal{K}$ is such a class provided in some richer signature there is an elementary class $\mathcal{L}$ so that $\mathcal{K}$ consists precisely of the "$\mathcal{K}$" parts of structures belonging to $\mathcal{L}$. More precisely a class $\mathcal{K}$ of structures of signature $\mathfrak{L}$ is a **relativized reduct of an elementary class** provided there is a signature $\mathfrak{L}^+$ expanding $\mathfrak{L}$ and a one-place relation symbol U of $\mathfrak{L}^+$ but not of $\mathfrak{L}$ and a set Λ of $\mathfrak{L}^+$ sentences such that

- The set Λ includes sentences that assert, in every $\mathfrak{L}^+$-structure, that U denotes a nonempty set that is closed under all the operations denoted by operation symbols of $\mathfrak{L}$.
- The class $\mathcal{K}$ consists precisely of those $\mathfrak{L}$-structures $\mathbf{A}$ obtained from $\mathfrak{L}^+$-structures $\mathbf{B}$ that are models of Λ by letting $U^{\mathbf{B}}$ be the universe of $\mathbf{A}$ and for each operation or relation symbol of $\mathfrak{L}$ by letting its interpretation in $\mathbf{A}$ be the restriction of its interpretation in $\mathbf{B}$ to $U^{\mathbf{B}}$.

In the last constraint above, we denote by $\mathbf{B}_U$ the $\mathfrak{L}$-structure obtained from the $\mathfrak{L}^+$-structure $\mathbf{B} \models \Lambda$. We call this the **relativized reduct over** U of $\mathbf{B}$ to $\mathfrak{L}$.

The Łoś-Tarski Theorem

In the mid-1950's Jerzy Łoś (1955a) and Alfred Tarski (1954), working independently, found versions the following theorem.

THEOREM 8.100 (The Łoś-Tarski Theorem). *Let $\mathcal{K}$ be a class of structures, all of the same signature. If $\mathcal{K}$ is a relativized reduct of an elementary class and $\mathcal{K}$ is closed under the formation of substructures, then $\mathcal{K}$ is an elementary class that can be axiomatized by some set of universal sentences.*

Proof. Let $\mathfrak{L}$ be the signature of $\mathcal{K}$ and let Ψ be the set of all universal $\mathfrak{L}$-sentences true in $\mathcal{K}$. Let $\mathfrak{L}^+$ be an expansion of $\mathfrak{L}$ and let Λ be a set of $\mathfrak{L}^+$-sentences that witness that $\mathcal{K}$ is a relativized reduct of an elementary class. Further let U be the new one-place relation symbol of $\mathfrak{L}^+$ that names the universes of structures in $\mathcal{K}$.

What we need is to prove that if $\mathbf{A} \models \Psi$, then $\mathbf{A} \in \mathcal{K}$. So suppose $\mathbf{A} \models \Psi$. Expand $\mathfrak{L}^+$ by adding new constant symbols c_a to name the elements of $a \in A$. Let Σ be the diagram of $\langle \mathbf{A}, a\rangle_{a\in A}$. (That is, Σ is the set of all atomic and negated atomic sentences holding in $\langle \mathbf{A}, a\rangle_{a\in A}$. See Definition 8.35 on page 336.) Observe that $\mathbf{C} \models \Sigma$ if and only if $\mathbf{A}$ is embeddable into $\mathbf{C} \restriction \mathfrak{L}$.

CLAIM 1.
The set $\Sigma \cup \{Uc_a \mid a \in A\} \cup \Lambda$ has a model.

Proof. Let us suppose our claim is false. Then there is a finite subset $\Phi \subseteq \Sigma$ and a finite $F \subseteq A$ so that $\Phi \cup \{Uc_a \mid a \in F\} \cup \Lambda$ that has no model. Let φ be the conjunction of the finitely many sentences in Φ. Notice that no quantifiers occur in φ. By enlarging F if necessary we can suppose that if c_a occurs in φ then $a \in F$. Let $F = \{a_0, a_1, \dots, a_{m-1}\}$. So we see

$$\Lambda \cup \{Uc_{a_0} \wedge Uc_{a_1} \wedge \cdots \wedge Uc_{a_{m-1}}\} \models \neg\varphi(c_{a_0}, \dots, c_{a_{m-1}}).$$

But then

$$\Lambda \models (Uc_{a_0} \wedge Uc_{a_1} \wedge \cdots \wedge Uc_{a_{m-1}}) \to \neg\varphi(c_{a_0}, \dots, c_{a_{m-1}}).$$

Since the new constants do not occur in Λ we have

$$\Lambda \models \forall \mathbf{y}((Uy_0 \curlywedge \cdots \curlywedge Uy_{m-1}) \to \neg\varphi(y_0, \dots, y_{m-1})).$$

Now suppose $\mathbf{B} \models \Lambda$. Then we see that

$$\mathbf{B} \models \forall \mathbf{y}((Uy_0 \curlywedge \cdots \curlywedge Uy_{m-1}) \to \neg\varphi(y_0, \dots, y_{m-1})).$$

But this means $\mathbf{B}_U \models \forall \mathbf{y} \neg\varphi(\mathbf{y})$. Since every structure in $\mathcal{K}$ is some such $\mathbf{B}_U$, we find that $\mathcal{K} \models \forall \mathbf{y}\neg\varphi(\mathbf{y})$. This means $\forall \mathbf{y}\neg\varphi(\mathbf{y}) \in \Psi$. But $\mathbf{A} \models \Psi$. This means

$$\mathbf{A} \models \forall \mathbf{y}\neg\varphi(\mathbf{y}), \text{ and}$$
$$\mathbf{A} \models \varphi(\mathbf{y})[a_0, \dots, a_{m-1}],$$

which is a contradiction. This establishes the claim. ■

So let $\mathbf{B} \models \Sigma \cup \{Uc_a \mid a \in A\} \cup \Lambda$. Now $\mathbf{B}$ is a structure for the signature $\mathfrak{L}^+$ expanded by the extra constant symbols. Let $\mathbf{B}_U^*$ be the relativized reduct of $\mathbf{B}$ to the signature $\mathfrak{L}$ expanded by the new constant symbols and let $\mathbf{B}_U$ be the relativized reduct of $\mathbf{B}$ to $\mathfrak{L}$. Since $\mathbf{B}_U^* \models \Sigma$, we see that $\mathbf{A}$ is isomorphic to a substructure of $\mathbf{B}_U$. Since $\mathbf{B}_U \in \mathcal{K}$, we finally obtain $\mathbf{A} \in \mathcal{K}$. This is what we wanted. It means that $\mathcal{K} = \mathbf{Mod}\,\Psi$. ■

A sentence φ is said to be **preserved under the formation of substructures** provided that if $\mathbf{A} \models \varphi$ and $\mathbf{B}$ is a substructure of $\mathbf{A}$, then $\mathbf{B} \models \varphi$.

THEOREM 8.101 (Sometimes also called the Łoś-Tarski Theorem). *A sentence is preserved under the formation of substructures if and only if it is logically equivalent to a universal sentence.*

Proof. It is evident that universal sentences are preserved under the formation of substructures. For the converse, suppose φ is preserved under substructures. Let $\mathcal{K} = \mathbf{Mod}\,\varphi$. This is an elementary class, so it is a relativized reduct of an elementary class. It is closed under the formation of substructures. By the Łos-Tarski Theorem it is axiomatized by some set Ψ of universal sentences. This means, in particular, that $\Psi \models \varphi$. By the Compactness Theorem there a finite $\Psi' \subseteq \Psi$ so that $\Psi' \models \varphi$. Let ψ be the conjunction of the finitely many universal sentences in Ψ'. Since a conjunction of universal sentences is logically equivalent to a universal sentence, we may suppose that ψ is itself universal. Also notice $\psi \in \Psi$. Altogether, this means $\psi \models \varphi$ and $\varphi \models \psi$. So φ is logically equivalent to the universal sentence ψ. ■

Theorems like the one above have the form "A sentence is preserved under the formation of some constructions if and only if the sentence is logically equivalent to one of a special syntactical form." Theorems like this are called **preservation theorems**. The reader is invited to formulate a preservation theorem that is a companion to the HSP Theorem.

Let us return to our example. Let n be a positive natural number and let $\mathcal{K}_n$ be the class of all groups that have faithful n-dimensional representations. We already convinced ourselves that $\mathcal{K}_n$ is a relativized reduct of an elementary class. But it is easy to see that any subgroup of a group with a faithful n-dimensional representation has itself a faithful n-dimensional representation. This means that $\mathcal{K}_n$ is actually an elementary class that can be axiomatized by a set of universal sentences.

Now a universal sentence has the form $\forall \mathbf{y}\theta(\mathbf{y})$ where no quantifiers occur in θ. Another way to view this, in fact our official view, is as $\neg\exists\mathbf{y}\neg\theta(\mathbf{y})$. That is, universal sentences are the negations of existential sentences: they assert that there are no elements that satisfy some kind of quantifier-free condition. Roughly speaking, each universal sentence can be understood to forbid some kind of finite configuration from occurring in any model of the sentence. So one way to view the Łoś-Tarski theorem is as asserting that certain kinds of classes of structures can be characterized by forbidding certain finite configurations, perhaps infinitely many such configurations.

(A. I. Mal'tsev, 1959a) has some related results.

Characterizations of Universal Classes

Shortly after he had published his version of the Łoś-Tarski theorem, Łoś found the following strengthening of it.

THEOREM 8.102 (The Łoś-Tarski Theorem, Ultraproduct Form). *Let $\mathcal{K}$ be a class of structures of the same signature. The class $\mathcal{K}$ is the class of all models of some set of universal sentences if and only if $\mathcal{K}$ is closed under the formation of ultraproducts and the formation of substructures.*

Proof. We already know that any elementary class is closed under the formation of ultraproducts and that any class axiomatized by a set of universal sentences must be closed under the formation of substructures. So we only concern ourselves with the converse. So we suppose that $\mathcal{K}$ is closed with respect to the formation of ultraproducts and substructures.

Let Ψ be the set of all universal sentences true in $\mathcal{K}$. All we need to do is show that if $\mathbf{A} \models \Psi$, then $\mathbf{A} \in \mathcal{K}$. So let $\mathbf{A} \models \Psi$. Expand the signature by adding a new constant c_a for every $a \in A$. Put $\mathbf{c} = \langle c_a \mid a \in A\rangle$. Let $\Sigma_{\mathbf{A}}$ be the diagram of $\mathbf{A}$.

For each finite subset $\Theta \subseteq \Sigma_{\mathbf{A}}$ we let θ be its conjunction. Let $\mathbf{y} = \langle y_a \mid a \in A\rangle$ be a system of distinct variables not occurring in $\theta(\mathbf{c})$. Let $\theta(\mathbf{y})$ be the formula obtained from θ by replacing, for all $a \in A$, the constant symbol c_a by the variable y_a. Let φ_Θ be the sentence $\forall\mathbf{y}\neg\theta(\mathbf{y})$. We see $\mathbf{A} \not\models \varphi_\Theta$. This means $\varphi_\Theta \notin \Psi$. So pick $\mathbf{B}_\Theta \in \mathcal{K}$ so that $\mathbf{B}_\Theta \models \neg\varphi_\Theta$. Untangling things a bit, we find that $\neg\varphi_\Theta$ is just another way to write $\exists\mathbf{y}\theta(\mathbf{y})$. This means that for each $a \in A$ so that the new constant symbol c_a occurs in $\theta(\mathbf{c})$ we can pick $b_\Theta(a) \in B_\Theta$ so that expanding $\mathbf{B}_\Theta$ by letting the new constant symbols name these elements leads to a model of Θ. Let us pick, arbitrarily, a default element $\infty_\Theta \in B_\Theta$. We can then extend b_Θ to be defined on A by setting $b_\Theta(a) = \infty_\Theta$ in case c_a does not occur in Θ.

Let $I = \{\Theta \mid \Theta(\mathbf{c}) \text{ is a finite subset of } \Sigma_{\mathbf{A}}\}$. For each $\Theta \in I$ let

$$E_\Theta = \{\Phi \mid \Phi \in I \text{ and } \Theta \subseteq \Phi\}.$$

Just as in the ultraproduct proof of the Compactness Theorem on page 334, the collection $\mathcal{C} = \{E_\Theta \mid \Theta \in I\}$ has the finite intersection property. Let $\mathcal{U}$ be an ultrafilter on I that extends $\mathcal{C}$.

Put $\mathbf{B} = \prod_I \mathbf{B}_\Theta/\mathcal{U}$. We see that $\mathbf{B} \in \mathcal{K}$, since $\mathcal{K}$ is closed under the formation of ultraproducts. It remains to prove that $\mathbf{A}$ can be embedded into $\mathbf{B}$.

Define $h : A \to \prod_I B_\Theta/\mathcal{U}$ by

$$h(a) := \langle b_\Theta(a) \mid \Theta(\mathbf{c}) \in T\rangle/\mathcal{U},$$

for all $a \in A$. The function h is our desired embedding. We need to show that it is one-to-one, that it preserves the basic operations, and that it preserves the basic relations as well as the failures of the basic relations. That is,

$$\mathbf{A} \models (\neg x_0 \approx x_1)[a,d] \text{ implies } \mathbf{B} \models (\neg x_0 \approx x_1)[h(a),h(d)]$$
$$\mathbf{A} \models (Qx_0 \cdots x_{r-1} \approx x_r)[a_0,\dots,a_r] \text{ implies } \mathbf{B} \models (Qx_0 \cdots x_{r-1} \approx x_r)[h(a_0),\dots,h(a_r)]$$
$$\mathbf{A} \models (Rx_0 \cdots x_{r-1})[a_0,\dots,a_{r-1}] \text{ implies } \mathbf{B} \models (Rx_0 \cdots x_{r-1})[h(a_0),\dots,h(a_{r-1})]$$
$$\mathbf{A} \models (\neg Rx_0 \cdots x_{r-1})[a_0,\dots,a_{r-1}] \text{ implies } \mathbf{B} \models (\neg Rx_0 \cdots x_{r-1})[h(a_0),\dots,h(a_{r-1})]$$

To say this another way, we need to show that h preserves the satisfaction of atomic formulas and their negations.

So let $\sigma(\mathbf{y})$ be any atomic formula. Taking into account the Fundamental Theorem of Ultraproducts, what we need to show is that for any assignment $\mathbf{a}$ from A,

$$\mathbf{A} \models \sigma(\mathbf{x})[\mathbf{a}] \text{ implies } \{\Theta(\mathbf{c}) \mid \mathbf{B}_\Theta \models \sigma(\mathbf{x})[b_\Theta(\mathbf{a})]\} \in \mathcal{U}$$

and

$$\mathbf{A} \models \neg\sigma(\mathbf{x})[\mathbf{a}] \text{ implies } \{\Theta(\mathbf{c}) \mid \mathbf{B}_\Theta \models \neg\sigma(\mathbf{x})[b_\Theta(\mathbf{a})]\} \in \mathcal{U}$$

Taking $\mathbf{a} = \langle a_0, a_1, a_2, \dots\rangle$, we see that $\mathbf{A} \models \sigma(\mathbf{x})[\mathbf{a}]$ means exactly the same as

$$\sigma(c_{a_0}, c_{a_1}, \dots) \in \Sigma_{\mathbf{A}}$$

and that $\mathbf{A} \models \neg\sigma(\mathbf{y})[\mathbf{a}]$ means $\neg\sigma(c_{a_0}, c_{a_1}, \dots) \in \Sigma_{\mathbf{A}}$. But recall that

$$E_{\{\sigma(c_{a_0},c_{a_1},\dots)\}} = \{\Theta \mid \Theta \in I \text{ and } \sigma(c_{a_0}, c_{a_1}, \dots) \in \Theta\} \in \mathcal{U}.$$

and a similar statement holds with $\neg\sigma(c_{a_0}, c_{a_1}, \dots)$ in place of $\sigma(c_{a_0}, c_{a_1}, \dots)$.

We have taken particular care in the definition of b_Θ so that if $\sigma(c_{a_0}, c_{a_1}, \dots) \in \Theta(\mathbf{c})$, then $\mathbf{B}_\Theta \models \sigma(\mathbf{x})[b_\Theta(\mathbf{a})]$. This means that

$$\text{If } \sigma(c_{a_0}, c_{a_1}, \dots) \in \Sigma_{\mathbf{A}}, \text{ then } E_{\{\sigma(c_{a_0},c_{a_1},\dots)\}} \subseteq \{\Theta(\mathbf{c}) \mid \mathbf{B}_\Theta \models \sigma(\mathbf{x})[b_\Theta(\mathbf{a})]\}.$$

Of course, the same applies with $\neg\sigma$ in place of σ. This completes the proof. ■

It is a straightforward exercise to establish that every relativized reduct of an elementary class is closed under the formation of ultraproducts. So the Łoś-Tarski Theorem is an immediate consequence of this ultraproduct version. Also, the ultraproduct version characterizes the elementary classes that can be axiomatized by sets of universal sentences. On the other hand, the version with relativized reducts of elementary classes is, in most cases, easier to apply.

Here is another characterization of universal classes. We call a class $\mathcal{K}$ of structures **locally finite** provided for all $\mathbf{A} \in \mathcal{K}$ we have that every finitely generated substructure of $\mathbf{A}$ is finite. The class $\mathcal{K}$ is **uniformly locally finite** provided there is a function $g : \omega \to \omega$ such that whenever $\mathbf{A} \in \mathcal{K}$ and n is a natural number, then every substructure of $\mathbf{A}$ generated by n or fewer elements has cardinality less than $g(n)$. The function g is called a **bounding function**. Finally, recall that the union of any chain of structures ordered by the substructure relation is again a structure (essentially, because the union of any chain of operations of the same rank is again an operation of that rank). The same applies to the union of any system of structures that is up-directed by the substructure relation—meaning that if $\mathbf{A}$ and $\mathbf{B}$ are in the system, then there is a structure $\mathbf{C}$ in the system so that $\mathbf{A}$ and $\mathbf{B}$ are substructures of $\mathbf{C}$. The union of such a system is called the *direct limit* of the system.

THEOREM 8.103 (Tarski's Characterization of Uniformly Locally Finite Universal Classes). *Let $\mathcal{K}$ be a uniformly locally finite class of structures all of the same finite signature. The following are equivalent:*

(i) *$\mathcal{K}$ is a universal class.*

(ii) *$\mathcal{K}$ is closed under*
- *the formation of isomorphic images,*
- *the formation of substructures, and*
- *the direct limit of any system of structures belonging to $\mathcal{K}$ that is up-directed by the substructure relation.*

(iii) *$\mathcal{K}$ is closed under the formation of isomorphic images, the formation of substructures, and for any structure $\mathbf{A}$ if every finitely generated substructure of $\mathbf{A}$ belongs to $\mathcal{K}$, then $\mathbf{A} \in \mathcal{K}$.*

Proof. (i) implies (ii)
Let $\mathcal{K} = \mathbf{Mod}\,\Gamma$, where Γ is a set of universal sentences. That $\mathcal{K}$ is closed under the formation of isomorphic images and substructures is evident. So let $\langle \mathbf{A}_i \mid i \in I\rangle$ be a system of structures, each belonging to $\mathcal{K}$, that is up-directed by the substructure relation. Let $\mathbf{A}$ be the limit of this system. To see that $\mathbf{A} \in \mathcal{K}$, let $\varphi \in \Gamma$ be chosen arbitrarily. So φ is $\forall \mathbf{y} \psi(\mathbf{y})$, where no quantifiers occur in ψ. We need to see that $\mathbf{A} \models \varphi$. To this end, let $\mathbf{a}$ be any assignment from A to the variables occurring in ψ. There are finitely many such variables, so using the up-directedness, we can pick $i \in I$ so that all the entries in $\mathbf{a}$ belong to A_i. Since $\varphi \in \Gamma$ and $\mathbf{A}_i \in \mathcal{K}$, we see that $\mathbf{A}_i \models \psi(\mathbf{y})[\mathbf{a}]$. But this means that $\mathbf{A} \models \psi(\mathbf{y})[\mathbf{a}]$ since ψ is quantifier-free and $\mathbf{A}_i$ is a substructure of $\mathbf{A}$. This means $\mathbf{A} \models \varphi$. So $\mathbf{A} \models \Gamma$, as desired.

(ii) implies (iii)
This follows immediately, since every structure is the limit of its up-directed system of finitely generated substructures, as the reader can easily verify.

(iii) implies (i)
Let $\Gamma = \{\varphi \mid \varphi$ is a universal sentence and $\mathcal{K} \models \varphi\}$. It only remains to prove that every model of Γ belongs to $\mathcal{K}$. So let $\mathbf{A} \models \Gamma$. We will argue that every finitely generated substructure of $\mathbf{A}$ belongs to $\mathcal{K}$ and then appeal to (iii). We need the following contention.

CLAIM 1.
The class $\mathbf{Mod}\,\Gamma$ *is uniformly locally finite.*

Proof of the claim. To see this, first observe that, up to isomorphism, for each n there are only finitely many structures generated by n or fewer elements that are substructures of structures belonging to $\mathcal{K}$. This follows since $\mathcal{K}$ is uniformly locally finite and since the signature is finite. Let g be a bounding function for $\mathcal{K}$. Let F_n be a set of distinct representatives of these isomorphism classes. Consider such a substructure $\mathbf{B} \in F_n$ and let $\{b_0, \ldots, b_{n-1}\}$ be a generating set of $\mathbf{B}$. Now B has no more than $g(n)$ elements, say $c_0, \ldots, c_{g(n)-1}$, repeating elements on this list as needed. For each $j < g(n)$ pick a term $t_j(x_0, \ldots, x_{n-1})$ so that

$$t_j^{\mathbf{B}}(b_0, \ldots, b_{n-1}) = c_j.$$

For each operation symbol Q, let its rank be r, we put into the set Φ all equations of the form

$$Qt_{j_0}\cdots t_{j_{r-1}} \approx t_{j_r}$$

whenever $Q^{\mathbf{B}}(c_{j_0},\dots,c_{j_{r-1}}) = c_{j_r}$. Also put into Φ all inequations $\neg t_j \approx t_k$ in the event that $c_j \neq c_k$, for $j,k < g(n)$. So Φ is a finite set of equations and inequations. Once an assignment of elements of some algebra has been made to the variables $x_0,\dots,x_{n-1}$, what Φ does is assure us that the set of values assigned to the representative terms is closed under the operations—as a consequence, the substructure generated by the assigned elements can have no more than $g(n)$ elements. Let $\varphi_{\mathbf{B},\mathbf{b}}$ denote the conjunction of the finite set Φ, where $\mathbf{b} = \langle b_0,\dots,b_{n-1}\rangle$. Now let ψ_n denote

$$\forall x_0, x_1,\dots,x_{n-1} \bigvee_{\substack{\mathbf{B}\in F_n\\ \mathbf{b}\text{ from }B}} \varphi_{\mathbf{B},\mathbf{b}}.$$

The sentence ψ_n asserts about a structure $\mathbf{A}$ that once any set of n or fewer elements of A is selected, then the substructure generated by these elements is isomorphic, as far as the basic operations are concerned, to one of the structures in F_n. Hence each ψ_n is true in $\mathcal{K}$. Since these sentences are universal, we conclude that $\psi_n \in \Gamma$ for each n. This, in turn, entails that $\mathbf{Mod}\,\Gamma$ is also uniformly locally finite—the same bounding function will do. This finishes the proof of the claim. ■

Returning to our proof that (iii) implies (i), let $\mathbf{A} \models \Gamma$. We argue that every finitely generated substructure of $\mathbf{A}$ belongs to $\mathcal{K}$.

So let $\mathbf{B}$ be a substructure of $\mathbf{A}$ generated by a finite set. Because $\mathbf{Mod}\,\Gamma$ is locally finite, we know that $\mathbf{B}$ is finite. Expand the signature by adding a new constant symbol for each element of B. Let $\delta(\mathbf{c})$ be the sentence resulting from forming the conjunction of the diagram of $\mathbf{B}$, which is finite since $\mathbf{B}$ is finite of finite signature. Observe that $\mathbf{A} \models \exists \mathbf{y}\delta(\mathbf{y})$. This means that $\mathbf{A} \not\models \forall \mathbf{y}\neg\delta(\mathbf{y})$. So $\forall\mathbf{y}\neg\delta(\mathbf{y}) \notin \Gamma$. Since this sentence is universal, we can pick $\mathbf{C} \in \mathcal{K}$ so that $\mathbf{C} \models \neg\forall\mathbf{y}\neg\delta(\mathbf{y})$. This is the same as $\mathbf{C} \models \exists\mathbf{y}\delta(\mathbf{y})$. So we see that $\mathbf{C}$ can be expanded to a model of the diagram of $\mathbf{B}$. This means that $\mathbf{B}$ is isomorphic to a substructure of $\mathbf{C}$. According to (iii), this entails that $\mathbf{B} \in \mathcal{K}$, just as desired. ■

Quasivarieties

A sentence φ is a **quasi-equation** provided it is of the form

$$\forall \mathbf{x}\,[\neg\epsilon_0 \vee \cdots \vee \neg\epsilon_{n-1} \vee \gamma\,]$$

where each ϵ_i as well as γ are equations. Here the list $\mathbf{x}$ of variables must include all the variables occurring in γ as well as in the ϵ_i's. We permit $n = 0$ above. In this case the sentence has the form $\forall\mathbf{y}\gamma$. Thus each equation is a quasi-equation. Casting quasi-equations in a more familiar form we have

$$\forall \mathbf{x}\,[(\epsilon_0 \wedge \cdots \wedge \epsilon_{n-1}) \implies \gamma\,].$$

Just as we did with equational logic, when dealing with quasi-equations we can suppress the leading universal quantifiers.

In this subsection we consider classes $\mathcal{K}$ of algebras, all of the same signature, such that $\mathcal{K} = \{\mathbf{A} \mid \mathbf{A} \models \Sigma\}$ for some set Σ of quasi-equations. Call such classes **quasivarieties**. Notice that every variety is a quasivariety.

The theory of quasivarieties, in some sense a step away from the theory of varieties of algebras, has taken its own place in the study of general algebraic systems. Here we provide the beginnings of the subject. A deeper investigation might begin with (Viktor A. Gorbunov, 1998) and continue with the recent monographs (Jennifer Hyndman and J. B. Nation, 2018) and (Kira Adaricheva, Jennifer Hyndman, J. B. Nation, and Joy Nishida, 2022).

Just as we did in §7.1 to illuminate the link between varieties and equational theories, we use a Galois connection to frame the connection between quasivarieties and quasi-equational theories.

The truth of quasi-equations in algebras imposes a two-place relation between the class of all algebras of our fixed signature and the set of all quasi-equations of the same signature. Of course $\models$ denotes this relation. Like any two-place relation, $\models$ gives rise to a Galois connection. The situation is displayed in Figure 8.12.

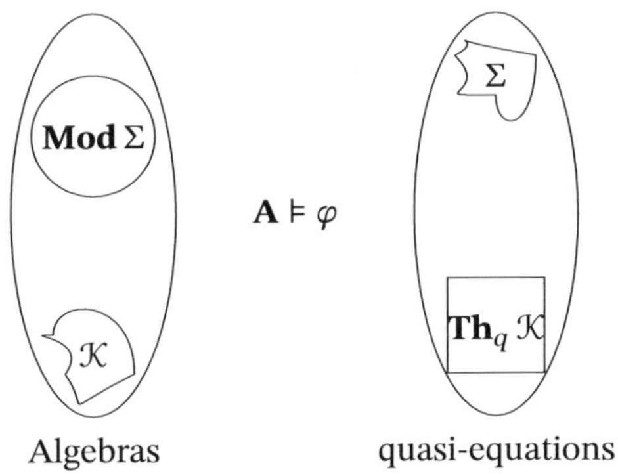

FIGURE 8.12. The Galois Connection Underlying the Study of Quasivarieties

The polarities of this connection are

$$\mathbf{Mod}\,\Sigma = \{\mathbf{A} \mid \mathbf{A} \models \varphi \text{ for all } \varphi \in \Sigma\},$$

where Σ is any set of quasi-equations
and

$$\mathbf{Th}_q\,\mathcal{K} = \{\varphi \mid \varphi \text{ is a quasi-equation and } \mathbf{A} \models \varphi \text{ for all } \mathbf{A} \in \mathcal{K}\},$$

where $\mathcal{K}$ is any class of algebras.

$\mathbf{Mod}\,\Sigma$ is called the the **quasivariety based** on Σ. It is the class of all models of Σ. The set $\mathbf{Th}_q\,\mathcal{K}$ is called the **quasi-equational theory** of $\mathcal{K}$. It is the set of all quasi-equations that are true in every algebra belonging to $\mathcal{K}$.

Just as with varieties, $\mathbf{Mod}\,\mathbf{Th}_q\,\mathcal{K}$ is the smallest quasivariety that includes the class $\mathcal{K}$. In this subsection, we characterize the closure operation $\mathbf{Mod}\,\mathbf{Th}_q$ in algebraic terms. We set aside the task of providing a syntactic description of $\mathbf{Th}_q\,\mathbf{Mod}$. This would amount to giving notion of formal deduction for quasi-equations and proving the corresponding completeness theorem. To see how this can be done in several ways, consult (A. Selman, 1972; David Kelly, 1984; George F. McNulty, 1977).

Since quasi-equations are universal sentences, it follows that quasivarieties are closed under the formation of subalgebras. But since quasi-equations are a restricted form of universal sentences, we can expect quasi-varieties to be closed not just with respect to ultraproducts, but under other constructions. In this subsection we will prove a sequence of theorems that play the role for quasivarieties that the HSP Theorem plays for varieties. In fact, it turns out that quasivarieties are exactly those classes of algebras closed with respect to the formation of (isomorphic copies) of subalgebras and the formation of (isomorphic copies) of arbitrary reduced products. It is interesting to note that while quasivarieties in general are not closed under the formation of homomorphic images, it is nevertheless true that each ultraproduct is a homomorphic image of a reduced product.

As a first step, we need the following theorem.

THEOREM 8.104. *Let $\mathcal{K}$ be a class of algebras, all of the same signature, so that $\mathbf{P}\,\mathcal{K} \subseteq \mathcal{K}$. Let θ be a universal sentence true in $\mathcal{K}$. Then there is a conjunction θ^* of quasi-equations so that $\models \theta^* \Rightarrow \theta$ and $\mathcal{K} \models \theta \Rightarrow \theta^*$.*

Proof. We may assume that θ is $\forall \mathbf{x}\theta'$ where θ' is in conjunctive-disjunctive normal form. That is θ' is $\varphi_0 \curlywedge \varphi_i \curlywedge \cdots \curlywedge \varphi_{n-1}$ for some natural number n and each φ_i has the form $\psi_{i,0} \curlyvee \cdots \curlyvee \psi_{i,m-1}$ for some natural number m and where each $\psi_{i,j}$ is either an equation or the negation of an equation. Notice that if exactly one of the disjuncts that make up φ_i is an equation, then $\forall \mathbf{x}\varphi_i$ is logically equivalent to a quasi-equation. If this happens to be true, at least in $\mathcal{K}$, for all $i < n$, then our proof can be completed with the observation that

$$\forall \mathbf{x}[\varphi_0 \curlywedge \cdots \curlywedge \varphi_{n-1}] \quad \text{is logically equivalent with} \quad \forall \mathbf{x}\varphi_0 \curlywedge \cdots \curlywedge \forall \mathbf{x}\varphi_{n-1}.$$

Since $\mathcal{K}$ is closed under the formation of arbitrary direct products (including the direct product of the empty system of algebras), we see that the 1-element algebra belongs to $\mathcal{K}$ and is therefore a model of θ. Consequently, the one-element algebra is also a model of $\forall \mathbf{x}\varphi_i$ for each $i < n$. This means that each φ_i has at least one disjunct that is an equation. It is harmless to suppose that for each $i < n$, the formula φ_i has been chosen (up to equivalence in $\mathcal{K}$) to have the smallest number of disjuncts that are equations.

In pursuit of a contradiction, suppose φ_i has at least two disjuncts that are equations. Indeed, let us take φ_i to be $\psi_0 \curlyvee \cdots \curlyvee \psi_j \curlyvee \psi_{j+1} \curlyvee \cdots \curlyvee \psi_{m-1}$, where $0 < j$ and $\psi_0, \dots, \psi_j$ are equations and the rest of the ψ_k's, should there be any, are negations of equations.

Let $\theta_{(0)}$ denote the sentence obtained from θ by deleting ψ_0 from φ_i within θ. Likewise, let $\theta_{(1)}$ denote the sentence obtained from θ by deleting ψ_1 from φ_i within θ. Evidently

$$\models \theta_{(\ell)} \implies \theta, \text{ for } \ell = 0, 1$$

but by the minimality imposed on the ψ_k's we have

$$\mathcal{K} \not\models \theta \implies \theta_{(\ell)} \text{ for } \ell = 0, 1.$$

So for each $\ell = 0, 1$ pick $\mathbf{B}_\ell \in \mathcal{K}$ so that $\mathbf{B}_\ell \models \theta$ but $\mathbf{B}_\ell \not\models \theta_{(\ell)}$. In particular, $\mathbf{B}_\ell \not\models \forall \mathbf{x}(\varphi_i)_{(\ell)}$. Let $\mathbf{B} := \mathbf{B}_0 \times \mathbf{B}_1$. Then $\mathbf{B} \in \mathcal{K}$ and so $\mathbf{B} \models \theta$. Thus, $\mathbf{B} \models \forall \mathbf{x}\varphi_i$. Since $\mathbf{B}_0 \not\models \forall \mathbf{x}(\varphi_i)_{(0)}$, pick an assignment $b_0 = \langle b_{0,0}, b_{0,1}, \dots \rangle$ of elements of B_0 to the variables so that b_0 fails to satisfy each of the disjuncts of $(\varphi_i)_{(0)}$. Likewise, let $b_1 = \langle b_{1,0}, b_{1,1}, \dots \rangle$ be an assignment of elements of B_1 that fails to satisfy each of the disjuncts of $(\varphi_i)_{(1)}$.

Let $b := \langle(b_{0,0}, b_{1,0}), (b_{0,1}, b_{1,1}), \dots\rangle$. So b is an assignment of elements of B to the variables. We see that b satisfies an equation $s \approx t$ in $\mathbf{B}$ if and only if b_0 satisfies $s \approx t$ in $\mathbf{B}_0$ and b_1 satisfies $s \approx t$ in $\mathbf{B}_1$. Note that each of the disjuncts of φ_i is either an equation or the negation of an equation. Now b_0 fails to satisfy each of the disjuncts of φ_i, with the exception of ψ_0, while b_1 fails to satisfy each of the disjuncts of φ_i, with the exception of ψ_1. Hence, b fails to satisfy each of the disjuncts of φ_i. But $\mathbf{B} \models \forall\mathbf{x}\varphi_i$. This is a contradiction. So each φ_i has exactly one disjunct that is an equation. This means $\forall\mathbf{x}\varphi_i$ is logically equivalent to a quasi-equation, as desired. ■

Theorems like Theorem 8.104 are sometimes called *interpolation theorems*.

THEOREM 8.105. *Let $\mathcal{K}$ be a class of algebras, all of the same signature. $\mathcal{K}$ is a quasivariety if and only if all of the following hold:*

(i) $\boldsymbol{S}\mathcal{K} \subseteq \mathcal{K}$,
(ii) $\boldsymbol{P_u}\mathcal{K} \subseteq \mathcal{K}$, *and*
(iii) $\boldsymbol{P}\mathcal{K} \subseteq \mathcal{K}$.

Proof. First suppose that $\mathcal{K}$ is a quasivariety. This means that $\mathcal{K}$ can be axiomatized by a set of quasi-equations. So conditions (i) and (ii) hold. For condition (iii) we need to show that quasi-equations are preserved with respect to the formation of arbitrary direct products. We leave this as an exercise for the reader.

Now suppose that $\mathcal{K}$ fulfills conditions (i),(ii), and (iii). Since (i) and (ii) hold we know that $\mathcal{K}$ can be axiomatized by a set Σ of universal sentences according the the ultraproduct version of the Łoś-Tarski Theorem. According to Theorem 8.104 for each sentence $\theta \in \Sigma$ there is a conjunction θ^* of quasi-equations such that $\mathcal{K} \models \theta \Leftrightarrow \theta^*$. Let

$$\Sigma^* = \{\varphi \mid \varphi \text{ is a conjunct of } \theta^* \text{ for some } \theta \in \Sigma\}.$$

We claim that $\mathcal{K} = \mathbf{Mod}\,\Sigma^*$. To see this observe that for each $\theta \in \Sigma$

$$\begin{aligned} \mathbf{A} \in \mathcal{K} &\Longrightarrow \mathbf{A} \models \theta \Leftrightarrow \theta^* \\ &\Longrightarrow \mathbf{A} \models \theta^* \\ &\Longrightarrow \mathbf{A} \models \varphi \text{ for each conjunct of } \theta^*. \end{aligned}$$

This means $\mathcal{K} \subseteq \mathbf{Mod}\,\Sigma^*$. On the other hand, Theorem 8.104 also gives to us $\models \theta^* \implies \theta$ for each $\theta \in \Sigma$, we see that $\mathbf{Mod}\,\Sigma^* \subseteq \mathbf{Mod}\,\Sigma = \mathcal{K}$. ■

COROLLARY 8.106. *Let $\mathcal{K}$ be a class of algebras, all of the same signature. $\mathcal{K}$ is a quasivariety if and only if all of the following hold:*

(i) $\boldsymbol{S}\mathcal{K} \subseteq \mathcal{K}$, *and*
(ii) $\boldsymbol{P_r}\mathcal{K} \subseteq \mathcal{K}$.

Proof. Suppose that $\mathcal{K}$ fulfills (i) and (ii) of this corollary. Since direct products and ultraproducts are each particular kinds of reduced products, we see that (i) and (ii) of this corollary entail (i), (ii), and (iii) of Theorem 8.105, we see that $\mathcal{K}$ must be a quasivariety.

For the converse, suppose $\mathcal{K}$ is a quasivariety. Then conditions (i), (ii), and (iii) of Theorem 8.105 hold for $\mathcal{K}$. According to property (v) of Theorem 8.48, we know that $\boldsymbol{P_r}\mathcal{K} \subseteq \boldsymbol{SPP_u}\mathcal{K}$. So we obtain $\boldsymbol{P_r}\mathcal{K} \subseteq \mathcal{K}$, as desired. ■

THEOREM 8.107. *A sentence in a signature for algebras is preserved under the formation of subalgebras and reduced products if and only if it is logically equivalent to a conjunction of quasi-equations.*

Proof. We already know that quasi-equations are preserved under the formation of subalgebras and reduced products. This secures the right-to-left implication of the theorem.

For the converse, suppose φ is a sentence preserved under the formation of subalgebras and reduced products. Let $\mathcal{K} = \mathbf{Mod}\,\varphi$. According to Corollary 8.106, the class $\mathcal{K}$ is a quasivariety. This means that there is a set Σ of quasi-equations that is logically equivalent to $\{\varphi\}$. By the Compactness Theorem, there is a finite set $\Gamma \subseteq \Sigma$ that is logically equivalent to $\{\varphi\}$. So the conjunction of the quasi-equations belonging to Γ is logically equivalent to φ. ■

Like Theorem 8.101, Theorem 8.107 is a preservation theorem.

THEOREM 8.108 (Mal'tsev's $\boldsymbol{SP_r}$ Theorem). *Let $\mathcal{K}$ be any class of algebras, all of the same signature. Then*

$$\mathbf{Mod\,Th}_q\,\mathcal{K} = \boldsymbol{SP_r}\mathcal{K} = \boldsymbol{SPP_u}\mathcal{K} = \boldsymbol{SP_uP}\,\mathcal{K}. \quad ■$$

This theorem is a more or less immediate consequence of Theorem 8.105 with the help of Theorem 8.48. Theorems 8.105–8.108 have a complex history. A. I. Mal'tsev (1966b), the great champion of the investigation of quasivarieties, deserves the leading place, see also (A. I. Mal'tsev, 1973), but G. Grätzer and H. Lakser (1973), Takao Kashiwagi (1972), and Tsuyoshi Fujiwara (1971) have added their efforts to these results.

In case $\mathcal{K}$ is a finite set of finite algebras of the same signature, we know that $\boldsymbol{P_u}\mathcal{K}$ is just the class of all algebras isomorphic with algebras in $\mathcal{K}$. So, in this case, $\boldsymbol{SP}\mathcal{K}$ is the smallest quasivariety that includes $\mathcal{K}$.

EXAMPLE 8.109. (Ralph McKenzie, 1977) There is a finite algebra **A** of finite signature so that $\boldsymbol{HSP}\,\mathbf{A}$ is finitely axiomatizable but $\boldsymbol{SP}\mathbf{A}$ is not finitely axiomatizable.

Proof. Let $\mathbf{G} = \langle V, E\rangle$ be a directed graph without loops. We devise an algebra $\mathbf{B_G}$ with universe $(V \times V \smallsetminus E) \cup V \cup \{0,1\}$ where 0 and 1 are new elements. This algebra has two constants, 0 and 1, and one two-place basic operation $*$ defined via

$$a * b = \begin{cases} (a,b) & \text{if } a,b \in V \text{ and } (a,b) \notin E \\ 1 & \text{if } (a,b) \in E \\ 0 & \text{otherwise.} \end{cases}$$

Now let **H** be the directed graph displayed in Figure 8.13.
Take $\mathbf{A} = \mathbf{B_H}$. By inspection the following equations hold in $\boldsymbol{HSP}\,\mathbf{A}$.

$$0 * x \approx x * 0 \approx 0 \qquad (x * y) * z \approx 0$$
$$1 * x \approx x * 1 \approx 0 \qquad x * (y * z) \approx 0$$

Observe that these equations entail that $t \approx 0$ whenever $*$ occurs at least twice in t or when 0 occurs in t or when both $*$ and 1 occur in t. So up to renaming variables,

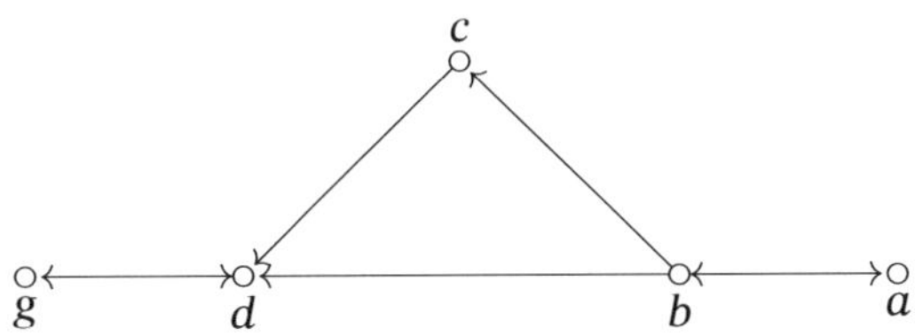

FIGURE 8.13. The directed graph **H**

this leaves only the following four terms: $1, x, x * x$, and $x * y$. So there are essentially nine nontautological equations with terms like one of these four on each side. None of them are true in **A**. So the equations displayed above constitute an equational basis for ***HSP*** **A**.

For contradiction, suppose that ***SP*A** were finitely axiomatizable. Then there would be finitely many quasi-equations that axiomatize it. The conjunction of these quasi-equations is logically equivalent to a universal sentence θ of the form

$$\forall x_0, \dots, x_{n-1} \varphi(\mathbf{x}),$$

where $\mathbf{x} = \langle x_0, \dots, x_{n-1} \rangle$ is the list of all variables occurring in the quantifier-free formula $\varphi(\mathbf{x})$. This entails that any algebra will belong to ***SP*A** as soon as all of its n-generated subalgebras belong to ***SP*A**.

We will invoke the algebras $\mathbf{B}_G$ for some further graphs **G**. For example, if **G** denotes an m-element cycle with $m > 2$, then we let $\mathbf{C}_m$ be the algebra obtained from the oriented cycle on m vertices and let $\mathbf{P}_m$ be the algebra obtained from the directed path on m vertices. The oriented cycle on 6 vertices and the directed path on 4 vertices are displayed in Figure 8.14.

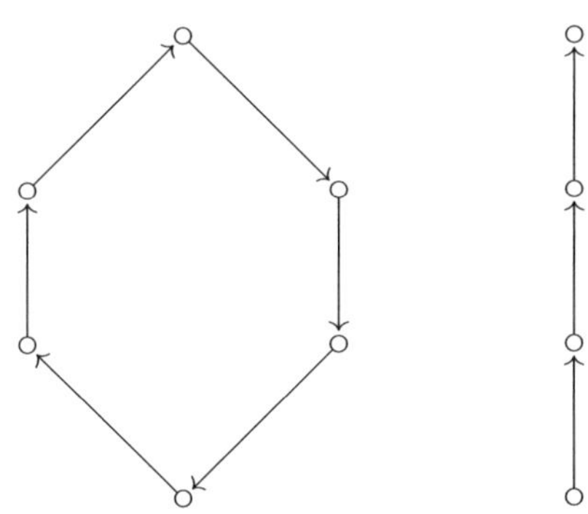

FIGURE 8.14. An oriented cycle and a directed path

After identifying appropriate vertices and edges, we can regard $\mathbf{P}_n$ as a subalgebra of $\mathbf{C}_{n+1}$—the reader can easily check that the operations work correctly. Picking n elements from $\mathbf{C}_{n+1}$ amounts to picking them from $\mathbf{P}_n$. So $\mathbf{C}_{n+1} \models \theta$ if and only if $\mathbf{P}_n \models \theta$. We will show below that $\mathbf{C}_{n+1} \notin$ ***SP*A** while $\mathbf{P}_n \in$ ***SP*A**. This will be our contradiction.

Observe that $\mathbf{B} \in$ ***SP*A** if and only if there is a system $\langle h_i \mid i \in I \rangle$ of homomorphisms from **B** to **A** that separates points. Suppose $\mathbf{C}_{n+1} \in$ ***SP*A** and let $h : \mathbf{C}_{n+1} \to \mathbf{A}$.

Let $c_0, \ldots, c_n$ denote the elements of $\mathbf{C}_{n+1}$ that constitute an oriented $n+1$-cycle. Then in $\mathbf{C}_{n+1}$ we have

$$c_i * c_j = 1 \text{ if } j \equiv i+1 \mod n+1.$$

Applying h we see that the equalities below hold in $\mathbf{A}$

$$h(c_i) * h(c_j) = 1 \text{ if } j \equiv i+1 \mod n+1.$$

So the system $\langle h(c_0), \ldots, h(c_n), h(c_0)\rangle$ must form a cycle in the graph $\mathbf{H}$. But the only cycles in that graph are 2-cycles. It follows that $h(c_0) = h(c_2)$. Since h was arbitrary, we see that no homomorphism from $\mathbf{C}_{n+1}$ to $\mathbf{A}$ can separate c_0 from c_2. So we find that $\mathbf{C}_{n+1} \notin \boldsymbol{SP}\mathbf{A}$.

Consider now the path algebra $\mathbf{P}_n$. Let $\langle p_0, \ldots, p_{n-1}\rangle$ be the underlying path. We contend that a map h from $\{p_0, \ldots, p_{n-1}\}$ into $\mathbf{A}$ that respects $*$ can be extended to a homomorphism from $\mathbf{P}_n$ to $\mathbf{A}$. This extension works by setting

$$\begin{aligned} h(p_i, p_j) &= h(p_i) * h(p_j) \quad \text{if } j \neq i+1 \\ h(1) &= 1 \\ h(0) &= 0 \end{aligned}$$

Evidently, in checking that there are enough homomorphisms to separate points, there will be a number of cases. We do enough to exhibit all the homomorphisms needed, but leave some cases to the reader.

Case: Separating vertices
Refer to Figure 8.13 for the labeling of the graph $\mathbf{H}$. Let $0 \leq i < j < n$.

If i is even let h map $p_0, p_2, \ldots, p_i$ to b, $p_1, p_3, \ldots, p_{i-1}$ to a, $p_{i+1}, p_{i+3}, \ldots$ to d, and $p_{i+2}, p_{i+4}, \ldots$ to g.

If i is odd let h map $p_0, p_2, \ldots, p_i$ to a, $p_1, p_3, \ldots, p_{i-1}$ to b, $p_{i+1}, p_{i+3}, \ldots$ to d, and $p_{i+2}, p_{i+4}, \ldots$ to g.

Case: Separating (p_i, p_j) and 1
First, suppose $i < j$. As in the case above, map p_i to b and then map p_j to g (if $j - i > 2$) using the vertex c in case $j - 1$ is even. To separate (p_i, p_{i+2}) from 1, map p_i to a and p_{i+2} to d (or c). If $j < i$, just map p_j to b and p_i to either d or g.

Case: Separating (p_i, p_j) and (p_k, p_ℓ)
There are a number of subcases that depend on the relative positions of the four vertices in the path. Here is an example: suppose $k < i < j$ and $j + 2 < \ell$. Then map p_k to a, p_i and p_j to a (or b) and p_ℓ to g. Then (p_i, p_j) is mapped to one of $1, (a, a)$, or (b, b), while (p_k, p_ℓ) is mapped to (a, g).

The remaining cases, such as separating a vertex from (p_i, p_j), can all be resolved in a manner similar to the cases handled above. ■

A. Ju. Ol'shanskiĭ (1974), using methods and results in group theory that lie outside the scope of our exposition, has characterized those finite groups $\mathbf{G}$ such that $\boldsymbol{SP}\,\mathbf{G}$ is finitely axiomatizable, as those groups all of whose Sylow subgroups are Abelian. This provides many more examples like the one above, since the equational theory of any finite group is finitely axiomatizable, according to the Oates-Powell Theorem.

Exercises 8.110

1. (A. I. Mal'tsev, 1941) Let n be a positive natural number. We say a group $\mathbf{G}$ has a *faithful n-dimensional representation* provided $\mathbf{G}$ is embeddable into the group of invertible $n \times n$ matrices over some field. Let $\mathbf{H}$ be a group so that every finitely generated subgroup of $\mathbf{H}$ has a faithful n-dimensional representation. Prove that $\mathbf{H}$ has a faithful n-dimensional representation.

2. Let T be an elementary theory that can be axiomatized by a set of universal sentences. Prove that $\mathbf{A} \models T$ if and only if every finitely generated substructure of $\mathbf{A}$ is a model of T.

3. Give an example of an elementary class $\mathcal{K}$ such that $\boldsymbol{P}\mathcal{K}$ is not elementary.

4. Let $\mathcal{K}$ be a class of algebras of the same signature. Prove $\boldsymbol{SP}\mathcal{K} = \boldsymbol{P_s S}\mathcal{K}$.

5. Give an example of a quasivariety that is not a variety.

6. Prove that a quasivariety is locally finite if and only if the variety it generates is locally finite.

7. Let $\mathbf{A}$ be a quasiprimal algebra (see Exercise 2 on page 18). Prove that $\boldsymbol{SP}\mathbf{A}$ is a variety.

*8. Let $\mathcal{K}$ be a class of algebras of some fixed signature. Suppose that $\boldsymbol{S}\mathcal{K} = \mathcal{K}, \boldsymbol{P_u}\mathcal{K} = \mathcal{K}$, the one-element algebra belongs to $\mathcal{K}$, and $\mathbf{A} \times \mathbf{B} \in \mathcal{K}$, whenever $\mathbf{A}, \mathbf{B} \in \mathcal{K}$. Prove that $\mathcal{K}$ is a quasivariety.

*9. (Robert L. Vaught, 1961) Let $\mathcal{K}$ be a class of algebras, all of the same signature, that is the relativized reduct of some elementary class. Prove each of the following:
   (i) $\boldsymbol{P_s}\mathcal{K}$ is the relativized reduct of some elementary class.
   (ii) $\boldsymbol{SP}\mathcal{K}$ is a universal class; indeed, $\boldsymbol{SP}\mathcal{K}$ is the smallest quasivariety that includes $\mathcal{K}$.
   (iii) Let $\mathcal{K}$ be a class of algebras, all of the same signature and suppose $\mathbf{A}$ is an algebra of the same signature. The $\mathcal{K}$-radical of $\mathbf{A}$ is the intersection of all the congruences θ of $\mathbf{A}$ such that $\mathbf{A}/\theta \in \mathcal{K}$. Suppose that $\mathcal{K}$ is a universal class of algebras and $\mathbf{A}$ is an algebra of the same signature. Let $a, b \in A$. In order that (a, b) is in the $\mathcal{K}$-radical of $\mathbf{A}$ it is necessary and sufficient that there is a quasi-equation of the form

$$(s_0(x, y, \mathbf{z}) \approx t_0(x, y, \mathbf{z}) \wedge\!\!\!\wedge \cdots \wedge\!\!\!\wedge s_{n-1}(x, y, \mathbf{z}) \approx t_{n-1}(x, y, \mathbf{z})) \Rightarrow x \approx y$$

that is true in $\mathcal{K}$ and a tuple $\mathbf{c}$ of elements of A so that

$$\mathbf{A} \models s_i(a, b, \mathbf{c}) \approx t_i(a, b, \mathbf{c}) \text{ for all } i < n.$$

8.8. Sentences True in Reduced Products

The Fundamental Theorem of Ultraproducts (Theorem 8.22) displays a strong connection between the satisfaction of a formula in the ultraproduct of structures and the corresponding satisfaction of the formula in the factor structures. While we know that the most obvious adaptation of this result to reduced products is false, it is still reasonable to look for some more subtle connection. Recall the proof of the Fundamental Theorem. The proof was a rather straightforward induction on formulas. The special feature of ultrafilters, among all filters, only played a role in handling negation (and disjunction, had we included it as primitive). The properties of filters generally were sufficient to take care of atomic formulas, conjunctions and the existential quantifier (and even the universal quantifier had we included it as primitive). This suggests that something like the Fundamental Theorem might be true for reduced products in general, but *only for a set of formulas in which negation and disjunction have a restricted role.* Actually, we can only deduce an implication rather than the full equivalence of the Fundamental Theorem. The reason is that our restricted class of formulas is not closed with respect to negation. The class of formulas is obtained by specifying certain basic formulas in which negation and disjunction play a very controlled part, and then closing this class of basic formulas under conjunction and both kinds of quantification.

DEFINITION 8.111. Let $\mathfrak{L}$ be an elementary language. An $\mathfrak{L}$-formula θ is a **basic Horn formula** if and only if θ is a disjunction $\varphi_0 \vee \cdots \vee \varphi_{n-1}$ for some $n > 0$ where each φ_i is either an atomic formula or the negation of an atomic formula and $\{\varphi_i \mid i < n\}$ contains at most one formula that is atomic; θ is a **strict basic Horn formula** if and only if θ is a basic Horn formula and one of its disjuncts is an atomic formula. Notice that while the negation of each atomic formula is a basic Horn formula, such formulas are not strict basic Horn formulas. The set of $\mathfrak{L}$-**(strict) Horn formulas** is smallest set H of $\mathfrak{L}$-formulas such that

(1) Every (strict) basic Horn formula belongs to H,
(2) if $\varphi, \psi \in H$, then $\varphi \wedge \psi \in H$, and
(3) if $\varphi \in H$ and x is a variable, then $\forall x\varphi, \exists x\varphi \in H$.

It is easy to see what strict basic Horn formulas are like if we write them using $\Rightarrow$. Then $\neg\psi_0 \vee \cdots \neg\psi_{n-2} \vee \psi_{n-1}$ can be recast as

$$(\psi_0 \wedge \cdots \wedge \psi_{n-2}) \Rightarrow \psi_{n-1},$$

where each ψ_i is an atomic formula. When $n = 1$ this just becomes ψ_0. Here are some familiar strict basic Horn formulas:

$$(x \leq y \wedge y \leq z) \Rightarrow x \leq z$$
$$(x \leq y \wedge u \leq v) \Rightarrow (x + u) \leq (y + v)$$
$$x + y \approx x + z \Rightarrow y \approx z$$

Of course, every atomic formula, including every equation, is a strict basic Horn formula. So is every quasi-equation, although strict basic Horn sentences might be composed of atomic formulas more generally, rather than just equations. Horn formulas are named after Alfred Horn who proved that a Horn sentence true in every factor structure is also true in their direct product, provided the index set is nonempty, see (Alfred Horn, 1951).

THEOREM 8.112 (C. C. Chang, see (T. Frayne, A. C. Morel, and D. S. Scott, 1962)). *Let $\mathfrak{L}$ be an elementary language and let I be any set. Let $\mathbf{A}_i$ be $\mathfrak{L}$-structures for all $i \in I$, let F be a filter on I and let $\mathbf{b}$ be a sequence of elements from $\prod_I \mathbf{A}_i$. Then*

(i) *If F is a proper filter and $\varphi(\mathbf{x})$ is a Horn formula such that $[\![\varphi(\mathbf{b})]\!] \in F$, then $\prod_i \mathbf{A}_i/F \models \varphi(\mathbf{b}/F)$.*
(ii) *If $\varphi(\mathbf{x})$ is a strict Horn formula such that $[\![\varphi(\mathbf{b})]\!] \in F$, then $\prod_i \mathbf{A}_i/F \models \varphi(\mathbf{b}/F)$.*

Proof. The proof of the Fundamental Theorem of Ultraproducts (page 330) proceeds by induction on the complexity of formulas. The proof of the theorem at hand departs from that proof only in how the base step in managed. So we restrict our attention to the base step in both parts (i) and (ii).

First, consider (ii). Let $\varphi(\mathbf{x})$ be a strict basic Horn sentence, which we render as $(\psi_0 \wedge\!\!\!\wedge \cdots \wedge\!\!\!\wedge \psi_{n-1}) \Rightarrow \theta$, where θ and each of the ψ_i's are atomic formulas. We take $[\![(\psi_0(\mathbf{b}) \wedge\!\!\!\wedge \cdots \wedge\!\!\!\wedge \psi_{n-1}(\mathbf{b})) \Rightarrow \theta(\mathbf{b})]\!] \in F$. Now if $[\![\theta(\mathbf{b})]\!] \in F$, then $\prod_I \mathbf{A}_i/F \models \theta(\mathbf{b})$, because θ is atomic, and so $\prod_I \mathbf{A}_i/F \models \varphi(\mathbf{b})$, as desired. So consider the case that $[\![\theta(\mathbf{b})]\!] \notin F$. Hence $[\![\psi_0(\mathbf{b}) \wedge\!\!\!\wedge \cdots \wedge\!\!\!\wedge \psi_{n-1}(\mathbf{b})]\!] \notin F$, since the intersection of this set with $[\![\varphi(\mathbf{b})]\!]$ is a subset of $[\![\theta(\mathbf{b})]\!]$. Since F is closed under finite intersections, we can also concluded that $[\![\psi_j(\mathbf{b})]\!] \notin F$, for some $j < n$. But then $\prod_I \mathbf{A}_i/F \models \neg\psi_j(\mathbf{b}/F)$, since ψ_j is atomic. Consequently, $\prod_I \mathbf{A}_i/F \models \varphi(\mathbf{b}/F)$. This supplies the base step for the inductive proof of (ii).

To get the base step for (i), suppose that F is a proper filter and that

$$\varphi \text{ is } \neg\psi_0 \vee\!\!\!\vee \cdots \vee\!\!\!\vee \neg\psi_{n-1},$$

where each ψ_i is atomic. Then

$$[\![\varphi(\mathbf{b})]\!] = [\![\neg\psi_0(\mathbf{b}) \vee\!\!\!\vee \cdots \vee\!\!\!\vee \neg\psi_{n-1}(\mathbf{b})]\!] = I \setminus [\![\psi_0(\mathbf{b}) \wedge\!\!\!\wedge \cdots \wedge\!\!\!\wedge \psi_{n-1}(\mathbf{b})]\!]$$

Since F is a proper filter, we see that $[\![\psi_0(\mathbf{b}) \wedge\!\!\!\wedge \ldots \wedge\!\!\!\wedge \psi_{n-1}(\mathbf{b})]\!] \notin F$. Now we can follow the last part of the argument above. ■

COROLLARY 8.113. *Horn sentences are preserved with respect to the formation of proper reduced products (i.e. the filter is a proper filter) and strict Horn sentences are preserved with respect to the formation of arbitrary reduced products.* ■

There is only one structure that arises from improper reduced products: the one-element structure (in which the basic relations hold for all tuples). Every strict Horn sentence is true in such a structure but no Horn sentence that is not strict can be true in such a structure. This is the root of the rather irritating intervention of the words "strict" and "proper".

The possible validity of the converse of Corollary 8.113 was immediately provocative. In fact, it turns out

THEOREM

An elementary sentence is preserved by proper reduced products if and only if it is logically equivalent to a Horn sentence.

This was first established by H. Jerome Keisler (1965) assuming the Continuum Hypothesis. Then Fred Galvin (1967) showed how the reliance on the Continuum Hypothesis could be removed, by a sophisticated metamathematical argument—establishing the displayed theorem in ZFC. Finally, Saharon Shelah (1972) obtained a direct proof without the Continuum Hypothesis. All these proofs require results and methods that lie beyond what we can include in this exposition.

The Fundamental Theorem of Ultraproducts says, roughly, that a sentence is true in an ultraproduct if and only if it is true in a large set of the factors. For reduced products more generally, Theorem 8.112 only says if a Horn sentence is true in a large set of the factors, then it is true in the reduced product. There may well be Horn sentences true in the reduced product that are true in none of the factors. For example, the sentence which expresses that there are at least 6 things is true in the 6-element Abelian groups but fails in each of its direct factors, namely the groups of order 2 and 3. This situation is unsatisfactory. What is the connection between the satisfaction of formulas in a reduced product and their satisfaction in the factor structures? Not all filters on the index set exhibit the same behavior in connection with this question. Ultraproducts and direct products (where the filter is $\{I\}$) seem to be very different. What is needed is some sort of machinery for distinguishing the effects of various filters. There has been an algebra which has gone unmentioned: the Boolean algebra of all subsets of the index set. This algebra offers the key to the problem.

Let $\mathfrak{L}$ be an elementary language and let $\mathbf{B} = \prod_I \mathbf{A}_i$, where each $\mathbf{A}_i$ is an $\mathfrak{L}$-structure. We denote the Boolean algebra of all subsets of I by $\mathbf{2}^I$. Now observe that $[\![\ \cdot\ ]\!]$ is a map from sentences, up to logical equivalence, of the diagram language of $\mathbf{B}$ into $\mathbf{2}^I$ that has the following properties:

$$\begin{aligned} [\![\varphi \mathbin{\&} \psi]\!] &= [\![\varphi]\!] \cap [\![\psi]\!] \\ [\![\varphi \vee \psi]\!] &= [\![\varphi]\!] \cup [\![\psi]\!] \\ [\![\neg\varphi]\!] &= I \setminus [\![\varphi]\!] \\ [\![\varphi]\!] = [\![\psi]\!] &\text{ if and only if } \mathbf{A}_i \models (\varphi \Leftrightarrow \psi) \text{ for all } i \in I. \end{aligned}$$

Hence $[\![\ \cdot\ ]\!]$ is a homomorphism from the Boolean algebra of sentences, up to logical equivalence, into $\mathbf{2}^I$. Now let F be a filter on I. This is the same as saying F is a filter of $\mathbf{2}^I$. As with any filter on a Boolean algebra, one can form the quotient algebra $\mathbf{2}^I/F$. Fortunately, this algebra is isomorphic to the reduced power, so our notational conventions are bent but not broken. The following are useful easy facts:

$$\begin{aligned} &[\![\varphi]\!] \in [\![\psi]\!]/F \text{ if and only if } [\![\varphi \Leftrightarrow \psi]\!] \in F \\ &I/F = X/F \text{ if and only if } X \in I/F \text{ if and only if } \mathbf{2}^I/F \models X/F \approx 1, \end{aligned}$$

where X is any subset of I.

In this setting, the Fundamental Theorem of Ultraproducts reads:

$$\prod_i \mathbf{A}_i/F \models \varphi(\mathbf{a}/F) \text{ if and only if } \mathbf{2}^I/F \models [\![\varphi(\mathbf{a})]\!]/F \approx 1,$$

where F is an ultrafilter on I (and so $\mathbf{2}^I/F \cong \mathbf{2}$). We take the basic operation symbols of the language of Boolean algebras to be $\vee, \wedge, ', 0$, and 1, a modest departure from the signature used in Volume I. In extending some form of this result to arbitrary proper reduced products, the basic insight is that we should try to associate with each formula φ some kind of sentence in the language of Boolean algebras that can play a role like

$$[\![\varphi]\!]/F \approx 1$$

does in the case of ultraproducts. More precisely, with each formula $\varphi(\mathbf{x})$ we will associate a sequence $\langle\sigma(y_0,\ldots,y_{n-1});\psi_0(\mathbf{x}),\ldots,\psi_{n-1}(\mathbf{x})\rangle$ where $\sigma(y_0,\ldots,y_{n-1})$ is a formula in the language of Boolean algebras and each $\psi_i(\mathbf{x})$ is an $\mathfrak{L}$-formula, and so that for any proper reduced product $\prod_I \mathbf{A}_i/F$ we have

$$\prod_I \mathbf{A}_i/F \models \varphi(\mathbf{a}/F) \text{ if and only if } \mathbf{2}^I/F \models \sigma([\![\psi_0(\mathbf{a})]\!]/F,\ldots,[\![\psi_{n-1}(\mathbf{a})]\!]/F).$$

This leads to the intricate definition of the determining sequence of a formula.

DEFINITION 8.114. For any $\mathfrak{L}$-formula $\varphi(\mathbf{x})$, where $\mathbf{x}$ is a tuple of variables including all those occurring freely in φ, the **determining sequence** is $\partial[\varphi(\mathbf{x})]$, where ∂ is the function on the set of formulas defined by the following recursion:

(1) $\partial[\varphi(\mathbf{x})] = \langle y \approx 1; \varphi(\mathbf{x})\rangle$ if $\varphi(\mathbf{x})$ is atomic,

(2) $\partial[\neg\varphi(\mathbf{x})] = \langle\neg\sigma(y_0',\ldots,y_{n-1}');\neg\psi_0(\mathbf{x}),\ldots,\neg\psi_{n-1}(\mathbf{x})\rangle$ where $\partial[\varphi(\mathbf{x})] = \langle\sigma(y_0,\ldots,y_{n-1});\psi_0(\mathbf{x}),\ldots,\psi_{n-1}(\mathbf{x})\rangle$,

(3) $\partial[\varphi(\mathbf{x}) \mathbin{\&} \theta(\mathbf{x})] = \langle\mu(y_0,\ldots,y_{m+n-1});\pi_0(\mathbf{x}),\ldots,\pi_{m+n-1}(\mathbf{x})\rangle$ where $\mu(y_0,\ldots,y_{m+n-1}) = \sigma(y_0,\ldots,y_{n-1}) \mathbin{\&} \tau(y_n,\ldots,y_{m+n-1})$ and $\pi_i(\mathbf{x}) = \psi_i(\mathbf{x})$ if $i < n$ and $\pi_i(\mathbf{x}) = \eta_i(\mathbf{x})$ if $n \le i < m+n$, where $\partial[\varphi(\mathbf{x})] = \langle\sigma(y_0,\ldots,y_{n-1});\psi_0(\mathbf{x}),\ldots,\psi_{n-1}(\mathbf{x})\rangle$ and $\partial[\theta(\mathbf{x})] = \langle\tau(y_0,\ldots,y_{m-1});\eta_0(\mathbf{x}),\ldots,\eta_{m-1}(\mathbf{x})\rangle$.

(4) $\partial[\exists u\pi(u,\mathbf{x})] = \langle\tau(y_0,\ldots,y_{m-1});\theta_0(\mathbf{x}),\ldots,\theta_{m-1}(\mathbf{x})\rangle$ where $\partial[\pi(u,\mathbf{x})] = \langle\sigma(y_0,\ldots,y_{n-1});\psi_0(u,\mathbf{x}),\ldots,\psi_{n-1}(u,\mathbf{x})\rangle$ and $m = 2^n - 1$ and $S_0 = \{0\}, S_1 = \{1\},\ldots,S_{n-1} = \{n-1\}, S_n,\ldots,S_{m-1}$ is a complete list of all the nonempty subsets of $\{0,1,\ldots,n-1\}$; moreover,

$$\theta_i(\mathbf{x}) \text{ is } \exists u \bigwedge_{j\in S_i} \psi_j(u,\mathbf{x}) \text{ for } i < m \text{ and } \tau(y_0,\ldots,y_{m-1}) \text{ is}$$

$$\exists z_0\ldots z_{m-1}[(\bigwedge_{j<m} z_j \le y_j) \mathbin{\&} (\bigwedge_{S_j\cup S_j=S_k} (z_i \wedge z_j \approx z_k)) \mathbin{\&} \sigma(z_0,\ldots,z_{n-1})].$$

This is a rather daunting definition. Reviewing a similar recursive definition like Definition 8.5 may help. We read $\partial(\varphi)$ as "the determining sequence of φ". Observe that ∂ is a function whose inputs are $\mathfrak{L}$ formulas and whose outputs are sequences made up of two parts: the first is a formula in the language of Boolean algebras with some number n of free variables and the second is an n-termed sequence of $\mathfrak{L}$ formulas all of whose free variables occur among the free variables of φ.

In view of the nice behavior of atomic formulas with respect to reduced products, (1) is a natural condition. Conditions (2) and (3), on inspection, are really transparent as well—although (2) seems, at first, to have some overly complicated double negation. It is condition (4) that is really the nonobvious part of this definition. The function ∂ is computable in the sense that if the underlying fundamental relation and operation symbols in $\varphi(\mathbf{x})$ are acceptable as inputs to a computable function, then ∂ itself is computable. The formulas in the sequence $\partial[\varphi]$ have essentially the same complexity as

φ; moreover, the formula $\sigma(\mathbf{y})$ in the language of Boolean algebras is monotone in the sense that

$$y_0 \leq w_0 \wedge \ldots \wedge y_{n_1} \leq w_{n-1} \wedge \sigma(y_0, \ldots, y_{n-1}) \Rightarrow \sigma(w_0, \ldots, w_{n-1})$$

is true in every Boolean algebra. This monotonicity has a role to play both in the proof of the Feferman-Vaught Theorem below and in some of its applications.

S. Feferman and R. L. Vaught (1959) proved a theorem more general than the one below, in the sense that it applies to a very general kind of product construction of which the reduced product is a special case. A. I. Mal'tsev (1959b) has a related result. Joseph M. Weinstein (1965) pointed out how to frame the version below. Fred Galvin (1967) has formulated a stronger version of the theorem below, see also (Fred Galvin, 1970), where the interconnections among these results are discussed in detail.

THEOREM 8.115 (The Feferman-Vaught Theorem for Reduced Products). *Let $\prod_i \mathbf{A}_i/F$ be any reduced product of $\mathfrak{L}$ structures. For any formula $\varphi(\mathbf{x})$, where $\mathbf{x} = \langle x_0, \ldots, x_{k-1}\rangle$, and any k-tuple $\mathbf{a}$ from $\prod_I \mathbf{A}_i$, , we have*

$$\prod_I \mathbf{A}_i/F \models \varphi(\mathbf{a}/F) \textit{ if and only if } \mathbf{2}^I/F \models \sigma(\llbracket \psi_0(\mathbf{a}) \rrbracket/F, \ldots, \llbracket \psi_{n-1}(\mathbf{a}) \rrbracket/F),$$

where $\partial[\varphi(\mathbf{x})] = \langle \sigma(y_0, \ldots, y_{n-1}); \psi_0(\mathbf{x}), \ldots, \psi_{n-1}(\mathbf{x})\rangle$.

Proof. The proof proceeds by induction on the complexity of $\varphi(\mathbf{x})$. If $\varphi(\mathbf{x})$ is atomic or the negation of a formula or the conjunction of two formulas, then the desired result emerges from unraveling the definitions involved. For example, knowing the result holds for $\varphi(\mathbf{x})$ we see

$$\begin{aligned}\prod_I \mathbf{A}_i/F \models \neg\varphi(\mathbf{a}/F) &\text{ iff } \prod_I \mathbf{A}_i/F \not\models \varphi(\mathbf{a}/F)\\ &\text{ iff } \mathbf{2}^I/F \not\models \sigma(\llbracket \psi_0(\mathbf{a}) \rrbracket/F, \ldots, \llbracket \psi_{n-1}(\mathbf{a}) \rrbracket/F)\\ &\text{ iff } \mathbf{2}^I/F \models \neg\sigma(\llbracket \psi_0(\mathbf{a}) \rrbracket/F, \ldots, \llbracket \psi_{n-1}(\mathbf{a}) \rrbracket/F)\\ &\text{ iff } \mathbf{2}^I/F \models \neg\sigma(I \setminus \llbracket \neg\psi_0(\mathbf{a}) \rrbracket/F, \ldots, I \setminus \llbracket \neg\psi_{n-1}(\mathbf{a}) \rrbracket/F)\\ &\text{ iff } \mathbf{2}^I/F \models \tau(\llbracket \theta_0(\mathbf{a}) \rrbracket/F, \ldots, \llbracket \theta_{n-1}(\mathbf{a}) \rrbracket/F)\end{aligned}$$

where

$$\partial[\varphi(\mathbf{x})] = \langle \sigma(\mathbf{y}); \psi_0(\mathbf{x}), \ldots, \psi_{n-1}(\mathbf{x}) \text{ and } \partial[\neg\varphi(\mathbf{x})] = \langle \tau(\mathbf{y}); \theta_0(\mathbf{x}), \ldots, \theta_{n-1}(\mathbf{x})\rangle.$$

Leaving the other two cases concerning atomic formulas and the connectives to the reader, we turn to the case of the existential quantifier. Let

$$\partial[\varphi(u, \mathbf{x})] = \langle \sigma(y_0, \ldots, y_{n-1}); \psi_0(u, \mathbf{x}), \ldots, \psi_{n-1}(u, \mathbf{x})\rangle.$$

Let $\prod_I \mathbf{A}_i/F$ be a reduced product and let $\mathbf{a}$ be an appropriate tuple from $\prod_I A_i$. Now

$$\begin{aligned}\prod_I \mathbf{A}_i/F \models \exists u\varphi(u, \mathbf{a}/F) &\text{ iff } \prod_I \mathbf{A}_i/F \models \varphi(b/F, \mathbf{a}/F) \text{ for some } b \in \prod_I A_i\\ &\text{ iff } \mathbf{2}^I/F \models \sigma(\llbracket \psi_0(b, \mathbf{a}) \rrbracket, \ldots, \llbracket \psi_{n-1}(b, \mathbf{a}) \rrbracket/F)\\ &\qquad \text{for some } b \in \prod_I A_i.\end{aligned}$$

Let m and S_i for $i < m$ be chosen in accordance with the definition of ∂, and let

$$\partial[\exists u\varphi(u,\mathbf{x})] = \langle \tau(y_0,\ldots,y_{n-1});\theta_0(\mathbf{x}),\ldots,\theta_{m-1}(\mathbf{x})\rangle.$$

What remains to be proven is

$$\mathbf{2}^I/F \models \sigma([\![\psi_0(b,\mathbf{a})]\!]/F,\ldots,[\![\psi_{n-1}(b,\mathbf{a})]\!]/F) \text{ for some } b \in \prod_I A_i$$
$$\text{if and only if}$$
$$\mathbf{2}^I/F \models \tau([\![\theta_0(\mathbf{a})]\!]/F,\ldots,[\![\theta_{m-1}(\mathbf{a})]\!]/F).$$

First suppose $\mathbf{2}^I/F \models \sigma([\![\psi_0(b,\mathbf{a})]\!]/F,\ldots,[\![\psi_{n-1}(b,\mathbf{a})]\!]/F)$ where $b \in \prod_I A_i$. Set

$$Y_k := [\![\theta_k(\mathbf{a})]\!] = [\![\exists u \bigwedge_{j\in S_k} \psi_j(u,\mathbf{a})]\!] \text{ and}$$
$$Z_k := [\![\bigwedge_{j\in S_k} \psi_j(b,\mathbf{a})]\!]$$

for all $k < m$. The following are clear

$$\begin{aligned} Z_k \subseteq Y_k \quad & \text{so } \mathbf{2}^I/F \models Z_K/F \le Y_k/F. \\ Z_i \cap Z_j = Z_k \text{ when } S_i \cup S_j = S_k \quad & \text{so } \mathbf{2}^I/F \models (Z_i/F \wedge Z_j/F) \approx Z_k/F \\ & \text{when } S_i \cup S_j = S_k. \\ Z_j = [\![\psi_j(b,\mathbf{a})]\!] \text{ when } j < n, \quad & \text{so } \mathbf{2}^I/F \models \sigma(Z_0/F,\ldots,Z_{n-1}/F). \end{aligned}$$

Consequently, $\mathbf{2}^I/F \models \tau([\![\theta_0(\mathbf{a})]\!]/F,\ldots,[\![\theta_{m-1}(\mathbf{a})]\!]/F)$, as desired.

For the converse, assume

$$\mathbf{2}^I/F \models \tau([\![\theta_0(\mathbf{a})]\!]/F,\ldots,[\![\theta_{m-1}(\mathbf{a})]\!]/F).$$

Again, let $Y_k = [\![\theta_k(\mathbf{a})]\!]$ for $k < m$. For each $k < m$, pick a set Z_k such that

$$\begin{aligned} &\mathbf{2}^I/F \models Z_k/F \le Y_k/F \\ &\mathbf{2}^I/F \models (Z_i/F \wedge Z_j/F) \approx Z_k/F \text{ when } S_i \cup S_j = S_k, \text{ and} \\ &\mathbf{2}^I/F \models \sigma(Z_0/F,\ldots,Z_{m-1}/F). \end{aligned}$$

Since F is a filter (and so closed under intersection), there is a set $X \in F$ such that

$$Z_k \cap X \subseteq Y_k \text{ for all } k < m \text{ and}$$
$$Z_i \cap Z_j \cap X = Z_k \cap X \text{ whenever } S_i \cup S_j = S_k.$$

For each $i \in I$ we pick $b_i \in A_i$ is the following manner. If $i \notin X$, then we make the choice arbitrarily. So suppose $i \in X$. Again, if $i \notin Z_k$ for all $k < m$, then pick b_i arbitrarily. In the remaining cases, let Z be the set among all the Z_k's smallest in the sense of set inclusion. The existence of such a Z follows for the connections stipulated between the Z_k's and the S_i's. Let S be the subset corresponding to Z. Thus

$$i \in Z \cap X \subseteq [\![\exists u \bigwedge_{j\in S} \psi_j(u,\mathbf{a}]\!].$$

This means that

$$\mathbf{A}_i \models \exists u \bigwedge_{j\in S} \psi_j(u,a_i).$$

So pick $b_i \in A_i$ such that

$$\mathbf{A}_i \models \psi_j(b_i, a_i) \text{ for all } j \in S.$$

Now put $b = (b_i \mid i \in I)$. For this choice of b we have

$$Z_k \cap X \subseteq [\![\psi_k(b, \mathbf{a})]\!] \text{ for all } k < n.$$

Thus $\mathbf{2}^I/F \models Z_k/F \leq [\![\psi_k(b, \mathbf{a})]\!]/F$ for all $k < n$. Since

$$\mathbf{2}^I/F \models \sigma(Z_0/F, \ldots, Z_{n-1}/F)$$

and since $\sigma(y_0, \ldots, y_{n-1})$ is monotone in every Boolean algebra, we can draw the desired conclusion:

$$\mathbf{2}^I/F \models \sigma([\![\psi_0(b, \mathbf{a})]\!]/F, \ldots, [\![\psi_{n-1}(b, \mathbf{a})]\!]/F) \text{ for some } b \in \prod_I A_i.$$

This completes the proof of the Feferman-Vaught Theorem for Reduced Products.

■

The Feferman-Vaught Theorem has a large number of significant consequences, notably ones that do not mention determining sequences. We conclude this section with several of these. Others can be found among the exercises, as well as in the literature.

COROLLARY 8.116. *Let I be any set and let F be any filter on I. Let $\mathbf{A}_i$ and $\mathbf{B}_i$ be structures for the same language for every $i \in I$. If $\{i \mid \mathbf{A}_i \preccurlyeq \mathbf{B}_i\} \in F$, then*

$$\prod_I \mathbf{A}_i/F \preccurlyeq \prod_I \mathbf{B}_i/F.$$

Proof. Let $X = \{i \mid \mathbf{A}_i \preccurlyeq \mathbf{B}_i\}$. Let

$$\mathbf{A} = \prod_i \mathbf{A}_i \text{ and } \mathbf{B} = \prod_I \mathbf{B}_i.$$

For any $\varphi(\mathbf{x})$ and any $\mathbf{a}$ from $\prod_I A_i$ let $[\![\varphi(\mathbf{a})]\!]^{\mathbf{A}} = \{i \mid \mathbf{A}_i \models \varphi(a_i)\}$ and define $[\![\varphi(\mathbf{a})]\!]^{\mathbf{B}}$ is a similar fashion.

For every formula $\psi(\mathbf{x})$ and every $\mathbf{a}$ from $\prod_I A_i$, our hypothesis yields

$$[\![\psi(\mathbf{a})]\!]^{\mathbf{A}} \cap X \subseteq [\![\psi(\mathbf{a})]\!]^{\mathbf{B}}$$

and so

$$\mathbf{2}^I/F \models [\![\psi(\mathbf{a})]\!]^{\mathbf{A}}/F \leq [\![\psi(\mathbf{a})]\!]^{\mathbf{B}}/F.$$

Now let $\varphi(\mathbf{x})$ be any formula and let $\mathbf{a}$ be any tuple from $\prod_I A_i$. Suppose that

$$\prod_i \mathbf{A}_i/F \models \varphi((\mathbf{a}/F).$$

By the Feferman-Vaught Theorem we get

$$\mathbf{2}^I/F \models \sigma([\![\psi_0(\mathbf{a})]\!]^{\mathbf{A}}/F, \ldots, [\![\psi_{n-1}(\mathbf{a})]\!]^{\mathbf{A}}/F)$$

where $\partial[\varphi(\mathbf{x})] = \langle \sigma(\mathbf{y}); \psi_0(\mathbf{x}), \ldots, \psi_{n-1}(\mathbf{x}) \rangle$. By the monotonicity of $\sigma(\mathbf{y})$, we obtain

$$\mathbf{2}^I/F \models \sigma([\![\psi_0(\mathbf{a})]\!]^{\mathbf{B}}/F, \ldots, [\![\psi_{n-1}(\mathbf{a})]\!]^{\mathbf{B}}/F).$$

Applying the Feferman-Vaught Theorem again, we conclude that

$$\prod_I \mathbf{B}_i \models \varphi(\mathbf{a}/F),$$

as needed. ■

COROLLARY 8.117 (Fred Galvin 1970). *Let I and J be any sets and let F be a filter on I and G be a filter on J. If $\mathbf{A} \equiv \mathbf{B}$ and $\mathbf{2}^I/F \equiv \mathbf{2}^J/G$, then $\mathbf{A}^I/F \equiv \mathbf{B}^J/G$.*

Proof. Let φ any sentence. Let $\partial[\varphi] = \langle\sigma(\mathbf{y}) : \psi_0, \ldots, \psi_{n-1}\rangle$. Observe that each ψ_i is a sentence. Therefore $\{i \mid i \in I \text{ and } \mathbf{A} \models \psi_k\}$ is either I or $\emptyset$; that is, it is either the greatest element of the power set of I or the smallest element of the power set of I. Likewise, $\{j \mid j \in J \text{ and } \mathbf{B} \models \psi_k\}$ is either the greatest or the smallest of the power set of J. Since $\mathbf{A} \equiv \mathbf{B}$, we know, for each $k < n$, that $\{i \mid i \in I \text{ and } \mathbf{A} \models \psi_k\}$ is empty if and only if $\{j \mid j \in J \text{ and } \mathbf{B} \models \psi_k\}$ is empty. Thus

$$\sigma([\![\psi_0]\!]/F, \ldots, [\![\psi_{n-1}]\!]/F) \text{ and } \sigma([\![\psi_0]\!]/G, \ldots, [\![\psi_{n-1}]\!]/G)$$

are exactly the same *sentence* (each $[\![\psi_k]\!]/F$ being either the Boolean 1 or the Boolean 0 and the same modulo G). The conclusion now follows immediately from $\mathbf{2}^I/F \equiv \mathbf{B}^J/G$ by the Feferman-Vaught Theorem. ■

The Feferman-Vaught Theorem establishes a connection between the elementary theory of a reduced product and the elementary theory of a related Boolean algebra. This opens the way for using the wealth of facts known about Boolean algebras and their elementary theories to obtain theorems of a quite general character about reduced products. See (Sabine Koppelberg, 1989; J. Donald Monk and Robert Bonnet, 1989b,a; Steven Givant and Paul Halmos, 2009) for a treasury of information about Boolean algebras. Here we present a theorem that is a consequence of the model completeness of the elementary theory of enhanced infinite atomic Boolean algebras, see Theorem 8.38.

We say a sequence $\langle\sigma^\circ(y_0, \ldots, y_{m-1}; \psi^\circ_0(\mathbf{x}), \ldots, \psi^\circ_{m-1}(\mathbf{x})\rangle$ is *partitioned* provided the formulas

$$\bigvee\!\!\!\bigvee_{j<m} \psi^\circ_j(\mathbf{x})$$

$$\neg(\psi^\circ_i(\mathbf{x}) \wedge\!\!\!\wedge \psi^\circ_j(\mathbf{x})) \text{ for } i \neq j$$

are valid—that is they hold in every structure.

LEMMA 8.118. *Let $\prod_I \mathbf{A}_i/F$ be any reduced product of $\mathfrak{L}$ structures. Let $\varphi(\mathbf{x})$ be any $\mathfrak{L}$-formula and let the free variables of φ be listed as $\mathbf{x}$. Let*

$$\partial[\varphi(\mathbf{x})] = \langle\sigma(y_0, \ldots, y_{n-1}); \psi_0(\mathbf{x}), \ldots, \psi_{n-1}(\mathbf{x})\rangle$$

be the determining sequence of $\varphi(\mathbf{x})$. There is a partitioned sequence

$$\langle\sigma^\circ(y_0, \ldots, y_{m-1}); \psi^\circ_0(\mathbf{x}), \ldots, \psi^\circ_{m-1}(\mathbf{x})\rangle,$$

where $m = 2^n$ such that for all tuples $\mathbf{a} \in \prod_I A_i$ we have

$$\mathbf{2}^I/F \models \sigma([\![\psi_0(\mathbf{a})]\!]/F, \dots, [\![\psi_{n-1}(\mathbf{a})]\!]/F)$$
if and only if
$$\mathbf{2}^I/F \models \sigma^\circ([\![\psi_0^\circ(\mathbf{a})]\!]/F, \dots, [\![\psi_{m-1}^\circ(\mathbf{a})]\!]/F)$$

Proof. Let $S_0, S_1, \dots, S_{m-1}$ be a list of all the subsets of $\{0, \dots, n-1\}$. To simplify notation, we insist that all the subsets that do not contain $n-1$ be listed first. So for the case when $n > 0$, putting $m' = 2^{n-1}$ we will have that $n-1 \notin S_j$ for all $j < m'$ and $n-1 \in S_j$ for all $j \geq m'$.

For $j < m = 2^n$ let

$$\psi_j^\circ(\mathbf{x}) = \bigwedge_{i \in S_j} \psi_i(\mathbf{x}) \wedge \bigwedge_{\substack{i \notin S_j \\ i<n}} \neg\psi_i(\mathbf{x})$$

where a conjunction over the empty set is simply dropped. Also let

$$\sigma^\circ = \sigma\left(\bigvee_{\substack{k<m \\ 0 \in S_k}} y_k, \dots, \bigvee_{\substack{k<m \\ n-1 \in S_k}} y_k\right).$$

A straightforward induction on n establishes that this is a partitioned sequence. In the base step of this induction we have $n = 0$ and $m = 1$. So there is only one set $S_0 = \emptyset$ and one $j < m$, namely $j = 0$, to consider. Thus ψ_0° is $\top$, a formula true in every structure. Thus in this base step we see that the sequence is partitioned. The distributive properties of the connectives comes into play in the inductive step, which we leave to the reader.

Now consider $\psi_i^\circ(\mathbf{x}) \wedge \psi_j^\circ(\mathbf{x})$ with $i \neq j$. So $S_i \neq S_j$. It is harmless to suppose there is $k \in S_i$ with $k \notin S_j$. Thus $\psi_k(\mathbf{x})$ is a conjunction of $\psi_i^\circ(\mathbf{x})$ while $\neg\psi_k(\mathbf{x})$ is a conjunct of $\psi_j^\circ(\mathbf{x})$. So we can regard $\psi_k(\mathbf{x}) \wedge \neg\psi_k(\mathbf{x})$ as a conjunct of $\psi_i^\circ(\mathbf{x}) \wedge \psi_k^\circ(\mathbf{x})$. Hence, this latter formula fails in every structure; its negation holds in every structure.

So our sequence is partitioned.

Now consider the equivalence to be proved. Observe that

$$\mathbf{2}^I/F \models \sigma^\circ([\![\psi_0^\circ(\mathbf{a})]\!]/F, \dots, [\![\psi_{m-1}^\circ(\mathbf{a})]\!]/F)$$
if and only if
$$\mathbf{2}^I/F \models \sigma(\bigcup_{\substack{k<m \\ 0 \in S_k}} [\![\psi_k^\circ(\mathbf{a})]\!]/F, \dots, \bigcup_{\substack{k<m \\ n-1 \in S_k}} [\![\psi_k^\circ(\mathbf{a})]\!]/F).$$

But for $\ell < n$ we have

$$\bigcup_{\substack{k<m \\ \ell \in S_k}} [\![\psi_k^\circ(\mathbf{a})]\!] = [\![\psi_\ell(\mathbf{a})]\!],$$

completing the proof. ■

THEOREM 8.119 (S. Feferman and R. L. Vaught 1959). *Let $\langle \mathbf{A}_i \mid i \in I \rangle$ be a system of $\mathfrak{L}$-structures and let φ be an $\mathfrak{L}$-sentence. If $\prod_J \mathbf{A}_i \models \varphi$ for each finite $J \subseteq I$, then $\prod_I \mathbf{A}_i \models \varphi$. Moreover, if $\prod_I \mathbf{A}_i \models \varphi$, then there is a finite $I' \subseteq I$ with such that $\prod_J \mathbf{A}_j \models \varphi$ for all finite J with $I' \subseteq J \subseteq I$.*

Proof. We may assume that I is infinite. Let $\langle \sigma(y_0, \dots, y_m); \psi_0, \dots, \psi_{m-1} \rangle$ be a partitioned sequence for φ. Let $\mathbf{A} = \prod_I \mathbf{A}_i$ and let

$$L := \{k \mid k < m \text{ and } [\![\psi_k]\!] \text{ is finite } \}.$$

Observe that the following informal statements can be formalized in the language of Boolean algebras, where s is some natural number:

- "$\langle y_0, \dots, y_{m-1} \rangle$ is a partition of the universe".
- "There are exactly $|[\![\psi_k]\!]|$ atoms below y_k". (Provided $|[\![\psi_k]\!]|$ is finite)
- "There are least s atoms below y_k".

Let θ_s be the following sentence in the language of Boolean algebras:

$$\forall y_0, \dots, y_{m-1} \big[(\langle y_0, \dots, y_{m-1} \rangle \text{ is a partitions of the universe } \mathbin{\&}$$
$$\bigwedge_{k \in L} \text{there are at exactly } |[\![\psi_k]\!]| \text{ atoms below } y_k \mathbin{\&}$$
$$\bigwedge_{\substack{k \notin L \\ k < m}} \text{there are at least } s \text{ atoms below } y_k) \Rightarrow \sigma(y_0, \dots, y_{m-1}) \big].$$

CLAIM 1. *There is a natural number s such that θ_s is true in every finite Boolean algebra.*

Proof of Claim 1. Suppose not. For each natural number s pick a finite Boolean algebra $\mathbf{B}_s$ so that θ_s fails in $\mathbf{B}_s$. Pick $b_0, \dots, b_{m-1} \in B_s$ so that the hypotheses of the implication in θ_s all hold, but $\sigma(b_0, \dots, b_{m-1})$ fails. Notice that since I is infinite there must be at least one $k < m$ so that $[\![\psi_k]\!]$ is infinite, since we are dealing with a partitioned sequence. In other words, there is at least one $k \notin L$. This means that $s \leq |B_s|$.

Let I_s be a finite subset of I that includes $[\![\psi_k]\!]$ for each $k \in L$ and so that $|I_s \cap [\![\psi_k]\!]|$ is at least as large as the number of atoms of $\mathbf{B}_s$ below b_k for each $k < m$ with $k \notin L$. Let $\mathbf{C}_s = \mathbf{2}^{I_s}$, which we regard as the Boolean algebra of all subsets of I_s.

There is an embedding $f : \mathbf{B}_s \rightarrowtail \mathbf{C}_s$ such that $f(b_k) = I_s \cap [\![\psi_k]\!]$ for all $k < m$.

So $\mathbf{B}_s \models \neg\sigma(b_0, \dots, b_{m-1})$ while $\mathbf{C}_s \models \sigma(f(b_0), \dots, f(b_{m-1}))$ by the hypothesis of our theorem and Theorem 8.115.

Let F be a nonprincipal ultrafilter on the set ω of natural numbers and put

$$\mathbf{B} := \prod_\omega \mathbf{B}_s/F \qquad \text{and} \qquad \mathbf{C} := \prod_\omega \mathbf{C}_s/F.$$

Now $\mathbf{B}_s$ and $\mathbf{C}_s$ are atomic Boolean algebras, for each $s \in \omega$. Since this notion is expressible by an elementary sentence, by the Fundamental Theorem of Ultraproducts, both $\mathbf{B}$ and $\mathbf{C}$ are (infinite) atomic Boolean algebras.

Now expand the language by adjoining a new one-place relation symbol A and add the sentence asserting $Ax \Leftrightarrow$ "x is an atom" to the elementary theory of infinite atomic Boolean algebras. By Theorem 8.38 we know that this expanded theory is model complete. Let $\mathbf{B}'$ and $\mathbf{C}'$ be the natural expansions of $\mathbf{B}$ and $\mathbf{C}$ to the expanded language. By Theorem 8.37 there is a universal formula $\tau(y_0, \dots, y_{m-1})$ equivalent, modulo this expanded theory, to $\sigma(y_0, \dots, y_{m-1})$. But the embedding f induces an embedding $f^* : \mathbf{B}' \rightarrowtail \mathbf{C}'$. Let $\mathbf{B}^*$ be the image of $\mathbf{B}'$ under this embedding. Then $\mathbf{B}^* \models \neg\sigma(f^*(b_0), \dots, f^*(b_{m-1})$. But we know as well that $\mathbf{B}^* \models \sigma(f^*(b_0), \dots, f^*(b_{m-1}))$ since $\mathbf{B}^*$ is a substructure of $\mathbf{C}'$ and $\sigma(f^*(b_0), \dots, f^*(b_{m-1}))$ is equivalent to $\tau(f^*(b_0), \dots,$

$f^*(b_{m-1})$), where τ is a universal formula (and so preserved under substructures). This is a contradiction and the Claim is established. ■

Now let s be a natural number so that θ_s is true in every finite Boolean algebra. By a familiar compactness argument, θ_s must hold in some infinite atomic Boolean algebra. But by Theorem 8.39 the elementary theory of infinite atomic Boolean algebras is complete. So θ_s holds in all infinite atomic Boolean algebras. So $\mathbf{2}^I \models \theta_s$. Therefore $\mathbf{2}^I \models \sigma([\![\psi_0]\!], \dots, [\![\psi_{m-1}]\!])$. By Theorem 8.115 we have $\prod_I \mathbf{A}_i \models \varphi$, as desired.

For the supplementary statement in the theorem, for each $k < m$ with $k \notin L$ select a subset $S_k \subseteq [\![\psi_k]\!]$ with $|S_k| = s$ and let

$$I' = \bigcup_{k \in L} [\![\psi_k]\!] \cup \bigcup_{\substack{k \notin L \\ k<m}} S_k.$$

Observe that θ_s holds in $\mathbf{2}^{I'}$. So

$$\begin{aligned} \mathbf{2}^{I'} \models \big[\, (\langle [\![\psi_0]\!], \dots, [\![\psi_{m-1}]\!] \rangle &\text{ is a partition of the universe} \\ &\wedge \bigwedge_{k \in L} \text{there are at exactly } |[\![\psi_k]\!]| \text{ atoms below } [\![\psi_k]\!] \\ &\wedge \bigwedge_{\substack{k \notin L \\ k<m}} \text{there are at least } s \text{ atoms below } [\![\psi_k]\!]\,) \Rightarrow \sigma([\![\psi_0]\!], \dots, [\![\psi_{m-1}]\!])\big]. \end{aligned}$$

Now I' (and hence for any finite J with $I' \subseteq J \subseteq I$) is rich enough so that the hypotheses of the implication above all hold. Thus $\mathbf{2}^J \models \sigma([\![\psi_0]\!], \dots, [\![\psi_{m-1}]\!])$. Invoking the Feferman-Vaught Theorem 8.115 completes the argument. ■

COROLLARY 8.120. $\mathbf{2}^I \equiv \mathbf{2}^J$ *if and only if either I and J are both infinite or* $|I| = |J|$. ■

COROLLARY 8.121. *Let $\mathcal{K}$ be an elementary class that contains the one-element structure with nonempty fundamental relations. Then*

$$\boldsymbol{P}\,\mathcal{K} = \mathcal{K} \text{ if and only if } \mathbf{A} \times \mathbf{B} \in \mathcal{K} \text{ whenever } \mathbf{A}, \mathbf{B} \in \mathcal{K}.$$

■

Exercises 8.122

1. Let I be any set and let F be any filter on I. Prove that if $\{i \mid \mathbf{A}_i \equiv \mathbf{B}_i\} \in F$, then $\prod_I \mathbf{A}_i/F \equiv \prod_I \mathbf{B}_i/F$.

2. Prove that if $\mathbf{A} \equiv \mathbf{B}$ and I and J are infinite, then $\mathbf{A}^I \equiv \mathbf{B}^J$.

3. Let $\mathcal{K}$ be any class of algebras of the same signature. Prove $\mathcal{K}$ is a variety if and only if each of the following conditions holds:
   (i) $\boldsymbol{H}\mathcal{K} = \mathcal{K}$,
   (ii) $\boldsymbol{S}\mathcal{K} = \mathcal{K}$,
   (iii) $\boldsymbol{P_u}\mathcal{K} = \mathcal{K}$,
   (iv) $\boldsymbol{P}_{\text{fin}}\mathcal{K} = \mathcal{K}$.

Here $\boldsymbol{P}_{\text{fin}}\mathcal{K}$ denotes the class of all algebras isomorphic to direct products of finite systems of algebras from $\mathcal{K}$. Since the empty system is finite, $\boldsymbol{P}_{\text{fin}}\mathcal{K}$ always contains the variety of trivial algebras.

***4.** (Fred Galvin, 1970) Prove that if $\mathbf{B} \equiv \mathbf{A}^I/F$ and $\mathbf{A} \equiv \mathbf{B}^J/G$, then $\mathbf{A} \equiv \mathbf{B}$.

***5.** (Fred Galvin, 1970) Prove that if $\mathbf{A} \equiv \mathbf{A} \times \mathbf{B} \times \mathbf{C}$, then $\mathbf{A} \equiv \mathbf{A} \times \mathbf{B}$. Compare this to Theorem 5.27 and its corollaries in Volume I beginning on page 331.

8.9. Elementary Chains, Amalgamation, and Interpolation

Reduced products—and especially ultraproducts—provide us with a way to put systems of structures together to make larger more complicated structures. Here we provide another method for putting structures together. Of course, we would like to know how the elementary properties of the less complicated structures relate to the elementary properties of the more involved structures.

Let $\langle \mathbf{A}_i \mid i \in I\rangle$ be a system of structures, all of the same signature. We say this system is **up-directed** by $\preccurlyeq$ provided for all $i, j \in I$ there is $k \in I$ so that

$$\mathbf{A}_i \preccurlyeq \mathbf{A}_k \text{ and } \mathbf{A}_j \preccurlyeq \mathbf{A}_k.$$

Of course, in a similar way we could make sense of what it means for the system to be up-directed with respect to some other binary relation in place of $\preccurlyeq$. The substructure relation and the relation of elementary embeddability are just two other interesting possibilities, although to include any kind of embeddability requires taking on some amalgamation property.

Let $\langle \mathbf{A}_i \mid i \in I\rangle$ be up-directed by $\preccurlyeq$. There is a transparent way to arrive at a limiting structure $\mathbf{A}$ of this system. Just let

$$\begin{aligned} A &= \bigcup_{i\in I} A_i \\ Q^{\mathbf{A}} &= \bigcup_{i\in I} Q^{\mathbf{A}_i} \text{ for each operation symbol } Q \\ R^{\mathbf{A}} &= \bigcup_{i\in I} R^{\mathbf{A}_i} \text{ for each relations symbol } R \end{aligned}$$

While it is evident that the union of a collection of r-place relations is itself an r-place relation, we might anticipate that the union of a collection of r-place operations could well fail to be an operation, unless a certain amount of compatibility is present. The up-directed character of the system ensures enough compatibility. Indeed, suppose Q is a 3-place operation symbol and $a, b, c \in A = \bigcup_{i\in I} A_i$. Using the fact that the system is up-directed, that must be a $k \in I$ so that $a, b, c \in A_k$. Let $d = Q^{\mathbf{A}_k}(a, b, c)$. Suppose $\ell \in I$ so that $a, b, c \in A_\ell$ and that $e = Q^{\mathbf{A}_\ell}(a, b, c)$. Using the up-directedness again, pick $m \in I$ so that $\mathbf{A}_k \preccurlyeq \mathbf{A}_m$ and $\mathbf{A}_\ell \preccurlyeq \mathbf{A}_m$. Then it follows that

$$d = Q^{\mathbf{A}_k}(a, b, c) = Q^{\mathbf{A}_m}(a, b, c) = Q^{\mathbf{A}_\ell}(a, b, c) = e.$$

This means $Q^{\mathbf{A}}$ really is a 3-place operation on A. A little reflection should convince you that $\mathbf{A}_k$ is a substructure of $\mathbf{A}$, for all $k \in I$. More is true.

THEOREM 8.123 (Tarski's Elementary Chain Theorem). *Let $\langle \mathbf{A}_i \mid i \in I\rangle$ be a system of structures, all of the same signature, that is up-directed by $\preccurlyeq$ and let $\mathbf{A}$ be the limiting structure of this system. Then $\mathbf{A}_k \preccurlyeq \mathbf{A}$ for all $k \in I$.*

Proof. What we have to prove is that for every formula $\phi(\mathbf{y})$ with, say, n free variables, for all $k \in I$, and all n-tuples $\mathbf{a}$ from A_k and all $k \in I$ we have

$$\mathbf{A}_k \models \phi[\mathbf{a}] \text{ if and only if } \mathbf{A} \models \phi[\mathbf{a}].$$

We do this by induction on the complexity of formulas. The proof is, except at one crucial point, almost identical to the proof of Tarski's Criterion. By now, we are able to see that the only step that is troublesome is
Inductive Step: ϕ is $\exists x \theta(x, \mathbf{y})$
Even here the implication

$$\mathbf{A}_k \models \phi[\mathbf{a}] \text{ implies } \mathbf{A} \models \phi[\mathbf{a}].$$

is entirely straightforward.

Here is the implication in the other direction.

$$\begin{aligned}
\mathbf{A} \models \phi[\mathbf{a}] &\text{ implies } \mathbf{A} \models \exists x\theta(x,\mathbf{y})[\mathbf{a}] \\
&\text{ implies } \mathbf{A} \models \theta(x,\mathbf{y})[b,\mathbf{a}] \text{ for some } b \in A \\
&\text{ implies } \mathbf{A} \models \theta(x,\mathbf{y})[b,\mathbf{a}] \text{ for some } b \in A_\ell \text{ for some } \ell \in I \\
&\text{ implies } \mathbf{A} \models \theta(x,\mathbf{y})[b,\mathbf{a}] \text{ for some } b \in A_m \text{ for some } m \in I \text{ so that } \mathbf{A}_k \preccurlyeq \mathbf{A}_m \\
&\text{ implies } \mathbf{A}_m \models \theta(x,\mathbf{y})[b,\mathbf{a}] \text{ for some } b \in A_m \text{ for some } m \in I \text{ so that } \mathbf{A}_k \preccurlyeq \mathbf{A}_m \\
&\text{ implies } \mathbf{A}_m \models \exists x\theta(x,\mathbf{y})[\mathbf{a}] \text{ for some } m \in I \text{ so that } \mathbf{A}_k \preccurlyeq \mathbf{A}_m \\
&\text{ implies } \mathbf{A}_m \models \phi[\mathbf{a}] \text{ for some } m \in I \text{ so that } \mathbf{A}_k \preccurlyeq \mathbf{A}_m \\
&\text{ implies } \mathbf{A}_k \models \phi[\mathbf{a}]
\end{aligned}$$

The fourth implication in this sequence relies on the up-directedness, the fifth invokes the induction hypothesis, and the last uses the elementary substructure relation. ∎

Tarski's Elementary Chain Theorem turns out to have a host of consequences. To use it, we need a method for assembling elementary chains or more elaborate systems of structures up-directed by the elementary substructure relation. For this purpose theorems like the next are well-suited. Roughly speaking, the next theorem specifies circumstances under which two structures that have some common overlap can be seen as part of a single more comprehensive structure—when this is possible the larger structure shows how to amalgamate the given structures over their common part. Essentially a concept from category theory, amalgamation can be framed by way of commutative diagrams. We have framed the next theorem using elementary equivalence, elementary embeddings, and the elementary substructure relation. The applications we have in mind actually use small variations of this theorem that, for example, might involve structures of several signatures or concern only elementary formulas of a particular form—leading to looser notions than elementary equivalence, elementary embeddings, and elementary substructure. The consequent adjustments to the statement and proof of the next theorem will require almost no effort.

THEOREM 8.124 (The Elementary Amalgamation Theorem). *Let* $\mathbf{A}$ *and* $\mathbf{B}$ *be structures of the same signature and let* $\mathbf{a}$ *and* $\mathbf{b}$ *be* I*-tuples of distinct elements of* A *and* B*, respectively. Let* $\mathbf{D}$ *be the substructure of* $\mathbf{A}$ *generated by the elements listed in the tuple* $\mathbf{a}$*. If* $\langle \mathbf{A}, \mathbf{a}\rangle \equiv \langle \mathbf{B}, \mathbf{b}\rangle$*, then there exist a structure* $\mathbf{C}$ *and an elementary embedding* g *so that*

- $\mathbf{A} \preccurlyeq \mathbf{C}$,
- $g : \mathbf{B} \to \mathbf{C}$ *such that* $g(\mathbf{b}) = \mathbf{a}$, *and*
- *There exists a unique embedding* f *such that* $f : \mathbf{D} \to \mathbf{B}$ *with* $f(\mathbf{a}) = \mathbf{b}$.

The situation described in this theorem is illustrated below.

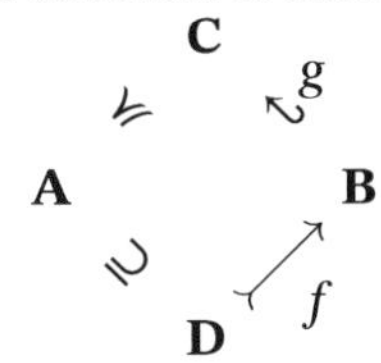

Proof. Definition 8.35 of the diagram of a single structure introduces new constants to name the elements of the structure. We are going to use the elementary diagrams of the two structures $\mathbf{A}$ and $\mathbf{B}$, but to keep things in good order we adopt the convention that the new constant symbols involved in these diagrams shall satisfy the following constraints:

- The same new constant symbol is associated with both a_i and b_i for all $i \in I$.
- Apart from the constant symbols in the line above, the constant symbols associated with the elements of A and B are entirely different.

Let $\Delta_{\mathbf{A}}$ be the elementary diagram of $\mathbf{A}$ and $\Delta_{\mathbf{B}}$ be the elementary diagram of $\mathbf{B}$. Notice that the structure $\mathbf{D}$ is fully described by sentences that belong to both $\mathbf{Th}\langle \mathbf{A}, \mathbf{a}\rangle$ and $\mathbf{Th}\langle \mathbf{B}, \mathbf{b}\rangle$. That justifies the bottom of the diagram. To simplify things, we assume the map f is just the inclusion map and, consequently, that $\mathbf{a} = \mathbf{b}$. We want a model of $\Delta_{\mathbf{A}} \cup \Delta_{\mathbf{B}}$. We'll use the Compactness Theorem.

Suppose no such model exists. Then there is some finite subset $\Gamma \subseteq \Delta_{\mathbf{B}}$ so that $\Delta_{\mathbf{A}} \cup \Gamma$ has no model. Γ is a finite set of sentences true in $\langle \mathbf{B}, B\rangle$. Let $\varphi(\mathbf{a}, \mathbf{d})$ be the conjunction of Γ, where $\mathbf{d}$ lists constant symbols for elements of B not listed in $\mathbf{a}$. Notice $\varphi(\mathbf{a}, \mathbf{d}) \in \Delta_{\mathbf{B}}$. So $\Delta_{\mathbf{A}} \cup \{\varphi(\mathbf{a}, \mathbf{d})\}$ has no model. Thus

$$\Delta_{\mathbf{A}} \models \neg\varphi(\mathbf{a}, \mathbf{d}) \text{ and so } \Delta_{\mathbf{A}} \models \forall \mathbf{y} \neg\varphi(\mathbf{a}, \mathbf{y})$$

since $\Delta_{\mathbf{A}}$ places no constraints on the constant symbols listed in $\mathbf{d}$. But $\forall \mathbf{y}\neg\varphi(\mathbf{a}, \mathbf{y})$ is just something written down in the language of $\langle \mathbf{A}, A\rangle$. So it is true in $\langle \mathbf{A}, A\rangle$. That is, it is true in $\langle \mathbf{A}, \mathbf{a}\rangle$. Since $\langle \mathbf{A}, \mathbf{a}\rangle$ and $\langle \mathbf{B}, \mathbf{a}\rangle$ are elementarily equivalent, it has to be true in $\langle \mathbf{B}, \mathbf{a}\rangle$:

$$\langle \mathbf{B}, \mathbf{a}\rangle \models \forall \mathbf{y}\neg\varphi(\mathbf{a}, \mathbf{y})$$

But we know that

$$\langle \mathbf{B}, \mathbf{a}\rangle \models \varphi(\mathbf{a}, \mathbf{d})$$

This is a contradiction. So our original supposition is wrong; hence there is a model of $\Delta_{\mathbf{A}} \cup \Delta_{\mathbf{B}}$. Let $\mathbf{C}'$ be a model of $\Delta_{\mathbf{A}} \cup \Delta_{\mathbf{B}}$ and let $\mathbf{C}$ be the reduct of $\mathbf{C}'$ back to the original signature. Now we know

- There is an elementary embedding g so that $g : \mathbf{B} \to \mathbf{C}$
- There is an elementary embedding h so that $h : \mathbf{A} \to \mathbf{C}$.

Since $\mathbf{C}' \models \Delta_{\mathbf{A}}$, it is harmless to suppose that $\mathbf{C}'$ has actually been chosen so that h is the inclusion map. Then $\mathbf{A} \preccurlyeq \mathbf{C}$. ■

There is a kind of hidden assumption in the statement of this theorem. In the event that I is empty, that is no elements are listed in $\mathbf{a}$ and $\mathbf{b}$, and that the signature provides no constant symbols, then we find that the empty set would be the universe of $\mathbf{D}$. But we have insisted that all our structures have nonempty universes. So the formulation of the Elementary Amalgamation Theorem excludes this possibility. Nevertheless, the proof still works. What it proves is

THEOREM 8.125 (The Elementary Joint Embedding Theorem). *If $\mathbf{A} \equiv \mathbf{B}$, then there are a substructure $\mathbf{C}$ and an elementary embedding g so that*

- $\mathbf{A} \preccurlyeq \mathbf{C}$,
- $g : \mathbf{B} \to \mathbf{C}$. ■

Multiple Signatures: Joint Consistency, Interpolation, and Definability

The next theorem has been extracted from a proof Abraham Robinson gave in the 1950's for his Joint Consistency Theorem. It is put here in a stand-alone form because it has become a paradigm for other similar constructions and because some of the consequences of the Joint Consistency Theorem are more readily seen as consequences of this theorem. While we gave it another name, reflecting its proof rather than its statement, it is a sort of two signature amalgamation theorem.

THEOREM 8.126 (Robinson's Two Signature Chain Construction Theorem (Abraham Robinson, 1956b)). *Let $\mathfrak{L}_0, \mathfrak{L}_1$, and $\mathfrak{L}_2$ be signatures where $\mathfrak{L}_0$ is the signature of symbols common to both $\mathfrak{L}_1$ and $\mathfrak{L}_2$. We assume the signatures $\mathfrak{L}_1$ and $\mathfrak{L}_2$ give their common symbols the same ranks. Let $\mathbf{A}$ be an $\mathfrak{L}_1$-structure and let $\mathbf{B}$ be an $\mathfrak{L}_2$-structure. Let $\mathbf{b}$ be a list of elements in both A and B. If $\langle \mathbf{A} \restriction_{\mathfrak{L}_0}, \mathbf{b}\rangle \equiv \langle \mathbf{B} \restriction_{\mathfrak{L}_0}, \mathbf{b}\rangle$, then there is some $\mathfrak{L}_1 \cup \mathfrak{L}_2$-structure $\mathbf{C}$ so that $\mathbf{A} \preccurlyeq C \restriction_{\mathfrak{L}_1}$ and there is an $\mathfrak{L}_2$-elementary embedding g so that*

$$g : \mathbf{B} \hookrightarrow \mathbf{C} \restriction_{\mathfrak{L}_2}$$

where $g(\mathbf{b}) = \mathbf{b}$.

Proof. The proof is almost fully displayed in this illustration:

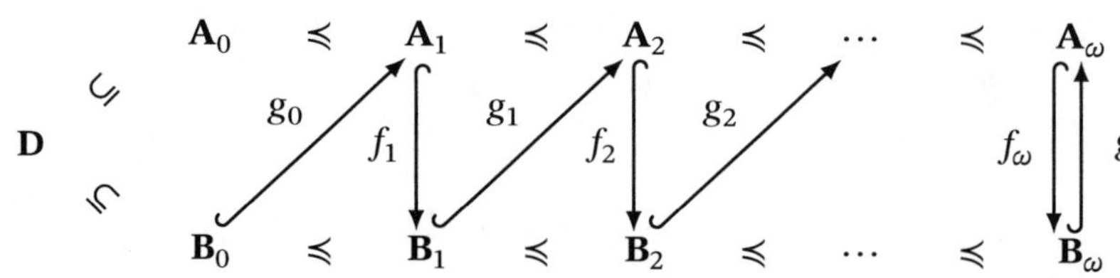

To start, we take $\mathbf{D}$ to be the $\mathfrak{L}_0$-substructure of $\mathbf{A} \restriction_{\mathfrak{L}_0}$ generated by the elements listed in the tuple $\mathbf{b}$. Notice that $\mathbf{D}$ is also an $\mathfrak{L}_0$-substructure of $\mathbf{B} \restriction_{\mathfrak{L}_0}$, since $\mathbf{A} \restriction_{\mathfrak{L}_0} \equiv \mathbf{B} \restriction_{\mathfrak{L}_0}$. We also put

$$\mathbf{A}_0 = \mathbf{A} \text{ and } \mathbf{B}_0 = \mathbf{B}.$$

To get $\mathbf{A}_1$ and g_0, we invoke a version of the Elementary Amalgamation Theorem. In this version, we use $\mathfrak{L}_1$ for the part involving $\mathbf{A}_0$ and $\mathfrak{L}_0 \cup B_0$ for the part involving

$\mathbf{B}_0$. This means that $\mathbf{A}_0 \preccurlyeq \mathbf{A}_1$ (in the appropriate, namely $\mathfrak{L}_1$, sense) but that g_0 is an elementary embedding of $\langle \mathbf{B}_0 \restriction_{\mathfrak{L}_0}, B_0 \rangle$ into $\langle \mathbf{A}_1 \restriction_{\mathfrak{L}_0}, g_0(B_0) \rangle$.

To get $\mathbf{B}_1$ and f_1, we invoke a version of the Elementary Amalgamation Theorem again. But in this version we use $\mathfrak{L}_2$ for the part involving $\mathbf{B}_0$ and $\mathfrak{L}_0$ for the part involving $\mathbf{A}_1$.

We continue in this way, taking a step for each natural number and alternating the use of $\mathfrak{L}_1$ and $\mathfrak{L}_2$. In the course of this construction, we build the $\mathfrak{L}_1$ elementary chain along the top of our illustration and the $\mathfrak{L}_2$ elementary chain along the bottom. The maps going up and down across the middle are elementary embeddings for $\mathfrak{L}_0$ enhanced by a growing supply of new constant symbols.

A simple but important property of our illustration is that it is a commutative diagram in the sense of category theory. That is that $f_{k+1}(g_k(b)) = b$ for all k and all $b \in B_k$ and likewise that $g_k(f_k(a)) = a$ for all $k > 0$ and all $a \in A_k$. The point is that these composite maps are elementary embeddings with respect to $\mathfrak{L}_0$ enhanced by enough new constants. For example, the formula $x \approx a$ is satisfied in $\mathbf{A}_k$ by a so the formula $x \approx a$ must be satisfied in $\mathbf{A}_k$ by $g_k(f_k(a))$. That is $g_k(f_k(a)) = a$.

At the limit, we just take unions, even for the functions. Ignoring all the new constants symbols accumulated in this construction, we find that f_ω is an $\mathfrak{L}_0$ isomorphism from $\mathbf{A}_\omega$ onto $\mathbf{B}_\omega$ and that g_ω is its inverse. To make the desired structure $\mathbf{C}$ we use these isomorphisms to impose on A_ω all the $\mathfrak{L}_2$ relations and operations not in $\mathfrak{L}_0$. For example, if R is a 3-place relation symbol of L_2 and $a, b, c \in A_\omega$ we put

$$(a, b, c) \in R^{\mathbf{C}} \text{ if and only if } (f_\omega(a), f_\omega(b), f_\omega(c)) \in R^{\mathbf{B}_\omega}.$$

In this way, g_ω will become an L_2 elementary embedding of $\mathbf{B}_\omega$ into $\mathbf{C} \restriction_{\mathfrak{L}_2}$ and $\mathbf{A}_\omega = \mathbf{C} \restriction_{\mathfrak{L}_1}$. ■

THEOREM 8.127 (Robinson's Joint Consistency Theorem (Abraham Robinson, 1956b)). *Let $\mathfrak{L}_0, \mathfrak{L}_1$, and $\mathfrak{L}_2$ be signatures where $\mathfrak{L}_0$ is the signature of symbols common to both $\mathfrak{L}_1$ and $\mathfrak{L}_2$. We assume the signatures $\mathfrak{L}_1$ and $\mathfrak{L}_2$ give their common symbols the same ranks. Let T be a complete $\mathfrak{L}_0$ theory, let T_1 be an $\mathfrak{L}_1$ theory with $T \subseteq T_1$, and let T_2 be an $\mathfrak{L}_2$ theory with $T \subseteq T_2$. If both T_1 and T_2 have models, then $T_1 \cup T_2$ has a model.*

Proof. Let $\mathbf{A} \models T_1$ and let $\mathbf{B} \models T_2$. We will use Robinson's Two Signature Chain Construction Theorem. We take $\mathbf{b}$ of that theorem to be the empty tuple. We see that $\mathbf{A} \restriction_{\mathfrak{L}_0} \models T$ and that $\mathbf{B} \restriction_{\mathfrak{L}_0} \models T$. Because T is a complete $\mathfrak{L}_0$ theory, we deduce that $\mathbf{A} \restriction_{\mathfrak{L}_0} \equiv \mathbf{B} \restriction_{\mathfrak{L}_0}$. By the Chain Construction Theorem there is an $\mathfrak{L}_1 \cup \mathfrak{L}_2$ structure $\mathbf{C}$ so that $\mathbf{A} \preccurlyeq \mathbf{C} \restriction_{\mathfrak{L}_1}$ and an $\mathfrak{L}_2$-elementary embedding $g : \mathbf{B} \hookrightarrow \mathbf{C} \restriction_{\mathfrak{L}_2}$. But this means $\mathbf{C} \models T_1 \cup T_2$. ■

THEOREM 8.128 (The Two Signature Interpolation Theorem). *Let $\mathfrak{L}_0, \mathfrak{L}_1$, and $\mathfrak{L}_2$ be signatures where $\mathfrak{L}_0$ is the signature of symbols common to both $\mathfrak{L}_1$ and $\mathfrak{L}_2$. We assume the signatures $\mathfrak{L}_1$ and $\mathfrak{L}_2$ give their common symbols the same ranks. Let T_1 be an $\mathfrak{L}_1$ theory and let T_2 be an $\mathfrak{L}_2$ theory. If $T_1 \cup T_2$ has no model, then there is an $\mathfrak{L}_0$ sentence ϕ so that $T_1 \models \phi$ and $T_2 \models \neg\phi$.*

Proof. Let $\Phi = T_1 \restriction_{\mathfrak{L}_0}$. By the Compactness Theorem, we only need to prove that $\Phi \cup T_2$ has no model. Because $T_1 \cup T_2$ has no models, there cannot be $\mathbf{A} \models T_1$ and $\mathbf{B} \models T_2$ so that $\mathbf{A} \restriction_{\mathfrak{L}_0} \equiv \mathbf{B} \restriction_{\mathfrak{L}_0}$, for otherwise the Chain Construction Theorem would provide a model of $T_1 \cup T_2$.

Suppose, for contradiction, that $\mathbf{B} \models \Phi \cup T_2$. We know that $\mathbf{Th}\,\mathbf{B} \restriction_{\mathfrak{L}_0} \cup T_1$ has no model, as noted above. So by the Compactness Theorem there must be $\theta \in \mathbf{Th}\,\mathbf{B} \restriction_{\mathfrak{L}_0}$ so that $\{\theta\} \cup T_1$ has no model. Therefore $T_1 \models \neg\theta$ and so that $\neg\theta \in \Phi$. Hence, we see that $\mathbf{B}$ is a model of both θ and $\neg\theta$, which cannot be. ■

Let $\mathfrak{L}$ be a language and let R be a relation symbol not belonging to $\mathfrak{L}$. Let r be the rank of R. A set Σ of $\mathfrak{L} \cup \{R\}$ sentences is said to **implicitly define** R provided that when Σ' is obtained from Σ by replacing each occurrence of R by R', where R' is a new relation symbol of rank r, then

$$\Sigma \cup \Sigma' \models \forall x_0, \dots, x_{r-1}[Rx_0 \dots x_{r-1} \Leftrightarrow R'x_0 \dots x_{r-1}].$$

We say that Σ **explicitly defines** R provided there is an $\mathfrak{L}$-formula $\varphi(x_0, \dots, x_{r-1})$ so that

$$\Sigma \models \forall x_0, \dots, x_{r-1}[Rx_0 \dots x_{r-1} \Leftrightarrow \varphi(x_0, \dots, x_{r_1})].$$

THEOREM 8.129 (Beth's Theorem (E. W. Beth, 1953)). *Let R be a relation symbol not in the language $\mathfrak{L}$. A set Σ of sentences in $\mathfrak{L} \cup \{R\}$ implicitly defines R if and only if Σ explicitly defines R.* ■

The proof of Beth's Theorem, as well as Craig's Interpolation Theorem just below, follows easily from the Two Signature Interpolation Theorem. These proofs are left as exercises. Historically, both Robinson's work and that of Craig arose from efforts to find a deeper understanding of Beth's Theorem.

THEOREM 8.130 (Craig's Interpolation Theorem (William Craig, 1957)). *Let ψ and θ be sentences so that $\psi \models \theta$. There is a sentence φ such that $\psi \models \varphi$ and $\varphi \models \theta$ and each relation symbol and each operation symbol that occurs in φ occurs also in both ψ and θ.* ■

This theorem, illuminating the word "interpolation", is just an instance of the Two Signature Interpolation Theorem in which T_1 is the set of all the logical consequences of ψ and T_2 is the set of all logical consequences of $\neg\theta$. William Craig gave the first demonstration of this theorem by means of proof theory.

The next theorem provides more information about interpolating sentences. For this theorem, we consider a different formulation of elementary logic. First we provide ourselves with two logical symbols $\top$ and $\bot$ that are always interpreted in every structure as true and false respectively. They count as atomic formulas (and could be regarded as denoting 0-place relations). In this formulation formulas are built up from atomic formulas with the help of the connectives to be $\vee$, $\wedge$, and $\neg$ and the quantifiers $\forall$ and $\exists$. (So we consider $\varphi \Rightarrow \psi$ as an abbreviation of $\neg\varphi \vee \psi$.) Every formula is logically equivalent to one built up from atomic and negated atomic formulas with the help of $\vee$, $\wedge$, $\forall$, and $\exists$—such a formula is said to be in **negation normal form**. A relation symbol, as well as the equality symbol $\approx$, is said to occur positively in a formula φ in negation normal form provided it occurs in an atomic subformula of φ that is not negated. The atomic formula $\top$ is regarded as positive while the atomic formula $\bot$ is negative (and is logically equivalent with $\neg\top$). Similarly, a relation symbol occurs negatively in φ if it occurs in a negated atomic subformula of φ. Notice that a relation symbol can have both positive and negative occurrences in φ.

THEOREM 8.131 (Lyndon's Interpolation Theorem (Roger C. Lyndon, 1959a; Arnold Oberschelp, 1968)). *Let Σ be a set of sentences in negation normal form and γ be a sentence in negation normal form such that $\Sigma \models \gamma$. There is sentence θ such that*

(i) *$\Sigma \models \theta$ and $\theta \models \gamma$;*
(ii) *Every relation symbol (excluding $\approx$) that occurs positively in θ also occurs positively in both Σ and γ;*
(iii) *Every relation symbol (excluding $\approx$) that occurs negatively in θ also occurs negatively in both Σ and γ.*
(iv) *If $\approx$ occurs positively in θ, then $\approx$ occurs positively in Σ;*
(v) *If $\approx$ occurs negatively in θ, then $\approx$ occurs negatively in γ.*

Moreover, If Σ has a model, then $\bot$ does not occur in θ, and if $\neg\gamma$ has a model then $\top$ does not occur in θ.

The statement above differs from that in Lyndon's original paper. Indeed, M. A. Taĭtslin (1960) in a review in the Journal of Symbolic Logic the next year pointed out the flaw and provided a counterexample. Leon Henkin (1963) provided a correct version, which, however did not allow for operation symbols. Finally, Arnold Oberschelp (1968) gave the version above. Tsuyoshi Fujiwara (1978) gave a slightly different version, see (Nobuyoshi Motohashi, 1984).

Proof. We make the harmless but convenient assumption that the sentences in Σ as well as γ are in negation normal form. We say that a sentence θ is a *Lyndon interpolant* between Σ and γ provided $\Sigma \models \theta \models \gamma$ and θ has properties (i)–(v). For a set Γ we say a sentence θ is a *Lyndon interpolant* between Σ and Γ provided θ has properties (i)–(v) (adjusted for Γ in place of γ) provided $\Sigma \models \theta$ and $\Gamma \models \neg\theta$. We will suppose that there is no Lyndon interpolant between Σ and γ and show that $\Sigma \not\models \gamma$.

Let κ be the cardinality of the set of formulas of our signature. We expand this signature by adjoining κ new constant symbols arranged in a well-ordered list $c_0, c_1, \ldots, c_\alpha, \ldots$ of order type κ. In this expanded signature Σ and γ will still fail to have an interpolant θ with the desired properties.

We will say that a sentence θ is *compatible* with Σ provided every relation symbol that occurs positively (negatively) in θ occurs positively (negatively) in Σ and if $\approx$ occurs positively in θ then it occurs positively in Σ. We also say that a sentence θ is *compatible* with γ provided every relation symbol that occurs positively (negatively) in θ occurs positively (negatively) in γ and if $\approx$ occurs negatively in θ then it occurs negatively in γ. Let Φ be the set of all sentences that are compatible with Σ and let Ψ be the set of all sentences compatible with γ. Then all the Lyndon interpolants between Σ and γ must lie in $\Phi \cap \Psi$. We use $\overline{\psi}$ to denote a negation normal form of $\neg\psi$ and $\overline{\Psi}$ to denote $\{\overline{\psi} \mid \psi \in \Psi\}$. We take $\overline{\top} = \bot$ and $\overline{\bot} = \top$.

We interrupt the proof of Theorem 8.131 for a lemma. That proof will resume on page 430 and finish on page 434.

Let χ be a formula. We use that notation $\chi\binom{c_a}{x}$ to denote the formula obtained from χ by replacing each free occurrence of the variable x with the constant symbol c_a.

LEMMA. *In the expanded signature there are sets Σ^* and Γ^* of sentences in negation normal form satisfying the following conditions:*

(i) $\Sigma \subseteq \Sigma^* \subseteq \Phi$ *and* $\overline{\gamma} \in \Gamma^* \subseteq \overline{\Psi}$;
(ii) Σ^* *and* $\overline{\Gamma^*}$ *are disjoint;*
(iii) *There is no Lyndon interpolant between* Σ^* *and* Γ^*;
(iv) *If* $\chi \in \Phi$ *and* $\chi \notin \Sigma^*$, *then there is a Lyndon interpolant between* $\Sigma^* \cup \{\chi\}$ *and* Γ^* *and if* $\chi \in \overline{\Psi}$ *and* $\chi \notin \Gamma^*$, *then there is a Lyndon interpolant between* Σ^* *and* $\Gamma^* \cup \{\chi\}$;
(v) *If* $\exists x\chi \in \Sigma^*$, *then there is a constant symbol* c_α *so that* $\chi\binom{c_\alpha}{x} \in \Sigma^*$ *and if* $\exists x\chi \in \Gamma^*$, *then there is a constant symbol* c_α *so that* $\chi\binom{c_\alpha}{x} \in \Gamma^*$. *If* $\forall x\chi \in \Sigma^*$, *then for every constant symbol* c_α *we have* $\chi\binom{c_\alpha}{x} \in \Sigma^*$ *and if* $\forall x\chi \in \Gamma^*$, *then for every constant symbol* c_α *we have* $\chi\binom{c_\alpha}{x} \in \Gamma^*$.
(vi) *For any sentence* χ *in negation normal form it is neither the case that* $\chi, \overline{\chi} \in \Sigma^*$ *nor* $\chi, \overline{\chi} \in \Gamma^*$, *where* $\overline{\chi}$ *is a negation normal form of* $\neg\chi$.
(vii) *For any sentence* $\chi \in \Phi$, *if* $\Sigma^* \models \chi$, *then* $\chi \in \Sigma^*$;
(viii) *For any sentences* $\chi \in \overline{\Psi}$, *if* $\Gamma^* \models \chi$, *then* $\chi \in \Gamma^*$;
(ix) *Let* χ *and* ρ *be any sentences in negation normal form. If* $\chi \vee \rho \in \Sigma^*$, *then* $\chi \in \Sigma^*$ *or* $\rho \in \Sigma^*$. *If* $\chi \vee \rho \in \Gamma^*$, *then* $\chi \in \Gamma^*$ *or* $\rho \in \Gamma^*$;
(x) *Let* χ *and* ρ *be any sentences in negation normal form. If* $\chi \wedge \rho \in \Sigma^*$, *then* $\chi \in \Sigma^*$ *and* $\rho \in \Sigma^*$. *If* $\chi \wedge \rho \in \Gamma^*$, *then* $\chi \in \Gamma^*$ *and* $\rho \in \Gamma^*$;

Proof of the Lemma. The sets Σ^* and Γ^* are built via a transfinite recursion. For this purpose we arrange a listing $\varphi_0, \varphi_1, \ldots, \varphi_\alpha, \ldots$ of all sentences in negation normal form. We begin by setting $\Sigma_0 = \Sigma$ and $\Gamma_0 = \{\overline{\gamma}\}$. For each ordinal $\alpha < \kappa$ we will define Σ_α and Γ_α so that the following five properties hold:

(1) If $\alpha' < \alpha$, then $\Sigma_{\alpha'} \subseteq \Sigma_\alpha \subseteq \Phi$ and $\Gamma_{\alpha'} \subseteq \Gamma_\alpha \subseteq \overline{\Psi}$;
(2) Σ_α and Γ_α have no Lyndon interpolant;
(3) In case $\alpha = \lambda + 3n + 1$ for some limit ordinal λ and some natural number n, for any $\eta < \lambda + n$, if $\varphi_\eta \in \Phi$ and $\varphi_\eta \notin \Sigma_\alpha$, then there is a Lyndon interpolant between $\Sigma_\alpha \cup \{\varphi_\eta\}$ and Γ_α;
(4) In case $\alpha = \lambda + 3n + 2$ for some limit ordinal λ and some natural number n and if $\varphi_\eta \in \overline{\Psi}$ and $\varphi_\eta \notin \Gamma_\alpha$, then there is a Lyndon interpolant between Σ_α and $\Gamma_\alpha \cup \{\varphi_\eta\}$.
(5) In case $\alpha = \lambda + 3n + 3$ for some limit ordinal λ and some natural number n, we have for all $\eta < \lambda + n$ if $\varphi_\eta \in \Sigma_\alpha$ and φ_η is $\exists x\rho$, then $\rho\binom{c}{x} \in \Sigma_\alpha$, for some new constant symbol c; likewise, if $\varphi_\eta \in \Gamma_\alpha$ and φ_η is $\exists x\rho$, then $\rho\binom{c}{x} \in \Gamma_\alpha$, for some new constant symbol c;
(6) In case $\alpha = \alpha' + 1$, there is at most one sentence in $\Sigma_\alpha \cup \Gamma_\alpha$ that is not in $\Sigma_{\alpha'} \cup \Gamma_{\alpha'}$.

To carry out the construction by recursion, we assume that for all $\alpha' < \alpha$ the construction have been carried out so that (1)–(5) are fulfilled. In the event that α is a limit ordinal we let

$$\Sigma_\alpha = \bigcup_{\alpha'<\alpha} \Sigma_{\alpha'} \text{ and } \Gamma_\alpha = \bigcup_{\alpha'<\alpha} \Gamma_{\alpha'}.$$

At limit ordinals only (1) and (2) are pertinent. Property (1) plainly holds, while property (2) follows from the fact that θ being a Lyndon interpolant only depends on finitely many sentences in each of Σ_α and Γ_α.

So consider the case when α is a successor ordinal. We break this into three subcases.

First, suppose $\alpha = \lambda + 3n + 1$ for some limit ordinal λ and some natural number n. If $\varphi_{\lambda+n}$ belongs to Φ and there is no Lyndon interpolant between $\Sigma_{\lambda+3n} \cup \{\varphi_{\lambda+n}\}$ and $\Gamma_{\lambda+3n}$, then put

$$\Sigma_\alpha = \Sigma_{\lambda+3n} \cup \{\varphi_{\lambda+n}\} \text{ and } \Gamma_\alpha = \Gamma_{\lambda+3n}.$$

In the remaining cases put

$$\Sigma_\alpha = \Sigma_{\lambda+3n} \text{ and } \Gamma_\alpha = \Gamma_{\lambda+3n}.$$

Once more, that properties (1)–(6) hold is transparent.

Next, suppose $\alpha = \lambda+3n+2$ for some limit ordinal λ and some natural number n. If $\varphi_{\lambda+n}$ belongs to $\overline{\Psi}$ and there is no Lyndon interpolant between $\Sigma_{\lambda+3n+1}$ and $\Gamma_{\lambda+3n+1} \cup \{\varphi_{\lambda+n}\}$ then put

$$\Sigma_\alpha = \Sigma_{\lambda+3n+1} \text{ and } \Gamma_\alpha = \Gamma_{\lambda+3n+1} \cup \{\varphi_{\lambda+n}\}.$$

In the remaining cases put

$$\Sigma_\alpha = \Sigma_{\lambda+3n+1} \text{ and } \Gamma_\alpha = \Gamma_{\lambda+3n+1}.$$

Once more, that properties (1)–(6) hold is transparent.

Now suppose $\alpha = \lambda + 3n + 3$, where λ is a limit ordinal and n is a natural number. In case it happened that $\varphi_{\lambda+n} \in \Sigma_{\lambda+3n+2}$ and $\varphi_{\lambda+n}$ is $\exists x \rho \in \Phi$ we put

$$\Sigma_\alpha = \Sigma_{\lambda+3n+2} \cup \left\{\rho\binom{c}{x}\right\} \text{ and } \Gamma_\alpha = \Gamma_{\lambda+3n+2}$$

where c is a constant symbols that has not been used. We need only show that there is no Lyndon interpolant between Σ_α and Γ_α. Suppose, to the contrary, that θ is such a Lyndon interpolant. Then

$$\Sigma_{\lambda+3n+2} \cup \{\rho\binom{c}{x}\} \models \theta \qquad \Gamma_{\lambda+3n} \models \neg\theta$$

$$\Sigma_{\lambda+3n+2} \models \neg\rho\binom{c}{x} \vee \theta \qquad \Gamma_{\lambda+3n+2} \models \neg\theta$$

$$\Sigma_{\lambda+3n+2} \models \forall x \neg\rho \vee \theta \qquad \Gamma_{\lambda+3n+2} \models \neg\theta$$

since c does not occur in $\Sigma_{\lambda+3n+2}$

$$\Sigma_{\lambda+3n+2} \models \neg\exists x \rho \vee \theta \qquad \Gamma_{\lambda+3n+2} \models \neg\theta$$

$$\Sigma_{\lambda+3n+2} \models \theta \qquad \Gamma_{\lambda+3n+2} \models \neg\theta$$

since $\exists x \rho \in \Sigma_{\lambda+3n+2}$.

But this means that θ is a Lyndon interpolant between $\Sigma_{\lambda+3n+2}$ and $\Gamma_{\lambda+3n+2}$, which is impossible.

At last, put

$$\Sigma^* = \bigcup_{\alpha<\kappa} \Sigma_\alpha \text{ and } \Gamma^* = \bigcup_{\alpha<\kappa} \Gamma_\alpha.$$

It is easy to check that properties (1) and (2) hold for Σ^* and Γ^*. This means parts (i) and (iii) of the Lemma are established.

For part (ii) observe that if $\theta \in \Sigma^* \cap \overline{\Gamma^*}$, then θ would be a Lyndon interpolant between Σ^* and Γ^*.

For part (iv) observe that all possible χ's have been considered during the course of the construction.

For part (v), the existential portion was accommodated at the "$3n+1$" and "$3n+2$" stages of the construction. Consider the case when $\forall x\chi \in \Sigma^*$. Suppose there is a constant c so that $\chi\binom{c}{x} \notin \Sigma^*$. Then there would be a Lyndon interpolant θ between $\Sigma^* \cup \{\chi\binom{c}{x}\}$ and Γ^*. This entails $\Sigma^* \models \neg\chi\binom{c}{x} \vee \theta$. But since $\forall x\chi \in \Sigma^*$, we also have $\Sigma^* \models \chi\binom{c}{x}$. This yields $\Sigma^* \models \theta$. So θ would be a Lyndon interpolant between Σ^* and Γ^*. The situation when $\forall x\chi \in \Gamma^*$ can be handled similarly.

For part (vi), observe that each of Σ^* and Γ^* must be consistent, lest $\bot$ (or $\top$) would be a Lyndon interpolant.

For parts (vii) and (viii), observe that χ would have been added since it would obstruct the Lyndon interpolants.

For part (ix) suppose $\chi \vee \rho \in \Sigma^*$ but that $\chi \notin \Sigma^*$ and $\rho \notin \Sigma^*$. Then there would be Lyndon interpolants θ_0 and θ_1 so that

$$\begin{aligned} &\Sigma^* \cup \{\chi\} \models \theta_0 && \Gamma^* \models \neg\theta_0 \\ &\Sigma^* \cup \{\rho\} \models \theta_1 && \Gamma^* \models \neg\theta_1 \end{aligned}$$

But then

$$\begin{aligned} &\Sigma^* \models \neg\chi \vee \theta_0 \\ &\Sigma^* \models \neg\rho \vee \theta_1 \end{aligned}$$

After some uses of distributivity and noting $\chi \vee \rho \in \Sigma^*$, we obtain

$$\Sigma^* \models \theta_0 \vee \theta_1 \qquad \Gamma^* \models \neg(\theta_0 \vee \theta_1),$$

yielding a Lyndon interpolant between Σ^* and Γ^*. The situation when $\chi \vee \rho \in \Gamma^*$ can be addressed in a similar fashion.

For part (x), suppose $\chi \wedge \rho \in \Sigma^*$ but $\chi \notin \Sigma^*$. Then there is a Lyndon interpolant θ so that $\Sigma^* \cup \{\chi\} \models \theta$ and $\Gamma^* \models \neg\theta$. But then $\Sigma^* \models \neg\chi \vee \theta$ and $\Sigma^* \models \chi$. Hence $\Sigma^* \models \theta$. This means that θ is a Lyndon interpolant between Σ^* and Γ^*. The case when $\chi \wedge \rho \in \Gamma^*$ is handled similarly.

In this way, all the parts of the Lemma have been established. ■

Returning to the proof of Theorem 8.131, remember that our aim is to show $\Sigma \not\models \gamma$. So it is enough to find a model of $\Sigma^* \cup \Gamma^*$, according to part (i) of the Lemma. As a first step in constructing our model, let $\mathbf{M}$ be a structure with universe the set of terms in which no variables occur. For an operation symbol Q and $t_0, \ldots, t_{r-1} \in M$, where r is the rank of Q, put

$$Q^{\mathbf{M}}(t_0, \ldots, t_{r-1}) = Qt_0 \ldots t_{r-1}.$$

For a relation symbol R and $t_0, \dots, t_{r-1} \in M$, where r is the rank of R put

$$\langle t_0, \dots, t_{r-1} \rangle \in R^{\mathbf{M}} \begin{cases} \text{if } Rt_0 \dots t_{r-1} \in \Sigma^* \text{ or} \\ \text{if } R_0 \dots t_{r-1} \in \Gamma^* \end{cases}$$

and put

$$\langle t_0, \dots, t_{r-1} \rangle \notin R^{\mathbf{M}} \begin{cases} \text{if } \neg Rt_0 \dots t_{r-1} \in \Sigma^* \text{ or} \\ \text{if } \neg Rt_0 \dots t_{r-1} \in \Gamma^* \end{cases}$$

These constraints are compatible by conditions (ii) and (vi) of the Lemma. Notice that these conditions define $R^{\mathbf{M}}$ provided each sequence $\langle t_0, \dots, t_{r-1} \rangle$ is placed either in $R^{\mathbf{M}}$ or in its complement. Observe by parts (vi) and (ii) of the Lemma, it is always possible to fulfill these constraints. This $\mathbf{M}$ is not yet the structure we need. Indeed, we desire a model of Σ^* in which all the sentence in Γ^* hold. So $\mathbf{M}$ has too many inequalities and not enough equalities.

We define a two-place relation $\sim$ on M via

$$s \sim t \text{ if and only if } \neg s \approx t \notin \Sigma^* \text{ and } \neg s \approx t \notin \Gamma^*.$$

CLAIM A.
The relation $\sim$ is a congruence relation on $\mathbf{M}$.

Proof of Claim A. To establish reflexivity, pick any $s \in M$. If $\neg s \approx s \in \Sigma^*$, then $\bot$ is a Lyndon interpolant between Σ^* and Γ^*, contrary to condition (iii) of the Lemma. Likewise, if $\neg s \approx s \in \Gamma^*$, then $\top$ is a Lyndon interpolant between Σ^* and Γ^*.

For symmetry of $\sim$ suppose $s \sim t$. Then $\neg s \approx t \notin \Sigma^*$ and $\neg s \approx t \in \Gamma^*$. From the first of these nonmemberships and part (iv) of the Lemma we have a Lyndon interpolant θ between $\Sigma^* \cup \{\neg s \approx t\}$ and Γ^*, since the negation of every equation belongs to Φ. But this means there is a Lyndon interpolant between $\Sigma^* \cup \{\neg t \approx s\}$ and Γ^*. By part (iii) of the Lemma we conclude that $\neg t \approx s \notin \Sigma^*$. Now consider that $\neg s \approx t \notin \Gamma^*$. Since every equation belongs to $\overline{\Psi}$, we have by part (iv) of the Lemma that there is a Lyndon interpolant between Σ^* and $\Gamma^* \cup \{\neg s \approx t\}$. But then the same interpolant serves between Σ^* and $\Gamma^* \cup \{\neg t \approx s\}$. Hence $\neg t \approx s \notin \Gamma^*$, as desired.

For transitivity, suppose $s \sim t \sim r$ but for contradiction suppose $s \sim r$ fails. So either $\neg s \approx r \in \Sigma^*$ or $\neg s \approx r \in \Gamma^*$. Then $\neg s \approx t \notin \Sigma^*$ and $\neg t \approx r \notin \Sigma^*$. So there is a Lyndon interpolant θ between $\Sigma^* \cup \{\neg s \approx t\}$ and Γ^*. Suppose first that $\neg s \approx r \in \Sigma^*$. So

$$\begin{aligned}
&\Sigma^* \models s \approx t \vee \theta \\
&\Sigma^* \models \neg s \approx r \\
&\Sigma^* \models \neg s \approx r \wedge (s \approx t \vee \theta) \\
&\Sigma^* \models (\neg s \approx r \wedge s \approx t) \vee (\neg s \approx r \wedge \theta) \\
&\Sigma^* \models \neg t \approx r \vee (\neg s \approx r \wedge \theta)
\end{aligned}$$

Since every inequality is in Φ we conclude by (ix) of the Lemma that $\neg t \approx r \in \Sigma^*$ or $\neg s \approx r \wedge \theta \in \Sigma^*$. The first alternative is impossible since $t \sim r$. But the other alternative is also impossible since the θ would be a Lyndon interpolant between Σ^* and Γ^*, violating part (iii) of the Lemma. So $\neg s \approx r \notin \Sigma^*$.

It remains to consider the case when $\neg s \approx r \in \Gamma^*$. An argument similar to the one above, but with Γ^* in the role of Σ^*, shows that $\neg s \approx r \notin \Gamma^*$, completing the proof of transitivity.

Now let Q be any operation symbol. To conserve notation, make the harmless assumption that the rank of Q is 3. Now suppose that $s_0 \sim t_0, s_1 \sim t_1$, and $s_2 \sim t_2$. We must prove that $Qs_0s_1s_2 \sim Qt_0t_1t_2$. Suppose otherwise. Then either $\neg Qs_0s_1s_2 \approx Qt_0t_1t_2 \in \Sigma^*$ or $\neg Qs_0s_1s_2 \approx Qt_0t_1t_2 \in \Gamma^*$. Consider the first of these alternatives. We have

$$\begin{array}{ll} \neg s_0 \approx t_0 \notin \Sigma^* & \neg s_0 \approx t_0 \notin \Gamma^* \\ \neg s_1 \approx t_1 \notin \Sigma^* & \neg s_1 \approx t_1 \notin \Gamma^* \\ \neg s_2 \approx t_2 \notin \Sigma^* & \neg s_2 \approx t_2 \notin \Gamma^* \end{array}$$

So there are Lyndon interpolants θ_0, θ_1, and θ_2 between $\Sigma^* \cup \{\neg s_0 \approx t_0\}$ and Γ^*, between $\Sigma^* \cup \{\neg s_1 \approx t_1\}$ and Γ^*, and between $\Sigma^* \cup \{\neg s_2 \approx t_2\}$ and Γ^* respectively. This entails

$$\Sigma^* \models (s_0 \approx t_0 \vee \theta_0) \wedge (s_1 \approx t_1 \vee \theta_1) \wedge (s_2 \approx t_2 \vee \theta_2).$$

After several uses of distributivity this yields

$$\begin{aligned} \Sigma^* \models (s_0 \approx t_0 \wedge s_1 \approx t_1 \wedge s_2 \approx t_2) \vee (s_0 \approx t_0 \wedge s_1 \approx t_1 \wedge \theta_2) \vee \\ \vee (s_0 \approx t_0 \wedge \theta_1 \wedge s_2 \approx t_2) \vee (\theta_0 \wedge s_1 \approx t_1 \wedge s_2 \approx t_2) \vee \\ \vee \cdots \vee (s_0 \approx t_2 \wedge \theta_1 \wedge \theta_2) \vee (\theta_0 \wedge \theta_1 \wedge \theta_2) \end{aligned}$$

Now we can weaken all but the first disjunct to obtain

$$\Sigma^* \models (s_0 \approx t_0 \wedge s_1 \approx t_1 \wedge s_2 \approx t_2) \vee (\theta_0 \vee \theta_1 \vee \theta_2).$$

To conserve notation let θ^* be $\theta_0 \vee \theta_1 \vee \theta_2$. So we have

$$\Sigma^* \models (s_0 \approx t_0 \wedge s_1 \approx t_1 \wedge s_2 \approx t_2) \vee \theta^*$$

Now we also have $\neg Qs_0s_1s_2 \approx Qt_0t_1t_2 \in \Sigma^*$. This yields

$$\Sigma^* \models \neg Qs_0s_1s_2 \approx Qt_0t_1t_2 \wedge \big((s_0 \approx t_0 \wedge s_1 \approx t_1 \wedge s_2 \approx t_2) \vee \theta^*\big).$$

Applying distributivity and weakening the last disjunct gives

$$\Sigma^* \models \big(\neg Qs_0s_1s_2 \approx Qt_0t_1t_2 \wedge s_0 \approx t_0 \wedge s_1 \approx t_1 \wedge s_2 \approx t_2\big) \vee \theta^*.$$

Observe that $\neg Qs_0s_1s_2 \approx Qs_0s_1s_2$ is a logical consequence of the first disjunct. So

$$\Sigma^* \models \big(\neg Qs_0s_1s_2 \approx Qs_0s_1s_2\big) \vee \theta^*.$$

Since the first disjunct above is always false we find that $\Sigma^* \models \theta^*$. But we already know that $\Gamma^* \models \neg\theta_0 \wedge \neg\theta_1 \wedge \neg\theta_2$. That is $\Gamma^* \models \neg\theta^*$. This means that θ^* is a Lyndon interpolant between Σ^* and Γ^*, which is impossible. So our conclusion is that $\neg Qs_0s_1s_2 \approx Qt_0t_1t_2 \notin \Sigma^*$. It remains to consider whether $\neg Qs_0s_1s_2 \approx Qt_0t_1t_2 \in \Gamma^*$. An argument similar to the one just given, but with the roles Γ^* and Σ^* reversed gives the conclusion that $\neg Qs_0s_1s_2 \approx Qt_0t_1t_2 \notin \Gamma^*$. Therefore $Qs_0s_1s_2 \sim Qt_0t_1t_2$.

So $\sim$ is a congruence relation, as desired and Claim A is established. ■

CLAIM B.

Suppose that the relation symbol R denotes a basic relation of $\mathbf{M}$. Let r be the rank of R. If $s_0, \ldots, s_{r-1}, t_0, \ldots, t_{r-1} \in M$ so that $s_0 \sim t_0, \ldots, s_{r-1} \sim t_{r-1}$, then

$$\langle s_0, \ldots, s_{r-1}\rangle \in R^{\mathbf{M}} \textit{ if and only if } \langle t_0, \ldots, t_{r-1}\rangle \in R^{\mathbf{M}}$$

Proof of Claim B. To conserve notation, we make the harmless assumption that $r = 3$. Since $\sim$ is symmetric, it is enough to prove one direction of the equivalence displayed above. So suppose that $\langle s_0, s_1, s_2\rangle \in R^{\mathbf{M}}$. This means that either $Rs_0s_1s_2 \in \Sigma^*$ or that $\neg Rs_0s_1s_2 \in \Gamma^*$. From the $\sim$ relations, we also have

$$\begin{array}{ll} \neg s_0 \approx t_0 \notin \Sigma^* & \neg s_0 \approx t_0 \notin \Gamma^* \\ \neg s_1 \approx t_1 \notin \Sigma^* & \neg s_1 \approx t_1 \notin \Gamma^* \\ \neg s_2 \approx t_2 \notin \Sigma^* & \neg s_2 \approx t_2 \notin \Gamma^* \end{array}$$

So there are Lyndon interpolants θ_0, θ_1, and θ_2 between $\Sigma^* \cup \{\neg s_0 \approx t_0\}$ and Γ^*, between $\Sigma^* \cup \{\neg s_1 \approx t_1\}$ and Γ^*, and between $\Sigma^* \cup \{\neg s_2 \approx t_2\}$ and Γ^* respectively. Reasoning just as in the proof of Claim A, we find

$$\Sigma^* \models (s_0 \approx t_0 \wedge s_1 \approx t_1 \wedge s_2 \approx t_2) \vee \theta^*$$

where θ^* is $\theta_0 \vee \theta_1 \vee \theta_2$. We also know that $\Gamma^* \models \neg\theta^*$. Now consider the alternative that $Rs_0s_1s_2 \in \Sigma^*$. As in the proof of Claim A, this yields

$$\Sigma^* \models (Rs_0s_1s_2 \wedge s_0 \approx t_0 \wedge s_1 \approx t_1 \wedge s_2 \approx t_2) \vee \theta^*.$$

But $Rt_0t_1t_2$ is a logical consequence of the first disjunct. So

$$\Sigma^* \models Rt_0t_1t_2 \vee \theta^*.$$

Observe that $Rt_0t_1t_2 \in \Phi$ since $Rs_0s_1s_2 \in \Phi$. This means that $Rt_0t_1t_2 \vee \theta^* \in \Phi$. By part (vii) of the Lemma, we obtain $Rt_0t_1t_2 \vee \theta^* \in \Sigma^*$. So by part (ix) of the Lemma $Rt_0t_1t_2 \in \Sigma^*$ or $\theta^* \in \Sigma^*$. The latter would mean that θ^* is a Lyndon interpolant between Σ^* and Γ^*. So we find that $Rt_0t_1t_2 \in \Sigma^*$. To conclude that $\langle t_0, t_1, t_2\rangle \in R^{\mathbf{M}}$ we will show that $\neg Rt_0t_1t_2 \notin \Sigma^*$ and $Rt_0, t_1, t_2 \notin \Gamma^*$. Now Σ^* must be consistent, lest $\bot$ become a Lyndon interpolant between Σ^* and Γ^*. So we have $\neg Rt_0t_1t_2 \notin \Sigma^*$. Finally $Rt_0t_1t_2 \notin \Gamma^*$, otherwise $\neg Rt_0t_1t_2$ would be a Lyndon interpolant between Σ^* and Γ^*. So we conclude that $\langle Rt_0t_1t_2\rangle \in R^{\mathbf{M}}$. ■

In view of Claims A and B, we can form the structure $\mathbf{N} = \mathbf{M}/\sim$, where we take $\langle t_0/\sim, \dots, t_{r-1}/\sim\rangle \in R^{\mathbf{N}}$ if and only if $\langle t_0, \dots, t_{r-1}\rangle \in R^{\mathbf{M}}$, for each basic relation $R^{\mathbf{M}}$—where r is the rank of R.

To complete the proof of Lyndon's Interpolation Theorem, we will show that

$$\mathbf{N} \models \Sigma^* \cup \Gamma^*.$$

First, consider an equation $s \approx t$ without variables, so that $s \approx t \in \Sigma^*$. Then $\neg s \approx t \notin \Sigma^*$. Suppose, for contradiction, that $\neg s \approx t \in \Gamma^*$. Pick a Lyndon interpolant θ so that $\Sigma^* \cup \{\neg s \approx t\} \models \theta$ and $\Gamma^* \models \neg\theta$. We obtain $\Sigma^* \models s \approx t \vee \theta$ and $\Gamma^* \models \neg\theta \wedge \neg s \approx t$. But the $s \approx t \vee \theta$ is a Lyndon interpolant between Σ^* and Γ^*, which is impossible. So $\neg s \approx t \notin \Gamma^*$. This means that $s \sim t$ and so $\mathbf{N} \models s \approx t$.

Now let $s \approx t \in \Gamma^*$. Then $\neg s \approx t \notin \Gamma^*$. For contradiction, suppose $\neg s \approx t \in \Sigma^*$. Pick a Lyndon interpolant θ so that $\Sigma^* \models \theta$ and $\Gamma^* \cup \{\neg s \approx t\} \models \neg\theta$. Then $\Sigma^* \models \neg s \approx t \wedge \theta$ and $\Gamma^* \models s \approx t \vee \neg\theta$. This means that $\neg s \approx t \wedge \theta$ is a Lyndon interpolant between Σ^* and Γ^*, which is impossible. So $\neg s \approx t \notin \Sigma^*$. Therefor $s \sim t$ and so $\mathbf{N} \models s \approx t$.

Next, consider $\neg s \approx t \in \Sigma^*$. Then $s \sim t$ fails and so $\mathbf{N} \not\models \neg s \approx t$. Likewise, if $\neg s \approx t \in \Gamma^*$ then $\mathbf{N} \not\models \neg s \approx t$.

Next, consider $Rs_0s_1 \dots s_{r-1} \in \Sigma^*$, where r is the rank of the relation symbol R. Then $\mathbf{M} \models Rs_0 \dots s_{r-1}$ and hence that $\mathbf{N} \models Rs_0 \dots s_{r-1}$. The same conclusion can be drawn from $Rs_0 \dots s_{r-1} \in \Gamma^*$.

Now, consider $\neg Rs_0 \dots s_{r-1} \in \Sigma^*$. Then $\mathbf{M} \models \neg Rs_0 \dots s_{r-1}$ and so $\mathbf{M} \models Rs_0 \dots s_{r-1}$. Likewise, if $\neg Rs_0 \dots s_{r-1} \in \Gamma^*$, then $\mathbf{N} \models \neg Rs_0 \dots s_{r-1}$.

At this stage we know that any atomic or negated atomic sentence belonging to $\Sigma^* \cup \Gamma^*$ holds in $\mathbf{N}$. This is the basis of an induction. Parts (ix) and (x) of the Lemma give us the induction steps associated with $\curlyvee$ and $\curlywedge$. So consider $\exists x \chi(x) \in \Sigma^* \cup \Gamma^*$. Then by part (v) of the Lemma, there is a constant symbol c_α so that $\chi(\substack{c_\alpha \\ x}) \in \Sigma^* \cup \Gamma^*$. Appealing to the inductive hypothesis, we see $\mathbf{N} \models \chi(\substack{c_\alpha \\ x})$. Therefore, $\mathbf{N} \models \exists x \chi(x)$. The universal quantifier is handled in a like manner, in view of the fact that every element of $\mathbf{N}$ is named by a constant symbol.

Therefore $\mathbf{N} \models \Sigma^* \cup \Gamma^*$. In particular, $\mathbf{N} \models \Sigma$ and $\mathbf{N} \models \neg\gamma$. This means $\Sigma \models \gamma$ fails, and the proof of the main part of the theorem in finished.

For the moreover statement in the theorem, we can assume that θ is in prenex form and the quantifier-free part of θ is a conjunction $\beta_0 \curlywedge \dots \curlywedge \beta_{n-1}$ where each β_i is a disjunction of atomic and negated atomic formulas. We can consider each occurrence of $\bot$ in each β_i. Each of these occurrences can be removed, except in the case that it is the only remaining disjunct. In that case β_i can be replaced by $\bot$. So we see that θ is logically equivalent to $\bot$. This is impossible if Σ has a model. So in that case θ is logically equivalent to a sentence in prenex-conjunctive-disjunctive normal form in which $\bot$ does not occur. Now consider in turn each occurrence of $\top$. If $\top$ is a disjunct of β_i, then β_i can be replaced by $\top$ and this whole conjunct can be deleted, unless it is the only remaining conjunct. In that case θ is logically equivalent to $\top$. This is impossible if $\neg\gamma$ has a model. In this case, θ is logically equivalent to a sentence in prenex-conjunctive-disjunctive normal form in which $\top$ does not occur. ■

Leon Henkin (1949) provided a proof of the the Completeness Theorem for elementary logic that is the clear ancestor of the proof we have given of Lyndon's Interpolation Theorem: we have constructed the desired model recursively, in stages, naming each element of the model with a new constant symbol. The intricate lemma inside the proof above ensures that this construction in stages can be carried out. The complexity of the lemma is a measure of the difficulty of this proof compared to Henkin's original construction.

Exercises 8.132

Beth's Definability Theorem, Craig's Interpolation Theorem, and Robinson's Joint Consistency Theorem, roughly speaking, lie close together. The next few exercises illuminate this remark.

1. Prove Beth's Theorem from the Two Signature Interpolation Theorem.

2. Prove Craig's Interpolation Theorem from the Two Signature Interpolation Theorem.

3. Prove Beth's Theorem from Craig's Interpolation Theorem.

4. Prove Craig's Interpolation Theorem from Robinson's Joint Consistency Theorem.

5. Prove Robinson's Joint Consistency Theorem from Beth's Definability Theorem.

* **6.** (Robert L. Vaught, 1961) An algebra **A** is called 0-**regular** provided there is a constant symbol c so that if $\theta, \varphi \in \operatorname{Con} \mathbf{A}$ and $c^{\mathbf{A}}/\theta = c^{\mathbf{A}}/\varphi$, then $\theta = \varphi$. The common congruence class of $c^{\mathbf{A}}$ is called a *normal* set.

Now let $\mathcal{K}$ be a finitely axiomatizable elementary class of algebras so that each algebra in $\mathcal{K}$ is 0-regular. Expand the signature to the elementary language $\mathfrak{L}$ by adjoining a one-place relation symbol U. Prove that there is an elementary sentence σ and an $\mathfrak{L}$-formula $\psi(x, y)$ so that for any $\mathbf{A} \in \mathcal{K}$ each of the following hold:

(i) A subset $N \subseteq A$ is normal for **A** if and only if $\langle \mathbf{A}, N \rangle \models \sigma$.

(ii) Let $\theta \in \operatorname{Con} \mathbf{A}$. Then for all $a, b \in A$ we have $(a, b) \in \theta$ if and only if $\langle \mathbf{A}, c^{\mathbf{A}}/\theta \rangle \models \psi(a, b)$.

8.10. Sentences Preserved under Homomorphic Images

Unlike classes closed under the formation of substructures (like varieties, universal classes, and quasivarieties) classes closed under the formation of subdirect products or under formation of homomorphic images need not be elementary classes. So we cannot expect the kind of axiomatizability results we obtained in those earlier situations. Nevertheless, we can still obtain preservation theorems in these cases.

A formula is **positive** provided it is built up from atomic formulas with the help of $\wedge$, $\vee$, and the quantifiers $\forall$ and $\exists$. The key result here is the theorem below, due to Roger C. Lyndon (1959c).

THEOREM 8.133. *Let* **A** *be an algebra and* $\mathcal{K}$ *be a nonempty elementary class of the same signature that is closed with respect to the formation of homomorphic images. The following are equivalent:*

(i) *There is* $\mathbf{B} \in \mathcal{K}$ *such that every positive sentence that holds in* **B** *must hold in* **A**;

(ii) **A** *has an elementary extension that is a homomorphic image of an algebra belonging to* $\mathcal{K}$.

Proof. We begin with the following lemma, which is of independent interest.

LEMMA 8.134. *Let* **A** *and* **B** *be algebras of the same signature. Let h be a homomorphism from* **B** *onto* **A**. *For every positive formula* $\varphi(\mathbf{x})$ *and any assignment* $\mathbf{b} = \langle b_0, b_1, \dots \rangle$ *of elements of B to the variables, we have*

if **b** *satisfies* $\varphi(\mathbf{x})$ *in* **B** *then* $h(\mathbf{b}) := \langle h(b_0), h(b_1), \dots \rangle$ *satisfies* $\varphi(\mathbf{x})$ *in* **A**.

Proof of Lemma 8.134. We proceed by induction on the complexity of $\varphi(\mathbf{x})$. The base case, when $\varphi(\mathbf{x})$ is just an equation, is something we already know. There are four inductive cases to consider.

CASE: $\varphi(\mathbf{x})$ is $\psi(\mathbf{x}) \wedge \theta(\mathbf{x})$
Here $\psi(\mathbf{x})$ and $\theta(\mathbf{x})$ are positive. So $\mathbf{b}$ satisfies both $\psi(\mathbf{x})$ and $\theta(\mathbf{x})$ in $\mathbf{B}$. By the inductive hypothesis, $h(\mathbf{b})$ satisfies both $\psi(\mathbf{x})$ and $\theta(\mathbf{x})$ in $\mathbf{A}$. Hence, $h(\mathbf{b})$ satisfies $\varphi(\mathbf{x})$ in $\mathbf{A}$.

CASE: $\varphi(\mathbf{x})$ is $\psi(\mathbf{x}) \vee \theta(\mathbf{x})$
This case is handled in a way similar to the argument above.

CASE: $\varphi(\mathbf{x})$ is $\exists y\psi(\mathbf{x})$
It is harmless to suppose $y = x_0$. So suppose $\mathbf{b}$ satisfies $\exists y\psi(\mathbf{x})$ in $\mathbf{B}$. This means that there is $b_0' \in B$ so that $\mathbf{b}' := \langle b_0', b_1, \dots \rangle$ satisfies $\psi(\mathbf{x})$ in $\mathbf{B}$. By the inductive hypothesis, $h(\mathbf{b}')$ satisfies $\psi(\mathbf{x})$ in $\mathbf{A}$. But this entails that $h(\mathbf{b})$ satisfies $\exists y\psi(\mathbf{x})$ in $\mathbf{A}$.

CASE: $\varphi(\mathbf{x})$ is $\forall y\psi(\mathbf{x})$
Again we assume $y = x_0$ and that $\mathbf{b}$ satisfies $\varphi(\mathbf{x})$ in $\mathbf{B}$. If d is any element of B, we know $\langle d, b_1, \dots \rangle$ satisfies $\psi(\mathbf{x})$ in $\mathbf{A}$. To see that $h(\mathbf{b})$ satisfies $\forall y\psi(\mathbf{x})$ in $\mathbf{A}$, let a be an arbitrary element of A. We want to show that $\langle a, h(b_1), h(b_2), \dots \rangle$ satisfies $\psi(\mathbf{x})$ in $\mathbf{A}$. Pick $d \in B$ so that $a = h(d)$. This is possible since h maps B onto A. Since $\langle a, h(b_1), h(b_2), \dots \rangle = \langle h(d), h(b_1), h(b_2), \dots \rangle$, we can appeal to the inductive hypothesis to secure the desired conclusion.

This completes the proof of the lemma. ■

Now suppose that property (ii) holds. Pick $\mathbf{B} \in \mathcal{K}$ and $\mathbf{A}'$ with $\mathbf{A} \preccurlyeq \mathbf{A}'$ and a homomorphism h from $\mathbf{B}$ onto $\mathbf{A}'$. Let σ be a positive sentence true in $\mathbf{B}$. By Lemma 8.134, σ must also be true in $\mathbf{A}'$. But every sentence true in $\mathbf{A}'$ must also be true in $\mathbf{A}$. So property (i) of the Theorem holds.

For the converse, suppose (i) holds. Pick $\mathbf{B} \in \mathcal{K}$ so that every positive sentence true in $\mathbf{B}$ is also true in $\mathbf{A}$. Our plan is to construct $\mathbf{B}_\omega$ and $\mathbf{A}_\omega$ so that these two algebras are, respectively, elementary extensions of $\mathbf{B}$ and $\mathbf{A}$, as well as a homomorphism from $\mathbf{B}_\omega$ onto $\mathbf{A}_\omega$. We use elementary chains to do this. Two more claims facilitate this construction.

CLAIM A.
Suppose that $\mathbf{C}$ *and* $\mathbf{D}$ *are algebras so that every positive sentence true in* $\mathbf{C}$ *is also true in* $\mathbf{D}$. *Then there is an elementary extension* $\mathbf{D}'$ *of* $\mathbf{D}$ *and a homomorphism* f *from* $\mathbf{C}$ *into* $\mathbf{D}'$ *so that every positive sentence true in* $(\mathbf{C}, c)_{c\in C}$ *is true in* $(\mathbf{D}', f(c))_{c\in C}$.

Proof of Claim A. Let Σ be the set of all positive sentences true in $(\mathbf{C}, c)_{c\in C}$ and let Δ be the set of all sentences true in $(\mathbf{D}, d)_{d\in D}$. So Δ is the elementary diagram of $\mathbf{D}$. Since every positive sentence true in $\mathbf{C}$ is also true in $\mathbf{D}$, it follows that $\Sigma \cup \Delta$ has a model $\mathbf{D}^* = (\mathbf{D}', c', d)_{c\in C, d\in D}$. Let $\mathbf{D}'$ be the reduct of $\mathbf{D}^*$ to the original signature. Since $\mathbf{D}^* \models \Delta$, we see that $\mathbf{D}'$ is an elementary extension of $\mathbf{D}$. For each $c \in C$ define $f(c) := c'$. So $f : C \to D'$. The map f is a homomorphism since for each operation symbol Q, say with rank r, and any $c_0, \dots, c_{r-1} \in C$ we can put $c_r = Q^{\mathbf{C}}(c_0, \dots, c_{r-1})$ and obtain a positive sentence $Q\hat{c}_0\hat{c}_1 \cdots \hat{c}_{r-1} \approx \hat{c}_r$ that belongs to Σ—here $\hat{c}_i$ is the constant symbol naming the element $c_i \in C$. So this sentence is true in $\mathbf{D}^*$, yielding $Q^{\mathbf{D}'}(f(c_0), \dots, f(c_{r-1})) = f(Q^{\mathbf{C}}(c_0, \dots, c_{r-1}))$. So f is a homomorphism. ■

CLAIM B.
Suppose that $\mathbf{C}$ *and* $\mathbf{D}$ *are algebras so that every positive sentence true in* $\mathbf{C}$ *is also true*

in $\mathbf{D}$. *Then there are an elementary extension* $\mathbf{C}'$ *of* $\mathbf{C}$ *and a function* g *from* $\mathbf{A}$ *into* $\mathbf{C}'$ *so that every positive sentence true in* $(\mathbf{C}', g(d))_{d\in D}$ *is true in* $(\mathbf{D}, d)_{d\in D}$. ■

Claim B has a proof resembling the proof of Claim A, but Δ should be the elementary diagram of $\mathbf{C}$ and Σ should be the set of sentences true in $(\mathbf{D}, d)_{d\in D}$ that are negations of positive sentences. Observe that we did not insist in Claim B that the function g be a homomorphism.

To complete the proof of Theorem 8.133, we now consider the diagram below. It is a guide to the elementary chain construction we will undertake. The heavy arrows, put in place by an invocation of Claim A, are homomorphisms. So $\mathbf{A}_1$ and the homomorphism f_0 are ensured by Claim A, as well as the fact that every positive sentence true in $(\mathbf{B}, b)_{b\in B}$ is true in $(\mathbf{A}_1, f_0(b))_{b\in B}$.

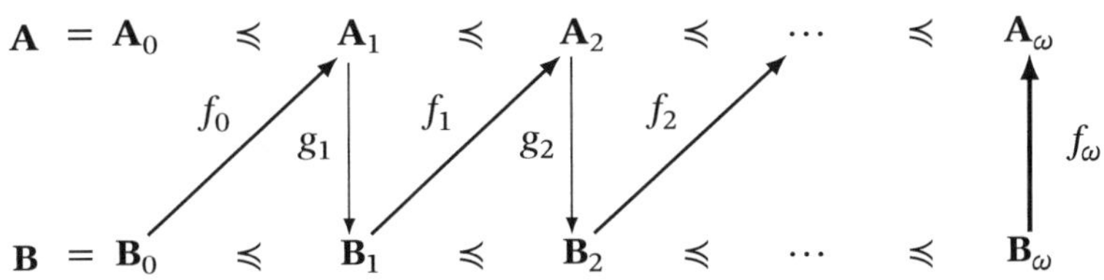

Once $\mathbf{A}_1$ and f_0 are in hand, along with information about those expansions, we can then apply Claim B to obtain $\mathbf{B}_1$ and the function g_1, as well as additional information about expansions. This in turn, sets up the next application of Claim A. And so it goes. We also note that f_1 extends f_0 and $g_1^{-1} \subseteq f_1$. These relations are repeated at every stage of the construction. Since the f_i's form an ascending chain, we can take f_ω to be the union of this chain of functions (recalling that functions are sets of ordered pairs...). It is easy to check that f_ω is a homomorphism from $\mathbf{B}_\omega$ into $\mathbf{A}_\omega$. Since $g_n^{-1} \subseteq f_\omega$ it also follows that f_ω maps $\mathbf{B}_\omega$ onto $\mathbf{A}_\omega$. By Tarski's Elementary Chain Theorem $\mathbf{B}_\omega$ and $\mathbf{A}_\omega$ are respectively elementary extensions of $\mathbf{B}$ and $\mathbf{A}$. Since $\mathbf{B} \in \mathcal{K}$, we see that $\mathbf{B}_\omega \in \mathcal{K}$. So we have in hand a member of $\mathcal{K}$ which has, as a homomorphic image, an elementary extension of $\mathbf{A}$. In this way property (ii) is established, given property (i). ■

THEOREM 8.135 (Lyndon's Preservation Theorem for Homomorphisms). *A sentence* σ *is preserved under the formation of homomorphic images if and only if* σ *is logically equivalent to a positive sentence.*

Proof. Lemma 8.134 gives us that every positive sentence is preserved under the formation of homomorphic images. For the converse, suppose that σ is a sentence preserved under homomorphic images. Let

$$\Theta := \{\varphi \mid \varphi \text{ is a positive sentence and } \sigma \models \varphi\}.$$

Evidently $\sigma \models \Theta$. If we could show that $\Theta \models \sigma$ then the Compactness Theorem would provide a finite subset $\Delta \subseteq \Theta$ so that Δ and σ are logically equivalent. Since Δ is a finite set of positive sentences then the conjunction of Δ will be a positive sentence logically equivalent to σ.

To prove that $\Theta \models \sigma$, let $\mathbf{A} \models \Theta$. Let $\mathcal{K} = \mathbf{Mod}\,\sigma$. We need to show that $\mathbf{A} \in \mathcal{K}$. Now $\mathcal{K}$ is an elementary class closed under the formation of homomorphic images. Next, let $\Gamma := \{\neg\psi \mid \mathbf{A} \models \neg\psi \text{ and } \psi \text{ is positive}\}$. Now $\{\sigma\} \cup \Gamma$ is consistent, for otherwise, by compactness, there are $\neg\psi_0, \dots, \neg\psi_{n-1} \in \Gamma$ so that $\sigma \models \neg(\neg\psi_0 \wedge \cdots \wedge \neg\psi_{n-1})$. That is, we would have $\sigma \models \psi_0 \vee \cdots \vee \psi_{n-1}$. Now $\psi_0 \vee \cdots \vee \psi_{n-1}$ is a positive sentence, so $\psi_0 \vee \cdots \vee \psi_{n-1} \in \Theta$. Hence $\mathbf{A} \models \psi_0 \vee \cdots \vee \psi_{n-1}$, which is impossible because $\neg(\psi_0 \vee \cdots \vee \psi_{n-1}) \in \Gamma$. So let $\mathbf{B} \models \{\sigma\} \cup \Gamma$. This means $\mathbf{B} \in \mathcal{K}$ and that every positive sentence true in **B** is also true in **A**. By Theorem 8.135, **A** has an elementary extension that is a homomorphic image of a member of $\mathcal{K}$. Since $\mathcal{K}$ is closed under the formation of homomorphic images, then the elementary extension of **A** belongs to $\mathcal{K}$ and hence **A** itself belongs to $\mathcal{K}$—just as desired. ∎

THEOREM 8.136. *Let $\mathcal{K}$ be an elementary class. $\mathcal{K}$ is closed under the formation of homomorphic images if and only if $\mathcal{K}$ can be axiomatized by some set of positive sentences.* ∎

A proof of this theorem can be made by letting Σ be a set of sentences so that $\mathcal{K} = \mathbf{Mod}\,\Sigma$, and then replacing σ everywhere in the proof above by Σ. In both proofs, a key property of positive sentences is used: the disjunction of two positive sentences is again positive.

We have framed the results in this section for algebras, our main focus, to simplify the presentation. But the results hold for structures more generally, see (C. C. Chang and H. J. Keisler, 1990) for the fuller account.

Exercises 8.137

1. Give a proof of Theorem 8.136.

2. Let Σ be a set of sentences so that $\mathbf{Mod}\,\Sigma$ is closed under the formation of homomorphic images. Let φ be any sentence. Prove that the following conditions are equivalent:
   (i) If $\mathbf{A} \models \Sigma$ and $\mathbf{A} \models \varphi$, then the sentence φ is true in every homomorphic image of **A**.
   (ii) There is a positive sentence ψ such that $\Sigma \models \varphi \Leftrightarrow \psi$.

8.11. Sentences Preserved under Subdirect Products

A formula φ is a **basic Lyndon formula** provided φ has the form $\pi \Rightarrow \alpha$ where π is a positive formula and α is an equation. A **Lyndon formula** is one that is built up from basic Lyndon formulas with the help of $\wedge$ and $\forall$.

THEOREM 8.138 (Roger C. Lyndon 1959b). *Let **A** be an algebra and $\mathcal{K}$ be an elementary class of the same signature. The following are equivalent:*
(i) *Every Lyndon sentence that holds in $\mathcal{K}$ holds in **A**;*
(ii) ***A** has an elementary extension that is a subdirect product of algebras belonging to $\mathcal{K}$.*

The following lemma secures the implication (ii)⇒(i) of the theorem above.

LEMMA 8.139. *Every Lyndon sentence is preserved under the formation of subdirect products.*

Proof. It is enough to prove this for Lyndon sentences of the form $\forall\mathbf{x}[\pi \Rightarrow \alpha]$, where the formula π is positive and the formula α is an equation. Now let I be a set and let $\mathbf{A}$ be an algebra and let $\langle \mathbf{B}_i \rangle_{i\in I}$ be a system of algebras and $\langle h_i \rangle_{i\in I}$ be system of homomorphisms $h_i : \mathbf{A} \twoheadrightarrow \mathbf{B}_i$ for $i \in I$ so that $\langle h_i \rangle_{i\in I}$ separates points. We argue that if $\forall\mathbf{x}[\pi \Rightarrow \alpha]$ fails in $\mathbf{A}$, then it fails in $\mathbf{B}_i$ for some $i \in I$. That the sentence fails in $\mathbf{A}$ means that there are $a_0, a_1, \cdots \in A$ so that $\pi(\mathbf{x})$ holds in $\mathbf{A}$ at $\langle a_0, a_1, \ldots \rangle$ but the $\alpha(\mathbf{x})$ fails in $\mathbf{A}$ at $\langle a_0, a_1, \ldots \rangle$. Since $\langle h_i \rangle_{i\in I}$ separates points we can pick $i \in I$ so that $\alpha(\mathbf{x})$ fails in $\mathbf{B}_i$ at $\langle h_i(a_0), h_i(a_1), \ldots \rangle$. But $\pi(\mathbf{x})$ is positive and positive formulas are preserved under homomorphic images. So $\pi(\mathbf{x})$ holds in $\mathbf{B}_i$ at $\langle h_i(a_0), h_i(a_1), \ldots \rangle$. Hence $\forall\mathbf{x}[\pi \Rightarrow \alpha]$ fails in $\mathbf{B}_i$. ■

Proof of Theorem 8.138. Let Γ be a set of elementary sentences so that $\mathcal{K} = \mathbf{Mod}\,\Gamma$ and assume that every Lyndon sentence that holds in $\mathcal{K}$ also holds in $\mathbf{A}$. Let Σ be the set of Lyndon sentences that hold in $\mathcal{K}$. So $\mathbf{A} \models \Sigma$.

Our object is to construct an elementary extension $\mathbf{A}'$ of $\mathbf{A}$ that is isomorphic to a subdirect product of algebras belonging to $\mathcal{K}$. Roughly speaking, what we will do is build an elementary chain of algebras, starting with $\mathbf{A}$, whose union will be $\mathbf{A}'$ and along the way we will build a system $\langle \theta_i \mid i \in I \rangle$ of congruences on $\mathbf{A}'$ that separates points and so that the quotient algebras all belong to $\mathcal{K}$.

For each ordinal μ the signature $\mathfrak{L}_\mu$ will be an expansion of the original signature $\mathfrak{L}$ by adjoining a two-place relation symbol θ_γ for each $\gamma < \mu$. We say that an algebra $\mathbf{A}_\mu$ of signature $\mathfrak{L}_\mu(A)$ satisfies *Lyndon's conditions* provided

(1) The reduct of $\mathbf{A}_\mu$ to $\mathfrak{L}(A)$ is an elementary extension of $\langle \mathbf{A}, a \rangle_{a\in A}$ and the reduct of $\mathbf{A}_\mu$ to the signature of $\mathbf{A}_\gamma$ is an elementary extension of $\mathbf{A}_\gamma$ for all $\gamma < \mu$;
(2) The relation $\theta_\gamma^{\mathbf{A}_\mu}$ is a congruence relation on $\mathbf{A}_\mu$, for each $\gamma < \mu$;
(3) The reduct of $\mathbf{A}_\mu/\theta_\gamma$ to L belongs to $\mathcal{K}$, for each $\gamma < \mu$.

To start the construction we let $\mathfrak{L}_0 = \mathfrak{L}(A)$ be the diagram language of $\mathbf{A}$ and we put $\mathbf{A}_0 = \langle \mathbf{A}, a \rangle_{a\in A}$. Observe that the Lyndon conditions hold for $\mathbf{A}_0$.

Let κ be the number of two-element subsets of A and let

$$\{a_0, a_0'\}, \{a_1, a_1'\}, \ldots, \{a_\gamma, a_\gamma'\}, \ldots$$

for $\gamma < \kappa$ be an enumeration of these subsets. For each ordinal $\mu < \kappa$ we construct an algebra $\mathbf{A}_\mu$ satisfying Lyndon's conditions so that the two-element set $\{a_\gamma, a_\gamma'\}$ is separated by $\theta^{\mathbf{A}_\mu}$ for each $\gamma < \mu$.

THE SUCCESSOR ORDINAL CASE

Suppose that $\mathbf{A}_\mu$ has been constructed. Let $\Gamma(\theta_\mu)$ be the set obtained from Γ by replacing each occurrence of $\approx$ by θ_μ. Let Δ be the elementary diagram of $\mathbf{A}_\mu$.

CLAIM A.

Suppose that $\langle \mathbf{A}, a \rangle_{a\in A} \models \neg a_\mu \approx a_\mu'$. Then the set $\Delta \cup \{\forall x, y[x \approx y \Rightarrow x\theta_\mu y]\} \cup \Gamma(\theta_\mu) \cup \{\neg a_\mu \theta_\mu a_\mu'\}$ has a model.

Proof of Claim A. Suppose otherwise. By the Compactness Theorem there is a finite subset $\Gamma' \subseteq \Gamma$ so that $\Delta \cup \{\forall x, y[(x \approx y) \Rightarrow x\theta_\mu y]\} \cup \Gamma'(\theta_\mu) \cup \{\neg a_\mu \theta_\mu a_\mu'\}$ has no model.

This means that

$$\Delta \cup \{\forall x, y[(x \approx y) \Rightarrow x\theta_\mu y]\} \models (\bigwedge\!\!\!\bigwedge \Gamma'(\theta_\mu)) \Rightarrow a_\mu \theta_\mu a'_\mu.$$

Now θ_μ is the only relation symbol occurring in $(\bigwedge\!\!\!\bigwedge \Gamma'(\theta_\mu)) \Rightarrow a_\mu \theta_\mu a'_\mu$ and θ_μ does not occur at all in Δ and occurs only positively in $\forall x, y[x \approx y \Rightarrow x\theta_\mu y]$. So according to Lyndon's Interpolation Theorem there is a positive sentence $\varphi(\theta_\mu)$ in which θ_μ is the only relation symbol to occur (not even $\approx$ occurs in φ) so that

$$\Delta \cup \{\forall x, y[(x \approx y) \Rightarrow x\theta_\mu y]\} \models \varphi(\theta_\mu) \text{ and } \varphi(\theta_\mu) \models (\bigwedge\!\!\!\bigwedge \Gamma'(\theta_\mu)) \Rightarrow a_\mu \theta_\mu a'_\mu.$$

Moreover, the only new constant symbols that can occur in φ are those associated with the elements a_μ and a'_μ. So we have

$$\Delta \cup \{\forall x, y[(x \approx y) \Rightarrow x\theta_\mu y]\} \models \varphi(\theta_\mu) \text{ and } (\bigwedge\!\!\!\bigwedge \Gamma'(\theta_\mu) \models \varphi(\theta_\mu) \Rightarrow a_\mu \theta_\mu a'_\mu.$$

Neither of the new constant symbols occur in Γ and so none occur in $(\bigwedge\!\!\!\bigwedge \Gamma'(\theta_\mu)$. So this sentence places no constraints on the new constant symbols that occur in $\varphi(\theta_\mu) \Rightarrow a_\mu \theta_\mu a'_\mu$. Therefore

$$(\bigwedge\!\!\!\bigwedge \Gamma'(\theta_\mu) \models \forall x, y[\varphi(\theta_\mu) \Rightarrow x\theta_\mu y].$$

Now, replace all the occurrences of θ_μ by $\approx$ to obtain

$$\Gamma' \models \forall x, y[\varphi \Rightarrow (x \approx y)].$$

So the Lyndon sentence $\forall x, y[\varphi \Rightarrow (x \approx y)]$ is also a consequence of Γ and so must hold in $\mathcal{K}$. Hence, $\mathbf{A} \models \forall x, y[\varphi \Rightarrow (x \approx y)]$. Since φ is a sentence, this means $\mathbf{A} \models \varphi \Rightarrow \forall x, y[x \approx y]$. But also since

$$\Delta \cup \{\forall x, y[(x \approx y) \Rightarrow x\theta_\mu y]\} \models \varphi(\theta_\mu),$$

we also obtain

$$\Delta \models \{\forall x, y[(x \approx y) \Rightarrow x\theta_\mu y]\} \Rightarrow \varphi(\theta_\mu).$$

Since the relation symbol θ_μ does not occur in Δ we arrive at

$$\Delta \models \{\forall x, y[(x \approx y) \Rightarrow x \approx y]\} \Rightarrow \varphi,$$

Of course, this means $\Delta \models \varphi$. So φ holds in $\langle \mathbf{A}, a\rangle_{a \in A}$ and hence also in $\mathbf{A}$. But we have both

$$\mathbf{A} \models \varphi \text{ and } \mathbf{A} \models \varphi \Rightarrow \forall x, y[(x \approx y)].$$

Therefore $\mathbf{A} \models \forall x, y[x \approx y]$. But then $\langle \mathbf{A}, a\rangle_{a \in A} \models a_\mu \approx a'_\mu$. This contradicts the hypothesis of Claim A. ■

Let $\mathbf{B}$ be a model of $\Delta \cup \{\forall x, y[(x \approx y) \Rightarrow x\theta_\mu y]\} \cup \Gamma(\theta_\mu) \cup \{\neg a_\mu \theta_\mu a'_\mu\}$ and let $\mathbf{A}_{\mu+1}$ be the reduct of $\mathbf{B}$ to the signature $\mathfrak{L}_{\mu+1}(A)$. We contend that Lyndon's conditions hold for $\mathbf{A}_{\mu+1}$. To see condition (1) just observe that $\mathbf{B} \models \Delta$ so $\mathbf{A}_{\mu+1}$ restricted to the signature of $\mathbf{A}_\mu$ is an elementary extension of $\mathbf{A}_\mu$. To see condition (2), there are two things to consider. First, for $\gamma < \mu$ the fact that $\theta_\gamma^{\mathbf{A}_\mu}$ is a congruence is recorded in Δ. Since $\mathbf{A}_{\mu+1} \models \Delta$ we conclude that $\theta_\gamma^{\mathbf{A}_{\mu+1}}$ is a congruence of $\mathbf{A}_{\mu+1}$. Second, to see that $\theta_\mu^{\mathbf{A}_{\mu+1}}$ is a congruence, observe that the fact that equality is a congruence is recorded in Γ. So that θ_μ denotes a congruence in $\mathbf{A}_{\mu+1}$ is supported by $\mathbf{A}_{\mu+1} \models \Gamma(\theta_\mu)$. Finally, to conclude that condition (3) holds for $\mathbf{A}_{\mu+1}$, observe that this is equivalent to the condition that $\mathbf{A}_{\mu+1} \models \Gamma(\theta_\gamma)$ for each $\gamma < \mu+1$. For $\gamma < \mu$ we see that $\Gamma(\theta_\gamma) \subseteq \Delta$ and so holds in $\mathbf{A}_{\mu+1}$.

For the remaining case, we have $\mathbf{A}_{\mu+1} \models \Gamma(\theta_\mu)$. Finally, we see that for each $\gamma < \mu + 1$ we have $\mathbf{A}_{\mu+1} \models \neg a_\gamma \theta_\gamma a'_\gamma$.

This concludes the successor ordinal case of the construction.

THE LIMIT ORDINAL CASE
Let μ be a limit ordinal and suppose that for all $\gamma < \mu$ the algebra $\mathbf{A}_\gamma$ has been constructed satisfying the Lyndon Conditions and so that $\neg a_\gamma \theta_\gamma a'_\gamma$ holds in $\mathbf{A}_{\gamma+1}$. Let $\mathbf{A}_\mu$ be the union of this chain of algebras. Now let $\rho < \gamma$. The algebras $\mathbf{A}_\gamma$ reduced to the signature $\mathfrak{L}_\rho(A)$ form an elementary chain whose union is $\mathbf{A}_\mu$ reduced to the signature of $\mathfrak{L}_\rho(A)$. So $\mathbf{A}_\mu$ satisfies the Lyndon Conditions and, moreover, for every $\gamma < \mu$ we have $\mathbf{A}_\mu \models \neg a_\gamma \theta_\gamma a'_\gamma$.

As a result, $\langle \mathbf{A}_\kappa, a\rangle_{a\in A}$ is an elementary extension of $\langle \mathbf{A}, a\rangle_{a\in A}$, each $\theta_\gamma^{\mathbf{A}_\mu}$ is a congruence on $\mathbf{A}_\mu$ so that $\mathbf{A}_\mu/\theta_\gamma^{\mathbf{A}_\mu} \in \mathcal{K}$, and the system $\langle \theta_\gamma^{\mathbf{A}_\mu}\rangle_{\gamma<\kappa}$ separates the points of A.

Now put $\mathbf{C}_0 = \langle \mathbf{A}, a\rangle_{a\in A}$ and let $\mathbf{C}_1 = \langle \mathbf{A}_\kappa, a\rangle_{a\in A_\kappa}$. Notice that the signature of $\mathbf{C}_1$ includes binary relation symbols that denote congruences of $\mathbf{C}_1$ that separate the points of C_0, and reduced to $\mathfrak{L}$ the quotient algebras all lie in $\mathcal{K}$. We have, as well, that $\mathbf{C}_1$ reduced to the signature of $\mathbf{C}_0$ is an elementary extension of $\mathbf{C}_0$. We iterate this construction ω times. That is, from $\mathbf{C}_k$ we construct $\mathbf{C}_{k+1}$ so that $\mathbf{C}_{k+1}$ reduced to the signature of $\mathbf{C}_k$ is an elementary extension of $\mathbf{C}_k$ and there are binary relations that name congruence relations on $\mathbf{C}_{k+1}$ that separate the points of C_k and, whose quotient algebras reduced to $\mathfrak{L}$ lie in $\mathcal{K}$. Form the union of this chain of structures to obtain $\mathbf{C}_\omega$. The reduct $\mathbf{A}^*$ of $\mathbf{C}_\omega$ to $\mathfrak{L}$ is an elementary extension of $\mathbf{A}$ by Tarski's Elementary Chain Theorem and $\mathbf{A}^*$ has a system of congruences that separate the points of A^* and whose images belong to $\mathcal{K}$. This means that $\mathbf{A} \prec \mathbf{A}^*$ and $\mathbf{A}^*$ is a subdirect product of algebras belonging to $\mathcal{K}$, establishing property (ii) of the theorem. ■

THEOREM 8.140 (Lyndon's Preservation Theorem for Subdirect Products (Roger C. Lyndon, 1959b)). *A sentence σ is preserved under the formation of subdirect products if and only if σ is logically equivalent to a Lyndon sentence.*

Proof. We have already proven that every Lyndon sentence is preserved under the formation of subdirect products. For the converse suppose that σ is a sentence preserved under subdirect products. Let

$$\Gamma = \{\varphi \mid \varphi \text{ is a Lyndon sentence and } \sigma \models \varphi\}.$$

We claim that $\Gamma \models \sigma$. To see this, let $\mathcal{K} = \mathbf{Mod}\,\sigma$ and let $\mathbf{A} \models \Gamma$. Then every Lyndon sentence true in $\mathcal{K}$ is true in $\mathbf{A}$. So there is $\mathbf{A}^*$, which is a subdirect product of models of σ so that $\mathbf{A} \prec \mathbf{A}^*$. But since σ is preserved under subdirect products we find that $\mathbf{A}^* \models \sigma$. Hence $\mathbf{A} \models \sigma$. Now this means that $\Gamma \models \sigma$. By Compactness there is a finite $\Gamma' \subseteq \Gamma$ so that $\Gamma' \models \sigma$. But then $\bigwedge\!\!\bigwedge \Gamma'$ and σ are logically equivalent. So just observe that $\bigwedge\!\!\bigwedge \Gamma'$ is a Lyndon sentence. ■

Exercises 8.141

1. Fix a finite signature which includes, among its operation symbols a binary symbol $+$ and a constant symbol 0 and let $\mathcal{V}$ be the variety of all algebras with zero in this signature so that the free algebra on one generator is finite. (See

§5.4 in Volume I or the last part of §8.6, in particular page 391.) Show that a subdirect product of centerless algebras in $\mathcal{V}$ is again centerless. (Compare Exercise 5.15(8) in Volume I.)

Bibliography

Kira Adaricheva, Jennifer Hyndman, J. B. Nation, and Joy Nishida

(2022) *A Primer of Subquasivariety Lattices.* CMS Books in Mathematics/Ouvrages de Mathématiques de la SMC. Springer, Cham. Forthcoming. 403

I. Š. Aliev

(1966) Smallest variety of symmetric algebras. *Algebra i Logika Sem.* **5**, 5–14. MR0205895. 141

Richard F. Arens and Irving Kaplansky

(1948) Topological representation of algebras. *Trans. Amer. Math. Soc.* **63**, 457–481. MR25453. 386

Tamás Bajusz, George McNulty, and Ágnes Szendrei

(1990) Lyndon's groupoid is not inherently nonfinitely based. *Algebra Universalis* **27**, 254–260. MR1037867. 194

Kirby A. Baker

(1971) Equational axioms for classes of lattices. *Bull. Amer. Math. Soc.* **77**, 97–102. MR288058. 368, 370

(1977) Finite equational bases for finite algebras in a congruence-distributive equational class. *Advances in Math.* **24**, 207–243. MR447074. 2, 191, 196, 203, 357, 363, 373

(1979) Congruence-distributive polynomial reducts of lattices. *Algebra Universalis* **9**, 142–145. MR523928. 134

Kirby A. Baker, George F. McNulty, and Ju Wang

(2004) An extension of Willard's finite basis theorem: congruence meet-semidistributive varieties of finite critical depth. *Algebra Universalis* **52**, 289–302. MR2161654. 28, 209

Kirby A. Baker, George F. McNulty, and Heinrich Werner

(1989) Shift-automorphism methods for inherently nonfinitely based varieties of algebras. *Czechoslovak Math. J.* **39(114)**, 53–69. MR983483. 191, 194, 195

Kirby A. Baker and Alden F. Pixley

(1975) Polynomial interpolation and the Chinese remainder theorem for algebraic systems. *Math. Z.* **143**, 165–174. MR371782. 130, 132

Kirby A. Baker and Ju Wang

(2002) Definable principal subcongruences. *Algebra Universalis* **47**, 145–151. MR1916612. 374, 375

John T. Baldwin and A. H. Lachlan

(1973) On universal Horn classes categorical in some infinte power. *Algebra Universalis* **3**, 98–111. MR351785. 141

Stefan Banach

(1935) Sur un théorème de M. Sierpiński. *Fund. Math.* **25**, 5–6. 229

Clifford Bergman

(2012) *Universal algebra*, volume 301 of *Pure and Applied Mathematics (Boca Raton)*. CRC Press, Boca Raton, FL. Fundamentals and selected topics. MR2839398. xiv, 16, 45, 81, 83

Joel Berman

(2005) The structure of free algebras. In *Structural theory of automata, semigroups, and universal algebra*, volume 207 of *NATO Sci. Ser. II Math. Phys. Chem.*. Springer, Dordrecht, pages 47–76. MR2210125. 66

Joel Berman, Paweł Idziak, Petar Marković, Ralph McKenzie, Matthew Valeriote, and Ross Willard

(2010) Varieties with few subalgebras of powers. *Trans. Amer. Math. Soc.* **362**, 1445–1473. MR2563736. 71

E. W. Beth

(1953) On Padoa's method in the theory of definition. *Nederl. Akad. Wetensch. Proc. Ser. A.* ***56*** = *Indagationes Math.* **15**, 330–339. MR0058537. 426

A. P. Birjukov

(1970) Varieties of idempotent semigroups. *Algebra i Logika* **9**, 255–273. MR0297897. 233

Garrett Birkhoff

(1935) On the structure of abstract algebras. *Proc. Cambridge Philos. Soc.* **31**, 433–454. 154, 157, 164, 171, 172

(1946) Universal algebra. In *Proc. First Canadian Math. Congress, Montreal, 1945.* University of Toronto Press, Toronto, pages 310–326. MR0019586. 233

A. D. Bol'bot

(1970) Varieties of Ω-algebras. *Algebra i Logika* **9**, 406–414. MR0286733. 233

William W. Boone

(1957) Certain simple, unsolvable problems of group theory. V, VI. *Nederl. Akad. Wetensch. Proc. Ser. A. 60* = *Indag. Math.* **19**, 22–27, 227–232. MR0098776. 256

Bruno Bosbach

(1970) Komplementäre Halbgruppen. Kongruenzen und Quotienten. *Fund. Math.* **69**, 1–14. MR277452. 19

Roger M. Bryant

(1982) The laws of finite pointed groups. *Bull. London Math. Soc.* **14**, 119–123. MR647192. 196

S. Bulman-Fleming, Alan Day, and Walter Taylor

(1974) Regularity and modularity of congruences. *Algebra Universalis* **4**, 58–60. MR382118. 97, 122

S. Bulman-Fleming and W. Taylor

(1976) Union-indecomposable varieties. *Colloq. Math.* **35**, 189–199, (Erratum ibid. bf 36 (1976) page 164). MR404100. 121, 134

S. Burris
(1975) Boolean powers. *Algebra Universalis* **5**, 341–360. MR447078. 386
(1979) On Baker's finite basis theorem for congruence distributive varieties. *Proc. Amer. Math. Soc.* **73**, 141–148. MR516452. 2, 205

S. Burris and H. P. Sankappanavar
(1981) *A course in universal algebra*, volume 78 of *Graduate Texts in Mathematics*. Springer-Verlag, New York-Berlin. Free online version, revised 2012, available at: `https://www.math.uwaterloo.ca/~snburris/htdocs/ualg.html`. MR648287. xiv, 19

Robert Cacioppo
(1987) *Finite Bases for Varieties and Pseudovarieties of Universal Algebra*. Ph.D. thesis, The University of Iowa. 194
(1993) Non-finitely based pseudovarieties and inherently non-finitely based varieties. *Semigroup Forum* **47**, 223–226. MR1230146. 178, 194

C. C. Chang and H. J. Keisler
(1990) *Model theory*, volume 73 of *Studies in Logic and the Foundations of Mathematics*. North-Holland Publishing Co., Amsterdam, third edition. MR1059055. 57, 313, 344, 355, 438

C. C. Chang and Anne C. Morel
(1958) On closure under direct product. *J. Symbolic Logic* **23**, 149–154. MR104606. 331

Alonzo Church
(1936) An Unsolvable Problem of Elementary Number Theory. *Amer. J. Math.* **58**, 345–363. MR1507159. 253

David M. Clark and Brian A. Davey
(1998) *Natural dualities for the working algebraist*, volume 57 of *Cambridge Studies in Advanced Mathematics*. Cambridge University Press, Cambridge. MR1663208. 379, 381, 382, 384

David M. Clark, Brian A. Davey, Ralph S. Freese, and Marcel Jackson
(2004) Standard topological algebras: syntactic and principal congruences and profiniteness. *Algebra Universalis* **52**, 343–376. MR2161658. 209

David M. Clark and Peter H. Krauss
(1976) Para primal algebras. *Algebra Universalis* **6**, 165–192. MR422113. 125

P. M. Cohn
(1995) *Skew fields: Theory of general division rings*, volume 57 of *Encyclopedia of Mathematics and its Applications*. Cambridge University Press, Cambridge. MR1349108. 256

A. L. S. Corner
(1969) On endomorphism rings of primary abelian groups. *Quart. J. Math. Oxford Ser. (2)* **20**, 277–296. MR258949. 394

William Craig
(1957) Three uses of the Herbrand-Gentzen theorem in relating model theory and proof theory. *J. Symbolic Logic* **22**, 269–285. MR104565. 426

Béla Csákány

(1964) Abelian properties of primitive classes of universal algebras. *Acta Sci. Math. (Szeged)* **25**, 202–208. MR169805. 142

(1976) Varieties whose algebras have no idempotent elements. *Colloq. Math.* **35**, 201–203. MR404109. 136

Gábor Czédli

(1981) A Mal'cev-type condition for the semidistributivity of congruence lattices. *Acta Sci. Math. (Szeged)* **43**, 267–272. MR640303. 33

(1983) A characterization for congruence semidistributivity. In *Universal algebra and lattice theory (Puebla, 1982)*, volume 1004 of *Lecture Notes in Math.*. Springer, Berlin, pages 104–110. MR716177. 28

(1994) The congruence variety of metaabelian [metabelian] groups is not self-dual. *Acta Math. Univ. Comenian. (N.S.)* **63**, 155–159. MR1342601. 101

Gábor Czédli and Ralph Freese

(1983) On congruence distributivity and modularity. *Algebra Universalis* **17**, 216–219. MR726275. 95

Gábor Czédli and Eszter K. Horváth

(2003) All congruence lattice identities implying modularity have Mal'tsev conditions. *Algebra Universalis* **50**, 69–74. MR2026828. 95, 111

Martin Davis

(1958) *Computability and unsolvability.* McGraw-Hill Series in Information Processing and Computers. McGraw-Hill Book Co., Inc., New York-Toronto-London. MR0124208. 251

Alan Day

(1969) A characterization of modularity for congruence lattices of algebras. *Canad. Math. Bull.* **12**, 167–173. MR248063. 40, 49

Alan Day and Ralph Freese

(1980) A characterization of identities implying congruence modularity. I. *Canadian J. Math.* **32**, 1140–1167. MR596102. 15, 95, 99, 101

Alan Day and Douglas Pickering

(1984) A note on the Arguesian lattice identity. *Studia Sci. Math. Hungar.* **19**, 303–305. MR874498. 104

Richard Dedekind

(1872) *Stetigkeit und irrationale Zahlen.* 6te unveränderte Aufl. Friedr. Vieweg & Sohn. Braunschweig. MR0114755. 151

(1888) *Was sind und was sollen die Zahlen?.* 8te unveränderte Aufl. Friedr. Vieweg & Sohn, Braunschweig. MR0106846. 151

(1900) Über die von drei Moduln erzeugte Dualgruppe. *Math. Ann.* **53**, 371–403. MR1511094. 13

(1963) *Essays on the theory of numbers. I:Continuity and irrational numbers. II:The nature and meaning of numbers.* Authorized translation by Wooster Woodruff Beman. Dover Publications, Inc., New York. MR0159773. 151

(1968) *Gesammelte mathematische Werke. Bände I–III.* Herausgegeben von Robert Fricke, Emmy Noether und Öystein Ore. Chelsea Publishing Co., New York. MR0237282. 151, 256

M. Dehn

(1911) Über unendliche diskontinuierliche Gruppen. *Math. Ann.* **71**, 116–144. MR1511645. 256

Dejan Delić

(2001) Finite bases for flat graph algebras. *J. Algebra* **246**, 453–469. MR1872631. 210

Topaz Dent, Keith A. Kearnes, and Ágnes Szendrei

(2012) An easy test for congruence modularity. *Algebra Universalis* **67**, 375–392. MR2965247. 69

W. Dyck

(1883) Gruppentheoretische studien. ii. ueber die zusammensetzung einer gruppe discreter operationen, über ihre primitivität und transitivität. *Math. Ann.* **22**, 70–108. 255

Wiesław Dziobiak

(1981) On infinite subdirectly irreducible algebras in locally finite equational classes. *Algebra Universalis* **13**, 393–394. MR631732. 278

Samuel Eilenberg and M. P. Schützenberger

(1976) On pseudovarieties. *Advances in Math.* **19**, 413–418. MR0401604. 178

Herbert B. Enderton

(2011) *Computability theory*. Elsevier/Academic Press, Amsterdam. An introduction to recursion theory. MR3292657. 251

P. Erdős, A. Hajnal, and R. Rado

(1965) Partition relations for cardinal numbers. *Acta Math. Acad. Sci. Hungar.* **16**, 93–196. MR202613. 55

P. Erdős and R. Rado

(1956) A partition calculus in set theory. *Bull. Amer. Math. Soc.* **62**, 427–489. MR81864. 55

Trevor Evans

(1951) The word problem for abstract algebras. *J. London Math. Soc.* **26**, 64–71. MR38958. 255

(1978) An algebra has a solvable word problem if and only if it is embeddable in a finitely generated simple algebra. *Algebra Universalis* **8**, 197–204. MR0472643. 259

(1981) Some remarks on the general theory of clones. In *Finite algebra and multiple-valued logic (Szeged, 1979)*, volume 28 of *Colloq. Math. Soc. János Bolyai*. North-Holland, Amsterdam-New York, pages 203–244. MR648617. 233

S. Feferman and R. L. Vaught

(1959) The first order properties of products of algebraic systems. *Fund. Math.* **47**, 57–103. MR108455. 414, 418

Charles Fennemore

(1970) All varieties of bands. *Semigroup Forum* **1**, 172–179. MR271257. 233

K. Fichtner

(1968) Varieties of universal algebras with ideals. *Mat. Sb. (N.S.)* **75 (117)**, 445–453. MR0222001. 58

(1972) Distributivity and modularity in varieties of algebras. *Acta Sci. Math. (Szeged)* **33**, 343–348. MR319858. 76

Alfred L. Foster

(1953a) Generalized "Boolean" theory of universal algebras. I. Subdirect sums and normal representation theorem. *Math. Z.* **58**, 306–336. MR57230. 386

(1953b) Generalized "Boolean" theory of universal algebras. II. Identities and subdirect sums of functionally complete algebras. *Math. Z.* **59**, 191–199. MR57231. 386

T. Frayne, A. C. Morel, and D. S. Scott

(1962) Reduced direct products. *Fund. Math.* **51**, 195–228. MR142459. 331, 335, 411

(1964) Correction to the paper "Reduced direct products". *Fund. Math.* **53**, 117. MR154807. 331

Ralph Freese

(1977) The structure of modular lattices of width four with applications to varieties of lattices. *Mem. Amer. Math. Soc.* **9**, vii+91. MR441803. 370, 372

(1980) Free modular lattices. *Trans. Amer. Math. Soc.* **261**, 81–91. MR576864. 256

(1994) Finitely based modular congruence varieties are distributive. *Algebra Universalis* **32**, 104–114. MR1287018. 109, 257

(1995) Alan Day's early work: congruence identities. *Algebra Universalis* **34**, 4–23. MR1344951. 115

(2013) Equations implying congruence n-permutability and semidistributivity. *Algebra Universalis* **70**, 347–357. MR3127978. 71

(2015) Notes on the Birkhoff construction of free algebras. Available at `uacalc.org`. 66

Ralph Freese, Jaroslav Ježek, and J. B. Nation

(1995) *Free lattices*, volume 42 of *Mathematical Surveys and Monographs*. American Mathematical Society, Providence, RI. MR1319815. 99

Ralph Freese and Bjarni Jónsson

(1976) Congruence modularity implies the Arguesian identity. *Algebra Universalis* **6**, 225–228. MR472644. 108

Ralph Freese, Emil Kiss, and Matthew Valeriote

(2008) Universal Algebra Calculator. Available at `uacalc.org`. www.uacalc.org. 20, 66

Ralph Freese, William A. Lampe, and Walter Taylor

(1979) Congruence lattices of algebras of fixed similarity type. I. *Pacific J. Math.* **82**, 59–68. MR549832. 219

Ralph Freese and Ralph McKenzie

(1981) Residually small varieties with modular congruence lattices. *Trans. Amer. Math. Soc.* **264**, 419–430. MR603772. 279

(1987) *Commutator theory for congruence modular varieties*, volume 125 of *London Mathematical Society Lecture Note Series*. Cambridge University Press,

Cambridge. Free online version available at: `http://math.hawaii.edu/~ralph/papers.html`. MR909290. 16, 46, 74, 95, 96

(2017) Maltsev families of varieties closed under join or Maltsev product. *Algebra Universalis* **77**, 29–50. MR3602782. 33

Ralph Freese, George F. McNulty, and J. B. Nation

(2002) Inherently nonfinitely based lattices. *Ann. Pure Appl. Logic* **115**, 175–193. MR1897025. 368

Ralph Freese and J. B. Nation

(1973) Congruence lattices of semilattices. *Pacific J. Math.* **49**, 51–58. MR332590. 38, 80, 114

Ralph Freese and Matthew Valeriote

(2009) On the complexity of some Maltsev conditions. *Internat. J. Algebra Comput.* **19**, 41–77. MR2494469. 4, 12, 17, 20

Tsuyoshi Fujiwara

(1971) On the construction of the least universal Horn class containing a given class. *Osaka Math. J.* **8**, 425–436. MR311546. 406

(1978) A variation of Lyndon-Keisler's homomorphism theorem and its applications to interpolation theorems. *J. Math. Soc. Japan* **30**, 287–302. MR494712. 427

Fred Galvin

(1967) *HORN SENTENCES*. Ph.D. thesis, University of Minnesota. 412, 414

(1970) Horn sentences. *Ann. Math. Logic* **1**, 389–422. MR325382. 414, 417, 421

O. C. Garcia and W. Taylor

(1984) The lattice of interpretability types of varieties. *Mem. Amer. Math. Soc.* **50**, v+125. MR749524. 141

E. Gedeonová

(1972) A characterization of p-modularity for congruence lattices of algebras. *Acta Fac. Rerum Natur. Univ. Comenian. Math.*, 99–106. MR313169. 96

David Geiger

(1974) Coherent algebras. *Notices Amer. Math. Soc.* **21**, 76T–A130. 57

J. A. Gerhard

(1970) The lattice of equational classes of idempotent semigroups. *J. Algebra* **15**, 195–224. MR263953. 233

J. A. Gerhard and Mario Petrich

(1989) Varieties of bands revisited. *Proc. London Math. Soc. (3)* **58**, 323–350. MR977480. 233

Steven Givant

(1979) A representation theorem for universal Horn classes categorical in power. *Ann. Math. Logic* **17**, 91–116. MR552417. 139, 141

Steven Givant and Paul Halmos

(2009) *Introduction to Boolean algebras*. Undergraduate Texts in Mathematics. Springer, New York. MR2466574. 417

Kurt Gödel

(1930) Die Vollständigkeit der Axiome des logischen Funktionenkalküls. *Monatsh. Math. Phys.* **37**, 349–360. MR1549799. 334

Viktor A. Gorbunov

(1998) *Algebraic theory of quasivarieties*. Siberian School of Algebra and Logic. Consultants Bureau, New York. Translated from the Russian. MR1654844. 403

Ewa Graczyńska

(1983) On regular identities. *Algebra Universalis* **17**, 369–375. MR729944. 167

Ronald L. Graham, Bruce L. Rothschild, and Joel H. Spencer

(1980) *Ramsey theory*. John Wiley & Sons Inc., New York. Wiley-Interscience Series in Discrete Mathematics, A Wiley-Interscience Publication. MR591457. 55

G. Grätzer

(1970a) Two Mal'cev-type theorems in universal algebra. *J. Combinatorial Theory* **8**, 334–342. MR279022. 51

(1970b) Universal algebra. In *Trends in Lattice Theory (Sympos., U.S. Naval Academy, Annapolis, Md., 1966)*. Van Nostrand Reinhold, New York, pages 173–210. MR0281674. 58

(1979) *Universal algebra*. Springer-Verlag, New York-Heidelberg, second edition. MR538623. xiv

G. Grätzer and H. Lakser

(1973) A note on the implicational class generated by a class of structures. *Canad. Math. Bull.* **16**, 603–605. MR369216. 406

G. Grätzer and E. T. Schmidt

(1963) Characterizations of congruence lattices of abstract algebras. *Acta Sci. Math. (Szeged)* **24**, 34–59. MR0151406. 211, 218

H. P. Gumm

(1978) Is there a Mal'cev theory for single algebras? *Algebra Universalis* **8**, 320–329. MR472647. 43, 46, 49

(1980) An easy way to the commutator in modular varieties. *Arch. Math. (Basel)* **34**, 220–228. MR590312. 43, 46

(1983) Geometrical methods in congruence modular algebras. *Mem. Amer. Math. Soc.* **45**, viii+79. MR714648. 4

Sergey Gusev, Edmond Lee, and Boris Vernikov

(2020) The lattice of varieties of monoids. ArXib:2011.03679v1. 234

J. Hagemann

(1973a) Mal'cev-conditions for algebraic relations. Preprint 75, Mathematical Preprints TU Darmstadt. 20, 71

(1973b) On regular and weakly regular congruences. Preprint 75, Mathematical Preprints TU Darmstadt. 51, 58, 73, 133

J. Hagemann and C. Herrmann

(1979) A concrete ideal multiplication for algebraic systems and its relation to congruence distributivity. *Arch. Math. (Basel)* **32**, 234–245. MR541622. 16

J. Hagemann and A. Mitschke

(1973) On n-permutable congruences. *Algebra Universalis* **3**, 8–12. MR330010. 13, 19, 20

Mark Haiman

(1991) Arguesian lattices which are not type-1. *Algebra Universalis* **28**, 128–137. MR1083826. 109

Marshall Hall, Jr.

(1949) The word problem for semigroups with two generators. *J. Symbolic Logic* **14**, 115–118. MR0030475. 266

T. O. Hand

(1976) Abstract 76t-a136. *Notices Amer. Math. Soc.* **23**, A–425. 166

William Hanf

(1957) On some fundamental problems concerning isomorphism of Boolean algebras. *Math. Scand.* **5**, 205–217. MR108451. 391

Junji Hashimoto

(1963) Congruence relations and congruence classes in lattices. *Osaka Math. J.* **15**, 71–86. MR151409. 58

Leon Henkin

(1948) *THE COMPLETENESS OF FORMAL SYSTEMS.* Ph.D. thesis, Princeton University. MR2937859. 335

(1949) The completeness of the first-order functional calculus. *J. Symbolic Logic* **14**, 159–166. MR33781. 335, 434

(1963) An extension of the Craig-Lyndon interpolation theorem. *J. Symbolic Logic* **28**, 201–216. MR171700. 427

Christian Herrmann

(1973) On the equational theory of submodule lattices. In *Proceedings of the University of Houston Lattice Theory Conference (Houston, Tex., 1973)*. MR0406882. 257

(1979) Affine algebras in congruence modular varieties. *Acta Sci. Math. (Szeged)* **41**, 119–125. MR534504. 45

(1983) On the word problem for the modular lattice with four free generators. *Math. Ann.* **265**, 513–527. MR721885. 256

Christian Herrmann and András Huhn

(1975) Zum Wortproblem für freie Untermodulverbände. *Arch. Math. (Basel)* **26**, 449–453. MR401562. 257

Edwin Hewitt

(1948) Rings of real-valued continuous functions. I. *Trans. Amer. Math. Soc.* **64**, 45–99. MR26239. 331

David Hobby and Ralph McKenzie

(1988) *The structure of finite algebras*, volume 76 of *Contemporary Mathematics*. American Mathematical Society, Providence, RI. MR958685. 33, 294

Wilfrid Hodges

(1993) *Model theory*, volume 42 of *Encyclopedia of Mathematics and its Applications*. Cambridge University Press, Cambridge. MR1221741. 313

John E. Hopcroft and Jeffrey D. Ullman

(1979) *Introduction to automata theory, languages, and computation.* Addison-Wesley Publishing Co., Reading, Mass. Addison-Wesley Series in Computer Science. MR645539. 251

Alfred Horn

(1951) On sentences which are true of direct unions of algebras. *J. Symbolic Logic* **16**, 14–21. MR40233. 411

Tah Kai Hu

(1969) Stone duality for primal algebra theory. *Math. Z.* **110**, 180–198. MR244130. 386, 389

A. P. Huhn

(1972) Weakly distributive lattices. Preprint, József Attila University, Szeged. 130

(1983) Schwach distributive Verbände. II. *Acta Sci. Math. (Szeged)* **46**, 85–98. MR739026. 130

George Hutchinson

(1973) Recursively unsolvable word problems of modular lattices and diagram-chasing. *J. Algebra* **26**, 385–399. MR332599. 256

(1977) Embedding and unsolvability theorems for modular lattices. *Algebra Universalis* **7**, 47–84. MR441804. 256

(1995) Lattice theory equations and interesting lattices. Available at: `http://www.math.u-szeged.hu/~czedli/m/research/hutchinson_lattice-theory-equations-and-interesting-lattices.pdf`. 257

George Hutchinson and Gábor Czédli

(1978) A test for identities satisfied in lattices of submodules. *Algebra Universalis* **8**, 269–309. MR469840. 118

Jennifer Hyndman and J. B. Nation

(2018) *The lattice of subquasivarieties of a locally finite quasivariety.* CMS Books in Mathematics/Ouvrages de Mathématiques de la SMC. Springer, Cham. MR3837544. 403

Thomas Ihringer

(2003) *Allgemeine Algebra*, volume 10 of *Berliner Studienreihe zur Mathematik* [*Berlin Study Series on Mathematics*]. Heldermann Verlag, Berlin. Mit einem Anhang über universelle Coalgebra von H. P. Gumm. [With an appendix on universal coalgebra by H. P. Gumm]. MR2002883. xiv, 45

I. M. Isaev

(1989) Inherently non-finitely based varieties of algebras. *Sibirsk. Mat. Zh.* **30**, 75–77. MR1043435. 196

John R. Isbell

(1966) On the problem of universal terms. *Bull. Acad. Polon. Sci. Sér. Sci. Math. Astronom. Phys.* **14**, 593–595. MR0207567. 229

Marcel Jackson and George F. McNulty

(2011) The equational complexity of Lyndon's algebra. *Algebra Universalis* **65**, 243–262. MR2793398. 177

Eugene Jacobs and Robert Schwabauer

(1964) The lattice of equational classes of algebras with one unary operation. *Amer. Math. Monthly* **71**, 151–155. MR162740. 250

N. Jacobson

(1945) Structure theory for algebraic algebras of bounded degree. *Ann. of Math. (2)* **46**, 695–707. MR0014083. 152

Jaroslav Ježek

(1969) Primitive classes of algebras with unary and nullary operations. *Colloq. Math.* **20**, 159–179. MR0246813. 250

(1970) On atoms in lattices of primitive classes. *Comment. Math. Univ. Carolinae* **11**, 515–532. MR0269571. 233, 251

(1976) Intervals in the lattice of varieties. *Algebra Universalis* **6**, 147–158. MR0419332. 234

(1981a) The lattice of equational theories. I. Modular elements. *Czechoslovak Math. J.* **31(106)**, 127–152, with a loose Russian summary. MR604120. 249

(1981b) The lattice of equational theories. II. The lattice of full sets of terms. *Czechoslovak Math. J.* **31(106)**, 573–603. MR631604. 249

(1982) The lattice of equational theories. III. Definability and automorphisms. *Czechoslovak Math. J.* **32(107)**, 129–164. MR646718. 249

(1985) Nonfinitely based three-element idempotent groupoids. *Algebra Universalis* **20**, 292–301. MR811690. 195

(1986) The lattice of equational theories. IV. Equational theories of finite algebras. *Czechoslovak Math. J.* **36(111)**, 331–341. MR831318. 249

Jaroslav Ježek and Ralph McKenzie

(1993) Definability in the lattice of equational theories of semigroups. *Semigroup Forum* **46**, 199–245. MR1200214. 250

Peter Jipsen and Henry Rose

(1992) *Varieties of lattices*, volume 1533 of *Lecture Notes in Mathematics*. Springer-Verlag, Berlin. MR1223545. 358

D. L. Johnson

(1997) *Presentations of groups*, volume 15 of *London Mathematical Society Student Texts*. Cambridge University Press, Cambridge, second edition. MR1472735. 256

Bjarni Jónsson

(1957) On isomorphism types of groups and other algebraic systems. *Math. Scand.* **5**, 224–229. MR108453. 391, 392

(1967) Algebras whose congruence lattices are distributive. *Math. Scand.* **21**, 110–121. MR237402. 2, 110, 333, 355, 372

(1968) Equational classes of lattices. *Math. Scand.* **22**, 187–196 (1969). MR246797. 372

(1972) The class of arguesian lattices is selfdual. *Algebra Universalis* **2**, 396–396. MR316325. 104

(1979) On finitely based varieties of algebras. *Colloq. Math.* **42**, 255–261. MR567561. 204, 205, 349, 362, 363

(1980) Congruence varieties. *Algebra Universalis* **10**, 355–394. MR564122. 33, 117

David Joyce

(1979) *An Algebraic Aproach To Symmetry With Applications To Knot Theory*. Ph.D. thesis, University of Pennsylvania. MR2628474. 166, 167

(1982) A classifying invariant of knots, the knot quandle. *J. Pure Appl. Algebra* **23**, 37–65. MR638121. 166, 167

Kalle Kaarli and Alden F. Pixley

(2001) *Polynomial completeness in algebraic systems*. Chapman & Hall/CRC, Boca Raton, FL. MR1888967. 132

Cornelia Kalfa

(1986) Decision problems concerning properties of finite sets of equations. *J. Symbolic Logic* **51**, 79–87. MR830075. 276

Jan Kalicki

(1955) The number of equationally complete classes of equations. *Nederl. Akad. Wetensch. Proc. Ser. A.* ***58*** = *Indag. Math.* **17**, 660–662. MR0074351. 233

Takao Kashiwagi

(1972) On $(\mathfrak{m}, A)$-implicational classes. *Math. Japon.* **17**, 1–12. MR327504. 406

Alexandr Kazda

(2019) Taylor term does not imply any nontrivial linear one-equality Maltsev condition. *Algebra Universalis* **80**, Art. 9, 9. MR3907941. 89

Keith A. Kearnes

(1993) Congruence permutable and congruence 3-permutable locally finite varieties. *J. Algebra* **156**, 36–49. MR1213783. 17

Keith A. Kearnes and Emil W. Kiss

(2013) The shape of congruence lattices. *Mem. Amer. Math. Soc.* **222**, viii+169. MR3076179. 33, 112

Keith A. Kearnes, Emil W. Kiss, and Ágnes Szendrei

(2016) Growth rates of algebras, I: Pointed cube terms. *J. Aust. Math. Soc.* **101**, 56–94. MR3518323. 164

Keith A. Kearnes and Ágnes Szendrei

(1998) The relationship between two commutators. *Internat. J. Algebra Comput.* **8**, 497–531. MR2000e:08001. 28, 112

Keith A. Kearnes, Ágnes Szendrei, and Ross Willard

(2016) A finite basis theorem for difference-term varieties with a finite residual bound. *Trans. Amer. Math. Soc.* **368**, 2115–2143. MR3449235. 196

(2022) Characterizing the commutator in varieties with a difference term. *Algebra Universalis*, to appear. arXiv:2112.00715. 196

Keith A. Kearnes and Ross Willard

(1999) Residually finite, congruence meet-semidistributive varieties of finite type have a finite residual bound. *Proc. Amer. Math. Soc.* **127**, 2841–2850. MR1636966. 279

H. Jerome Keisler

(1961) Ultraproducts and elementary classes. *Nederl. Akad. Wetensch. Proc. Ser. A 64* = *Indag. Math.* **23**, 477–495. MR0140396. 344

(1965) Reduced products and Horn classes. *Trans. Amer. Math. Soc.* **117**, 307–328. MR170812. 412

David Kelly

(1972) *Basic Equations, Free Algebras, Word Problems, and Mal'cev Conditions.* Ph.D. thesis, Queen's University. 75, 164, 165

(1973) Basic equations: word problems and mal'cev conditions. *Notices Amer. Math. Soc.* **20**, abstract 701-08-4. 75, 76

(1984) Complete rules of inference for universal sentences. *Studia Sci. Math. Hungar.* **19**, 347–361. MR874503. 403

Jussi Ketonen

(1978) The structure of countable Boolean algebras. *Ann. of Math.* (2) **108**, 41–89. MR491391. 391, 394

Emil Kiss

(1992) Three remarks on the modular commutator. *Algebra Universalis* **29**, 455–476. MR1201171. 46

Emil Kiss and Matthew Valeriote

(1991) Strongly abelian varieties and the Hamiltonian property. *Canad. J. Math.* **43**, 331–346. MR1113758. 136

L. Klukovits

(1975) Hamiltonian varieties of universal algebras. *Acta Sci. Math. (Szeged)* **37**, 11–15. MR401611. 135

J. Kollár

(1979) The category of unary algebras, containing a given subalgebra. I. *Acta Math. Acad. Sci. Hungar.* **33**, 407–417. MR542491. 136, 138

Sabine Koppelberg

(1989) *Handbook of Boolean algebras. Vol. 1.* North-Holland Publishing Co., Amsterdam. Edited by J. Donald Monk and Robert Bonnet. MR991565. 417

Marcin Kozik, Andrei Krokhin, Matthew Valeriote, and Ross Willard

(2015) Characterizations of several Maltsev conditions. *Algebra Universalis* **73**, 205–224. MR3350327. 76, 79, 80

Robert L. Kruse

(1973) Identities satisfied by a finite ring. *J. Algebra* **26**, 298–318. MR325678. 196

A. V. Kuznetsov

(1956) O problema toždestva i funcional'noĭ polnoty dlya algebraičeskih sistem. In *Trudy 3-go Vsesojuznog Mat. S"ezda (Moscow) 1956*, volume 2. pages 145–146, Akad. Nauk SSSR vol. 2. 258

H. Lakser

(1982) Principal congruences in *n*-permutable varieties. *Algebra Universalis* **14**, 64–67. MR634417. 21

William A. Lampe

(1982) Congruence lattices of algebras of fixed similarity type. II. *Pacific J. Math.* **103**, 475–508. MR705246. 225

(1986) A property of the lattice of equational theories. *Algebra Universalis* **23**, 61–69. MR873596. 246

(1991) Further properties of lattices of equational theories. *Algebra Universalis* **28**, 459–486. MR1128384. 249

F. William Lawvere

(1963) Functorial semantics of algebraic theories. *Proc. Nat. Acad. Sci. U.S.A.* **50**, 869–872. MR158921. xii

Edmond W. H. Lee

(2009) Lyndon's groupoid generates a small almost Cross variety. *Algebra Universalis* **60**, 239–246. MR2491425. 173, 194

F. W. Levi

(1944) Notes on group theory. *J. Indian Math. Soc.* **8**, 1–9. MR10590. 166

Paolo Lipparini

(1998) A characterization of varieties with a difference term. II. Neutral = meet semi-distributive. *Canad. Math. Bull.* **41**, 318–327. MR1637665. 28

(2020) The Gumm level equals the alvin level in congruence distributive varieties. *Arch. Math. (Basel)* **115**, 391–400. MR4145765. 12

(2021) The Tschantz and the alvin higher condtions are equivalent in congruence distributive varieties. *Algebra Universalis* **82**. 12

L. Lipshitz

(1974) The undecidability of the word problems for projective geometries and modular lattices. *Trans. Amer. Math. Soc.* **193**, 171–180. MR364040. 256

Jerzy Łoś

(1950) Un théorème sur les superpositions des fonctions définies dans les ensembles arbitraires. *Fund. Math.* **37**, 84–86. MR0041202. 229

(1955a) On the extending of models. I. *Fund. Math.* **42**, 38–54. MR72071. 397

(1955b) Quelques remarques, théorèmes et problèmes sur les classes définissables d'algèbres. In *Mathematical interpretation of formal systems*. North-Holland Publishing Co., Amsterdam, pages 98–113. MR0075156. 330, 331

I. V. L'vov

(1973) Varieties of associative rings. I, II. *Algebra i Logika* **12**, 269–297, 363; ibid. 12 (1973), 667–688, 735. MR0389973. 196

Roger C. Lyndon

(1954) Identities in finite algebras. *Proc. Amer. Math. Soc.* **5**, 8–9. MR60482. 173, 177

(1959a) An interpolation theorem in the predicate calculus. *Pacific J. Math.* **9**, 129–142. MR106825. 427

(1959b) Properties preserved in subdirect products. *Pacific J. Math.* **9**, 155–164. MR107606. 438, 441

(1959c) Properties preserved under homomorphism. *Pacific J. Math.* **9**, 143–154. MR108441. 435

Roger C. Lyndon and Paul E. Schupp

(2001) *Combinatorial group theory*. Classics in Mathematics. Springer-Verlag, Berlin. Reprint of the 1977 edition. MR1812024. 256

Angus Macintyre

(1973) The word problem for division rings. *J. Symbolic Logic* **38**, 428–436. MR0337554. 256

Roberto Magari

(1969) Una dimostrazione del fatto che ogni varietà ammette algebre semplici. *Ann. Univ. Ferrara Sez. VII (N.S.)* **14**, 1–4. MR262146. 138, 139

Wilhelm Magnus, Abraham Karrass, and Donald Solitar

(2004) *Combinatorial group theory*. Dover Publications, Inc., Mineola, NY, second edition. Presentations of groups in terms of generators and relations. MR2109550. 256

A. I. Mal'tsev

(1936) Untersuchungen aus dem Gebiete der mathematischen Logik. *Rec. Math. [Mat. Sbornik] N.S.* **43**, 332–336. 334

(1941) On a general method for obtaining local theorems in group theory. *Ivanov. Gos. Ped. Inst. Uč. Zap. Fiz.-Mat. Fak.* **1**, 3–9. MR0075939. 335, 409

(1954) On the general theory of algebraic systems. *Mat. Sbornik* **77**, 3–20, in Russian, see Mal'tsev 1963 for an English translation. MR0065533. 1, 24, 74, 164

(1958) O gomomorfizmah na konečnye gruppy. *Učemye Zapiski Ivanov, Ped. Inst.* **28**, 49–60. 257

(1959a) Model correspondences. *Izv. Akad. Nauk SSSR. Ser. Mat.* **23**, 313–336. MR0120152. 399

(1959b) Regular products of models. *Izv. Akad. Nauk SSSR Ser. Mat.* **23**, 489–502. MR0124221. 414

(1963) On the general theory of algebraic systems. *Amer. Math. Soc. Transl. (2)* **27**, 125–142, translation of (Mal'tsev 1954). MR0151416. 24

(1966a) Identity relations on manifolds of quasigroups. *Mat. Sb. (N.S.)* **69 (111)**, 3–12. MR0200162. 266

(1966b) Several remarks on quasivarieties of algebraic systems. *Algebra i Logika Sem.* **5**, 3–9. MR0205902. 331, 406

(1968) Some borderline problems of algebra and logic. In *Proc. Internat. Congr. Math. (Moscow, 1966)*. Izdat. "Mir", Moscow, pages 217–231. MR0233751. 233

(1973) *Algebraic systems*. Springer-Verlag, New York-Heidelberg. Posthumous edition, edited by D. Smirnov and M. Taĭclin, Translated from the Russian by B. D. Seckler and A. P. Doohovskoy, Die Grundlehren der mathematischen Wissenschaften, Band 192. MR0349384. 406

David Marker

(2002) *Model theory*, volume 217 of *Graduate Texts in Mathematics*. Springer-Verlag, New York. An introduction. MR1924282. 313

A. Markov

(1947) On the impossibility of certain algorithms in the theory of associative systems. *C. R. (Doklady) Acad. Sci. URSS (N.S.)* **55**, 583–586. MR0020528. 256, 266, 276

Peter Mayr and Nik Ruškuc

(2018) Finiteness properties of direct products of algebraic structures. *J. Algebra* **494**, 167–187. MR3723176. 276

Ralph McKenzie

(1970) Equational bases for lattice theories. *Math. Scand.* **27**, 24–38. MR0274353. 196

(1971) Definability in lattices of equational theories. *Ann. Math. Logic* **3**, 197–237. MR0280349. 249, 251, 365

(1972) Equational bases and nonmodular lattice varieties. *Trans. Amer. Math. Soc.* **174**, 1–43. MR313141. 99

(1975) On spectra, and the negative solution of the decision problem for identities having a finite nontrivial model. *J. Symbolic Logic* **40**, 186–196. MR0376323. 276

(1977) A finite algebra **A** with ***SP***(**A**) not elementary. *Algebra Universalis* **8**, 5–7. MR491409. 406

(1978) Para primal varieties: A study of finite axiomatizability and definable principal congruences in locally finite varieties. *Algebra Universalis* **8**, 336–348. MR0469853. 210, 375, 376

(1983) Finite forbidden lattices. In R. Freese and O. Garcia, editors, *Universal Algebra and Lattice Theory*. Springer-Verlag, New York, pages 176–205. Lecture notes in Mathematics, vol. **1004**. MR716183. 246

(1984) A new product of algebras and a type reduction theorem. *Algebra Universalis* **18**, 29–69. MR743456. 222, 246, 311, 312

(1987) Finite equational bases for congruence modular varieties. *Algebra Universalis* **24**, 224–250. MR931614. 196

(1996a) The residual bound of a finite algebra is not computable. *Internat. J. Algebra Comput.* **6**, 29–48. MR1371733. 311

(1996b) The residual bounds of finite algebras. *Internat. J. Algebra Comput.* **6**, 1–28. MR1371732. 191, 279, 282, 294, 362

(1996c) Tarski's finite basis problem is undecidable. *Internat. J. Algebra Comput.* **6**, 49–104. MR1371734. 311

Ralph McKenzie and Richard J. Thompson

(1973) An elementary construction of unsolvable word problems in group theory. In *Word problems: decision problems and the Burnside problem in group theory (Conf., Univ. California, Irvine, Calif. 1969; dedicated to Hanna Neumann)*, volume 71 of *Studies in Logic and the Foundations of Math.*. North Holland, pages 457–478. MR0396769. 256

George F. McNulty

(1972) *The Decision Problem for Bases of Equational Theories of Algebras*. Ph.D. thesis, University of California, Berkeley. MR2940149. 269

(1976a) The decision problem for equational bases of algebras. *Ann. Math. Logic* **10**, 193–259. MR0432440. 269

(1976b) Undecidable properties of finite sets of equations. *J. Symbolic Logic* **41**, 589–604. MR0485307. 273, 274

(1977) Fragments of first order logic. I. Universal Horn logic. *J. Symbolic Logic* **42**, 221–237. MR498081. 403

George F. McNulty, Zoltán Székely, and Ross Willard

(2008) Equational complexity of the finite algebra membership problem. *Internat. J. Algebra Comput.* **18**, 1283–1319. MR2483124. 191

George F. McNulty and Ross Willard

(2018) Congruence meet-semidistributive locally finite varieties and a finite basis theorem. *Algebra Universalis* **79**, Art. 44, 20. MR3802279. 28, 209, 210

Peter Mederly

(1975) Three Maltsev type theorems and their application. *Math. Časopis Sloven. Akad. Vied.* **25**, 83–95. MR384650. 96

Aleit Mitschke

(1971) Implication algebras are 3-permutable and 3-distributive. *Algebra Universalis* **1**, 182–186. MR309828. 19

(1978) Near unanimity identities and congruence distributivity in equational classes. *Algebra Universalis* **8**, 29–32. MR453609. 129

J. Donald Monk and Robert Bonnet, editors

(1989a) *Handbook of Boolean algebras. Vol. 2.* North-Holland Publishing Co., Amsterdam. MR991595. 417

(1989b) *Handbook of Boolean algebras. Vol. 3.* North-Holland Publishing Co., Amsterdam. MR991607. 417

Nobuyoshi Motohashi

(1984) Equality and Lyndon's interpolation theorem. *J. Symbolic Logic* **49**, 123–128. MR736607. 427

V. L. Murskiĭ

(1965) The existence in the three-valued logic of a closed class with a finite basis having no finite complete system of identities. *Dokl. Akad. Nauk SSSR* **163**, 815–818. MR0186539. 192

(1971) Unrecognizable properties of finite systems of identity relations. *Dokl. Akad. Nauk SSSR* **196**, 520–522. MR0286877. 269, 276

(1979) The number of k-element algebras with a binary operation which do not have a finite basis of identities. *Problemy Kibernet.*, 5–27, 208. MR539884. 191, 192

J. B. Nation

(1974) Varieties whose congruences satisfy certain lattice identities. *Algebra Universalis* **4**, 78–88. MR354501. 71, 95

(1996) A counterexample to the finite height conjecture. *Order* **13**, 1–9. MR1402646. 358, 362

G. C. Nelson

(1984) Boolean powers, recursive models, and the Horn theory of a structure. *Pacific J. Math.* **114**, 207–220. MR755490. 386

Hanna Neumann

(1967) *Varieties of groups.* Springer-Verlag New York, Inc., New York. MR0215899. 196

P. S. Novikov

(1955) *Ob algoritmičeskoĭ nerazrešimosti problemy toždestva slov v teorii grupp.* Trudy Mat. Inst. im. Steklov. no. 44. Izdat. Akad. Nauk SSSR, Moscow. MR0075197. 256

Anvar M. Nurakunov

(2008) Equational theories as congruences of enriched monoids. *Algebra Universalis* **58**, 357–372. MR2415287. 247

(2017) Ultraproducts preserve finite subdirect reducibility. *Algebra Universalis* **78**, 181–192. MR3697188. 350, 351, 355, 365

Sheila Oates and M. B. Powell

(1964) Identical relations in finite groups. *J. Algebra* **1**, 11–39. MR161904. 196

Arnold Oberschelp

(1968) On the Craig-Lyndon interpolation theorem. *J. Symbolic Logic* **33**, 271–274. MR239937. 427

A. Ju. Ol'shanskĭĭ

(1974) Conditional identities in finite groups. *Sibirsk. Mat. Ž.* **15**, 1409–1413, 1432. MR0367068. 408

Miroslav Olšák

(2017) The weakest nontrivial idempotent equations. *Bull. Lond. Math. Soc.* **49**, 1028–1047. MR3743486. 90

O. Ore

(1942) Theory of equivalence relations. *Duke Math. J.* **9**, 573–627. MR7388. 13

Kathryn S. Owens

(2010) Every shift automorphism variety has an infinite subdirectly irreducible member. *J. Aust. Math. Soc.* **88**, 231–238. MR2629932. 191

P. P. Pálfy and P. Pudlák

(1980) Congruence lattices of finite algebras and intervals in subgroup lattices of finite groups. *Algebra Universalis* **11**, 22–27. MR593011. 219

P. P. Pálfy and C. Szabó

(1995) An identity for subgroup lattices of abelian groups. *Algebra Universalis* **33**, 191–195. MR1318983. 109

E. A. Palyutin

(1975) Description of categorical quasivarieties. *Algebra i Logika* **14**, 145–185, 240. MR0398821. 139, 141

Dona Papert

(1964) Congruence relations in semi-lattices. *J. London Math. Soc.* **39**, 723–729. MR0166123. 32, 33

Robert E. Park

(1976) *Equational Classes of Non-associative Ordered Algebras.* Ph.D. thesis, University of California, Los Angeles. MR2626431. 194, 294

(1980) A four-element algebra whose identities are not finitely based. *Algebra Universalis* **11**, 255–260. MR588218. 194

Peter Perkins

(1966) *Decision Problems for Equational Theories of Semigroups and General Algebras.* Ph.D. thesis, Thesis (Ph.D.)–University of California, Berkeley. MR2615990. 195, 273

(1967) Unsolvable problems for equational theories. *Notre Dame J. Formal Logic* **8**, 175–185. MR0236012. 273, 276

(1969) Bases for equational theories of semigroups. *J. Algebra* **11**, 298–314. MR233911. 195

(1984) Basic questions for general algebras. *Algebra Universalis* **19**, 16–23. MR748904. 195

Don Pigozzi

(1976) Base-undecidable properties of universal varieties. *Algebra Universalis* **6**, 193–223. MR0434793. 276

Jane Pitkethly and Brian Davey

(2005) *Dualisability*, volume 9 of *Advances in Mathematics (Springer)*. Springer, New York. Unary algebras and beyond. MR2161626. 379

Alden F. Pixley

(1970) Functionally complete algebras generating distributive and permutable classes. *Math. Z.* **114**, 361–372. MR262148. 279, 341

(1972) Local Malcev conditions. *Canad. Math. Bull.* **15**, 559–568. MR309837. 95, 109

J. Płonka

(1967) On a method of construction of abstract algebras. *Fund. Math.* **61**, 183–189. MR0225701. 167

S. V. Polin

(1976) Identities of finite algebras. *Sibirsk. Mat. Ž.* **17**, 1356–1366, 1439. MR0439715. 196

(1977) Identities in congruence lattices of universal algebras. *Mat. Zametki* **22**, 443–451, transl. Math. Notes, **22**(1977), 737–742. MR491407. 15, 100

Emil L. Post

(1920) *Introduction to a General Theory of Elementary Proposition.* Ph.D. thesis, Columbia University. 378

(1921) Introduction to a General Theory of Elementary Propositions. *Amer. J. Math.* **43**, 163–185. MR1506440. 378

(1941) *The Two-Valued Iterative Systems of Mathematical Logic.* Annals of Mathematics Studies, no. 5. Princeton University Press, Princeton, N. J. MR0004195. xi

(1947) Recursive unsolvability of a problem of Thue. *J. Symbolic Logic* **12**, 1–11. MR0020527. 254, 256, 266, 276

Pavel Pudlák

(1976) A new proof of the congruence lattice representation theorem. *Algebra Universalis* **6**, 269–275. MR0429699. 218

Pavel Pudlák and J. Tůma

(1980) Every finite lattice can be embedded in the lattice of all equivalences over a finite set. *Algebra Universalis* **10**, 74–95. MR552159. 219

Robert W. Quackenbush

(1971) Equational classes generated by finite algebras. *Algebra Universalis* **1**, 265–266. MR294222. 279, 294, 342

(1982) Varieties with n-principle compact congruences. *Algebra Universalis* **14**, 292–296. MR654398. 142

F. P. Ramsey

(1930) On a problem of formal logic. *Proc. London Math. Soc.* **30**, 264–286. MR1576401. 55

Abraham Robinson

(1956a) *Complete theories*. North-Holland Publishing Co., Amsterdam. MR0075897. 337

(1956b) A result on consistency and its application to the theory of definition. *Nederl. Akad. Wetensch. Proc. Ser. A.* ***59*** = *Indag. Math.* **18**, 47–58. MR0078307. 424, 425

Hartley Rogers, Jr.

(1987) *Theory of recursive functions and effective computability*. MIT Press, Cambridge, MA, second edition. MR886890. 251

Ivo Rosenberg

(1970) Über die funktionale Vollständigkeit in den mehrwertigen Logiken. Struktur der Funktionen von mehreren Veränderlichen auf endlichen Mengen. *Rozpravy Československé Akad. Věd Řada Mat. Přirod. Věd* **80**, 93. MR0292647. xi

Mark V. Sapir

(1987) Inherently non-finitely based finite semigroups. *Mat. Sb. (N.S.)* **133(175)**, 154–166, 270. MR905002. 196

(1988) Sur la propriété de base finie pour les pseudovariétés de semigroupes finis. *C. R. Acad. Sci. Paris Sér. I Math.* **306**, 795–797. MR949035. 178, 194

E. T. Schmidt

(1969) *Kongruenzrelationen algebraischer Strukturen*. Mathematische Forschungsberichte, XXV. VEB Deutscher Verlag der Wissenschaften, Berlin. MR0255474. 13

(1970) Über reguläre Mannigfaltigkeiten. *Acta Sci. Math. (Szeged)* **31**, 197–201. MR281679. 125

(1972) On n-permutable equational classes. *Acta Sci. Math. (Szeged)* **33**, 29–30. MR302539. 17

Dana Scott

(1956) Equationally complete extensions of finite algebras. *Nederl. Akad. Wetensch. Proc. Ser. A.* ***59*** = *Indag. Math.* **18**, 35–38. MR0082960. 251

A. Selman

(1972) Completeness of calculi for axiomatically defined classes of algebras. *Algebra Universalis* **2**, 20–32. MR313170. 403

Saharon Shelah

(1971) Every two elementarily equivalent models have isomorphic ultrapowers. *Israel J. Math.* **10**, 224–233. MR297554. 344

(1972) Uniqueness and characterization of prime models over sets for totally transcendental first-order theories. *J. Symbolic Logic* **37**, 107–113. MR316239. 412

L. N. Shevrin, B. M. Vernikov, and M. V. Volkov

(2009) Lattices of semigroup varieties. *Izv. Vyssh. Uchebn. Zaved. Mat.*, 3–36. MR2581451. 234

W. Sierpiński

(1935) Sur les suites infinies de fonctions définies dans les ensembles quelconques. *Fund. Math.* **24**, 209–212. 229

Michael Sipser

(2012) *Introduction to the Theory of Computation, 3rd Edition.* Cenage Learning. 251

Thoralf Skolem

(1934) Über die Nicht-charakterisierbarkeit der Zahlenreihe mittels endlich oder abzählbar unendlich vieler Aussagen mit ausschliesslich Zahlenvariablen. *Fundamenta Mathematicae* **23**, 150–161. 331

Douglas D. Smith

(1972) Non-recursiveness of the set of finite sets of equations whose theories are one-based. *Notre Dame J. Formal Logic* **13**, 135–138. MR0299477. 276

M. H. Stone

(1936) The theory of representations for Boolean algebras. *Trans. Amer. Math. Soc.* **40**, 37–111. MR1501865. 333, 378, 381

(1937) Applications of the theory of Boolean rings to general topology. *Trans. Amer. Math. Soc.* **41**, 375–481. MR1501905. 381

W. Stormquist

(1974) Solution to elementary problem #2411. *Amer. Math. Monthly* **81**, 410. 166

Alfred Tarski

(1930) Une contribution à la théorie de la mesure. *Fundamenta Mathematicae* **15**, 42–50. MR1549800. 333

(1949) *Cardinal Algebras. With an Appendix: Cardinal Products of Isomorphism Types, by Bjarni Jónsson and Alfred Tarski.* Oxford University Press, New York, N. Y. MR0029954. 394

(1954) Contributions to the theory of models. I. *Nederl. Akad. Wetensch. Proc. Ser. A.* **57**, 572–581 = Indagationes Math. 16, 572–581 (1954). MR0066301. 397

(1968) Equational logic and equational theories of algebras. In *Contributions to Math. Logic (Colloquium, Hannover, 1966).* North-Holland, Amsterdam, pages 275–288. MR0237410. 165

Alfred Tarski and Robert L. Vaught

(1958) Arithmetical extensions of relational systems. *Compositio Math.* **13**, 81–102. MR95121. 322

M. A. Taĭtslin

(1960) Review of Roger C. Lyndon. An interpolation theorem in the predicate calculus. Pacific Journal of Mathematics, vol. 9 (1959), pp. 129–142. *Journal of Symbolic Logic* **25**, 273–274. 427

Walter Taylor

(1972) Residually small varieties. *Algebra Universalis* **2**, 33–53. MR0314726. 277, 281, 343

(1973) Characterizing Mal'cev conditions. *Algebra Universalis* **3**, 351–397. MR0349537. 80

(1974) Uniformity of congruences. *Algebra Universalis* **4**, 342–360. MR376483. 56, 59, 61, 120, 124

(1977a) Equational logic. In *Contributions to universal algebra*. North-Holland Publishing Co., Amsterdam, pages 465–501. Proceedings of the Colloquium held in Szeged, 1975. Colloquia Mathematica Societatis János Bolyai, vol. 17. MR0472645. 59

(1977b) Varieties obeying homotopy laws. *Canad. J. Math.* **29**, 498–527. MR434928. 83

(1977c) Varieties of topological algebras. *J. Austral. Math. Soc. Ser. A* **23**, 207–241. MR447077. 22

(1981) Some universal sets of terms. *Trans. Amer. Math. Soc.* **267**, 595–607. MR626492. 227, 230

(1982) Some applications of the term condition. *Algebra Universalis* **14**, 11–24. MR634412. 43, 139

(2009) Simple equations on real intervals. *Algebra Universalis* **61**, 213–226. MR2564320. 62, 63

A. M. Turing

(1936) On Computable Numbers, with an Application to the Entscheidungsproblem. *Proc. London Math. Soc. (2)* **42**, 230–265. MR1577030. 253, 254

M. Valeriote and R. Willard

(2014) Idempotent n-permutable varieties. *Bull. Lond. Math. Soc.* **46**, 870–880. MR3239624. 71

J. Varlet

(1972) A regular variety of type $< 2, 2, 1, 1, 0, 0, 0 >$. *Algebra Universalis* **2**, 218–223. MR325477. 58

Robert L. Vaught

(1961) The elementary character of two notions from general algebra. In *Essays on the foundations of mathematics*. Magnes Press, Hebrew Univ., Jerusalem, pages 226–233. MR0170816. 51, 409, 435

(1965) The Löwenheim-Skolem theorem. In *Logic, Methodology and Philos. Sci. (Proc. 1964 Internat. Congr.)*. North-Holland, Amsterdam, pages 81–89. MR0210574. 57

Ju Wang

(1988) *Projectivities and minimal blocks in algebraic systems*. Ph.D. thesis, University of California, Los Angeles. 209

(1990) A proof of the Baker conjecture. *Acta Math. Sinica* **33**, 626–633. MR1082596. 209

Joseph M. Weinstein

(1965) *First Order Properties Preserved by Direct Products*. Ph.D. thesis, The University of Wisconsin - Madison. 414

P. Whitman

(1946) Lattices, equivalence relations and subgroups. *Bull. Amer. Math. Soc.* **52**, 507–522. MR16750. 218

Ross Willard

(2000) A finite basis theorem for residually finite, congruence meet-semidistributive varieties. *J. Symbolic Logic* **65**, 187–200. MR1782114. 28, 196, 203, 204

Rudolf Wille

(1969) Primitive Länge und primitive Weite bei modularen Verbänden. *Math. Z.* **108**, 129–136. MR241332. 119

(1970) *Kongruenzklassengeometrien.* Lecture Notes in Mathematics, Vol. 113. Springer-Verlag, Berlin-New York. MR0262149. 4, 20, 51, 95, 109

Index

270 **Leslie Hogben, Jephian C.-H. Lin, and Bryan L. Shader,** Inverse Problems and Zero Forcing for Graphs, 2022

269 **Ralph S. Freese, Ralph N. McKenzie, George F. McNulty, and Walter F. Taylor,** Algebras, Lattices, Varieties: Volume III, 2022

268 **Ralph S. Freese, Ralph N. McKenzie, George F. McNulty, and Walter F. Taylor,** Algebras, Lattices, Varieties: Volume II, 2022

267 **Dragana S. Cvetković Ilić,** Completion Problems on Operator Matrices, 2022

266 **Kate Juschenko,** Amenability of Discrete Groups by Examples, 2022

265 **Nabil H. Mustafa,** Sampling in Combinatorial and Geometric Set Systems, 2022

264 **J. Scott Carter and Seiichi Kamada,** Diagrammatic Algebra, 2021

263 **Vugar E. Ismailov,** Ridge Functions and Applications in Neural Networks, 2021

262 **Ragnar-Olaf Buchweitz,** Maximal Cohen–Macaulay Modules and Tate Cohomology, 2021

261 **Shiri Artstein-Avidan, Apostolos Giannopoulos, and Vitali D. Milman,** Asymptotic Geometric Analysis, Part II, 2021

260 **Lindsay N. Childs, Cornelius Greither, Kevin P. Keating, Alan Koch, Timothy Kohl, Paul J. Truman, and Robert G. Underwood,** Hopf Algebras and Galois Module Theory, 2021

259 **William Heinzer, Christel Rotthaus, and Sylvia Wiegand,** Integral Domains Inside Noetherian Power Series Rings, 2021

258 **Pramod N. Achar,** Perverse Sheaves and Applications to Representation Theory, 2021

257 **Juha Kinnunen, Juha Lehrbäck, and Antti Vähäkangas,** Maximal Function Methods for Sobolev Spaces, 2021

256 **Michio Jimbo, Tetsuji Miwa, and Fedor Smirnov,** Local Operators in Integrable Models I, 2021

255 **Alexandre Boritchev and Sergei Kuksin,** One-Dimensional Turbulence and the Stochastic Burgers Equation, 2021

254 **Karim Belabas and Henri Cohen,** Numerical Algorithms for Number Theory, 2021

253 **Robert R. Bruner and John Rognes,** The Adams Spectral Sequence for Topological Modular Forms, 2021

252 **Julie Déserti,** The Cremona Group and Its Subgroups, 2021

251 **David Hoff,** Linear and Quasilinear Parabolic Systems, 2020

250 **Bachir Bekka and Pierre de la Harpe,** Unitary Representations of Groups, Duals, and Characters, 2020

249 **Nikolai M. Adrianov, Fedor Pakovich, and Alexander K. Zvonkin,** Davenport–Zannier Polynomials and Dessins d'Enfants, 2020

248 **Paul B. Larson and Jindrich Zapletal,** Geometric Set Theory, 2020

247 **István Heckenberger and Hans-Jürgen Schneider,** Hopf Algebras and Root Systems, 2020

246 **Matheus C. Bortolan, Alexandre N. Carvalho, and José A. Langa,** Attractors Under Autonomous and Non-autonomous Perturbations, 2020

245 **Aiping Wang and Anton Zettl,** Ordinary Differential Operators, 2019

244 **Nabile Boussaïd and Andrew Comech,** Nonlinear Dirac Equation, 2019

243 **José M. Isidro,** Jordan Triple Systems in Complex and Functional Analysis, 2019

242 **Bhargav Bhatt, Ana Caraiani, Kiran S. Kedlaya, Peter Scholze, and Jared Weinstein,** Perfectoid Spaces, 2019

241 **Dana P. Williams,** A Tool Kit for Groupoid C^*-Algebras, 2019

For a complete list of titles in this series, visit the AMS Bookstore at **www.ams.org/bookstore/survseries/**.